FORENSIC DNA APPLICATIONS
AN INTERDISCIPLINARY PERSPECTIVE

Edited by
Dragan Primorac, M.D., Ph.D.
Adjunct Professor of Forensic Science, Eberly College of Science,
Pennsylvania State University, University Park, PA, USA, and
Henry C. Lee College of Criminal Justice and Forensic Sciences,
University of New Haven, West Haven, CT, USA
Professor of Pediatric Medicine, School of Medicine, University of Split,
Split, Croatia and School of Medicine, University of Osijek, Osijek, Croatia

Moses Schanfield, Ph.D.
Professor of Forensic Science and Anthropology,
Department of Forensic Science,
George Washington University, Mount Vernon Campus,
Washington, D.C., USA

CRC Press is an imprint of the
Taylor & Francis Group, an **informa** business

CRC Press
Taylor & Francis Group
6000 Broken Sound Parkway NW, Suite 300
Boca Raton, FL 33487-2742

First issued in paperback 2020

© 2014 by Taylor & Francis Group, LLC
CRC Press is an imprint of Taylor & Francis Group, an Informa business

No claim to original U.S. Government works

ISBN-13: 978-1-4665-8022-0 (hbk)
ISBN-13: 978-0-367-77848-4 (pbk)

This book contains information obtained from authentic and highly regarded sources. Reasonable efforts have been made to publish reliable data and information, but the author and publisher cannot assume responsibility for the validity of all materials or the consequences of their use. The authors and publishers have attempted to trace the copyright holders of all material reproduced in this publication and apologize to copyright holders if permission to publish in this form has not been obtained. If any copyright material has not been acknowledged please write and let us know so we may rectify in any future reprint.

Except as permitted under U.S. Copyright Law, no part of this book may be reprinted, reproduced, transmitted, or utilized in any form by any electronic, mechanical, or other means, now known or hereafter invented, including photocopying, microfilming, and recording, or in any information storage or retrieval system, without written permission from the publishers.

For permission to photocopy or use material electronically from this work, please access www.copyright.com (http://www.copyright.com/) or contact the Copyright Clearance Center, Inc. (CCC), 222 Rosewood Drive, Danvers, MA 01923, 978-750-8400. CCC is a not-for-profit organization that provides licenses and registration for a variety of users. For organizations that have been granted a photocopy license by the CCC, a separate system of payment has been arranged.

Trademark Notice: Product or corporate names may be trademarks or registered trademarks, and are used only for identification and explanation without intent to infringe.

Visit the Taylor & Francis Web site at
http://www.taylorandfrancis.com

and the CRC Press Web site at
http://www.crcpress.com

We bring this volume to the forensic community in honor and memory of the innocent who fell victim to violence and brutality around the world and for their loved ones who suffered the pain of loss. We dedicate this volume to all forensic scientists and practitioners whose work is essential to uphold justice in the hope that the world will become a better and less violent place.

I dedicate my contribution to this volume to the memory of my father Marinko, and to my mother Dragica, wife Jadranka, daughters Lara and Matea, brother Damir and his family, and to all my friends and colleagues who have been the source of love, inspiration, and strength.

—**Dragan Primorac**

I dedicate my contribution to my mother Fannie, wife Patricia, daughters Amanda and Samantha, brother David, sister Miriam, and their families for all the support over the years. I thank all my colleagues, living and those who passed on, for their support, inspiration, and ideas that have contributed in their own way to this book.

—**Moses Schanfield**

Contents

Foreword	xi
Preface	xiii
Acknowledgments	xix
Authors	xxi
Contributors	xxiii

Section I
GENERAL BACKGROUND AND METHODOLOGICAL CONCEPTS

1 Basic Genetics and Human Genetic Variation — 3
DRAGAN PRIMORAC, MOSES S. SCHANFIELD, AND DAMIR MARJANOVIĆ

2 Forensic DNA Analysis and Statistics — 55
MOSES S. SCHANFIELD, DRAGAN PRIMORAC, AND DAMIR MARJANOVIĆ

3 Forensic Aspects of mtDNA Analysis — 85
MITCHELL M. HOLLAND AND GORDAN LAUC

4 Y Chromosome in Forensic Science — 105
MANFRED KAYSER AND KAYE N. BALLANTYNE

5 Forensic Application of X Chromosome STRs — 135
TONI MARIE DIEGOLI

6 Low Copy Number DNA Profiling — 171
THERESA CARAGINE, KRISTA CURRIE, AND CRAIG O'CONNOR

7 Forensic DNA Mixtures, Approaches, and Analysis — 189
THERESA CARAGINE, ADELE MITCHELL, AND CRAIG O'CONNOR

8 Forensic DNA Typing and Quality Assurance — 205
DANIEL VANEK AND KATJA DROBNIČ

Section II
USES AND APPLICATIONS

9 Collection and Preservation of Physical Evidence — 253
HENRY C. LEE, TIMOTHY M. PALMBACH, DRAGAN PRIMORAC, AND ŠIMUN ANĐELINOVIĆ

10 Identification of Missing Persons and Mass Disaster Victim Identification by DNA — 277
BARBARA A. BUTCHER, FREDERICK R. BIEBER, ZORAN M. BUDIMLIJA, SHEILA M. DENNIS, AND MARK A. DESIRE

11 Bioterrorism and Forensic Microbiology — 293
ALEMKA MARKOTIĆ, JAMES W. LE DUC, AND JENNIFER SMITH

12 Forensic Animal DNA Analysis — 317
MARILYN A. MENOTTI-RAYMOND, VICTOR A. DAVID, STEPHEN J. O'BRIEN, SREE KANTHASWAMY, PETAR PROJIĆ, VEDRANA ŠKARO, GORDAN LAUC, AND ADRIAN LINACRE

13 Application of DNA-Based Methods in Forensic Entomology — 353
JEFFREY D. WELLS AND VEDRANA ŠKARO

14 Forensic Botany: Plants as Evidence in Criminal Cases and as Agents of Bioterrorism — 367
HEATHER MILLER COYLE, HENRY C. LEE, AND TIMOTHY M. PALMBACH

Section III
RECENT DEVELOPMENTS AND FUTURE DIRECTIONS IN HUMAN FORENSIC MOLECULAR BIOLOGY

15 Forensic Tissue Identification with Nucleic Acids — 385
DMITRY ZUBAKOV AND MANFRED KAYSER

16 Evolving Technologies in Forensic DNA Analysis — 417
CASSANDRA D. CALLOWAY AND HENRY ERLICH

| 17 | Prediction of Physical Characteristics, such as Eye, Hair, and Skin Color Based Solely on DNA | 429 |

ELISA WURMBACH

| 18 | Molecular Autopsy | 453 |

GRACE AXLER-DIPERTE, FREDERICK R. BIEBER,
ZORAN M. BUDIMLIJA, ANTTI SAJANTILA, DONALD SIEGEL,
AND YINGYING TANG

| 19 | Genetic Genealogy in the Genomic Era | 483 |

JAKE K. BYRNES, NATALIE M. MYRES, AND PETER A. UNDERHILL

Section IV
LAW, ETHICS, AND POLICY

| 20 | DNA as Evidence in the Courtroom | 509 |

DAVID H. KAYE, FREDERICK R. BIEBER, AND DAMIR PRIMORAC

| 21 | Some Ethical Issues in Forensic Genetics | 527 |

ERIN D. WILLIAMS AND DAVID H. KAYE

| 22 | DNA in Immigration and Human Trafficking | 539 |

SARA HUSTON KATSANIS AND JOYCE KIM

| 23 | DNA Databases | 557 |

CHRISTOPHER ASPLEN

| **Author Index** | 571 |
| **Subject Index** | 603 |

Foreword

I appreciate the opportunity to introduce the most comprehensive volume on forensic and ancestral DNA analysis yet published. Drs. Primorac and Schanfield deserve considerable credit for having brought this wide-ranging material together under one cover. The coverage is comprehensive in terms of both the topics and of its international scope. The book grew out of the Sixth International Society of Applied Biological Sciences on Human Genome Project Based Applications in Forensic Science, Anthropology and Individualized Medicine in Split, Croatia, in 2009.

Four logical sections comprise the book: General Background and Methodological Concepts; Uses and Applications; Recent Developments and Future Directions in Human Forensic Molecular Biology, and Law, Ethics, and Policy.

The first section reviews the history and development of DNA typing and profiling for criminal and civil purposes. It includes chapters on the statistical interpretation of results with case examples, mitochondrial DNA testing, Y single nucleotide polymorphisms (SNPs) and short tandem repeats (STRs), X SNP and STR testing, low copy number DNA typing, mixtures, and quality assurance/quality control.

The second section, Uses and Applications, has chapters on collection and preservation of biological evidence under a variety of different circumstances, identification of human remains including at mass disaster settings, applications to bioterrorism investigations, animal DNA testing in criminal cases, pedigree questions and wildlife forensic problems, applications in forensic entomology, and forensic botany.

The third section, which focuses on recent developments, contains chapters on developing technologies, including the rigorous identification of tissue of origin (e.g., saliva, vaginal, skin), mtDNA profiling using immobilized probe strips (HLA-DQA1 strips) as well as chips and next-generation sequencing, use of SNPs to ascertain phenotypic characteristics such as eye color and skin color, and finally the "molecular autopsy" that looks at aspects of toxicogenetics and pharmacogenetics.

The fourth and final section on law, ethics, and policy has chapters on the use of DNA evidence in the criminal justice system in both the United States and Europe, ethical issues in forensic laboratory practices, familial searches, DNA databases, ancestry searches, physical phenotyping, and report writing. In addition, there are chapters on DNA applications in immigration and human trafficking cases, and international perspectives on DNA databases.

The beauty and value of this book are that it treats and tries to connect all the disparate dots in present-day forensic molecular biology, taken in its broadest possible sense.

Human identification by an agreed-upon set of anonymous STRs is now a robust and reliable technology, but there are still issues surrounding mixture interpretation and obtaining reliable types with less than optimal quantities of target DNA (at least using CE-laser dye techniques). The level of sensitivity of DNA typing is a two-sided coin, at times helpful to sorting out the case, and at other times confusing as to what is signal and what is noise. The introduction of Y STRs holds promise for sexual assault evidence analysis, because the commonly used differential extraction procedure is not always completely successful. However, these markers are not sufficiently polymorphic in most populations to be very useful for individualization. Coverage of DNA "statistics" is welcome. The probability of chance match is key to evaluating DNA results, and there are still arguments about different approaches.

Applications to human identification in criminal cases and to resolving disputed parentage cases coalesce in the identification of mass disaster victims. Similarly, animal DNA techniques have helped in cases involving pet hair, and are obviously the *sine qua non* of game regulation, poaching, and endangered species law enforcement. Plant DNA technologies can help controlled substance investigations. DNA techniques applied to insect classes that are forensically important can assist forensic entomologists in estimating time since death. Microbial DNA analysis is perhaps the only fast method for diagnosing bioterrorist toxins or organisms. In a large-scale attack or epidemic, these techniques provide perhaps the only hope of containing the problem.

Forensic tissue of origin identification has been a long-standing problem in forensic biology. RNA methods may finally provide the unique tissue identification markers needed to do a thorough criminalistics workup in criminal cases. Methods for predicting ancestry may be useful in investigations but must be carefully interpreted at this time. If robust, reliable techniques for common descriptive phenotypic traits (eye color, hair color, etc.) are developed, they would be very helpful in criminal investigations and considerably more reliable than eyewitnesses. As the genetic variation in CYP genes is unraveled, genetic typing could become important in the diagnosis of drug-induced death, thus bringing genetic typing into the realm of forensic toxicology.

It is widely accepted that the law is slow to catch up with fast-moving science and technology. DNA technologies have proved no exception. Thus, coverage of this important subject is key to societal acceptance and use of these technologies in democratic countries. The public must accept that the technologies serve a compelling societal good, or their use will be curtailed or prohibited.

DNA techniques and resulting technologies have spread their wings into the obvious traditional forensic areas, but also into many new and unusual ones. It is possible today to think about solving forensic problems that were simply intractable with the techniques available even thirty years ago. This book is the only one I have seen that pulls all the material together in a detailed, organized, and comprehensive manner. For this, the editors and contributing authors deserve our gratitude.

R. E. Gaensslen
Prof. Emeritus, Forensic Science
University of Illinois, Chicago

Preface

Forensic DNA Applications: An Interdisciplinary Perspective was created to fill a void we perceived in the literature for a book that could be a textbook for forensic molecular biology students, and a reference book for practitioners of forensic molecular biology, as well as lawyers and judges dealing with civil and criminal cases involving DNA technology. The book provides up-to-date coverage of the ever-broadening field of forensic DNA testing. The individual chapters written by multiple authors, all of whom are experts in their particular areas, provide a compact review of the start of the art for that particular topic. Every effort has been made to have each topic current at the time of submission.

The book was developed as an outgrowth of the Sixth International Society of Applied Biological Sciences (ISABS) Conference on Human Genome Project Based Applications in Forensic Science, Anthropology, and Individualized Medicine, Split, Croatia, June 1–5, 2009, biannual educational conference of the ISABS held in Croatia, whereupon the organizers and speakers at the meeting felt that there was a need for a book that could be used as a textbook but also as a reference book for people working in the field of forensic molecular biology as well as individuals investigating and adjudicating cases involving DNA evidence, whether they be civil or criminal cases. The approach is international, so some things may not be relevant to analysts working only in the United States or in Europe, but as the title states the approach is "interdisciplinary."

There are many texts on forensic molecular biology, some authored by a single author, others edited volumes. This particular volume is unique in the sense that the 51 authors who provided the 23 chapters are experts in their specific areas. The authors of this book span those who worked during the period before DNA testing was done through the present, and those now working in developing some of the newer technologies mentioned and who therefore can provide a unique perspective on the history and practice of forensic DNA testing. The authors come from Australia, Bosnia and Herzegovina, Croatia, the Czech Republic, Finland, the Netherlands, Slovenia, and the United State of America.

This edition has four sections and 23 chapters, representing all aspects of forensic DNA methodology, ethics, law, and policy.

Section I, General Background and Methodological Concepts, consists of eight chapters reviewing the overall background of the field.

Chapter 1 (Basic Genetics and Human Genetic Variation) is a review of the history of forensic DNA testing, including screening tests for biological fluids, a review of DNA methods from the introduction of restriction fragment length polymorphisms (RFLP) testing, through the beginning of the use of the polymerase chain reaction (PCR) to the use of modern PCR for both small tandem repeat (STR) testing, single nucleotide polymorphism (SNP) testing, and DNA sequencing procedures.

The next chapter, Forensic DNA Analysis and Statistics, is a review of forensic DNA analysis and statistical analysis of the data generated including parentage testing, with both complete cases and the different forms of partial cases, identification of remains, forensic identification using nuclear markers, as well as X and Y markers and mtDNA. This chapter includes many examples taken from actual cases that allow the reader to see how all of the calculations are performed.

Chapter 3 (Forensic Aspects of mtDNA Analysis) provides a review of mitochondrial DNA (mtDNA) testing, how it is performed, and the pitfalls and difficulties with interpretation. The chapter covers the sequencing of the hypervariable regions I and II as well as the SNP markers in the coding region of mtDNA that define the population-associated haplogroups, and their statistical interpretation.

The fourth chapter, Y Chromosome in Forensic Science, looks at the use of Y SNPs and Y STRs to define haplogroups and haplotypes, as well as the calculation of match likelihoods, which is also discussed in Chapter 2. The chapter also includes information on commercially available Y STR and Y SNP kits, and combining the information in Y chromosome markers with nuclear markers.

Chapter 5 (Forensic Application of X Chromosome STRs) looks at the relatively new field of STR and SNP markers on the X chromosome and the unique statistical calculations for this haplodiploid system, as males are haploid and do not recombine and females are diploid and are the only source of recombination for these markers.

The next chapter focuses on low copy number DNA testing (Low Copy Number DNA Profiling), which can occur when a single source sample is very dilute or very degraded, or when there is a mixture of two or more samples and there is either degradation, or dilution, or both. LCN situations occur when the amount of amplifiable DNA is less than 100 pg. The chapter is written by staff of the Office of the Chief Medical Examiner of New York City, one of the first laboratories to validate LCN DNA testing in the United States.

Chapter 7 (Forensic DNA Mixtures, Approaches, and Analysis) is on the testing, and analysis of DNA mixtures, which is a logical extension of LCN DNA testing, since as soon as the mixture is less than a 1:1 ratio it is possible for the minor component to be less than 100 pg of amplifiable DNA. Mixture interpretation requires the inclusion of LCN DNA analysis with the guidelines established by the Technical Working Group on DNA Analysis Methods (TWGDAM), European DNA Profiling Group (EDNAP), and the various regulatory bodies propounding guidelines and standards for the interpretation of mixtures.

The final chapter of this section, Forensic DNA Typing and Quality Assurance, covers issues of quality assurance in the forensic DNA typing laboratory. This chapter covers both the old U.S. American Society of Crime Laboratory Directors-Laboratory Accreditation Board (ASCLD-LAB) standards and the new ISO 17025 standards initially implemented in Europe and more recently the United States, as well as some additional European standards.

Section II, Uses and Applications, focuses on a multitude of applications of the DNA technologies described in the previous chapters, ranging from the collection and preservation of biological evidence, identification of missing persons in single and mass disasters, bioterrorism, animal DNA analysis, forensic entomology, and forensic botany.

Chapter 9 (Collection and Preservation of Physical Evidence) is an international approach to the collection and preservation of physical evidence, but provides basic guidelines on the preservation of biological evidence, although the title says physical evidence. The chapter also goes into the various forms of crime scene reconstruction that aid in the interpretation of the aggregate physical evidence collected at that scene.

Chapter 10 (Identification of Missing Persons and Mass Disaster Victim Identification by DNA) deals with the seemingly unending job of identifying remains, either of a single missing person or of a mass disaster. There seems to be an ever-increasing number of mass casualty events, from the 9/11 terrorism attack in the United States, to the mass graves of the Balkan War, to the long list of countries around the world with large numbers of missing individuals. This chapter relies on LCN type data, as well as computer programs and other ways of linking DNA profiles to identify victims of these events.

The next chapter, Bioterrorism and Forensic Microbiology, provides a history of bioterrorism, which is older than one would think, and includes a classification of these "select agents." Forensic microbiology encompasses the tools used to identify these agents in a timely fashion so that an appropriate response can be initiated. The areas of biosafety and biosecurity are also included.

Chapter 12 (Forensic Animal DNA Analysis) covers the subject of forensic animal DNA testing. This happens to be a very broad subject and only a limited number of areas are covered. The earliest use of animal DNA was the case involving "Snowball" the cat, so cat forensic DNA applications are discussed first followed by dog forensics. As a significant portion of the world populations has pets including cats or dogs, their hairs are often found at crime scenes. Bovine (cattle) DNA is included next as cattle are often involved because of theft, as well as occasional fraud. In the United States, all pedigreed cattle are DNA tested. The final area covered in this chapter is wildlife forensic DNA testing. This field is extremely broad and includes poaching of nonthreatened animals such as elk or deer, to the illegal importation of protected/endangered species such as sea mammals, elephants, and tigers.

Next is a chapter on the Application of DNA-Based Methods in Forensic Entomology, which presents two sides to this topic. The first is the identification of forensically useful insect species, as maggots are not readily identifiable. This is critical in trying to assess the postmortem interval on a decomposed body. The second area of interest is the testing of human DNA retrieved from insects to aid in identification of the victim, as that may be the only surviving human tissue on a decomposed body.

Chapter 14 (Forensic Botany: Plants as Evidence in Criminal Cases and as Agents of Bioterrorism) concentrates on the subject of forensic botany, which has several areas. One of the most widely used controlled substances is marijuana, which in all U.S. jurisdictions is the number two controlled substance submitted for analysis. This chapter goes into the physical as well as DNA methods for the identification of plant material containing controlled substances. The rest of the chapter discusses the use of identification of pollen and fungus to help locate where a body has been, as well as the toxicology and possible bioterrorism applications of plant materials such as "ricin," as well as drug enforcement issues regarding plant material.

Section III, Recent Developments and Future Directions in Human Forensic Molecular Biology, looks at the evolving technologies in forensic DNA testing that may not be widely used at the present but are being developed as future technologies, including improved identification of specific tissues; new applications of mtDNA testing; the future of next generation sequencing in forensic analysis; the use of SNPs to predict physical characteristics such as eye, hair, and skin color; and finally the area of molecular autopsy.

Chapter 15 (Forensic Tissue Identification with Nucleic Acids) reviews the classical tests for the identification of forensically relevant tissues such as peripheral blood, menstrual blood, parturient blood, fetal and neonatal blood, nasal blood, saliva, semen, vaginal secretions, and skin. The chapter then goes on to discuss the use of RNA to identify these tissues, as well as looking at DNA methylation, and the use of bacterial DNA and RNA for forensic tissue identification.

Chapter 16 (Evolving Technologies in Forensic DNA Analysis) primarily deals with testing for mtDNA using immobilized probe strips, similar to the DQA1 and Polymarker tests described in Chapter 1. The chapter then describes the use of chip technology and Next Generation Sequencing (NGS) technology for HLA and mtDNA as well as other genetic marker systems.

Chapter 17 (Prediction of Physical Characteristics, such as Eye, Hair, and Skin Color Based Solely on DNA) discusses the use of SNPs, some of the complexities of the inheritance of these traits, and the future directions of the use of these assays.

The next chapter, Molecular Autopsy, concentrates on the use of molecular tools to look at unexplained deaths. This is a rapidly growing area of interest that involves the interaction of a large family of genes that have to do with the metabolism of drugs, or in some cases the lack of metabolism. Although not normally a cause of death, adverse reactions to the anticoagulant warfarin can be life-threatening. They usually occur in individuals taking a normal dose of the drug, but they have a genetic variant of one of the two genes that metabolize the drug that is defective so the drug levels are higher than normal, leading to an adverse drug reaction. This chapter looks at the various aspects of these genes and how they can lead to unexplained deaths. The chapter has two major sections, the genomics of sudden natural death, and the use of toxicogenetics/pharmacogenetics in cause of death investigations.

Chapter 19 (Genetic Genealogy in the Genomic Era) concentrates on the field of genetic genealogy. Although this has limited application to forensic DNA analysis, this area is getting a lot of attention in the popular media, and provides useful background information in this area of public excitement. Ancestry determination has historically been based on the use of mtDNA and Y chromosome markers; however, nuclear ancestry informative SNPs have recently become available, improving our ability to identify ancestry in broad categories.

Section IV, Law, Ethics, and Policy, helps to set this book aside from others as it deals with the legal, ethical, and policy issues on the use of DNA evidence. As a practitioner involved in some of the earliest RFLP-DNA cases, this section is very useful for me to refresh my memory on the many arguments that have gone before. This section is a good refresher for lawyers and judges as well as practitioners who are new to the field and us old hands.

Chapter 20 (DNA as Evidence in the Courtroom) provides some insight to both the European and U.S. systems of jurisprudence. The chapter covers the legal standards for acceptance of scientific evidence, a history of the acceptance of DNA evidence in courts, some of the newer methods of DNA analysis, postconviction relief, legal procedures, and other relevant material.

The next chapter (Some Ethical Issues in Forensic Genetics) deals with ethical issues in forensic genetics. Given the recent spate of crime laboratories in the media for various infractions of quality assurance standards and other problems, this is a critical chapter. It covers the general concepts of bioethics, and then goes on to deal with ethical issues in acquiring DNA samples, DNA databanks, phenotyping and ancestry identification, identification of remains, as well as, the ethics of report writing and testimony. This chapter provides the ethical framework and background for many of the previous chapters.

Chapter 22 (DNA in Immigration and Human Trafficking) delves into the area of DNA testing in immigration cases and human trafficking. As more and more immigrants try to enter the United States, European Union, and other countries, governments are requiring that the parties prove their claimed relationships via DNA testing. Similar programs are developing to address human trafficking, by establishing the relationship of victims to possible parents. The chapter also deals with the ethical, legal, and social considerations of this type of DNA identification testing.

The final chapter (DNA Databases) covers international perspectives on forensic DNA databases. This chapter is up-to-date and inclusive of data on a worldwide basis, including U.S. Supreme Court decisions in June of 2013. The chapter reviews the size and scope of DNA databases and their use from a global perspective.

Each of the chapters includes an extensive list of references for the reader.

Dragan Primorac
Moses Schanfield

Acknowledgments

We acknowledge the International Society of Applied Biological Sciences for support essential in bringing this volume to the light of day. We thank Ana Banić Göttlicher and Maša Vukmanović/offstudio, Zagreb, Croatia, for the cover art.

Authors

Dragan Primorac, M.D., Ph.D., is a pediatrician, forensic expert, and geneticist. Currently he serves as an adjunct professor at Eberly College of Science, The Pennsylvania State University, and the Henry C. Lee College of Criminal Justice and Forensic Sciences, University of New Haven in the United States and at medical schools in Split and Osijek in Croatia. Professor Primorac is a pioneer in the application of DNA analysis for identification of bodies in mass graves. He has authored close to 200 scientific papers, book chapters, and abstracts in areas of forensic science, clinical medicine, molecular genetics, population genetics, genetic legacy of *Homo sapiens sapiens*, education, science, and technology policy. His papers have been cited more than 1600 times.

Professor Primorac received the Young Investigator Award of the American Society for Bone and Mineral Research in 1992, the Michael Geisman Fellowship Award of the Osteogenesis Imperfecta Foundation in 1993, the Life Time Achievement Award by the Henry C. Lee's Institute of Forensic Science in 2002, the Award of the Italian Region Veneto for Special Achievements in Promoting Science in the EU in 2007, and the University of New Haven's International Award for Excellence in 2010. Professor Primorac is the co-founder of the International Society of Applied Biological Sciences (ISABS). He has been an invited speaker to 70 national and international scientific meetings. Several renowned media outlets, both electronic and print, have reported on his work, such as the *New York Times, JAMA, Science, Profiles in DNA, Die Presse, Haaretz, Kleine Zeitung* and many others. Furthermore, Connecticut TV Station Channel 8 filmed a television serial on forensic work in Bosnia and Herzegovina and the Republic of Croatia performed by Professor Primorac and his colleagues.

From 2003 to 2009, he served as the Minister of Science, Education, and Sports of the Republic of Croatia. According to the International Republican Institute survey of October 1, 2007, he was rated as the most successful minister in the Croatian government with 31% approval rate. An award for numerous efforts made by Professor Primorac and his team on the Croatian educational system was a survey by *Newsweek* (appearing in the August 16, 2010 issue) in which the Croatian education system was rated 22nd in the world, ahead of 12 countries from the G20 group.

Moses S. Schanfield, Ph.D., has undergraduate and master's degrees in anthropology and a Ph.D. degree in human genetics. He has been involved full time in forensic testing for over 25 years. His career in forensic biology began before DNA testing was done. He was involved in some of the earliest forensic DNA cases and in some famous forensic cases including the O.J. Simpson case and the JonBenet Ramsey case. Since 1995, Professor Schanfield has been involved in repatriation of remains in the Balkans and the identification of remains in mass casualty events. This led to the creation of the first intensive course in forensic genetics, which has now been replaced by the biannual education conference of the International Society of Applied Biological Sciences (ISABS), in which Professor

Schanfield is one of the permanent organizers. Professor Schanfield has been to Croatia many times as part of his ISABS and other duties. He was involved in four of the first eight PCR cases admitted in court and reviewed at the appellate level in the United States, and testified in many of the early DNA admissibility hearings. Professor Schanfield has been involved in the development of some of the critical functions of modern forensic DNA testing including the in-lane size ladder, which is used in all human identification testing. He has published extensively on forensic DNA testing and has testified in state (39 states), federal, and military courts in the United States and Canada, Puerto Rico, and Barbados on forensic cases over 100 times. Professor Schanfield has lectured all over the world on blood transfusion, forensic science, and anthropological issues. He is currently a professor of forensic science and anthropology at George Washington University in Washington, D.C. Professor Schanfield has published over 145 articles and books in areas of anthropology, forensic science, immunology, and other scientific areas.

Contributors

Šimun Anđelinović
Department of Forensic Sciences
Split Medical School
University of Split
Split, Croatia

Christopher Asplen
Asplen and Associates
Chalfont, Pennsylvania, USA

Grace Axler-DiPerte
Office of Chief Medical Examiner of the City of New York
New York, New York, USA

Kaye N. Ballantyne
Office of the Chief Forensic Scientist
Victoria Police Forensic Services Department
Macleod, Australia

Frederick R. Bieber
Harvard Medical School
and
Brigham and Women's Hospital
Boston, Massachusetts, USA

Zoran M. Budimlija
Office of Chief Medical Examiner of the City of New York
New York, New York, USA

Barbara A. Butcher
Office of Chief Medical Examiner of the City of New York
New York, New York, USA

Jake K. Byrnes
Ancestry.com DNA
San Francisco, California, USA

Cassandra D. Calloway
Children's Hospital Oakland Research Institute
Oakland, California, USA

Theresa Caragine
Office of Chief Medical Examiner of the City of New York
New York, New York, USA

Heather Miller Coyle
Henry C. Lee College of Criminal Justice and Forensic Science
University of New Haven
West Haven, Connecticut, USA

Krista Currie
Office of Chief Medical Examiner of the City of New York
New York, New York, USA

Victor A. David
Laboratory of Genomic Diversity
National Cancer Institute
Frederick, Maryland, USA

Sheila M. Dennis
Office of Chief Medical Examiner of the City of New York
New York, New York, USA

Mark A. Desire
Department of Forensic Biology
New York City Office of Chief Medical Examiner
New York, New York, USA

Toni Marie Diegoli
Armed Forces DNA Identification Laboratory
Armed Forces Medical Examiner System
Dover Air Force Base
Dover, Delaware, USA

and

American Registry of Pathology
Camden, Delaware, USA

and

School of Biological Sciences
Flinders University
South Australia, Australia

Katja Drobnič
National Forensic Laboratory
Police
Ministry of the Interior
Ljubljana, Slovenia

and

Faculty of Criminal Justice and Security
University of Maribor
Maribor, Slovenia

Nanette Elster
Spence & Elster, P.C.
Chicago, Illinois, USA

Henry Erlich
Roche Molecular Systems
Pleasanton, California, USA

and

Children's Hospital Oakland Research Institute
Oakland, California, USA

Mitchell M. Holland
Pennsylvania State University
University Park, Pennsylvania, USA

Sree Kanthaswamy
University of California–Davis
Davis, California, USA

Sara Huston Katsanis
Duke Institute for Genome Sciences and Policy
Duke University
Durham, North Carolina, USA

David H. Kaye
Pennsylvania State University
University Park, Pennsylvania, USA

Manfred Kayser
Erasmus MC University Medical Center Rotterdam
Rotterdam, the Netherlands

Joyce Kim
Duke Institute for Genome Sciences and Policy
Duke University
Durham, North Carolina, USA

Gordan Lauc
Genos, DNA Laboratory
Zagreb, Croatia

James W. Le Duc
Galveston National Laboratory
University of Texas Medical Branch
Galveston, Texas, USA

Henry C. Lee
Forensic Science Department
College of Criminal Justice and Forensic Science
University of New Haven
West Haven, Connecticut, USA

Adrian Linacre
School of Biological Sciences
Flinders University
South Australia, Australia

Damir Marjanović
Institute for Genetic Engineering and Biotechnology (INGEB)
Sarajevo, Bosnia and Herzegovina

and

Genos, DNA Laboratory
Zagreb, Croatia

Alemka Markotić
Department for Research
University Hospital for Infectious Diseases
Zagreb, Croatia

Marilyn A. Menotti-Raymond
Laboratory of Genomic Diversity
National Cancer Institute
Frederick, Maryland, USA

Adele Mitchell
Office of Chief Medical Examiner of the City of New York
New York, New York, USA

Natalie M. Myres
Ancestry.com DNA
Provo, Utah, USA

Stephen J. O'Brien
Laboratory of Genomic Diversity
National Cancer Institute
Frederick, Maryland, USA

Craig O'Connor
Office of Chief Medical Examiner of the City of New York
New York, New York, USA

Contributors

Timothy M. Palmbach
Connecticut Forensic Science Laboratory
Meriden, Connecticut, USA

Damir Primorac
Department for Forensic Sciences
University of Split
Split, Croatia

Dragan Primorac
Department of Forensic Sciences
Split Medical School
University of Split
Split, Croatia

and

University of Osijek Medical School
Osijek, Croatia

and

Pennsylvania State University
University of New Haven
New Haven, Connecticut, USA

Petar Projić
Genos, DNA Laboratory
Zagreb, Croatia

Antti Sajantila
Department of Forensic Medicine
University of Helsinki
Helsinki, Finland

Moses S. Schanfield
Department of Forensic Science
George Washington University at Mount Vernon College
Washington, District of Columbia, USA

Donald Siegel
Office of Chief Medical Examiner of the City of New York
New York, New York, USA

Vedrana Škaro
Genos, DNA Laboratory
Zagreb, Croatia

Jennifer Smith
Pennsylvania State University
University Park, Pennsylvania, USA

Yingying Tang
Department of Forensic Biology
New York City Office of Chief Medical Examiner
New York, New York, USA

Peter A. Underhill
Department of Genetics
Stanford University School of Medicine
Stanford, California, USA

Daniel Vanek
Forensic DNA Service
Prague, Czech Republic

Jeffrey D. Wells
Department of Biological Sciences
and
International Forensic Research Institute
Florida International University
Miami, Florida, USA

Erin D. Williams
Center for Transforming Health
The MITRE Corporation
McLean, Virginia, USA

Elisa Wurmbach
Office of Chief Medical Examiner of the City of New York
New York, New York, USA

Dmitry Zubakov
Erasmus MC University Medical Center Rotterdam
Rotterdam, the Netherlands

General Background and Methodological Concepts

I

Basic Genetics and Human Genetic Variation

1

DRAGAN PRIMORAC
MOSES S. SCHANFIELD
DAMIR MARJANOVIĆ

Contents

1.1	Introduction	4
1.2	Historical Overview of DNA Research	5
	1.2.1 Introduction to Human Genetics	6
	1.2.2 Genome Structure	6
	1.2.3 Chromosomes and Genes	8
	1.2.4 Deoxyribonucleic Acid	10
	1.2.5 Genetic Diversity	10
	1.2.6 Variability of DNA	11
	1.2.7 Structure and Nomenclature of STR Markers	12
	1.2.8 Analysis of Sex Chromosomes	15
	1.2.8.1 Y Chromosome DNA Testing	15
	1.2.8.2 X Chromosome DNA Testing	17
	1.2.9 Mitochondrial DNA	18
	1.2.10 RNA Profiling	19
	1.2.11 Application of New Molecular Markers	19
1.3	Potential Biological Sources of DNA	21
	1.3.1 Basic Models and Steps of Forensic DNA Analysis	22
	1.3.2 Collecting and Storing Samples	22
	1.3.3 Determination of Biological Evidence	23
	1.3.3.1 Blood	23
	1.3.3.2 Semen	24
	1.3.3.3 Vaginal Body Fluid	25
	1.3.3.4 Saliva	25
	1.3.3.5 Urine	26
	1.3.3.6 Feces	26
1.4	Methods of DNA Isolation	26
	1.4.1 DNA Isolation with Organic Solvents	26
	1.4.2 Other DNA Isolation Methods	28
	1.4.2.1 Chelex®100	28
	1.4.2.2 Qiagen DNA Isolation Procedure	28
	1.4.2.3 DNA IQ System	28
	1.4.2.4 FTA®	28

1.5	DNA Quantification	29
	1.5.1 Quantifying DNA Using Method of Spectrophotometry	29
	1.5.2 "Yield" Gel Method	29
	1.5.3 Hybridization (Slot-Blot) Method	30
	1.5.4 AluQuant® Human DNA Quantitation System	30
	1.5.5 Quantitative RT-PCR Quantification Technology	31
1.6	Methods for Measuring DNA Variation	32
	1.6.1 Restriction Fragment Length Polymorphism	32
	1.6.2 Polymerase Chain Reaction	33
1.7	PCR Methods	34
	1.7.1 Early PCR Amplification and Typing Kits	34
	1.7.2 Multiplex STR Systems	35
	1.7.3 PowerPlex™ 16 System	38
	1.7.4 AmpFLSTR® Identifiler™ PCR Amplification Kit	38
	1.7.5 PowerPlex® ESX and ESI Systems	40
	1.7.6 AmpFℓSTR® NGM™ PCR Amplification Kit	41
	1.7.7 Investigator ESSplex Plus Kit and Investigator IDplex Plus Kit	42
	1.7.8 PLEX-ID SNP Assay	42
1.8	Detection of PCR Products	43
	1.8.1 Analytical Thresholds and Sensitivity for Forensic DNA Analysis	44
	1.8.2 Sequencing	44
1.9	Forensic Analysis of Plant DNA	46
1.10	Forensic Analysis of Animal DNA	46
1.11	Contamination Issues in Field and in Laboratory	47
References		47

1.1 Introduction

Nowadays, deoxyribonucleic acid (DNA) undoubtedly has an inimitable role in forensic science. Since 1985, when Alec Jeffreys and colleagues first applied DNA analysis to solve forensic problems, numerous medico-legal cases have been won based on this method (Jeffreys et al. 1985a,b). It is beyond a doubt that DNA analysis has become "a new form of scientific evidence" that is being constantly evaluated by both the public and professionals. More and more courts around the world accept the results of DNA analysis, and nowadays this technology is almost universally accepted in most legal systems. The foremost applications of DNA analysis in forensic medicine include criminal investigation, personal identification, and paternity testing. According to some sources, more than 300,000 DNA analyses in the different areas of expertise are performed annually in the United States. The fact that more than 30% of men who were identified as possible fathers are excluded using DNA technology articulates the importance of DNA analysis. Through the "DNA Innocence project," launched in the United States to acquit wrongfully convicted people, more than 300 persons have been exonerated by DNA testing (February 2013), including several individuals who were sentenced to death. DNA played an extremely important role in projects involving identification of war victims in Croatia and surrounding countries (Primorac et al. 1996; Primorac 2004; Džijan et al. 2009; Gornik et al. 2002; Alonso et al. 2001). Using this powerful "molecular" tool, identities were determined for thousands of skeletal remains, and their families had the opportunity to bury their loved ones with dignity (Anđelinović et al. 2005). Although the development of DNA typing in forensic science was extremely fast, the process is still not finished, and today we are witnessing a new era in the development of DNA technology that involves the introduction of automation and "chip" technology. In this chapter, we will explain the structure of DNA, principles of inheritance of genetic information, technological advances in DNA analysis, as well as the common application of mathematical methods in forensic practice.

1.2 Historical Overview of DNA Research

Since the time of ancient Greeks, there has been a need to explain the cause of diseases that repeatedly occur in certain families in obvious patterns. The Augustinian friar Gregor Mendel, who is often called the father of modern genetics, laid the foundations of genetics as we know it today. As a result of his experiments with garden pea, Mendel devised the fundamental principles of inheritance, which will be discussed later. Although Mendel published his first results in 1866, they were completely forgotten until 1900, when they were confirmed by scientists in three different laboratories. The next important discovery came in 1944 when Oswald Avery and his colleagues showed that genes are composed of the basic hereditary material—DNA. In 1994, O. Avery and collaborators conclusively showed that DNA was the genetic material, whereas the structure of DNA was first presented by Nobel Prize winners James Watson and Francis Crick in 1953, the groundwork for development of modern molecular genetics (Tamarin 2002). The knowledge of the number of chromosomes in humans came in 1956, when it was confirmed that the entire human genetic material is organized in 23 pairs or a total of 46 chromosomes. In 1980, David Botstein and his colleagues demonstrated that there are small variations in genetic material, which differ from person to person, and Alec Jeffrey in 1985 showed that certain regions of DNA contain repetitive sequences that are variable in different individuals. It is this fact that was critical in resolving the first forensic case using DNA analysis (Butler 2010). After the murder of two girls, Lynda Mann and Dawn Ashworth, in 1983 and 1986, respectively, the police organized the testing of more than 4000 men and eventually found the killer (Wambaugh 1989). However, the discovery of the polymerase chain reaction (PCR) method in 1983 certainly determined the future of DNA analysis, both in clinical and forensic medicine. In particular, with the discovery of this technology, it has become possible to analyze biological samples containing minuscule amounts of DNA. In 2001, the world witnessed the revolutionary discovery of the human genome structure, which was made by two separate groups of researchers: the International Human Genome Sequencing Consortium, led by Eric Lander, and Celera Genomics, which was spearheaded by Craig Venter (International Human Genome Sequencing Consortium 2001; Venter et al. 2001). The aim of both groups was to determine the complete nucleotide sequence of the human haploid genome, which contains about 3.0×10^9 base pairs (3.0 Gbp). However, they estimated that the size of the human diploid male genome is 6.294×10^9 bp, whereas for the human diploid female genome the size is 6.406×10^9 bp. The real surprise came with the finding that the human genome has only 20,000 to 25,000 genes, and not—as previously predicted—100,000 genes. But apparently, because of alternative splicing, it is certain that individual genes give more than one mRNA, resulting in the fact that the specified number of genes, according to some predictions, code for more than 100,000 proteins.

Particularly surprising was the fact that just more than 1% of the DNA contains protein-coding sequences (Cooper and Hausman 2004). On the other hand, non-protein–coding regions (ncDNA) are still poorly understood, and at least four hypotheses explained the role of these nonfunctional regions of the genome (Wagner 2013). The *selectionist hypothesis* suggested that ncDNA regulate gene expression, whereas the *neutralist hypothesis* (junk DNA) implied that these regions are without function but are transmitted passively as relics of evolutionary processes. The *intragenomic selection hypothesis* (selfish DNA) posited that ncDNA stimulate their own transmission and accumulate because of their prominent reproduction rate compared to protein-coding regions. Finally, *nucleotypic hypothesis* posited that ncDNA regions act to preserve the organizational/structural integrity of genome (Graur and Li 2000). Table 1.1 shows a summary of non-protein–coding genomic elements (Wagner 2013).

Doležel et al. (2003) proposed that 1 pg of DNA would represent 0.978×10^9 bp, and therefore diploid human female and male nuclei in G_1 the phase of the cell cycle should contain 6.550 and 6.436 pg of DNA, respectively. In the future, a great task lies before scientists: to discover the functions and interactions of many genes, as well as to exploit the possibility of gene and cell therapy. All advances in the field of molecular genetics will have an important role in the development of new technologies and methods in forensic DNA analysis (Primorac 2009).

1.2.1 Introduction to Human Genetics

All living creatures, including humans, are built of cells, which represent the smallest structural and functional units of our body.

1.2.2 Genome Structure

There is an essential difference between the two existing groups of cells: prokaryotic and eukaryotic. Prokaryotic cells have no nucleus or other cellular organelles, their average size is about 1 μm, and they contain one circular DNA molecule. Prokaryotic cells are mainly bacteria and contain the so-called extrachromosomal DNA (plasmid), which in bacteria can create resistance to certain antibiotics. Eukaryotic cells are up to 10 times larger than prokaryotic cells, have a nucleus and a number of cellular organelles, and their DNA is mainly located in one or more linear formations, called chromosomes, located in the nucleus. The somatic cells of eukaryotes commonly have two sets of genes derived from both parents, and these cells are called diploids. Cells (the ovum and the sperm) are having only one set of chromosomes, and these cells are called haploids. Most prokaryotes are haploids, as are the germ cells of eukaryotes. Based on their cellular structure, fungi, plants, animals, and humans are eukaryotic organisms. It is assumed that the human organism is made up of 200 different cell types and that the total number of cells is about 100 trillion. Although human cells are specialized for different functions, their very structure is fundamentally similar, as they are built from the nucleus, cytoplasm, and membranes. The basic genetic information is located in the nucleus, whereas the cytoplasm contains numerous cellular structures that keep the cell alive. Figure 1.1 shows the cellular organization of eukaryotic cell as well as the structure and location of the nuclear and mitochondrial DNA (mtDNA). One of these cellular structures is the mitochondrion; mitochondria act as a "cellular power plant" because they generate most of the cell's supply

Basic Genetics and Human Genetic Variation

Table 1.1 Summary of Nonprotein-Coding Genomic Elements

Nonprotein-Coding Genomic Element		Brief Description
Transcription regulatory elements		Molecular elements considered typical of gene structure, such as promoters, enhancers and intronic splicing signals (Slack 2006)
Introns		Segments of DNA located within genes that interrupt or separate exons from one another
5′ and 3′ untranslated regions	UTRs	Transcribed DNA sequences preceding (5′ UTR) and following (3′ UTR) coding sequences containing regulatory elements, such as binding sites for microRNAs (miRNAs) and polyadenylation signals (Maroni 2001)
RNA-specifying genes	MicroRNAs miRNAs	Destabilize or inhibit the translation of targeted mRNAs; 19–25 nucleotides in length (Neilson and Sandberg 2010)
	Transfer RNAs tRNAs	Facilitate translation by transporting specific amino acids to the ribosome; ca. 80 nucleotides in length (Neilson and Sandberg 2010)
	Ribosomal RNAs rRNAs	Facilitate the movement of tRNAs along the mRNA during translation; four types (18S, 28S, 5.8S, and 5S) (Neilson and Sandberg 2010)
	Spliceosomal RNAs snRNAs	Facilitate the processing of pre-mRNAs (i.e., help splice introns that are not self-splicing); five types (U1, U2, U4, U5, and U6) (Neilson and Sandberg 2010)
	Small nucleolar RNAs snoRNAs	Facilitate post-transcription modifications of rRNAs, tRNAs, and snRNAs; two types (H/ACA box and C/D box) (Neilson and Sandberg 2010)
	Piwi-interacting RNAs piRNAs	Protect the integrity of the genome in germline cells during spermatogenesis; 25–33 nucleotides in length (Neilson and Sandberg 2010)
	RNAse P/MRP genes	Process tRNA and rRNA precursors (Neilson and Sandberg 2010)
	Long non-coding RNAs lncRNAs	About 200+ nucleotides in length, such as *XIST*, which silences an X chromosome during X inactivation (Neilson and Sandberg 2010)
Repeat elements	Satellite DNA	DNA sequences often near centromeres and telomeres α-Satellite or alphoid DNA, a 171-bp sequence that is repeated in tandem and clustered at the centromeres of all chromosomes Repeat size of satellite DNA may be between 2 and 2000 bp and the size of the repeat array may be greater than 1000 bp (10,21)
	Minisatellites or variable number tandem repeats VNTRs	Repeat units of 10–200 bp clustered into repeat arrays of 10–100 units Found near the telomeres (the terminal ends of chromosomes), but are also distributed across the chromosomes (Graur and Li 2000; Slack 2006)
	Microsatellites or short tandem repeats STRs	Repeat units of 2–5 bp arranged in arrays of 10–100 units (Graur and Li 2000; Slack 2006)
	Short interspersed nucleotide elements SINEs	About 1,500,000 copies of SINEs present in the genome account for more than 10% of the genome (Graur and Li 2000; Slack 2006)
	Long interspersed nucleotide elements LINEs	About 850,000 copies of LINEs present in the genome, account for roughly one-fifth of the genome (Graur and Li 2000; Slack 2006)
	Retrovirus-like elements	About 450,000 copies present in the genome (Slack 2006)
	Transposons	About 300,000 copies present in the genome (Slack 2006)
Pseudogenes		Exhibit similarity to genes but lack introns and promoters and contain poly-A tails Most pseudogenes have lost the ability to be transcribed (Graur and Li 2000; Slack 2006; Jobling et al. 2004)

Source: Wagner, J.K. 2013. *J Forensic Sci* 58:292–294.

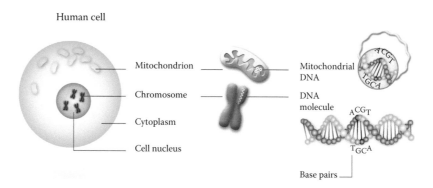

Figure 1.1 Sources of DNA in nucleated human cells include the nucleus (2n) and mitochondrion (100–1000 copies of mitochondrial DNA).

of adenosine triphosphate (ATP), which transports chemical energy within cells for metabolism. From the forensic perspective, mitochondria are important because they contain mtDNA, which is inherited directly through the maternal line (mtDNA will be discussed in greater detail in Chapter 3). Briefly, mtDNA does not contain introns, is not subject to classical recombination, and has a higher mutation rate than nuclear DNA. A human cell has between 100 and 1000 mitochondria, each of them containing multiple copies of circular mtDNA, which is 16,569 bp long. Mitochondria comprise one of the basic lines of evidence for the symbiogenetic theory of evolution, because mitochondria and mtDNA have a number of features that are characteristic to prokaryotic organisms. This suggests that this organelle derived from an ancient, engulfed (endosymbiotic) prokaryotic organism, which was one of the most important moments for the development of the first unicellular and later multicellular eukaryotic organisms. Although the average cell size corresponds to about one-tenth of the diameter of a hair, finding biological evidence that contains a sufficient numbers of cells can undoubtedly be crucial in solving forensic cases (Primorac and Marjanović 2008).

1.2.3 Chromosomes and Genes

Given the fact that every person inherits half of the genetic material from their father and half from their mother (excluding mtDNA, which is normally inherited exclusively from the mother, and the Y chromosome, which is only inherited paternally), DNA testing can be used to determine the genetic relationship between individuals. Another important feature is that DNA is an extremely stable molecule, and that over time, if properly stored *in vitro*, the order of its constituent units does not change, so a suitably deposited sample can be used to compare its DNA profile with the profile of another sample taken years later. As noted above, a small portion of nuclear DNA carries the genetic message, encoded in the *genes*. In other words, genes are active segments located at specific places (*loci*) in DNA strands. Because of the normal functioning of genes, each individual has certain parameters of growth and development of the organism (e.g., it is genetically specified that a person has two hands and two feet), but also some other traits, such as eye color and height. Thus, it is obvious that DNA represents the central molecule of life, which—as the primary (main) carrier of hereditary information—controls the growth and development of every living being. Total nuclear DNA is almost 2 m long and is located in structures called *chromosomes*. The word chromosome comes from two Greek words—*chromos* meaning color and *soma* meaning body. From a total of 46 chromosomes found in the somatic cells of each individual, 23 is inherited from the mother and 23 from the father. In particular, during the formation of zygote, 23 chromosomes come from the egg, and 23 from the sperm cell. Chromosomes are found in the nucleus in pairs, and humans have a total of 23 pairs of chromosomes. One of these pairs is sex chromosomes (X, Y), which are important in determining the gender of each individual and how they differ from one another. The remaining 22 pairs are called *autosomes*, and they are homologous in both sexes. Excluding possible chromosomal abnormalities, in general, the rule is that two X chromosomes (XX) determine the female, and the combination of X and Y chromosomes (XY) determines the male sex. It is important to note that all somatic cells (e.g., bone cells, skin cells, white blood cells) contain 46 chromosomes (23 pairs), whereas gametes (sperm and ovum) contain half as much, or more precisely, 23 chromosomes. A large number of scientific studies have shown that there are certain genes located in mtDNA, so the study of their functions has become very intense in recent years (Report of the Committee on the Human Mitochondrial Genome 2009). The physical location of a gene on a chromosome is called a *locus*. Two autosomes that make a pair match each other both by structure and function. This is why two such chromosomes, which are similar in structure and carry the same genes, are called *homologous chromosomes*. Furthermore, variants of genes that occupy the same position or locus on homologous chromosomes and determine different forms of the same genetic trait are called *alleles*. Alleles are, in fact, alternative forms of the same gene or genetic loci with differences in sequence or length (Primorac and Marjanović 2008; Primorac and Paić 2004). The following terms, often used in genetics, including forensic medicine, are *homozygote* and *heterozygote*. Homozygote indicates the genotype (genetic version) that has identical alleles or variants of the same gene at a particular locus on a pair of homologous chromosomes. On the contrary, the heterozygote genotype indicates the person or genotype that has two different alleles at a particular locus on a pair of homologous chromosomes (Figure 1.2).

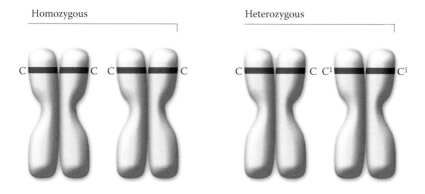

Figure 1.2 Example of homozygous and heterozygous loci on chromosomes. Note: Chromosomes are represented as sister chromatid before cell division, so each chromosome is represented twice, and as such, each allele is represented twice. Once division has occurred there will be only two copies of each nuclear gene.

1.2.4 Deoxyribonucleic Acid

DNA molecule is constructed in the form of double-stranded helix, and it is constituent of all 46 chromosomes. DNA consists of units called *nucleotides*, and the human genome (haploid) contains approximately 3.2×10^9 of such units. The nucleotide itself is composed of three subunits: a five-carbon sugar, a phosphate group and the nitrogen containing nucleotide bases: *adenine* (A), *guanine* (G), *cytosine* (C), and *thymine* (T). It is good to remember that in a double helix adenine always pairs with thymine (A–T), and cytosine with guanine (C–G). Under normal circumstances, base mating under another scheme is not possible. This occurs because adenine and guanine are a chemical structure called a *purine*, whereas thymine and cytosine are a chemical structure called a *pyrimidine*. The purine and pyrimidine can form complementary structure held together by hydrogen bonds. There are two hydrogen bonds between adenine and thymine and three between guanine and cytosine holding the two strands of nucleotides in DNA together (Figure 1.3) and provide exceptional stability to this molecule. Every single contact between these units is called a *base pair* (bp), and the entire human genome (haploid) has about 3.2 billion bp. Changes in these bases or variations in the number of base pair repetitions are the basis for personal identification (Marjanović and Primorac 2013).

1.2.5 Genetic Diversity

Genes (protein coding sequence) make up only about 2% of the human genome. However, the rest (noncoding DNA sequence) do have important biological functions including the transcriptional and translational regulation of protein-coding sequences. The Encyclopedia of DNA Elements (ENCODE) underlined that gene's regulation is influenced by multiple stretches of regulatory DNA located both near and far from the gene itself and by strands of RNA not translated into proteins, the so-called noncoding RNA (Pennisi 2012). In many ways, recently published data are changing genomics concepts that are written in current textbooks. However, just as genes can contain genetic variation, non-coding DNA contains many types of genetic variation. These include differences in a single nucleotide referred to as *single nucleotide polymorphisms* (SNPs), repeated sequences of DNA that can range in size from 1 to 70 bp called *variable number of tandem repeats* (VNTR; Inman and Rudin 1997; Budowle et al. 1991, 1994; FBI 1990), and insertion and deletion polymorphisms (InDels) that can vary between 1 and thousands of bp. VNTR loci have proven to be the most useful forensically. One of the methods for analyzing VNTR is through *amplified fragment length*

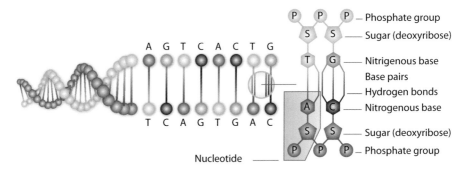

Figure 1.3 Structure of DNA molecule from double helix to base pairs.

Basic Genetics and Human Genetic Variation

polymorphism (Latorra et al. 1994; Budowle et al. 1992), but one can also use other methods (which will be explained later). Some VNTR regions code for three base pairs and are the same as the sequence of three bases coding for an amino acid in proteins. The three base pair code for amino acids is called a *codon*. If the number or order of these sequences is disrupted in a gene, the structure can be disrupted, which is one of the prerequisites for the development of genetic diseases. Examples of two diseases that involve trinucleotide repeats are mental retardation associated with the X chromosome (fragile X) and Huntington's disease. Most of the VNTR loci are located in the parts of DNA that do not encode proteins or in *introns*—noncoding parts of genes. Given the high degree of molecular diversity, VNTR loci have long been used in forensic analysis, in which one or more of these loci are examined. Initially, the larger VNTR loci (30–70 bp repeats) were used forensically; however, because the procedure was not sufficiently sensitive, could not be done on degraded samples, and had many technical problems that affected the final results, these markers were replaced by shorter VNTRs detected by the PCR reaction in forensic DNA analysis (Goodwin et al. 2011).

Discovery and application of new, smaller molecular markers have significantly shortened the process. These markers are called *short tandem repeats* (STRs), and they would be explained in more detail in following sections of this chapter (Butler 2011).

1.2.6 Variability of DNA

Analysis of DNA in forensic science is based on the fact that only 0.5% of the DNA varies in every person. Nevertheless, that small part of DNA contains a great number of so-called *polymorphisms* (Greek: *poly* for many, *morfoma* for form), or differences in DNA sequence among individuals, and therefore we can almost certainly claim that each of us has a unique genetic material, with the exception of identical twins. The fact that we are genetically different is the basis of analysis of evidence found at the crime scene, identification of the victims, identification of rape offenders, or routine determination of kinship. The majority of loci that encode proteins have only one form of the gene. This is because most genes are not tolerant of mutations. Those genes that tolerate mutations have more than one form, that is, have their alleles. Loci containing alleles with relatively high frequency are called *polymorphic loci*. Genetic variation in blood groups, serum proteins, and transplantation antigens on the protein level is a reflection of that particular polymorphism, namely, the variability at the DNA level. Advances in DNA technology enabled the detection of variability (polymorphisms) in specific DNA sequences. Figure 1.4 shows an example of gene polymorphisms (i.e., sequential polymorphism) and displays the existence of two alleles for one gene where the change in a base pair is framed.

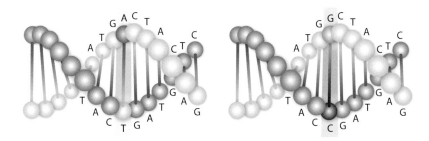

Figure 1.4 Example of single nucleotide polymorphism (SNP) on DNA double helix.

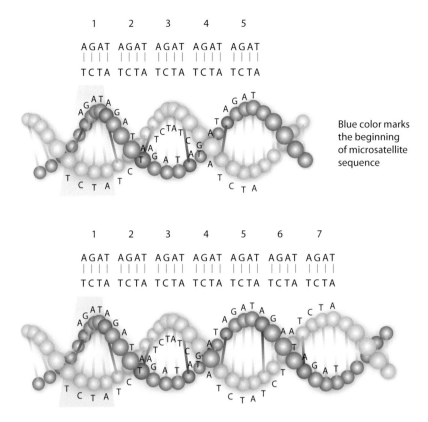

Figure 1.5 Example of heterozygote for short tandem repeat (STR) variable number tandem repeat (VNTR) with five and seven repeats respectively on DNA double helixes.

Other forms of polymorphisms are changes in the length of DNA between two homologous DNA segments (Figure 1.5).

During laboratory testing, the differences in alleles within a specific segment of DNA must be shown with a plain and understandable method that will clearly demonstrate the difference in the length of DNA between the two microsatellite alleles.

1.2.7 Structure and Nomenclature of STR Markers

The development of the Human Genome Project, which marked the beginning of the comprehensive studying and mapping of human genes, has imposed a need for developing guidelines on the identification and designation of new genes, and the inclusion of old, existing systems. The very beginning was the establishment of the International System of Gene Nomenclature in 1987 (Shows et al. 1987; Primorac et al. 2000). Meanwhile, the DNA Commission of the International Society for Forensic Human Genetics issued specific recommendations regarding the use of *restriction fragment length polymorphism* (RFLP) and PCR (1991 Report of the DNA Commission; Mayr 1993).

There are thousands of STRs that can be used for forensic DNA analysis. STR markers consist of sequences of length 2–7 (according to some sources, 1–10) bp that are repeated numerous times at the locus. However, the majority of loci used in forensic genetics are tetranucleotides repeats (Figure 1.6), which include a 4-bp repeat motif (Goodwin et al.

Basic Genetics and Human Genetic Variation 13

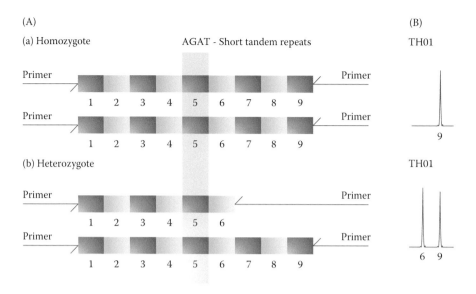

Figure 1.6 (A) Amplification and (B) electropherograms of the STR TH01 in homozygous and heterozygous samples.

2011). Figure 1.6 shows the homozygote genotype (a) for TH01 locus with two identical alleles with nine repeats, whereas the heterozygote genotype (b) indicates that the person has two different alleles, one with six and second with nine repeats. The number of repeats varies from person to person in a pseudo-normal distribution with usually one common allele and increasing and decreasing numbers of repeats moving away from the most common allele. Because of this, it is not uncommon for two individuals to share the same allele at the observed STR locus or even two alleles at a locus, but the least theoretical probability that there are two unrelated people with identical profile at say 15 STR loci represented in the PowerPlex 16 system for Caucasian population is $1/1.83 \times 10^{17}$ (Sprecher et al. 2000).

However, the real value of the application of these markers lies in the simplicity and rapidity of the process and the possibility of simultaneously testing of a large number of STR markers in multiplex STR systems, enabling an extremely high degree of individualization in identifying biological evidence. In addition to its wide application in forensic DNA analysis, STRs have become very attractive as a subject of genetic research from a medical point of view, because it was shown that the trinucleotide STR loci, in the case of their hyperexpansion, are associated with certain genetic disorders (Read and Donnai 2011).

Tetranucleotide STR loci, that is, those STR markers that have four bases in the repeat sequence are the best studied and most frequently used in the analysis of individual and population diversity. Also, several trinucleotide and pentanucleotide systems have found practical applications. The tetra and penta loci are often included in commercial multiplex analysis systems. Such systems provide the results with high index of exclusion. Desirable characteristics of a STR locus are

1. High heterozygosity
2. Clearly defined repeat sequences
3. Clearly defined allelic variants
4. Simple and reliable amplification

There are several types of STRs (Urquhart et al. 1994):

1. Simple consisting—one repeating sequence (*D5S818*, *D13S317*, *D16S539*, *TPOX* [Figure 1.7a], *CSF1PO*, etc.)
2. Simple with nonconsensus alleles—one type of repeating sequence (*TH01* [Figure 1.7b], *D18S51*, *D7S820*, etc.)
3. Compound consisting—two or more different repeat sequences (*GABRB15*)
4. Compound with nonconsensus alleles—two or more different repeat sequences (*D3S1358*, *D8S1179*, *FGA*, *vWA* [Figure 1.7c])
5. *Complex repeats*—more repetitive sequences with the presence of DNA insertions (*D21S11*, etc.) (Figure 1.7d)
6. Hypervariable repeats (*SE33*)

In October 1993, The DNA Commission of the International Society of Forensic Haemogenetics (now called the International Society of Forensic Genetics [ISFG]) recommended a nomenclature of STR loci and the allelic variants that is still in use today. The name of an STR contains the number of chromosome on which it was found; so, for portions of noncoding DNA, such as D21S11, the name indicates that it is a DNA sequence (*D*) placed on the 21st chromosome that is a "single copy" or a sequence present in only one copy, at only one site in the genome (*S*), and it is the 11th discovered and categorized marker on the 21st chromosome. Instead of the letter S that stands when the sequence occurs in only

(a) TPOX
6    $[AATG]_6$
7    $[AATG]_7$
8    $[AATG]_8$
9    $[AATG]_9$

(b) TH01
8      $[AATG]_8$
8.3 $[AATG]_5 ATG [AATG]_3$
9      $[AATG]_9$
9.3 $[AATG]_6 ATG [AATG]_3$
10     $[AATG]_{10}$
10.3 $[AATG]_6 ATG [AATG]_4$

(c) VWA03
10   TCTA TCTG TCTA $[TCTG]_4 [TCTA]_3$
11   TCTA$[TCTG]_3 [TCTA]_7$
12   TCTA$[TCTG]_4 [TCTA]_7$
13   $[TCTA]_2 [TCTG]_4 [TCTA]_3$ TCCA$[TCTA]_3$

(d) D21S11
29     $[TCTA]_4 [TCTG]_6 [TCTA]_3$ TA $[TCTA]_3$ TCA $[TCTA]_2$ TCCA TA $[TCTA]_{11}$
29     $[TCTA]_6 [TCTG]_5 [TCTA]_3$ TA $[TCTA]_3$ TCA $[TCTA]_2$ TCCA TA $[TCTA]_{10}$
29.2 $[TCTA]_5 [TCTG]_5 [TCTA]_3$ TA $[TCTA]_3$ TCA $[TCTA]_2$ TCCA TA $[TCTA]_{10}$ TA TCTA

Figure 1.7 (a) Simple repeat sequences in loci such as TPOX, CSF1PA, D5S818, D13S317, D16S539, Penta D, and Penta E. (b) Simple repeat sequences with nonconsensus repeats in loci such as TH01, D18S51, and D7S820. (c) Compound consisting of two or more different repeat sequences with nonconsensus repeats in loci such as VWA03, D3S1358, D8S1179, and FGA. (d) Complex repeats consisting of two or more different repeat sequences with nonconsensus repeats and DNA insertions found in loci such as D21S11. (From Van Kirk, M.E., McCary, A., and Podini, D., 2009. In *Proceedings of the American Academy of Forensic Sciences, Annual Scientific Meeting*, February 16–21, Denver, CO, Abstract A134, p. 118. With permission.)

one place, the letter *Z* is the label used if the sequences occur in several places (Primorac and Marjanović 2008). Nomenclature of allelic variants is based on the number of their repeat units. If allelic variant contains nine repeating sequences, then it is indicated by the number 9. If allele has no common repeating motif, that is, if it contains nine repetitive sequences and another series of, for example, three bases, then the allelic variant is marked *n.m*, that is, 9.3, where *n* [9] is the number of complete repetitive sequences and *m* [3] is the number of bases of incomplete repeating sequence that allelic variant includes.

STR loci located within introns of coding regions are given the name of the gene such that the CODIS (Combined DNA Indexing System) loci CSF1PO, FGA, TH01, TPOX and vWA. For example, one of the highly informative STR markers TH01 was named after the gene in which the system belongs: tyrosine hydroxylase gene, which is positioned on the 11th chromosome, which cannot be seen from its name. Designation 01 comes from the fact that was the first variant found at this locus. Often used in the literature is its supplemented name with prefix "HUM" (full name: HUMTH01), which further clarifies that it is part of the gene for *human* tyrosine hydroxylase. It should be noted that vWA is an incorrect identifier, and the correct designation should be VWA03, as it was the third STR found at the VWA locus. According to the system of gene nomenclature, abbreviations such as VWA are all in capital letters (Shows et al. 1987).

1.2.8 Analysis of Sex Chromosomes

Analysis of the sex chromosomes is important in the determination of gender and instantly excludes 50% of the population. The human gene usually analyzed to determine gender is the Amelogenin (AMEL) locus, after its amplification using PCR DNA fragments of different lengths can be generated. The sequence on the *X chromosome* is shorter by 6 bp compared to the allele on the *Y chromosome* (male sex). The results of gender determination using Amelogenin are presented in Figure 1.8 (Primorac and Marjanović 2008).

1.2.8.1 Y Chromosome DNA Testing

According to the criteria of Denver convention, the human *Y chromosome* belongs to group G chromosomes, that is, the category of the shortest chromosomes in human genome, in which the 21st and 22nd chromosomes also belong. The Y chromosome contains about 50 million bp, representing about 1.8% of the entire human genome. Y chromosome can provide important information when it comes to determining the lineages of a specific man. This is possible because the Y chromosome contains highly polymorphic regions (Jobling

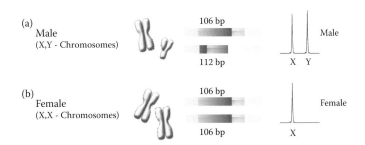

Figure 1.8 Schematic drawing of (a) male and (b) female Amelogenin locus at the chromosome, amplicon, and electropherogram levels.

et al. 1999, 2004). The human Y chromosome is present in normal males in a single copy and is passed from father to son; 95% of the chromosome is not subject to recombination and is referred to as NRY. Only 5% of the chromosome can potentially recombine with the X chromosome, and this region is called the pseudo-autosomal region of the X and Y chromosomes. Forensic analysis of the Y chromosome can play an important role in rape cases, particularly those involving more than one male (Kayser and Sajantila 2000; Prinz and Sansone 2001). These markers can also be useful in cases of paternity testing of male children, and in the process of identification, when relatives only from the father's side are present. Additionally, the Y chromosome is increasingly used in determining the migration routes of some people in the past, because the Y chromosome do not undergo recombination during the generational transfer of genetic material (Semino et al. 2000). In recent years, a considerable number of population studies have been conducted using the NRY of the Y chromosome (Y Chromosome Consortium 2002), both worldwide and in Europe. Besides offering very interesting models and scenarios of human history in this region (Barać et al. 2003; Marjanović et al. 2005; Primorac et al. 2011), these studies promoted the utilization of *SNP markers*, which are based on substitution, at only 1 bp. SNPs continue to be explored as potential supplements to STR markers already in use but will probably not replace STRs in the near future. These markers will be discussed further in subsequent sections of this chapter. Furthermore, analysis of Y STR loci plays an important role in rare but important cases where there is a lack of Amelogenin gene in man (Prinz and Sansone 2001).

From the standpoint of jurisprudence, it is important to emphasize that the identification and possible matching of Y STR DNA profiles from two evidence samples, regardless of how many molecular markers were analyzed, does not mean complete individualization. That is, the Y chromosome is inherited by the male line—from father to son—so every male related through "father parental line" will share the same profile. Markers found on the Y chromosome exhibit less diversity than those on other chromosomes because 95% of this chromosome is not subject to recombination. Consequently, STR diversity on Y chromosomes results solely from mutation. Therefore, the most common interpretation of the results in court, in a case when the Y STR profile of two samples match, is that this person cannot be excluded as a potential biological source of the analyzed evidence material. Usually, the DNA profile report contains the relative frequency of occurrence of the observed Y STR profile (i.e., haplotype) in a specific, regional, or global population. Statistical interpretations of Y chromosome markers are discussed in Chapter 2.

One indicator of seriousness and importance of the Y chromosome analysis in forensics is a presence of several commercial multiplex systems that allow simultaneous investigation of more than 10 STR loci, such as PowerPlex® Y with 12 and the recently released Prototype PowerPlex® Y23 System with 23 loci (Davis et al. 2013), both manufactured by Promega Corporation. The PowerPlex® Y23 System provides all materials necessary to amplify Y-STR regions of human genomic DNA. Furthermore, the PowerPlex® Y23 System allows coamplification and four-color fluorescent detection of 23 loci, including DYS576, DYS389I, DYS448, DYS389II, DYS19, DYS391, DYS481, DYS549, DYS533, DYS438, DYS437, DYS570, DYS635, DYS390, DYS439, DYS392, DYS643, DYS393, DYS458, DYS385a/b, DYS456, and Y-GATA-H4 (Figure 1.9). The AmpFLSTR®Yfiler PCR Amplification Kit, manufactured by Life Technologies (Applied Biosystems), allows typing of 17 Y STR loci, and Investigator Argus Y-12 QS Kit (made by Qiagen) allows typing of 12 Y STR loci (only available in Europe).

Basic Genetics and Human Genetic Variation

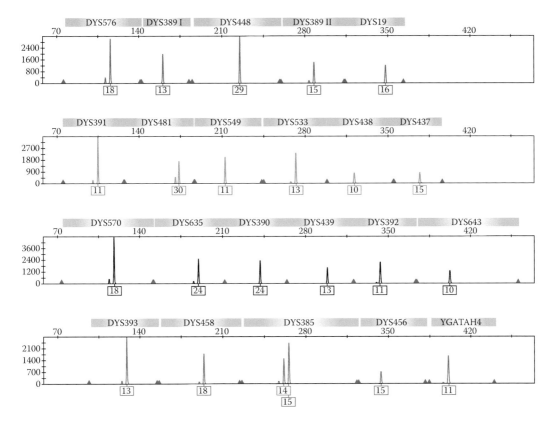

Figure 1.9 Electropherogram of Powerplex Y23 (23 Y chromosome SNPs) PCR amplification kit (Promega Corporation). Note: Five-color detection system (four colors for loci and fifth color for sizing ladder, not shown).

1.2.8.2 X Chromosome DNA Testing

In recent years, more importance has been given to the application of the human X chromosome, both in forensic and population genetics (Szibor et al. 2003). Unlike the Y chromosome, such an approach can be effective in cases of paternity testing of female offspring in conditions where the potential father is not present and results can be obtained by testing his relatives. In addition, the results can be achieved in the analysis of motherhood, and most effectively in the analysis of mother–son relationship, which really is rare, but is still carried out for different purposes.

Application of this marker is still relatively limited, and it is mainly based on the combined use of X-related markers with common autosomal loci (Pereira et al. 2007). One reason for this limitation was the insufficient number of molecular markers examined on the X chromosome, as well as limited population data and calculation of forensic statistics. An increasing number of research groups are looking into this problem, so it is logical to predict that the application of this intriguing chromosome will soon become much more intense. However, it has been showed that X STRs can distinguish pedigrees that are otherwise indistinguishable using only unlinked autosomal markers (Pinto et al. 2011). (See Chapter 2 for comments on forensic calculations using the X chromosome markers and Chapter 5 for a review of X chromosome markers.)

1.2.9 Mitochondrial DNA

mtDNA—or, as some call it, "cytoplasmic chromosome"—is also a form of DNA that is analyzed and used in forensics and makes up approximately 1% of the total cellular DNA. The mtDNA will be discussed in further detail in the next section, but here we will briefly point to some of its most important characteristics. An mtDNA molecule is 16,569 bp long, circular in shape, and does not contain intron sequences. Two highly polymorphic noncoding regions are especially important for forensic analysis: HV1 and HV2. Unlike nuclear DNA, mitochondria and their DNA come from the cytoplasm of an oocyte that contributed to the development of zygote and therefore are of maternal origin. Therefore, mtDNA is inherited from the mother and indicates the female ancestors of an individual. In contrast to nuclear DNA, mtDNA may be present in a few thousand copies depending on the energy requirements of the tissue. Because mtDNA molecules replicate independently of each other, unlike nuclear chromosomes in which the replication of individual chromosomes occurs first, while pairing and recombination of genes happens in meiosis 1, there is no mechanism for recombination of mtDNA. Mutations that can lead to changes in the sequence of base pairs are the only source of variability in mtDNA. Because mtDNA is inherited from the mother, a person cannot be heterozygote, and this fact is useful in determining maternal lineages within families and populations. mtDNA is mainly used in forensic cases where it is necessary to analyze the evidence that does not contain enough nuclear DNA (Holland et al. 1993; Coble et al. 2004). mtDNA is also found in tissues that do not contain nuclei, such as a strand of hair. However, one of the biggest problems in working with mtDNA from hair is the presence of two or more subpopulations of mtDNA in one individual (heteroplasmy). Point heteroplasmy (PHP) is observed in approximately 6% of blood and saliva samples and more frequently in hair and metabolically active tissues such as muscle (Irwin et al. 2009). Heteroplasmy is most likely the result of frequent errors or mutations in mtDNA replication, which are rarer in nuclear DNA. The reason may be that the mtDNA molecules replicate independently of each other and are not strictly related to meiotic or mitotic cell division, which has proofreading functions. Since each cell contains a population of mtDNA molecules, a single cell may contain some molecules that have a particular mtDNA mutation that others do not have. It is possible that this phenomenon is responsible for the expression of different diseases that are inherited through the mitochondria. However, it is also possible that the number of mutated mtDNA molecules changes during the segregation in cell division, while the cells are multiplying and the number of mitochondria increases. Heteroplasmy is important for forensic purposes because it can help in the forensic examination of identity, but it can also make the analysis very complex. Still, it is obvious that heteroplasmy represents a new level of variability, which in most cases can increase the accuracy of mtDNA testing. Additional information on this issue may be found in an excellent paper by Holland and Parsons (1999). MtDNA has an important role in identifying human remains, especially skeletal, as well as in cases of decomposed bodies. One of the most famous cases solved by using this technology was the identification of Czar Nicholas II, when it was confirmed that he had the same heteroplasmy as the remains of his brother, Georgij Romanov, Grand Duke of Russia (Ivanov et al. 1996). Analysis of mtDNA is nowadays performed using the common method of sequencing, and the obtained results are compared with so-called revised Anderson sequence (International Society for Forensic Haemogenetics; Andrews et al. 1999). In recent years, *sequence-specific oligonucleotide probe* (SSOP) analysis, in which the amplified DNA hybridizes with the existing probes previously bound to the nylon or another membrane, has been applied to mtDNA testing. Previously, it had been used for DQA1 and Polymarker typing (see below). Using this analysis method, the time needed to examine a large number of specimens is significantly shortened (Gabriel et al. 2001); however, the level of individualization is lower than that generated by DNA sequencing, because a limited numbers of polymorphic sites are surveyed (Škaro et al. 2011). Given the large number of copies of mtDNA in the cell, the sample can easily be contaminated with foreign mtDNA if handled recklessly.

1.2.10 RNA Profiling

RNA differs from DNA in several respects: it has the sugar ribose in place of deoxyribose, it has the base uracil (U) instead of thymine (T), and it usually occurs in a single-stranded form (Tamarin 2002). Messenger RNA (mRNA) is a large family of RNA molecules that convey genetic information from DNA to ribosome. During transcription, the DNA serves as a template and an enzyme called RNA polymerase II catalyzes the formation of a pre-mRNA molecule, which is then by a process called splicing processed to form mature mRNA. Ribosomes link amino acids together in the order specified by mRNA and generate new proteins. Eukaryotes have segments of DNA within genes (introns) that are transcribed into mRNA but never translated into protein. The segments of the gene between introns that are transcribed and translated and hence exported to the cytoplasm and expressed to proteins are called exons. The normal mechanism by which introns are excised from unprocessed RNA and exons spliced together to form a mature mRNA is dependent on particular nucleotide sequences located at intron–exon (acceptor site) and exon–intron (donor site). It has been known for years that either improper splicing or inadequate mRNA transport can cause cancer or numerous diseases (Kaida et al. 2012; Stover et al. 1994; Primorac et al. 1994, 1999; Johnson et al. 2000). However, alternative splicing can play a role even before life and after death (Kelemen et al. 2013). Identification of the tissue of origin for biological stains continues to be pursued using RNA profiling with reverse transcription (RT) and point PCR and real-time PCR (Brettell et al. 2011). Recently, the European DNA Profiling Group (EDNAP) performed a collaborative exercise involving RNA/DNA coextraction and showed the potential use of an mRNA-based system for the identification of saliva and semen in forensic casework that is compatible with current DNA analysis methodologies (Haas et al. 2013).

1.2.11 Application of New Molecular Markers

Forensic genetics is an extremely dynamic scientific discipline, and one of its basic features is certainly the almost daily evolution in terms of discoveries of new procedures, molecular markers, or improvement of existing systems so as to enhance their utility value. One of the newest approaches that is slowly finding its application in forensic DNA analysis is the use of the SNP *molecular markers* (Figure 1.10). These are the most abundant forms of DNA polymorphisms that occur within the human genome. In recent years, especially with the discovery and use of the new technologies, these features of nucleic acids have achieved full recognition not only in population-genetic research, but also in medical diagnostics and testing of identity. The mainstream of SNP usage in forensic purposes could be recognized in analysis of highly degraded DNA and novel phenotyping approach.

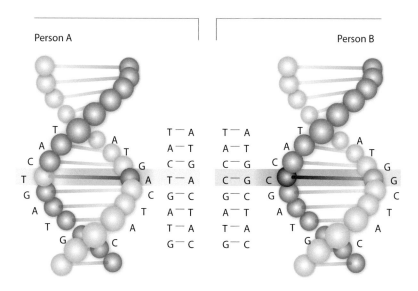

Figure 1.10 Example of DNA sequence variation between two different individuals A and B.

As already noted, this polymorphism is based primarily on the substitution of only one base in a standard order of observed sequences. Although less polymorphic compared to the now widely applied STR markers, their presence in almost every 1000 bp makes them extremely informative. One of the primary directions for development of testing procedures of these markers is directed to an additional analysis of mtDNA (Parsons and Coble 2001). In fact, as already mentioned in previous sections, mtDNA has three main features that give it the status of an extremely favorable molecule for use in DNA identification: (1) successful isolation from very degraded samples, (2) significant degree of informativeness when the reference samples are only those from the maternal side, and (3) application, when the possibility of using nuclear molecular markers does not exist. However, recently, The Spanish and Portuguese Speaking Working Group of the International Society for Forensic Genetics (GHEP-ISFG) during collaborative exercise in order to monitor the current practice of mtDNA reporting, noticed a 10-fold range of reported likelihood ratio (LR) values, mainly due to the selection of different reference datasets in EDNAP Forensic mtDNA Population Database (EMPOP) but also due to different applied formulae. Prieto et al. (2013) suggested that more standardization and harmonization of the mtDNA reporting is needed. On the other hand, its application can be limited because of the extremely low discriminatory power. The analysis of a large number of SNP markers positioned not only on the hypervariable region but also in the constant region can increase the discriminatory power, and therefore, may give a whole new dimension to the application of mtDNA in forensic genetics.

Also, one of the new interesting approaches in forensic genetics, which is still on its preliminary stages, is using InDels. These genetic markers are mostly considered short amplicons, and their usage could be suited for degraded DNA. Another relatively new approach in forensic DNA testing already provides extremely useful and promising results. The focus is not on the application of some new molecular markers, but rather on a different approach to the analysis of already existing ones. The so-called miniSTR molecular markers represent the modified version of existing standard STR DNA sequences, which are primarily based on the moving the forward and reverse PCR primers in closer to the

STR polymorphic region. In this way the total molecular mass, that is, the size of these markers, is reduced so there is a higher likelihood for the successful amplification of samples in cases when analyzing highly degraded DNA. These markers have already been used in the analysis and identification of victims of terrorist attacks on the World Trade Center in New York. The current research objective is to develop and optimize a miniSTR multiplex system that will enable the simultaneous analysis of a large number of these markers, which has yielded the Minifiler kit (Life Technologies) (Coble and Butler 2005).

A final area of research is the identification of SNPs that can be used as *Ancestry Informative Markers* (AIMs) and *Phenotype Informative Markers* (PIMs). AIMs provide information about the geographic population of origin, such as Europeans, Africans, and East Asians, of a DNA profile that does not match any samples in the DNA databases. This is of less use in ethnically homogeneous populations but is very useful in countries with highly diverse populations. Similarly, phenotypic informative markers provide information about skin, eye, and hair color, as well as some other traits. Our recent study on over 5000 individuals from Europe and China revealed very strong association of some IgG glycans with age and currently we are validating IgG glycosylation analysis for possible forensic applications (Primorac et al. 2013). At present, these markers and therefore multiplexes are still the subject of research. (See Chapter 16 for information on these markers.)

1.3 Potential Biological Sources of DNA

Forensic laboratories receive different types of biological evidence for testing. Evidence that can be tested using some of the methods of DNA analysis (with the exception of the mtDNA) is limited to those containing cells with a nucleus. In this sense, it is possible to successfully isolate and analyze DNA from the following biological sources:

1. Whole blood and blood cells
2. Semen and sperm cells
3. Tissues and organs
4. Bones and teeth
5. Hair roots and dandruff
6. Saliva, urine, feces, and other bodily secretions
7. Epithelial cell found on clothes

Biological evidence that lacks nuclei such as a sweat (if there are no epithelial cells), tears, or hair shafts cannot be tested using standard DNA analysis. DNA can be extracted even from materials such as gastric juice and feces. However, it is sometimes difficult to obtain sufficient amount of DNA from these sources. It should be also noted that, although the good results of DNA analysis can be obtained from the listed evidence, in many cases the quality and/or quantity of the sample turn out to be unsuitable for DNA analysis. The success of the detection of a DNA profile from a certain item of evidence primarily depends on three basic conditions:

1. Amount of sample. Methods of DNA analysis, especially PCR are very sensitive, but still have limitations.
2. The level of degradation of DNA. If an item of evidence, even a large blood stain, is exposed to the harsh outdoor conditions for a prolonged time there can be enough

DNA degradation due to environmental insult or bacterial contamination that makes the sample unsuitable for further analysis.
3. Purity of the sample. Sometimes, dirt, grease, some fabric colors, and similar inhibitors of certain phases of DNA analysis (mainly amplification) can seriously affect the performance of DNA analysis.

1.3.1 Basic Models and Steps of Forensic DNA Analysis

Forensic DNA analysis takes place in several successive, interrelated steps, which are defined by clearly specified procedures.

1.3.2 Collecting and Storing Samples

The development of modern DNA methods made it possible to analyze biological evidence with a small quantity of DNA or highly degraded DNA; however, extraordinary precautions must be taken while handling the evidence because of the possibility of contamination. The collection, storage, and transfer of such biological evidence are the initial—but also the most critical—phases of a successful implementation of the DNA testing. If the evidence is not properly documented, collected, and stored, it probably will not be accepted in court. The documentation of evidence must be detailed and must follow all guidelines for the use of these materials in such purposes. All collected samples must have a label with its number, date, time, place, and name of the person who collected the sample. If applicable, the case identification number should also be recorded on the evidence. A detailed description of how to collect and preserve evidence for DNA testing can be found in the work of Lee and colleagues (1998). Thus, we will now mention only a few basic rules that must be followed when collecting biological evidence intended for DNA analysis (Primorac and Marjanović 2008):

- Any biological evidence found in liquid or moist condition must first be dried, then packed, and transported to the place of its analysis.
- Biological evidence should never be permanently packed in plastic (PVC) packaging. This form of packaging may be used only for short-term transport to the place where it will be dried. Such transport must be quick to avoid accelerating the process of (irreversible) degradation (progressive destruction) of the evidence.
- In the process of collecting biological evidence, it is of the utmost importance to use sterile (latex, powder-free) gloves. Gloves must be changed when collecting more than one item of evidence, following the principle: one pair of gloves—one biological evidence! This way, the possible contamination of collected biological material will be avoided.
- During the collection of biological evidence, it is recommended (implied as required) to wear face masks covering mouth and nose. In some cases, it is necessary to wear a jumpsuit with a hood to prevent any potential contamination of biological evidence (e.g., by hair), but also for the personal safety of crime scene technicians and investigators from a possible source of infection.
- If one does not have a mask (for objective or subjective reasons), possible contamination of the collected evidence can be well avoided by eliminating the talking, coughing, and preventing other nonessential personnel from entering the scene during the sample collection.

- When collecting a larger number of biological evidence samples from a single location, care must be taken to process each item of evidence individually, that is, each of them should be collected using new, sterile material and then packaged separately.
- When sending biological evidence to the laboratory, it is recommended to also provide, if possible, a sample of the adjacent material (or part of the material) from which the evidence was collected (clothes, smaller objects, etc.) to provide substrate controls. Substrate controls aid in the interpretation of results as they can verify that the DNA profile came from the stain and not the substrate.
- Packaging in which collected evidence is placed must be clearly and legibly marked with understandable labels that are more extensively described and recorded in the accompanying transmittal forms. Labels on the packaging and labels in the accompanying communication must match!
- When collecting samples it is obligatory to use a sample collection kit.
- The collection, packaging, and storage of these and other biological evidence will be further discussed in Chapter 9.

1.3.3 Determination of Biological Evidence

Forensic biological evidences may appear in different forms and states. Nearly every evidence sample has to be seen as an individual phenomenon and carefully examined before it is submitted for DNA analysis. There are a number of different methods for the isolation of DNA from a wide range of biological evidences. Most of these procedures are adapted and optimized for a particular type of evidence, for example, blood stain, semen stain, saliva from a cigarette butt or chewing gum, epithelial cells collected from the clothing.

In order to select an appropriate procedure for DNA isolation, it is necessary to assess the type of biological evidence in question. It is not rare that evidence, collected at the crime scene, is submitted to the laboratory without prior determination whether it is of biological origin, and if so, what type of evidence is it. Such pieces of evidence are usually described in the related documentation as evidences that resemble the blood, semen, saliva, human hair, etc. Sometimes, biological evidences are invisible to the naked eye, and their existence can only be assumed. In rape cases, parts of clothing items that can potentially hold the evidence of semen are often examined. Given that it would be extremely expensive, complicated, and time-consuming to cut the entire garment into small fragments and process them separately, it is essential to preliminarily examine the item in order to verify and locate potential biological evidences and determine their origin. Similar procedures are conducted for those items and scenes believed to have been cleaned after a crime in order to conceal the evidence (usually blood).

1.3.3.1 Blood

Blood is a liquid tissue composed of watery plasma and cells immersed in it: red blood cells (erythrocytes), white blood cells (leukocytes), and platelets (thrombocytes). From the standpoint of DNA analysis, red blood cells have no value because they do not contain DNA. However, they are extremely important in testing for the presence of hemoglobin, which is the basis of most tests for the detection of blood as biological evidence.

The fastest method—but not the most reliable one considering the high possibility of false positive tests—is the detection of blood with specially designed strips that detect the presence of hemoglobin, that is, iron in hemoglobin. This method is commonly used only as a preliminary test because of its nonspecificity. Phenolphtalein (Kastel–Meyer) and Leucomalachite green are generally considered most specific, whereas *ortho*-toluidine and tetramethylbenzidine are the most sensitive (Fisher and Fisher 2012).

One of the most widespread and—thanks to a great number of CSI-type movies and TV series—the most popular method is the luminol (3-aminophthalhydrazide) test. This test allows the rapid investigation of large areas and the possibility of detection of blood that was diluted up to 10 million times (Saferstein 2001). It is only used in the detection of latent (nonvisible) bloodstains. Luminol reacts with the peroxidase activity in hemoglobin, which is manifested as a yellow green to blue glow visible to the naked eye in dim light. This presumptive test indicates the presence/absence of biological evidence that could be further analyzed (Gunn 2006). Luminol does not damage the genetic material and does not inhibit any of the stages of DNA analysis. Today, one can find a large number of commercial kits based on chemiluminescence and on luminol (e.g., Bluestar®Forensic, Monte Carlo, Monaco), which extremely simplified the procedures and does not have any damaging consequences on DNA and allows the naked eye to detect bloodstains down to 1:10,000 dilutions, such as minute traces or droplets that have been washed off, with or without detergent. The luminol method, as it can be concluded from the above, is mostly used when analyzing small or large, but concealed, traces of blood. When it comes to visible evidence that could potentially represent the bloodstains, there are a number of simpler, faster, and cheaper methods for their preliminary testing, such as Kastle–Mayer (phenolphtalein-based) reaction and a commercial test, Heglostix.

The application of Kastle–Mayer reaction involves treating stains with a mixture of ethanol (improves the sensitivity), phenolphthalein (a color indicator), and hydrogen peroxide (an oxidant) (Gunn 2006). The hydrogen peroxide interacts with the heme molecule of hemoglobin and is broken down into water plus free oxygen radicals, which then interact with the phenolphthalein resulting in the solution changing from colorless to pink. Positive (known bloodstain) and negative (distilled water) probes are always used in such analyses.

It must be borne in mind that in certain cases a false positive result can occur. Specifically, hydrogen peroxide can decompose in reaction with other chemicals that can be present, for example, in potato. Hence, modern-day tests are increasingly based on the specific reactions that not only confirm the presence of blood, but successfully differentiate its human origin. These tests can be generally divided into two basic categories: diffusion reactions and electrophoretic methods. It is important to remember that the preceding chemical reagents (Leucocrystal violet and Leucomalachite green) do not allow for DNA typing when blood is found in small amounts (Geberth 2006).

1.3.3.2 Semen

Semen is a complex gelatinous mixture produced in male sexual organs and is ejaculated as a result of sexual stimulation. At least four male urogenital glands produce seminal fluid: seminal vesicle gland, the prostate, the epididymis, and bulbourethral glands. It consists of male sex cells (spermatozoa), amino acids, sugars, salts, ions, and other components produced by sexually mature males. The volume of ejaculate varies from 2 to 6 mL and typically contains between 100 and 150 million spermatozoa/mL (Houck and Seigel 2006). Sperm cells are an interesting biological structure, approximately 55 μm in length, with a head containing DNA and a mobile tail that enables movement.

Semen contains an acid phosphatase, a common enzyme in nature that occurs at a very high level in semen. This feature is a basis of a number of commercial kits for the detection of semen in biological evidence. One of the most common tests is Brentamine Fast Blue B applied to the sample on an alpha-naphthyl phosphate substrate. A piece of sterile filter paper is moistened with distilled water and applied to the examined stain. The Brentamine Fast Blue B reagent is added, and if an intense purple color is visible, the test is positive; if no color reaction occurs within 2 minutes, there is no semen is present in the questioned sample (Houck and Seigel 2006). If a sample is old, the intensity of the color can be decreased or a false negative reaction can occur because acid phosphatase becomes less active over time. On the other hand, false positive reactions are possible since acid phosphatase is not exclusive for semen, but can be found in some amount even in vaginal secretions.

Semen can be visualized by special lamps (wavelength of 450 nm) regardless of whether the stain is on a light or dark surface. This method is performed in the dark, and its major advantage is that it allows quick examination of a relatively large areas.

Confirmatory tests for semen are based either on microscopic visualization of sperm or on the detection of prostate specific antigen (PSA) or p30. A method called ELISA (enzyme-linked immunosorbent assay) was previously used. This method is based on the antibody–antigen reaction and is very sensitive (the detectable level of p30 as low as 0.005 ng/mL) (Gunn 2006). However, nowadays, an increasing number of laboratories use straightforward and rapid commercial immunochromatography assay kits.

From the standpoint of DNA analysis, the most conclusive and the only usable component of the semen for generating DNA profile are certainly spermatozoa. Previous experience has shown that the presence of semen does not always mean that the DNA profile will be obtained from the examined evidence. In particular, the lack of spermatozoa (azoospermia) or their low concentration (oligospermia) can cause the absence or insufficient presence of DNA in the analyzed semen sample.

The most common test for sperm identification is Christmas Tree Stain (CTS) method that stains the tip of the sperm's head pink, the bottom of the head dark red, the middle portion blue, and the tail yellowish-green. An extract of the semen stain is dissolved in distilled water and applied to a microscopic slide. It is then treated with CTS reagents and examined under the microscope at magnifications of 400 to 1000.

A very important parameter in the analysis of semen evidences is the time since ejaculation. Sperm ejaculated into a woman's vagina can live in a woman's cervical mucus or upper genital tract for 3 to 5 days. The life span of sperm after ejaculation depends on the circumstances. The presence of sperm in the vagina, under some circumstances, can be demonstrated after 26 hours, whereas the heads alone can be detected even after 3 days. However, the persistence is greater in the cervix. Intact sperm can be found in the rectum after 65 hours, but rarely survive more than 6 hours in the mouth (heavily dependent on whether the victim is dead or alive) (Marjanović and Primorac 2013).

1.3.3.3 *Vaginal Body Fluid*

Vaginal body fluid can be detected and separated from various other body fluids by microbial (bacteria of the female genital tract) signature detection using a multiplex real-time PCR assay (Giampaoli et al. 2012).

1.3.3.4 *Saliva*

Saliva is commonly found biological evidence. It can be recovered from cigarette butts, chewing gums, stamps, envelopes, bottles, glasses, etc. (Abaz et al. 2002). Detection of saliva is based on the presence of enzyme amylase. The problem is that amylase can occur in many other body fluids, so the test is not nearly specific as for blood and semen. Saliva contains large numbers of epithelial cells from the buccal mucosa and therefore is easy to type for DNA analysis. Today, the most widely used commercial test for detection of saliva is Phadebas amylase test.

1.3.3.5 Urine

Urine is the waste fluid produced by the kidneys through excretion. It is presumptively tested based on the presence of urea (using enzyme urease) or creatine (using picric acid). Urine has few epithelial cells so DNA analysis has to be optimized for potentially minute amounts of sample evidence.

1.3.3.6 Feces

Feces is a waste product from digestive tract expelled through the annus or cloaca during defecation. However, identification of feces is very important in variety of crime investigations. Recently, novel fecal identification method by detection of the gene sequences specific to fecal bacteria (*Bacteroides uniformis*, *Bacteroides vulgatus*, and *Bacteroides thetaiotaomicron*) in various body (feces, blood, saliva, semen, urine, vaginal fluids, and skin surfaces) and forensic (anal adhesions) specimens have been developed (Nakanishi et al. 2013).

1.4 Methods of DNA Isolation

In cells, DNA is not in a pure form, but is associated with many other molecules such as proteins, lipids, and many other contaminating substances. The first step in the isolation of DNA requires that the cell membranes are broken down and the cellular content with molecules of carbohydrate, proteins, lipids, etc., is released. Given that these molecules can interfere with PCR, they must be removed using special techniques. After the cell lysis, the DNA is initially released from the nucleus, and in the next step large quantities of protein are removed. Today, several methods of DNA extraction are routinely implemented, but before giving a brief listing of these methods, it is essential to point out several important facts.

A large number of cases coming to the laboratory for analysis contain minimal amounts of DNA, and any unnecessary manipulation can have a direct impact on the quality and quantity of DNA. All persons involved in the process of DNA analysis must be made aware of the possibility of contamination of the existing DNA with some other DNA. Some of the substances used in the isolation of DNA are potentially harmful to the DNA if not removed completely and as soon as possible. Figure 1.11 shows the schematic representation of some DNA extraction methods applied in forensic genetics.

1.4.1 DNA Isolation with Organic Solvents

In addition to DNA, there are many other molecules in cells, many of which can inhibit DNA process, such as amplification of DNA during PCR. This method removes a large amount of proteins and other molecules released during the extraction of DNA from the cells. Isolated DNA is preserved in large pieces. At the end of the procedure, the DNA is precipitated or concentrated in special tubes.

This method is highly universal, meaning that it can be applied for different types of samples. All the phases of the protocol for organic extraction can be modified, but its

Figure 1.11 Four different methods of DNA extraction: (a) The classic phenol–chloroform method; (b) Chelex®100 resin based extraction; (c) DNA extraction from FTA paper; (d) solid phase DNA extraction methods.

"skeleton" is repeated over and over again through each of these variants. Specifically, in the first phase of this protocol, different amounts of digestive buffer are added. The amount of buffer added depends on the nature and quantity of the sample being extracted.

The composition of digestive buffer can vary, but is primarily based on the presence of chemicals such as TRIS, NaCl, sodium dodecyl sulfate (SDS), and ethylenediaminetetraacetic acid (EDTA). The role of the buffer, as well as all others, comes down to creating conditions in which generalized enzyme proteinase K can act (e.g., perform its function of protein digestion). This way leads to degradation of membrane complexes within cells and releasing the total genomic DNA to "swim freely" in the new suspension. Subsequent mixing of the aqueous phase and the organic phase (phenol–chloroform), followed by centrifugation results in the upper layer (aqueous phase) containing the "purified DNA." The solution of phenol–chloroform–isoamyl alcohol denatures protein and lipid, which becomes an interface between the upper aqueous layer and the lower organic layer (phenol–chloroform). Further washes with chloroform remove an excess of phenol and other contaminants that remain. The next step is the recovery of DNA from the aqueous phase by precipitation. Depending on the quantity and quality of DNA, the precipitation step may be performed using various methods.

1.4.2 Other DNA Isolation Methods

1.4.2.1 Chelex®100

DNA extraction using Chelex®100 is very effective in isolating DNA from samples in which the existence minimal amounts of DNA is expected, and only PCR-based testing will be used on the extracted DNA. It is based on heating the sample in the presence of commercial chelating resin called Chelex®100. After heating, the proteins and fats have been denatured, the DNA is released, and Chelex®100 binds the ions (e.g., Mg^{2+}) present with the DNA in the cell. During the next step, all the molecules associated with Chelex®100 are removed and only pure DNA remains in aqueous solution. This method denatures the DNA double helix, and the method can be applied only if the PCR amplification will be performed afterward.

1.4.2.2 Qiagen DNA Isolation Procedure

The Qiagen DNA isolation procedure is one of the most widely used approaches in the forensic community. It is produced in several varieties optimized for different types of samples. Its working principle is based on the use of chemicals supplied as part of a kit and down to the basic stages of digestion in the presence of proteinase K, transferring the samples to a special silica membrane, purification of bound DNA molecules and, at the end, its elution into special collection tubes. It is noteworthy is that every kit is supplied with a comprehensive handbook containing protocols for DNA isolation from a number of sample sources. These procedures, often with none or some minor changes (including a very recent QIAmp DNA Investigator Kit), give a very good results (high molecular weight DNA that is cleaner than DNA from Chelex®100 extractions) for wide range of sample types.

1.4.2.3 DNA IQ System

The DNA IQ System (Promega Corporation) employs a technology with paramagnetic particles. Those resins are designed to remove PCR inhibitors and contaminants frequently encountered in casework samples. This kit contains the proprietary resin and several specialized buffers, including lysis, wash, and elution. The method demonstrates efficient DNA extraction from small samples and the removal of inhibiting compounds.

1.4.2.4 FTA®

The FTA® concept involves applying a weak base, chelating agent, anionic surfactant or detergent, and uric acid to a cellulose-based matrix (filter paper). A sample containing

DNA could then be applied to the treated filter paper for preservation and long-term storage. There are a number of methods of DNA extraction, which is used in the analysis of known specimens (buccal mucosa or blood stains) collected on an FTA® card. One of the fastest, easiest, and low-priced options consists of successive rinsing with DNA-free water, centrifugation, and shaking of isolated parts of the card. One of the peculiarities of this extremely inexpensive method is its wide application in the analysis of known evidence using some of the automated isolation units. This approach allows simultaneous processing of a large number of samples, which significantly improves and accelerates the analysis of the known samples, especially those deposited in national databases.

Nowadays, numerous different DNA isolation methods exist, and many more new approaches are developing almost by daily rate. Moreover, automatization of DNA extraction procedures, as one of the most intensive current process in forensic genetics, is initiating extensive research work in finding new solutions for successful massive DNA isolation methods.

1.5 DNA Quantification

DNA quantification is a required step in forensic PCR-based testing in the Federal Bureau of Investigation (FBI) DNA standards. The outcome of DNA analysis depends on its quality and quantity. PCR is the most common method used for the analysis of evidence. Since the optimized methods require specific amounts of DNA, it is extremely important to establish methods of quantifying the exact concentration of DNA, but also to determine its origin, that is, whether the DNA found is derived from human source. Also, one of the key parameters for successful application of PCR is the purity of the sample. Specifically, the samples arriving for the processing are very often contaminated with bacterial DNA or containing large amounts of so-called PCR inhibitors. Hence, the methods used in the quantification of DNA will be explained in more detail in the following section.

1.5.1 Quantifying DNA Using Method of Spectrophotometry

Large amounts of DNA are rarely available for sample analysis in forensic medicine, so the standard method of spectrophotometric determination of DNA concentration in the sample is very rarely applied. This method is based on the measurement of light transmission through the liquid (at wavelengths of 260 and 280 nm, 1 nm = 10^{-9} m), on what basis the concentration of existing DNA can be determined. By measuring the absorption at wavelengths of 260 and 280 nm the purity of DNA can be determined, whereas the same measurements at a wavelength of 230 nm reveals the existence of peptides, phenols, carbohydrates, and other compounds that contaminate DNA. However, the lack of specificity of this method, that is, the fact that in this way the total DNA is measured, regardless of its origin, makes it highly inadequate in the process of forensic DNA analysis.

1.5.2 "Yield" Gel Method

The first nonspectrophotometric method used forensically used "yield" gel. It is performed by adding purified DNA onto a gel, usually agarose, electrophoresing the DNA and comparing it with fragments of high-molecular-weight DNA of known size tested on the same gel. In the United States, the standards can be obtained from the National Institute of Standards and Technology (U.S. Department of Commerce) in Standard Reference Material 2390 (SRM 2390). This method was developed for RFLP technology and quantitates all DNA in a sample, which is only relevant to RFLP testing and is of limited use for PCR-based testing. Therefore, its application in modern forensic DNA analysis is extremely limited and overshadowed by newer methods.

1.5.3 Hybridization (Slot-Blot) Method

The next innovation was the application of initially the dot blot and later the slot blot to the quantification of purified DNA samples. Commercial kits were developed based on hybridization of the human-specific probes with human DNA potentially present in biological evidence. The benefit of using this method (commonly detected using either chemiluminescence or colorimetric method) is that it may determine low levels of DNA in a sample while only using a small amount of extract. Denatured DNA from the samples is first captured and immobilized on the membrane and subsequently bound to the 40-bp DNA probe specific for the human DNA, but also for the DNA of higher primates (chimpanzees and gorillas). The strength of the detected band is compared with the quantitative controls, of known concentrations of DNA. Previously, one of the most widely used commercial kit was QuantiBlot Human DNA Quantitation Kit (Applied Biosystems), which uses sequence from the human chromosome 17 as a DNA probe. It is an extremely sensitive method, and it successfully detects the amount of DNA up to 150 pg (picogram = 10^{-12} g). On the other hand, the procedure is very complex, time-consuming, and depends on the ability of analysts who performs it. Furthermore, it was proved that sometimes samples that cannot be quantified by other methods can be successfully analyzed using QuantiBlot. This fact suggests the need for quantification in order to optimize the next phase of amplification, not a decision on continuation or termination of the process. Nowadays, this technology is no longer widely used, especially if Q-PCR is available.

1.5.4 AluQuant® Human DNA Quantitation System

This system is not in use anymore. Approximately 10 years ago, Aluquant® was introduced as "the next generation" of tests for the quantification of human DNA before PCR amplification. After DNA is denatured, specially designed probes, specific for human DNA, bind to highly repetitive sequences in the genome. The method was strictly primate-specific because it does not measure the existence of some foreign DNA (e.g., bacterial DNA). Quantitation of human DNA by this system was provided by a series of reactions. Following an initial denaturation, DNA samples are incubated with polymerase, kinase, and human-specific probe. This coupled enzymatic reaction produces ATP relative to the amount of human DNA present. In a second incubation, ATP produced in the first reaction is used by Luciferase to produce a proportional and measurable amount of light. Detection of the results was based on system utilizing luciferase-produced light. The amount of DNA is determined indirectly through the measured intensity of light that emerges from the release of energy-rich compounds. According to preliminary results, this method was successfully used to determine the amount of DNA that is 0.1 to 50 ng (1 ng = 10^{-9} g) (Mandrekar et al. 2001).

1.5.5 Quantitative RT-PCR Quantification Technology

The most recent improvement for the quantification of human DNA is the use of quantitative PCR reactions in a real-time (quantitative real-time PCR [QRT-PCR]) instrument. The real-time PCR scan is done on a Cepheid Smart Cycler (Figure 1.12). Currently, the most widely used commercial kits are Quantifiler® Human DNA Quantification Kit and Quantifiler® Y Human Male DNA Quantification Kit (Applied Biosystems 2012), The Investigator® Quantiplex Kit and The Investigator® Quantiplex HYres Kit (Qiagen 2012), and The Plexor® HY System (Promega Corporation 2013).

Both Quantifiler (Applied Biosystems) kits are based on TaqMan (5′ nuclease assay) technology with probes labeled with two different fluoroscent dyes (Reporter dye attached on the 5′ end of probe and Quencher dye located on the 3′ end). If hybridization between probe and DNA target occurs, a reporter dye will be released and start to fluoresce. This reaction will be used for detection of the presence of the DNA target. Also, IPC (internal PCR control) is included within each reaction to verify the reaction setup and/or to point presence of PCR inhibitors.

The Plexor HY (Promega Corporation) is based on the interaction between two modified nucleotides to achieve quantitative PCR analysis. One of the PCR primers contains a modified nucleotide (iso-dC) linked to a fluorescent label at the 5′ end. The second PCR primer is unlabeled. The reaction mix includes deoxynucleotides and iso-dGTP modified with the quencher, which is incorporated opposite the iso-dC residue in the primer. The incorporation of the quencher-iso-dGTP at this position results in quenching of the

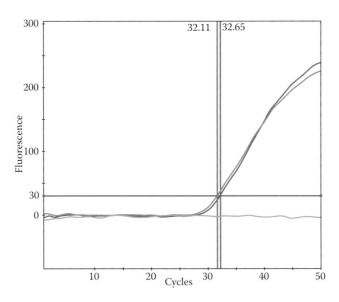

Figure 1.12 The real-time PCR scan done on a Cepheid Smart Cycler at the Department of Forensic Sciences, George Washington University, Washington, DC, show two samples going through the cycle threshold (CT) between cycles 32 and 33, and a third negative sample showing a flat baseline that never rises to the CT cutoff height. (Used with permission from Forensic Science, Department of Forensic Sciences, George Washington University, Washington DC.)

fluorescent dye on the complementary strand and a reduction in fluorescence, which allows quantitation during amplification. IPC control is added to each reaction.

Finally, Qiagen Quantiplex kits (Qiagen) are based on the detection of amplification using "Scorpion" primers and fast PCR chemistry. Scorpion primers are bifunctional molecules containing a PCR primer covalently linked to a probe. The fluorophore in this probe interacts with a quencher, also incorporated into the probe, which reduces fluorescence. During PCR, when the probe binds to the PCR products, the fluorophore and quencher become separated. This leads to an increase in fluorescence. This kit contains reagents and a DNA polymerase for specific amplification of proprietary region present on several autosomal chromosomes of the human genome.

As could be seen, different companies developed different methods, but with one main goal: to improve sensitivity and stability of DNA quantification procedure.

1.6 Methods for Measuring DNA Variation

1.6.1 Restriction Fragment Length Polymorphism

The first method that has been used in forensic DNA analysis was analysis of RFLP. In this method, the double-stranded DNA is cleaved in the presence of a restriction endonuclease, after which DNA fragments are separated using electrophoresis on agarose gel. Using special techniques, DNA fragments are transferred to a nylon membrane and finally detected using radioactive or chemiluminescent probes that exclusively bind to particular fragments of genomic DNA (Jeffreys et al. 1985a,b; NRC 1996). Restriction enzymes are generally bacterial products that can cut foreign DNA in the bacterial cell. Previously, these enzymes were widely applied in medical diagnostics. It should be noted that each of these many enzymes cut the DNA at specific restriction site in the nucleotide sequence. For example, the enzyme *Hae*III cuts DNA whenever there is a sequence GGCC within the DNA. In the human genome, there are approximately 12 million loci with this sequence and at each of these loci the DNA will be cut.

The basic type of DNA polymorphism being studied using RFLP technology was minisatellite VNTR loci that consist of short series of bases repeated different number of times. These loci vary in length because of the different number of repetitive DNA sequences at each locus. Large VNTR loci have repeat sequences of length 15–70 bp. Standardization of testing in forensic science led to the use of two restriction enzymes: *Hae*III in the United States and *Hin*fI in Europe, whereas the rest of the countries have used one of these two enzymes.

However, this technique required a relatively large amount of intact DNA, and the process was extremely time-consuming and laborious, and originally required the use of radioactive materials. Another problem in utilizing this method was the fact that the amount of DNA needed for RFLP methods is 20–500 ng, whereas PCR methods requires only 0.1–1 ng of DNA (Butler 2010).

All these facts explain why the RFLP method, which was extensively used during the late 1980s and 1990s, was completely abandoned at the end of the twentieth century.

An elaborate historical overview of the possibilities of using RFLP technology and its forensic applications can be found in the study of by Waye (2000).

1.6.2 Polymerase Chain Reaction

PCR has literally revolutionized the field of molecular genetics and biology in general. The starting amount of DNA, which for decades has been a limiting factor in research, is now reduced to the existence of only one well-preserved molecule, and the arduous procedures of DNA isolation and purification have become simpler. Suddenly, whole new opportunities opened up before the researchers. There is almost no social activity that the technique has not directly or indirectly touched and changed forever. The main "culprits" for this situation are Kary Mullis and his colleagues from the Department of Human Genetics, Cetus Corporation. In 1985, the journal *Science* published the research study by R. Saiki, K. Mullis and H. Erlich, with the first description of *in vitro* amplification of specific DNA fragments, catalyzed by DNA polymerase, the enzyme isolated from *Escherichia coli* (Saiki et al. 1985). However, because the *E. coli* enzyme is inherently degraded by heating to denature the DNA, a new enzyme was needed. In 1988, the same team of scientists published the paper in which they introduced heat-stable DNA polymerase isolated from bacterium *Thermus aquaticus* (*Taq*), instead of the previously used *E. coli* polymerase (Saiki et al. 1988). Despite numerous modifications, various applications, and process automation, the basic aspects of this technique, established in 1985 and 1988, have not been significantly changed.

The basic premises of the PCR are as follows—in the native state DNA is a double helix consisting of two antiparallel polynucleotide chains that are interconnected by hydrogen bonds: the nucleotides in one polynucleotide chain are joined by a covalent bond between the sugar and the phosphate group of adjacent nucleotides, and hydrogen bonds that hold two polynucleotide chains together are always formed between complementary bases, that is, A (adenine) always binds to T (thymine) and C (cytosine) with G (guanine). Hydrogen bonds are weak and easily disrupted by heating (denaturation), whereas the covalent bonds persist; the cooling leads to a reestablishment of hydrogen bonds between complementary bases (renaturation).

Based on the explanation above, the basic model of PCR reaction consists of three principal steps: denaturation (separation of polynucleotide strands caused by heating to 95°C), hybridization (annealing of artificially synthesized DNA primers that can be fluorescently labeled and flank the DNA fragments to be amplified at 50°C–65°C), and chain elongation (binding of complementary bases on the free sites, in the presence of denatured matrix, DNA primers, and *Taq* polymerase activity at 72°C). Under ideal conditions and 100% amplification efficiency (which certainly do not exist in nature) by cycle 30, approximately a billion copies of the target area on the DNA template have been produced (Butler 2011).

Therefore, PCR is the method by which the fragment of DNA multiplies to produce billions of copies (depending on the number of reaction cycles) that are completely identical to the DNA fragment of interest (Figure 1.13). This technique has several key advantages over RFLP: it requires a smaller amount of DNA; it is faster (results are obtained in a matter of hours); and it does not involve radioactive isotopes (NRC 1996). However, there are two fundamental drawbacks in the use of this method: (1) because of the exceptional sensitivity, there is a possibility of contamination; (2) many loci analyzed by PCR, especially STR molecular markers, have fewer alleles than minisatellite VNTR loci analyzed by RFLP. However, these two problems can be relatively easily overcome by conducting this method in strictly controlled conditions and using multiplex STR systems that, for example, nowadays allow the simultaneous analysis of up to15 STR loci.

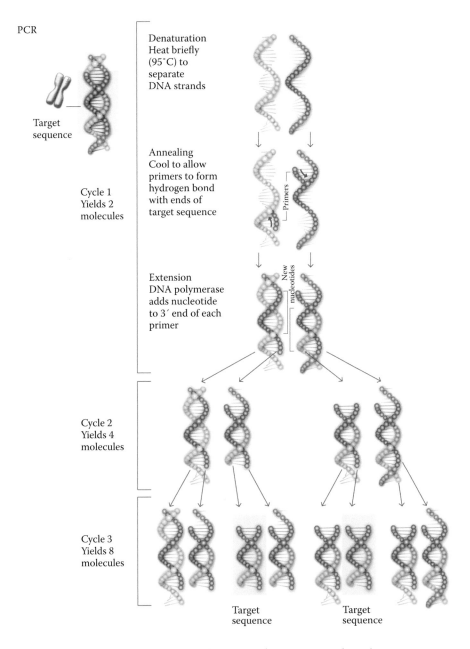

Figure 1.13 Schematic drawing of the polymerase chain reaction (PCR) process.

1.7 PCR Methods

1.7.1 Early PCR Amplification and Typing Kits

The first PCR system to be used in forensics analyzed the HLA-DQA1 locus, which is a part of the human histocompatibility region. Amplitype DQA1 was a typing system that contained probes for detection of DNA sequence variations that were originally identified using antibodies and live cells. Initially, the original system contained six alleles. When testing the evidence, in approximately 16% of cases two people had the same genotype, so it was necessary to develop additional PCR systems for typing. The next system that was developed using a similar principle of dot blot analysis was Amplitype Polymarker. This method further increased the number of analyzed genes, and its biggest contribution was that it actually increased the power of exclusion (discrimination power) during testing. This kit is not in forensic use anymore.

The next system was AmpliFLP™ D1S80 PCR Amplification Kit, commonly known as D1S80. It was used to detect genetic variations at the polymorphic VNTR locus, D1S80. The basic repeat sequence at the D1S80 locus was 16 bp long, with the number of repeats varying between 14 and 41 (350–1000 bp). Alleles were characterized by the number of repeats, so a person who has the alleles 18 and 22 is described as D1S80 18, 22. D1S80 was originally detected as an RFLP locus, which was converted to PCR to increase the information content. As other RFLP loci, DNA fragments of different lengths will be observed in different people. Unlike RFLP loci, where you only had the approximate length of the fragment, with PCR the number of repeats can be determined increasing the information content significantly. However, because this system was analyzing only one gene locus, the power of exclusion (discrimination power) was not great; nevertheless, this system in combination with others showed to be quite useful in the "premultiplex-STR era." Typically, when using this method, bands were visualized by staining the gel with silver nitrate.

1.7.2 Multiplex STR Systems

Like the RFLP system, the first tests for DNA analysis in forensic medicine were based on the variability of DNA sequences (sequential polymorphism). Researchers soon discovered another class of polymorphic loci with short repetitive blocks (two to nine nucleotide repeats). This class of loci is called STR or *microsatellites*; these are in contrast to systems such as D1S80, which are referred to as long tandem repeats or *minisatellites*. These molecular markers have already been discussed in previous text, and it has been mentioned that one of their main advantages is the possibility of relatively simple, rapid, and simultaneous processing of more than 10 STR loci.

Almost all systems used in the forensic testing have repeat units of 4 bp (tetranucleotide) or 5 bp (pentanucleotide), which occur from 5 to about 50 times, depending on the locus (Hammond et al. 1994; Alford et al. 1994). This type of molecular markers is very short (100–400 bp) and very useful in the analysis of degraded DNA, in contrast to the RFLP fragments whose length can vary from 500 to 12,000 bp. The advantage of these systems is the fact that the occurrence of so-called *stutter* bands is much less pronounced than in loci with one to three nucleotide repeats, making the interpretation of results much easier. The term *stutter* band is used for DNA fragments that are one repeat unit shorter than the expected band. Stutter occurs relatively frequently in tetranucleotide repeat sequences, as an amplicon that is for one 4-bp repeat unit shorter than expected allele. Stutter bands can make the interpretation of mixtures difficult, when it is important to determine whether the minor fragment that appears in a mixture is a stutter band or belongs to a different source. As the number of nucleotides in a repeat increase, the formation of stutter bands decreases; thus, the frequency of stutter bands for pentanucleotide repeats in PowerPlex 16 is much lower (Primorac 2001). The United States has adopted 13 STR loci for a national database of convicted offenders (McEwen 1995). This database is called CODIS. This system was initially limited to convicted rapists and murderers, but was later extended to other crimes (Hares 2011). Thirteen CODIS loci were selected in a manner to coincide with other loci chosen by Forensic Science Services in Great Britain for their national database and organizations such as INTERPOL. Two facts are related with CODIS loci: none of the CODIS markers are linked within exons, and no CODIS genotype is associated with known phenotypes (Katsanis and Wagner 2013). Genomic characterization of CODIS markers and phenotypic relevance of genomic regions of CODIS markers are shown in the Tables 1.2 and 1.3, respectively (Katsanis and Wagner 2013).

The new STR typing systems have several advantages over the old ones. The first advantage, already mentioned, is that more loci can be amplified simultaneously (multiplex reaction). Other features important in development of STR multiplexes includes: discrete and distinguishable alleles, amplification of the locus should be robust, a high

Table 1.2 Genomic Characterization of CODIS Markers

	CODIS Marker	Cytogenetic Location	Intragenic or Distance from Nearest Gene	Included in Marshfield Human Genetic Linkage Maps	Number of (#) SNPs (dbSNP build 132) within 1 kbp
1	D18S51	18q21.33	Intron 1	Included	13
2	FGA	4q28	Intron 3		4
3	D21S11	21q21.1	>100 kb from nearest gene	Removed	7
4	D8S1179	8q24.13	>50 kb from nearest gene	Included	14
5	VWA[a]	12p13.31	Intron 40		27
6	D13S317	13q31.1	>100 kb from nearest gene	Included	10
7	D16S539	16q24.1	~10 kb from nearest gene	Removed	29
8	D7S820	7q21.11	Intron 1	Included	7
9	TH01	11p15.5	Intron 1		8
10	D3S1358	3p21.31	Intron 20		11
11	D5S818	5q23.2	>100 kb from nearest gene	Removed	8
12	CSF1PO	5q33.1	Intron 6		15
13	D2S1338	2q35	~20 kb from nearest gene	Included	11
14	D19S433	19q12	Intron 1	Included	10
15	D1S1656	1q42	Intron 6	Removed	22
16	D12S391[a]	12p13.2	~40 kb from nearest gene	Included	18
17	D2S441	2p14	~30 kb from nearest gene	Removed	13
18	D10S1248	10q26.3	~3 kb from nearest gene	Included	16
19	Penta E	15q26.2	within uncharacterized EST; ~50 kb from nearest gene		19
20	DYS391	Yq11.21	~5 kb from nearest gene		0
21	TPOX	2p25.3	Intron 10		33
22	D22S1045	22q12.3	Intron 4	Included	19
23	SE33	6q14	~30 kb from nearest gene, pseudogene		8
24	Penta D	21q22.3	Intron 4		6

Source: Katsamis, S.H. Duke Institute for Genome Sciences and Policy, Duke University, Durham, NC. With permission.

Note: Markers are shown in their relative rank according to Hares (2011).

[a] VWA and D12S391 are colocated on 12p13 within 6 Mbp.

Table 1.3 Reported Phenotypic Relevance of Genomic Regions of CODIS Markers

	CODIS Marker	Gene Name	Disorder(s) Caused by Gene Mutations	Number of (#)Phenotypes Associated within 1 kbp	Predicted DNA Elements
1	D18S51	*BCL2* (B-cell CLL/lymphoma 2)	Leukemia/lymphoma, B-cell	11	ELAV1 binding site
2	FGA	*FGA* (fibrogen alpha chain)	Congenital afibrinogenemia; hereditary renal amyloidosis; dysfibrinogenemia (alpha type)	17	PABPC1 binding site
3	D21S11	None		1	None
4	D8S1179	None		17	None
5	VWA[a]	*VWF* (von Willebrand factor)	von Willebrand disease	12	ELAV1 binding site
6	D13S317	None		5	None
7	D16S539	None		8	None
8	D7S820	*SEMA3A* (sema domain, immunoglobulin domain, short basic domain, secreted (semaphoring) 3A)		8	CELF1, ELAV1 and PABPC1 binding site
9	TH01	*TH* (tyrosine hydroxylase)	Segawa syndrome, recessive	18	ELAVL1, PABPC1 and SLBP binding site
10	D3S1358	*LARS2* (leucyl-tRNA synthetase 2, mitochondria)		15	None
11	D5S818	None		5	None
12	CSF1PO	*CSF1R* (colony stimulating factor 1 receptor)	Predisposition to myeloid malignancy	15	eGFP-GATA2 transcription factor; PABPC1 binding site
13	D2S1338	None		9	None
14	D19S433	*C19orf2* (uncharacterized gene)		7	DNase I hypersensitivity site; SLBP binding site
15	D1S1656	*CAPN9* (calpain 9)		10	PABPC1 binding site
16	D12S391[a]	None		6	None
17	D2S441	None		6	None
18	D10S1248	None		6	DNase I hypersensitivity site
19	Penta E	EST: BG210743 (uncharacterized EST)		8	None
20	DYS391	None		1	None
21	TPOX	*TPO* (thyroid peroxidase)	Thyroid dyshormonogenesis 2A	5	PABPC1 and SLBP binding site
22	D22S1045	*IL2RB* (interleukin 2 receptor, beta)		11	None
23	SE33	None		9	None
24	Penta D	*HSF2BP* (heat shock factor 2-binding protein)		6	PABPC1 and SLBP binding site

Source: Katsamis, S.H. Duke Institute for Genome Sciences and Policy, Duke University, Durham, NC. With permission.

Note: Markers are shown in their relative rank according to Hares (2011).

[a] VWA and D12S391 are colocated on 12p13 within 6 Mbp.

power of discrimination, an absence of genetic linkage with other loci being analyzed, and low levels of artifact formation during the amplification (Goodwin et al. 2011). Since one primer of each pair has a fluorescent tag, it is possible to differentiate different loci by both size and color after separation by electrophoresis (Buel et al. 1998).

The most common STR systems that are now used in routine forensic work are manufactured by three companies: Promega, Life Technologies, and Qiagen.

A detailed overview of the basic PCR technology, the STR-PCR technology, and their applications in forensic investigations can be found in supplementary literature (McEwen 1995; Schanfield 2000; Butler 2010). Here, we will only look at the most used multiplex STR systems.

1.7.3 PowerPlex™ 16 System

PowerPlex™ 16 (PP16) allows coamplification and three-dye detection of 16 loci (15 STR loci and Amelogenin) (Promega Corporation 2013). Each of these loci is specifically located in the genome and has defined repetitive units. The kit contains 13 STR tetranucleotide (also known as standard CODIS loci) and two pentanucleotide loci. PP16 system is based on the use of three different dyes to label STR primers:

1. Primers for the loci Penta E, D18S51, D21S11, TH01, D3S1358, labeled by fluorescein (FL—nominal blue color)
2. Primers for the loci FGA, TPOX, D8S1179, VWA, Amelogenin, labeled with carboxytetramethylrhodamine (TMR—nominal yellow color, printed in black on computer output)
3. Primers for the loci Penta D, CSF1PO, D16S539, D7S820, D13S317, D5S818, labeled by 6-carboxy-4′,5′-dichloro-2′,7′-dimethoxyfluorescein (JOE—nominal green color)

The postamplification part of this kit contains all the chemicals necessary for the automated detection and analysis of the results. Internal Lane Standard (ILS) is required for DNA sizing analysis. It consists of 22 artificially synthesized DNA fragments ranging in size from 60 to 600 bp. Each fragment is labeled with carboxy-x-rhodamine (red dye). In this standard, the relative migration position of each fragment depends only on the length of the fragments, and not on their sequence.

PowerPlex® 16 Allelic Ladder Mix is a mixture of amplified DNA fragments representing allelic variants that were discovered in the Caucasian population in previous studies. As a result of combined use of ILS and allelic ladders, and application of appropriate software, each allelic variant is automatically identified, that is, the number of repeats is determined.

It is worth repeating the probability that there are two unrelated people with identical allelic variants at 15 STR loci, represented in the PowerPlex 16 kit, for the Caucasian population is 1.83×10^{-17} (Sprecher et al. 2000). It speaks volumes about the level of discrimination and individualization provided by the application of this kit.

1.7.4 AmpFLSTR® Identifiler™ PCR Amplification Kit

The AmpFLSTR® Identifiler™ PCR Amplification Kit (Applied Biosystems 2005) is also a DNA identification kit that can simultaneously analyze 15 STR loci, but, unlike the previous kits, all loci are tetranucleotides. It contains all 13 CODIS loci, but also two additional loci (D2S133 and D19S433). This system is based on the use of four different dyes to label the STR primers:

1. Loci D8S1179, D21S11, D7S820, CSF1PO labeled with the dye 6-FAM (nominal blue color)
2. Loci D3S1358, TH01, D13S317, D16S539, D2S1338 labeled with the dye VIC nominal green color)
3. Loci D19S433, TPOX, vWA, D18S51 labeled with the dye NED (nominal yellow color, printed in black on printout)
4. Amelogenin, D5S818, FGA labeled with the dye PET (nominal red color)

In comparison with previous kits, besides different colored labels, Identifiler has five-dye chemistry, with new orange dye LIZ that labels Size Standard, which consists of fragments of lengths up to 500 bp.

The probability that there are two unrelated people with identical allelic variants at 15 STR loci, represented in the Identifiler kit, for the Caucasian population of the United States is 5.01×10^{-19} (Life Technologies 2010), which clearly indicates the level of discrimination and individualization provided by the application of this reagent kit. The improvements in the Identifiler® Plus Kit formulation allow for more rapid amplification—eliminating approximately 1 hour from the amplification step, when compared to the current Identifiler® Kit. In addition, the Identifiler® Plus Master Mix includes the *Taq* polymerase, which eliminates two tubes and an extra box from the kit packaging, and enables streamlined reaction setup (Figure 1.14).

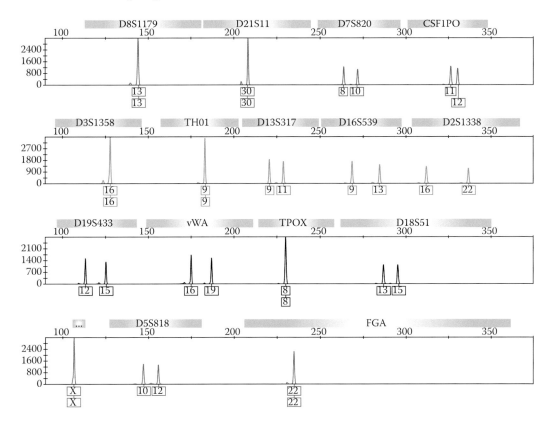

Figure 1.14 Electropherogram of AmpFlSTR Identifiler Plus PCR amplification kit (Life Technologies). Note: Five-color detection system (four colors for loci and fifth color for sizing ladder, not shown).

1.7.5 PowerPlex® ESX and ESI Systems

Nowadays, in response to the European Network of Forensic Science Institutes (ENFSI) and EDNAP groups' call for new STR multiplexes for Europe, Promega developed a suite of four new DNA profiling kits: PowerPlex® ESX 16 and PowerPlex® ESI 16 Systems (Promega 2013), and PowerPlex® ESI 17 Systems with SE33 (Promega 2013). These new multiplexes each contain the five new loci selected by ENFSI and EDNAP (D10S1248, D22S1045, D2S441, D12S391 and D1S1656) as well as 11 other loci commonly used throughout Europe (D8S1179, D18S51, D21S11, FGA, TH01, vWA, D2S1338, D3S1358, D16S539, D19S433 and Amelogenin). The PowerPlex® ESX Systems deliver D10S1248, D22S1045, and D2S441 as miniSTRs, and D12S391 and D1S1656 as midiSTRs (Figure 1.15). On the other hand, the PowerPlex® ESI Systems include the new ENFSI/EDNAP loci but focus on miniaturization of current ESS loci (D16S539, TH01 and D8S1179). Both formats are available with and without SE33. One primer for each locus is labeled with fluorescein, JOE, TMR-ET, or CXR-ET and can be resolved in the blue, green, yellow, or red channel, respectively, on the different Genetic Analyzers. Also included in each kit is CC5-labeled ILS 500 (CC5 ILS 500), which is resolved in the orange channel.

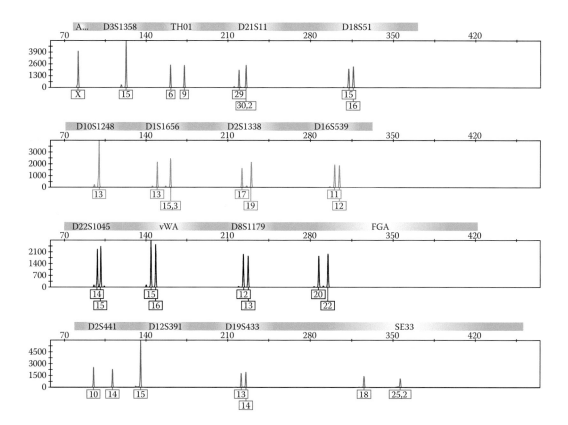

Figure 1.15 Electropherogram of Powerplex ESX PCR amplification kit (Promega Corporation). Note: Five-color detection system (four colors for loci and fifth color for sizing ladder, not shown).

Basic Genetics and Human Genetic Variation

1.7.6 AmpFℓSTR® NGM™ PCR Amplification Kit

The NGM™ Kit has been also designed specifically to address the requirements stipulated by the ENFSI/EDNAP groups. This kit amplifies the 10 loci previously included in SGM Plus Kit (D3S1358,vWA, D16S539, D2S1338, D8S1179, D21S11, D18S51, D19S433, TH01, and FGA), which include the seven ESS loci, together with the five additional loci recommended by ENFSI and EDNAP (D10S1248, D22S1045, D2S441, D1S1656, and D12S391) (Figure 1.16). These additional loci have been recommended because of their discrimination power and ability to be engineered to produce amplicons in the shorter size range to facilitate the analysis of degraded samples.

This system is based on the use of four different dyes to label the STR primers:

1. Loci D10S1248, vWA, D16S539, D2S1338 labeled with the dye 6-FAM (nominal blue color)
2. Loci Amelogenin, D8S1179, D21S11, D18S51 labeled with the dye VIC (nominal green color)

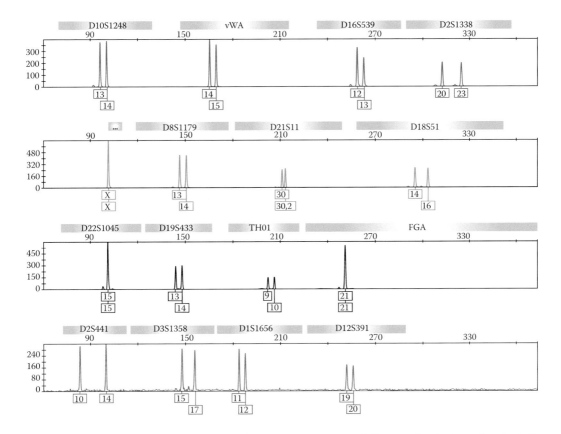

Figure 1.16 Electropherogram of AmpFlSTR NGM PCR amplification kit (Life Technologies). Note: Five-color detection system (four colors for loci and fifth color for sizing ladder, not shown).

3. Loci D22S10453, D19S433, TH01, FGA labeled with the dye NED (nominal yellow color, printed in black on printout)
4. Loci D2S441, D3S1358, D1S1656, D12S391 labeled with the dye PET (nominal red color)

1.7.7 Investigator ESSplex Plus Kit and Investigator IDplex Plus Kit

Qiagen, as the new company in the field of the manufacturing of multiplex STR systems, released more than 10 different human identification assays within 2011 and 2012. Two of them (Investigator ESSplex Plus Kit and Investigator IDplex Plus Kit) could be recognized as the current standard approach for the forensic genetics community.

The Investigator ESSplex Plus Kit is assembling the 15 polymorphic STR markers recommended by the ENFSI and the European DNA Profiling Group (EDNAP) as the new European Standard Set of loci (D1S1656, D2S441, D2S1338, D3S1358, D8S1179, D10S1248, D12S391, D16S539, D18S51, D19S433, D21S11, D22S1045, FGA, TH01, and vWA), and the gender-specific Amelogenin.

The primers are fluorescence-labeled with the following dyes:

1. Loci Amelogenin, TH01, D3S1358, vWA, D21S11 labeled with the dye 6-FAM (nominal blue color)
2. Loci D16S539, D1S1656, D19S433, D8S1179, D2S1338 labeled with the dye BTG (nominal green color)
3. Loci D10S1248, D22S1045, D12S391, FGA labeled with the dye BTY (nominal yellow color)
4. Loci D2S441, D18S51 with the dye BTR (nominal red color)

On the other hand, the Investigator IDplex Plus Kit is a multiplex application for human identification covering the 13 CODIS (Combined DNA Index System) core loci, D2S1338, D19S433, and Amelogenin.

The primers are fluorescence-labeled with the following dyes:

1. Loci Amelogenin, TH01, D3S1358, vWA, D21S11 labeled with the dye 6-FAM (nominal blue color)
2. Loci TPOX, D7S820, D19S433, D5S818, D2S1338 labeled with the dye BTG (nominal green color)
3. Loci D16S539, CSF1PO, D13S317, FGA labeled with the dye BTY (nominal yellow color)
4. Loci D18S51, D8S1179 labeled with the dye BTR (nominal red color)

1.7.8 PLEX-ID SNP Assay

Autosomal SNP testing could be used on highly degraded samples. Recently, Abbott introduced the PLEX-IS SNP, the panel of 40-loci based on the study by Pakstis et al. (2007) that identified SNP markers with the most evenly distributed heterozygosity and the least amount of population bias across populations globally. Kiesler and Valone (2013) showed that the combined random match probabilities for the 40 SNP assay (PLEX-ID SNP40) ranged from 10^{-16} to 10^{-21}.

1.8 Detection of PCR Products

The next step after amplification is detection of PCR products. It is conducted on one of the analytical machines, using different software to generate final genetic profiles. Detection, as the final phase of the process of determining the genetic identity of the person or biological evidence, is an automated procedure that is most susceptible to frequent modifications and innovations. These changes essentially follow the trend of progress in computational and fluorescent detection technology and aim for complete automation of the process. Because of the necessity that all laboratories should produce results that can be shared, all testing is done with standardized forensically validated kits. At this time, the shift from gel to capillary electrophoresis has been completed.

Unlike RFLP testing, which was hampered by the inability to identify and specify alleles but rather a DNA fragment that had significant measurement errors and limited reproducibility, STR-based testing produces measured DNA fragments, which—using the in lane size ladder and comparing to a known "allele ladder"—generates specific alleles based on repeat number. The allele ladder contains the majority of commonly detected allelic variants at a given locus and allows detection and analysis of particular allelic variants. This makes the technology extremely useful in parentage as well as forensic testing. Because only the length of the fragment is at issue, and the DNA migration process of PCR amplified DNA is quite different than restriction digested genomic DNA, many different detection methods have been used, with the initial technology using radioactivity and silver staining, followed by the development of fluorescent detection and the use of fluorescent labeled primers. This was further enhanced by the development of charge-coupled detectors and computer software that could simultaneously unscramble multiple colors allowing for single reaction, single read detection of multiple colors. In most cases, a number of STR loci are used, some of which match in size and position on the gel. To avoid misinterpretation, four—and recently, five—dyes are used to distinguish individual loci. The loci with considerably different molecular weights are labeled in the same color, and different colors are used for the loci whose allele ranges overlap. Since alleles from different loci emit fluorescent light at different wavelengths, genetic analyzers are able to distinguish each allele as they simultaneously pass through the detection system. Thus, after the data are collected and digitally processed, the resulting peaks do not represent individual bases, but rather allelic variants, which are indicators of how many repeats are present in a given fragment. Consequently, the term "sequencing," which is often used in laboratory jargon for this process, is inappropriate for this type of testing, which is normally referred to as "DNA profiling or genotyping."

Although some of the loci used forensically contain either sequence variation in the repeats or a combined pattern of repeat sequence variation and repeat order variation, these are not detected, such that for two individuals with the type D21S11 29, one could be homozygous for the same allele for repeat sequence variation, whereas another individual could have two different repeat sequence alleles. However, to determine this, each D21S11 allele would have to be sequenced. Although this would increase the information content, it is not practical in routine forensic testing. Sequencing of DNA has been a critical part of the evolution of molecular genetics tools. Although it has limited application forensically, it will become more important as sequencing technology improves and becomes less expensive.

1.8.1 Analytical Thresholds and Sensitivity for Forensic DNA Analysis

Heterozygous STR loci for forensic casework and database samples produce two peaks in the profile. In an ideal situation, the ratio between these two peaks will be 1:1 in terms of peak height and area (Goodwin et al. 2011). Unfortunately, it does not happen commonly and variations in peak height are common. However, in good-quality DNA extracts, smaller peak is on average approximately 90% the size of the larger peak (Gill et al. 1997). During forensic DNA analysis when the quantity of analyzed DNA is too low, it may be difficult to distinguish true low-level peaks from technical artifacts, including noise. While working with low amounts of DNA, degraded DNA, or DNA mixtures, most forensic scientists experienced the so-called "stochastic effect during PCR amplification of low level DNA" because of a random primer banding in the low copy number milieu. In that case during the detection process, several outcomes are possible: *allele drop-in*, *allele drop-out*, *increased stutter production*, *locus drop-out*, and *heterozygote peak imbalance* (Butler 2011; National Forensic Science Technology Center 2013). Allele-drop in (additional allele is observed) could be of unknown origin or from a variety of intralaboratory sources including consumable items (Gill 2010) and personnel, whereas a single allelic drop out produces false homozygous profiles, because of stochastic effects. On the other hand, stutter peaks are formed because of strand slippage during the extension of nascent DNA strand during PCR amplification (Schlotterer and Tautz 1992). Stutter peaks usually appear one repeat unit before true alleles but can occur one repeat unit after true alleles, and most of the stutters (di- and trinucleotide repeats are more prone to stutter than are tetra- and pentanucleotid repeats) are less than 15% of the main peak (Goodwin et al. 2011). Heterozygote peak imbalance also happens because of the stochastic PCR effect, where one of the alleles is amplified preferentially by chance. In order to distinguish a real peak from noise it is necessary to establish minimum distinguishable signal that may be considered a relative fluorescent unit (RFU) or analytical threshold (AT) (Bregu 2013). The RFU peak height mainly depends on the amount of DNA being analyzed. Recently, the Scientific Working Group on DNA Analysis Methods (SWGDAM) published Interpretation Guidelines for Autosomal STR Typing by Forensic DNA Testing Laboratories, where they underline that in general, nonallelic data such as stutter, nontemplate-dependent nucleotide addition, disassociated dye, and incomplete spectral separation are reproducible, whereas spikes and raised baseline are nonreproducible (Scientific Working Group on DNA Analysis Methods 2010). Bregu and his group just published an excellent paper stating that determination of ATs should be widely implemented. However, they observed that when a substantial mass of DNA (>1 ng) was amplified the baseline noise increased, and that the number and intensity of noise peaks increased with increasing injection times (Bregu et al. 2013).

1.8.2 Sequencing

DNA sequencing is the process of determining the order of bases within the DNA strand or fragment, and it represents one of the basic methods underlying molecular genetics. Knowing the DNA sequence is a precondition for any manipulation of the target segment of the hereditary material. For example, a computer search of sequences of all known restriction (endonuclease) sites can result in the creation of complete and accurate restriction maps. There are several techniques of sequencing, which differ in their basic principles:

1. Maxam–Gilbert (M-G) method
2. Sanger method
3. Automated sequencing

Simplicity, rapidity, and a wide range of applicability promoted the automated sequencing analysis currently in use. Manual sequencing of DNA fragments, represented by the first two approaches (M-G and Sanger), includes four separate PCR reactions whose products are then run in four individual lanes on a gel, usually on polyacrylamide gels, and detected by autoradiography. This makes these methods extremely laborious, complicated, and demanding from the standpoint of the needs of forensic DNA analysis to process as many samples in the shortest possible period. The Sanger sequencing method is based on stopping the enzymatic DNA synthesis by using dideoxynucleotide triphosphates as chain terminators. Frederick Sanger received the Nobel Prize for his discovery.

The initial requirement for a successful automatic sequencing is the simultaneous separation of fragments using electrophoresis and the detection of labeled fragments. In automated detection, fluorescent DNA labeling has the advantage over other detection methods. However, no matter what generation of the automated sequencing is used, the principals of each technique are based on Sanger method. Initially, fluorescent tags were linked to the primers. Four separate reactions were conducted each with a different fluorescent dye on the same primer and each with 10% of a different chain terminating dideoxynucleotide (ddNTP). DNA fragments were detected on the basis of their fluorescence during their passage through the detector region. Horizontal or vertical scanning by the detector made it possible to simultaneously observe several different colors, one sequence per lane.

The next generation of fluorescence techniques is based on the use of fluorescent markers linked to the chain-terminator ddNTPs, which implies a different labeling of each of the bases (C, G, T, A). Each of these four ddNTPs fluoresces in a different spectrum. The labeled ddNTP is incorporated into the DNA molecule by using *Taq* polymerase and has two functions: to stop further synthesis of DNA chain and to bind fluorophore to the end of the molecules. The main advantage of this method is that the reaction does not have to be conducted in four separate tubes. Even with this method, the final detection occurs during electrophoresis, when labeled fragments pass through the detector's reading region. Using this method, 3–500 bp can be generated per lane in either a lane of an acrylamide gel or in a capillary. What is important to note is the fact that this approach to testing of hereditary material in forensic DNA analysis is primarily applied in the analysis of mtDNA and its hypervariabile regions.

Automated sequencing is designed to sequence large pieces of DNA such at the control regions of mtDNA where information is distributed throughout the region or if you are looking for specific substitutions (SNPs). As the number of AIMs and PIMs are developed and validated as forensic investigative tools for the majority of DNA profiles that do not match DNA database files, the need for rapid SNP based testing will increase, and commercial kits will be developed. Minisequencing multiplexes have the ability to amplify multiple regions in a single reaction, and then with a second reaction detect specific alleles (SNPs) at these regions. The system that appears to be the most widely used and is compatible with the capillary electrophoresis DNA analyzers currently in use is the Life Technologies SNAPSHOT™ single base extension assay or "minisequencing" system. The system consists of multiplexes of unlabeled primers that amplify one or more loci. The amplicon is cleaned with ExoSap (a mixture of Exonuclease and Shrimp Alkaline Phosphatase that simultaneously destroys the old primers and spent nucleotides), and then added to a new reaction mixture that contains unlabeled primers for the targeted SNP and the four ddNTPs each labeled with a different fluorophore. The material is cleaned up with ExoSAP and detected on a standard capillary electrophoresis DNA analyzer. The SNP primers have been designed so that different loci have different sized fragments, and each allele has a different color.

1.9 Forensic Analysis of Plant DNA

Analysis of DNA isolated from plants has a primary role in connecting individuals with the crime scenes, or linking the evidence to the specific geographic locations and, since recently, it is possible to monitor the movement of certain drugs from breeding sites to the places of consumption. The first forensic case that involved analysis of DNA from plants, which demonstrated the importance and potential use of plant DNA testing, occurred in 1992. In Arizona, the body of a female was found under a Palo Verde tree. Lying close to the body, there was a pager that could be traced to its owner. During the investigation, police found a few seed pods from a Palo Verde tree in the suspect's truck. Police wanted to know if the seeds found in a truck could be associated with the Palo Verde tree next to the body establishing the suspect's tie to the crime scene, which the suspect denied. A study was made of DNA from Palo Verde trees in the area and they were all found to be different. The DNA from the seeds in the truck was identical to the DNA from the tree next to the body. This was the first use of plant DNA to link a suspect to a crime scene and was critical in finding the suspect guilty of murder (Miller Coyle et al. 2001).

Another area of interest is the possible identification of origin of plant material. Marijuana plants propagate clonally so all descendants of a parent plant will have the same genetic profile, although there is variation among clones. Because of this, it is now possible to monitor the distribution of seeds and leaves, connecting paths of the drugs distribution, as well as drug dealers (Shirley et al. 2013). Surely, in the near future, it will be required to implement a number of population studies on different plants in order to accurately determine the occurrence of particular genes in specific types of plants within a certain geographic area, which will ultimately have a major impact on the acceptance of the results of plant DNA analysis in court (Zeller et al. 2001). This issue is so topical that it is separately discussed in more detail in Chapter 14.

1.10 Forensic Analysis of Animal DNA

Many studies have clearly shown that the animal DNA has a unique genetic code, making it possible to distinguish individual animals even within the same species and subspecies (breed). Recently, a new SNP assay was developed in order to identify individual horses from urine samples that are submitted for postracing doping tests (Kakoi et al. 2013). It is often necessary to determine the origin of animal hairs found at a crime scene or on the clothes of the victim or suspect. Usually, it is a dog or cat hair, or hair from other pets. So far, numerous research studies have been carried out that clearly showed the specificity of nuclear DNA or mtDNA isolated from various evidence of animal origin (Savolainen et al. 2000; Fridez et al. 1999; Padar et al. 2001; Roney et al. 2001; Raymond-Menotti 2001). This topic is discussed further in Chapter 12.

1.11 Contamination Issues in Field and in Laboratory

Because of the increasing number of cases, contamination is becoming one of the major problems in forensic casework analysis. The appropriate collection of biological evidence in order to prevent initial contamination is imperative (for further reading, please check Chapter 9). Contamination of forensic evidence with foreign DNA can result in misidentification and mixed DNA profiles, which can possibly lead to a loss of crucial evidence and unsuccessful case solving. Even when working in the field during the remains recovery process, preventative measures are needed in order to eliminate the possibility of contamination by exogenous DNA (Edson and Christensen 2013). In a recent paper Pilli et al. (2013) showed that teeth are less prone to contamination than the other skeletal areas and may be considered a material of choice for classical ancient DNA (aDNA) studies. In addition, they showed that bones can also be a good candidate for human aDNA analysis if they come directly from the excavation site and are accompanied by a clear taphonomic history (Pilli et al. 2013). Also, it has been known that while working with samples recovered from mass graves or mass disasters, forensic scientists are unfortunately facing serious issues related to DNA quality, DNA degradation, DNA contamination, and presence of inhibitors including humic acid (Sutlović 2005). Surprisingly, it has been shown that even during the manufacturing process, contamination of "sterile" materials can occur (Gill et al. 2010). In the same report, the authors published the joint statement by the ENFSI, the United States of America SWGDAM, and Australia/New Zealand Biology Specialist Advisory Group (BSAG) on how to prevent DNA contamination of disposable plastic ware and other reagents by manufacturers. However, PCR amplification of minute quantities of degraded DNA is a challenge and the problem of contamination is underestimated. The most serious threats related to contamination are coming from carryover contamination (e.g., products of previous PCR amplification), laboratory surface, and reagent contamination. A multistrategy procedure for decontamination of PCR reagents, involving gamma and UV irradiation and treatment with a heat-labile double-strand specific DNase was recently described (Champlot et al. 2010). Results of our recent study pointed out the inefficiency of 10% sodium hypochlorite and 20% ethanol for decontamination of working surfaces. High UVC doses are necessary for complete decontamination of working surfaces with inflicted nondiluted blood and semen controls (Gršković et al. 2013).

References

1991 Report concerning recommendations of the DNA Commission of the International Society for Forensic Haemogenetics relating to the use of DNA polymorphism. 1992. *Vox-Sanguinis* 63:70–73.

Abaz, J., S.J. Walsh, J.M. Curran et al. 2002. Comparison of variables affecting the recovery of DNA from common drinking containers. *Forensic Sci Int* 126:233–240.

Alford, R.L., H.A. Hammond, I. Coto, and C.T. Caskey. 1994. Rapid and efficient resolution of parentage by amplification of short tandem repeats. *Am J Hum Genet* 55:190–195.

Alonso, A., S. Andjelinovic, P. Martin et al. 2001. DNA typing from skeletal remains: Evaluation of multiplex and megaplex STR systems on DNA isolated from bone and teeth samples. *Croat Med J* 42(3):260–266.

Andjelinović, Š., D. Sutlović, I. Erceg-Ivkošić et al. 2005. Twelve-year experience in identification of skeletal remains from mass graves. *Croat Med J* 46(4):530–539.

Andrews, R.M., I. Kubacka, P.F. Chinnery, R.N. Lightowlers, D.M. Turnbull, and N. Howell. 1999. Reanalysis and revision of the Cambridge reference sequence for human mitochondrial DNA. *Nat Genet* 23:147.

Applied Biosystems. 2005. *AmpFLSTR® Identifiler® PCR Amplification Kit User's Manual.*

Applied Biosystems. 2012. *Quantifiler® Human DNA Quantification Kit and Quantifiler® Y Human Male DNA Quantification Kit User's Manual.*

Barać, L., M. Peričić, I.M. Klarić et al. 2003. Y chromosomal heritage of Croatian population and its island isolates. *Eur J Hum Genet* 11:535–542.

Bregu, J., D. Conklin, E. Coronado, M. Terrill, W.R. Cotton, and M.C. Grgicak. 2013. Analytical thresholds and sensitivity: Establishing RFU thresholds for forensic DNA analysis. *J Forensic Sci* 58(1):120–129.

Brettell, T.A., J.M. Butler, and J.R. Almirall. 2011. Forensic science. *Anal Chem* 83:4539–4556.

Budowle, B., R. Chakraborty, A.M. Guisti, A.J. Eisenberg, and R.C. Allen. 1991. Analysis of the VNTR locus DlS80 by the PCR followed by high resolution PAGE. *Am J Hum Genet* 48:137–144.

Budowle, B., F.S. Baechtel, and C.T. Comey. 1992. Some considerations for use of AMP-FLPs for identity testing. In *Advances in Forensic Haemogenetics*, ed. C. Ritter and P.M. Schneider, 11–17. New York: Springer Verlag.

Budowle, B., K.L. Monson, A.L. Guisti, and B.L. Brown. 1994. The assessment of frequency estimates of Hae-III generated VNTR profiles in various reference databases. *J Forensic Sci* 39:319–352.

Buel, E., M.B. Schwartz, and M.J. LaFountain. 1998. Capillary electrophoresis STR analysis: Comparison to gel-based systems. *J Forensic Sci* 43:164–170.

Butler, J.M. 2010. *Fundamentals of Forensic DNA Typing*. San Diego: Elsevier Academic Press.

Butler, J.M. 2011. *Advanced Topics in Forensic DNA Typing: Methodology*. San Diego: Elsevier Academic Press.

Champlot, S., C. Berthelot, M. Pruvost, E.A. Bennett, T. Grange, and E.M. Geigl. 2010. An efficient multistrategy DNA decontamination procedure of PCR reagents for hypersensitive PCR applications. *PloS ONE* 5(9):e13042.

Coble, M.D., and J.M. Butler. 2005. Characterization of new miniSTR loci to aid analysis of degraded DNA. *J Forensic Sci* 50:43–53.

Coble, M.D., R.S. Just, J.E. O'Callaghan et al. 2004. Single nucleotide polymorphisms over the entire mtDNA genome that increase the power of forensic testing in Caucasians. *Int J Legal Med* 118:137–146.

Cooper, G.M., and R.E. Hausman. 2004. *The Cell: A Molecular Approach*, 3rd edn. Washington, DC: ASM Press.

Davis, C., J. Ge, and C. Sprecher. 2013. Prototype PowerPlex Y23 System: A concordance study. *Forensic Sci Int* 7:204–208.

Doležel, J., J. Bartoš, H. Voglmayr, and J. Greilhuber. 2003. Nuclear DNA content and genome size of trout and human. *Cytometry A* 51(2):127–128.

Džijan, S., G. Ćurić, D. Pavlinić, M. Marcikić, D. Primorac, and G. Lauc. 2009. Evaluation of the reliability of DNA typing in the process of identification of war victims in Croatia. *J Forensic Sci* 54(3):608–609.

Edson, S.M., and A.F. Christensen. 2013. Field contamination of skeletonized human remains with exogenous DNA. *J Forensic Sci* 58:206–209.

Federal Bureau of Investigation. 1990. *The Application of Forensic DNA Testing to Solve Violent Crimes*. Washington, DC: US Department of Justice.

Fisher, A.B., and D.R. Fisher. 2012. *Techniques of Crime Scene Investigation*. Boca Raton, FL: Taylor and Francis.

Fridez, F., S. Rochat, and R. Coquoz. 1999. Individual identification of cats and dogs using mitochondrial DNA tandem repeats. *Sci Justice* 39:167–71.

Gabriel, N.M., D.C. Calloway, L.R. Rebecca, Š. Andjelinović, and D. Primorac. 2001. Population variation of human mitochondrial DNA hypervariable regions I and II in 105 Croatian individuals demonstrated by immobilized sequence specific oligonucleotide probe analysis. *Croat Med J* 42:328–335.

Giampaoli, S., A. Berti, F. Valeriani et al. 2012. Molecular identification of vaginal fluid by microbial signature. *Forensic Sci Int Genet* 6:559–564.

Gill, P., R. Sparkes, and C. Kimpton. 1997. Development of guidelines to designate alleles using an STR multiplex system. *Forensic Sci Int* 89:185–197.

Gill, P., D. Rowlands, G. Tully, I. Bastisch, T. Staples, and P. Scott. 2010. Manufacturer contamination on disposable plastic-ware and other reagents—an agreed position statement by ENFSI, SWGDAM and BSAG. *Forensic Sci Int Genet* 4:269–270.

Geberth, J.V. 2006. *Practical Homicide Investigation: Tactics, Procedures and Forensic Techniques*. Boca Raton, FL: Taylor and Francis.

Goodwin, W., A. Linacre, and S. Hadi. 2011. *An Introduction to Forensic Genetics*. Hoboken, NJ: Wiley-Blackwell.

Gornik, I., M. Marcikić, M. Kubat, D. Primorac, and G. Lauc. 2002. The identification of war victims by reverse paternity is associated with significant risks of false inclusion. *Int J Legal Med* 116:255–257.

Graur, D., and W.H. Li. 2000. *Fundamentals of Molecular Evolution*, 2nd edn. Sunderland, MA: Sinauer, pp. 14, 274–275, 386–387, 392–394.

Gršković, B., D. Zrnec, M. Popovic, M.J. Petek, D. Primorac, and G. Mrsic. 2013. Effect of ultraviolet C radiation on biological samples. *Croat Med J* 54:263–271.

Gunn, A. 2006. *Essential Forensic Biology*. Chichester: John Willey & Sons.

Haas, C., E. Hanson, M.J. Anjos et al. 2013. RNA/DNA co-analysis from human saliva and semen stains. Results of a third collaborative EDNAP exercise. *Forensic Sci Int Genet* 7:230–239.

Hammond, H.A., L. Jin, Y. Hong, C.T. Caskey, and R. Chakraborty. 1994. Evaluation of 13 short tandem repeat loci for use in personal identification applications. *Am J Hum Genet* 55:175–189.

Hares, D.R. 2011. Expanding the CODIS core loci in the United States. *Forensic Sci Int Genet* 6(1):e52–e54.

Holland, M.M., D.L. Fisher, L.G. Mitchell et al. 1993. Mitochondrial DNA sequence analysis of human skeletal remains: Identification of remains from the Vietnam War. *J Forensic Sci* 38:542–53.

Holland, M.M., and T.J. Parsons. 1999. Mitochondrial DNA sequence analysis—validation and use for forensic casework. *Forensic Sci Rev* 11:22–50.

Houck, M.M., and J.A. Siegel. 2006. *Fundamentals of Forensic Science*. London: Elsevier Academic Press.

Inman, K., and N. Rudin. 1997. *An Introduction to Forensic DNA Analysis*. New York: CRC Press.

International Human Genome Sequencing Consortium. 2001. Initial sequencing and analysis of the human genome. *Nature* 409:860–921.

Irwin, J.A., J.L. Saunier, Niederstatter et al. 2009. Investigation of heteroplasmy in the human mitochondrial DNA Control Region: A synthesis of observations from more than 5000 global population samples. *J Mol Evol* 68:16–527.

Ivanov, P.L., M.J. Wadhams, R.K. Roby, M.M. Holland, V.W. Weedn, and T. Parsons. 1996. Mitochondrial DNA sequence heteroplasmy in the Grand Duke of Russia Georgij Romanov, establishes the authenticity of the remains of Tzar Nicholas II. *Nat Genet* 12:417–420.

Jeffreys, A.J., V. Wilson, and S.L., Thein. 1985a. Hypervariable "minisatellite" regions in human DNA. *Nature* 314:67–73.

Jeffreys, A.J., S.L. Thein, and V. Wilson. 1985b. Individual specific "fingerprints" of human DNA. *Nature* 316:76–79.

Jobling, M.A., E. Heyer, P. Dieltjes, and P. de Knijff. 1999. Y-chromosome-specific microsatellite mutation rates re-examined using a minisatellite, MSY1 *Hum Mol Gen* 8:2117–2120.

Jobling, M.A., M.E. Hurles, and C. Tyler-Smith. 2004. *Human Evolutionary Genetics: Origins, Peoples and Disease*. New York: Garland Science, Taylor & Francis Group.

Johnson, C.V., D. Primorac, D.M. McKinstry, D.W. Rowe, and J.B. Lawrence. 2000. Tracking COL1A1 RNA in osteogenesis imperfecta: Splice-defective transcripts initiate transport from the gene but are retained within the SC35 domain. *J Cell Biol* 150:417–432.

Kaida, D., T. Schneider-Poetsch, and M. Yoshida. 2012. Splicing in oncogenesis and tumor suppression. *Cancer Sci* 103:1611–1616.

Kakoi, H., I. Kijima-Suda, H. Gawahara et al. 2013. Individual identification of racehorses from urine samples using a 26-plex single-nucleotide polymorphism assay. *J Forensic Sci* 58(1):21–28.

Katsanis, H.S., and J.K. Wagner. 2013. Characterization of the standard and recommended CODIS markers. *J Forensic Sci* 58:S169–S171.

Kayser, M., and A. Sajantila. 2000. Mutations at Y-STR loci: Implications for paternity testing and forensic analysis. *Forensic Sci Int* 118:116–121.

Kelemen, O., P. Convertini, Z. Zhang et al. 2013. Function of alternative splicing. *Gene* 514:1–30.

Kiesler, K.M., and P.M. Vallone. 2013. 40 Autosomal SNP loci typed for U.S. African American, Caucasian and Hispanic samples using electrospray ionization mass spectrometry. *Croat Med J* (in press).

Latorra, D., M.C. Stern, and M.S. Schanfield. 1994. Characterization of human AFLP systems apolipoprotein B, Phenylalanine hydroxylase and D1S80. *Genome Res* 3:351–358.

Lee, H.C., C. Ladd, C.A. Scherczinger, and M.T. Bourke. 1998. Forensic applications of DNA typing: Part 2. Collection and preservation of evidence. *Am J For Med Pathol* 19:10–18.

Life Technologies. 2008. *Quantifiler Human DNA Quantification Kit, User's Manual*. Applied Biosystems.

Life Technologies. 2010. *AmpFlSTR Identifiler PCR Amplification Kit User's Manual*. Applied Biosystems.

Mandrekar, N.M., M.A. Erickson, K. Kopp et al. 2001. Development of a human DNA quantitation system. *Croat Med J* 42:336–9.

Marjanović, D., and D. Primorac. 2013. DNA Variability and molecular markers in forensic genetics. In *Forensic Genetics: Theory and Application*, ed. D. Marjanović, D. Primorac, L.L. Bilela et al., 75–98. Sarajevo: Lelo Publishing (Bosnia and Herzegovina's edition).

Marjanovic, D., S. Fomarino, S. Montagna et al. 2005. The peopling of modern Bosnia-Herzegovina: Y-chromosome haplogroups in the three main ethnic groups. *Ann Hum Genet* 69:757–763.

Maroni, G. 2001. *Molecular and Genetic Analysis of Human Traits*. Malden, MA: Blackwell Science.

Mayr, W.R. 1993. Recommendations of the DNA Commission of the International Society for Forensic Haemogenetics relating to the use of PCR-based polyrmorphisms. *Vox Sanguinis* 64:124–126.

McEwen, J.E. 1995. Forensic DNA data banking by state crime laboratories. *Am J Hum Genet* 56:1487–1492.

Miller Coyle, H., C. Ladd, T. Palmbach, and H.C. Lee. 2001 The green revolution: Botanical contributions to forensics and drug enforcement. *Croat Med J* 42:340–345.

Nakanishi, H., H. Shojo, T. Ohmori et al. 2013. *Forensic Sci Int* 7:176–179.

National Forensic Science Technology Center. 2013. Stochastic Effects of LCN DNA Analysis http://www.nfstc.org/pdi/Subject09/pdi_s09_m01_03_b.htm (accessed March 25, 2013).

National Research Council, National Academy of Sciences. 1996. *The Evaluation of Forensic DNA Evidence*. Washington, DC: National Academy Press.

Neilson, J.R., and R. Sandberg. 2010. Heterogeneity in mammalian RNA 3' end formation. *Exp Cell Res* 316(8):1357–1364.

Padar, Z., B. Egyed, K. Kontadakis et al. 2001. An importance of canine identification in Hungarian forensic practice. In *Proceedings of The Second European–American Intensive Course in Clinical and Forensic Genetics*, ed. D. Primorac, I. Erceg, and A. Ivkošić, September 3–14, 2001, Dubrovnik, Croatia. Zagreb, Croatia: Studio HRG.

Pakstis, A.J., W.C. Speed, J.R. Kidd et al. 2007. Candidate SNPs for a universal individual identification panel. *Hum Genet* 121(3–4):305–317.

Parsons, T.J., and M.D. Coble. 2001. Increasing the forensic discrimination of mitochondrial DNA testing through analysis of the entire mitochondrial DNA genome. *Croat Med J* 42:304–309.

Pennisi, E. 2012. Genomics ENCODE Project writes eulogy for junk DNA. *Science* 337:1159–1161.

Pereira, R., I. Gomes, A. Amorim, and L. Gusmão. 2007. Genetic diversity of 10 X chromosome STRs in northern Portugal. *Int J Legal Med* 121:192–197.

Pilli, E., A. Modi, C. Serpico et al. 2013. Monitoring DNA contamination in handled vs. directly excavated ancient human skeletal remains. *PLoS ONE* 8(1):e52524. doi:10.1371/journal.pone.0052524.

Pinto, N., L. Gusmao, and A. Amorim. 2011. X-chromosome markers in kinship testing: A generalisation of the IBD approach identifying situations where their contribution is crucial. *Forensic Sci Int Genet* 5(1):27–32.

Prieto, L., C. Alves, B. Zimmermann et al. 2013. GHEP-ISFG proficiency test 2011: Paper challenge on evaluation of mitochondrial DNA results. *Forensic Sci Int: Genet* 7:10–15.

Primorac, D. 2001. Validation of PP16 System on teeth and bone samples. *Proceedings of the Promega's STR Educational Forum*, June 5–6. Zagreb, Croatia: Promega Corporation.

Primorac, D. 2004. The role of DNA technology in identification of skeletal remains discovered in mass graves. *Forensic Sci Int* 146(Suppl.):S163–S164.

Primorac, D. 2009. Human Genome Project-based applications in forensic sciences. *Croat Med J* 50(30):2005–2006.

Primorac, D., and F. Paić. 2004. DNA Analysis in forensic medicine. In *Medical and Diagnostic Biochemistry in Clinical Practice*, ed. E. Topic, D. Primorac, and S. Jankovic, 365–384. Zagreb: Medicinska Naklada (Croatian edition).

Primorac, D., and D. Marjanović. 2008. DNA Analysis in forensic medicine and legal system. In *DNA Analysis in Forensic Medicine and Legal System*, ed. D. Primorac, 1–59. Zagreb: Medicinska Naklada (Croatian edition).

Primorac, D., M.L. Stover, S.H. Clark, and D.W. Rowe. 1994. Molecular basis of nanomelia, a heritable chondrodystrophy of chicken. *Matrix Biol* 14:297–305.

Primorac, D., S. Andelinovic, M. Definis-Gojanovic et al. 1996. Identification of war victims from mass graves in Croatia and Bosnia and Herzegovina trough the use of DNA typing and standards forensic methods. *J Forensic Sci* 41:891–894.

Primorac, D., C.V. Johnson, J.B. Lawrence et al. 1999. Premature termination codon in the aggrecan gene of nanomelia and its influence on mRNA transport and stability. *Croat Med J* 40:528–532.

Primorac, D., S.M. Schanfield, and D. Primorac. 2000. Application of forensic DNA in the legal system. *Croat Med J* 41:33–47.

Primorac, D., D. Marjanović, P. Rudan, R. Villems, and P. Underhill. 2011. Croatian genetic heritage: Y chromosome story. *Croat Med J* 52:225–234.

Primorac, D., G. Lauc, and I. Rudan. 2013. Predicting age from biological markers in forensic traces. *The Eight ISABS Conference on Forensic, Anthropologic and Medical Genetics and Mayo Clinic Lectures in Translational Medicine*, June 24–28 (Book of abstracts, p. 66). Split, Croatia.

Prinz, M., and M. Sansone. 2001. Y Chromosome-specific short tandem repeats in forensic casework. *Croat Med J* 42:288–91.

Promega Corporation. 2013. *Plexor® HY System for the Applied Biosystems 7500 and 7500 FAST Real-Time PCR Systems – Technical Manual* (Revised 5/13). Madison, WI: Promega Corporation.

Promega. 2013. *PowerPlex® ESX 16 System – Technical Manual-Instruction for use of products DC6770 and DC6771* (Revised 5/13). Madison, WI: Promega Corporation.

Promega. 2013. *PowerPlex® ESI 17 Pro System – Technical Manual-Instruction for use of products DC6780 and DC7781* (Revised 5/13). Madison, WI: Promega Corporation.

Qiagen. 2012. *Investigator® Quantiplex Handbook.*

Qiagen. 2012. *Investigator® Quantiplex HYres Handbook.*

Raymond-Menotti, M. 2001. Genetics of coat patterns in the domestic cat. In: *Proceedings of The Seventh ISABS Conference in Forensics, Anthropologic and Medical Genetics and Mayo Clinic Lectures in Translational Medicine*, ed. D. Primorac, M. Schanfield, and S. Vuk-Pavlovic, June 20–24, 2011. Bol, Croatia: ISABS.

Report of the Committee on the Human Mitochondrial Genome. http://www.mitomap.org/report (accessed July 7, 2009).

Read, A., and D. Donnai. 2011. *New Clinical Genetics.* Banbury: Scion Publishing Ltd.

Roney, C., A. Spriggs, C. Tsang, and J. Wetton. 2001. A DNA test for the identification of tiger bone. In: *Proceedings of The Second European–American Intensive Course in Clinical and Forensic Genetics*, ed. D. Primorac, I. Erceg, A. Ivkošić, September 3–14, 2001, Dubrovnik, Croatia. Zagreb, Croatia: Studio HRG.

Saferstein, R. 2001. *Criminalistics: An Introduction to Forensic Science*, 7th edn. New Jersey: Prentice-Hall.

Saiki, R.K., D.H. Gelfand, S. Stoffel et al. 1988. Primer-directed enzymatic amplification of DNA with a thermostable DNA polymerase. *Science* 2239:487–491.

Saiki, R.K., S. Scharf, F. Falooona et al. 1985. Enzymatic amplification of beta-globin sequences and restriction site analysis for diagnosis of sickle cell anemia. *Science* 230(4372):1350–1354.

Savolainen, P., L. Arvestad, and J. Lundeberg. 2000. A novel method for forensic DNA investigations: repeat-type sequence analysis of tandemly repeated mtDNA in domestic dogs. *J Forensic Sci* 45:990–999.

Schanfield, M. 2000. DNA: PCR-STR. In: *Encyclopedia of Forensic Sciences*, J.A. Siegel, P.J. Saukko, and G.C. Knupfer, 526–535. London: Academic Press.

Schlotterer, C., and D. Tautz. 1992. Slippage synthesis of simple sequence DNA. *Nucleic Acids Res* 20:211–215.

Scientific Working Group on DNA Analysis Methods (SWGDAM). 2010. http://www.fbi.gov/about-us/lab/biometric-analysis/codis/swgdam.pdf (accessed March 25, 2013).

Semino, O., G. Passarino, J.P. Oefner et al. 2000. The genetic legacy of paleolithic *Homo sapiens sapiens* in extant Europeans: A Y Chromosome perspective. *Science* 290:155–159.

Shirley, N., L. Allgeier, T. LaNier, and H. Miller Coyle. 2013. Analysis of the NMI01 marker for a population database of Cannabis seeds. *J Forensic Sci* 58:S176–S182.

Shows, T.B., P.J. McAlpine, C. Boucheix et al. 1987. Guidelines for human gene nomenclature. An international system for human gene nomenclature (ISGN, 1987). *Cytogenet Cell Genet* 46:11–28.

Škaro, V., C.D. Calloway, S.M. Stuart et al. 2011. Mitochondrial DNA polymorphisms in 312 individuals of Croatian population determined by 105 probe panel targeting 61 hypervariable and coding region sites. Abstract. *Proceedings of the Seventh ISABS Conference in Forensic, Anthropologic and Medical Genetics and Mayo Clinic Lectures in Translational Medicine.* http://www.isabs.hr/PDF/ISABS_2011_abstract_book_web.pdf (accessed February 24, 2013).

Slack, F.J. 2006. Regulatory RNAs and the demise of "junk" DIVA. *Genome Biol* 7(9):328.

Sprecher, C., B. Krenke, B. Amiott, D. Rabbach, and K. Grooms. 2000. The PowerPlex™ 16 System. *Profiles DNA* 4:3–6.

Stover, M.L., D. Primorac, S.C. Liu, M.B. McKinstry, and D.W. Rowe. 1994. Defective splicing of mRNA from one COL1A allele of Type I collagen in nondeforming (Type I) osteogenesis imperfecta. *J Clin Invest* 92:1994–2002.

Sutlović, D., M. Definis-Gojanović, S. Anđelinović, D. Gugić, and D. Primorac. 2005. Taq polymerase reverses inhibition of quantitative real time polymerase chain reaction by humic acid. *Croat Med J* 46(4):556–562.

Szibor, R., M. Krawczak, S. Hering, J. Edelmann, E. Kuhlisch, and D. Krause. 2003. Use of X-linked markers for forensic purpose. *Int J Legal Med* 117:67–74.

Tamarin, H.R. 2002. *Principles of Genetics*. New York: McGraw-Hill.
Urquhart, A., C.P. Kimpton, T.J. Downes, and P. Gill. 1994. Variation in short tandem repeat sequences—a survey of twelve microsatellite loci for use as forensic identification markers. *Int J Leg Med* 107:13–20.
Van Kirk, M.E., A. McCary, and D. Podini. 2009. The molecular basis of microvariant STR alleles at the D21S11 locus using forensic DNA identification. In *Proceedings of the American Academy of Forensic Sciences, Annual Scientific Meeting*, February 16–21, Denver, CO, Abstract A134, p. 118.
Venter, J.C., M.D. Adams, E.W. Myers et al. 2001. The sequence of the human genome. *Science* 291:1304–1351.
Wagner, J.K. 2013. Out with the "junk DNA" phrase. *J Forensic Sci* 58:292–294.
Wambaugh, J. 1989. *The Blooding*. New York: Bantam Books.
Waye, J. 2000. DNA: RFLP. In: *Encyclopedia of Forensic Sciences*, ed. J.A. Siegel, P.J. Saukko, and G.C. Knupfer. London: Academic Press.
Y Chromosome Consortium. 2002. A nomenclature system for the tree of human Y-chromosomal binary haplogroups. *Genome Res* 12:339–348.
Zeller, M., D.H. Wehner, and V. Hemleben. 2001. DNA fingerprinting of plants. In: *Proceedings of The Second European–American Intensive Course in Clinical and Forensic Genetics*, ed. D. Primorac, I. Erceg, and A. Ivkošić, September 3–14, 2001, Dubrovnik, Croatia. Zagreb, Croatia: Studio HRG.

Forensic DNA Analysis and Statistics

2

MOSES S. SCHANFIELD
DRAGAN PRIMORAC
DAMIR MARJANOVIĆ

Contents

2.1	Introduction	56
	2.1.1 Genetic and Statistical Principles in Forensic Genetics	56
	2.1.2 Principles of Parentage Testing	56
	2.1.3 Hardy–Weinberg Equilibrium	56
	2.1.4 Linkage Equilibrium	58
2.2	DNA Evidence in Court	58
2.3	Forensic Identification	59
	2.3.1 Correction for Substructuring	60
	2.3.2 Individualization and Identification	61
	2.3.3 Analysis of Mixed Samples	62
	2.3.4 Low Copy Number Calculations	64
	2.3.5 Presentation of Y-STR Analysis Results	65
	2.3.6 Mitochondrial DNA Testing	66
	2.3.7 X Chromosome STR Testing	66
	2.3.8 Parentage Testing	67
	2.3.9 Paternity Index or Combined Paternity Index	68
	2.3.10 Probability of Paternity	69
	2.3.11 Random Man Not Excluded	71
	2.3.12 Motherless Paternity Testing	72
	2.3.13 Effect of Mutations	74
	2.3.14 Maternity Testing	74
	2.3.15 Parentage Testing with Mixed Populations	76
	2.3.16 Using Y and X Chromosome Markers in Cases of Disputed Parentage	76
2.4	Identification of Human Body Remains	77
	2.4.1 Victim Identification Using Parental DNA	78
	2.4.2 Victim Identification Using Child's DNA	79
	2.4.3 Parentage Testing versus Forensic Identification	80
References		81

2.1 Introduction

2.1.1 Genetic and Statistical Principles in Forensic Genetics

The distribution (segregation) of parental genotypes in offspring depends on the combination of alleles in the parents. Mendelian laws of inheritance determine the expected distribution of alleles in offspring of various matings. The principles of Mendelian inheritance are formulated in the form of three laws.

1. *Monohybrid crossing:* if a father who is homozygote (A,A) mates with a mother who is also homozygous (B,B), all offspring will be heterozygous, that is, A,B.
2. *Segregation:* If two heterozygous A,B parents mate, the following genotypes are possible: A,A, A,B, and B,B at a ratio of 1:2:1. This example shows that there is segregation of alleles at the same locus, such that on average 50% of gametes will carry the A allele, whereas the other 50% of gametes will carry the B allele.
3. *Independent assortment:* When individuals are segregating at more than one unlinked loci—that is, each locus segregates independently of the other—then, the segregation of alleles at one locus is independent of the segregation of alleles of the other loci during meiosis.

2.1.2 Principles of Parentage Testing

Based on Mendelian laws, the four rules for paternity/maternity testing are established:

1. A child, with the exception of a mutation, cannot have a marker (allele) that is not present in one of its parents (direct exclusion).
2. A child must inherit one marker (allele) from a pair of genetic markers from each parent (direct exclusion).
3. A child cannot have a pair of identical genetic markers, unless both parents have the same marker (indirect exclusion).
4. A child must have a genetic marker that is present as an identical pair in both parents (indirect exclusion).

2.1.3 Hardy–Weinberg Equilibrium

Gregor Mendel described the behavior of alleles in various matings. Similarly, in 1908, Hardy and Weinberg, independently, explained the behavior of alleles in the population. The Hardy–Weinberg equilibrium (HWE) describes the expected relationship between allele frequencies and genotype frequencies at a single locus (Weir 1993; Primorac et al. 2000). Thus, for a locus with two alleles A and B, in which A has a frequency of p and B has a frequency q, the HWE expected frequencies of genotypes from these allele frequencies is the binomial expansion of $p + q$, such that

$$(p + q)^2 = p^2 + 2pq + q^2 = 1.00,$$

where

$$p + q = 1.00,$$

where p^2 is the frequency of the homozygous A,A individuals; q^2 is the frequency of the homozygous B,B individuals; and $2pq$ is the frequency of the heterozygous A,B individuals.

This can be expanded to any number of alleles in the multinomial example. Thus, for three alleles, A, B, and C, with frequencies of p, q, and r, respectively, and $p + q + r = 1.000$. Then, the HWE expected values for the genotypes will be the multinomial expansion of the allele frequencies, such that the HWE expected distribution of $(p + q + r)^2$ will be

$$p^2 + 2pq + 2pr + q^2 + 2qr + r^2 = A,A + A,B + A,C + B,B + B,C + C,C.$$

The HWE relates to the constancy of the frequency of genotypes over generations, and since allele frequencies are generated from the genotype frequencies, then the constancy of allele frequencies over generations is also implied. The HWE has several assumptions: that the population is very large, that individuals are mating at random, that the allele frequencies in males and females are the same, that there are nonoverlapping generations, and that there is no mutation and no migration (gene flow) into the population or natural selection. For human populations, many of these assumptions are or can be violated.

However, deviations from HWE may occur for several reasons:

1. Nonrandom formation of reproductive couples violates the assumption of random mating. In human populations, this can occur in two ways: *inbreeding* (mating of genetically related individuals) and *outbreeding* (mating of individuals known not to be related); only inbreeding violates HWE. Inbreeding—because the parents are closely related—can lead to a reduction in heterozygosity; in contrast, outbreeding increases heterozygosity in a parent population. *Inbreeding only affects genotype frequencies and not allele frequencies, whereas inbreeding affects all loci.*
2. A population that is not infinitely large can lead to g*enetic drift*, which is a random change in allele frequency within smaller populations when the next generation is not a random sample of the genotypes in the previous generation. Under certain conditions, some alleles may entirely disappear (loss of alleles) or be present in every individual within that population (fixation). Thus, genetic drift will lead to changes in allele and genotype frequencies over time. Genetic drift affects individual loci.
3. Mutation or change in alleles, which may affect the allele frequency; however, it acts very slowly as most mutation rates are very low.
4. Gene flow between populations changes the frequency alleles and impacts the HWE until it reaches a new equilibrium. Continuous gene flow will alter the genetic structure by changing allele frequencies and genotype frequencies over generations. The affect of gene flow will depend on the differences in allele frequencies between the two mixing populations.
5. Natural selection will lead to increased frequency of the favorable alleles over time since children with the favored allele are more likely to survive than children with other alleles. Examples of G6PD deficiency or thalassemia are found in areas with malaria. Note: In adulthood, the genotype distribution will not be HWE; it will only be seen in newborns or before the natural selection occurs.

6. Random mating also implies that the population is not substructured or divided into multiple breeding units that individually may be in HWE, but across the populations isolated by distance or physical barriers there will be an overall reduction in heterozygosity in the total population.

Although many of the parameters of HWE are often violated, it is very difficult to establish that a population is out of HWE. To do so, you would have to show significant deviations from HWE statistical tests at multiple loci. It is possible to indicate significant substructuring at the population over large distances and sometimes within a population that has had obvious recent mixing, in the absence of gene flow.

2.1.4 Linkage Equilibrium

Two loci close to each other on a chromosome will not recombine because they are physically too close together for sister chromosomes to exchange DNA. This is referred to as genetic linkage. The further apart two loci are, the higher the likelihood of recombination. At its maximum, recombination will occur in 50% of meiosis. Loci on different chromosomes are not physically linked, but through segregation will recombine 50% of the time. Linkage equilibrium denotes concepts by which loci on the same chromosome that are physically unlinked and loci on different chromosome should behave as independent loci. This is the HWE equivalent of Mendel's law of independent assortment. *Linkage disequilibrium* can occur when a new mutation occurs in a region of DNA that was previously in linkage equilibrium, but now the mutation is associated with a single region. Until recombination moves the new mutation to all of the alternative sequences for that region, the region will be in linkage disequilibrium. The amount of time it will take the region to come into linkage equilibrium is inversely related to the genetic distance between the markers—that is, the closer together they are on the same DNA segment, the longer it will take to reach equilibrium. Regions of DNA that are in linkage disequilibrium are inherited in blocks referred to either as "haplogroups" or "haplotypes" depending on the number and type of markers that define the region. For markers on different chromosomes, the maximum rate of recombination due to segregation is 0.5. *Gametic disequilibrium* occurs when individuals from two different populations mix, and when alleles on different chromosomes have markedly different frequencies then certain unlinked alleles may occur together more often than predicted by chance because of the recent mixing. Gametic disequilibrium will disappear in a mixed population more quickly than linkage disequilibrium. Both types of equilibrium can be detected at the population level and at the segregation level in families. Because these states of disequilibrium violate the law in independent assortment, they have an impact on the calculation of match likelihood and parentage statistics.

2.2 DNA Evidence in Court

Chapters 20, 21, 22, and 23 describe in great detail DNA in the courtroom, ethics in the laboratory and the courtroom, DNA immigration and human trafficking, and DNA data banking. In many countries, presentation of scientific evidence belongs to the testimony of experts (expert witness). But the fact that the experts carried out analyses does not mean that such evidence will be automatically accepted. Under normal circumstances, it must be proven that the presented evidence material is reliable and informative. The Council of Europe and the Committee of Ministers, issued on February 10, 1992, during its 470th meeting, the recommendation No. 92 on the use of DNA analysis in the criminal justice system (Council of Europe 1992). Some of the most important guidelines are the following:

1. Samples collected for DNA analysis and the information derived from such analysis for the purpose of the investigation and prosecution of criminal offences must not be used for other purposes. However, where the individual from whom the samples have been taken so wishes, the information should be given to him.
2. Samples collected from living persons for DNA analysis for medical purposes, and the information derived from such samples, may not be used for the purposes of investigation and prosecution of criminal offences unless in circumstances laid down expressly by the domestic law.
3. Samples or other body tissues taken from individuals for DNA analysis should not be kept after the rendering of the final decision in the case for which they were used, unless it is necessary for purposes directly linked to those for which they were collected.

The FBI launched the National DNA Index System (NDIS) in 1998—along with the Combined DNA Index System (CODIS) software to manage the program—and since that time it has become the world's largest repository of known offender DNA records. CODIS includes a Convicted Offender Index (containing profiles of offenders submitted by states) and a Forensic Index (containing DNA profiles of evidence related to unsolved crimes). Last year, in partnership with local, state, and federal crime laboratories and law enforcement agencies, CODIS aided nearly 25,000 criminal investigations (FBI 2011).

Chris Asplen (personal communication) provides the following for U.S. courts: Samples or other body tissues taken from individuals for DNA analysis should not be kept after the rendering of the final decision in the case for which they were used, unless it is necessary for purposes directly linked to those for which they were collected. Sample retention in the United States is a state-specific policy particular to each State's legislation. However, Federal legislation does require that all states have legislation requiring the destruction of sample and removal of profile, subsequent to a finding of "not guilty" or the dropping of charges.

2.3 Forensic Identification

The aim of the forensic identification is to compare evidence (blood, body fluids, or tissue) with a victim or suspect. Within a given population, each person has a unique genetic record and, except in case of identical twins (identical twins, are genetically identical except for mutations that occur after separation of the twins), DNA results will normally unambiguously link or exclude a suspect/victim from an item of evidence. When evidence and a possible donor are the same, then the match probability is calculated to provide additional weight of the evidence. For example, formerly, in sexual assault cases ABO blood group system was used. If the victim has blood group O, and the evidence belongs to blood group A, then the suspect must have blood type A. If the person does have blood group A, then it cannot be excluded from the list of suspects. Is this information useful and does it serves as a proof in the legal sense? Given that approximately 40% of European populations belong to blood group A, such information is not too helpful. When it comes to DNA analysis, only a small portion of the population can be randomly matched. In determining kinship, the frequency of allele donors is an important value. For identification, the statistics based on the expected frequency of genotype are used. As mentioned above, the expected frequencies of genotypes are determined by HWE.

In forensic identification, two basic approaches to communication of results can be applied: the first involves informing the results in the form of frequency of the assessed DNA profile in a specific population (Marjanović et al. 2006). Thus, the final formulation may contain the fact that the frequency of the assessed DNA profile within the Croatian population is 1.33×10^{-9}. This is often referred to as the random match probability (RMP). This formulation clearly and accurately displays the results in mathematical terms. In contrast, a likelihood ratio (LR), which is simply the inverse of the RMP, denotes how many times more likely it is that the tested person contributed to the evidence sample than someone at random from the population. For example, if the RMP is 0.001, the LR says that the person tested is 1000 times as likely to be the donor of the sample as a random individual. Note: This is not the probability that the person tested is the donor. In this way, the common stand of the prosecution (evidence originates from a suspect) whose probability is 1, and the stand of the defense (evidence comes from someone else), where the probability for homozygotes is p^2, and for heterozygotes is $2pq$, are compared. In the forensic genetics statistics, this concept is called a *likelihood ratio* (Butler 2010). LR for one, say, heterozygous locus, equals $1/2pq$.

Table 2.1 shows the likelihood of a random match (RMP) is 4.214×10^{-18} or the LR originated from the suspect 237,303,870 billion times greater than an unrelated random individual.

2.3.1 Correction for Substructuring

In the example above, only the HWE estimators of genotype frequencies were used to estimate the RMP and LR values. In the United States where forensic DNA testing was initially regulated by the DNA Advisory Board and later the U.S. Federal Bureau of Investigation, calculations were influenced by the results of the second National Research Council (NRC 1996). In that report, the NRC suggested that homozygous genotypes can significantly decrease the likelihood of an RMP. They suggested that a conservative approach was to increase the frequency of the homozygous genotype $\left(p_i^2\right)$ by the correction for substructuring, such that the following formula applied.

$$\text{Corrected homozygous genotype frequency} = p_i^2 + \theta \times p_i \times (1 - p_i),$$

where θ is the degree of substructuring and p_i are the allele frequencies. For areas with large populations such as the United States, Europe, and Croatia, θ is set at 0.01, whereas for smaller endogamous populations such as Roma or religous isolates θ is set at 0.03.

If the corrected value is used for the homozygous genotype, whereas the uncorrected value is for heterozygous genotypes, the following inequality will always hold, and the results will always be the most conservative value for the RMP.

Forensic DNA Analysis and Statistics

Table 2.1 Example of Forensic Identification of a Blood Stain by Analyzing 15 STR Loci

STR Locus	Known Sample of a Suspect	Question Blood Evidence	p_i	p_j	Formula	Frequency
D3S1358	16, 18	16, 18	0.265	0.155	$2p_ip_j$	0.0821
TH01	6, 9.3	6, 9.3	0.225	0.330	$2p_ip_j$	0.1485
D21S11	28, 32.2	28, 32.2	0.165	0.085	$2p_ip_j$	0.0281
D18S51	12, 17	12, 17	0.080	0.100	$2p_ip_j$	0.0160
PENTA E	7, 13	7, 13	0.190	0.155	$2p_ip_j$	0.0589
D5S818	12, 12	12, 12	0.340	0.340	p_i^2	0.1156
D13S317	12, 13	12, 13	0.270	0.075	$2p_ip_j$	0.0405
D7S820	10, 11	10, 11	0.290	0.185	$2p_ip_j$	0.1073
D16S539	11, 12	11, 12	0.340	0.280	$2p_ip_j$	0.1904
CSF1PO	11, 12	11, 12	0.245	0.345	$2p_ip_j$	0.1691
PENTA D	9, 9	9, 9	0.245	0.245	p_i^2	0.0600
VWA	16, 19	16, 19	0.205	0.065	$2p_ip_j$	0.0267
D8S1179	12, 15	12, 15	0.165	0.080	$2p_ip_j$	0.0264
TPOX	8, 8	8, 8	0.570	0.570	p_i^2	0.3249
FGA	21, 22	21, 22	0.155	0.190	$2p_ip_j$	0.0589
Amelogenin	X Y	X Y				
Combined frequency (CF)						4.214×10^{-18}
Likelihood 1/CF						237,303,870,652,031,000

Note: p_i and p_j denote frequency of allele variants in Bosnia-Herzegovinian population (Marjanovic et al. 2006).

$$\text{NRC2 inequality: } p_i^2 + \theta \times p_i \times (1 - p_i) + 2p_ip_j$$
$$> p_i^2 + \theta \times p_i \times (1 - p_i) + 2p_ip_j > 2\theta \times p_i \times (1 - p_i)$$

For the example in Table 2.1 (the RMP), if $\theta = 0.01$ is used, the RMP goes from 4.214×10^{-18} to 4.462×10^{-18} or a 5.6% decrease in the RMP, indicating that it is a more conservative number.

2.3.2 Individualization and Identification

Individualization is based on a fact that a particular person is characterized by that individual's genetic information. The example of that is the identification through fingerprints, where it is possible to distinguish that a certain fingerprint belongs to the exact person. Even identical twins do not have identical fingerprints, although they carry the same genetic structure. Today, with the number of available markers, it is possible to achieve practically any degree of individualization. The degree of individualization required for identification, if there is no international agreement, must be established at the local level. In order to do that, it is important to clearly determine the frequency of allele variation within the studied population, as well as its total size. The resulting probability in Table 2.1 is more than sufficient for both the Bosnia-Herzegovinian and the Croatian populations, but even much lower values obtained by DNA profiling of an individual at a smaller number of STR loci in some cases may be sufficient for the Croatian population. Specifically, the probability of an accidental match of, for example, 1 in 23 billion Croats or population of Bosnia and Herzegovina, is solid evidence that the sample originated from a common source. The definition of individualization is an administrative one. Such a rule originally proposed by the FBI was approximately 100 times the U.S. population, such that any value less than 1 in 30 billion was considered individualized, such that an expert could report that "to a reasonable degree of scientific knowledge the profile had to originate with the suspect or their identical twin" (Budowle et al. 2000).

2.3.3 Analysis of Mixed Samples

One of the important aspects of DNA analysis is certainly the possibility of determining the DNA profile from the mixed evidence and prospect of their analysis. Such things could have not been achieved using any prior methods. Mixtures arise when two or more individuals are contributors to the sample being tested, for example, they are biological sources of the sample (Primorac 2008). In contrast to the DNA profiles of "common biological evidence," DNA profiles of mixed samples often show more than two prominent peaks at the analyzed loci, and/or apparent discordance within those loci that have two or only one peak. According to Clayton et al. (1998), there are six basic steps in interpreting the results of DNA analysis of mixed evidence: (1) identifying the presence of mixed evidence, (2) precise discrimination of all allelic variants present in the mixture, (3) determining the number of potential contributors, (4) determining the quantitative ratio of contribution of individual biological evidence of each of the donors who form the mixture, (5) determining all possible genotypic combinations, and (6) comparison with the referential known samples.

Mixed samples are not uncommon in forensics, particularly in cases of rape, where a large number of evidence (vaginal swab, evidence of semen on the underwear, evidence of sperm from the victim's body, etc.) are mixed, with a perpetrator and a victim as donors (contributors). Analysis of this evidence implies additional statistical tests that first try to distinguish clearly between each profile that makes up the mixture. The circumstances that can facilitate such analysis are: the number of contributors in a mixture is the lowest possible number (two contributors); the quantity of each component is different, that is, the dominant and minor fraction can be distinguished; and contributors have as many unshared allelic variants at one locus. In fact, when it comes to two contributors who are heterozygotes at the same locus with unshared allelic variants, and if the two fractions are represented in different quantities, then the analysis of mixed evidence, primarily based on analysis of peak area, is relatively simple (Evett and Weir 1998). On the other hand, if there are multiple fractions, and if they are all more or less equally represented, the analysis of such evidence is much more complicated.

Mixtures fall into two categories. In sexual assault cases you can have a mixture of a known individual (the victim) and an unknown individual(s) (the perpetrator[s]). This is very different from mixtures of blood or touch DNA from two or more unknown individuals. In the case of sexual assault evidence, it appears to be permissible to use the known DNA profiles to deduce perpetrator profile; however, this may not be totally satisfactory as there can be situations with overlapping alleles. Thus, if at a given locus there are four peaks, with two matching the victim, then the other two peaks have to come from the semen donor. However, if there are less than four peaks, this can have a more complex interpretations. A quick look at 10 European paternity cases tested for eight loci (Powerplex 1.1) on the average 3.6 loci shared no alleles between the male and female, 3.6 loci shared one allele between the male and female, and in 0.8 cases the male and female shared two alleles (AGTC unpublished data). For other types of mixtures, the interpretation is more complex. To aid in the interpretation of these mixtures, various guidelines have been put forward.

Since the detection of mixtures is becoming more common as the types of evidence being tested are expanded. To deal with the interpretation of mixture evidence, both the International Society of Forensic Genetics (ISFG) (Schneider et al. 2006; Gill et al. 2006) and the German Stain Commission (Schneider et al. 2009) have put forward guidelines for interpreting mixtures. John Buckleton, a member of the ISFG committee that propounded recommendations for the interpretation of mixed stains, was concerned enough to do additional work on the interpretation of mixed stains (Curran and Buckleton 2010). As of the present time, there has not been a forensic validation study on the error rates for the different guidelines. The guidelines identify three classes of stains referred to as A, B and C. Class A stains are a mixture in which all peaks represent large amounts of DNA, such that it is recommended that an RMP be calculated for all of the bands together (formula below). The FBI/SWGDAM also issued a report on the interpretation of mixtures (Budowle et al. 2009).

The RMP for a mixture is based on the combined HWE estimates of all of the possible genotypes, for example, if there are four alleles found such as D8S1179 9,10,13 and 14, then the possible genotypes are

$$2 \times 9,10 + 2 \times 9,13 + 2 \times 9,14 + 2 \times 10,13 + 2 \times 10,14 + 2 \times 13,14 + 9^2 + 10^2 + 13^2 + 14^2.$$

The number in the formula such as $2 \times 9,10$ are 2 times the frequency of the 9 allele and the 10 allele. The numbers represent the frequency of all of the different genotypes in the mixture, which is the same as P^2 where P is the sum of the allele frequencies of all of the "alleles" in the mixture.

Fortunately, this is the same as P^2, where P is the sum of the allele frequencies observed in the Class A array. Furthermore, all of the individual RMP for the loci can be multiplied to obtain the combined RMP for the mixture.

Class B is described as a single group in which there are dominant peaks representing a lot of DNA, and minor peaks, but representing enough DNA to generate peaks that are readily identified. In reality, class B must be subdivided into the former group that will be referred to as B1 and a second group that will be referred to as B2, in which the low level minor component can have stochastic effects and not all alleles present are detected. For class B1, it was recommended that only the major peaks should be used to calculate an RMP, with the calculation performed as in class A mixtures. However, in class B2 this often leads to loci in which the perpetrator's alleles are not detected. To solve this problem, some individuals eliminated loci from the calculation that do not include the suspect. Curran and Buckleton (2010) indicate that this procedure has not only not been forensically validated, but in a simulation performed with known pairs indicates that a high percentage of uninvolved third-party individuals can be included with a high degree of statistical significance, creating what is considered to be the worst type of statistical error, the false inclusion of individuals that are not involved.

Class C mixtures are those in which all levels of DNA are present and stochastic effects have occurred (see low copy number calculations). In this case, no calculations should be made because of the uncertainty.

Two basic principles can be clearly distinguished when interpreting the results. One of them is directed to the generalized conclusion that the examinee cannot be excluded as a potential contributor. Such an approach is usually applied in cases where the fractions are equally represented and when each individual contributor cannot be clearly defined. Then, on the basis of comparison of DNA profile of a known sample and the DNA profile of a mixture, only the (im)possibility of exclusion can be discussed. The second is more exact and gives statistically stronger results, and is applied in cases where the fractions in the mixture can quantitatively be clearly distinguished, that is, when it can be confirmed with a great certainty which specific allelic variant belongs to which contributor. Then, different statistical approaches can be preformed to calculate different levels of probability that, as a final result, give the number that indicates how likely it is that the examinee is a mixture contributor or how likely that he is not. See Chapter 7 for a more detailed discussion of testing and interpreting mixtures.

2.3.4 Low Copy Number Calculations

As DNA technology attempts to resolve greater numbers of forensic issues, the complexity of the statistics increases. Low copy number (LCN) DNA testing is presented in Chapter 6; however, it is necessary to include it here because it has an impact on the calculation of mixtures, which follows. The presence of ambiguous results when doing LCN DNA complicates the calculation of a coincidental match probability. The calculation of routine random match probabilities is described above, and in its simplest form consists of the product of all of the homozygous genotype frequencies times the product of all of the heterozygous genotype frequencies. In the case of LCN testing, there is a third phenotype, the ambiguous result indicated by the presence of a "Z", for example, D7S820 10,Z (Caragine et al. 2009). The expected value for "Z" phenotypes (note that this is a phenotype, because the true genetic nature is not known). The frequency of "Z" is taken to be $2p_i$.

Therefore, the RMP for an LCN DNA profile will be

$$\text{RMP} = \prod p_i^2 \times \prod 2 \times p_i \times p_j \times \prod 2p_i,$$

where Π represents the product of the homozygous, heterozygous, and ambiguous types detected in the LCN profile. An example of a calculation is found in Table 2.2.

Table 2.2 Comparison of RMP Values for Complete Profiles and LCN Profiles

Locus	Genotype	freq 1	freq 2	HWE	RMP	LCN	Likelihood	RMP
D13S317	11, 12	0.379	0.238	$2p_ip_j$	0.18	11, Z	$2p_i$	0.757
D16S539	11, 12	0.332	0.272	$2p_ip_j$	0.181	11, Z	$2p_i$	0.663
D8S1179	10, 15	0.073	0.136	$2p_ip_j$	0.02	10, 15	$2p_ip_j$	0.02
CSF1PO	10, 12	0.223	0.356	$2p_ip_j$	0.159	10, Z	$2p_i$	0.446
D21S11	30, 30	0.214		p_i^2	0.046	30, 30	p_i^2	0.046
D7S820	8, 10	0.178	0.223	$2p_ip_j$	0.079	10, Z	$2p_i$	0.446
					3.7×10^{-7}			9.0×10^{-5}

Note: Profiles are taken from Figures 8, 9, and 10 of Caragine et al. (2009). Allele frequencies are taken from Schanfield et al. (2002).

In Table 2.2, the LCN RMP is 9.0×10^{-5} whereas the complete profile is 3.7×10^{-7}, indicating that the LCN profile is 2 orders of magnitude less individualizing than the complete profile RMP, highlighting the loss of information when LCN is needed.

2.3.5 Presentation of Y-STR Analysis Results

As mentioned above, molecular markers positioned on the Y chromosome—besides their scientific importance in population genetics—can provide very significant results in the field of applied forensic genetics. (See Chapter 4 for more information on Y chromosome analysis in forensic sciences.) Their application was met with widespread approval and acceptance throughout the forensic community.

However, one of the unsolved parameters of their application is statistical presentation of the obtained results. Specifically, as previously stated, these markers can help in the so-called partial identification and individualization, because of their exclusive paternal inheritance and high degree of conservatism. In the absence of mutations, the Y-STR profile is identical in the male line through generations, that is, it is passed from father to all sons, and they pass it to their male offspring, etc. In fact, this very feature is the basic limit to the use of these markers in court.

One of the most widely accepted method of presenting results is determining the frequency of certain haplotypes within a specific population. In this regard, the forensic genetics community has settled on the so-called *minimal haplotype* (minHt) consisting of seven single copy Y-STR loci and one multicopy locus. Today, systems with significantly increased number of Y-STR loci are used, for example, PowerPlex® Y (Promega Corporation) with 11 loci (10 single copy loci and one polymorphic multicopy locus) and Yfiler™ (Applied Biosystems) with 16 loci (15 single copy loci and one polymorphic multicopy locus). Promega has announced the release of PowerPlex® Y23 system, which contains 23 loci in a five-color system.

When the Y-STR profiles of the examined evidence do not match, then one can conclude that they do not originate from the same evidence, and the statistical representation is usually not needed. But in situations where the Y-STR profiles of examined evidence match, then it is advised to use data from the world's largest online Y database, www.yhrd.org, which at the time this chapter is being written, contained 107,152 seven-locus haplotypes from (as of February 23, 2013). The simplest approach is to enter a specific haplotype in the database and to determine its prevalence in the world and the local population, on the basis of a formula $p = X/N$, where p is the frequency of a haplotype, X represents the number of times the analyzed haplotype is observed in the population, and N is the total number of all submitted haplotypes, that is, individuals in a given population database. However, a statistically more correct approach is to take into account the fact that the database contains only a limited sample of the population, so that the above formula should be corrected to $p = X/N + 1.96\{[(p)(1-p)]/N\}^{1/2}$. This correction places 95% *upper bound confidence interval*, and its application takes into account the fact that the number of rare haplotypes is not included in the database. For example, if some haplotype appears only five times in the YHRD database, its expected frequency in world population would be $p = 5/86{,}568 + 1.96\{[(5/86{,}568)(1 - 5/86{,}568)]/86{,}568\}^{1/2}$, which is 0.0000506 or 0.00506%. Of course, that number increases by reducing the population, that is, if the data for regional subpopulations is taken into the calculation, but it is still more than sufficient, especially if used along with the results of the standard DNA profiling. As the Y markers are independent of autosomal markers, the RMP can be multiplied by the corrected Y haplotypes frequency. According to Butler, when a Y-STR haplotype (or mtDNA sequence) has not been observed in a database of size N, the 95% confidence interval is $1 - (0.05)^{1/N}$ (Butler 2011).

2.3.6 Mitochondrial DNA Testing

The forensic aspects of mitochodrial (mtDNA) testing are covered in detail in Chapter 3; however, as with Y haplotypes frequencies, it is being included in this section on statistics. Since mtDNA is inherited in haplotypes or haplogroups, its frequency estimates are the same as for the Y haplotypes. All mtDNA haplotypes are based on SNPs and will vary in nature from those only based on two hyper-variable coding regions HVI or HVI, and HVII or total mtDNA sequence, and each will yield slightly different haplotypes or haplogroups affecting frequencies. Unfortunately, unlike the Y haplotypes registry for STR loci, there is no single registry for mtDNA haplotypes (haplogroups); many are commercial enterprises and some are public domain. In any case, if a frequency can be obtained, although they represent maternal lineages, they can be combined with other data as the Y haplotypes.

2.3.7 X Chromosome STR Testing

STR loci on the X chromosome are the newest addition to forensic and parentage testing (see Chapter 5 for more information on X chromosome STR loci). Loci on the X chromosome in mammals XX and XY have the same pattern of inheritance as all genes in haplodiploid organisms such as bees, ants, and wasps. In both types of organisms, there is an initial difference in allele frequencies in the two sexes, with ultimate convergence on an equilibrium frequency after several generations. For our purposes, we will assume that females are homogametic XX and the males will be heterogametic XY in this case. Thus, females are diploid, whereas males are haploid. This ultimately complicates the calculation of RMP for males and females. For female profiles, the calculation of random match likelihoods is the same as for any other diploid locus with heterozygotes equal to $2p_ip_j$ and homozygotes p_i^2, with or without the substructuring correction. In contrast, hemizygous males will always be estimated by the allele frequency p_i.

Unlike the autosomal markers used forensically and in parentage testing, which are known to be unlinked and scattered throughout the genome, all of the X-linked makers are on single chromosome and can be within a measureable linkage distance. Closely linked markers often are in linkage disequilibrium, which precludes simple multiplication and requires that haplotypes frequencies are used instead of allele frequencies for those linked markers. The total RMP will be the product of the respective individual loci, or individual loci and haplotypes if there are linkage groups included.

At present, there are a limited number of kits commercially available. The two that the authors are aware of are Mentype Argus X-8 from Biotype Diagnostic, GmbH (Dresden, Germany) (see Chapter 5 for more information) and GenePhile X-Plex from GenePhile Bioscience Co., Ltd. (Taipei, Taiwan). Both are single multiplexes with eight and 13 loci, respectively, which have had analysis of linkage disequilibrium performed. The Argus X-8 kit was extensively studied by Tillmar et al. (2008) and further discussed by Machado and Medina-Acosta (2009). The four pairs of markers in each of the four linkage groups on the X chromosome are in marked linkage disequilibrium and require the generation of

Table 2.3 Distribution of X-STR Loci in Three Sets of Kits for Use in Forensic Testing

Linkage Group	Map	Mentype Argus 8	Genphile	Diegoli and Coble
1	Xpter-22.2		DXS6807	
1	Xp22.4 (22.1)		DXS9902	DXS9902
1	Xp22.32	DXS10135		
1	Xp22.31	DXS8378	DXS8378	DXS8378
1	Xp22.1			DXS6795
2	Xq11.2	DXS7132	DXS7132	DXS7132
2	Xq12	DXS10074		
2	Xq21.1		DXS9898	
2	Xq21.1			DXS6803
2	Xq21.2		DXS6809	
2	Xq21.3		DXS6789	DXS6789
2	Xq22		DXS7424	DXS7424
2	Xq22		DXS101	DXS101
2	Xq23 (24)		GATA172D05	GATA172D05
2	Xq24			DXS7130
2	Xq25			GATA165B12
3	Xq26.2	HPRTB	HPRTB	HPRTB
3	Xq26.2	DXS10101		
4	Xq27			GATA31E08
4	Xq28			DXS10147
4	Xq28		DXS8377	
4	Xq28	DXS10134		
4	Xq28	DXS7423	DXS7423	DXS7423

Note: Map locations taken from publications, disagreements indicated in parentheses. Linked loci require the calculation of haplotype frequencies (the frequency of joint markers, e.g., HPRTB and DXS10101).

two locus haplotypes for each of the linkage groups. In contrast, Hwa et al. (2009) did not find a significant linkage disequilibrium across the 13 loci in the multiplex they evaluated, allowing them to simply multiply the individual loci. Recently, Diegoli and Coble (2010) developed two mini-multiplexes for forensic applications at the U.S. Armed Forces DNA Identification Laboratory under the guidance of the U.S. National Institute of Justice. They found no linkage disequilibrium among the 15 loci in Europeans, African Americans, U.S. Hispanics, and U.S. Asians. A comparison of the loci used in the three sets of kits is found in Table 2.3. See Chapter 5 for frequency data and more information on kits and other studies.

2.3.8 Parentage Testing

The need for determining the identity of an individual person or parentage has been around since biblical days. Even King Solomon encountered this problem, when two women came before him, each claiming to be the true mother of a child, and asking him to decide which one was the true mother. Realizing that he would not be able to resolve this based on their testimony, as only one could be speaking the truth, and trusting the fact that the real mother would do anything to save her child, Solomon suggested that the child should be cut in half and that each mother will be given a half. Solomon recognized the real mother as the woman who fiercely opposed the decision and assigned the child to her (Kaštelan and Duda 1987). Today, modern forensic laboratories perform parentage testing in less painful and more efficient ways, thanks to the development of DNA technology.

The primary function of all the forensic testing, regardless of whether it is parentage testing or personal identification, is to exclude the maximum number of individuals. In cases of parentage testing, this is done by identifying the obligate allele(s) and determining if the presumed or alleged parent also carries this allele(s) (in cases of disputed parentage, the disputed parent is always referred to as the *alleged parent*) (Schanfield 2000). The obligate allele is the allele that originated from the biological parent. For example, if we assume that the mother is AA, and the child is AB, then child's B allele had to arise from the biological father. In a case of disputed paternity, it is always assumed that the mother is truly the mother of the child, unless the test proves disagreement with Mendelian laws of inheritance. If the alleged father is homozygous (two B alleles) or heterozygous (one B allele) for the B allele, he cannot be excluded as the biological parent. If the alleged father does not have the B allele, that is, a child cannot have inherited it from him, he must be excluded. The exception to this rule is when there is a relatively common mutation, A to B. Therefore, it is generally accepted that in the case when at least three genetic inconsistencies are found with the child being tested, the alleged father is automatically excluded as the biological father of the child. If both the mother and child are A,B, then the obligate alleles are A and B, and either allele, A or B could have come from the biological father. It is obvious that in this case there are two obligate alleles. Every alleged father who is the carrier of A or B allele cannot be excluded. However, if the alleged father has C,C or any other non-A or non-B genotype, he can be excluded.

2.3.9 Paternity Index or Combined Paternity Index

If the alleged father is not excluded as a possible biological father, one method of evaluating the weight of the evidence is equal to the relative probability that the alleged father passed the obligate allele to the child versus an unrelated individual from the same population. The ratio between the relative probability that the alleged father passed the obligate allele(s) and the probability that a randomly selected unrelated man from the same population could pass the allele to the child is called *paternity index* (PI). The PI is an LR or an "odds ratio," that is, the ratio of probabilities that the alleged father, and not another man from the population, passed the obligate allele. Mendelian laws determine probability or possibility that the alleged father can pass the gene. If the father is homozygous for obligate allele or has both obligate alleles, then the likelihood that he passed the obligate allele is 1.0 (2/2). If the father has only one copy of obligate allele or only one of the two obligate alleles, then the probability that he passed obligate allele(s) equals to 0.5 (1/2). The value of PI for a particular locus is equal to the probability that the alleged father can pass allele, divided by the frequency of obligate allele(s). Therefore, PI will be $1/p$ or $0.5/p$, depending on the number of obligate alleles, which the alleged father carries. If the genotypes of mother and child suggest two obligate alleles, p will be calculated using the following formula: $p = p_1 + p_2$, and the PI will be:

$$PI = 1/(p_1 + p_2) \text{ or } PI = 0.5/(p_1 + p_2).$$

PI is calculated for each locus tested.

The rules of probability theory determine how to combine multiple PIs to determine the combined probability of paternity. (Note: There are two rules for combining probabilities, the "and" rule and the "or" rule. The expected value of the heterozygote $2pq$ uses both rules. The probability [or likelihood] of getting a "p" and a "q" is the product of p and q, whereas the probability [or likelihood] of getting a p from the mother and q from the father, or a p from the father and a q from the mother, is the sum of the two p,q's.) To combine all of the individuals' paternity indices, each of the individual PIs is multiplied together. This calculation is called *combined paternity index* (CPI). CPI is the LR or odds ratio, that is, the CPI is the likelihood that the alleged father transmitted all of the obligate alleles to the child versus an unrelated male from the same population. This is expressed as the alleged father is X times as likely to have transmitted the obligate alleles as a random man. Rules for combining probabilities are quite simple. In this case, if you want to know the combined probability of the traits A and B and C, it is necessary to multiply the individual probabilities of A and B and C. Therefore, the likelihood that the alleged father passed alleles 1*A, 2*C and locus 3*E to the child equals to the product of probabilities 1*A2*C3*E. Simply put, no matter how many loci are used for the analysis of contested paternity, the CPI is determined by multiplying all calculated PIs, that is, the paternity indices calculated for each of the observed loci. In the beginning, when a relatively small number of loci were used for these purposes (3–5 loci), the acceptable value of CPI, say, for the German judiciary was 1000, whereas in the United States this value was initially 100. In other words, in these cases, CPI expresses the fact that the alleged father is 1000 or 100 times more likely to be biological father than any other nonrelated man from the same population. Nowadays, in the use of, say, 15 STR loci represented in most commercial kits (PowerPlex 16 or Identifiler), it is not unusual for the value of the CPI in the standard paternity test (mother–child–presumed father) to be in the range of several hundred thousand up to several million, whereas in the case of the motherless paternity testing, this value is somewhat lower. It is important to note that fluctuations in these values primarily depend on the frequency of alleles that biological father passes to a child. Since many have difficulties in understanding and interpreting the likelihood ratios, an alternative way of showing data was suggested that is based on a conversion of probability ratio in the probability of paternity.

2.3.10 Probability of Paternity

In the 18th century, the mathematician Bayes developed a theorem to assess the probability that an event occurred, even when this event cannot be directly measured. This is the basis of Bayesian statistics that is widely used, especially in areas such as genetic counseling for genetic risk assessment. Bayesian formula to estimate the possibility for an event to happen is:

$$\frac{X \times p}{X \times p + Y \times (1-p)},$$

where X is the probability that some event will occur, Y is the probability that it will not occur, p is the prior probability that X will happen, and $1 - p$ is the prior probability that it will not happen.

Bayes' formula has several limitations. The formula is exhaustive, that is, it includes all possibilities of the event space. In the case of paternity testing, X is the combined probability that the alleged father passed all the obligate alleles, and consists of the product of all 0.5 and 1.0 values for each locus, and Y is the combined probability that a nonrelated individual in the population is the biological father and the product of frequencies of all obligate alleles. The prior probability is the probability that this event could have happened without the knowledge of the current results. In this case, it represents the probability that the alleged father is the biological father before any laboratory tests were performed. There are different ways to calculate the prior probability that the alleged father is also the biological father. It can be assumed that it is a laboratory inclusion rate that is approximately 70%. It can also be assumed that the prior probability is the number of males that may be taken into account within the time when fertilization occurred or, more precisely, men from the area, but who are of reproductive age. However, it is customary to assume that it is the same probability that the alleged father is or is not the biological father. This is the so-called neutral prior probability. Therefore, the prior probability equals to 0.5. Some statisticians argue that the prior probability should be calculated by dividing the desired result by the total number of possible outcomes, for example, the prior probability to get "one" when throwing dice (half pair of dice) is equal to 1/6, whereas the probability of getting "heads" when tossing a coin is equal to 1/2. On the occasion of paternity testing, only two outcomes are possible—he is or he is not the biological father—the starting probability is equal to 1 for the two options, or 0.5, which means that it is possible that the "neutral starting probability" is the possible correct starting probability. If the value 0.5 is chosen as the initial probability, the formula reduces to the one that uses the CPI.

Probability of paternity (if prior probability = 0.5) = PP = $1/[1 + (1/CPI)]$

This formula can be further simplified to

Probability of paternity = $CPI/(CPI + 1)$.

All prosecutors, judges, and lawyers who read this formula will see clearly why it is not possible to obtain a score of 100%. In fact, no matter how large the CPI number is, it must always be divided by a number that is greater than 1 and so this ratio will never equal 1, and hence the probability of paternity, put in percentage, will never be 100%. Therefore, when the score is, say, 99.999999678998%, it is ridiculous to ask for absolute certainty. Unfortunately, however, DNA experts are often asked: "Are you 100% sure that the person concerned is the child's biological father?"

Thus, the probability of paternity is a possibility that the alleged father is the biological father of the child. Therefore, if a hypothetical CPI equals, say 6,477,698, then the probability that the alleged father is the biological father equals 0.999999846 or 99.9999846% [6,477,698/(6,477,698 + 1)]. In contrast, the probability that the alleged father is not the biological father is 0.000000154 or 0.0000154%.

The logical question that can be asked is, what is the lowest value of the probability of paternity, so that it can be fully accepted? In general, the rules indicate that in the standard testing, when the mother is present, it is desirable to get at least 99%, and if 13 to 15 loci are tested the value should be significantly greater than 99%. In cases of motherless paternity, because of the lack of certain information, the number must be greater than 0.9999, or

99.99%. Thus a high percentage of probability is determined by using a larger number of molecular markers, which is nowadays seldom less than 13. A large number of laboratories suggest analysis of additional molecular markers if they do not obtain these values, regardless of whether the father's identified allelic variants match with the obligate alleles at all loci tested.

2.3.11 Random Man Not Excluded

PI and probability of paternity have a common assumption with Mendelian likelihood that the alleged father is also the biological father of the child. To assess the same data in a different way, it is possible to ask how much genetic information is present in the mother–child pair, that is, what is the discrimination power of a test at preventing erroneous inclusion of alleged father. This is similar to the concept of "the power of the test" in statistics. Ideally, the test should be strong and practical and should exclude all falsely accused alleged fathers. Today, by the method of DNA analysis, approximately 30% of accused alleged fathers are excluded worldwide (Allen 1999).

The power of the paternity test can be determined via the so-called *random man not excluded* (RMNE). This statistical test can be compared with the calculation of a random match probability in the population, which is applied in forensic identification (see below). RMNE is the proportion of the population that can provide all obligate alleles and cannot be excluded. This term is used to express the frequency of individuals that cannot be excluded as potential donors of genetic material.

The formula for RMNE for one locus is

$$p^2 + 2p(1-p) \text{ or, simply, } 1 - (1-p)^2.$$

Combined random man not excluded (CRMNE) is analogous to the CPI, that is, it is equal to the product of individual values. The value of CRMNE is usually very low. However, it is easier to use the reciprocal value of formula (1 − CRMNE), for which the term "probability of exclusion" is commonly used. But since it is possible to mix the concept of "probability of exclusion" with the notion of "probability of paternity," the less confusing term "exclusionary power" (EP) or "power of exclusion" (PE) is used. EP or PE is the probability of excluding a man falsely accused (1 − CRMNE). When an *a priori* probability of 0.5 is used, PE equals the Bayesian probability of paternity in the so-called nonexclusion model. To avoid confusion with the term "probability of paternity," which derives from the model of inheritance, one should use the term "nonexcluded probability of paternity" or "Weiner's probability of paternity" by the scientist who first proposed the concept in 1976 (Weiner 1976). The term nonexcluded probability of paternity has several advantages over the standard probability of paternity (Li and Chakravarti 1988). The biggest advantage of this method is that it always increases by increasing the number of conducted tests and is completely analogous to the concept of population frequency or random match probability, which is used in forensic identifications. The probability of paternity (the term traditionally used for the probability of paternity with respect to the transfer of alleles) can be reduced if the alleged father is heterozygote and the mother and child are carrying the same heterozygous genotype containing two common alleles, where $p_1 + p_2$ is greater than 0.5.

Table 2.4 shows the general formulas used in paternity testing, whereas Table 2.5 shows an example of testing. The fourth column in Table 2.5 displays all of the obligate alleles of

Table 2.4 Formulas for Calculating Paternity Index

Alleles of a Mother	Alleles of a Child	Alleles of a Father	PI
AB	AA	AA	$1/a$
BC or BB	AB	AA	$1/a$
AA	AA	AA	$1/a$
BB or BC	AB	AB or AC	$1/2a$
AB	AA	AB or AC	$1/2a$
AA	AA	AB	$1/2a$
AB	AB	AB	$1/(a+b)$
AB	AB	AA	$1/(a+b)$
AB	AB	AC	$1/2(a+b)$

Source: More details can be found at www.DNA-view.com.
Note: A, B, C—alleles; a, b, c—frequency of A, B, C alleles.

the alleged father. The column p denotes the frequency of each allele, which is based on the Croatian database (Projić et al. 2007). RMNE is calculated by the above-mentioned formula. X indicates the probability that alleged father can pass the obligate allele, and it is determined by the father's genotype (if the father has only one copy of obligate allele or only one of two obligate alleles, then the probability [X] that he passed the obligate allele to a child equals to 0.5). PI is calculated as X/p or, in the case that there are two obligate alleles at the same locus, as $X/(p_1 + p_2)$. From the combined values, shown at the bottom of the table, it is evident that the probability of the alleged father is almost 790 million times higher than of any other unrelated male in the Croatian population. This value indicates the probability of paternity of 99.99999987%. RMNE shows that the probability to exclude a falsely accused father is 99.99999973%. Thus, in this case, the probability of paternity based on the passage of alleles is approximate to the likelihood based on the method of nonexclusion.

2.3.12 Motherless Paternity Testing

Occasionally, it is necessary to conduct a paternity testing when the mother is unavailable for analysis. In Germany, these cases are labeled as motherless or deficient test, because not all parties are present. In such cases, there is a considerable loss of information. The formula used for calculating the frequency of potential fathers in a given population is as follows:

$$\text{RMNE} = p^2 + q^2 + 2pq + 2p(1 - p - q) + 2q(1 - p - q) = (p + q)^2 + 2(p + q)(1 - p - q)$$

Accordingly, the PI is calculated differently. If one parent is missing, a special formula for each individual phenotype of an alleged parent is used. Formulas taken from Brenner (Brenner 1993) and further elaborated based on the data from www.DNA-view.com are shown in Table 2.6. To demonstrate the loss of information, as well as differences in the results, the paternity test, shown in Table 2.5, the CPI and exclusion probability were recalculated without the data on the mother, using appropriate formulas from Table 2.6. The results are shown in Table 2.7. The first thing that one would notice is that the power of exclusion decreases approximately 20,525-folds, whereas the CPI drops only

Forensic DNA Analysis and Statistics

Table 2.5 Example of Complete Paternity Testing

STR Loci	Mother	Child	Obligate Allele	Alleged Father	p(Y)	RMNE	X	PI
D8S1179	13, 14	11, 13	11	11, 13	0.0564	0.1096	0.5	8.8652
D21S11	30, 32.2	31, 32.2	31	30, 2 31	0.0590	0.1145	0.5	8.4746
D7S820	10, 12	9, 10	9	9, 9	0.1795	0.3268	1	5.5710
CSF1PO	12, 12	10, 12	10	10, 12	0.2538	0.4432	0.5	1.9701
D3S1358	15, 15	15, 18	18	15, 18	0.1487	0.2753	0.5	3.3625
TH01	6, 9.3	6, 8	8	6, 8	0.1308	0.2445	0.5	3.8226
D13S317	8, 10	10, 11	11	11, 11	0.3308	0.5522	1	3.0230
D16S539	9, 11	11, 13	13	9, 13	0.1718	0.3141	0.5	2.9104
D2S1338	17, 18	17, 23	23	23, 25	0.1026	0.1947	0.5	4.8733
D19S433	14.2, 16	13, 16	13	12, 13	0.2128	0.3803	0.5	2.3496
VWA	15, 16	16, 17	17	17, 17	0.2256	0.4003	1	4.4326
TPOX	8, 9	9, 9	9	8, 9	0.0846	0.1620	0.5	5.9102
D18S51	16, 17	15, 17	15	14, 15	0.1359	0.2533	0.5	3.6792
D5S818	11, 12	11, 11	11	11, 12	0.3256	0.5452	0.5	1.5356
FGA	20, 21	20, 25	25	20, 25	0.1000	0.1900	0.5	5.0000
Amelogenin	X X	X Y		X Y				
Combined						2.67×10^{-9}		790, 169, 857
Power of exclusion	1 − CRMNE					99.9999997%		
Likelihood	1/CRMNE					374, 531, 835		
Probability of paternity	CPI/CPI + 1					99.9999973%		99.9999987%

Note: $p(Y)$, frequency of each allele or each obligate allele based on Croatian database [62]; MNE, random man not excluded; PI, paternity index; X, the probability that alleged father can pass obligate allele.

Table 2.6 Formulas for Calculating Index of Paternity or Maternity in Absence of the Other Parent

Alleles of a Child	Alleles of a Tested Parent	PI
AA	AA	$1/a$
AB	AA	$1/2a$
AA	AB	$1/2a$
AB	AB	$a + b/4ab$
AB	AC	$1/4a$

Source: More details are available on www.DNA-view.com.
Note: A, B, C—alleles; a, b, c—frequency of A, B, C alleles.

about 760-fold; this illustrates the differences between the two methods for calculating the probability of parentage. In these instances, the CPI tends to overestimate the likelihoods. Based on the exclusion probability, incomplete cases have a significantly greater loss in exclusionary power than that indicated by the CPI. As more genetic markers are used, this difference increases. If the exclusionary power method is used, there is no discrepancy that the probability of paternity reflects the exclusionary power of the test.

2.3.13 Effect of Mutations

One of the open, but very important, issues is a potential occurrence of mutations, which can lead to discrepancies between a DNA profile of the alleged father and a specified set of obligate alleles, on one but sometimes even on two molecular markers. This issue points to the fact that paternity/maternity may not be automatically rejected if the inconsistency in allele variants is observed on one out of 15 tested molecular markers. In this case, complex statistical forms are applied that take into account the frequency of mutations at specific loci (more information can be found on www.DNA-view.com). At present, there does not appear to be a published consensus minimum exclusionary power on incomplete paternity cases. When this does not happen, and there is a likelihood of mutations, it is customary to request further testing on the X or Y related markers (depending on the child's gender). It is important to note that as more markers are used in the analysis, the bigger the chance becomes to detect the mutations at these loci. However, these cases are not too frequent and are treated differently in different countries. Previous AABB guidelines for RFLP testing indicated that two or more exclusions were adequate to exclude an alleged father. However, as more and more parentage cases are tested with increasing numbers of loci, it has been observed that two exclusions can occur by chance due to mutations, and it is necessary to carefully evaluate cases with only two or three exclusions, as most nonparents have an average of four to five exclusions (Promega 1996).

2.3.14 Maternity Testing

Sometimes, there is a need to find out who the biological mother of an abandoned or a deceased child is. In this case, unlike paternity testing, there is no knowledge about the child's parents. Therefore, as with any other forensic evidence, the genetic profile of the child is considered a proof. If a child is a heterozygote at a particular locus, then there are two obligate alleles. Conversely, when the child is homozygote, there is only one obligate allele.

Forensic DNA Analysis and Statistics

Table 2.7 Example of Motherless Paternity Testing

STR Loci	Child	Alleged Father	Obligate Alleles*	p(Y)	RMNE	PI
D8S1179	11, 13	11, 13	11, 13	0.0564 / 0.3487	0.6461	5.1496
D21S11	31, 32.2	30.2, 31	31, 32.2	0.0590 / 0.1128	0.3141	4.2373
D7S820	9, 10	9, 9	9, 10	0.1795 / 0.2590	0.6847	2.7855
CSF1PO	10, 12	10, 12	10, 12	0.2538 / 0.3538	0.8460	1.6916
D3S1358	15, 18	15, 18	15, 18	0.1487 / 0.2538	0.6430	2.6663
TH01	6, 8	6, 8	6, 8	0.1308 / 0.2641	0.6338	2.8580
D13S317	10, 11	11, 11	10, 11	0.3308 / 0.0615	0.6307	1.5115
D16S539	11, 13	9, 13	11, 13	0.1718 / 0.2795	0.6990	1.4552
D2S1338	17, 23	23, 25	17, 23	0.1026 / 0.2154	0.5348	2.4366
D19S433	13, 16	12, 13	13, 16	0.2128 / 0.0641	0.4772	1.1748
VWA	16, 17	17, 17	16, 17	0.2256 / 0.2205	0.6932	2.2163
TPOX	9, 9	8, 9	9	0.0846 / 0.0846	0.3097	5.9102
D18S51	15, 17	14, 15	15, 17	0.1359 / 0.0897	0.4003	1.8396
D5S818	11, 11	11, 12	11	0.3256 / 0.3256	0.8784	1.5356
FGA	20, 25	20, 25	20, 25	0.1000 / 0.1231	0.3965	2.5000
Amelogenin	X Y	X Y				
Combined					1.61×10^{-4}	826 693
Power of exclusion	1− CRMNE				99.9839%	
Likelihood	1/CRMNE				6170	
Probability of paternity	CPI/CPI + 1				99.983795%	99.9997808%

Note: In the absence of the mother, there is no identification of the paternal alleles; thus, the biological father could contribute either allele. p(Y), the frequency of each allele or each obligate allele based on Croatian database (Projic et al. 2007); PI, paternity index; RMNE, random man not excluded.

If there is no information about the mother, the formula for calculating *random female not excluded* (RFNE) is the same as one used for computing RMNE. The mother is excluded from further analysis if she does not have either of the child's two alleles. But a mother cannot be excluded if she has only one of the child's two alleles. If this counts for all loci, the alleged mother cannot be excluded. *Combined random female not excluded* (CRFNE) is the product of all individual RFNEs and displays the percentage of women in the population that cannot be excluded at random or may be falsely included. The opposite is 1/CRFNE, showing the number of women who should be tested in order to reach the matching result. 1/CRFNE is also Bayesian probability of maternity. Conversely, it is possible to calculate the index of maternity based on the passage of alleles, similar to the aforementioned PI, and convert it into Bayes' probability. So, in case of lack of information about the father, maternity index is calculated by the formulas shown in Table 2.3. The basis of this is identical to that of motherless cases found in Table 2.6, where the obligate alleles are defined by the child's DNA profile, and are compared with allelic in the mother's DNA profile.

2.3.15 Parentage Testing with Mixed Populations

When calculating the parentage index, it is common to use the allele frequencies for the alleged fathers or mother depending on the question. If the alleged father or mother originates from the mixture of two populations, such as Croatian and sub-Saharan Africa, then the frequency of the obligate allele, if possible, should be the average of the two frequencies. On the other hand, if the alleged parent is Italian, and the only database available is the one in Croatia, it will not drastically affect the results as long as populations from the same geographical area are used.

In reality, almost all human populations are mixed to some extent. Alleles specific to Central Asia, northern Asia, and Africa may be found in European populations, as there is historical evidence of admixture with the Mongols, the Avars, and African slaves throughout history. It is not required to make any special adjustments for these historical events, because they will be included in the database of a population. On the other hand, if the person that we are testing is a tourist, for example, from Puerto Rico, it would be very desirable, if possible, to use the database from that country. However, if a specific database is not available, the analysis may lead to certain errors. If the genetic structure of the population in question is known, the frequency can be calculated using the allele frequencies proportional to the corresponding subpopulation.

Precisely because of these reasons, journals such as *Forensic Science International*, *Forensic Science International: Genetics*, *Journal of Forensic Science*, *Internal Journal of Legal Medicine*, as well as *Croatian Medical Journal* often publish population data for a large number of the world's populations. This approach allows comparative analysis in order to investigate if there are statistically significant differences between the populations, and to enable the accurate calculation of certain forensic genetics parameters.

2.3.16 Using Y and X Chromosome Markers in Cases of Disputed Parentage

The addition of Y and X markers to parentage testing and in inheritance cases has been quite useful. Each has its own unique applications. Y chromosome markers can only be used with male children or descendants in a male line, such that it is an easy matter of verifying male children, or male siblings in an inheritance case. In contrast, for a paternity case, X chromosome markers can only be used to study the inheritance in daughters; however, for mothers, children of either gender can be studied. For Y chromosome markers, it must be remembered that you are looking at the inheritance of a single haplotypes so that there will only be a single PI value or RMNE. In contrast, if unlinked X STRs are used, there will be a PI or RMNE for each unlinked locus. An example is provided in Table 2.8 using data from a paternity trio taken from Schanfield, Diegoli, and Coble

Table 2.8 Paternity Case Using X STR Loci

Sample Name	Mother 002-M	Child 002-C	AF 002-F	Obligate Allele	Frequency	RMNE	SI
DXS8378	11, 12	9, 11	9	9	0.017	0.017	58.824
DXS9902	10, 10	10, 10	10	10	0.305	0.305	3.279
DXS6795	9, 11	11, 11	11	11	0.161	0.161	6.211
DXS7132	12, 14	12, 15	15	15	0.229	0.229	4.367
DXS6803	11, 13	13, 13	13	13	0.093	0.093	10.753
DXS6789	20, 20	20, 21	21	21	0.22	0.22	4.545
DXS7424	15, 15	15, 16	16	16	0.136	0.136	7.353
DXS101	24, 25	25, 25	25	25	0.119	0.119	8.403
GATA172D05	6, 11	6, 12	12	12	0.025	0.025	40.000
DXS7130	13, 16.3	13, 15.3	15.3	15.3	0.237	0.237	4.219
GATA165B12	9, 10	10, 10	10	10	0.297	0.297	3.367
HPRTB	11, 11	11, 12	12	12	0.263	0.263	3.802
GATA31E08	12, 14	12, 12	12	12	0.246	0.246	4.065
DXS10147	9, 9	8, 9	8	8	0.407	0.407	2.457
DXS7423	16, 16	15, 16	15	15	0.356	0.356	2.809
Combined						1.044×10^{-12}	9.577×10^{11}
Exclusionary power						99.9999999999%	9.9999999999%
Probability paternity							

Source: Data taken from Schanfield, Diegoli and Coble (unpublished) from a U.S. European family and European allele frequencies.

(unpublished). (Note: The RMNE for a haploid male is the allele frequency, and the SI = 1/allele frequency.)

2.4 Identification of Human Body Remains

Identification of the human remains may be achieved by a number of methods, depending on the circumstances, as well as the condition of human remains. If the decedent is recovered before decomposition, visual identification of the remains by a close relative is possible as is identification of the victim using scars or other physical marks (tattoos), fingerprints (if any), dental record or fractures, and medical implants as well as DNA analysis.

In wartime, the identification of victims is often difficult, because a large number of bodies are often buried in mass graves, and *premortem* data are not always available as well (Gunby 1994). Development of new technologies for the analysis of genomic (autosomal as well as sex) and mitochondrial DNA provided forensic experts with new tools for identification. Furthermore, DNA analysis has become a standard method that is used for the confirmation/rejection of the results of previous analyses, even in cases where the level of recognition and possibilities for body identification are relatively good. Because of the rapid decay of soft tissue of bodies in war or in mass catastrophes, forensic experts often can only perform analysis of DNA from the skeletal and dental remains for identification purposes (Gaensslen and Lee 1990). Several authors emphasize that DNA extraction methods that precede the amplification of DNA are of major importance for a higher percentage of successful identifications (Fisher et al. 1993; Ortner et al. 1992; Hochmeister et al. 1991; Burgi 1997). Furthermore, it is stated that the analysis of mtDNA is the best choice if working with highly degraded material. However, using modified standard procedures of DNA extraction and purifying the isolated DNA (repurification) with NaOH or other chemicals, it was found that the success rate of identification by genomic DNA may reach more than 96% (Andjelinović et al. 2005; Primorac 1999; Primorac et al. 1996; Keys et al. 1996; Alonso et al. 2001). It was also noted that the analysis of dental remains in comparison with the results of the analysis of long bones gives better results in 20%–30% of all cases analyzed (Primorac 1999).

2.4.1 Victim Identification Using Parental DNA

In the previous section, we discussed identification of the father or mother of a living person. Now, we will talk about the identification of remains, which is also possible if either parents or children of a presumed victim are available for testing. Sometimes, in order to determine the identity of the victim, reconstruction is necessary. The first case involves a missing person who is assumed to be dead, but whose body is yet to be found (e.g., in one's home blood was found). Do the blood stains found at the scene belong to the person who lived there and whose parents are available for testing? In the second case, the human remains are found in mass grave or they are from the murder case, which cannot be identified in any other way than through genetic analysis.

In the process of identifying the remains, the calculation is slightly different from that in paternity testing. In this case, instead of determining the frequency of an obligate allele in a population of potential fathers or mothers (RMNE or RFNE), it is necessary to compute the likelihood of the allele in common in each of the parents. If we go back to the product rule for combining probabilities, the probability of finding a parent with allele A and the other parent with allele B is the frequency of potential parents with allele A multiplied by the frequency of potential parents with allele B. In practice, it is calculated by computing the RMNE/RFNE for each allele, which is then multiplied by the specific RMNE for first allele multiplied by RFNE for the second allele to obtain a probability of two parents who have the same allele. This method is called *random parents not excluded* (RPNE). Therefore, the formula for calculating RPNE for any two alleles, regardless of whether they are similar or different, is as follows: $(p^2 + 2p[1 - p]) \times (q^2 + 2q[1 - q])$. For example, using the locus D3S1358 and results in an unidentified stain of D3S1358 15,17 with an incidence of 0.2425 and 0.1875 for each allele, the probability of finding two parents, one with allele *D3S1358 * 15* and the other with allele *D3S1358 * 17,* would be

$$(0.2425^2 + 2 \times 0.2425 \times 0.7575) \times (0.1875^2 + 2 \times 0.1875 \times 0.8125) = 0.4262 \times 0.3398 = 0.1448.$$

Accordingly, 14.5% of parents in Croatia would not be excluded as possible parents. On the other hand, 85.5% of parents in Croatia would be excluded by this simple test. By multiplying all RPNEs for each locus, a combined RPNE or *combined parents not excluded* (CRPNE) is created, generating a relative frequency or proportion of the population of possible parents. Similarly, the probability of parents who are not excluded from the analysis

Table 2.9 Identification of Remains of Child Using Parents

Locus	Mother	Bone	Father	Obligate Alleles	p	RMNE	RPNE
LDLR	B	A,B	A	A	0.410	0.6519	
				B	0.590	0.8319	0.5423
GYPA	A,B	A,B	A,B	A	0.560	0.8064	
				B	0.440	0.6864	0.5535
HBGG	A	A,B	B	A	0.530	0.7791	
				B	0.470	0.7191	0.5603
D7S8	A	A,B	A,B	A	0.653	0.8796	
				B	0.347	0.5736	0.5045
GC	A,C	A,C	A,C	A	0.279	0.4802	
				C	0.595	0.8360	0.4014
DQA1	2,4	3,4	2,3	3	0.105	0.1990	
				4	0.340	0.5644	0.1123
Combined							0.0038
Likelihood of parenthood							261
Probability of parenthood							99.62%

Source: Primorac, D. et al., *J Forensic Sci*, 41, 891–894, 1996. With permission.

is 1/CRPNE. Bayes' probability of parentage, utilizing the method of nonexclusion and *a priori* probability of 0.5, is 1 − CRPNE.

Identification of remains found in Croatia and Bosnia during and after the Balkan War was carried out before the STR technique became available (Primorac et al. 1996). In Table 2.9, we show an example of the identification of remains of a missing individual, with both parents available for testing. In this case, the population of potential parents was computed for each allele, which means that the alleles shown in the RMNE column were multiplied with each other, and the product was RPNE for each locus. In that time, the data were based on an analysis of DQA1 and Polymarker, which are less informative than STR (Keys et al. 1996; Gill and Evett 1995; Budowle et al. 1999), and the values are lower than those shown in previous tables. In this case, only 0.4% of couples in Croatia could be possible parents of the found victims, or 99.62% of couples would be excluded from further analysis. This means that the chances were 260 to 1 that some of the couples are parents of the found victims, or that there is a 99.62% probability to determine the parentage by the method of nonexclusion.

2.4.2 Victim Identification Using Child's DNA

If the missing person is a father whose children are available, identification is usually carried out following the procedure described above. In Table 2.10, data are presented for the identification of NN, who was assumed to be the father of the child. The circumstances are more favorable if the mother is also available for analysis. The testing results of mother and child indicate that 99.9969% of unrelated males would be excluded; however, the remains of NN were not excluded. Calculating the SI and PI suggests that it is approximately 45,000 times more likely that the NN is the father of a boy than another unrelated man, or more precisely, the probability that it is the boy's father is 99.9978%. The method of nonexclusion exhibited the probability of almost 32,000 to 1, that is, the probability of paternity

Table 2.10 Identification of Remains Using a Child of the Decedent

Loci	Wife	Child	Obligate Allele	Remains NN	p	RMNE	X	SI
D3S1358	14,16	15,16	15	15,17	0.2425	0.4262	0.50	2.062
VWA	15,16	16,19	19	19	0.0900	0.1719	1.00	11.111
FGA	21,22	21,22	21	21,22	0.2125			
			22		0.1575	0.6031	1.00	2.703
THO1	7,8	7,8	7	8,9	0.1724	FBI		
			8		0.1150	0.4922	0.50	1.740
TPOX	11	10,11	10	9,10	0.0625	0.1211	0.50	8.000
CSF1PO	10,12	10	10	10,11	0.2750	0.4744	0.50	1.818
D5S818	11,12	10,12	10	10,12	0.0775	0.1490	0.50	6.452
D13S317	8,12	8,11	11	11,12	0.3350	0.5578	0.50	1.493
D7S820	8,11	8	8	8,10	0.1650	0.3028	0.50	3.030
AMEL	X,X	X,Y		X,Y				
Combined						3.14×10^{-5}		45,720
Exclusion Power						99.9969%		
Likelihood						31,812		
Probability of Paternity						99.99781%		

is 99.998%. The two methods coincide with each other, and it is fairly certain that the remains found are of those of the boy's father.

It is easy to notice that Table 2.10 is based on the use of only nine STR loci (based on the kit available at the time), and on the frequencies that are not entirely a realistic presentation of the situation. Such an approach was deliberately designed to show how—despite all problems that were being solved along the way—good results were achieved during the past 15 years in the DNA identification of Balkan War victims (Primorac and Marjanović 2008; Marjanović and Primorac 2013). As it is likely in a situation such as this that only a partial profile is obtained, a statistically valid identification can be accomplished.

Identification of war victims in Croatia was one of the greatest scientific challenges, not only for Croatians, but also for world's forensic scientists. The experiences gained through this project, as well as in a similar project that was implemented in Bosnia and Herzegovina, under the coordination of the International Commission on Missing Persons, have found a global application. One very interesting project, whose first phase has been completed, was the identification of the victims from a mass grave in Slovenia dated to the end of World War 2 (Marjanovic et al. 2007). The success of this analysis gives us hope that in the near future, this method will allow for the identification of victims of that conflict, who died during WW2 and in the postwar years.

2.4.3 Parentage Testing versus Forensic Identification

In normal parentage testing, the frequency of possible parents (mothers or fathers) is determined by the frequency of RM/FNE or population of potential allele donors. In forensic identification, the frequency of potential phenotype (genotype) donors is relevant (Schanfield 2000). Table 2.11 shows a comparison of the results of parentage testing (RMNE = $(p + q)^2 + 2(p + q)(1 - p - q)$) and the results of forensic identification ($2pq$). In this case, the paternity donor population (RMNE) is approximately seven times bigger

Table 2.11 Comparison of Calculating Population Frequencies at One Locus in the Case of Paternity Testing versus Forensic Identification

	RMNE (Paternity)	Forensic Evidence
Allele	D3S1358 15, 17	D3S1358 15, 17
Frequency	$(p + q)2 + 2(p + q)(1 - p - q)$	$2pq$
	0.6751	0.0909

Note: $p(15) = 0.2425$, $q(17) = 0.1875$.

Table 2.12 Comparison of Calculating Population Frequencies at One Locus in the Case of Paternity Testing versus Forensic Identification with U.S. National Research Council Recommendations

	RMNE (Paternity)	Forensic Evidence
Allele	D3S1358 15, 15	D3S1358 15, 15
Frequency	$p^2 + 2p(1 - p)$	$p^2 + 2p(1 - p)(0.01)$
	0.4262	0.0625

Note: $p(15) = 0.2425$.

than the frequency obtained from forensic identification. Thus, there is almost an entire order of magnitude difference at one locus between the two methods of identification.

The estimates shown in Table 2.11 are based on the heterozygous phenotype. If the results were for a homozygote, p^2 would serve for the assessment of HWE, assuming there is no significant deviation from HWE due to population substructuring, or some other event. The U.S. National Research Council recommends that substructuring correction factor should be taken into account for all populations, even those in which substructuring was not observed, such as Croatia. In this case, to assess the HWE, the formula $p^2 + p(1 - p) \times (0.01)$ is used (NRC 1996). The example in Table 2.12 shows that in that case, for a homozygote sample at D3S1358 (15, 15) the calculation for the expected match likelihood would be $0.2425^2 + 2 \times 0.2425 \times 0.7575 \times 0.01 = 0.0625$, because the RMP for a homozygous person is $p^2 + 2p(1 - p) 0.01$. Note: The substructuring value of (θ or F) for large populations is taken to be 0.01; for small or inbred populations, such as Native Americans or Roma, the value of 0.03 is used.

It is clear that it is much easier to obtain very low match likelihoods for forensic evidence rather than a parentage situation. The final step in forensic identification is to unite any individual measures in order to obtain the frequencies of the multilocus genotype or profile. This is achieved by multiplying the frequencies of the genotypes of each locus.

References

Allen, R. 1999. The American Association of Blood Banks (AABB) Annual Meeting. Paternity testing, specialist interested group. *Annual Report*. San Francisco: AABB.

Alonso, A., Š. Andjelinović, P. Martin et al. 2001. DNA typing from skeletal remains: Evaluation of multiplex and megaplex systems on DNA isolated from bone and teeth samples. *Croat Med J* 42:260–266.

Andjelinović, Š., D. Sutlović, I.Erceg-Ivkošić et al. 2005. Twelve-year experience in identification of skeletal remains from mass graves. *Croat Med J* 46:530–539.

Brenner, C. 1993. A note on paternity computation in cases lacking a mother. *Transfusion* 33:51–54.

Budowle, B., T.R. Moretti, A.L. Baumstark et al. 1999. Population data on the thirteen CODIS core short tandem repeat loci in African Americans, U.S. Caucasians, Hispanics, Bahamians, Jamaicans, and Trinidadians. *J Forensic Sci* 44:1277–1286.

Budowle, B., R. Chakraborty, G. Carmody et al. 2000. Source attribution of a forensic DNA profile. *Forensic Sci Commun* 2(3).

Budowle, B., A.J. Onorato, T.F. Callaghan et al. 2009. Mixture interpretation: Defining the relevant features for guidelines for the assessment of mixed DNA profiles in forensic casework. *J Forensic Sci* 54(4), doi: 10.1111/j.1556-4029.2009.1046.x (online).

Burgi, S.B. editor. 1997. *First European–American intensive Course in PCR Based Clinical and Forensic Testing. Laboratory Manual*; September 23–October 3 1997; Split, Croatia: Laboratory for Clinical and Forensic Genetics.

Butler, J.M. 2010. *Fundamentals of Forensic DNA Typing*. San Diego: Elsevier Academic Press.

Butler, J.M. 2011. *Advanced Topics in Forensic DNA Typing: Methodology*. San Diego: Elsevier Academic Press.

Caragine, T., R. Mikulasovich, J. Tamariz et al. 2009. Validation of testing and interpretation protocols for low template DNA samples using AmpFlSTR® Identifiler®. *Croat Med J* 50:250–267.

Clayton, T.M., P. Gill, R. Sparkes et al. 1998. Analysis and interpretation of mixed stains using DNA STR profiling. *Forensic Sci Int* 91:55–70.

Council of Europe. 1992. Committee of Minister's. Recommendation No.R (92) 1 on the use of analysis of deoxyribonucleic acid (DNA) within the framework of the criminal justice system. 470th meeting of the Minister's Deputies.

Curran, J.M., and J. Buckleton. 2010. Inclusion probabilities and dropout. *J Forensic Sci* 55(5):1171–1173.

Diegoli, T.M., and M.D. Coble. 2010. Development and characterization of two mini_X chromosomal short tandem repeat multiplexes. *Forensic Sci Int Genet* 5:415–421, doi 10.1016/j.fsigen.2010.08.019 (online).

Evett, I.W., and B.S. Weir. 1998. *Interpreting DNA Evidence: Statistical Genetics for Forensic Scientists*. Sunderland MA: Sinaure Associates Inc.

FBI. 2011. The FBI and DNA. http://www.fbi.gov/news/stories/2011/november/dna_112311 (accessed March 25, 2013).

Fisher, D.L., M.M. Holland, L. Mitchell et al. 1993. Extraction, evaluation and amplification of DNA from decalcified and un-decalcified United States civil war bone. *J Forensic Sci* 38:60–68.

Gaensslen, R.E., and C.H. Lee. 1990. Genetic markers in human bone tissue. *Forensic Sci Rev* 2:126–146.

Gill, P., and I. Evett. 1995. Population genetics of short tandem repeat (STR) loci. In: *Human Identification: The Use of DNA Markers*, ed. B. Weir, 69–87. Dordrecht: Kluwer Publishing.

Gill, P., C.H. Brenner, J.S. Buckleton et al. 2006. DNA commission of the International Society of Forensic Genetics: Recommendations on the interpretations of mixtures. *For Sci Int* 160:90–101.

Gunby, P. 1994. Medical team seeks to identify human remains from mass graves of war in former Yugoslavia. *JAMA* 272:1804–1806.

Hochmeister, M.N., B. Budowle, U.V. Borer et al. 1991. Typing of deoxyribonucleic acid (DNA) extracted from compact bone from human remains. *J Forensic Sci* 36:1649–1661.

Hwa, H.L., Y.Y. Chang, J.C. Lee et al. 2009. Thirteen X-chromosomal short tandem repeat loci multiplex data from Taiwanese. *Int J Legal Med* 123:263–269.

Kaštelan, J., and B. Duda. 1987. *Bible—The Old and New Testaments*. Zagreb: Kršćanska sadašnjost (Croatian edition).

Keys, K.M., B. Budowle, Š. Andjelinović et al. 1996. Northern and Southern Croatian population data on seven PCR-based loci. *Forensic Sci Int* 81:191–199.

Li, C.C., and A. Chakravarti. 1988. An expository review of two methods of calculating a paternity probability. *Am J Hum Genet* 43:197–205.

Machado, F.B., and E. Medina-Acosta. 2009. Genetic map of human X-linked microsatellites used in forensic practice *For Sci Int: Genet* 3:202–204.

Marjanović, D., N. Bakal, N. Pojskić et al. 2006. Allele frequencies for 15 short tandem repeat loci in a representative sample of Bosnians and Herzegovinians. *Forensic Sci Int* 156:79–81.

Marjanović, D., A. Durmić-Pasić, N. Bakal et al. 2007. DNA Identification of skeletal remains from the Second World War mass graves uncovered in Slovenia—first results. *Croat Med J* 48:520–527.

Marjanović, D., and D. Primorac. 2013. DNA Variability and molecular markers in forensic genetics. In *Forensic Genetics: Theory and Application,* ed. D. Marjanović, D. Primorac, L.L. Bilela et al., pp. 75–98. Sarajevo: Lelo Publishing (Bosnia and Herzegovina's edition).

NRC. 1996. *The Evaluation of Forensic DNA Evidence*, Committee on Forensic DNA: An update. Washington, DC: National Research Council, National Academy Press.

Ortner, D.J., N. Tuross, and A.I. Stix. 1992. New approach to the study of disease in archeological New World populations. *Hum Biol* 64:337–360.

Primorac, D. 1999. Identification of human remains from mass graves found in Croatia and Bosnia and Herzegovina. In *Proceedings of the 10th International Symposium on Human Identification*; September 29–October 2, 1999; Orlando, Florida. Madison, USA: Promega Corporation.

Primorac, D., and D. Marjanović. 2008. DNA Analysis in forensic medicine and legal system. In *DNA Analysis in Forensic Medicine and Legal System*, ed. D. Primorac, pp. 1–59. Zagreb: Medicinska Naklada (Croatian edition).

Primorac, D., Š. Andjelinović, M. Definis-Gojanović et al. 1996. Identification of war victims from mass graves in Croatia, Bosnia and Herzegovina by the use of standard forensic methods and DNA typing. *J Forensic Sci* 41:891–894.

Primorac, D., S.M. Schanfield, and D. Primorac. 2000. Application of Forensic DNA in the legal system. *Croat Med J* 41:33–47.

Promega Statistics Workshop. 1996. September 16–18, 1996, Scottsdale, AZ.

Projić, P., V. Škaro, I. Šamija et al. 2007. Allele frequencies for 15 short tandem repeat loci in representative sample of Croatian population. *Croat Med J* 48:473–477.

Schanfield, M. 2000. DNA: Parentage testing. In: *Encyclopedia of Forensic Sciences*, ed. J.A. Siegel, P.J. Saukko, and G.C. Knupfer, pp. 504–515. London: Academic Press.

Schanfield, M.S., M.N. Gabriel, Š. Andjelinović et al. 2002. Allele frequencies for the 13 CODIS STR loci in a sample of Southern Croatians. *J Forensic Sci* 47:669–670.

Schneider, P.M., P. Gill, and A. Carracedo. 2006. Editorial on the recommendations of the DNA commission of the ISFG on the interpretation of mixtures. *Forensic Sci Int* 160:89.

Schneider, P.M., M. Keil, H. Schmitter et al. 2009. The German Stain Commission: Recommendations for the interpretation of mixed stains. *Int J Legal Med* 123:1–5.

Tillmar, A.O., P. Mostad, T. Egeland et al. 2008. Analysis of linkage and linkage equilibrium for 8 X-STR markers. *For Sci Int Genet* 3:37–41.

Weiner, A.S. 1976. Likelihood of parentage. In *Paternity Testing in Blood Grouping*, ed. L.N. Sussman, pp. 124–131. Springfield, VA: Charles C. Thomas.

Weir, B.S. 1993. Population genetics of DNA profiles. *J Forensic Sci* 33:218–225.

Forensic Aspects of mtDNA Analysis

3

MITCHELL M. HOLLAND
GORDAN LAUC

Contents

3.1	Introduction	85
3.2	Mitochondrion and mtDNA Genome Structure	85
3.3	mtDNA Copy Number	89
3.4	mtDNA Inheritance	90
3.5	mtDNA Heteroplasmy	91
3.6	Application of mtDNA Analysis to Forensic Cases	96
3.7	Genetic Variability and Random Match Probabilities	97
References		101

3.1 Introduction

Numerous features of the human mitochondrial DNA (mtDNA) genome have been studied in great detail, such as structure, function, copy number, mutation rate, and inheritance patterns. These studies have revealed that mtDNA is a circular genome of approximately 16,569 nucleotides in length. A well-defined, noncoding portion of the genome tolerates the accumulation of mutations, which can be queried to develop a forensic mtDNA sequence profile. The profile information will often be helpful when identifying missing individuals and solving crime. For example, a comparison of mtDNA profiles frequently allows biological evidence to be associated with one specific individual in a forensic case, while excluding unrelated individuals in the same case as a source of DNA recovered from the crime scene. Unfortunately, the maternal inheritance pattern of the mitochondrial genome reduces the overall discrimination potential of the mtDNA testing system, as relatives from the same maternal lineage typically have identical profiles. However, the higher mutation rate of the genome can result in the occurrence of heterogeneous pools of mtDNA profiles that can significantly increase the power of discrimination and can be transmitted in differing ratios across maternal lines. The material presented in this chapter provides an overview of the forensically relevant characteristics of the mtDNA genome, and how these characteristics can be used to answer questions raised in forensic investigations.

3.2 Mitochondrion and mtDNA Genome Structure

Molecules of DNA located in the cell nucleus, in the form of chromosomes, are not the only source of DNA in modern eukaryotic cells. Mitochondria, organelles found in the cytoplasm of most cell types, contain a second intracellular DNA genome. According to the widely accepted endosymbiotic theory of mitochondrial origin, mitochondria were derived from α-proteobacteria that lived approximately 2 billion years ago inside pre-eukaryotic cells resembling ancient protists (Gray 1992; Lang et al. 1997; Gray et al. 2004). Over the course of evolutionary time, the endosymbiont lost its ability to survive outside the eukaryotic cell, and portions of its associated DNA were retained (Andersson et al. 1998; Dolezal et al. 2006). The endosymbiont became the mitochondrion, and its genome became mtDNA. Interestingly, the mitochondrion continues to play a vital role in the production of cellular energy by means of oxidative phosphorylation and the synthesis of ATP molecules. During the evolutionary process, the mitochondrion maintained a number of genes involved in this important biochemical process, while losing many of the activities needed to survive as an independent organism. Instead, the mtDNA gene products work together with the products of hundreds of genes located in the nuclear DNA (nucDNA) that are necessary for organelle function, and therefore, are directly linked to the survival of the cell.

Early molecular surveys suggested that mtDNA from different species had very similar nucleotide base composition. Speculation persisted for many years that mtDNA genomes from yeast, fungi, plants, and animals were quite similar in both enzyme content and function (Haldar et al. 1966). Although many of the similarities in gene content have fundamentally held true, the size and structure of mtDNA genomes has revealed significant differences between species (Borst and Grivell 1981; Marienfeld et al. 1999; Wu et al. 2009). The first complete mtDNA genome sequences reported in the literature came, perhaps surprisingly, from mouse and humans (Bibb et al. 1981; Anderson et al. 1981). The publication of numerous genome sequences from other sources has followed, including the sequence of a legume genome, *Vigna radiate* (mung bean) (Alverson et al. 2011). As expected, the mouse and human genomes are quite similar in size and structure; approximately 16,295 nucleotides in length versus 16,569, respectively, with identical gene organization. However, the legume genome is approximately 25 times larger in size—more than 400,000 nucleotides. In fact, most mtDNA genomes of lower organisms include the characteristic of being large in size, reaching as high as 1–3 million bp in plants (Lilly and Havey 2001). This is consistent with the phenomenon that as the proteobacterium took up residence in the primitive pre-eukaryotic cell, portions of the mtDNA genome were transferred to the nuclear genome or became obsolete over evolutionary time, considerably reducing the size of the genome with the emergence of animals and humans.

The human mtDNA genome is a circular piece of double-stranded DNA of approximately 16,569 bp in length; a schematic representation of the genome can be found in Figure 3.1. Detailed information about the genome is available on the Internet in a variety of general and specialized databases (MITOMAP; Lupi et al. 2010). The complementary strands of mtDNA sequence are quite different in their composition. The "heavy strand" is rich in purine nucleosides (adenosine and guanosine), whereas the "light strand" is subsequently rich in pyrimidine nucleosides (thymidine and cytidine). The coding region encompasses approximately 93% of the genome, with the gene sequences densely arranged. There are 37 genes found in the coding region sequence: 22 genes for transport RNAs (tRNA), two genes for ribosomal RNAs (rRNA, the 12S and 16S subunits), and 13 genes for enzymes in the respiratory chain involved in the process of oxidative phosphorylation and ATP production. Function, mutation, hereditary disorders, and population variability related to the genome sequence have been studied in great detail (Arnheim and Cortopassi 1992; Wallace 2010; Lodeiro et al. 2010; Irwin et al. 2011; Federico et al. 2012; Holland and

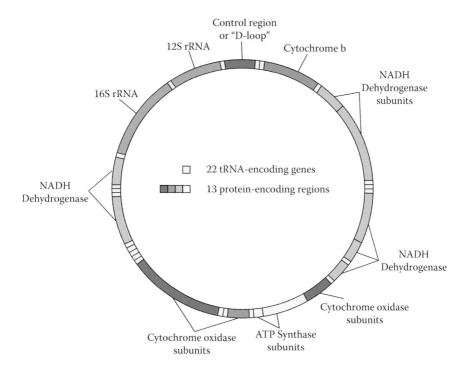

Figure 3.1 Mitochondrial DNA genome. All portions of the genome not colored blue represent coding regions, with a total of 37 genes that include 22 genes for transport RNAs (tRNA), two genes for ribosomal RNAs (rRNA, the 12S and 16S subunits), and 13 genes for enzymes in the respiratory chain involved in the process of oxidative phosphorylation and ATP production. The portion colored in blue represents the control region, and is primarily where forensic scientists focus their DNA tests.

Parsons 1999 for an expanded list of citations). Specific deletions, duplications, and point mutations can lead to a variety of pathophysiologic syndromes. In addition, somatic deletions and mutations that are acquired and accumulated during the life of an individual are held to be one of the most important causes for degeneration of cells and tissues in the aging process (Arnheim and Cortopassi 1992). Work on patients exposed to nucleoside analogue antiretroviral drugs to combat the HIV virus has support these assertions by connecting the progressive accumulation of somatic mtDNA mutations to premature aging in these patients (Payne et al. 2011).

Beyond the densely structured portions of the mtDNA genome that code for genes, there is a small noncoding segment of the genome called the control region (CR), accounting for approximately 7% of the genome sequence. The CR is sometimes referred to as the displacement loop (D-loop), as during the initiation of transcription and replication, the DNA in the CR forms a three-stranded loop structure visible through an electron microscope and by atomic force microscopy (Brown and Clayton 2006). The range of sequence encompassing the CR is approximately 1125 bp in length, and includes promoters for transcription of polycystronic messages of the heavy and light strands. The initiation of heavy strand replication can also be found in the CR, highlighting the fact that although no gene sequences exist in this portion of the mtDNA genome, it still contains vital regulatory elements for genome function. Subsequent initiation of light strand replication occurs at one or more sites in the coding region, from a primary initiation site or from multiple alternative initiation sites. Nonetheless, and not surprisingly, the CR readily tolerates changes to its sequence. Therefore, unlike the coding regions of mtDNA, sequence within the CR is quite variable between individuals in the population because of reduced evolutionary pressures on the noncoding sequence.

The numbering system of the circular mtDNA genome begins in the middle of the CR (i.e., position 1) and proceeds in a clockwise fashion until it reunites with position 1 at nucleotide position 16,569 (Anderson et al. 1981). Therefore, the CR is typically defined as the sequence between nucleotide position 16,024 (immediately following the tRNA for proline) and position 576 (immediately before the tRNA for phenylalanine), having crossed over position 1 in the genome's numbering system. As noted above, the sequence between 16,024 and 576 contains no coding information, but does include regulatory sites such as the

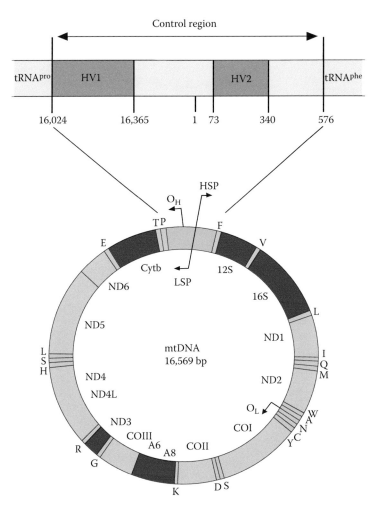

Figure 3.2 Mitochondrial DNA control region. There are two hypervariable (HV) segments located in the control region that contain a high percentage of the polymorphic content of the genome. Hypervariable region 1 (HV1) begins at nucleotide position 16,024 and ends at 16,365, and hypervariable region 2 (HV2) starts at 73 and ends at 340. There are two variable regions (VRs) of DNA sequence that span the remaining portions of the control region, i.e., VR1 encompasses sequence positions 16,366–16,372, and VR2 encompasses sequence positions 341–576.

origin of heavy strand replication around nucleotide position 191. There are two hypervariable (HV) segments located in the CR that contain a high percentage of the polymorphic content of the genome. Hypervariable region 1 (HV1) begins at nucleotide position 16,024 and ends at 16,365, and hypervariable region 2 (HV2) starts at 73 and ends at 340. Although these are the most often defined ranges for HV1 and HV2, the actual starting and ending points may vary depending on laboratory or study design. However, finding polymorphic sequence before position 16,024 and after 576 is highly unlikely without considerable effort. Therefore, because of the high levels of variability observed within HV1 and HV2, DNA sequencing of these two segments of mtDNA is most often exploited for forensic purposes (Holland and Parsons 1999). There are two variable regions (VRs) of DNA sequence that span the remaining portions of the CR: VR1 encompasses sequence positions 16,366–16,372 and VR2 encompasses sequence positions 341–576 (Figure 3.2). Although these regions have less discrimination potential, the polymorphic sequence information found in VR1 and VR2 can be used to distinguish between individuals with the same HV sequences (Lutz et al. 2000). Lastly, wobble positions in codons of gene sequences in the coding region of the mtDNA genome have been used to help differentiate between individuals with the same CR sequence (Coble et al. 2006; Parsons and Coble 2001). A comprehensive review of alternative approaches for analyzing mtDNA is available in the literature (Melton et al. 2012).

3.3 mtDNA Copy Number

There are numerous mitochondria located inside a single eukaryotic cell, and inside a single mitochondrion there are multiple copies of the mtDNA genome. The total number is typically dependent on the cell's energy requirements. For example, the average somatic cell may contain 200 to more than 1700 copies of mtDNA, whereas the higher energy requirements of a mature oocyte demands the need for at least 100,000–150,000 copies (Bogenhagen and Clayton 1974; Robin and Wong 1988; Piko and Matsumoto 1976; Michaels et al. 1982). Regardless of the precise number of copies per cell, the elevated amount enables the isolation of suitable quantities of mtDNA from poor-quality forensic samples for sequence analysis. In comparison to the two copies of each forensic autosomal locus found in a single cell, or the single copy of each Y chromosome locus in males, this phenomenon allows mtDNA analysis to remain accessible when nucDNA testing has failed. As an example, given that at least 200 pg of total genomic DNA is typically required for short tandem repeat (STR) analysis in forensic casework, this translates into approximately 34 cells worth of DNA. Within the DNA recovered from these 34 cells, there are 68 copies of STR alleles (two copies per locus), but 34,000 copies of mtDNA (assuming a per cell copy number of 1000). To further enhance the sensitivity of mtDNA testing, since allelic dropout is not an issue with mtDNA, the analysis typically involves a higher number of polymerase chain reaction (PCR) cycles for amplification; for example, 38 cycles for mtDNA amplification versus 28 cycles for STR amplification. Therefore, one can reliably obtain an mtDNA sequencing result from DNA samples containing femtogram (subpicogram) levels of total genomic DNA. Historically, the types of evidence where this concept has most often been successfully applied is for the analysis of hair shafts and older, poorly preserved bones and teeth (Hopgood et al. 1992; Melton et al. 2005; Fisher et al. 1993; Ivanov et al. 1996; Ginther et al. 1992). In recent years, the development of new DNA extraction techniques have made it possible to generate nucDNA testing results on a higher percentage of older bones, and to successfully perform mtDNA analysis on the most highly degraded samples (Loreille et al. 2007). However, to date, nucDNA analysis has remained an elusive and ineffective technique when analyzing hair shafts (Muller et al. 2007), unless scientists are able to combine hair material from many individual hairs and take samplings on the proximal end of the hair shaft near the root (Amory et al. 2007).

3.4 mtDNA Inheritance

A second important characteristic of the mtDNA genome, especially for forensic purposes, is the expected pattern of maternal inheritance. All relatives within the same maternal lineage will typically share the same mtDNA sequence; for example, all siblings, male and female, should have the same mtDNA sequence as their mother's. The reason for this mode of inheritance appears to be simple, at first glance. The mature ovum has several thousand mitochondria with approximately 100,000–150,000 copies of mtDNA, whereas the sperm cell entering the egg has only a few copies of mtDNA (Chen et al. 1995). The total number of mtDNA copies found in the midsection of the mature sperm cell has been reported to range from about 2 to 100 copies (Wai et al. 2010). Therefore, after fertilization—and assuming the mtDNA from the sperm has entered the egg cell—there is certainly an excess of maternal mtDNA in the zygote in relation to paternal copies of the genome. In addition to the disparity in relative copy number, there is evidence for a specific mechanism that recognizes and eliminates the several copies of paternal mtDNA as they enter the cytoplasm of the mature oocyte (Sutovsky et al. 2000; Kaneda et al. 1995) However, a small number of studies have revealed the presence of paternal copies of mtDNA in mature adults (Schwartz and Vissing 2002; Gyllensten et al. 1991). Therefore, the elimination of sperm cell mtDNA does seem to be a leaky process at some level. Studies have also revealed that in certain species (e.g., blue mussels), paternal mtDNA can be inherited and maintained (Sano et al. 2011). Therefore, it is still unknown to what extent paternal copies of mtDNA remain in adult humans, but examples of paternal inheritance at the germ cell level have not been reported in the literature.

Evidence for the control of paternal inheritance can be found in experiments that have introduced foreign mitochondria and mtDNA into germ-line or somatic cells, and evaluated the outcomes. For example, in one study researcher found that immediately after inserting foreign sperm mitochondria into somatic cells lacking mtDNA, 10%–20% of the cells had functional sperm mitochondria, but only one in 10,000 of those cells survived more than 48 hours (Manfredi et al. 1997). In a second study, when mitochondria from somatic cells were introduced into the cytoplasm of foreign somatic cells, a rapid replacement of the original mtDNA with newly introduced molecules was observed (King and Attardi 1988). This study also suggested that selective pressures can have a significant and direct impact on the proliferation of specific copies of mtDNA. Elimination of paternal mtDNA from a fertilized ovum is also species specific. In animal experiments, it was shown that molecules of mtDNA originating from males of the same species are eliminated much faster (Sutovsky et al. 2000; Kaneda et al. 1995; Gyllensten et al. 1991). However, the precise mechanism responsible for the elimination of paternal mtDNA from the zygote and early embryo could not be explained. Therefore, it remains unclear how this process functions to completely eliminate paternal copies of mtDNA originating from sperm cells.

The studies presented in this section illustrate that mechanisms exist to eliminate mtDNA originating from the male gamete; in control studies, somatic cells can accept, proliferate, and permanently maintain foreign copies of mtDNA, and paternal mtDNA has been rarely identified in humans. Therefore, the literature supports the contention that for practical purposes of mtDNA analysis in forensic investigations, it is correct to assume that mtDNA will typically be inherited through the maternal line. Support for this assertion has been documented, including by researchers who were not able to identify paternal mtDNA in the somatic cells of offspring by means of routine methods used in forensic laboratories (Parsons et al. 1997). However, with the emergence of quantitative PCR methods and deep DNA sequencing (second or third generation DNA sequencing), the validity of the paradigm that mtDNA is "strictly" maternal inheritance could be challenged in the future (Tang and Huang 2010; Li et al. 2010; Holland et al. 2011; Van Wormhoudt et al. 2011). The ability for these technologies to query pools of sequence types or assess each copy of mtDNA from a cell or cluster of cells, may allow researcher to determine whether paternal copies of mtDNA persist in adult humans, albeit at a low level. More importantly, these methods may allow researchers to determine whether copies of paternal mtDNA can be found in female germ cells.

3.5 mtDNA Heteroplasmy

Heteroplasmy is defined as a mixture of more than one mtDNA genome sequence within a cell or between cells of a single individual. As they accumulate, these heteroplasmic variants can lead to disease and can have a profound impact on the aging process (Arnheim and Cortopassi 1992; Wallace 2010; Payne et al. 2011). In a forensic context, these variants can provide additional layers of discrimination potential to help identify human remains (Ivanov et al. 1996), and have become a routine part of the interpretation process when conducting analysis on samples in forensic casework (Melton et al. 2005; Melton 2004). Therefore, because the presence of heteroplasmy can impact the interpretation of mtDNA comparisons between evidentiary and known samples encountered in forensic casework, it is important to understand the biological basis of heteroplasmy, how often it is observed in relation to the sequence detection method being used, and how best to interpret the results.

Since most eukaryotic cells contain hundreds to thousands of copies of mtDNA, it is possible and indeed very likely that random mutations exist in multiple copies of the mitochondrial genome throughout an individual. The mtDNA CR has a 10-fold higher mutation rate than the nucDNA genome. Although it is often speculated that the high mutation rate is due to low fidelity in the γ-polymerase used for replicating the genome, it turns out that this version of the replicative DNA polymerase is quite faithful (Longley et al. 2001). Instead, the likely source of mutation is from DNA damage induced by reactive oxygen species (ROSs) generated during oxidative phosphorylation, followed by the actions of a less-than-adequate repair system (Yakes and VanHouten 1997). The repair of ROS-induced damage occurs primarily through the base-excision repair pathway (Driggers et al. 1993), as nucleotide excision repair appears to be missing in the mitochondrion, and double-stranded break repair systems such as homologous recombination and nonhomologous end joining pathways are either absent or highly attenuated (Sykora et al. 2011). In general, base-excision repair involves the removal of the damaged base by the action of a specific

DNA glycosylase, removal of the ribose moiety at the apurinic or apyrimidinic site, and replicative insertion of the original nucleotide to repair the lesion. Although this is an effective process, cells struggle to keep up with the rate of ROS-induced DNA damage in the mitochondrial genome. It is this constant challenge of combating mtDNA damage, and the failure ultimately to do so, that allows for the accumulation of mtDNA heteroplasmic variants over the lifetime of an individual.

In addition to single nucleotide heteroplasmy (SNH), there is considerable length heteroplasmy (LH) observed in the CR of the mtDNA genome, specifically within long stretches of GC base pairs (Forster et al. 2010). Whereas the majority of SNH is the result of ROS-induced mutations and the lack of their repair, LH is typically caused by slippage of the replicative polymerase, similar to the type of slippage that occurs at STR loci to create stutter artifacts. Two primary regions of the CR are hot spots for LH: 16,184–16,193 in HV1 when there is a T to C transition at position 16,189 in comparison to the Cambridge reference sequence (Anderson et al. 1981), and 303–315 in HV2. When the number of homopolymeric nucleotides reaches eight, γ-polymerase struggles to faithfully replicate these stretches of DNA and will create a set of length variants. In fact, many polymerases struggle to faithfully replicate these homopolymeric stretches (e.g., *Taq* polymerase; unpublished data). The resulting family of length variants will produce uninterpretable and out-of-reading-frame data when performing conventional, Sanger-based DNA sequencing. Therefore, alternative sequencing strategies are required when faced with LH, for example, using sequencing primers that sit immediately downstream of the homopolymeric stretch.

A genetic bottleneck theory was proposed in the early 1980s to explain the rapid segregation of heteroplasmic sequence variants in Holstein cows, providing a mechanism for the transmission of heteroplasmy from cell to cell or from mother to child (Hauswirth and Laipis 1982; Hauswirth et al. 1984). The theory purported a reduction in the number of mtDNA molecules during the development of germ cells. Only a small number of mtDNA molecules pass through the bottleneck and are subsequently replicated to an amplified number of more than 100,000 in the mature ovum. Since these first observations in cows, researchers have been attempting to identify where in the developmental process the bottleneck occurs, and elucidate the mechanism of the bottleneck (Jenuth et al. 1996; Cao et al. 2007; Cree et al. 2008). Reports by Jenuth et al. (1996) and Cree et al. (2008) both support a reduction in the number of mtDNA molecules found in primordial germ cells of mice before the rapid expansion that leads to the high copy number state in mature oocytes. However, the findings reported by Cao et al. (2007) support a different theory. Instead of a reduction in copy number, these authors provided evidence to suggest that a bottleneck occurs through the selection of specific replicating units, potentially in conjunction with protein-rich structures called nucleoids. Given the increasing knowledge of the structure and function of nucleoids within a mitochondrion (Bogenhagen et al. 2008; Prachar 2010; Bogenhagen 2011), the theory put forth by Cao and coauthors is certainly a plausible explanation.

The mtDNA genome does not associate with proteins such as histones that are required for packaging of the nucDNA genome into nucleosomes. Instead, similar to bacterial genomes such as that found in *Escherichia coli*, the mtDNA genome is coated with proteins that allow for packaging into nucleoids (i.e., nucleosome-like structures). The nucleoids are composed of a number of protein activities, for example, single-stranded binding proteins, transcription factors, and replication complexes. Early studies suggested that there are approximately 3–4 nucleoids per mitochondrion, and 4–10 copies of mtDNA per nucleoid

(Satoh and Kuroiwa 1991). However, approximately 30% of mitochondria had initially been reported to lack nucleoids, so it was unclear how packaging and mtDNA function were related. One of the most abundant proteins associated with the mtDNA genome is *transcription factor A for the mitochondrion* (TFAM). Although the principle role of TFAM appears to be twofold, packaging and transcription regulation (Kanki et al. 2004; Kukat et al. 2011), what is clear is that the presence of TFAM is essential for the maintenance of the mtDNA genome. Researchers who suppressed the expression of TFAM through RNA interference have found that mtDNA content decreases with decreasing levels of TFAM (Kanki et al. 2004). Of course, one possible and plausible explanation for this observation is the tightly linked processes of transcription and replication in the displacement loop (CR) of the genome. More recent and advanced studies have revealed that on average more than 1000 molecules of TFAM are bound to a single copy of the mtDNA genome, as few as 1–2 copies of mtDNA are found in mammalian nucleoids, and more than 1000 nucleoids are typically found in a single cell (Lodeiro et al. 2012). In addition, it appears that the CR lacks bound TFAM, increasing the accessibility of ROSs. Along with a greater tolerance for sequence variation, these observations may help to explain the higher mutation rate in the mtDNA CR.

An important question regarding mtDNA inheritance and the nature of developmental bottlenecks is whether only a subset of nucleoids are participating in replication, with others remaining relatively inactive. If replication of mtDNA were restricted to only a few nucleoids, or if replication centers within the nucleoids restricted which genomes were being replicated, ratios of mtDNA heteroplasmy could shift rapidly and in an isolated manner, supporting the theories of Cao et al. (2007). Researchers have tried to address this question in lower eukaryotes. Studies on yeast have revealed that there is an mtDNA-separable, self-replicating unit that is faithfully inherited during cell division (Meeusen and Nunnari 2003). This two-membrane protein structure can exist with or without the presence of an mtDNA molecule, functions as a replisome, and is essential to both mtDNA inheritance and maintenance. Although these data may support the theories of Cao, it is clear that the mechanism of genetic bottlenecks at both the somatic and germ cell levels is still not well understood, especially in higher eukaryotes and humans. Nor is it evident whether mutated copies of the mtDNA genome readily propagate or are transmitted to dividing cells.

When occurring in germ cells, *de novo* mutations in mtDNA are the basis for potential changes in lineage-based CR haplotypes over evolutionary time. However, complete fixation of these changes is most certainly preceded by a heteroplasmic state, which may persist for many generations. The frequency of observing heteroplasmy in humans, detected by conventional Sanger sequencing methods, ranges from 2% to 8% of the population, including examples of SNH at two or three sites within the CR (Melton 2004; Irwin et al. 2009). This relatively low rate is primarily due to the lack of sensitivity of the Sanger method in detecting minor sequence variants. At best, the Sanger method can detect minor variants at a ratio of 1:10 to 1:20, or 5%–10% of the major component. Low-level variants are masked by the predominant sequence, and even higher level variants are not routinely reported by operating forensic laboratories. Alternative methods such as denaturing gradient gel electrophoresis (DGGE) and denaturing high-performance liquid chromatography (dHPLC) are more sensitive for detecting minor variants (down to 1%–5% of the major component), but do not have the ability to directly detect the nucleotide position with sequence variation (Tully et al. 2000; Kristinsson et al. 2009). The dHPLC method allows for automated fractionation of the variants, which can be individually sequenced, but still lacks sufficient sensitivity to uncover lower levels of mtDNA heteroplasmy. Deep sequencing is an exciting emerging technology that may allow for the detection of low-level heteroplasmy (<1%) in support of forensic investigations.

Of the deep sequencing or second-generation sequencing (SGS) platforms available as of mid-2012 (e.g., the 454 Life Science FLX/Junior from Roche, the SOLiD and the Ion Torrent/Ion Proton Sequencer from LifeTechnologies, the HiSeq/MiSeq from Illumina, and the Heliscope Single Molecule Sequencer from Helicos), the one best suited for targeted sequencing of forensic loci is the 454 Life Sciences system, as it can directly sequence amplicons of 400–500 bp in length (Rothberg and Leamon 2008; Hert et al. 2008). A small bench-top instrument (the GS Junior) makes the technology more accessible to forensic laboratories, at a cost of less than $100,000 to purchase and a cost per run of less than $1000. The sequencing chemistry is based on a pyrosequencing approach (http://www.454.com/; GS Junior Titanium Series Amplicon Library Preparation Method Manual 2010). The four nucleotides flow sequentially across a picotiter plate containing ~100,000 reaction wells (~1/10th the capacity of the larger 454 FLX instrument). Each well is ~44 μm in diameter and will accommodate a single DNA capture bead of ~28 μm coated with amplified copies of a single amplicon molecule. This second round of amplification occurs through a unique emulsion PCR (emPCR) technique, where each bead attaches to a single amplicon molecule from an initial PCR reaction. The emPCR occurs in an aqueous microdroplet containing the PCR reagents and surrounded by an oil-based emulsion. During the subsequent pyrosequencing, if a nucleotide is incorporated, a coupled reaction between sulfurylase and luciferase will generate photons of light that can be captured by a charge-coupled device (CCD) camera. When multiple nucleotides are incorporated into a homopolymeric stretch, the signal intensity increases, and is proportional to the total number of incorporations. Therefore, the pyrosequencing approach and the GS Junior technology allows for the parallel sequencing of ~100,000 amplicons in a single 8-hour run, or is comparable to running approximately one thousand 96-well plates of Sanger sequencing simultaneously. The result is a level of sequencing depth only seen in cloning experiments, at a small fraction of the cost and time necessary to complete such a study.

The 454 Life Sciences GS Junior instrument produces reliable and reportable results for the analysis of markers of interest to the forensic community, including the mtDNA CR (Holland et al. 2011). For example, when compared to conventional Sanger dye-terminator results, the 454 sequence of HV1 was correctly determined for 25 different maternal lineages, across 30 different individuals. When analyzing the Sanger data, only one of the 25 lineages exhibited reportable heteroplasmy (4%). However, analysis of the 454 data revealed heteroplasmy in 11 of the 25 lineages; a 44% rate of identifiable and reportable heteroplasmy. These 11 lineages exhibited low-level heteroplasmy, between 0.33% and 4.50% of the major component sequence, with the vast majority at less than 3%. In addition, many of the samples had multiple positions of heteroplasmy. Therefore, the effective rate of heteroplasmy was 19 observations across the 25 lineages, or approximately 0.253% of the sites exhibiting heteroplasmy (19/~300 sites per lineage × 25 lineages), compared to a single observation in the Sanger data (0.013%). That single observation displayed 18.40% and 20.14% of the minor component sequence in the Sanger and 454 data, respectively. Of course, a question that must be answered with deep sequencing data is whether low-level variants are true heteroplasmy, or instead, are artifacts of the PCR and sequencing process.

The nucleotide positions where heteroplasmy was observed in the study described above (Holland et al. 2011) were consistent with HV1 mutational hot spots, and sites where forensic polymorphisms and heteroplasmy have been observed in past studies (Parsons et al. 1997; Tully et al. 2000). In particular, when the sensitivity level of heteroplasmy detection is elevated (e.g., when using a DGGE approach), many of the same sites of heteroplasmy have been identified (Tully et al. 2000). Concerns regarding the reliability of the 454 sequence data were addressed through reproducibility studies. Multiple samples were run in either duplicate or triplicate confirmed positions of heteroplasmy, with very similar minor component percentages. Finally, the coverage rates and total number of reads for all heteroplasmic variants were typically high. Reported instances of low-level heteroplasmy required at least 40 reads of sequence (most with more than 100 reads) and a balanced ratio of forward and reverse reads when compared to the total read ratio. Therefore, PCR and sequencing artifacts were ruled out as the source of the relatively high rate of reported heteroplasmy when performing amplicon-based 454 deep sequencing.

Given a standard of at least 40 required reads of deep sequence data, and when the ratio of forward to reverse reads is consistent with the total, mixtures of 1:100 (1%), 1:250 (0.4%), 1:500 (0.2%), and 1:1000 (0.1%) would need total coverage of at least 4000, 10,000, 20,000, and 40,000 reads, respectively. This level of coverage is well within the capability of the GS Junior instrument. Although increasing coverage rates would reduce throughput and drive up costs, the significant increase in heteroplasmy detection is of sufficient value to make the 454 approach an attractive method when attempting to increase discrimination potential or differentiate between individuals with the same primary haplotype, including maternal relatives. In addition, a deep sequencing approach would allow for the reporting of heteroplasmy on a routine basis, significantly increasing the value of the testing method and potentially expanding the use of mtDNA analysis in forensic casework.

A second operational system has emerged that can resolve the variants of mtDNA heteroplasmy (Oberacher et al. 2006; Hall et al. 2009; Howard et al. 2011). The electrospray ionization mass spectrometry (ESI-MS) approach from IBIS Biosciences is capable of resolving mtDNA mixtures with component ratios as low as 1:20 and has been used in a number of forensic laboratories performing mtDNA analysis across the United States. The principal differences between the ESI-MS and 454 SGS systems are the inability of the ESI-MS to identify the location where a sequence difference exists, and the inability of the 454 SGS system to fully resolve homopolymeric stretches and LH. The ESI-MS system generates the precise molecular mass of each pool of amplicons to identify species with different sequence content. However, the location of sequence differences can only be implied. In contrast, the 454 pyrosequencing approach results in primary sequence information but provides poor resolution of homopolymeric sequences, and reveals PCR and sequencing artifacts that will require a filtering mechanism to eliminate their consideration during analysis, similar to filters needed for STR stutter and spectral bleed through. Therefore, neither system is currently capable of providing complete sequence information across the entire mtDNA CR, nor do they allow for the deconvolution of mtDNA mixtures. Further work will be necessary to identify which system is most desirable for routine mtDNA sequence analysis in forensic laboratories. Complicating this challenge is the pace of technology development and the newly emerging third-generation DNA sequencing techniques (Stoddart et al. 2009; Lieberman et al. 2010).

When heteroplasmy is observed in a forensic case using conventional Sanger sequencing, it may actually complicate the interpretation process, and as noted above, in most instances will not be reported. For example, hairs from the same individual or those collected from different maternal relatives can have differing ratios of heteroplasmic variants, including shifts that result in apparent substitutions (Parsons et al. 1997; Melton 2004). Empirical evidence shows that the concept of maternal inheritance is correct in almost all cases, and that children inherit the same mtDNA sequence from their mother (Schwartz and Vissing 2002; Parsons et al. 1997). If different mtDNA sequences were to be transmitted to the offspring in an uncontrollable manner, the concept of maternal mtDNA inheritance would simply not be plausible. Nonetheless, detecting and reporting heteroplasmy is an important consideration, as it is clear that the presence of heteroplasmy will increase the significance of a match, so if ignored will lessen the value of the testing results. Heteroplasmy occurs at only a small number of positions in the mtDNA sequence, so when the occurrence of heteroplasmy in the evidence sample is the same as that found in a reference, this leads to additional narrowing of the possible donor list.

One of the best known cases in which heteroplasmy was observed and reported is the case of the Russian royal family in which skeletal remains of Tsar Nicolas II were identified (Ivanov et al. 1996). This case exemplifies the manner in which heteroplasmy results can be reported, and provides an approach to assess the statistical significance of a match involving heteroplasmy. The subsequent sections will address the details of this case.

3.6 Application of mtDNA Analysis to Forensic Cases

The analysis of mtDNA sequence has been used in thousands of forensic cases to help exonerate individuals who were falsely convicted of crimes they did not commit, to assist the trier of fact in criminal investigations by identifying potential perpetrators of crime, to assist with the identification of military personnel and historical figures, and has withstood the test of time. The possible impact of mtDNA recombination or paternal inheritance, the efficacy and reliability of population databases, and the frequency and proper interpretation of heteroplasmy in hairs has not diminished the usefulness and reliability of mtDNA sequence analysis when applied to forensic casework. As the forensic community continues to expand the use of mtDNA analysis, it will be important to address the new challenges ahead. Larger and well-maintained databases will be needed, analysis and interpretation of mtDNA mixtures when working with touch evidence will be required, and enhancing the discrimination potential of mtDNA will need to be considered.

As noted in the previous sections, there are certain advantages of mtDNA analysis that have ensured its entry into everyday practice in forensic cases and for human identification. One of the main advantages of mtDNA is the concept that every somatic cell with aerobic metabolism has several hundred copies of mtDNA. The segments of the mtDNA genome that are studied are only about 300 bp long, which augments the possibility of "survival" of the targeted molecules. The other important characteristic of mtDNA is the manner in which it is inherited. Maternal inheritance allows for reference samples to be used from relatives who may be distantly related to the person who is being identified, but are connected through their maternal line either horizontally (relatives with same grandmother or even great grandmother) or vertically (grandmother, mother, siblings). This is not possible when using nucDNA for identification, as relatives from the first inheritance line (parents, children) are generally used for positive identification.

Procedures involved in mtDNA analysis are identical to those used in nucDNA analysis in the first few steps (Holland and Parsons 1999). All procedures performed on human remains or on trace evidence must be performed in a laboratory separated from reference samples, and reference samples should be analyzed after the analysis of evidence, when possible, to minimize the chance of contamination. The PCR product generated from extracted DNA is subjected to Sanger sequencing. After the data have been confirmed by analysis of the complementary strands of mtDNA sequence, and in most cases through duplicate analysis, the results are compared to the standard reference sequence (Anderson et al. 1981), and the differences are examined. The list of differences becomes the mtDNA "profile" of the individual associated with the sample. Only then are different sample profiles compared to determine whether matches have occurred. In the case of a mismatch, a person can be rejected as a potential donor of the sample or as a potential relative of the person who is being identified. Inconclusive results may also emerge, especially when heteroplasmy or only a single sequence difference exists, as an apparent difference can be the result of a severe shift in heteroplasmic variant ratios. In the case of a match, the profile must be compared to a population database and the probability of that match can be calculated.

The main disadvantage of mtDNA analysis is its limited ability for positive identification. When analyzing the HV regions, an absolute match between samples enables positive identification with a probability of 0.995 or less (Holland and Parsons 1999). The reason for such a low probability, despite the high variability of the HV regions, is in the fact that the sequence is inherited together as a single locus (linked). In the case of a mismatch, a person considered as a potential source of a sample can be ruled out with absolute certainty. With an increase in the number of analyzed population samples and the availability of an international human mtDNA database, it will be possible to calculate the probabilities of positive matches more accurately, but the probabilities themselves will not improve considerably until the database sizes increase dramatically.

The identification of the last Russian Tsar, Nicholas Romanov, and his family has been the subject of great interest and controversy for decades (Ivanov et al. 1996; Gill et al. 1994; Zhivotovsky 1999; Coble et al. 2009). Sequence heteroplamy was observed at position 16,169 in the remains of Nicholas, a reference sample from his living maternal relative (Princess Xenia Cheremeteff Sfiri), and the skeletal remains of his brother, the Grand Duke Georgij Romanov. Heteroplasmy in the sample profile from the living relative was confirmed using DGGE analysis, as the Sanger sequencing results showed apparent homoplasmy. Remarkably, the findings in the Tsar's case were the first documented report of mtDNA heteroplasmy in a forensic or identification case in the scientific literature. Before 1994, many articles had been written on observed deletions in the mtDNA genome that cause disease, and some related to SNH that cause disease through a dosing effect.

3.7 Genetic Variability and Random Match Probabilities

The strengths and limitations of the discrimination potential of conventional mtDNA testing have been well documented (Holland and Parsons 1999; Melton et al. 2012). Increasing the discrimination level has been attempted by looking outside of the HV1 and HV2 regions, and expanding the scope of analysis to the coding region (Lutz et al. 2000; Coble et al. 2006; Parsons and Coble 2001). However, it has also been well documented that tapping into the underlying levels of heteroplasmy in the CR is an excellent method for enhancing the discrimination of the typing system (Ivanov et al. 1996). In the identification case of Nicholas Romanov, the presence of heteroplasmy in the Romanov family allowed for the calculation of a likelihood ratio (LR) that took both the haplotype frequency and the probability of observing heteroplasmy at any one position along the mtDNA CR into account. The LR for identity of the Tsar when based strictly on the haplotype of the skeletal remains was calculated as 150, and the LR for the presence of a heteroplasmic sequence shared by two brothers was calculated as 2500 (Nicholas and Georgij Romanov). Therefore, since these two events were considered to be independent of one another, the LR for the mtDNA evidence was 375,000 more likely if the remains were, in fact, those of Nicholas Romanov. Similar calculations could be generated in forensic cases to assess the weight of observing low-level heteroplasmic variants that are shared by the evidence and a reference profile.

For proper interpretation of mtDNA results in forensic science, it is necessary to know the variability of mtDNA profiles in a population groups. Tens of thousands of mtDNA sequences have been reported in the literature and through online sources. Since the sources of information regarding mtDNA variability are numerous and the examined populations are different, it is challenging to use global data as a reference in forensic practice. There are several databases commercially available for forensic use and also several scientific databases that are mostly available for free (e.g., Mitochondrial DNA Control Region Database, http://empop.org/). The main disadvantage of commercial databases is the limited populations examined for database formation. Although such databases can be used in forensic and research laboratories examining nonidentical databases, limitations of such an approach are clear. It is therefore necessary for laboratories to use databases of the populations routinely investigated and if the population is racially and ethnically inhomogeneous, a database of the specific ethnic group or subgroup could be used. One of the best examples of the use of different databases is the system used by the Armed Forces DNA Identification Laboratory and the Federal Bureau of Investigation. Their database is used by other laboratories that do not have their own databases.

Since mtDNA is inherited along maternal lines, all relatives in the same line should have identical mtDNA, it is clear that mtDNA cannot be used for positive identification. In fact, persons that may appear unrelated may share the same mtDNA sequence, inherited by a distant maternal ancestor. It is therefore necessary to be familiar with the frequencies of different mtDNA alleles in a population before making conclusions about the significance of a positive match between an investigated sample and a reference sample. All investigated populations share some common characteristics: there are a small number of sequences that appear more frequently, whereas the majority of appear once or twice in the database. For example, in a database of North American Caucasians (604 persons), the most frequent profile appears 26 times (4.3%), whereas there are 390 appearing only once (Holland and Parsons 1999). In the African-American database (149 people) the most frequent profile appears in only 2.7% of the individuals, whereas there are 118 sequences that appear only once. In addition, there is only one match between these two databases, which illustrates the need for separate databases of different ethnic groups.

The size of the database is the principal limitation in calculating the frequency of a particular mtDNA sequence profile, especially for those with very low frequencies. Profiles appearing in a database only once can have their frequency greatly overestimated. A simple experiment can illustrate that fact: increasing of the number of people in the database results in a majority of newly discovered sequences, whereas only a few profiles reoccur. In one such experiment (Holland and Parsons 1999), the number of people in a database was increased from 700 to 800 with only six profiles reoccurring and 66 new profiles being observed. An increase in database size raised the frequency for only six profiles, whereas the frequency was lowered for the 445 unique profiles. Therefore, it is clear that the number of people in the databases must be increased greatly before the real frequencies of different profiles are known with a higher degree of certainty.

Sequencing of mtDNA in forensic cases should offer the answer to two fundamental questions: "Is it possible to exclude a certain person as the possible source of the sample examined?" and "If it is not possible to exclude a person as a source, what is the probability that the person is the source of the sample?" The importance of a positive answer to the first question is obvious. Every DNA analysis, including mtDNA, is reliable in relation to the reporting of exclusion. If it is not possible to exclude the person, the second question must be asked. A positive match without an objective and explicit calculation of probability that the person investigated is in fact the source of the sample is meaningless. When significance of a match is investigated it is important to present the results in a scientifically based way and with mathematical accuracy. Although the presentation of results is based on those two premises and on knowledge of potentials and limitations of statistical methods, forensic scientist should be allowed to choose the best method available.

The significance of a match in mtDNA analysis depends on the specific case and on the profiles in question. Definite results of identification are possible only in cases with a closed population, for example, a traffic accident in which it is sufficient to identify which body is which person from the list of people involved in the accident. If a match is made between a reference sample (from maternal relative) and only one of the investigated samples, a qualified positive identification can be established. Such certainty does not rely on calculation of absolute probabilities, but on the fact that the sample in question could only have originated from a small number of people.

Most forensic cases do not involve a closed group of people, and thus must consider the entire population as a potential source. In that case, a positive match cannot be considered as a definite result. The significance of such a match must be calculated in all such cases. For every population with known frequencies of mtDNA profiles, it is possible to calculate the general significance of any result by calculating the probability of the event that two randomly chosen people in a database will have the same profile. If, for example, profiles of all the people ($N = 604$) in the database of American Caucasian database are compared, 669 matches can be found after comparing 182,106 comparisons. This gives the general probability that two randomly chosen people will have the same CR profile of 0.36% or 1 in 272. This fact is valuable in the context of evaluating general weights of mtDNA testing results, but cannot support the significance of a match in a specific case. A more precise indicator for the interpretation of a particular match is derived from knowing the relative frequency (relative rarity) of the profile in question and calculating the probability that a random person has that same profile. The simplest way to determine relative rarity of a single profile is the "counting method." This method expresses only the absolute observations of the profile in different databases. The calculations based on the counting method have little value for very rare alleles, so in the judicial practice of the United States only absolute numbers are quoted and not the frequencies resulting from them.

For relatively frequent profiles, the frequency can be determined with greater certainty. The most frequent profile in the American Caucasian database (263G, 315.1C) appears 26 times in the database of 604 people. The relative frequency of this profile is 0.043. Assuming a normal distribution and a 95% confidence interval, the frequency of the profile in the population can be calculated according to this formula:

$$p = p' \pm 1.96 \, (p \times q/n)^{1/2},$$

where p' is the frequency of the allele in the database, $q = 1 - p'$, and n is the number of people in the database.

For the common profile mentioned above, the 95% confidence interval is 0.027–0.059 (0.043 ± 0.016). Thus, the highest frequency of the profile (with 95% confidence) is 0.059, and 94.1% of the population can be excluded as potential donors with 95% certainty. On the other hand, one in every 17 people (5.9%) in the population will have that particular profile.

For rare mtDNA profiles, including the newly found sequences not in the database, a confidence limit from zero proportion is applied:

$$p = 1 - \alpha^{1/N},$$

where N is the size of a database and $\alpha = 1 -$ confidence level (for 95% level of confidence, $\alpha = 0.05$).

For example, the value calculated for a new allele in the database of 604 American Caucasians would be 0.005, which means that 99.5% of the population can be excluded as a potential source of the sample with 95% certainty.

Another approach for expressing the significance of a match is calculating an odds ratio (Council 1996; Evett and Wier 1998). An odds ratio is a relative value of mtDNA evidence compared to other, contrary hypotheses. A typical example would be a trial in which the prosecution's hypothesis that the defendant really is the source of the sample given as evidence is confronted by the defense's hypothesis that the real source of the evidence is a random person from the population. An odds ratio higher than 1 would suggest a higher probability that the identity determined by mtDNA analysis is real. The higher the odds ratio is, the higher that probability would be in support of the prosecution's hypothesis. A simple example of a match without heteroplasmy would be that the value of the evidence that the person is a source of the evidence is 1.0 (complete match is assumed if the person is really the source). On the other hand, the value of the evidence that an unknown person from the population is the source of the evidence is the relative frequency of the profile under consideration in the population. An odds ratio in such a case would be the inverted value of the profile frequency in the population. For a newly discovered profile in the American Caucasian database, the odds ratio would be 1:200, whereas for the most frequent allele in the same database it would be 1:17. The advantage of expressing an odds ratio is that the chance of a person being the source of the evidence may be different from 1.0, which enables taking into consideration the effects of heteroplasmy and mutations.

A similar approach to calculating an odds ratio is the application of Bayes' theorem, which takes into account "existing chance," the probability of the person being the source given by all other non-mtDNA evidence (Council 1996; Evett and Wier 1998). A practical problem in the application of this method is the inability to mathematically express the values of most evidence. Anglo-American judicial practice allows the jury to evaluate relative values of the evidence submitted. The role of the mtDNA evidence is only to present a factor by which all other evidence is augmented or divided. Given the relatively low power of mtDNA discrimination (without heteroplasmy), evidence in some cases must rely on other evidence.

References

Alverson, A.J., S. Zhuo, D.W. Rice, D.B. Sloan, and J.D. Palmer. 2011. The mitochondrial genome of the legume *Vigna radiata* and the analysis of recombination across short mitochondrial repeats. *PLoSOne* 20:16404–12.

Amory, S., C. Keyser, E. Crubezy, and B. Lubes. 2007. STR typing for ancient DNA extracted from hair shafts of Siberian mummies. *Forensic Sci Int* 166:218–29.

Anderson, S., A.T. Bankier, B.G. Barrell et al. 1981. Sequence and organization of the human mitochondrial genome. *Nature* 290:457–65.

Andersson, S.G., A. Zomorodipour, J.O. Andersson et al. 1998. The genome sequence of *Rickettsia prowazekii* and the origin of mitochondria. *Nature* 396:133–40.

Arnheim, N., and G. Cortopassi. 1992. Deleterious mitochondrial DNA mutations accumulate in aging human tissues. *Mutat Res* 275:157–67.

Bibb, M.J., R.A. Van Etten, C.T. Wright, M.W. Walberg, and D.A. Clayton. 1981. Sequence and gene organization of mouse mitochondrial DNA. *Cell* 26:167–80.

Bogenhagen, D., and D.A. Clayton. 1974. The number of mitochondrial deoxyribonucleic acid genomes in mouse L and human HeLa cells: Quantitative isolation of mitochondrial deoxyribonucleic acid. *J Biol Chem* 249:7991–5.

Bogenhagen, D.F. 2011. Mitochondrial DNA nucleoid structure. *Biochem Biophys Acta* 1819:914–20.

Bogenhagen, D.F., D. Rousseau, and S. Burke. 2008. The layered structure of human mitochondrial DNA nucleoids. *J Biol Chem* 283:3665–75.

Borst, P., and L.A. Grivell. 1981. Small is beautiful—Portrait of a mitochondrial genome. *Nature* 290:443–4.

Brown, T.A., and D.A. Clayton 2006. Genesis and wanderings: Origins and migrations in asymmetrically replicating mitochondrial DNA. *Cell Cycle* 5:917–21.

Cao, L., H. Shitara, T. Horii et al. 2007. The mitochondrial bottleneck occurs without reduction of mtDNA content in female mouse germ cells. *Nat Genet* 39:386–90.

Chen, X., R. Prosser, S. Simonetti, J. Sadlock, G. Jaqiello, and E.A. Schon. 1995. Rearranged mitochondrial genomes are present in human oocytes. *Am J Hum Genet* 57:239–47.

Coble, M.D., P.M. Vallone, R.S. Just, T.M. Diegoli, B.C. Smith, and T.J. Parsons. 2006. Effective strategies for forensic analysis in the mitochondrial DNA coding region. *Int J Legal Med* 120:27–32.

Coble, M.D., O.M. Loreille, M.J. Wadhams et al. 2009. Mystery solved: The identification of the two missing Romanov children using DNA analysis. *PlosOne* 4:4838–46.

Council, N. 1996. *The Evaluation of Forensic DNA Evidence*. Washington, DC: National Academy Press.

Cree, L.M., D.C. Samuels, S.C. de Sousa Lopes et al. 2008. A reduction of mitochondrial DNA molecules during embryogenesis explains the rapid segregation of genotypes. *Nat Genet* 40:249–54.

Dolezal, P., V. Likic, J. Tachezy, and T. Lithgow. 2006. Evolution of the molecular machines for protein import into mitochondria. *Science* 313:314–18.

Driggers, W.J., S.P. LeDoux, and G.L. Wilson. 1993. Repair of oxidative damage within the mitochondrial DNA RINr 38 cells. *J Biol Chem* 268:22042–45.

Evett, I., and B.S. Wier. 1998. *Interpreting DNA Evidence*. Sunderland, MA: Sinauer Associates.

Federico, A., E. Cardaioli, P. DaPozzo, P. Formichi, G.N. Gallus, and E. Radi. 2012. Mitochondria, oxidative stress and neurodegeneration. *J Neuro Sci*, in press.

Fisher, D.L., M.M. Holland, L. Mitchell et al. 1993. Extraction, evaluation, and amplification of DNA from decalcified and undecalcified United States Civil War bone. *J Forensic Sci* 38:60–8.

Forster, L., P. Forster, S.M.R. Gurney et al. 2010. Evaluating length heteroplasmy in human mitochondrial DNA control region. *Int J Legal Med* 124:133–42.

Gill, P., P.L. Ivanov, C. Kimpton et al. 1994. Identification of the remains of the Romanov family by DNA analysis. *Nat Genet* 6:130–35.

Ginther, C., L. Issel-Tarver, and M.C. King. 1992. Identifying individuals by sequencing mitochondrial DNA from teeth. *Nat Genet* 2:135–8.

Gray, M.W. 1992. The endosymbiont hypothesis revisited. *Int Rev Cytol* 141:233–357.

Gray, M.W., B.F. Lang, and G. Burger. 2004. Mitochondria of protists. *Annu Rev Genet* 38:477–524.

GS Junior Titanium Series Amplicon Library Preparation Method Manual, Sequence emPCR Amplification Method Manual, and Sequencing Method Manual. 2010 (May). 454 Sequencing. Available at: http://www.454.com/.

Gyllensten, U., D. Wharton, A. Josefsson, and A.C. Wilson. 1991. Paternal inheritance of mitochondrial DNA in mice. *Nature* 352:255–7.

Haldar, D., K. Freeman, and T.S. Work. 1966. Biogenesis of mitochondria. *Nature* 211:9–12.

Hall, T.A., K.A. Sannes-Lowery, L.D. McCurdy et al. 2009. Base composition profiling of human mitochondrial DNA using polymerase chain reaction and direct automated electrospray ionization mass spectrometry. *Anal Chem* 81:7515–26.

Hauswirth, W.W., and P.J. Laipis. 1982. Mitochondrial DNA polymorphism in a maternal lineage of Holstein cows. *Proc Natl Acad Sci USA* 79:4686–90.

Hauswirth, W.W., M.J. Van de Walle, P.J. Laipis, and P.D. Olivo. 1984. Heterogeneous mitochondrial DNA D-loop sequence in bovine tissue. *Cell* 37:1001–7.

Hert, D.G., C.P. Fredlake, and A.E. Barron. 2008. Advantages and limitations of next-generation sequencing technologies: A comparison of electrophoresis and non-electrophoresis methods. *Electrophoresis* 29:4618–26.

Holland, M.M., and T.J. Parsons. 1999. Mitochondrial DNA sequence analysis—Validation and use for forensic casework. *Forensic Sci Rev* 11:21–50.

Holland, M.M., M.R. McQuillan, and K.A. O'Hanlon. 2011. Second generation sequencing allows for mtDNA mixture deconvolution and high resolution detection of heteroplasmy. *Croat Med J* 52:299–313.

Hopgood, R., K.M. Sullivan, and P. Gill. 1992. Strategies for automated sequencing of human mitochondrial DNA directly from PCR products. *Biotechniques* 13:82–92.

Howard, R., V. Encheva, J. Thomson et al. 2011. Comparative analysis of human mitochondrial DNA from World War I bone samples by DNA sequencing and ESI-TOF mass spectrometry. *Forensic Sci Int Genetics*, Epub ahead of print.

Irwin, J.A., W. Parson, M.D. Coble, and R.S. Just. 2011. mtGenome reference population databases and the future of forensic mtDNA analysis. *Forensic Sci Int Genet* 5:222–5.

Irwin, J.A., J.L. Saunier, H. Niederstatter et al. 2009. Investigation of heteroplasmy in the human mitochondrial DNA control region: A synthesis of observations from more than 5000 global population samples. *J Mol Evol* 68:516–27.

Ivanov, P.L., M.J. Wadhams, R.K. Roby, M.M. Holland, V.W. Weedn, and T.J. Parsons. 1996. Mitochondrial DNA sequence heteroplasmy in the Grand Duke of Russia Georgij Romanov establishes the authenticity of the remains of Tsar Nicholas II. *Nat Genet* 12:417–20.

Jenuth, J.P., A.C. Peterson, K. Fu, and E.A. Shoubridge. 1996. Random genetic drift in the femail germline explains the rapid segregation of mammalian mitochondrial DNA. *Nat Genet* 14:146–51.

Kaneda, H., J. Hayashi, S. Takahama, C. Taya, K.F. Lindahl, and H. Yonekawa. 1995. Elimination of paternal mitochondrial DNA in intraspecific crosses during early mouse embryogenesis. *Proc Natl Acad Sci* 92:4542–6.

Kanki, T., K. Ohgaki, M. Gaspari et al. 2004. Architectural role of TFAM in maintenance of human mitochondrial DNA. *Mol Cell Biol* 24:9823–34.

King, M.P., and G. Attardi. 1988. Injection of mitochondria into human cells leads to a rapid replacement of the endogenous mitochondrial DNA. *Cell* 52:811–9.

Kristinsson, R., S.E. Lewis, and P.B. Danielson. 2009. Comparative analysis of the HV1 and HV2 regions of human mitochondrial DNA by denaturing high-performance liquid chromatography. *J Forensic Sci* 54:28–36.

Kukat, C., C.A. Wurm, H. Spahr, M. Falkenberg, N.G. Larsson, and S. Jakobs. 2011. Super-resolution microscopy reveals that mammalian mitochondrial nucleoids have a uniform size and frequently contain a single copy of mtDNA. *Proc Natl Acad Sci* 108:13534–39.

Lang, B.F. et al. 1997. An ancestral mitochondrial DNA resembling a eubacterial genome iminiature. *Nature* 387:493–7.

Li, M., A. Schonberg, M. Schaefer, R. Schroeder, I. Nasidze, and M. Stoneking. 2010. Detecting heteroplasmy from high-throughput sequencing of complete human mitochondrial DNA genomes. *Am J Hum Genet* 87:237–49.

Lieberman, K.R., G.M. Cherf, M.J. Doody, F. Olasagasti, Y. Kolodji, and M. Akeson. 2010. Processive replication of single DNA molecules in a nanopore catalyzed by phi29 DNA polymerase. *J Am Chem Soc* 132:17961–72.

Lilly, J.W., and M.J. Havey. 2001. Small, repetitive DNAs contribute significantly to the expanded mitochondrial genome of cucumber. *Genetics* 159:317–28.

Lodeiro, M.F., A. Uchida, M. Bestwick et al. 2012. Transcription from the second heavy-strand promoter of human mtDNA is repressed by transcription factor A in vitro. *Proc Natl Acad Sci* 109:6513–18.

Lodeiro, M.F., A.U. Uchida, J.J. Arnold, S.L. Reynolds, I.M. Moustafa, and C.E. Cameron. 2010. Identification of multiple rate-limiting steps during the human mitochondrial transcription cycle in vitro. *J Biol Chem* 285:16387–402.

Longley, M.J., D. Nguyen, T.A. Kunkel, and W.C. Copeland. 2001. The fidelity of human DNA polymerase gamma with and without exonucleolytic proofreading and the p55 accessory subunit. *J Biol Chem* (Epub) 276:38555–62.

Loreille, O.M., T.M. Diegoli, J.A. Irwin, M.D. Cobleand, and T.J. Parsons. 2007. High efficiency DNA extraction from bone by total demineralization. *Forensic Sci Int Genet* 1:191–5.

Lupi, R., P.D. de Meo, E. Picardi et al. 2010. MitoZoa: A curated mitochondrial genome database of metazoans for comparative genomics studies. *Mitochondrion* 10:192–9.

Lutz, S., H. Wittig, H.J. Weisser et al. 2000. Is it possible to differentiate mtDNA by means of HVIII in samples that cannot be distinguished by sequencing the HVI and HVII regions? *Forensic Sci Int* 113:97–101.

Manfredi, G., D. Thyagarajan, L.C. Papadopoulou, F. Pallotti, and E.A. Schon. 1997. The fate of human sperm-derived mtDNA in somatic cells. *Am J Hum Genet* 61:953–60.

Marienfeld, J., M. Unseld, and A. Brennicke. 1999. The mitochondrial genome of Arabidopsis is composed of both native and immigrant information. *Trends Plant Sci* 4:495–502.

Meeusen, S., and J. Nunnari. 2003. Evidence for a two membrane-spanning autonomous mitochondrial DNA replisome. *J Cell Biol* 163:503–10.

Melton, T. 2004. Mitochondrial DNA heteroplasmy. *Forensic Sci Rev* 16:1–20.

Melton, T., G. Dimick, B. Higgins, L. Lindstrom, and K. Nelson. 2005. Forensic mitochondrial DNA analysis of 691 casework hairs. *J Forensic Sci* 50:73–80.

Melton, T., C. Holland, and M. Holland. 2012. Forensic mitochondrial DNA—Current practice and future potential. *Forensic Sci Rev* 10:101–22.

Michaels, G.S., W.W. Hauswirth, and P.J. Laipis. 1982. Mitochondrial DNA copy number in bovine oocytes and somatic cells. *Dev Biol* 94:246–51.

Mitochondrial DNA Control Region Database. Available at: http://empop.org/.

MITOMAP. A human mitochondrial genome database. Available at: http://www.mitomap.org (accessed on June 30, 2012).

Muller, K., R. Klein, E. Miltner, and P. Wiegand. 2007. Improved STR typing of telogen hair root and hair shaft DNA. *Electrophoresis* 28:2835–42.

Oberacher, H., H. Niederstatter, C.G. Huber, and W. Parson. 2006. Accurate determination of allelic frequencies in mitochondrial DNA mixtures by electrospray ionization time-of-flight mass spectrometry. *Anal Bioanal Chem* 384:1155–63.

Parsons, T.J., and M.D. Coble. 2001. Increasing the forensic discrimination of mitochondrial DNA testing through analysis of the entire mitochondrial DNA genome. *Croat Med J* 43:304–9.

Parsons, T.J., D.S. Muniec, K. Sullivan et al. 1997. A high observed substitution rate in the human mitochondrial DNA control region. *Nat Genet* 15:363–8.

Payne, B.A., I.J. Wilson, C.A. Hateley et al. 2011. Mitochondrial aging is accelerated by anti-retroviral therapy through the clonal expansion of mtDNA mutations. *Nat Genet* 43:806–10.

Piko, L., and L. Matsumoto. 1976. Number of mitochondria and some properties mitochondrial DNA in the mouse egg. *Dev Biol* 49:1–10.

Prachar, J. 2010. Mouse and human mitochondrial nucleoid – detailed structure in relation to function. *Gen Physiol Biophys* 29:160–74.

Robin, E.D., and R. Wong. 1988. Mitochondrial DNA molecules and virtual number of mitochondria per cell in mammalian cells. *J Cell Physiol* 136:507–13.

Rothberg, J.M., and J.H. Leamon. 2008. The development and impact of 454 sequencing. *Nature Biotechnology* 26:1117–24.

Sano, N., M. Obata, Y. Ooie, and A. Komaru. 2011. Mitochondrial DNA copy number is maintained during spermatogenesis and in the development of male larvae to sustain the doubly uniparental inheritance of mitochondrial DNA system in the blue mussel *Mytilus galloprovincialis*. *Dev Growth Differ* 53:816–21.

Satoh, M., and T. Kuroiwa. 1991. Organization of multiple nucleoids and DNA molecules in mitochondria of a human cell. *Exp Cell Res* 196:137–40.

Schwartz, M., and J. Vissing. 2002. Paternal inheritance of mitochondrial DNA. *N Engl J Med* 347:576–80.

Stoddart, D., A.J. Heron, E. Mikhailova, G. Maglia, and H. Bayley. 2009. Single-nucleotide discrimination in imobilized DNA oligonucleotides with a biological nanopore. *Proc Natl Acad Sci* 106:7703–07.

Sutovsky, P., R.D. Moreno, J. Ramalho-Santos, T. Dominko, C. Simerly, and G. Schatten. 2000. Ubiquitinated sperm mitochondria, selective proteolysis, and the regulation of mitochondrial inheritance in mammalian embryos. *Biol Reprod* 63:582–590.

Sykora, P., D.M. Wilson III, and V.A. Bohr. 2011. Repair of persistent strand breaks in the mitochondrial genome. *Mech Aging Dev* 133:169–75.

Tang, S., and T. Huang. 2010. Characterization of mitochondrial DNA heteroplasmy using a parallel sequencing system. *Biotechniques* 48:287–96.

Tully, L.A., T.J. Parsons, R.J. Steighner, M.M. Holland, M.A. Marino, and V.L. Prenger. 2000. A sensitive denaturing gradient-gel electrophoresis assay reveals a high frequency of heteroplasmy in hypervariable region I of the human mtDNA control region. *Am J Hum Genet* 67:432–43.

Van Wormhoudt, A., V. Roussel, G. Courtois, and S. Huchette. 2011. Mitochondrial DNA introgression in the European abalone *Haliotis tuberculata tuberculata*: Evidence for experimental mtDNA paternal inheritance and a natural hybrid sequence. *Mar Biotechnol* 13:563–74.

Wai, T., A. Ao, X. Zhang, D. Cyr, D. Dufort, and E.A. Shoubridge. 2010. The role of mitochondrial DNA copy number in mammalian fertility. *Biol Reprod* 83:52–62.

Wallace, D.C. 2010. Mitochondrial DNA mutations in disease and aging. *Environ Mol Mutagen* 51:440–50.

Wu, Y., J. Yang, F. Yang, T. Liu, W. Leng, Y. Chu, and Q. Jin. 2009. Recent dermatophyte divergence revealed by comparative and phylogenetic analysis of mitochondrial DNA genomes. *BMC Genomics* 10:238–51.

Yakes, F.M., and B. VanHouten 1997. Mitochondrial DNA damage is more extensive and persists longer than nuclear DNA damage in human cells following oxidative stress. *Proc Natl Acad Sci* 94:514–19.

Zhivotovsky, L.A. 1999. Recognition of the remains of Tsar Nicholas II and his family: A case of premature identification? *Ann Hum Biol* 26:569–77.

Y Chromosome in Forensic Science

4

MANFRED KAYSER
KAYE N. BALLANTYNE

Contents

4.1	Introduction	105
4.2	Sex Determination	106
4.3	Paternal Lineage Differentiation and Identification	107
	4.3.1 Y-STR Markers Currently Used in Forensic Science	109
	4.3.2 Interpretation of Y-STR Profiles	112
	4.3.3 Additional Y-STR Markers for Improving Male Lineage Resolution	115
	4.3.4 Combining Autosomal and Y-STR Evidence	118
4.4	Paternal Male Relative Differentiation and Identification	118
4.5	Paternity Testing and DVI	120
4.6	Paternal Biogeographic Origin Inferences	122
	4.6.1 Y-SNP Haplogroups	122
	4.6.2 Geographic Information from Y-STR Haplotypes?	126
	4.6.3 Haplogroups from Haplotypes?	126
	4.6.4 Y-SNP Typing Technologies in Forensics	127
Acknowledgments		128
References		128

4.1 Introduction

The Y chromosome has long been regarded as the poor cousin of the human genome. At only ~60 MB in size it is the second smallest human chromosome (after chromosome 21), contains the lowest number of genes, and—in contrast to all other human chromosomes—is extremely rich in repetitive DNA sequences of all kinds. Only the euchromatic region, covering about half of the Y, is genetically active, consisting of Y-specific single-copy regions, Y-specific repetitive regions, and X–Y homologous regions, as well as the repetitive centromere region of the Y. The heterochromatic part consists of only repetitive sequences and does not harbor any genes. Unlike any other human chromosome, most of the Y chromosome (~95%) does not undergo homologous recombination during meiosis (termed the nonrecombining portion of the Y [NRY]). Recombination with homologous regions on the X chromosome only occurs at the pseudoautosomal regions, located at the tips of the Y (Tilford et al. 2001). The most important gene functions on the Y are those involved in male sex determination (e.g., *SRY* gene) and spermatogenesis (e.g., *AZF*), although genes with other functions are also found, often with homolog partners on the X chromosome (such as *AMELY/AMELX*). The latter indicates the shared evolutionary history of Y and X going back to a homolog pair of autosomes in early mammalian history. The relative dearth of coding genes, combined with the Y's largely haploid nature, has resulted in this chromosome displaying genetic features and encoding human evolutionary history unlike any other human chromosome. For the most part, forensic science has been relatively slow in taking full advantage of this remarkable chromosome, although this is changing as our knowledge increases and new markers become available.

In order to fully utilize the Y chromosome for forensic purposes, it is necessary to understand precisely what makes it such a unique chromosome. The usual absence of Y chromosomes in females allows the use of the Y chromosome as a marker for human sex identification, which can add helpful information in forensic investigations. The strict male-specific inheritance of the NRY provides opportunities to specifically analyze DNA components of a crime scene sample that were provided by males only, and differentiate them from those provided by females, which can be highly important in mixed stain analysis in forensics such as in cases of sexual assault. At the same time, recombination-free inheritance from fathers to sons, combined with low to moderate mutation rates of most NRY-DNA polymorphisms, means that male relatives usually share the same NRY polymorphisms. This feature has both advantages and disadvantages for forensic applications of Y chromosome DNA. Disadvantages come in the way that conclusions from Y-chromosome DNA analysis usually cannot be made on an individual level, as desired in forensic investigation. This is because in the event of a matching Y-DNA profile between samples from a suspect and a crime scene the hypotheses that either the suspect or alternatively, any of his paternal male relatives, has left the crime scene sample have the same estimated probability (but see below for potential solution). Advantages are that because of shared Y-DNA profiles between male relatives, a close paternal male relative of a deceased alleged father can be used to replace the father in paternity testing of a male offspring using Y-DNA analysis in deficiency cases, where autosomal DNA profiling often is not informative. The same principle can also be used in disaster victim identification (DVI) of males using close or distant paternal male relatives in cases where autosomal DNA profiling fails.

The haploid nature of the NRY also leads to the Y-chromosome having a lower effective population size than the autosomes, with four copies of autosomal loci relative to each Y locus (Jobling and Tyler-Smith 2003). This lower effective population size results in the Y chromosome displaying the lowest genetic diversity of any chromosome (International SNP Map Working Group 2001). As consequence of the lower effective population size, Y polymorphisms can be more strongly affected by genetic drift or population-level events such as bottlenecks or founder effects than autosomal loci. In addition, the asymmetrical spread of distinct polymorphisms is aided by the patrilineal transmission of the Y mirroring certain cultural practices, such as patrilocality (where males retain their familial lands, with females relocating), or polygyny (low numbers of males having the highest reproductive success) (Oota et al. 2001; Seielstad et al. 1998). These features in part explain the relatively strong geographic information content provided with some Y-chromosomal DNA polymorphisms, as further outlined below.

4.2 Sex Determination

The use of the Y chromosome for male sex determination in forensic applications started about 40 years ago when luminescence microscopy was applied for detecting Y chromosomes in cells from cadaver material (Radam and Strauch 1973). In the late 1980s/early

1990s, specific Y chromosome DNA sequences were used for this purpose (Ebensperger et al. 1989; Fattorinil et al. 1991). However, analyzing only Y-specific DNA for the purpose of male sex determination is semioptimal, as the absence of the signal in principle can mean either the presence of female material or a negative result due to technical reasons. Therefore, systems have been developed that take advantage of the homologous nature of the human X and Y chromosomes, targeting sites that display sequences with length polymorphisms between the copies. Since the early 1990s (Akane et al. 1992), the amelogenin system has been used for human sex determination in forensics and other applications such as in paleogenetics, and is part of many commercial kits for human identification. The polymerase chain reaction (PCR) primers most often used amplify a 112-bp Y fragment together with a 106-bp X fragment (Sullivan et al. 1993), where observing a fragment of 106 bp indicates the presence of female DNA with two X copies of the same length, whereas two fragments of 106 and 112 bp indicates male DNA. However, this system is not free of error as it was observed that some men can carry a Y chromosome deletion that includes the *AMELY* gene locus and consequently appear as females in the test results. Although the frequency of the respective deletion is low (<1%) in many geographic regions such as Europe (Mitchell et al. 2006; Steinlechner et al. 2002), it can be as high as 3% in some populations such as from India or Sri Lanka (Chang et al. 2003; Thangaraj et al. 2002). To make DNA-based sex determination more reliable, proposals have been made to combine the *AMELY/AMELX* system with other X–Y differential markers or with Y-specific markers that are more distant to the *AMELY* region (Santos et al. 1998). This has recently occurred with the release of PowerPlex Fusion Kit (Promega) containing both *AMELY/AMELX* and the male-specific Y-STR (Y-chromosome short tandem repeat) DYS391, as well as the GlobalFiler Kit (Life Technologies) containing *AMELY/AMELX*, DYS391, and a Y-chromosomal insertion/deletion (indel) marker.

4.3 Paternal Lineage Differentiation and Identification

In principle, any NRY-DNA marker with a low or medium mutation rate is suitable for characterizing groups of male relatives belonging to the same paternal lineage, especially when multiple markers are combined to create compound haplotypes. However, if the mutation rate of a NRY marker is too low, as it for instance is for Y-chromosomal single nucleotide polymorphisms (Y-SNPs) with a mutation rate per site per generation of about 10^{-8} (Xue et al. 2009), the markers will not be practically useful for forensic applications (besides biogeographic ancestry inference, see below). Although such a low mutation rate ensures that all males carrying a particular Y-SNP mutation can be linked back to a common ancestor, this timing is expected to be long if the Y-SNP is frequent enough. Consequently, close but also very distantly related males (so distant that it usually escapes family knowledge) will carry such Y-SNP mutations, and as such the level of male lineage identification is extremely low. Therefore, more polymorphic NRY markers, that is, those with a higher underlying mutation rate such as Y-STRs that have an average mutation rate about 100,000 times higher than Y-SNPs (Goedbloed et al. 2009), are the preferred choice for male lineage differentiation for forensic purposes either alone as usually applied or in combination with Y-SNPs (but see below for a different forensic application of Y-SNPs).

The introduction of Y-STRs for paternal lineage identification was relatively straightforward for forensic biology, as they are biologically and analytically similar to autosomal STRs, although they are haploid, rather than diploid. However, in the early days of forensic STR analysis, the knowledge about Y-STR markers was lagging far behind that of autosomal STRs. This was because autosomal STRs were mainly identified in systematic studies with considerable funding, aiming to provide polymorphic markers for gene mapping purposes and disease gene identification. However, because of the nonrecombining nature of most of the Y chromosome, the principle of linkage mapping does not work in practice; hence, the Y chromosome was left out in the search for STRs. It was not until the early 1990s that the first human polymorphic male-specific Y-STR, DYS19, was identified (Roewer and Epplen 1992a) and, because of the NRY features described above, was immediately applied to a sexual assault case where it provided an exclusion constellation (Roewer and Epplen 1992b). The coming years saw only a very minor increase in Y-STR markers, so that in the late 1990s less than 20 Y-STRs were known. In a milestone study published in 1997, 14 Y-STRs were analyzed in a multicenter approach involving many colleagues from the forensic genetic community interested in the future application of Y-STRs to forensic casework. The considerable population data generated resulted in the recommendation of seven Y-STRs for forensic application, the so-called Minimal Haplotype (MH), and three extra Y-STRs for supplementation (Kayser et al. 1997). This publication marks the beginning of Y-STR implementation in forensic case work, so that today after only 16 years Y-STR profiling has become a routine application in most forensic laboratories worldwide. Although about 30 additional Y-STRs were identified in the following years, it was not until 2004 that a systematic search for polymorphic STRs on the Y chromosome was published, which provided 166 new useful Y-STRs (Kayser et al. 2004) as valuable resource for forensic and other applications. The International Society of Forensic Genetics has issued guidelines on the use of Y-STRs for forensic purposes (Gill et al. 2001; Gusmao et al. 2006).

Since their introduction to forensic science, Y-STRs have been used for one main purpose—to identify male lineages for the purpose of identifying and excluding suspects (Roewer 2009; Kayser 2007). In particular, Y-specific DNA amplification is useful in mixtures where female cells are present in substantially higher quantities than the male contribution. Although this is most commonly observed in sexual assault cases, any mixed sample (such as blood/saliva or skin/skin mixtures) can benefit from the male-specific amplification (Dekairelle and Hoste 2001; Sibille et al. 2002). In particular, Y-STR testing of sperm-negative sexual assault samples, where differential DNA extraction cannot be applied, can provide informative profiles in ~45% of cases where autosomal profiling would be unsuccessful (Olofsson et al. 2011).

Y-STR profiling can also be highly useful in DNA dragnets or mass screenings, if legally allowed, in cases where the true perpetrator escapes from voluntarily participation. Autosomal STRs may still be informative as long as very close relatives of the perpetrator take part and may provide hints via allele sharing ("familial searching"). However, Y-STR profiling usually is more informative because it allows identifying close and distant relatives of a nonparticipating perpetrator and thus provides direct leads for further investigation to find the true perpetrator, as shown in practice (Dettlaff-Kakol and Pawlowski 2002; Huang et al. 2011). For instance, Y-STR profiling was used in 2012 to finally solve the Marianne Vaastra murder case in the Netherlands that was unsolved for more than one decade (even though the true perpetrator did participate in the Y-STR dragnet together

4.3.1 Y-STR Markers Currently Used in Forensic Science

Until now, the commonly used core set of Y-STRs used in forensic investigations consists of up to 17 markers, which display low- to midrange diversity values and mutation rates (Table 4.1). Commercial multiplexes such as AmpFlSTR® Yfiler™ (Applied Biosystems; Figure 4.1) (Mulero et al. 2006) and PowerPlex Y® (Promega) (Krenke et al. 2005) provide highly sensitive methods of amplifying 12 or 17 overlapping markers, respectively, with template requirements of only 0.25–0.5 ng, although full Y-STR profiles can reliably be obtained from only 125 pg of DNA (Mulero et al. 2006; Krenke et al. 2005; Gross et al. 2008; Sturk et al. 2009). Analytically, the methodology is identical to that used for autosomal STRs—highly multiplexed single-reaction PCRs followed by capillary electrophoresis and semiautomatic software-supported allele scoring are used to genotype samples (Figure 4.1). Loci range in size from 90 to 330 bp, allowing amplification in high-quality or moderately degraded DNA samples (Table 4.1). Both Yfiler and PowerPlex Y are able to specifically amplify Y-STRs in the presence of overwhelming quantities of female DNA, with reported successful amplification at 1:2000 ratios of male/female DNA (Krenke et al. 2005; Mulero et al. 2006). Full male profiles are obtainable from 1:10 mixtures of male DNA, with partial profiles routinely seen from 1:20 and lower (Gross et al. 2008; Mulero et al. 2006). Y-STR results can be obtained from sperm-negative (as measured with Sperm Hy-liter, microscopic evaluation, or prostate-specific antigen detection) sexual assault samples (Dekairelle and Hoste 2001; Sibille et al. 2002; Olofsson et al. 2011), or in fingernail scrapings that show a single female profile with autosomal STR typing (Malsom et al. 2009). The full set of 17 core Y-STRs provide high levels of discrimination between paternal lineages in outbred populations, with haplotype resolution (a measure of the ability of the set of markers to discriminate between unrelated males) reaching 0.989 in Europeans, 0.889 in sub-Saharan Africa, and 0.905 in East Asia (Table 4.1) (Ballantyne et al. 2011). The high discrimination power ensures a high probability of differentiating between male lineages in a population. However, the high discrimination power is not seen in all populations (but see below for solution).

To facilitate the ability to amplify Y-STRs in highly degraded samples, mini-Y-STRs have been developed recently to complement the autosomal mini-STR multiplexes. Primers were redesigned for eight of the core 17 Y-STR loci to reduce the amplicon size by 49–197 bp, resulting in an increase in amplification success from degraded material, such as skeletal remains or enzymatically degraded DNA (Park et al. 2007). However, because of the large repeat size of some of the core Y-STRs, it is not possible to reduce all amplicons below 200 bp in size. Therefore, various panels of Y-STRs with amplicon sizes of only 91–151 bases were designed, covering a selection of both Yfiler loci and novel Y-STRs (Asamura et al. 2007, 2008; Park et al. 2007). Although the mini-Y-STRs do not provide an increase in haplotype resolution or discrimination capacity relative to Yfiler (Asamura et al. 2008), the increased amplification success from degraded samples suggests they will be useful for DVI or analysis of skeletal remains, where both autosomal and conventional Y-STR analysis may fail.

Table 4.1 Locus Information of Current Forensically Applied Y-STRs

Marker[a]	Repeat Motif	Alleles	Mutation Rate (95% Credible Interval; Number of Meiotic Transfers Investigated)	Gene Diversity
DYS19[b,c,d]	$(TAGA)_3(TAGG)_1(TAGA)_{6-16}$	9–19	2.6×10^{-3} (1.6×10^{-3}–3.9×10^{-3}; 11,900)	0.758
DYS389I[b,c,d]	$(TCTG)_3(TCTA)_{6-14}$	9–17	2.4×10^{-3} (1.1×10^{-3}–4.0×10^{-3}; 10,103)	0.691
DYS389II[b,c,d]	$(TCTG)_{4-5}(TCTA)_{10-14} N_{28}(TCTG)_3(TCTA)_{6-14}$	24–36	3.0×10^{-3} (1.8×10^{-3}–4.5×10^{-3}; 10,079)	0.646
DYS390[b,c,d]	$(TCTG)_8(TCTA)_{9-14}(TCTG)_1(TCTG)_4$	17–29	1.9×10^{-3} (8.0×10^{-4}–3.2×10^{-3}; 11,385)	0.774
DYS391[b,c,d]	$(TCTG)_3(TCTA)_{6-15}$	5–16	2.8×10^{-3} (1.6×10^{-3}–4.1×10^{-3}; 11,336)	0.474
DYS392[b,c,d]	$(TAT)_{4-20}$	4–20	4.0×10^{-4} (1.0×10^{-5}–1.1×10^{-3}; 11,268)	0.669
DYS393[b,c,d]	$(AGAT)_{7-18}$	7–18	9.0×10^{-4} (3.0×10^{-4}–1.7×10^{-3}; 10,079)	0.676
DYS385a/b[b,c,d]	$(AAGG)_4N_{14}(AAAG)_3 N_{12}(AAAG)_3N_{29}(AAGG)_{6-7}(GAAA)_{7-23}$	6–28	2.0×10^{-3} (1.3×10^{-3}–2.9×10^{-3}; 19,108 joined analysis)	0.968
DYS438[c,d]	$(TTTTC)_{7-16}$	7–18	5.0×10^{-4} (1.0×10^{-4}–1.3×10^{-3}; 6947)	0.664
DYS439[c,d]	$(GATA)_3N_{32}(GATA)_{5-19}$	5–19	5.5×10^{-3} (3.5×10^{-3}–7.9×10^{-3}; 6908)	0.699
DYS437[c,d]	$(TCTA)_{4-12}(TCTG)_2(TCTA)_4$	10–18	1.1×10^{-3} (30×10^{-4}–2.3×10^{-3}; 6919)	0.506
DYS448[d]	$(AGAGAT)_{11-13}N_{42}(AGAGAT)_{8-9}$	14–24	2.0×10^{-4} (2.0×10^{-5}–8.0×10^{-4}; 3531)	0.748
DYS456[d]	$(AGAT)_{11-23}$	5–23	4.3×10^{-3} (1.7×10^{-3}–9.5×10^{-3}; 3384)	0.597
DYS458[d]	$(GAAA)_{11-24}$	11–24	6.5×10^{-3} (2.3×10^{-3}–1.3×10^{-2}; 3382)	0.795
DYS635[d]	$(TCTA)_4(TGTA)_2(TCTA)_2(TGTA)_2(TCTA)_2(TATG)_{0-2}(TCTA)_{4-17}$	16–30	3.7×10^{-3} (1.5×10^{-3}–6.6×10^{-3}; 4349)	0.791
Y-GATA-H4[d]	$(TAGA)_3N_{12}(TAGG)_3(TAGA)_{8-15}N_{22}(TAGA)_4$	8–15.1	2.9×10^{-3} (1.3×10^{-3}–5.5×10^{-3}; 4534)	0.614

Note: Mutation rates are Median rates from Bayesian approach using summarized family data (Goedbloed et al. 2009), whereas global diversity values are calculated from the HGDP-CEPH panel of 604 males (Ballantyne et al. 2011).

[a] The recently released PowerPlex® Y23 System (Promega) includes all 17 Y-STRs considered in the AmpFLSTR® Yfiler® PCR Amplification Kit, and additionally six highly lineage differentiating simple single-copy Y-STRs DYS481, DYS549, DYS533, DYS643, DYS570, DYS576 (the latter two being RM Y-STRs).
[b] Minimal Haplotype (MH).
[c] Included in the commercial PowerPlex® Y System (Promega).
[d] Included in the commercial AmpFLSTR® Yfiler® PCR Amplification Kit (Applied Biosystems/Life Technologies).

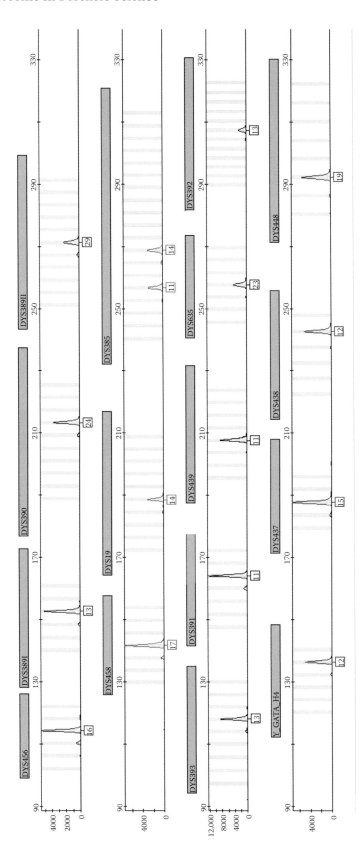

Figure 4.1 Y-STR profile. Representative electropherogram of a 17-locus Y-STR haplotype profile that was generated with the AmpFLSTR® Yfiler® PCR Amplification Kit (Applied Biosystem/Life Technologies).

4.3.2 Interpretation of Y-STR Profiles

Although Y-STRs may be similar analytically to autosomal STRs, the interpretation of Y-STR profiles differs in several key aspects from that of autosomal STR profiles. Allele calling is generally less complex than for autosomal STRs, because of the haploid status. This also simplifies the separation of the haplotypes in mixed samples, with mixtures containing two to four males being relatively easily genotyped, provided there are measurable differences in relative amounts of all the contributors (Cerri et al. 2003; Parson et al. 2001; Prinz et al. 1997). However, some of the current Y-STRs (such as DYS385) have several male-specific copies, and cases with multicopy Y-STR status were observed for almost all markers that usually are present in single-copy ones (see www.yhrd.org for details). Although multicopy Y-STRs can be difficult to interpret in some situations, such as mixed samples, the multiple copies give greater diversity within the marker, and have been shown to be the most informative markers in differentiating between haplotypes (Ballantyne et al. 2011).

When it comes to the way of estimating the strength of evidence, Y-STRs are conceptionally different from autosomal STRs in two ways. First, as the NRY is inherited intact from father to son, any statement of nonexclusion of a suspect from being the donor of a crime stain must also include all the suspects' paternal relatives in the nonexclusion (de Knijff 2003). In the absence of mutations, relatives separated by as many as 20 generations have been shown to share identical 17 locus haplotypes, and the nonexclusion of these tens or hundreds of male relatives must be conveyed in the evidentiary statement (Ballantyne et al. 2011). Second, because of the complete genetic linkage of all NRY markers, Y-STR frequencies have to be collected and used for statistical interpretation on the level of complete haplotypes instead of single loci. As such, the statistical interpretation of Y-STRs does not use the product rule with multiplication of the individual allele frequencies, but is instead most commonly done by estimating the frequency of the entire haplotype within the population of interest using representative databases.

Because compound haplotypes are much more variable than single independent STRs ever can be, the population databases used to derive Y-STR haplotype frequency estimates have to be much larger than for autosomal STRs. The larger the number of individuals in the frequency database, the more accurate the frequency estimate of common haplotypes will be, and the greater the chance of observing rare haplotypes (see Table 4.2). There are currently three Y-STR haplotype frequency databases available that were created for forensic purposes—YHRD, US Y-STR, and the Yfiler Haplotype Database. As of February 2013, the largest is YHRD (www.yhrd.org), with 108,949 seven-locus haplotypes and 49,782 seventeen-locus haplotypes from eight metapopulations across 108 countries. Any laboratory worldwide can submit data to this public database, provided quality control measures are met. The US Y-STR database (http://usystrdatabase.org/) contains 25,787 eleven-locus profiles and 15,616 seventeen-locus profiles, whereas the Applied Biosystems 17 locus Yfiler Haplotype database (http://www6.appliedbiosystems.com/yfilerdatabase/) lists 11,393 haplotypes.

When the questioned profile is contained in the database, the frequency of the haplotype can be calculated as x observations in a sample of N haplotypes (Gill et al. 2001). Because this is an estimate from a sample, rather than the entire population, upper and lower bounds of the estimate should be presented with the frequency. Although the confidence interval was previously calculated using the normal approximation, there has recently been a shift to using the binomial Clopper–Pearson confidence interval calculation

Table 4.2 Comparison of Frequency Estimates within the Western European Population of Haplotype Shown in Figure 4.1 across Different Y-STR Sets and Different Frequency Databases

Y-STR Set	YHRD				US Y-STR			Yfiler		
	Frequency (Normal 95% CI)	Frequency (Clopper–Pearson 95% CI)	Frequency Surveying Mean (Mode)	Haplotype Not Present (Upper 95% CI)	Frequency (Normal 95% CI)	Frequency (Clopper–Pearson 95% CI)	Haplotype Not Present	Frequency (Normal 95% CI)	Frequency (Clopper–Pearson 95% CI)	Haplotype Not Present
Minimal Haplotype (7 loci)	7.09×10^{-2} (6.74×10^{-2}–7.43×10^{-2})	7.09×10^{-2} (6.75×10^{-2}–7.44×10^{-2})	4.17×10^{-2} (4.16×10^{-2})	0 (1.39×10^{-4})	5.00×10^{-3} (8.44×10^{-2}–9.80×10^{-2})	5.00×10^{-3} (8.45×10^{-2}–9.82×10^{-2})	0 (4.17×10^{-4})	6.08×10^{-2} (5.35×10^{-2}–6.81×10^{-2})	6.08×10^{-2} (5.37×10^{-2}–6.85×10^{-2})	0 (5.87×10^{-4})
PowerPlex Y (12 loci)	9.46×10^{-3} (7.22×10^{-3}–1.17×10^{-2})	9.46×10^{-3} (7.36×10^{-3}–1.20×10^{-2})	4.17×10^{-2} (4.16×10^{-2})	0 (4.30×10^{-4})	1.07×10^{-2} (8.0×10^{-3}–1.33×10^{-2})	1.07×10^{-2} (8.16×10^{-3}–1.37×10^{-2})	0 (5.23×10^{-4})	5.80×10^{-3} (3.51×10^{-3}–8.16×10^{-3})	5.80×10^{-3} (3.74×10^{-3}–8.67×10^{-3})	0 (9.17×10^{-4})
Yfiler (17 loci)	7.84×10^{-4} (1.6×10^{-5}–1.55×10^{-3})	7.84×10^{-4} (2.14×10^{-5}–2.01×10^{-3})	4.17×10^{-2} (4.16×10^{-2})	0 (7.28×10^{-4})	3.06×10^{-4} (0–9.07×10^{-4})	3.06×10^{-4} (7.76×10^{-6}–1.71×10^{-3})	0 (7.28×10^{-4})	2.43×10^{-4} (0–7.19×10^{-4})	2.43×10^{-4} (6.15×10^{-6}–1.35×10^{-3})	0 (7.28×10^{-4})

Note: Although the two confidence interval estimation methods are similar when databases are large (>5000 haplotypes), the Clopper Pearson method produces more conservative estimates when database size decreases. When the profile is not observed in the database, the upper bound of the confidence interval is approximately equal to 3/N.

as being more conservative for rare haplotypes (Table 4.2) (Buckleton et al. 2011). Despite the large sizes of current databases, it is estimated that 95% of all 17-locus Y-STR profiles will not be represented (Butler 2011). Because a frequency estimate cannot be obtained for these singleton haplotypes, there are several options for estimating evidentiary value. A 95% confidence interval, approximately equal to $3/N$, may be used as a conservative estimate of the upper bound of the frequency (Buckleton et al. 2011). An alternative approach using the number of singletons already existing in the database has recently been proposed as a method to estimate the probability of an innocent suspect matching a previously unobserved haplotype (Brenner 2010). Although promising, this approach is alleged to be anticonservative in its estimation of frequencies, and has yet to be adopted (Buckleton et al. 2011). A further method to estimate the frequency, and therefore the evidentiary value of a haplotype, is termed frequency surveying (Roewer et al. 2000; Willuweit et al. 2011). Because only mutation (and not recombination) generates new Y-STR haplotypes, it logically follows that clusters of closely related haplotypes will occur within populations. Indeed, strong population clustering is observed throughout Europe, and follows historical geopolitical lines, indicating a distant level of relatedness between males in a given population group. By utilizing knowledge of this clustering, the frequency surveying method uses a Bayesian approach to estimate the frequency of a given haplotype on the basis of both the number of observations of the haplotype within the database and the number of close mutational neighbors. In this way, rare haplotypes, not observed in the database, can still have meaningful frequency estimates generated based on the known mutational processes of Y-STR alleles, rather than a $0/N$ frequency as in the counting method (Willuweit et al. 2011). An observed haplotype that is genetically distant from the modal haplotype cluster of a population will be given more statistical weight than a genetically close haplotype, even if the two have identical counts within the database. Recently, a new method for estimating trace–suspect match probabilities for singleton Y-STR haplotypes using coalescent theory was published (Andersen et al. 2013). A first analysis revealed that this coalescent-based approach is characterized by lower bias and lower mean squared error than the uncorrected count estimator and the surveying estimator. However, as currently developed, this method is highly computational intensive, which needs to be improved before it can be considered for practical applications. Thus far, none of the mentioned methods have reached the consensus status of being universally accepted and applied, which clearly marks a disadvantage of the forensic use of Y-STRs compared with autosomal STRs. All currently proposed approaches come with advantages and disadvantages, and the statistical forensic genetic community—perhaps with further progress in enlarging size and structure of Y-STR haplotype frequency databases—is encouraged to develop a consensus approach in the interest of Y-STR acceptance in the courtroom.

A general problem of currently unknown dimension is posed by the fact that all currently available Y-STR haplotype frequency databases for forensic use were generated from unrelated individuals only. Thus, in principle, the derived frequency estimates are underestimating the true frequencies in the population where close and distant relatives are often living in the same population. Without empirical data available, it is difficult to know how problematic this is for final conclusions in forensic case work, but it can be expected that the difference between estimated frequencies from such databases and the true frequencies are larger in rural areas and are smaller in metropolitan areas. Ideally, Y-STR haplotype frequency databases should be established from randomly chosen men, including related and unrelated individuals, to reflect the amount of male population substructure in a region.

4.3.3 Additional Y-STR Markers for Improving Male Lineage Resolution

Although the currently used Y-STRs, in particular the complete set of 17 Yfiler markers, provide high haplotype resolution in many populations, they show reduced diversity in certain populations (such as Finns, Xhosa, and Polynesians; D'Amato et al. 2010; Hedman et al. 2011; Kayser et al. 2000a) that have experienced population bottlenecks or sex-biased migration. In addition, as the number of males in Y-STR frequency databases expands, the number of unrelated individuals sharing haplotypes is growing. Thus, to ensure that the evidentiary value of Y-STRs remains high, it will be necessary to expand the core set of loci in the future, in the same manner as the autosomal core set has recently been increased in Europe and the United States. The current set of Y-STRs, with their low- to midrange diversities, was selected from a limited panel of known markers, as only 30 Y-STRs were described by 2002. Notably, the vast majority of currently known Y-STRs were identified after the commercial Y-STR kits were developed, and hence, could not be considered. When creating new STR panels, the key for successful discrimination between unrelated individuals is the selection of sufficient numbers of markers that show high levels of diversity within the population of interest. The inclusion of highly diverse markers within the haplotype ensures that few males will carry identical haplotypes by chance (referred to as Identity-by-State [IBS]), rather than by a shared origin (Identity-by-Descent [IBD]).

The Y-STRs showing the highest levels of diversity within and between worldwide populations generally share key characteristics that generate increased mutation rates, and therefore increased allelic diversity within the locus. These molecular features include the number of repeats within a locus (the higher the number of repeats, the greater the mutation rate), the complexity of the STR sequence (the more complex a sequence in terms of numbers of repeat blocks the greater the rate), and the repeat size (mutation rate decreasing as repeat size increases from tri- to tetra- to pentanucleotide repeats, etc.) (Ballantyne et al. 2010). Furthermore, multicopy status of Y-STRs usually provides enhanced value in differentiating between haplotypes (Ballantyne et al. 2011). The improved knowledge regarding both the number and characteristics of diverse Y-STRs has allowed the selection of candidate loci to improve current Y-STR testing capabilities. There are now mutation rates and sequence data available for 186 Y-STRs (Ballantyne et al. 2010; Goedbloed et al. 2009), and limited population data for ~110 of these (D'Amato et al. 2009, 2010; Ehler et al. 2010; Hanson et al. 2006; Hedman et al. 2011; Leat et al. 2007; Lessig et al. 2009; Lim et al. 2007; Maybruck et al. 2009; Redd et al. 2002; Rodig et al. 2008; Xu et al. 2010; Geppert et al. 2009; Hanson and Ballantyne 2007; Palha et al. 2011). To date, various combinations of 41 distinct Y-STR markers have been proposed to either supplement or replace the current set (for a recent review of additional Y-STRs in forensics, see Ballantyne and Kayser 2012). These new panels of 7–21 markers can increase the power of discrimination by 1.25% in American Africans and Europeans (Hanson and Ballantyne 2007), 11.1% in South Africans (D'Amato et al. 2011), and 28% in Finns (Hedman et al. 2011). The panel of novel Y-STRs with the highest increase in resolution on a global scale is a set of 13 Y-STRs characterized by high mutation rates (1×10^{-2} or higher), called rapidly mutating (RM) Y-STRs, as will be further described below (Table 4.3, Figure 4.2) (Ballantyne et al. 2010, 2011).

In 2012, the first new-generation commercial Y-STR kit has been released that took advantage of the recent scientific developments in the field of Y-STRs. The PowerPlex Y23 kit (Promega) contains all 17 Yfiler Y-STRs, plus an additional six Y-STR loci. These six Y-STRs, DYS481, DYS549, DYS533, DYS643, DYS570, and DYS576 (the latter two

Table 4.3 RM Y-STR Marker Information

Marker	Copy Number	Repeat Motif	Mutation Rate	Diversity (Global)
DYS449	1	$(TTCT)_{13-19}N_{22}(TTCT)_3 N_{12}(TTCT)_{13-19}$	1.22×10^{-2} $(7.54 \times 10^{-3}$–$1.85 \times 10^{-2}, 1617)$	0.88
DYS518	1	$(AAAG)_3(GAAG)_1(AAAG)_{14-22}(GGAG)_1(AAAG)_4N_6(AAAG)_{11-19}$ $N_{27}(AAGG)_4$	1.84×10^{-2} $(1.25 \times 10^{-2}$–$2.60 \times 10^{-2}, 1556)$	0.87
DYS526 a/b	2	$(CCTT)_{10-17}$ (a) $(CCCT)_3N_{20}(CTTT)_{11-17}(CCTT)_{6-10}N_{113}(CCTT)_{10-17}$ (b)	2.72×10^{-3} $(9.52 \times 10^{-4}$–$5.97 \times 10^{-3}, 1716)$ (a) 1.25×10^{-2} $(7.88 \times 10^{-3}$–$1.87 \times 10^{-2}, 1651)$ (b)	0.88
DYS547	1	$(CCTT)_{9-13}T(CTTC)_{4-5}N_{56}(TTTC)_{10-22}N_{10}(CCTT)_4(TCTC)_1(TTTC)_{9-16}$ $N_{14}(TTTC)_3$	2.36×10^{-2} $(1.70 \times 10^{-2}$–$3.18 \times 10^{-2}, 1679)$	0.87
DYS570[a]	1	$(TTTC)_{14-24}$	1.24×10^{-2} $(7.52 \times 10^{-3}$–$1.91 \times 10^{-2}, 1426)$	0.83
DYS576[a]	1	$(AAAG)_{13-22}$	1.43×10^{-2} $(9.41 \times 10^{-3}$–$2.07 \times 10^{-2}, 1727)$	0.83
DYS612	1	$(CCT)_5(CTT)_1(TCT)_4(CCT)_1(TCT)_{19-31}$	1.45×10^{-2} $(9.61 \times 10^{-3}$–$2.09 \times 10^{-2}, 1767)$	0.85
DYS626	1	$(GAAA)_{14-23}N_{24}(GAAA)_3 N_6(GAAA)_5(AAA)_1(GAAA)_{2-3}(GAAAG)_1(GAAA)_3$	1.22×10^{-2} $(7.70 \times 10^{-3}$–$1.82 \times 10^{-2}, 1689)$	0.85
DYS627	1	$(AGAA)_3N_{16}(AGAG)_3(AAAG)_{12-24}N_{81}(AAGG)_3$	1.23×10^{-2} $(7.80 \times 10^{-3}$–$1.81 \times 10^{-2}, 1766)$	0.85
DYF387S1	2	$(AAAG)_3(GTAG)_1(GAAG)_4N_{16}(GAAG)_9(AAAG)_{13}$	1.59×10^{-2} $(1.08 \times 10^{-2}$–$2.24 \times 10^{-2}, 1804)$	0.95
DYF399S1	3	$(GAAA)_3N_{7-8}(GAAA)_{10-23}$	7.73×10^{-2} $(6.51 \times 10^{-2}$–$9.09 \times 10^{-2}, 1794)$	0.99
DYF403S1a/b	4	$(TTCT)_{10-17}N_{2-3}(TTCT)_{3-17}$ (a) $(TTCT)_{12}N_2(TTCT)_8(TTCC)_9(TTCT)_{14} N_2(TTCT)_3$ (b)	3.10×10^{-2} $(2.30 \times 10^{-2}$–$4.07 \times 10^{-2}, 1504)$ (a) 1.19×10^{-2} $(7.05 \times 10^{-3}$–$1.86 \times 10^{-2}, 1402)$ (b)	0.89 (a)/ 0.99 (b)
DYF404S1	2	$(TTTC)_{10-20}N_{42}(TTTC)_3$	1.25×10^{-2} $(7.92 \times 10^{-3}$–$1.84 \times 10^{-2}, 1739)$	0.92

Note: Mutation rates are from family data (Ballantyne et al. 2010), whereas global diversity values are calculated from the HGDP-CEPH panel (Ballantyne et al. 2011).

[a] Included in the PowerPlex® Y23 System (Promega).

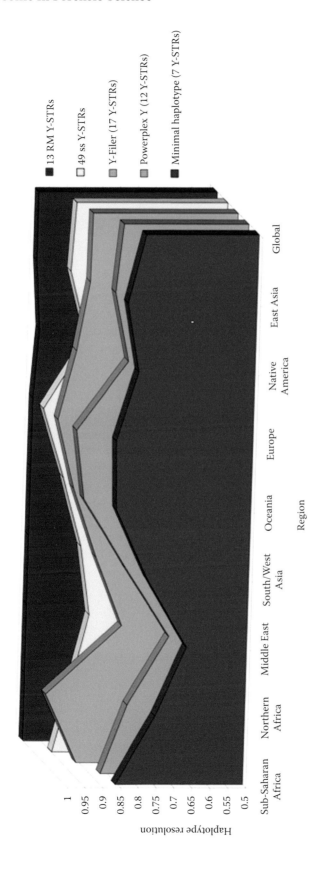

Figure 4.2 Male lineage differentiation using different Y-STR sets. Haplotype resolution of different Y-STR sets shown for larger geographic regions as well as globally based on the HGDP-CEPH samples. Color coding is according to Y-STR sets (see inbuilt legend). Note that only the RM Y-STRs provide near-complete resolution in all global populations in this dataset. (From Ballantyne, K.N. et al., *Forensic Sci Int Genet*, 6, 208–218, 2011; Vermeulen, M. et al., *Forensic Sci Int Genet*, 3, 205–213, 2009. With permission).

representing RM Y-STRs), were shown in a previous worldwide population study to be most informative—together with the commonly used Y-STRs—in discriminating paternal lineages out of a large number of simple single-copy Y-STRs tested (Vermeulen et al. 2009). As expected, PPY23 provides additional discrimination power as demonstrated already for U.S. populations (Davis et al. 2012), and this new kit comes with improved sensitivity and PCR inhibition resistance (Thompson et al. 2012). Although this new Y-STR panel will aid in resolving some IBS nonexclusions, greater benefit is obtained from applying both the 17 Yfiler Y-STRs and the 13 RM Y-STRs. Furthermore, Applied Biosystems/Life Technologies has already announced their next-generation Y-STR kit based on 27 Y-STRs, including more than two RM Y-STRs, which is expected to be commercially available in the end of 2013.

4.3.4 Combining Autosomal and Y-STR Evidence

At present, guidelines on Y-STR interpretation allow for the combination of Y-STR haplotype frequency estimates and autosomal STR match probabilities (Gusmao et al. 2006; Walsh et al. 2008; Gill et al. 2001). However, there has been some criticism of this, because of the different assumptions used in each weight-of-evidence calculation (Amorim 2008). In particular, the frequencies of current Y-STR haplotypes do not exclude an individual's paternal relatives, whereas the autosomal probabilities generally do. This leads to each calculation addressing different, and exclusive, hypotheses, leading to joint probabilities being factually incorrect. To counter these arguments, it has been stressed that the nonexclusion of relatives is a case-specific problem, and may be discounted in some instances. Alternatively, any combined likelihood ratio should include the caveat that it may not be relevant for excluding relatives of the suspect (Buckleton et al. 2011).

4.4 Paternal Male Relative Differentiation and Identification

The strongest limitation of Y-STR sets currently in use for forensic applications is that conclusions cannot be drawn to a single individual because paternal male relatives of a suspect usually share the same Y-STR haplotype due to relatively low mutation rates of the Y-STR loci involved (10^{-4} to 10^{-3}, see below; Goedbloed et al. 2009). Notably, most of the currently used Y-STR sets were ascertained at times where only limited resources regarding the number of Y-STRs were available. Hence, a systematic ascertainment procedure aiming for the most suitable Y-STRs for forensic applications considering all types of genetic and population parameters, as ideally is desirable, could not be carried. However, it could be speculated that if Y-STRs with a considerably higher mutation rate than the ones currently in forensic use were available, paternal male relatives may be differentiable, and thus male individuals may be identifiable by means of mutations observed. Recently, this speculation was replaced by real findings with the first systematic study investigating a large number of Y-STRs at a reasonably large number of DNA-confirmed father–son pairs became available (Ballantyne et al. 2010). Not only did this study provide locus-specific mutation rate estimates for almost 190 Y-STRs, in many cases for the first time ever, but it indeed identified 13 Y-STRs with mutation rates of 1×10^{-2} and higher. This set of 13 RM Y-STRs, in addition to providing increased differentiation between unrelated males of between 1% and 16% on a continental scale (see above), is also able to differentiate between close paternal relatives.

This has been demonstrated thus far based on theoretical grounds but also empirically with paternal male relatives separated between 1 and 20 generations—notably being independent samples from those used for mutation rate estimation (Ballantyne et al. 2010, 2011). Although many more paternal male relatives need to be investigated to further establish the power of RM Y-STRs in male relative differentiation and thus identification for future forensic case work, the available data are very promising. Nearly 50% of the father–sons, 60% of brothers, and 75% of cousins could be distinguished with RM Y-STRs. It should, however, be noted that these observed rates, particularly for the father–son and brother pairs, exceed the theoretical estimates generated from mutation rates (19.5% and 39%, respectively) and it may be that they represent overestimates due to relatively small sample size. More data are therefore required to resolve the true rate of differentiation with the RM Y-STR set for these relative pairs. From nine and more separating meiotic transfers, all paternal male relatives tested were differentiable with the RM Y-STR set, in line with

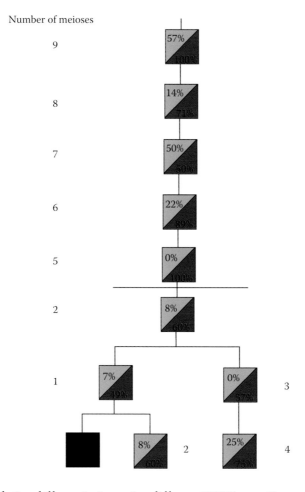

Figure 4.3 Male relative differentiation using different Y-STR sets. Example of a male multigeneration pedigree with observed rates of male relative differentiation using the Yfiler Y-STR set (above the diagonal) and the RM Y-STR set (below the diagonal), with respect to the level of relationship with the proband (black square). Full data and sample descriptions can be found in our previous work. (From Ballantyne, K.N. et al., *Forensic Sci Int Genet*, 6, 208–218, 2011. With permission.)

theoretical estimates from mutation rates. These results sharply contrast to Yfiler findings in the very same samples providing separation of only 7.7% of the father–sons, 8% of the brothers, 25% of cousins, and even at nine and more separating generations only 44% on average (range 0%–57%) were differentiable. Overall, 66% of paternal male relatives were distinguishable by means of mutations with the RM Y-STR set compared with 15% with Yfiler. Hence, the set of 13 RM Y-STRs identified carries a remarkable value for differentiating paternal male relatives and thus individualizing males by means of Y chromosome analysis (Figure 4.3). We propose that RM Y-STRs should not only be applied to those forensic cases with an *a priori* hypothesis of male relatives being involved, but moreover to all forensic cases where Yfiler analysis has revealed a match between suspect and crime scene materials to test for involvement of paternal males relatives (Ballantyne et al. 2011). The application of RM Y-STRs allows, at least in many cases, male individualization, by using the general advantage of Y-chromosome DNA analysis in specifically targeting male DNA components and clearly differentiate them from female components. In the absence of commercial kits for RM Y-STR typing, the 13 RM Y-STRs can be analyzed with three multiplex reactions (Ballantyne et al. 2011); companies are encouraged to develop and provide a commercial RM Y-STR product for forensic applications. The recently announced next-generation Y-STR kit by Life Technologies, which will include more RM Y-STRs than the two in the PowerPlex Y23 kit, is the first step in this direction.

4.5 Paternity Testing and DVI

The use of the human Y chromosome in paternity testing of male offspring goes back 47 years ago where for the first time whole Y-chromosome length differences were applied to conclude an exclusion from paternity (Nuzzo et al. 1966). The true value of applying NRY-DNA polymorphisms to paternity/family testing lies in the ability to replace unavailable persons such as deceased alleged fathers with paternal relatives (Kayser et al. 1998; Junge et al. 2006). In particular, such deficiency cases can only be solved with autosomal DNA analysis in cases where both parents of the deceased alleged father are available for DNA analysis, which is often not the case. Deficiency paternity cases involving male offspring can also be addressed with NRY-DNA analysis when the paternity issue occurred many generations ago, if paternal male relatives of both the alleged father and the male child are available for analysis. Obviously, such cases are impossible to solve with autosomal DNA analysis because of recombination issues. The most famous such case is that of U.S. president Thomas Jefferson (1743–1826), who was speculated to have sired Eston Hemings Jefferson (born 1808), the son of his African American slave Sally Hemings (1773–1835) (Foster et al. 1998). NRY-DNA analysis of a modern direct male descendant of Eston Hemings Jefferson and of several modern direct male descendants of Field Jefferson, the brother of Thomas Jefferson's father, revealed complete identity in the haplogroup characterized by several Y-SNPs and in 11 Y-STRs. This result is consistent with Thomas Jefferson's fathering of Easton Hemings Jefferson. However, given the nature of NRY-DNA, any one of Thomas Jefferson's contemporary paternal relatives, including his brother Randolph, could have been the true father with the same probability as Thomas Jefferson himself.

One of the key requirements to use NRY markers for paternity testing in deficiency cases involving male offspring, and to reconstruct more comprehensive paternal family relationships, is that the markers applied display a sufficiently low mutation rate, since mutations usually complicate the statistical interpretations. By now, a relatively large amount of family-based mutation rate data has been gathered for the particular Y-STR markers used in forensics. A recent study investigating the mutation rates of all 17 Yfiler Y-STRs in close to 2000 father–son pairs, also considering studies for the same loci published before, provides a comprehensive overview (Goedbloed et al. 2009) (Table 4.1). For the seven Y-STRs of the minimal haplotype and DYS385, more than 10,000 male meiotic transfers have been studied at each locus, resulting in somewhat reliable locus-specific mutation rate estimates between 4×10^{-4} (DYS392) and 3×10^{-3} (DYS389II). Since other forensically used Y-STRs were introduced to the community later, the number of meiotic transfers investigated is smaller, hence the mutation rate estimates are somewhat less reliable, such as about 7000 meioses for DYS437, -38, and -39 with estimated mutation rates between 5×10^{-4} (DYS438) and 5×10^{-3} (DYS439) and about 3–4000 meiotic transfers for the remaining Yfiler markers with estimated mutation rates between 2×10^{-4} (DYS448) and 6.5×10^{-3} (DYS458). The average mutation rate of all 17 Y-STRs currently used in forensics (Yfiler) was estimated as 2.2×10^{-3} based on more than 135,000 meiotic transfers (Goedbloed et al. 2009). Such relatively low rates established from large amounts of family data practically mean that in most cases of Y-STR applications to paternity testing, a true biological father will show the same alleles as his son.

If allelic differences are observed between a son and his putative father (or replacing paternal relative), they need to be considered in the paternity probability estimation (Rolf et al. 2001). It was previously recommended that an exclusion from paternity should be based on exclusion constellations at the minimum of three Y-STR loci, requiring the analysis of a sufficiently large number (≥9) of Y-STRs (Kayser and Sajantila 2001). This conclusion was based on the observation of two DNA-confirmed father–son pairs that both showed mutation at two out of nine Y-STRs tested (Kayser et al. 2000b). Expectedly, this knowledge, and thus the recommendation, may be biased by the relatively small number of father–son pairs investigated at several Y-STRs in parallel. Indeed, more recently a larger study involving close to 2000 father–son pairs and analyzing 17 Y-STRs (Yfiler) found one DNA-confirmed father–son pair with mutations at three Y-STRs (Goedbloed et al. 2009). Hence, based on the previous recommendation, this individual would have been excluded as the true father if only Y-STR data were available. However, not only did autosomal DNA analysis of the trio provided clear evidence in favor of paternity, but extended analysis of 157 Y-STRs showed no further allelic differences, clearly confirming the paternity (Ballantyne and Kayser 2012). Such an observation led us to recommend recently the use of probabilistic estimates of the likelihood of observing X number of mutations in a given meiotic transfer, where the locus-specific mutation rates of the (Y) STRs applied should be considered (as well as the number of meiotic transfers) (Ballantyne et al. 2011). A further solution for the issue of mutations in paternity testing of male offspring would be to use Y-STRs with an even lower mutation rate than the markers currently in use. Many of such slowly mutating (SM) yet polymorphic Y-STRs are available from the recently published Y-STR mutation survey (Ballantyne et al. 2010), although no specific SM Y-STR set has yet been developed for paternity testing.

As with paternity testing, Y-STRs are also useful for DVI in cases where all other means of investigation including autosomal STR (or SNP) profiling were not informative. In particular, Y-STRs are suitable in DVI cases of males where only distant paternal male relatives are available for DNA testing that usually cannot be linked via autosomal STR profiling because of recombination events that have occurred. This can be important in disaster events involving entire families (hence all close relatives) such as airplane disasters to/from holiday destinations, or disasters involving the destruction of residential areas including holiday resorts such as the 2004 Tsunami disaster in Southeast Asia.

4.6 Paternal Biogeographic Origin Inferences

4.6.1 Y-SNP Haplogroups

Although currently used Y-STRs are suitable for differentiating paternal lineages, their mutation rates limit their applicability for longer time scale analyses. Instead, Y-SNPs are the marker of choice to illuminate paternal aspects of human demographic including migration history. Currently, the Y-chromosome Consortium Phylogenetic tree lists ~600 Y-SNPs defining 316 haplogroups, with each being defined by 1–25 characteristic Y-SNPs (Karafet et al. 2008). Across the globe, there are 22 main branches of the Y chromosome phylogeny referred to as major haplogroups and labeled A–T (Figure 4.4) (Karafet et al. 2008; Y Chromosome Consortium 2002). The formation and spread of each major haplogroup reflects the establishment and expansion of major population groups, and can give an indication of the time scales and the route of major migration events. At the same time, major Y haplogroups can reveal information about paternal biogeographic origins. For example, the oldest and most variable haplogroups are the African haplogroups A and B, which is in line with the Out-of-Africa hypothesis of modern human origins. Distributions of other major haplogroups have been used as evidence for the spread of populations out of Africa and major migration routes throughout the globe such as haplogroup J in Northern African, European, and Central/Southern Asian populations, or haplogroup Q within the Americas (Chiaroni et al. 2009).

Consequently, at least some major haplogroups can be used directly to infer the geographic region of paternal genetic origin such as A and B indicating a sub-Saharan African origin, H indicating an S/C/W Asian (and Roma) origin, M indicating an origin in Oceania. However, most major haplogroups do not allow geographic inferences because they include subhaplogroups of diverse geographic origins; hence, additional Y-SNP typing is needed for determining the specific subhaplogroup to infer geographic information correctly. For instance, the major haplogroup C is found throughout N Asia, SE Asia, C Asia, Oceania, and Australia, and in Native Americans. However, certain C subhaplogroups specify particular geographic subregions such as C1 (M8) for Japan, C2 (M208) for Near Oceania, C2a1 (P33) for Remote Oceania, C3b (P39) for Native Americans, C4 (M347) for Australia, and C5 (M356) for South Asia. With some haplogroups, the specific geographic information can only be derived after high-resolution Y-SNP typing and subhaplogrouping, such as for the major haplogroup E, which occurs in Europe, Africa, and W Asia. However, certain E subhaplogroups indicate particular geographic subregions such as E2 (M75) for sub-Saharan Africa, E1b1b1a1b (V13) for Europe, and E1b1b1b1 (M81) for Northern Africa/Europe and E1b1b1c1 (M34) for Middle East/Anatolia/Northern Africa/Europe. Another example for a major haplogroup with various derived subhaplogroups carrying different geographic signatures is the major haplogroup R such as R2a (M124) for South Asia, R1b1c (V88) for Africa/Middle East, R1a1a1g (M458) for (Eastern) Europe, and R1a1a1f (M434) for West Asia. These examples demonstrate that geographic inferences from NRY DNA analysis must be accompanied by complete knowledge about the (sub)

Y Chromosome in Forensic Science

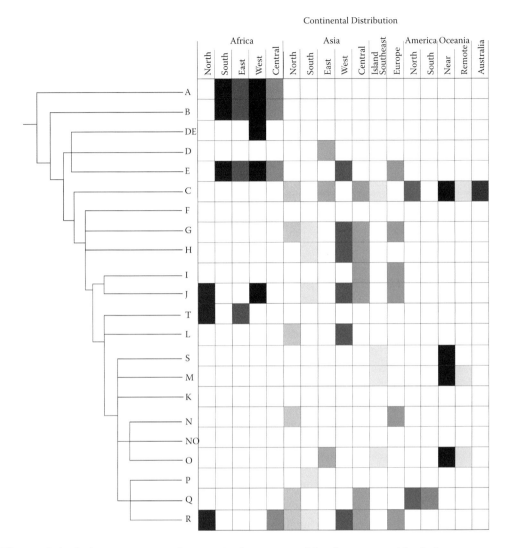

Figure 4.4 Phylogenetic tree of major Y-chromosomal haplogroups with their occurrences around the world. Colour coding is according to geographic regions. Haplogroups without geographic information are either found in all regions outside of Africa in low frequencies (i.e., F and K), or have insufficient data available thus far (i.e., NO). For more information, see Karafet et al. (2008) and Chiaroni et al. (2009).

haplogroup in question and its geographic distribution from appropriate reference data, which do not yet fully exist for all subhaplogroups.

The predominant use for Y-SNPs in forensic science will most likely be for inference of paternal biogeographic ancestry in search for unknown suspects, as detailed above. However, the success rate will be strongly dependent on the population in question, and the haplogroup of the individual as indicated above. Therefore, when designing Y-SNP multiplex tools for paternal biogeographic ancestry inferences, it is necessary to account for the distributions of all subhaplogroups within each major clade in case they reveal different geographic information. Recently, a hierarchical Y-SNP typing strategy has been described to screen for all major worldwide haplogroups, genotyping 28 SNPs for 29 branches of the phylogenetic tree (van Oven et al. 2011). This new tool provides a method to rapidly determine the major haplogroup of a sample. Some of the revealed haplogroups already allow a broad geographic inference such as A and B for sub-Saharan Africa, H for S/C/W Asia, and P for S Asia, whereas most major haplogroups identified are shared between larger geographic regions, and additional Y-SNP typing is needed to accurately focus on the correct subhaplogroups with the respective geographic information associated. Figure 4.5 shows examples of two individuals genotyped with Multiplex 1 and 2 (van Oven et al. 2011), where the results show that they belong to haplogroups I and O, respectively. Haplogroup I has a largely European distribution (rare occurrences in Northern Africa, Central and Western Asia) with highest frequencies in Scandinavia and the northwestern Balkans. Hence, haplogroup I most likely indicates a European paternal ancestry. Haplogroup O is the major Y haplogroup in East/Southeast Asia and also occurs in those regions settled by E/SE Asians such as Madagascar and parts of Near and all of Remote Oceania; hence, it indicates paternal ancestry from such geographic regions. To provide finer resolution, Y-SNP multiplexes have also been developed for European-specific haplogroups (Brion et al. 2004), East African (Gomes et al. 2010), and South America (Geppert et al. 2011), to list just a few examples. The new YHRD.org 3.0 database currently considers 138 Y-SNPs defining 127 haplogroups (R42, 2013). Although the current YHRD data content on Y-STR haplotypes with Y-SNP information is still relatively small (10,321, R42), this is expected to grow in the near future to serve as convenient resource for geographic inference of paternal genetic ancestry.

Y-SNPs can also be extremely useful for detecting admixture within individuals and populations. The former is of high relevance when performing forensic casework in regions of the world where genetic admixture between different continental groups is known to have played a substantial role in human population history. For instance, Y-SNP haplogroup data usually show high proportions of European ancestry in many urban populations in South America believed to be of mainly European ancestry (Corach et al. 2010). However, mtDNA and suitable autosomal markers analyzed in the same individuals revealed much less European and much more Native American and African admixture components (Corach et al. 2010). This usually is explained by sex-biased genetic admixture starting with the European occupation of South America dominated by males and continuing with the transfer of African slaves. Furthermore, this is similarly important in any region of the world where different continental groups are nowadays living next to each other due to more recent migrations, such as in many metropolitan areas in Europe or North America, because admixture is possible and cannot be excluded *per se* when investigating unknown samples. For instance, Y-SNP haplogrouping in self-declared U.S. African Americans revealed considerably high proportions of European admixture, and such proportions were much smaller when detected with suitable mtDNA and autosomal DNA markers in the same individuals (Lao et al. 2010; Vallone and Butler 2004). An example for very distant admixture detected with Y-SNP analysis is that of an indigenous British male with a rare east Yorkshire surname carrying a haplogroup A1 Y chromosome, indicating an African origin (King et al. 2007). Some other men with the same rare surname also carried the African A1 Y chromosomes, suggesting that the A1 lineage need not have been a founding type at the time of surname establishment in Britain about 700 years ago, or before (King et al. 2007). Clearly, an admixture event that old would not be manifested in any appearance traits anymore and is unlikely to be known from family records, but can still be detected with Y-SNP haplogroup analysis. These examples illustrate on one hand how powerful in general Y-SNPs are for geographic origin inference, but on the

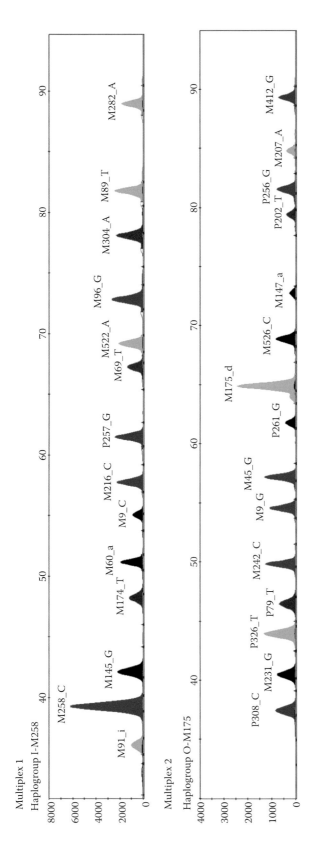

Figure 4.5 Y-SNP haplogrouping for inferring paternal biogeographic ancestry. Electropherograms for an individual belonging to Y chromosome haplogroup I as obtained with "Multiplex 1" and for an individual belonging to haplogroup O as obtained with "Multiplex 2" described in our previous work (van Oven et al. 2011). For each peak, the marker name and the observed allele is indicated. Haplogroup I strongly indicates a European paternal ancestry (it occurs also in Northern Africa, Central and Western Asia but much less frequently than in Europe). Haplogroup O strongly indicates a paternal origin in East/Southeast Asia (or in regions harboring populations originating from E/SE Asia such as Madagascar and parts of Oceania). (Adapted with additional information from van Oven, H. et al., *Int J Legal Med*, 125, 879–885, 2011.)

other hand demonstrate that solely analyzing Y-SNPs to infer the geographic origin of an individual can be misleading in cases of admixture involving ancestors from different geographic regions. Hence, Y-SNP analysis should always be combined with suitable mtDNA and autosomal markers for accurate inference of a person's geographic origin from DNA (Corach et al. 2010; Lao et al. 2010). Whereas autosomal ancestry-informative markers are most sensitive to reveal recent admixture, nonrecombining Y-SNPs (and mtDNA SNPs) are able to detect recent as well as ancient admixture and additionally allow inferring the parental side (paternal vs. maternal) through which an admixture was introduced.

On a further note, the ability to detect admixture with Y-SNPs may be useful for creating population-specific Y-STR haplotype frequency databases. The National Research Council recommended the use of separate databases for major population groups within a population of heterogeneous ancestry, to ensure that frequency estimates were not biased for rare alleles or haplotypes of minor population groups (National Research Council 1996). However, in admixed societies, it is likely that some individuals who self-declare as one group may carry a Y chromosome of another group as a result of paternal admixture at any time as described above. By genotyping Y-SNPs for these individuals, it would be possible to obtain accurate frequency estimates for the parental population for Y-STR databases (Hammer et al. 2006).

4.6.2 Geographic Information from Y-STR Haplotypes?

Although most Y-STR haplotypes do not carry any geographic information, some particular ones indeed do. Direct geographic inference from Y-STR haplotypes is possible in cases where a Y-SNP haplogroup that contains geographic information harbors a relatively frequent Y-STR haplotype, for instance, because of population effects such as bottlenecks or founders. Examples of particular Y-STR haplotypes include DYS19–DYS389I–DYS389II–DYS390–DYS391–DYS392–DYS393–DYS385a,b 17-13-30-25-10-11-13-10,14, indicating an Eastern European origin due to association with haplogroup R1a, haplotype 14-13-29-24-11-13-13-11,14 indicating a Western European origin due to association with haplogroup R1b, haplotype 13-13-30-24-10-11-13-16,18 indicating a Southwestern European origin due to association with haplogroup E3b, haplotype 15-13-31-21-10-11-13-16,17 indicating a sub-Saharan African origin due to association with haplogroup E*(xM35), and haplotype 15-12-28-24-10-13-13-12,16 indicating an origin in East/Southeast Asia or Oceania due to association with haplogroup O3. The reader is invited to search the named haplotypes in YHRD.org to see their geographic distribution and grasp the geographic value attached. However, as explained above, Y-SNPs are usually the better choice for inferring paternal geographic origins compared with Y-STRs.

4.6.3 Haplogroups from Haplotypes?

Because only very few Y-STR haplotypes allow geographic inference directly, and because Y-SNP typing is not yet as widespread among forensic laboratories as Y-STR analysis is, approaches were developed to predict the Y-SNP haplogroup from Y-STR haplotype data. There may also be cases where the evidential material is too limited for an additional Y-SNP analysis. Programs such as HaplogroupPredictor (Athey 2005) use a training database of Y-STR haplotypes within known Y-SNP based haplogroup data to predict the unknown haplogroup based on individual STR allele frequencies within each haplogroup. An alternative approach involves a tandem analysis strategy combining decision trees, Bayesian models and support vector machines to provide a 98.8% accuracy of haplogroup prediction (Schlecht et al. 2008). However, such prediction models have been criticized for being biased toward European haplogroups (because of skewed database compositions), and for failing when recurrent mutation generates identical Y-STR haplotypes on different haplogroup backgrounds (Muzzio et al. 2011a,b). The latter problem can only be overcome by including a large number of Y-STRs in the underlying reference databases, whereas the former may be rectified by the addition of more worldwide Y-STR/Y-SNP datasets. In the absence of these, however, it is generally considered more reliable to simply genotype the specific Y-SNPs required for haplogroup assignment in the population of interest.

4.6.4 Y-SNP Typing Technologies in Forensics

There now is a wide range of technologies and platforms available to genotype SNPs rapidly and reliably depending on the degree of sample as well as marker throughput from singleplex assays suitable for parallel genotyping of large number of samples such as with TaqMan to parallel genotyping of up to 1 million of SNPs in single samples that can be upgraded to large numbers of samples as well (Ding and Jin 2009). However, a small selection is usually applied in forensic laboratories (Sobrino et al. 2005) mostly concentrating on low-throughput approaches that can make use of already existing equipment such as DNA analyzers. The SNP-typing platform commonly used by the forensic community currently is SNaPshot from Applied Biosystems, a single-base extension assay. An initial PCR amplifies a short segment (50–100 bp) around the target SNP, followed by a second PCR, where a fluorescently labeled dideoxy nucleotide, complementary to the SNP of interest, is added to a single primer annealed immediately adjacent to the SNP. Multiplexes of 20–30 SNPs can be developed but become more demanding in their design the more SNPs that are added. Nonoverlap of extension fragment lengths during capillary electrophoresis can be achieved by added nonhuman sequence "tails" to the end of the extension primer. There are SNaPshot-based assays for continental-level resolution of Y-SNP haplogroups (Figure 4.5) (van Oven et al. 2011) as well as for Y-SNPs informative within continental levels such as Europe (Brion et al. 2004), East Africa (Gomes et al. 2010), and South America (Geppert et al. 2011). Alternatively, the concept of allele-specific hybridization is sometimes being used in forensics to distinguish between SNP alleles, by utilizing two specific probes differing in sequence at only the target site. A signal will only be generated if the oligonucloetide probe matches the target (and thus the SNP) exactly. Attaching different color fluorescent dyes to each allele-specific probe allows the genotype to be determined. Such probes are generally detected using real-time PCR methods (such as TaqMan (Applied Biosystems), LightCycler (Roche), or Molecular Beacon probes (Invitrogen), which limits the multiplexing ability. Because the PCR products are not separated by size, as with capillary electrophoresis, the number of SNPs that can be analyzed in a single assay is limited by the number of dyes available (at present, five colors simultaneously, or two SNPs). Another limitation of this approach is that accurate allele calling requires reference DNA samples of known genotypes, which are not always easily available. Other methods, although only rarely applied in forensic laboratories for SNP typing, include direct DNA sequencing, restriction enzyme digestion, or matrix-assisted laser desorption/ionization–time of flight (MALDI-TOF) mass spectrometry (Sequenom)–based technologies. Notably, there currently is no commercial kit available for Y-SNP analysis; all available Y-SNP typing protocols were developed in forensic (and other) laboratories and therefore usually miss the high degrees of robustness often available with commercial assays. However, until such commercial products are available for Y-SNP analysis, for example, to infer paternal geographic origins in cases with unknown crime scene sample donors, the published tools will reveal useful results in such cases.

Acknowledgments

We thank Mannis van Oven and Arwin Ralf for providing Figure 4.5. We are grateful to the numerous colleagues who published on the Y chromosome and their application to forensic science, whose work we had the privilege to partly summarize here. MK was supported by funding from the Netherlands Forensic Institute (NFI) and by a grant from the Netherlands Genomics Initiative (NGI)/Netherlands Organization for Scientific Research (NWO) within the framework of the Forensic Genomics Consortium Netherlands (FGCN).

References

Akane, A., S. Seki, H. Shiono et al. 1992. Sex determination of forensic samples by dual PCR amplification of an X–Y homologous gene. *Forensic Science International* 52: 143–148.

Amorim, A. 2008. A cautionary note on the evaluation of genetic evidence from uniparentally transmitted markers. *Forensic Science International Genetics* 2: 376–378.

Andersen, M.M., A. Caliebe, A. Jochens, S. Willuweit, and M. Krawczak 2013. Estimating trace-suspect match probabilities for singleton Y-STR haplotypes using coalescent theory. *Forensic Science International Genetics* 7: 264–271.

Asamura, H., H. Sakai, M. Ota, and H. Fukushima. 2007. MiniY-STR quadruplex systems with short amplicon lengths for analysis of degraded DNA samples. *Forensic Science International Genetics* 1: 56–61.

Asamura, H., S. Fujimori, M. Ota, T. Oki, and H. Fukushima. 2008. Evaluation of miniY-STR multiplex PCR systems for extended 16 Y-STR loci. *International Journal of Legal Medicine* 122: 43–49.

Athey, W. 2005. Haplogroup prediction from Y-STR values using an allele-frequency approach. *Journal of Genetic Genealogy* 1: 1–7.

Ballantyne, K.N., and M. Kayser. 2012. Additional Y-STRs in forensics: Why, which and when. *Forensic Science Review* 24: 63–78.

Ballantyne, K.N., M. Goedbloed, R. Fang et al. 2010. Mutability of Y-chromosomal microsatellites: Rates, characteristics, molecular bases, and forensic implications. *American Journal of Human Genetics* 87: 341–353.

Ballantyne, K.N., V. Keerl, A. Wollstein et al. 2011. A new future of forensic Y-chromosome analysis: Rapidly mutating Y-STRs for differentiating male relatives and paternal lineages. *Forensic Science International Genetics* 6: 208–218.

Brenner, C.H. 2010. Fundamental problem of forensic mathematics—The evidential value of a rare haplotype. *Forensic Science International Genetics* 4: 281–291.

Brion, M., B. Sobrino, A. Blanco-Verea, M.V. Lareu, and A. Carracedo. 2004. Hierarchical analysis of 30 Y-chromosome SNPs in European populations. *International Journal of Legal Medicine* 119: 10–15.

Buckleton, J.S., M. Krawczak, and B.S. Weir. 2011. The interpretation of lineage markers in forensic DNA testing. *Forensic Science International Genetics* 5: 78–83.

Butler, J.M. 2011. *Advanced Topics in Forensic DNA Typing: Methodology*. Waltham, MA: Academic Press.

Cerri, N., U. Ricci, I. Sani, A. Verzeletti, and F. De Ferrari. 2003. Mixed stains from sexual assault cases: autosomal or Y-chromosome short tandem repeats? *Croatian Medical Journal* 44: 89–292.
Chang, Y.M., L.A. Burgoyne, and K. Both. 2003. Higher failures of amelogenin sex test in an Indian population group. *Journal of Forensic Science* 48: 1306–1313.
Chiaroni, J., P.A. Underhill, and L.L. Cavalli-Sforza. 2009. Y chromosome diversity, human expansion, drift, and cultural evolution. *Proceedings of the National Academy of Sciences of the United States of America* 106: 20174–20179.
Corach, D., O. Lao, C. Bobillo et al. 2010. Inferring continental ancestry of Argentineans from autosomal, Y-chromosomal and mitochondrial DNA. *Annals of Human Genetics* 74: 65–76.
D'Amato, M.E., M. Benjeddou, and S. Davison. 2009. Evaluation of 21 Y-STRs for population and forensic studies. *Forensic Science International Genetics Supplement Series* 2: 446–447.
D'Amato, M.E., L. Ehrenreich, K. Cloete, M. Benjeddou, and S. Davison. 2010. Characterization of the highly discriminatory loci DYS449, DYS481, DYS518, DYS612, DYS626, DYS644 and DYS710. *Forensic Science International Genetics* 4: 104–110.
D'Amato, M.E., V.B. Bajic, and S. Davison. 2011. Design and validation of a highly discriminatory 10-locus Y-chromosome STR multiplex system. *Forensic Science International Genetics* 5: 122–125.
Davis, C., J. Ge, C. Sprecher et al. 2013. Prototype PowerPlex Y23 System: A concordance study. *Forensic Science International Genetics* 7: 204–208.
de Knijff, P. 2003. Son, give up your gun: Presenting Y-STR results in court. *Profiles in DNA* 7: 3–5.
Dekairelle, A.F., and B. Hoste. 2001. Application of a Y-STR pentaplex PCR (DYS19, DYS389I and II, DYS390 and DYS393) to sexual assault cases. *Forensic Science International* 118: 122–125.
Dettlaff-Kakol, A., and R. Pawlowski. 2002. First Polish DNA "manhunt"—an application of Y-chromosome STRs. *International Journal of Legal Medicine* 116: 289–291.
Ding, C., and S. Jin. 2009. High-throughput methods for SNP genotyping. *Methods in Molecular Biology* 578: 245–254.
Ebensperger, C., R. Studer, and J.T. Epplen. 1989. Specific amplification of the ZFY gene to screen sex in man. *Human Genetics* 82: 289–290.
Ehler, E., R. Marvan, and D. Vanek. 2010. Evaluation of 14 Y-chromosomal short tandem repeat haplotype with focus on DYS449, DYS456, and DYS458: Czech population sample. *Croatian Medical Journal* 51: 54–60.
Fattorinil, P., S. Cacció, S. Gustincich, J. Wolfe, B.M. Altamura, and G. Graziosil. 1991. Sex determination and species exclusion in forensic samples with probe cY97. *International Journal of Legal Medicine* 104: 247–250.
Foster, E.A., M. Jobling, P.G. Taylor et al. 1998. Jefferson fathered slave's last child. *Nature* 6706: 27–28.
Geppert, M., J. Edelmann, and R. Lessig. 2009. The Y-chromosomal STRs DYS481, DYS570, DYS576 and DYS643. *Legal Medicine* 11: S109–S110.
Geppert, M., M. Baeta, C. Nunez et al. 2011. Hierarchical Y-SNP assay to study the hidden diversity and phylogenetic relationship of native populations in South America. *Forensic Science International Genetics* 5: 100–104.
Gill, P., C.H. Brenner, B. Brinkmann et al. 2001. DNA Commission of the International Society of Forensic Genetics: Recommendations on forensic analysis using Y-chromosome STRs. *Forensic Science International* 124: 5–10.
Goedbloed, M., M. Vermeulen, R.N. Fang et al. 2009. Comprehensive mutation analysis of 17 Y-chromosomal short tandem repeat polymorphisms included in the AmpFlSTR Yfiler PCR amplification kit. *International Journal of Legal Medicine* 123: 471–482.
Gomes, V., P. Sanchez-Diz, A. Amorim, A. Carracedo, and L. Gusmao. 2010. Digging deeper into East African human Y chromosome lineages. *Human Genetics* 127: 603–613.
Gross, A.M., A.A. Liberty, M.M. Ulland, and J.K. Kuriger. 2008. Internal validation of the AmpFlSTR Yfiler amplification kit for use in forensic casework. *J Forensic Sci* 53: 125–134.

Gusmao, L., J.M. Butler, A. Carracedo et al. 2006. DNA Commission of the International Society of Forensic Genetics (ISFG): An update of the recommendations on the use of Y-STRs in forensic analysis. *Forensic Science International* 157: 187–197.

Hammer, M.F., V.F. Chamberlain, V.F. Kearney et al. 2006. Population structure of Y chromosome SNP haplogroups in the United States and forensic implications for constructing Y chromosome STR databases. *Forensic Science International* 164: 45–55.

Hanson, E., and J. Ballantyne. 2007. Population data for 48 'non-core' Y chromosome STR loci. *Legal Medicine* 9: 221–231.

Hanson, E.K., and J. Ballantyne. 2007. An ultra-high discrimination Y chromosome short tandem repeat multiplex DNA typing system. *PLoS One* 2: e688.

Hanson, E.K., P.N. Berdos, and J. Ballantyne. 2006. Testing and evaluation of 43 "noncore" Y chromosome markers for forensic casework applications. *Journal of Forensic Science* 51: 1298–1314.

Hedman, M., A.M. Neuvonen, A. Sajantila, and J.U. Palo. 2011. Dissecting the Finnish male uniformity: The value of additional Y-STR loci. *Forensic Science International Genetics* 5: 199–201.

Huang, D., S. Shi, C. Zhu et al. 2011. Y-haplotype screening of local patrilineages followed by autosomal STR typing can detect likely perpetrators in some populations. *Journal of Forensic Science* 56: 1340–1342.

International SNP Map Working Group. 2001. A map of human genome sequence variation containing 1.42 million single nucleotide polymorphisms. *Nature* 409: 928–933.

Jobling, M.A., and C. Tyler-Smith. 2003. The human Y chromosome: An evolutionary marker comes of age. *Nature Reviews Genetics* 4: 598–612.

Junge, A., B. Brinkmann, R. Fimmers, and B. Madea. 2006. Mutations or exclusion: An unusual case in paternity testing. *International Journal of Legal Medicine* 120: 360–363.

Karafet, T.M., F.L. Mendez, M.B. Meilerman, P.A. Underhill, S.L. Zegura, and M.F. Hammer. 2008. New binary polymorphisms reshape and increase resolution of the human Y chromosomal haplogroup tree. *Genome Research* 18: 830–838.

Kayser, M. 2007. Uni-parental markers in human identity testing including forensic DNA analysis. *Biotechniques* 43: Sxv–Sxxi.

Kayser, M., and A. Sajantila. 2001. Mutations at Y-STR loci: Implications for paternity testing and forensic analysis. *Forensic Science International* 118: 116–121.

Kayser, M., A. Caglia, D. Corach et al. 1997. Evaluation of Y-chromosomal STRs: A multicenter study. *International Journal of Legal Medicine* 110: 125–133, 141–149.

Kayser, M., C. Krüger, M. Nagy, G. Geserick, and L. Roewer. 1998. Y-chromosomal DNA-analysis in paternity testing: Experiences and recommendations. In *Progress in Forensic Genetics*, edited by B. Olaisen and B. Brinkmann. Amsterdam: Elsevier.

Kayser, M., S. Brauer, G. Weiss et al. 2000a. Melanesian origin of Polynesian Y chromosomes. *Current Biology* 10: 1237–1246.

Kayser, M., L. Roewer, M. Hedman et al. 2000b. Characteristics and frequency of germline mutations at microsatellite loci from the human Y chromosome, as revealed by direct observation in father/son pairs. *American Journal of Human Genetics* 66: 1580–1588.

Kayser, M., R. Kittler, A. Erler et al. 2004. A comprehensive survey of human Y-chromosomal microsatellites. *American Journal of Human Genetics* 74: 1183–1197.

King, T.E., E.J. Parkin, G. Swinfield et al. 2007. Africans in Yorkshire? The deepest-rooting clade of the Y phylogeny within an English genealogy. *European Journal of Human Genetics* 15: 288–293.

Krenke, B.E., L. Viculis, M.L. Richard et al. 2005. Validation of male-specific, 12-locus fluorescent short tandem repeat (STR) multiplex. *Forensic Science International* 151: 111–124.

Lao, O., P.M. Vallone, M.D. Coble et al. 2010. Evaluating self-declared ancestry of U.S. Americans with autosomal, Y-chromosomal and mitochondrial DNA. *Human Mutation* 31: E1875–E1893.

Leat, N., L. Ehrenreich, M. Benjeddou, K. Cloete, and S. Davison. 2007. Properties of novel and widely studied Y-STR loci in three South African populations. *Forensic Science International* 168: 154–161.

Lessig, R., J. Edelmann, J. Dressler, and M. Krawczak. 2009. Haplotyping of Y-chromosomal short tandem repeats DYS481, DYS570, DYS576 and DYS643 in three Baltic populations. *Forensic Science International Genetics Supplement Series* 2: 429–430.

Lim, S.K., Y. Xue, E.J. Parkin, and C. Tyler-Smith. 2007. Variation of 52 new Y-STR loci in the Y Chromosome Consortium worldwide panel of 76 diverse individuals. *International Journal of Legal Medicine* 121: 124–127.

Malsom, S., N. Flanagan, C. McAlister, and L. Dixon. 2009. The prevalance of mixed DNA profiles in fingernail samples taken from couples who co-habit using autosomal and Y-STRs. *Forensic Science International Genetics* 3: 57–62.

Maybruck, J.L., E. Hanson, J. Ballantyne, B. Budowle, and P.A. Fuerst. 2009. A comparative analysis of two different sets of Y-chromosome short tandem repeats (Y-STRs) on a common population panel. *Forensic Science International Genetics* 4: 11–20.

Mitchell, R.J., M. Kreskas, E. Baxter, L. Buffalino, and R.A. van Oorschot. 2006. An investigation of sequence deletions of amelogenin (AMELY), a Y-chromosome locus commonly used for gender determination. *Annals of Human Biology* 33: 227–240.

Mulero, J.J., C.W. Chang, L.M. Calandro et al. 2006. Development and validation of the AmpFlSTR Yfiler PCR amplification kit: A male specific, single amplification 17 Y-STR multiplex system. *Journal of Forensic Science* 51: 64–75.

Muzzio, M., V. Ramallo, J.M.B. Motti, M.R. Santos, J.S. Lopez Camelo, and G. Bailliet. 2011a. About the letter "Comments on the article, 'Software for Y-Haplogroup predictions, a word of caution.'" *International Journal of Legal Medicine* 125: 905–906.

Muzzio, M., V. Ramallo, J.M.B. Motti, M.R. Santos, J.S. Lopez Camelo, and G. Bailliet. 2011b. Software for Y-haplogroup predictions: A word of caution. *International Journal of Legal Medicine* 125: 143–147.

National Research Council. 1996. *The Evaluation of Forensic DNA Evidence*, edited by Committee on DNA Forensic Science: An Update, National Research Council. Washington DC: The National Academies Press.

Nuzzo, F., F. Caviezel, and L. de Carli. 1966. Y chromosome and exclusion of paternity. *Lancet* 2: 260–262.

Olofsson, J., H.S. Mogensen, B.B. Hjort, and N. Morling. 2011. Evaluation of Y-STR analyses of sperm cell negative vaginal samples. *Forensic Science International Genetics Supplement Series* 3: e141–e142.

Oota, H., W. Settheetham-Ishida, D. Tiwawech, T. Ishida, and M. Stoneking. 2001. Human mtDNA and Y-chromosome variation is correlated with matrilocal versus patrilocal residence. *Nature Genetics* 29: 20–21.

Palha, T., E. Ribeiro-Rodrigues, A. Ribeiro-dos-Santos, and S. Santos. 2011. Fourteen short tandem repeat loci Y chromosome haplotypes: Genetic analysis in populations from Northern Brazil. *Forensic Science International Genetics* 6: 413–418.

Park, M.J., H.Y. Lee, U. Chung, S.-C. Kang, and K.-J. Shin. 2007. Y-STR analysis of degraded DNA using reduced size amplicons. *International Journal of Legal Medicine* 121: 152–157.

Parson, W., H. Niederstätter, S. Kochl, M. Steinlechner, and B. Berger. 2001. When autosomal short tandem repeats fail: Optimized primer and reaction design for Y-chromosome short tandem repeat analysis in forensic casework. *Croatian Medical Journal* 42: 285–287.

Prinz, M., K. Boll, H.J. Baum, and B. Shaler. 1997. Multiplexing of Y chromosome specific STRs and performance for mixed samples. *Forensic Science International* 85: 209–218.

Radam, G., and H. Strauch. 1973. Lumineszenzmikroskopischer Nachweis des Y Chromosoms in Knochenmarkszellen – Eine neue Methode zur Geschlechtserkennung an Leichenmaterial. *Kriminalistik und Forensische Wissenschaften* 6: 149–151.

Redd, A.J., A.B. Agellon, V.A. Kearney et al. 2002. Forensic value of 14 novel STRs on the human Y chromosome. *Forensic Science International* 130: 97–111.

Rodig, H., L. Roewer, A.M. Gross et al. 2008. Evaluation of haplotype discrimination capacity of 35 Y-chromosomal short tandem repeat loci. *Forensic Science International* 174: 182–188.

Roewer, L. 2009. Y chromosome STR typing in crime casework. *Forensic Science Medicine and Pathology* 5: 77–84.

Roewer, L., and J.T. Epplen. 1992a. Simple repeat sequences on the human Y chromsome are equally polymorphic as their autosomal counterparts. *Human Genetics* 89: 389–394.

Roewer, L., and J.T. Epplen. 1992b. Rapid and sensitive typing of forensic stains by PCR amplification of polymorphic simple repeat sequences in case work. *Forensic Science International* 53: 163–171.

Roewer, L., M. Kayser, P. de Knijff et al. 2000. A new method for the evaluation of matches in non-recombining genomes: Application to Y-chromosomal short tandem repeat (STR) haplotypes in European males. *Forensic Science International* 114: 31–43.

Rolf, B., W. Keil, B. Brinkmann, L. Roewer, and R. Fimmers. 2001. Paternity testing using Y-STR haplotypes: Assigning a probability for paternity in cases of mutations. *International Journal of Legal Medicine* 115: 12–15.

Santos, F.R., A. Pandya, and C. Tyler-Smith. 1998. Reliability of DNA-based sex tests. *Nature Genetics* 18: 103.

Schlecht, J., M.E. Kaplan, K. Barnard, T. Karafet, M.F. Hammer, and N.C. Merchant. 2008. Machine-learning approached for classifying haplogroup from Y chromosome STR data. *PLoS Computational Biology* 4: e1000093.

Seielstad, M.T., F. Minch, and L.L. Cavalli-Sforza. 1998. Genetic evidence for a higher female migration rate in humans. *Nature Genetics* 20: 278–280.

Sibille, I., C. Duverneuil, G. Lorin de Grandmaison et al. 2002. Y-STR DNA amplification as biological evidence in sexually assaulted female victims with no cytological detection of spermatozoa. *Forensic Science International* 125: 212–216.

Sobrino, B., M. Brion, and A. Carracedo. 2005. SNPs in forensic genetics: A review on SNP typing methodologies. *Forensic Science International* 154: 181–194.

Steinlechner, M., B. Berger, H. Niederstätter, and W. Parson. 2002. Rare failures in the amelogenin sex test. *International Journal of Legal Medicine* 116: 117–120.

Sturk, K.A., M.D. Coble, S.M. Barritt, and J.A. Irwin. 2009. Evaluation of modified Yfiler amplification stategy for comprimised samples. *Croatian Medical Journal* 50: 228–238.

Sullivan, K.M., A. Mannucci, C.P. Kimpton, and P. Gill. 1993. A rapid and quantitative DNA sex test: Fluorescence-based PCR analysis of X–Y homologous gene amelogenin. *Biotechniques* 15: 636–638, 640–641.

Thangaraj, K., A.G. Reddy, and L. Singh. 2002. Is the amelogenin gene reliable for gender identification in forensic casework and prenatal diagnosis? *International Journal of Legal Medicine* 116: 121–123.

Thompson, J.M., M.M. Ewing, W.E. Frank et al. 2013. Developmental validation of the PowerPlex Y23 System: A single multiplex Y-STR analysis system for casework and database samples. *Forensic Science International Genetics* 7: 240–250.

Tilford, C.A., T. Kuroda-Kawaguchi, H. Skaletsky et al. 2001. A physical map of the human Y chromosome. *Nature* 409: 943–945.

Vallone, P.M., and J.M. Butler. 2004. Y-SNP typing of U.S. African American and Caucasian samples using allele-specific hybridization and primer extension. *Journal of Forensic Science* 49: 723–732.

van Oven, M., A. Ralf, and M. Kayser. 2011. An efficient multiplex genotyping approach for detecting the major worldwide human Y-chromosome haplogroups. *International Journal of Legal Medicine* 125: 879–885.

Vermeulen, M., A. Wollstein, K. van der Gaag et al. 2009. Improving global and regional resolution of male lineage differentiation by simple single-copy Y-chromosomal short tandem repeat polymorphisms. *Forensic Science International Genetics* 3: 205–213.

Walsh, B., A.J. Redd, and M.F. Hammer. 2008. Joint match probabilities for Y chromosomal and autosomal markers. *Forensic Science International* 174: 234–238.

Willuweit, S., A. Caliebe, M.M. Andersen, and L. Roewer. 2011. Y-STR frequency surveying method: A critical reappraisal. *Forensic Science International Genetics* 5: 84–90.

Xu, Z., H. Sun, Y. Yu et al. 2010. Diversity of five novel Y-STR loci and their application in studies of north Chinese populations. *Journal of Genetics* 89: 29–36.

Xue, Y., Q. Wang, Q. Long et al. 2009. Human Y chromosome base-substitution mutation rate measured by direct sequencing in a deep-rooting pedigree. *Current Biology* 19: 1453–1457.

Y Chromosome Consortium. 2002. A nomenclature system for the tree of human Y-chromosomal binary haplogroups. *Genome Research* 12: 339–348.

Forensic Application of X Chromosome STRs

5

TONI MARIE DIEGOLI

Contents

5.1	Introduction	135
5.2	Commonly Used X STR Markers and Assays	138
5.3	Reference Population Databases	140
5.4	Mutation Rates	144
5.5	Requirements for Statistical Interpretation of a Match	150
5.6	Exchange and Comparability of Data	153
5.7	Future Directions	154
Acknowledgments		154
Appendix		155
References		161

5.1 Introduction

Autosomal short tandem repeat (STR) testing has been the cornerstone of forensic identity testing since the early 1990s, when the first fluorescently labeled STR markers were described. Since then, thousands of polymorphic STR markers have been characterized, and today combinations of autosomal markers allow the identification of individuals to the degree of one in trillions. In missing persons and deficiency paternity cases, Y chromosomal STRs expand the pool of family reference samples that can be used to confirm identity, and have proven useful in situations where the male portion of a mixture is of interest, often in the presence of large excesses of female DNA, as would be the case with a vaginal swab from a sexual assault.

More recently, STR markers located on the X chromosome have emerged as additional tools in this forensic arsenal. X chromosomal STRs can be used to supplement traditional STR typing because of their unique inheritance pattern and, correspondingly, the breadth of published literature on the subject has expanded greatly in recent years. STR markers on the X chromosome may be useful in several forensic contexts. To begin, missing persons cases usually require the analysis of relatives because of a lack of direct reference material. Oftentimes, mitochondrial DNA (mtDNA) typing can be used to address the potential for degraded or low quantity samples such as skeletal remains, particularly in closed populations and when a direct maternal reference is available, because of its relatively high copy number and protected location within the mitochondria of the cell. However, mtDNA is maternally inherited; therefore, where maternal references are unavailable or where the unidentified individual matches one of the most common mtDNA haplotypes, mtDNA testing alone may be inadequate. In such cases, markers on the X chromosome may provide additional information (Silva et al. 2009; Andreaggi et al. 2010), offering the potential to both augment traditional STR testing and mtDNA sequencing for human remains identification as well as differentiate pedigrees that would be otherwise indistinguishable with unlinked autosomal STRs (Pinto et al. 2011).

X chromosomal STRs can be particularly useful for any parent–child relationship that involves at least one female (e.g., father–daughter, mother–son, or mother–daughter) (Krawczak 2007). For example, Figure 5.1 shows the relationship pedigree of parents with one son who has a daughter. In this scenario, if the son and his wife are unavailable for testing, it may be necessary to use the grandparents' DNA profiles to reassociate their granddaughter. In this specific scenario, autosomal STRs generally give a low likelihood ratio of a relationship since there is on average only one-quarter sharing of alleles between a grandparent and a grandchild. X chromosomal STRs, on the other hand, prove to be more useful since the X chromosome of the son was inherited entirely from his mother's genome with no contribution from his father (in the nonrecombining region). This X chromosome was then passed in full to the granddaughter. In this example, one would expect to see one allele from each X STR marker of the grandmother present in the granddaughter's X STR profile; therefore, X chromosomal STRs will most likely outperform autosomal STRs. Other maternally related scenarios, such as identifying cousins or aunt–niece relationships using X chromosomal STRs to augment or replace autosomal STR testing, have been proposed (Szibor 2007).

In criminal incest investigations, the use of X chromosomal STRs could be potentially more informative than any other marker system. In the published example illustrated in Figure 5.2, X chromosomal STRs are used to exclude incest in a case of questioned paternity without involving either potential father (Schmidtke et al. 2004). A pregnant female (proband) presented with a question as to whether her father (putative father 1) or an unrelated man (putative father 2) was the true father of her unborn daughter. A combination of 21 autosomal STRs was uninformative as to the likely father since only the proband, her

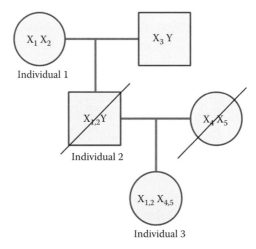

Figure 5.1 Example of X chromosomal STR (short tandem repeat) use in relationship testing. In this scenario, the matriarch (individual 1) passes a combination of her X chromosomes (X_1, X_2) to her son (individual 2). The son passes this combination ($X_{1,2}$) in its entirety to his daughter (individual 3) without recombination. Without knowing the X STR profile of either parent, the granddaughter will share at least one allele at each X STR marker (50% allele sharing) with her paternal grandmother. Using autosomal STRs, the grandmother and granddaughter share only one-quarter of their alleles.

Forensic Application of X Chromosome STRs

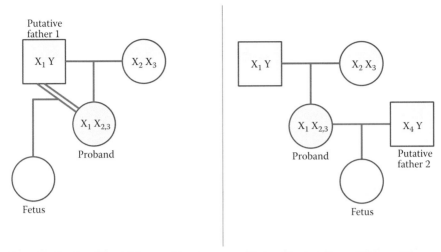

H_1: no introduction of new alleles possible H_2: introduction of new alleles possible

Figure 5.2 Example of X chromosomal STR use in a paternity scenario involving incest. In this scenario, there are two hypotheses being investigated. In hypothesis 1 (H_1), the father of the proband (yellow) is the putative father of her unborn child. Here, the female child (fetus) would inherit one half of her X STR alleles from the father (X_1), and a number of X_1 alleles would be transmitted to her from her mother (who also inherited the same X_1 chromosome from the father). A portion of the alleles transmitted to the child from her mother (proband) will contain the alleles from the proband's mother (present on chromosome $X_{2,3}$). Therefore, under H_1, it is expected that only alleles present in the mother's profile will be present in the fetus' profile and that there would be a high degree of homozygosity. Under the second hypothesis (H_2), the X_4 chromosome of the alternate putative father would be transmitted in its entirety to the fetus. Here, the fetus will have a number of alleles that are not present in her mother. The X chromosomal STRs can provide additional evidence to the investigator, whereas autosomal STRs provided little evidence given the limited number of individuals available for testing (see Schmidtke et al. 2004 for details).

mother, and her fetus were available for testing. Additionally, Y chromosomal STRs and mtDNA analysis are both useless in this case. X chromosomal STRs, however, provide convincing evidence since the fetus would not possess any alleles not present in the proband's profile if that fetus indeed resulted from the incestuous relationship. That is, if the father of the proband was the true father of the proband's child, then the child would exhibit 100% allele sharing with the father's X chromosome (X_1) along with a combination of alleles from the X chromosomes of the proband (X_1 from the father and $X_{2,3}$ inherited from the grandmother). Alternatively, if the unrelated man was the true father, then the child would show additional alleles at some or all of the X STR markers tested due to the contribution from his X chromosome.

It has been demonstrated that X STRs can differentiate pedigrees that are otherwise indistinguishable using only unlinked autosomal markers (Pinto et al. 2011). For example, using autosomal STRs, the avuncular, half-sibling, and grandparent–grandchild relationships cannot be differentiated (Pinto et al. 2010). However, Figure 5.3 depicts just one example of how X STR markers could distinguish the avuncular relationship from the other two in the case of a questioned relationship between two females. In the grandmother–granddaughter pedigree, similar to Figure 5.1, the grandmother passes on a combination of her X

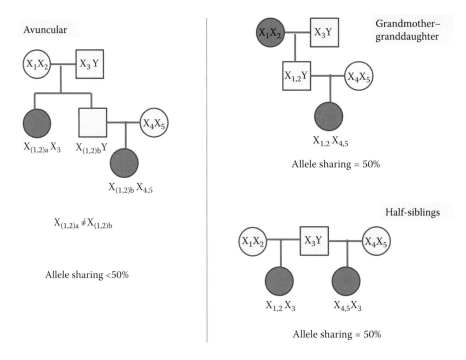

Figure 5.3 Example of X chromosomal STR use in a XX. In this scenario, the relationship between two women is questioned; no additional relatives are available for testing. Autosomal STRs cannot distinguish between the three relationship scenarios depicted here, and unlinked X chromosomal markers are utilized to further discriminate between the avuncular relationship and the grandparent–grandchild and half-sibling relationships. Although exactly 50% of X STR alleles are expected to be shared by a grandmother and granddaughter or half-sisters, this sharing will be less than 50% between an aunt and her niece due to recombination.

chromosomes (X_1, X_2) to her son, who passes along this X chromosome ($X_{1,2}$) in its entirety to the granddaughter. The grandmother and granddaughter, then, can be expected to share exactly one allele at every X STR locus. Similarly, in the case of half-siblings, the common father passes his X chromosome (X_3) in its entirety without recombination to each of his daughters. The half-sisters therefore can be expected to share exactly one allele at every X locus. However, a closer look at the aunt–niece relationship depicted here reveals that the two women would not be expected to share an allele at each locus, and could potentially share none. Because of recombination of the unlinked X markers, the combination of X_1 and X_2 that each female receives will be different (barring unlikely recombination events generating two identical X chromosomes). Unlinked X chromosomal markers are similarly useful in the same pedigree scenarios involving one male and one female as well as two males (Pinto et al. 2011).

5.2 Commonly Used X STR Markers and Assays

For both the Y chromosome and the autosomes, commercial kits are available that probe a wide variety of genetic markers, and commonly the work completed during development of these multiplexes is published; PowerPlex® 16 (Promega Corporation) (Krenke et

Forensic Application of X Chromosome STRs

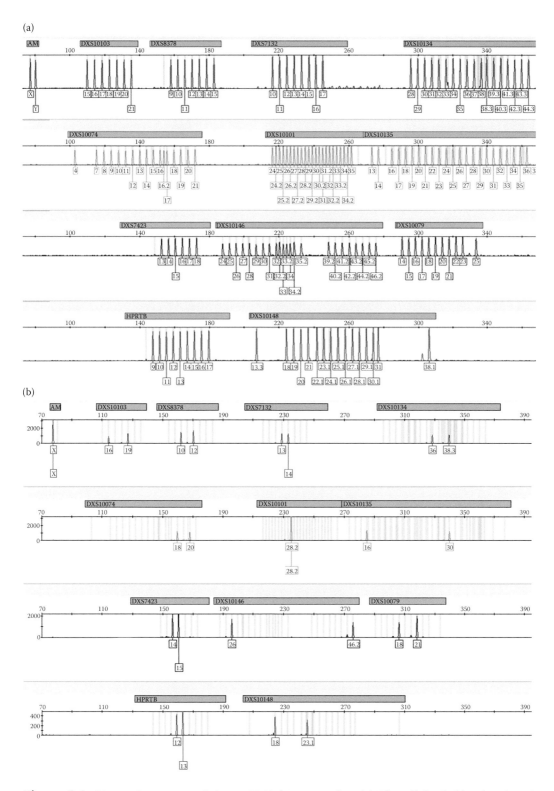

Figure 5.4 Qiagen Investigator® Argus X-12 kit examples. (a) The allelic ladder for the 12 X STR loci and sex-typing locus Amelogenin included in the commercially available kit. (b) Investigator® Argus X-12 profile for 500 pg of control DNA XX28 (included with the kit).

al. 2002), AmpFLSTR® Minifiler™ (Life Technologies) (Mulero et al. 2008), PowerPlex® Y (Promega Corporation) (Krenke et al. 2005), AmpFLSTR® Yfiler® (Life Technologies) (Mulero et al. 2006), PowerPlex® ESX/ESI (Promega Corporation) (Tucker et al. 2010, 2011), and AmpFLSTR® Identifiler® (Life Technologies) (Collins et al. 2004) are all examples of commercial kits used by forensic laboratories with published validation studies. For X STRs, there have been several iterations of one commercial kit. The Mentype® Argus X-UL kit simultaneously amplified four X STR loci (DXS8378, DXS7423, HPRTB, and DXS7132) plus amelogenin in a single dye channel (Biotype AG 2006). The Mentype® Argus X-8 kit expanded the number of loci to eight by adding four additional STRs (DXS10134, DXS10074, DXS10101, and DXS10135) (Biotype AG 2007). The current iteration of this multiplex is produced by Qiagen and is known as the Investigator® Argus X-12 kit, which includes yet another four loci (DXS10103, DXS10079, DXS10146, and DXS10148) present in four dye channels (Qiagen 2010) (Figure 5.4). Because of patent and intellectual property issues between United States and European STR kit manufacturers, these kits cannot be sold or marketed in the United States at this time. It is believed that going forward, as more laboratories demand X STRs, additional X STR kits will be manufactured for the forensic community. For now, most X STR typing currently relies upon noncommercial multiplex assays, amplifying two to 12 loci in a single reaction, that have been designed and published by individual laboratories interested in X STR typing (Table 5.1).

When specifically targeting DNA that is known to be degraded, short amplicon sizes are favored with the goal of recovering the maximum number of alleles (Wiegand and Kleiber 2001). Genotyping using reduced-size amplicons for the 13 core autosomal loci used in the United States produced an increase in the ability to recover information from compromised samples while maintaining concordant profiles (Butler et al. 2003), and additional mini-STR loci were characterized to further increase the information that could be obtained from degraded samples (Coble and Butler 2005). In an X STR study examining degraded samples, Asamura et al. (2006a) demonstrated the success of two quadruplex reactions consisting of amplicons ranging from 76 to 169 base pairs (bp) in length, and reduced size amplicons have since been designed for additional X STRs (Diegoli and Coble 2011; Castaneda et al. 2013).

5.3 Reference Population Databases

Population databases must be generated as part of the validation of a specific marker system in order to ensure the ability to quantify the value of a "match" between two samples. These reference databases should contain at least 100–250 unrelated individuals from relevant populations (Schneider 2007; ISFH 1992a) and are used to calculate population-specific allele or haplotype frequencies. Observed and expected genotype frequencies are used to confirm that the population is in Hardy–Weinberg equilibrium, and additional parameters such as expected heterozygosity, power of discrimination, polymorphism information content, and mean exclusion chance are also calculated to evaluate the usefulness of the selected set of markers. The mean exclusion chance and power of discrimination are calculated differently for X chromosomal markers than for autosomal markers, and depending on sex and/or kinship situation; that is, trios involving a daughter are treated differently than father–daughter duos (Szibor 2007). The Forensic ChrX Research website's

Forensic Application of X Chromosome STRs

Table 5.1 Published Multiplex PCR Assays Targeting X Chromosomal STR Loci

Reference	N (loci)	Populations Studied	N (tot)	DXS101	HPRTB	DXS7423	DXS6789	DXS8377	DXS7132	DXS7424	DXS8378	DXS7133	GATA172D05	DXS6809	DXS6807	DXS9898	GATA31E08	DXS6800	DXS981	GATA165B12	DXS10011	DXS9902	DXS6803	HumARA	DXS6797	DXS7130	DXS6801	DXS9895	DXS10074	DXS10079	DXS6799	DXS6854	DXS6804	DXS10075	DXS6795	DXS6808	DXS10147	DXS10078	DXS6810	Additional Studies Using Multiplex
Zarrabeitia et al. 2002b	5	Cantabria	131	×	×	×		×																×																Cantabria and Basque (Zarrabeitia et al. 2004)
Athanasiadou et al. 2003	4	Germany	120	×	×			×											×																					Tuscany (Toni et al. 2003); case example (Toni et al. 2006)
Bini et al. 2004	7	Italians	556	×	×	×	×								×						×			×																
Lee et al. 2004	4	Koreans	300	×	×					×				×	×	×					×																			
Shin et al. 2004	4	Koreans	300	×	×			×					×																											
Poetsch et al. 2005	10	NE Germans	205	×	×	×	×	×		×	×	×				×					×			×																Latvia (Poetsch et al. 2006)
Tabbada et al. 2005	5	China, Japan, Thai, German, Philippines	562	×	×		×	×											×																					Kurd and German (Fracasso et al. 2008)
Asamura et al. 2006a	8	Japan	333	×		×	×	×	×		×	×					×			×																				
Asamura et al. 2006b	8	Japan	401		×			×	×	×	×		×		×	×					×				×															
Coletti et al. 2006	6	Italy	100	×	×	×	×	×							×									×																Italy (Massetti et al. 2008)
Gomes et al. 2006	4	N. Spain	65	×	×	×	×	×																																
Robino et al. 2006	12	NW Italian	160	×	×		×	×	×	×	×	×	×	×	×	×		×			×						×													Algeria (Bekada et al. 2009); Ivory Coast (Pasino et al. 2011); case example (Inturri et al. 2011)

(continued)

Table 5.1 Published Multiplex PCR Assays Targeting X Chromosomal STR Loci (Continued)

Reference	N (loci)	Populations Studied	N (tot)	DXS101	HPRTB	DXS7423	DXS6789	DXS8377	DXS7132	DXS7424	DXS8378	DXS7133	GATA172D05	DXS6809	DXS6807	DXS9898	GATA31E08	DXS6800	DXS981	GATA165B12	DXS10011	DXS9902	DXS6803	HumARA	DXS6797	DXS7130	DXS6801	DXS9895	DXS10074	DXS10079	DXS6799	DXS6854	DXS6804	DXS10075	DXS6795	DXS8098	DXS10147	DXS10078	DXS6810	Additional Studies Using Multiplex
Gomes et al. 2007b	10	United States, 3 groups	377	×	×	×	×	×	×	×	×	×	×	×		×																								Spain (Aler et al. 2007); Columbia (Pico et al. 2008); Angola (Gomes et al. 2007a); Mozambique (Gomes et al. 2007a); Uganda (Gomes et al. 2007a); N. Portugal (Pereira et al. 2007)
Turrina et al. 2007	12	N. Italy	200	×	×	×	×		×	×	×	×	×	×	×		×																							India (Mukerjee et al. 2010)
Cybulska et al. 2008	4	Belarus, Slovakia	296	×	×	×		×																																
Liu et al. 2008	5	Han, China	827	×			×		×					×					×																					
Martinez et al. 2008	4	Peru	172		×								×					×						×																
Martins et al. 2008	4	Bauru, Brazil	90							×	×							×							×								×							Brazil (Martins et al. 2010)
Ribeiro Rodrigues et al. 2008	11	Brazil Amazon	324	×		×	×		×		×	×	×	×		×	×				×					×		×												Brazil (Rodrigues et al. 2010; Ribeiro-Rodrigues et al. 2011)
Tavares et al. 2008	6	Brazil	263			×			×	×	×	×			×																									Paternity testing (Silva et al. 2009; Aquino et al. 2009)
Gusmao et al. 2009	10	Iberian/Latin American	2959	×		×	×		×	×	×	×	×	×		×	×						×																	Azores (Silva et al. 2010); Uganda (Gomes et al. 2009b); paternity testing (Toscanini et al. 2009a; Serra et al. 2008); Brazil (Martins et al. 2009; Caine et al. 2010); Macau (Gomes et al. 2009a); Portuguese Gypsies (Valente et al. 2009); Argentina (Bobillo et al. 2011; Toscanini et al. 2009b); prostate cancer study (Carvalho et al. 2008); Spain (Garcia et al. 2012; Illescas et al. 2011a, 2011b, 2012)

(*continued*)

Table 5.1 Published Multiplex PCR Assays Targeting X Chromosomal STR Loci (Continued)

Reference	N (loci)	Populations Studied	N (tot)	DXS101	HPRTB	DXS7423	DXS6789	DXS8377	DXS7132	DXS7424	DXS8378	DXS7133	GATA172D05	DXS6809	DXS6807	DXS9898	GATA31E08	DXS6800	DXS981	GATA165B12	DXS10011	DXS9902	DXS6803	HumARA	DXS6797	DXS7130	DXS6801	DXS9895	DXS10074	DXS10079	DXS6799	DXS6854	DXS6804	DXS10075	DXS6795	DXS6808	DXS10147	DXS10078	DXS6810	Additional Studies Using Multiplex
Hwa et al. 2009	13	Taiwan	221	×	×	×	×	×	×	×	×	×	×	×	×	×						×																		Taiwan (Hwa et al. 2011)
Li et al. 2009	4	Beijing Han	400						×										×	×		×									×									
Nadeem et al. 2009	5	Pakistan	212									×							×	×					×															
Poetsch et al. 2009	11	Ghana	243	×	×	×	×	×	×		×	×		×	×	×		×																						Morocco (Poetsch et al. 2011); Madagascar (Poetsch et al. 2011)
Tariq et al. 2009	5	Pakistan	302	×		×	×			×					×												×													
Wu et al. 2009	7	Chinese Han	696	×	×	×	×			×							×			×																				
Zeng et al. 2009	10	Chinese Daur	138	×	×	×	×	×	×		×	×														×	×													China (Sun et al. 2012)
Ferreira da Silva et al. 2010	5	Brazil	746							×																			×	×	×			×						
Nakamura and Minaguchi 2010	16	Japan	512	×	×	×	×	×	×	×	×	×	×	×	×	×	×	×					×					×												
Diegoli and Coble 2011	15	United States, 4 groups	1363			×		×	×	×		×	×				×	×	×	×		×			×	×	×		×	×					×		×			Bosnia and Herzegovina (Diegoli et al. 2011b)
Li et al. 2011	11	Tibet	664	×	×	×	×	×	×	×		×													×		×								×					China (Li et al. 2012)
Liu et al. 2011	9	China, 3 groups	890	×	×	×	×	×	×	×	×					×						×										×								
Zeng et al. 2011a	12	Chinese Han	360	×		×	×		×	×	×	×	×	×	×	×		×							×	×											×			
Liu et al. 2012	15	China, 3 groups	830	×		×	×	×	×	×	×	×	×	×	×	×	×	×	×	×					×	×	×			×				×					×	×

Note: Thirty-five published X STR multiplexes have been established in the forensic community and used for population studies, paternity investigations, and other purposes. Thirty-six of the most commonly used markers were included in the various multiplexes.

"Evaluate & Calculate" section can be used to automatically calculate these (and other) forensic efficiency parameters specifically for X STR markers, and contains allele frequency information for selected markers and published populations (Szibor et al. 2006).

There are a number of population studies that have been performed with the various iterations of the commercial X STR, and these are summarized in Table 5.2. Also, a number of the noncommercially produced published multiplexes have been used by additional laboratories to increase the population data available for a particular marker set and are noted in the right-most column of Table 5.1. Appendix Tables 5.5–5.8 present allele frequencies for 23 commonly used X STRs in four U.S. population groups.

Ideally, during the development of a multiplex, multiple flanking region sequences would be examined in an alignment to look for variation that could affect primer binding before inclusion of the particular primer pair into a multiplex (Schoske et al. 2003). However, polymorphisms that result in reduced primer binding efficiency are often only recognized when a large sample set, such as that required for a population database, is typed with the multiplex. In one example, an excess of homozygosity at D8S1179 was observed in a population study of Asian samples using AmpFLSTR® Profiler Plus®, revealing a population-specific mutation under the reverse primer binding site (Leibelt et al. 2003). In this case, a degenerate primer was designed and added to the primer mix for two commercial kits (AmpFLSTR® Profiler Plus® and AmpFLSTR® Identifiler®), and previously null alleles were recovered in the population samples. Similarly, null alleles have been detected for X STR loci, such as the null allele caused by a U.S. Hispanic-specific G to A substitution under the reverse primer binding site for GATA172D05 observed in two different multiplexes using the same primers (Gomes et al. 2007b; Diegoli and Coble 2011). This particular null allele was observed only twice in a total of 998 samples. In contrast to the D8S1179 strategy of adding an additional primer, no changes were made to either multiplex because of the rarity of the GATA172D05 mutation. This was also the case for additional null alleles observed at markers DXS7132, GATA165B12, and GATA31E08 (Diegoli and Coble 2011) as well as certain markers within the commercial kit.

5.4 Mutation Rates

According to the 1991 report of the International Society for Forensic Genetics (ISFG; formerly International Society for Forensic Haemogenetics [ISFH]) relating to the use of DNA polymorphisms in paternity testing, mutation rates must be known in order to adequately address possible mismatches attributable to mutational events (ISFH 1992a). Mutations typically occur as a result of strand slippage during DNA replication, and are the major mechanism of the high degree of polymorphism seen in human microsatellites (Levinson and Gutman 1987). Single-step mutations (insertions or deletions of one repeat unit) are most common, affecting longer alleles more frequently than shorter ones (Brinkmann et al. 1998). This trend is observed for markers on the Y chromosome (Ge et al. 2009), and no difference is expected on the X chromosome. Mutation rates can differ, however, between males and females, with one estimate of the ratio of paternal to maternal mutations at 17:3 (Brinkmann et al. 1998). Mutation rates can also vary with population, as was demonstrated in several studies of Y chromosomal markers where members of the African American population had a slightly higher rate than other populations (Ge et al. 2009; Decker et al. 2008). This variation is most likely related to the length of the most frequently observed alleles in each population.

Table 5.2 Summary of Published Studies Using Commercial X STR Multiplexes

Reference	Kit	Population Studied	N
Pepinski et al. 2005	X-UL	Poland	240
Cerri et al. 2006	X-UL	Italy	90
Oguzturun et al. 2006	X-UL	United Kingdom	200
Oguzturun et al. 2006	X-UL	Ireland	200
Oguzturun et al. 2006	X-UL	South Asia	200
Tang and To 2006	X-UL	China	500
Caine et al. 2007b	X-UL	Brazil	184
Caine et al. 2007a	X-UL	Portugal	200
Pepinski et al. 2007	X-UL	Poland, 2 minority groups	420
Zalan et al. 2007	X-UL	Hungary[a]	384
Barbaro et al. 2008	X-8	Italy	100
Becker et al. 2008	X-8	Germany	259
Becker et al. 2008	X-8	Ghana	59
Becker et al. 2008	X-8	Japan	93
Cerri et al. 2008	X-8	Italy	131
Hashiyada et al. 2008	X-8	Japan	258
Thiele et al. 2008	X-8	Ghana	182
Zalan et al. 2008	X-8	Hungary[a]	384
Acar et al. 2009	X-8	Turkey	100
Hedman et al. 2009	X-8	Finland	300
Hedman et al. 2009	X-8	Somalia	300
Lim et al. 2009	X-8	Korea	300
Turrina et al. 2009	X-8	NE Italy	176
Tie et al. 2010	X-8	Japan	492
Luczak et al. 2011	X-8	Poland	311
Luo et al. 2011	X-8	China	303
Yoo et al. 2011	X-8	Korea	138
Zhang et al. 2011	X-8	China	198
Bekada et al. 2010	X-8+4	Algeria	210
Pasino et al. 2011	X-8+4	Ivory Coast	125
Bentayebi et al. 2012	X-12	Morocco	145
Diegoli et al. 2011a	X-12	United States, 4 groups	853
Edelmann et al. 2012	X-12	Germany	1037
Horvath et al. 2012	X-12	Hungary[a]	407
Turrina et al. 2011	X-12	NE Italy	207
Zeng et al. 2011b	X-12	Chinese Han	272
Samejima et al. 2012	X-12	Kuala Lumpur	283
Tillmar 2012	X-12	Sweden	652
Tomas et al. 2012	X-12	Greenland	198
Tomas et al. 2012	X-12	Denmark	210
Tomas et al. 2012	X-12	Somalia	441

Note: "X-UL" refers to the Mentype® Argus X-UL; "X-8" refers to the Mentype® Argus X-8 kit; and "X-12" refers to the Qiagen Investigator® Argus X-12 kit. "X-8+4" indicates that the same 12 markers in the Qiagen Investigator® Argus X-12 kit were typed but that the Mentype® Argus X-8 kit was used in combination with a non-commercial 4plex. N denotes the total number of individuals included in study.

[a] The three Hungarian populations are identical, but were processed separately with all three kits.

Table 5.3 Summary of Published Overall X Chromosomal STR Mutation Rates

Reference	Population	Markers (N)	Mutations (N)	Meioses (N)	Overall mutation rate (%)	95% confidence interval (%)
Szibor et al. 2000	Germany	1	2	580	0.34	0.042–1.24
Edelmann and Szibor 2001	Germany	1	0	340	0.00	0.00–1.08
Hering, Kuhlisch, and Szibor 2001	Germany	1	0	404	0.00	0.00–0.91
Edelmann et al. 2002b	Germany	1	0	300	0.00	0.00–1.22
Zarrabeitia et al. 2002a	Spain	2	1	214	0.47	0.012–2.58
Zarrabeitia et al. 2002b	Spain	5	0	125	0.00	0.00–2.91
Athanasiadou et al. 2003	Germany	4	0	372	0.00	0.00–0.99
Huang et al. 2003	China	2	0	312	0.00	0.00–1.17
Szibor et al. 2003b	Germany	16	16	7658	0.21	0.12–0.34
Wiegand et al. 2003	Austria and Germany	3	0	834	0.00	0.00–0.44
Shin et al. 2004	Korea	5	1	180	0.56	0.014–3.01
Turrina and De Leo 2004	Italy	3	0	240	0.00	0.00–1.52
Pepinski et al. 2005	Poland	4*	0	320	0.00	0.00–1.15
Poetsch et al. 2005	Germany	10	0	500	0.00	0.00–0.73
Tabbada et al. 2005	Philippines	5	1	445	0.22	0.006–1.25
Pepinski et al. 2006	Poland	4*	0	264	0.00	0.00–1.39

Forensic Application of X Chromosome STRs

Study	Country	Markers	Mutations	Meioses	Rate	95% CI
Tang and To 2006	China	4*	0	424	0.00	0.00–0.87
Pepinski et al. 2007	Poland	4*	0	600	0.00	0.00–0.61
Turrina et al. 2007	Italy	12	0	1080	0.00	0.00–0.34
Zalan et al. 2007	Hungary	4*	1	768	0.13	0.003–0.72
Becker et al. 2008	Germany	8*	1	2800	0.04	0.001–0.20
Hundertmark et al. 2008	Germany	3	3	1029	0.29	0.06–0.85
Liu et al. 2008	China	5	8	4295	0.19	0.08–0.37
Pico et al. 2008	Columbia	10	4	1460	0.27	0.07–0.70
Tariq et al. 2008	Pakistan	13	0	1300	0.00	0.00–0.28
Glesmann et al. 2009	Argentina	7	1	1015	0.10	0.02–0.55
Nadeem et al. 2009	Pakistan	5	0	840	0.00	0.00–0.44
Poetsch et al. 2009	Ghana	11	0	198	0.00	0.00–1.85
Castaneda et al. 2012	Spain	6	4	1164	0.34	0.09–0.88
Liu et al. 2012	China	15	13	11,850	0.11	0.06–0.19
Tetzlaff et al. 2012	Germany	8	8	1680	0.48	0.21–0.94
Tomas, Pereira, and Morling 2012	Greenland, Denmark, Somalia	12*	20	6156	0.33	0.20–0.50
Nishi et al. 2013	Japan	12	0	648	0.00	0.00–0.57
Totals			84	50,395	0.17%	0.13–0.21

* Indicates that a commercial kit was used in this study (Mentype® Argus X-UL (4 markers), Mentype® Argus X-8 kit (8 markers), or Qiagen Investigator® Argus X-12 kit (12 markers).

Table 5.4 Summary of Published X Chromosomal STR Mutation Rates by Marker

Marker	Mutations (N)	Meioses (N)	Pooled Mutation Rate (%)	95% confidence interval (%)	References
ARA	4	673	0.59	0.16–1.51	Szibor et al. 2003; Shin et al. 2004; Poetsch et al. 2005; Zarrabeitia et al. 2002b
DXS8377	10	1702	0.59	0.28–1.08	Szibor et al. 2003; Shin et al. 2004; Poetsch et al. 2005; Zarrabeitia et al. 2002a; Pico et al. 2008; Tabbada et al. 2005; Wiegand et al. 2003; Athanasiadou et al. 2003; Tariq et al. 2008; Poetsch et al. 2009
DXS10135	8	1416	0.56	0.24–1.11	Becker et al. 2008; Hundertmark et al. 2008; Tomas, Pereira, and Morling 2012; Tetzlaff, Wegener, and Lindner 2012
DXS10079	7	1497	0.47	0.19–0.96	Tomas, Pereira, and Morling 2012; Liu et al. 2012; Castaneda et al. 2012
DXS10103	2	513	0.39	0.047–1.40	Tomas, Pereira, and Morling 2012
DXS10146	2	513	0.39	0.047–1.40	Tomas, Pereira, and Morling 2012
DXS10134	4	1127	0.35	0.097–0.91	Becker et al. 2008; Tomas, Pereira, and Morling 2012; Tetzlaff, Wegener, and Lindner 2012; Nishi et al. 2013
DXS10148	3	856	0.35	0.072–1.02	Hundertmark et al. 2008; Tomas, Pereira, and Morling 2012
DXS10075	3	984	0.30	0.063–0.89	Liu et al. 2012; Castaneda et al. 2012
DXS7132	10	4064	0.25	0.12–0.45	Szibor et al. 2003; Pico et al. 2008; Tariq et al. 2008; Poetsch et al. 2009; Becker et al. 2008; Tomas, Pereira, and Morling 2012; Tetzlaff, Wegener, and Lindner 2012; Liu et al. 2012; Nishi et al. 2013; Turrina and De Leo 2004; Zalan et al. 2007; Pepinski et al. 2005; Liu et al. 2008; Pepinski et al. 2007; Turrina et al. 2007; Tang and To 2006; Pepinski et al. 2006
DXS10074	5	2111	0.24	0.077–0.55	Becker et al. 2008; Tomas, Pereira, and Morling 2012; Tetzlaff, Wegener, and Lindner 2012; Liu et al. 2012; Castaneda et al. 2012; Nishi et al. 2013
DXS6803	2	1015	0.20	0.024–0.71	Liu et al. 2008; Huang et al. 2003
DXS6809	4	2133	0.19	0.051–0.48	Pico et al. 2008; Liu et al. 2012; Castaneda et al. 2012; Nishi et al. 2013; Liu et al. 2008; Turrina et al. 2007
HPRTB	6	3627	0.17	0.061–0.36	Szibor et al. 2003; Shin et al. 2004; Poetsch et al. 2005; Zarrabeitia et al. 2002b; Pico et al. 2008; Tabbada et al. 2005; Athanasiadou et al. 2003; Tariq et al. 2008; Poetsch et al. 2009; Becker et al. 2008; Tomas, Pereira, and Morling 2012; Tetzlaff, Wegener, and Lindner 2012; Zalan et al. 2007; Pepinski et al. 2005; Pepinski et al. 2007; Turrina et al. 2007; Tang and To 2006; Pepinski et al. 2006; Szibor et al. 2000; Glesmann et al. 2009
DXS7424	2	1805	0.11	0.013–0.40	Szibor et al. 2003; Poetsch et al. 2005; Poetsch et al. 2009; Liu et al. 2012; Turrina et al. 2007; Glesmann et al. 2009; Edelmann et al. 2002
DXS8378	3	2982	0.10	0.021–0.29	Szibor et al. 2003; Poetsch et al. 2005; Pico et al. 2008; Tariq et al. 2008; Poetsch et al. 2009; Becker et al. 2008; Hundertmark et al. 2008; Tomas, Pereira, and Morling 2012; Tetzlaff, Wegener, and Lindner 2012; Nishi et al. 2013; Zalan et al. 2007; Pepinski et al. 2005; Pepinski et al. 2007; Turrina et al. 2007; Tang and To 2006; Pepinski et al. 2006; Glesmann et al. 2009

Marker		N		Range	References
DXS10101	1	1073	0.093	0.002–0.52	Becker et al. 2008; Tomas, Pereira, and Morling 2012; Tetzlaff, Wegener, and Lindner 2012
GATA31E08	1	1127	0.089	0.002–0.49	Tariq et al. 2008; Liu et al. 2012; Nishi et al. 2013; Turrina et al. 2007; Glesmann et al. 2009
DXS6789	3	3478	0.086	0.018–0.25	Szibor et al. 2003; Pico et al. 2008; Tabbada et al. 2005; Tariq et al. 2008; Liu et al. 2012; Castaneda et al. 2012; Nishi et al. 2013; Liu et al. 2008; Turrina et al. 2007; Hering, Kuhlisch, and Szibor 2001
DXS7423	2	2614	0.077	0.009–0.28	Szibor et al. 2003; Zarrabeitia et al. 2002b; Pico et al. 2008; Tariq et al. 2008; Poetsch et al. 2009; Becker et al. 2008; Tomas, Pereira, and Morling 2012; Tetzlaff, Wegener, and Lindner 2012; Nishi et al. 2013; Zalan et al. 2007; Pepinski et al. 2005; Turrina et al. 2007; Tang and To 2006; Pepinski et al. 2006; Glesmann et al. 2009
DXS9898	1	1936	0.052	0.001–0.29	Szibor et al. 2003; Poetsch et al. 2005; Pico et al. 2008; Poetsch et al. 2009; Liu et al. 2012; Nishi et al. 2013; Glesmann et al. 2009
DXS101	1	2534	0.039	0.001–0.22	Szibor et al. 2003; Shin et al. 2004; Poetsch et al. 2005; Zarrabeitia et al. 2002b; Pico et al. 2008; Tabbada et al. 2005; Wiegand et al. 2003; Athanasiadou et al. 2003; Poetsch et al. 2009; Liu et al. 2012; Turrina et al. 2007; Glesmann et al. 2009; Edelmann and Szibor 2001
DXS10011	0	50	0.00	0.00–7.11	Poetsch et al. 2005
DXS10147	0	54	0.00	0.00–6.60	Nishi et al. 2013
DXS6810	0	100	0.00	0.00–3.62	Tariq et al. 2008
DXS6793	0	100	0.00	0.00–3.62	Tariq et al. 2008
DXS6797	0	168	0.00	0.00–2.17	Nadeem et al. 2009
DXS9902	0	458	0.00	0.00–0.80	Szibor et al. 2003; Tariq et al. 2008; Nishi et al. 2013
DXS6807	0	598	0.00	0.00–0.62	Szibor et al. 2003; Poetsch et al. 2005; Poetsch et al. 2009; Turrina et al. 2007
GATA172D05	0	876	0.00	0.00–0.42	Szibor et al. 2003; Shin et al. 2004; Pico et al. 2008; Tariq et al. 2008; Nishi et al. 2013; Turrina and De Leo 2004; Turrina et al. 2007
DXS9895	0	917	0.00	0.00–0.40	Szibor et al. 2003; Huang et al. 2003
GATA165B12	0	958	0.00	0.00–0.38	Liu et al. 2012; Nadeem et al. 2009
DXS6801	0	1084	0.00	0.00–0.34	Tariq et al. 2008; Liu et al. 2012; Castaneda et al. 2012
DXS7133	0	1459	0.00	0.00–0.25	Szibor et al. 2003; Poetsch et al. 2005; Poetsch et al. 2009; Liu et al. 2012; Turrina and De Leo 2004; Turrina et al. 2007; Nadeem et al. 2009
DXS6800	0	1694	0.00	0.00–0.22	Szibor et al. 2003; Wiegand et al. 2003; Poetsch et al. 2009; Liu et al. 2012; Nadeem et al. 2009
DXS981	0	2099	0.00	0.00–0.18	Tabbada et al. 2005; Athanasiadou et al. 2003; Tariq et al. 2008; Liu et al. 2012; Liu et al. 2008; Nadeem et al. 2009

Typical mutation rates for autosomal and Y-chromosomal markers are in the range of 0.1–0.5% (Schneider 2007). Initial reports of X chromosomal mutation rates fall within this range, although many are based on a relatively small number of meiotic events (see Table 5.3 for a summary of published X chromosomal mutation rates). As evident from Table 5.3, there is a need to investigate mutation rates in populations outside of Europe in general and Germany specifically. Currently, there are no studies in the literature that investigate mutation rates among U.S. African American or U.S. Hispanic groups, or from the continent of Australia, for example. Additionally, marker- and even allele-specific mutation rates should be investigated. A summary of markers with published observed mutation rates are presented in Table 5.4. Note that the two markers with the highest mutation rate are ARA and DXS8377, both of which contain trinucleotide repeats (Edelmann et al. 2001). In addition, both have very long repeat stretches (on the order of 10–50 repeat units) that are highly polymorphic (>20 alleles), a characteristic that favors higher mutation rates (Dupuy et al. 2004). The ARA locus is no longer considered suitable for forensic use because it falls within the coding region of a gene in which a mutation would give rise to X-linked spinal and bulbar muscular atrophy (La Spada et al. 1991).

5.5 Requirements for Statistical Interpretation of a Match

Standard 8.1.3.2 of the U.S. DNA Advisory Board's Quality Assurance Standards for Forensic DNA Testing Laboratories states that laboratories shall establish and document match criteria on the basis of empirical data (DNA Advisory Board 2000). Another requirement of the 1991 ISFG report relating to the use of DNA polymorphisms is that questions of independent assortment and linkage disequilibrium be addressed (ISFH 1992a). For autosomal STRs, this ensures that the product rule can be used to multiply individual marker frequencies together to determine the overall rarity of a profile. It does not preclude the use of linked markers, however. Y chromosomal STRs, for example, are linked to one another and are considered together as a group called a haplotype. Haplotype frequencies are measured directly from population data, and the counting method is used to determine the rarity of the profile. It follows that X chromosomal STRs may be a combination of the two techniques: the organization of several physically close markers into linkage groups, forming haplotypes, whose frequencies could then be multiplied together once independent assortment of the groups was established.

From the ISFG report, it is clear that both linkage and linkage disequilibrium must be studied. Linkage refers to the cosegregation of closely located markers within a pedigree and can be measured by calculating the recombination fraction (RF) from family samples. Families in which the gametic phase of the parents is known (requires grandparental profiles) and which have multiple offspring are typically used to directly observe recombination events using informative meioses. A meiosis is informative for linkage if it can be unambiguously determined that the gamete is recombinant. Homozygous markers, for example, will prevent a meiosis from being informative. A statistical test (logarithm of the odds) is then applied to determine if the observed RF value is significantly different from that expected for independent assortment (0.5). Family types other than the three-generation pedigree have also been investigated, however. Of note, one novel approach specific to the analysis of X chromosome markers incorporates the mutation rate within a maximum likelihood estimation of the recombination rate (Nothnagel et al. 2012). With this model, both three-generation and two-generation pedigrees can be addressed.

A set of families satisfying the requirements of linkage study (multiple generations and offspring) have been established at the National Institute of General Medical Sciences repository from lymphoblastic cell lines donated by the Centre d'Etudes du Polymorphisme Humaine (CEPH). The collection includes families from Utah, France, Venezuela, and Amish country, and is an important resource for the characterization of DNA polymorphisms and the construction of the human genetic map. In addition, any research effort that requires access to a common dataset can find value in the use of the CEPH reference families, as evidenced by the large amount of data that has already been collected from them and shared in a database made available to contributing researchers. Several reliable linkage maps of the human genome have resulted from such collaborations (Anonymous 1992b; Weissenbach et al. 1992; Broman et al. 1998). In fact, most forensic publications refer to the location of markers on the X chromosome according to the Marshfield map (Broman et al. 1998), which is based on analysis of recombination rates in a subset of eight CEPH families. The CEPH families were also used to create the Rutgers combined linkage-physical map of the human genome, which is a denser map incorporating both sequence-based positional information as well as recombination-based data (Matise et al. 2007). This Rutgers map has been used to generate a consolidated list of physical and genetic distances between 39 commonly used forensic X chromosomal markers (Machado and Medina-Acosta 2009), but because the entire set of 39 markers had not been directly measured, some genetic distances were interpolated. Additionally, certain X chromosomal markers are missing from this analysis (e.g., DXS6795). Therefore, although the physical map of the X chromosome may be well established, further study of chosen markers to create a genetic map through family studies is required. See Figure 5.5 for the physical map of 24 commonly used STR markers on the X chromosome.

Linkage disequilibrium measures the nonrandom association of two or more alleles that are not necessarily closely located on a chromosome, and is estimated from allele and haplotype frequencies. Statistical tests designed to indicate linkage disequilibrium across the genome can potentially highlight cosegregating markers on the X chromosome that may be physically or genetically linked. However, large sample sizes are required to obtain reliable estimates, and this measure alone cannot establish groups of markers that should be considered as haplotypes. Although most population studies include a test for linkage disequilibrium, little scrutiny of linkage on the X chromosome has been performed. Early linkage studies performed with 182 mother–multiple son constellations produced a map of the X chromosome that divided 16 X chromosomal STRs into four linkage groups (Szibor et al. 2003b) (Figure 5.5). The hypothesis stated that alleles at linked markers combine to form haplotypes that could recombine during meiosis as "blocks." In the same study, linkage disequilibrium was estimated from a population of more than 200 males and showed association between only two markers: DXS7424 and DXS101. Since this initial work, various sets of physically close markers have been studied, demonstrating that alleles do indeed cosegregate as stable haplotypes (Hundertmark et al. 2008; Hering et al. 2006b; Szibor et al. 2005), especially markers located around the centromere where recombination rates are reduced (Edelmann et al. 2009a, 2010).

In a study of the Argus X-8 kit, Tillmar et al. (2008) were able to use observed haplotype frequencies to reveal linkage disequilibrium between markers within the same linkage group, but not between markers located in different linkage groups. The study showed that the paternity index would be significantly influenced if this observed linkage disequilibrium was not taken into account. Additionally, 32 families were studied and the RF between linkage groups 3 and 4 was found to be approximately 25%, indicating non-random assortment.

In another recombination report, three-generation pedigrees were analyzed at 39 X STR markers (Hering et al. 2010). Previous studies were confirmed, including a loose association between groups 3 and 4, which could be potentially misleading in certain kinship cases. The need for larger, collaborative recombination studies was emphasized.

The X chromosomal STR commercial kits Argus X-UL, X-8, and X-12 were designed with the specialized linkage situation on the X chromosome in mind. Because the chosen

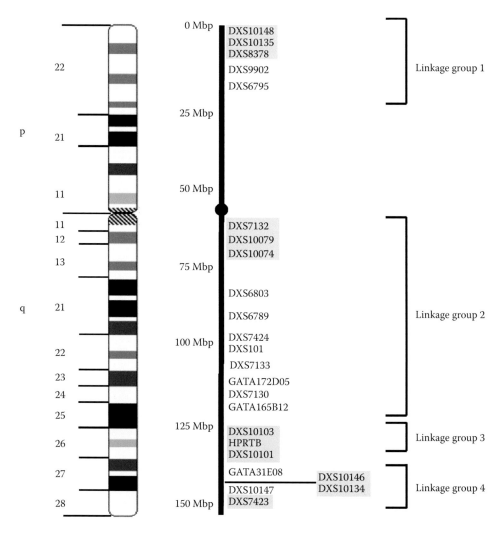

Figure 5.5 Physical map of the X chromosome. The physical location of 24 X STR markers and the four proposed linkage groups are shown. Markers included in the commercial kit Investigator® Argus X-12 are highlighted in yellow.

markers reside on a single chromosome, the initial four markers were chosen because they were physically far apart and unlinked. Additional loci were chosen based upon their reported linkage to the original markers, creating four linkage trios that could be viewed as haplotypes: DXS8378–DXS10135–DXS10148, DXS7132–DXS10079–DXS10074, DXS10103–HPRTB–DXS10101, and DXS10146–DXS10134–DXS7423 (Qiagen 2010). The first trio was proposed for inclusion into the commercial multiplex after confirmation of heterogeneity in a German population and a small recombination study (89 informative meioses) in which the stability of the region containing the three proposed markers was assessed (Hundertmark et al. 2008). DXS10074 and DXS10079 were validated for forensic use through a study of their allele structure and recombination rate in a German population (Hering et al. 2006a,b). Although no recombination was observed during this study of approximately 92 informative meioses, linkage disequilibrium was established. In comparison, a study of two Brazilian populations revealed ambiguous results for DXS10079 and DXS10074 and a third marker within a 280-kb region of Xq12, where significant linkage disequilibrium was confirmed in the absence of an indication of significant linkage (Ferreira da Silva et al. 2010). Evaluation of DXS10103 and DXS10101 for acceptable heterozygosity and reliable amplification was performed as part of a study of the 133.14–133.45 Mb region of the X chromosome surrounding HPRTB (Rodig et al. 2010). The haplotype stability of the trio DXS10146–DXS10134–DXS7423 was assessed in a recombination study of less than 109 informative meioses (Edelmann et al. 2008). Although linkage disequilibrium was not tested because of the small sample size, two crossing over events were observed between DXS10146 and DXS10135 in 80 informative meioses. The authors still recommended these two markers for inclusion into the new commercial kit because of their high degree of polymorphism, pending a more accurate estimation of the genetic distance between them. A recent study, however, found evidence for less-than-free recombination between these groups as well as substantial intra-group recombination rates (Nothnagel et al. 2012), indicating that further study is necessary before adoption of a robust statistical approach.

5.6 Exchange and Comparability of Data

As the number of new markers described in the literature increases, the potential for nomenclature differences also increases. There are 36 X STRs that are commonly used by different forensic laboratories (Table 5.1), and in some cases, differences in allele nomenclature make comparisons of published population data between the laboratories at best tedious and at worst impossible. Although several publications have explicitly addressed these nomenclature, discrepancies for specific X markers such as DXS7423, DXS8377 (Szibor et al. 2003a), and HPRTB (Szibor et al. 2009), other instances still exist. DXS6803, GATA172D05, and GATA31E08 are all examples of X STR markers for which nomenclature variations between publications affects the allele designation. The standard repeat unit defined for GATA31E08 is the AGAT repeat unit. However, allele sequencing results revealed additional variation resulting from AGGG repeats immediately before this AGAT repeat, changing the allele designation by +2 (Diegoli and Coble 2011; Gomes et al. 2009c). Depending on the repeat unit used for GATA172D05 nomenclature, TAGA (Edelmann et al. 2002a) or GATA (Edelmann et al. 2001), the allele designations can vary by one.

Additionally, DXS9902 and DXS7424 are examples of markers for which the nomenclature differences exist despite attempts at adhering to the ISFG guidelines. Publication of corresponding allele sequencing data and/or definition of the repeat structure used in a particular study can ease a portion of this confusion.

5.7 Future Directions

As for the autosomes and Y chromosome, other non-STR markers on the X chromosome have been investigated for forensic use. These include single-nucleotide polymorphisms (Li et al. 2010; Pereira et al. 2010; Tomas et al. 2008, 2010) and insertion/deletion markers (Ribeiro-Rodrigues et al. 2009; Freitas et al. 2010; Edelmann et al. 2009b). Research has also been undertaken on additional species such as dogs (van Asch et al. 2010). However, although forensic use of autosomal and Y chromosomal markers is well documented and accepted by both the scientific community and courts of law, the routine use of markers on the X chromosome is less common, especially in the United States, and much of the groundwork necessary is still in progress. Through the collaborative process of scientific research that is already ongoing, X chromosomal markers will offer the alternative perspective necessary to answer some of the more challenging kinship questions often faced by forensic scientists in any laboratory.

Acknowledgments

The authors thank Dr. Michael Coble for discussion and support; Dr. Adrian Linacre for manuscript review; and the AFDIL Emerging Technologies section, James Canik, Col. Louis Finelli, Lt. Col. Laura Regan, Dr. Timothy McMahon and the Armed Forces Medical Examiner System for logistical and administrative support. The opinions or assertions presented hereafter are the private views of the author and should not be construed as official or as reflecting the views of the Department of Defense, its branches, the U.S. Army Medical Research and Material Command or the Armed Forces Medical Examiner System.

Appendix: Allele Frequency Data

Table 5.5 X STR Allele Frequencies for Four U.S. Populations

Marker	DXS10135				DXS10148				DXS101				DXS10103				DXS10079				DXS6789			
Pop.	AF	AS	CN	Hisp	AF	AS	CN	Hisp	AF	AS	CN	Hisp	AF	AS	CN	Hisp	AF	AS	CN	Hisp	AF	AS	CN	Hisp
N	261	238	337	140	261	238	336	140	349	501	268	245	261	238	336	140	262	238	337	140	349	501	268	245
Allele																								
13																	0.015							
13.3					0.103																			
14							0.003		0.002								0.008							
15	0.011		0.003	0.021					0.004		0.041	0.008		0.013	0.021	0.021		0.004	0.006	0.064	0.006	0.005	0.002	0.005
16	0.023		0.009	0.007	0.004				0.004				0.138	0.265	0.116	0.243	0.008	0.025	0.024	0.007	0.212	0.156	0.048	0.060
17	0.031	0.017	0.030	0.029		0.004	0.003		0.002		0.005		0.069	0.105	0.140	0.114	0.084	0.055	0.065	0.050	0.115	0.363	0.019	0.063
17.1	0.019			0.007																	0.011	0.031	0.005	0.005
18	0.065	0.025	0.039	0.014	0.123	0.143	0.134	0.157	0.063		0.097	0.057	0.203	0.181	0.164	0.136	0.126	0.105	0.166	0.107	0.017			0.003
18.1	0.019		0.009	0.007																				
19	0.096	0.088	0.062	0.079	0.050	0.059	0.015	0.007	0.075	0.009	0.046	0.027	0.444	0.353	0.423	0.386	0.221	0.223	0.217	0.250	0.052	0.025	0.022	0.035
19.1	0.073		0.012	0.021																				
20	0.050	0.097	0.050	0.071	0.008	0.025		0.014	0.075	0.005	0.024	0.025	0.134	0.076	0.122	0.071	0.260	0.261	0.300	0.271	0.207	0.194	0.357	0.401
20.1	0.034		0.012		0.011																			
21	0.069	0.122	0.059	0.086		0.008	0.006	0.007	0.115	0.009	0.029	0.033	0.011	0.008	0.015	0.029	0.168	0.139	0.128	0.157	0.207	0.139	0.319	0.256
21.1	0.057	0.134	0.006	0.021	0.011	0.008	0.015	0.014	0.067	0.055	0.005	0.038												
22	0.073	0.092	0.068	0.107	0.008	0.042	0.057	0.050	0.078	0.119	0.068	0.060					0.084	0.164	0.065	0.043	0.11	0.07	0.14	0.12
22.1	0.011		0.021		0.011		0.071	0.036																
23	0.046	0.092	0.089	0.114	0.038	0.092	0.012	0.014	0.103	0.278	0.191	0.251					0.019	0.025	0.006	0.021	0.055	0.014	0.070	0.041
23.1	0.011				0.034	0.155	0.158	0.064																
24	0.054	0.109	0.098	0.071	0.034	0.134	0.161	0.229	0.098	0.223	0.181	0.158					0.008				0.008	0.002	0.014	0.008
24.1	0.008		0.003	0.014	0.019	0.097	0.185	0.200												0.002				0.003
25	0.038	0.050	0.107	0.071	0.054				0.098	0.174	0.147	0.213												
25.1	0.004		0.003		0.015																			
26	0.042	0.021	0.086	0.071	0.034																			
26.1					0.004																			
26.3																								

(*continued*)

Table 5.5   X STR Allele Frequencies for Four U.S. Populations (Continued)

Marker	DXS10135				DXS10148				DXS101				DXS10103				DXS10079				DXS6789			
Pop.	AF	AS	CN	Hisp	AF	AS	CN	Hisp	AF	AS	CN	Hisp	AF	AS	CN	Hisp	AF	AS	CN	Hisp	AF	AS	CN	Hisp
27	0.031	0.067	0.059	0.100	0.008				0.117	0.075	0.065	0.074												
27.1	0.008		0.003		0.031	0.105	0.098	0.093																
27.2					0.008	0.004																		
28	0.015	0.055	0.047	0.043	0.023				0.061	0.039	0.051	0.030												
28.1	0.004				0.023	0.050	0.045	0.036																
29	0.027	0.042	0.036	0.021	0.019				0.023	0.010	0.029	0.014												
29.1					0.027	0.038	0.033	0.036																
29.2						0.004																		
30	0.015	0.025	0.042	0.014	0.023				0.010	0.005	0.019	0.011												
30.1					0.011	0.017	0.006	0.014																
31	0.019	0.021	0.024		0.019				0.004		0.002													
31.1						0.008		0.007																
32		0.021	0.009		0.008																			
32.2						0.004																		
33		0.008	0.009						0.004			0.003												
34			0.003	0.007																				
34.1	0.004																							
35	0.004	0.004																						
35.1	0.004				0.004																			
35.2	0.011																							
36.1	0.004																							
38			0.003																					
38.2	0.008																							
39.2	0.011																							
Null					0.234			0.014																

Note: AF, African American; AS, U.S. Asian; CN, U.S. Caucasian; Hisp, U.S. Hispanic; N, number of alleles. For further study details, see Diegoli and Coble (2011) for markers DXS6789, DXS7130, GATA31E08, DXS7424, GATA165B12, DXS101, DXS6795, GATA172D05, DXS10147, DXS8378, DXS7132, DXS6803, HPRTB, DXS7423, and DXS99902; and see Diegoli et al. (2011a) for markers DXS10135, DXS10148, DXS10103, DXS10079, DXS10134, DXS10101, DXS10074, and DXS10146.

Table 5.6 X STR Allele Frequencies for Four U.S. Populations

Marker	DXS10146				DXS10134				DXS10101			
Pop.	AF	AS	CN	Hisp	AF	AS	CN	Hisp	AF	AS	CN	Hisp
N	261	238	336	140	261	238	337	140	261	238	336	141
Allele												
23	0.004								0.004			
24	0.004	0.034	0.006	0.014								
24.2				0.007								0.035
25	0.019	0.055	0.060	0.036					0.038			
25.2									0.004		0.009	
26	0.065	0.092	0.134	0.079					0.015			
26.2									0.019		0.015	0.007
27	0.080	0.218	0.128	0.164					0.008	0.004	0.003	0.050
27.2									0.027	0.008	0.051	0.028
28	0.080	0.218	0.164	0.121	0.004				0.073	0.034	0.018	
28.1					0.004							
28.2									0.031	0.017	0.095	0.014
29	0.069	0.160	0.140	0.250	0.011			0.007	0.096	0.034	0.030	0.028
29.2	0.008								0.038	0.084	0.119	0.057
30	0.073	0.084	0.119	0.064	0.073			0.014	0.092	0.122	0.071	0.057
30.1					0.004							
30.2	0.004								0.073	0.084	0.158	0.156
31	0.069	0.063	0.039	0.050	0.019	0.004	0.003	0.014	0.134	0.164	0.137	0.113
31.2	0.004				0.038				0.038	0.092	0.122	0.121
32	0.073	0.029	0.006	0.014	0.042	0.017	0.024	0.014	0.138	0.130	0.074	0.156
32.1					0.004							
32.2	0.008	0.013	0.003		0.084	0.034	0.047	0.050	0.015	0.097	0.024	0.050
33	0.069					0.004			0.130	0.084	0.045	0.071
33.2	0.011				0.004				0.004	0.017	0.009	0.007
33.3												
34	0.011	0.008			0.084	0.071	0.125	0.050	0.023	0.025	0.018	0.043
34.2	0.054											
35	0.008				0.142	0.181	0.151	0.150		0.004		0.007
35.1					0.004		0.003	0.007				

(*continued*)

Table 5.6 X STR Allele Frequencies for Four U.S. Populations (Continued)

Marker	DXS10146				DXS10134				DXS10101			
Pop.	AF	AS	CN	Hisp	AF	AS	CN	Hisp	AF	AS	CN	Hisp
35.2	0.042			0.014	0.004							
36					0.199	0.223	0.211	0.250				
36.1								0.007				
36.2	0.011			0.007								
37					0.176	0.210	0.160	0.171			0.003	
37.2							0.003					
37.3					0.004	0.008	0.006	0.014				
38					0.103	0.134	0.080	0.121				
38.2	0.038		0.006	0.021								
38.3					0.004	0.025	0.021	0.007				
39					0.015	0.063	0.024	0.021				
39.1	0.008											
39.2	0.011		0.039	0.029				0.007				
39.3					0.004	0.004	0.024	0.029				
40							0.009	0.007				
40.2	0.034	0.004	0.030	0.029		0.004	0.033	0.021				
40.3						0.008						
41			0.003									
41.2	0.011	0.004	0.012	0.014	0.011	0.004	0.030	0.014				
41.3												
42.2	0.027	0.013	0.012	0.007		0.004	0.039	0.007				
42.3	0.004											
43.2	0.015		0.048	0.007								
43.3							0.003	0.014				
44.2	0.008		0.033	0.036								
44.3							0.006					
45.2	0.004		0.009	0.007								
46.2		0.004	0.012									
47.2				0.007								
Null	0.073			0.021								

Note: AF, African American; AS, U.S. Asian; CN, U.S. Caucasian; Hisp, U.S. Hispanic; N, number of alleles. For further study details, see Diegoli and Coble (2011) for markers DXS6789, DXS7130, GATA31E08, DXS7424, GATA165B12, DXS101, DXS6795, GATA172D05, DXS10147, DXS8378, DXS7132, DXS6803, HPRTB, DXS7423, and DXS9902; and see Diegoli et al. (2011a) for markers DXS10135, DXS10148, DXS10103, DXS10079, DXS10134, DXS10101, DXS10074, and DXS10146.

Table 5.7 X STR Allele Frequencies for Four U.S. Populations

Marker	DXS10074				DXS9902				DXS7424				GATA165B12				DXS7423				DXS10147				HPRTB			
Pop.	AF	AS	CN	Hisp	AF	AS	CN	Hisp	AF	AS	CN	Hisp	AF	AS	CN	Hisp	AF	AS	CN	Hisp	AF	AS	CN	Hisp	AF	AS	CN	Hisp
N	261	238	336	140	349	501	268	245	349	501	268	245	349	501	268	245	349	501	268	245	349	501	268	245	349	501	268	245
Allele																												
5																					0.002	0.001						
6																					0.130	0.277	0.220	0.341				
7	0.115		0.083	0.036	0.002	0.004															0.281	0.199	0.039	0.060				
8	0.015	0.146	0.057			0.033		0.003	0.008				0.038	0.004	0.012	0.014					0.363	0.476	0.300	0.447				0.002
9	0.004		0.009	0.036	0.075	0.017	0.034	0.027			0.002		0.170	0.255	0.314	0.245					0.174	0.047	0.413	0.139	0.002			
10	0.019				0.285	0.422	0.336	0.373	0.011	0.001	0.005		0.331	0.564	0.314	0.450					0.044		0.024	0.014	0.038		0.002	0.003
10.1								0.005																				
11	0.034				0.369	0.335	0.386	0.351	0.059	0.010	0.005	0.003	0.359	0.151	0.333	0.256					0.006		0.005		0.013			0.011
11.1					0.010		0.022	0.033																				
11.3		0.004																										
12	0.077				0.224	0.212	0.215	0.202	0.055	0.011	0.046	0.019	0.098	0.026	0.027	0.035									0.296	0.258	0.341	0.256
12.1								0.005																				
13	0.096	0.004	0.009	0.007	0.004	0.009	0.007		0.216	0.066	0.058	0.125	0.004				0.082	0.005	0.097	0.030					0.296	0.258	0.341	0.256
14	0.111	0.008	0.015	0.057		0.001			0.222	0.134	0.200	0.180					0.482	0.383	0.350	0.300					0.289	0.409	0.319	0.395
14.2				0.007																					0.174	0.186	0.162	0.199
14.3	0.008																											
15	0.180	0.076	0.077	0.171					0.193	0.322	0.254	0.234					0.327	0.558	0.382	0.480					0.073	0.052	0.046	0.049
15.3	0.004	0.004																										
16	0.153	0.143	0.196	0.150					0.172	0.370	0.246	0.278					0.076	0.052	0.126	0.095					0.017	0.005	0.010	0.014
16.3	0.011																											
17	0.111	0.336	0.199	0.243					0.050	0.062	0.138	0.123					0.019	0.001	0.046	0.095								
18	0.034	0.244	0.167	0.107					0.015	0.021	0.039	0.025																
19	0.023	0.160	0.086	0.071					0.002	0.002	0.005	0.005																
19.2			0.003																									
20		0.017	0.006	0.014					0.002		0.002																	
21	0.004		0.003																									
22		0.004																										

Note: AF, African American; AS, U.S. Asian; CN, U.S. Caucasian; Hisp, U.S. Hispanic; N, number of alleles. For further study details, see Diegoli and Coble (2011) for markers DXS6789, DXS7130, GATA31E08, DXS7424, GATA165B12, DXS101, DXS6795, GATA172D05, DXS10147, DXS7132, DXS8378, DXS7423, DXS6803, HPRTB, DXS7423, and DXS9902; and see Diegoli et al. (2011a) for markers DXS10135, DXS10148, DXS10103, DXS10079, DXS10134, DXS10101, DXS10074, and DXS10146.

Table 5.8 X STR Allele Frequencies for Four U.S. Populations

Marker	DXS7132				GATA31E08				DXS6795				DXS8378				GATA172D05				DXS6803				DXS7130			
Pop.	AF	AS	CN	Hisp	AF	AS	CN	Hisp	AF	AS	CN	Hisp	AF	AS	CN	Hisp	AF	AS	CN	Hisp	AF	AS	CN	Hisp	AF	AS	CN	Hisp
N	349	501	268	245	349	501	268	245	349	501	268	245	349	501	268	245	349	501	268	245	349	501	268	245	349	501	268	245
Allele																												
6																	0.185	0.055	0.179	0.112								
7					0.021				0.002								0.038	0.009	0.002	0.008				0.005				
8					0.029	0.001	0.002	0.003					0.010	0.001	0.002	0.003	0.178	0.186	0.162	0.134								
9					0.140	0.099	0.196	0.177	0.117	0.041	0.302	0.183	0.011	0.021	0.007	0.008	0.277	0.094	0.051	0.071	0.002				0.008		0.002	
10		0.001			0.147	0.027	0.019	0.087	0.268	0.171	0.017	0.109	0.281	0.563	0.355	0.381	0.161	0.383	0.297	0.313	0.025		0.002	0.025	0.023	0.007		0.003
10.3																					0.143	0.011	0.043	0.003				
11	0.011	0.004	0.014	0.005	0.061	0.194	0.222	0.180	0.195	0.312	0.471	0.223	0.359	0.267	0.345	0.316	0.109	0.227	0.203	0.251	0.367	0.195	0.263	0.256	0.076	0.212	0.036	0.057
11.3																					0.010	0.069	0.010	0.057				
12	0.101	0.080	0.097	0.082	0.249	0.240	0.205	0.319	0.088	0.032	0.029	0.147	0.306	0.131	0.258	0.270	0.052	0.046	0.099	0.109	0.233	0.142	0.263	0.349	0.193	0.181	0.101	0.213
12.3																					0.069	0.442	0.109	0.131				
13	0.258	0.172	0.309	0.275	0.262	0.321	0.261	0.185	0.098	0.419	0.179	0.300	0.027	0.011	0.027	0.019		0.007		0.003	0.080	0.040	0.128	0.063	0.149	0.067	0.058	0.079
13.3																					0.036	0.094	0.155	0.112	0.023	0.005	0.043	0.019
14	0.371	0.370	0.360	0.322	0.082	0.105	0.111	0.033	0.052	0.021		0.019	0.006	0.004	0.005	0.003					0.023	0.002	0.010		0.044	0.005	0.017	0.016
14.3																						0.005		0.017	0.189	0.080	0.225	0.223
15	0.218	0.273	0.152	0.251	0.008	0.012	0.010	0.016	0.170	0.002	0.002	0.019	0.001															
15.3																									0.222	0.337	0.341	0.332
16	0.031	0.082	0.063	0.054	0.002	0.001			0.008												0.002				0.002			
16.3	0.002																								0.061	0.096	0.145	0.041
17	0.004	0.017	0.002	0.011					0.002																			
17.3																									0.008	0.009	0.031	0.016
18	0.004		0.002																									
18.3																									0.002			

Note: AF, African American; AS, U.S. Asian; CN, U.S. Caucasian; Hisp, U.S. Hispanic; N, number of alleles. For further study details, see Diegoli and Coble (2011) for markers DXS6789, DXS7130, GATA31E08, DXS7424, GATA165B12, DXS101, DXS6795, GATA172D05, DXS10147, DXS8378, DXS7132, DXS6803, HPRTB, DXS7423, and DXS9902; and see Diegoli et al. (2011a) for markers DXS10135, DXS10148, DXS10103, DXS10079, DXS10134, DXS10101, DXS10074, and DXS10146.

References

Anonymous. 1992. A comprehensive genetic linkage map of the human genome. NIH/CEPH collaborative mapping group. *Science (New York, N.Y.)* 258(5079): 67–86.

Acar, E., O. Bulbul, G. Rayimoglu et al. 2009. Optimization and validation studies of the Mentype® Argus X-8 kit for paternity cases. *Forensic Sci. Int. Genet. Suppl. Ser.* 2: 47–8.

Aler, M., P. Sanchez-Diz, I. Gomes et al. 2007. Genetic data of 10 X-STRs in a Spanish population sample. *Forensic Sci. Int.* 173(2–3): 193–6.

Andreaggi, K. S., R. S. Just, T. M. Diegoli et al. 2010. Application of modified STR amplification protocols to commingled remains from the USS *Oklahoma*. Abstract.

Aquino, J., C. Peixe, D. Silva, C. Tavares, and E. F. de Carvalho. 2009. A X-chromosome STR hexaplex as a powerful tool in deficiency paternity cases. *Forensic Sci. Int. Genet. Suppl. Ser.* 2: 45–6.

Asamura, H., H. Sakai, K. Kobayashi, M. Ota, and H. Fukushima. 2006a. MiniX-STR multiplex system population study in Japan and application to degraded DNA analysis. *Int. J. Legal Med.* 120(3): 174–81.

Asamura, H., H. Sakai, M. Ota, and H. Fukushima. 2006b. Japanese population data for eight X-STR loci using two new quadruplex systems. *Int. J. Legal Med.* 120(5): 303–9.

Athanasiadou, D., B. Stradmann-Bellinghausen, C. Rittner, K. W. Alt, and P. M. Schneider. 2003. Development of a quadruplex PCR system for the genetic analysis of X-chromosomal STR loci. *Int. Congr. Ser.* 1239: 311–4.

Barbaro, A., P. Cormaci, S. Votano, and A. Barbaro. 2008. Population data of 8 X-STRs in south Italy (Calabria) using the Mentype® Argus X-8 PCR amplification kit (Biotype). *Forensic Sci. Int. Genet. Suppl. Ser.* 1: 135–9.

Becker, D., H. Rodig, C. Augustin et al. 2008. Population genetic evaluation of eight X-chromosomal short tandem repeat loci using Mentype® Argus X-8 PCR amplification kit. *Forensic Sci. Int. Genet.* 2(1): 69–74.

Bekada, A., S. Benhamamouch, A. Boudjema et al. 2009. Analysis of 12 X-chromosomal STRs in an Algerian population sample. *Forensic Sci. Int. Genet. Suppl. Ser.* 2: 400–1.

Bekada, A., S. Benhamamouch, A. Boudjema et al. 2010. Analysis of 21 X-chromosomal STRs in an Algerian population sample. *Int. J. Legal Med.* 124(4): 287–94.

Bentayebi, K., A. Picornell, M. Bouabdeallah et al. 2012. Genetic diversity of 12 X-chromosomal short tandem repeats in the Moroccan population. *Forensic Sci. Int. Genet.* 6(1): e48–9.

Bini, C., S. Ceccardi, G. Ferri et al. 2004. Development of a heptaplex PCR system to analyse X-chromosome STR loci from five Italian population samples. A collaborative study. *Int. Congr. Ser.* 1261: 272–4.

Biotype AG. 2006. Mentype® Argus X-UL PCR amplification kit manual (December 2006).

Biotype AG. 2007. Mentype® Argus X-8 PCR amplification kit manual (August 2007).

Bobillo, C., A. Sala, L. Gusmao, and D. Corach. 2011. Genetic analysis of 10 X-STRs in Argentinian population. *Forensic Sci. Int. Genet.* 5(1): e14–6.

Brinkmann, B., M. Klintschar, F. Neuhuber, J. Huhne, and B. Rolf. 1998. Mutation rate in human microsatellites: Influence of the structure and length of the tandem repeat. *Am. J. Hum. Genet.* 62(6): 1408–15.

Broman, K. W., J. C. Murray, V. C. Sheffield, R. L. White, and J. L. Weber. 1998. Comprehensive human genetic maps: Individual and sex-specific variation in recombination. *Am. J. Hum. Genet.* 63(3): 861–9.

Butler, J. M., Y. Shen, and B. R. McCord. 2003. The development of reduced size STR amplicons as tools for analysis of degraded DNA. *J. Forensic Sci.* 48(5): 1054–64.

Caine, L. M., L. Pontes, D. Abrantes, G. Lima, and F. Pinheiro. 2007a. Genetic data of 4 X-chromosomal short tandem repeats in a north of Portugal population. *J. Forensic Sci.* 52(2): 500–1.

Caine, L. M., L. Pontes, D. Abrantes, G. Lima, and F. Pinheiro. 2007b. Genetic data of four X-chromosomal STRs in a population sample of Santa Catarina, Brazil. *J. Forensic Sci.* 52(2): 502–3.

Caine, L. M., M. T. Zarrabeitia, J. A. Riancho, and M. F. Pinheiro. 2010. Genetic data of a Brazilian population sample (Santa Catarina) using an X-STR decaplex. *J. Forensic Legal Med.* 17(5): 272–4.

Carvalho, R., M. F. Pinheiro, and R. Medeiros. 2008. Study of 16 X-STRs in a prostate cancer population sample (preliminary results). *Forensic Sci. Int. Genet. Suppl. Ser.* 1: 142–4.

Castaneda, M., V. Mijares, J. A. Riancho, and M. T. Zarrabeitia. 2012. Haplotypic blocks of X-linked STRs for forensic cases: Study of recombination and mutation rates. *Journal of Forensic Sciences* 57(1) (Jan): 192–5.

Castaneda, M., A. Odriozola, J. Gomez, and M. T. Zarrabeitia. 2013. Development and validation of a multiplex reaction analyzing eight miniSTRs of the X chromosome for identity and kinship testing with degraded DNA. *Int. J. Legal Med.* 127: 735–9.

Cerri, N., A. Verzeletti, F. Gasparini, B. Bandera, and F. De Ferrari. 2006. Population data for four X-chromosomal STR loci in a population sample from Brescia (northern Italy). *Int. Cong. Ser.* 1288: 286–8.

Cerri, N., A. Verzeletti, F. Gasparini, A. Poglio, E. Mazzeo, and F. De Ferrari. 2008. Population data for 8 X-chromosome STR loci in a population sample from northern Italy and from the Sardinia island. *Forensic Sci. Int. Genet. Suppl. Ser.* 1: 173–5.

Coble, M. D., and J. M. Butler. 2005. Characterization of new miniSTR loci to aid analysis of degraded DNA. *J. Forensic Sci.* 50(1): 43–53.

Coletti, A., L. Lottanti, M. Lancia, G. Margiotta, E. Carnevali, and M. Bacci. 2006. Allele distribution of 6 X-chromosome STR loci in an Italian population. *Int. Cong. Ser.* 1288: 292–4.

Collins, P. J., L. K. Hennessy, C. S. Leibelt, R. K. Roby, D. J. Reeder, and P. A. Foxall. 2004. Developmental validation of a single-tube amplification of the 13 CODIS STR loci, D2S1338, D19S433, and amelogenin: The AmpFlSTR® Identifiler® PCR amplification kit. *J. Forensic Sci.* 49(6): 1265–77.

Cybulska, L., J. Wysocka, K. Rebała et al. 2008. Polymorphism of four X-chromosomal STR loci in Belarusians and Slovaks. *Forensic Sci. Int. Genet.* 1: 145–6.

Decker, A. E., M. C. Kline, J. W. Redman, T. M. Reid, and J. M. Butler. 2008. Analysis of mutations in father–son pairs with 17 Y-STR loci. *Forensic Sci. Int. Genet.* 2(3): e31–5.

Diegoli, T. M., and M. D. Coble. 2011. Development and characterization of two mini-X chromosomal short tandem repeat multiplexes. *Forensic Sci. Int. Genet.* 5(5): 415–21.

Diegoli, T. M., A. Linacre, P. M. Vallone, J. M. Butler, and M. D. Coble. 2011a. Allele frequency distribution of twelve X-chromosomal short tandem repeat markers in four U.S. population groups. *Forensic Sci. Int. Genet. Suppl. Ser.* 3: e481–3.

Diegoli, T. M., L. Kovacevic, N. Pojskic, M. D. Coble, and D. Marjanovic. 2011b. Population study of fourteen X chromosomal short tandem repeat loci in a population from Bosnia and Herzegovina. *Forensic Sci. Int. Genet.* 5: 350–1.

DNA Advisory Board. 2000. Quality assurance standards for forensic DNA testing laboratories. *Forensic Sci. Commun.* 2(3).

Dupuy, B. M., M. Stenersen, T. Egeland, and B. Olaisen. 2004. Y-chromosomal microsatellite mutation rates: Differences in mutation rate between and within loci. *Hum. Mutat.* 23(2): 117–24.

Edelmann, J., and R. Szibor. 2001. DXS101: A highly polymorphic X-linked STR. *Int. J. Legal Med.* 114(4–5): 301–4.

Edelmann, J., S. Hering, M. Michael et al. 2001. 16 X-chromosome STR loci frequency data from a German population. *Forensic Sci. Int.* 124(2–3): 215–8.

Edelmann, J., D. Deichsel, S. Hering, I. Plate, and R. Szibor. 2002a. Sequence variation and allele nomenclature for the X-linked STRs DXS9895, DXS8378, DXS7132, DXS6800, DXS7133, GATA172D05, DXS7423 and DXS8377. *Forensic Sci. Int.* 129(2): 99–103.

Edelmann, J., S. Hering, E. Kuhlisch, and R. Szibor. 2002b. Validation of the STR DXS7424 and the linkage situation on the X-chromosome. *Forensic Sci. Int.* 125(2–3): 217–22.

Edelmann, J., S. Hering, C. Augustin, and R. Szibor. 2008. Characterisation of the STR markers DXS10146, DXS10134 and DXS10147 located within a 79.1 kb region at Xq28. *Forensic Sci. Int. Genet.* 2(1): 41–6.

Edelmann, J., S. Hering, C. Augustin, U. Immel, and R. Szibor. 2009a. Chromosome X centromere region—Haplotype frequencies for different populations. *Forensic Sci. Int. Genet. Suppl. Ser.* 2: 398–9.

Edelmann, J., S. Hering, C. Augustin, and R. Szibor. 2009b. Indel polymorphisms—An additional set of markers on the X-chromosome. *Forensic Sci. Int. Genet. Suppl. Ser.* 2: 510–2.

Edelmann, J., S. Hering, C. Augustin, S. Kalis, and R. Szibor. 2010. Validation of six closely linked STRs located in the chromosome X centromere region. *Int. J. Legal Med.* 124(1): 83–7.

Edelmann, J., S. Lutz-Bonengel, J. Naue, and S. Hering. 2012. X-chromosomal haplotype frequencies of four linkage groups using the Investigator™ Argus X-12 kit. *Forensic Sci. Int. Genet.* 6(1): e24–34.

Ferreira da Silva, I. H., A. G. Barbosa, D. A. Azevedo et al. 2010. An X-chromosome pentaplex in two linkage groups: Haplotype data in Alagoas and Rio de Janeiro populations from Brazil. *Forensic Sci. Int. Genet.* 4(4): e95–100.

Fracasso, T., M. Schurenkamp, B. Brinkmann, and C. Hohoff. 2008. An X-STR meiosis study in Kurds and Germans: Allele frequencies and mutation rates. *Int. J. Legal Med.* 122(4): 353–6.

Freitas, N. S., R. L. Resque, E. M. Ribeiro-Rodrigues et al. 2010. X-linked insertion/deletion polymorphisms: Forensic applications of a 33-markers panel. *Int. J. Legal Med.* 124(6): 589–93.

Garcia, B., M. Crespillo, M. Paredes, and J. L. Valverde. 2012. Population data for 10 X-chromosome STRs from north-east of Spain. *Forensic Sci. Int. Genet.* 6(1): e13–5.

Ge, J., B. Budowle, X. G. Aranda, J. V. Planz, A. J. Eisenberg, and R. Chakraborty. 2009. Mutation rates at Y chromosome short tandem repeats in Texas populations. *Forensic Sci. Int. Genet.* 3(3): 179–84.

Glesmann, L. A., P. F. Martina, R. L. Vidal, and C. I. Catanesi. 2009. Mutation rate of 7 X-STRs of common use in population genetics. *J. Basic Appl. Genet.* 20(2): 37–41.

Gomes, I., A. Amorim, V. Pereira, A. Carracedo, and L. Gusmao. 2009a. Genetic patterns of 10 X chromosome short tandem repeats in an Asian population from Macau. *Forensic Sci. Int. Genet. Suppl. Ser.* 2: 402–4.

Gomes, I., A. Carracedo, A. Amorim, and L. Gusmao. 2006. A multiplex PCR design for simultaneous genotyping of X chromosome short tandem repeat markers. *Int. Congr. Ser.* 1288: 313–5.

Gomes, I., C. Alves, K. Maxzud et al. 2007a. Analysis of 10 X-STRs in three African populations. *Forensic Sci. Int. Genet.* 1(2): 208–11.

Gomes, I., V. Pereira, V. Gomes et al. 2009b. The Karimojong from Uganda: Genetic characterization using an X-STR decaplex system. *Forensic Sci. Int. Genet.* 3(4): e127–8.

Gomes, I., M. Prinz, R. Pereira et al. 2009c. X-chromosome STR sequence variation, repeat structure, and nomenclature in humans and chimpanzees. *Int. J. Legal Med.* 123(2): 143–9.

Gomes, I., M. Prinz, R. Pereira et al. 2007b. Genetic analysis of three U.S. population groups using an X-chromosomal STR decaplex. *Int. J. Legal Med.* 121(3): 198–203.

Gusmao, L., P. Sanchez-Diz, C. Alves et al. 2009. A GEP–ISFG collaborative study on the optimization of an X-STR decaplex: Data on 15 Iberian and Latin American populations. *Int. J. Legal Med.* 123(3): 227–34.

Hashiyada, M., Y. Itakura, and M. Funayama. 2008. Polymorphism of eight X-chromosomal STRs in a Japanese population. *Forensic Sci. Int. Genet. Suppl. Ser.* 1: 150–2.

Hedman, M., J. U. Palo, and A. Sajantila. 2009. X-STR diversity patterns in the Finnish and the Somali population. *Forensic Sci. Int. Genet.* 3(3): 173–8.

Hering, S., E. Kuhlisch, and R. Szibor. 2001. Development of the X-linked tetrameric microsatellite marker HumDXS6789 for forensic purposes. *Forensic Sci. Int.* 119(1): 42–6.

Hering, S., C. Augstin, J. Edelmann, M. Heidel, J. Drebler, and R. Szibor. 2006a. A cluster of six closely linked STR-markers: Recombination analysis in a 3.6-mb region at Xq12–13.1. *Int. Cong. Ser.* 1288: 289–91.

Hering, S., C. Augustin, J. Edelmann et al. 2006b. DXS10079, DXS10074 and DXS10075 are STRs located within a 280-kb region of Xq12 and provide stable haplotypes useful for complex kinship cases. *Int. J. Legal Med.* 120(6): 337–45.

Hering, S., J. Edelmann, C. Augustin, E. Kuhlisch, and R. Szibor. 2010. X chromosomal recombination—A family study analysing 39 STR markers in German three-generation pedigrees. *Int. J. Legal Med.* 124(5): 483–91.

Horvath, G., A. Zalan, Z. Kis, and H. Pamjav. 2012. A genetic study of 12 X-STR loci in the Hungarian population. *Forensic Sci. Int. Genet.* 6(1): e46–7.

Huang, D., Q. Yang, C. Yu, and R. Yang. 2003. Development of the X-linked tetrameric microsatellite markers HumDXS6803 and HumDXS9895 for forensic purpose. *Forensic Sci. Int.* 133(3): 246–9.

Hundertmark, T., S. Hering, J. Edelmann, C. Augustin, I. Plate, and R. Szibor. 2008. The STR cluster DXS10148–DXS8378–DXS10135 provides a powerful tool for X-chromosomal haplotyping at Xp22. *Int. J. Legal Med.* 122(6): 489–92.

Hwa, H. L., Y. Y. Chang, J. C. Lee et al. 2009. Thirteen X-chromosomal short tandem repeat loci multiplex data from Taiwanese. *Int. J. Legal Med.* 123(3): 263–9.

Hwa, H. L., J. C. Lee, Y. Y. Chang et al. 2011. Genetic analysis of eight population groups living in Taiwan using a 13 X-chromosomal STR loci multiplex system. *Int. J. Legal Med.* 125(1): 33–7.

Illescas, M. J., J. M. Aznar, S. Cardoso et al. 2011a. Genetic diversity of 10 X-STR markers in a sample population from the region of Murcia in Spain. *Forensic Sci. Int. Genet. Suppl. Ser.* 3: e437–8.

Illescas, M. J., J. M. Aznar, A. Odriozola, D. Celorrio, and M. M. de Pancorbo. 2011b. X-STR admixture analysis of two populations of the Basque Diaspora. *Forensic Sci. Int. Genet. Suppl. Ser.* 3: e441–2.

Illescas, M. J., A. Perez, J. M. Aznar et al. 2012. Population genetic data for 10 X-STR loci in autochthonous Basques from Navarre (Spain). *Forensic Sci. Int. Genet.* 6: e146–8.

Inturri, S., C. Robino, I. Carboni, U. Ricci, and S. Gino. 2011. An Italian Jean Jacques Rousseau: A complex kinship case. *Forensic Sci. Int. Genet. Suppl. Ser.* 3: e520–1.

ISFH. 1991 report concerning recommendations of the DNA commission of the international society for forensic haemogenetics relating to the use of DNA polymorphisms. 1992a. *Forensic Sci. Int.* 52(2): 125–30.

Krawczak, M. 2007. Kinship testing with X-chromosomal markers: Mathematical and statistical issues. *Forensic Sci. Int. Genet.* 1(2): 111–4.

Krenke, B. E., A. Tereba, S. J. Anderson et al. 2002. Validation of a 16-locus fluorescent multiplex system. *J. Forensic Sci.* 47(4): 773–85.

Krenke, B. E., L. Viculis, M. L. Richard et al. 2005. Validation of a male-specific, 12-locus fluorescent short tandem repeat (STR) multiplex. *Forensic Sci. Int.* 148(1): 1–14.

La Spada, A. R., E. M. Wilson, D. B. Lubahn, A. E. Harding, and K. H. Fischbeck. 1991. Androgen receptor gene mutations in X-linked spinal and bulbar muscular atrophy. *Nature* 352(6330): 77–9.

Lee, H. Y., M. J. Park, C. K. Jeong et al. 2004. Genetic characteristics and population study of 4 X-chromosomal STRs in Koreans: Evidence for a null allele at DXS9898. *Int. J. Legal Med.* 118(6): 355–60.

Leibelt, C., B. Budowle, P. Collins et al. 2003. Identification of a D8S1179 primer binding site mutation and the validation of a primer designed to recover null alleles. *Forensic Sci. Int.* 133(3): 220–7.

Levinson, G., and G. A. Gutman. 1987. Slipped-strand mispairing: A major mechanism for DNA sequence evolution. *Mol. Biol. Evol.* 4(3): 203–21.

Li, C., T. Ma, S. Zhao et al. 2011. Development of 11 X-STR loci typing system and genetic analysis in Tibetan and northern Han populations from China. *Int. J. Legal Med.* 125(5): 753–6.

Li, C., J. Xu, S. Zhao et al. 2012. Genetic analysis of the 11 X-STR loci in Uigur population from China. *Forensic Sci. Int. Genet.* 6: e139–40.

Li, H., H. Tang, Q. Zhang, Z. Jiao, J. Bai, and S. Chang. 2009. A multiplex PCR for 4 X chromosome STR markers and population data from Beijing Han ethnic group. *Legal Med. (Tokyo, Japan)* 11(5): 248–50.

Li, L., C. Li, S. Zhang, S. Zhao, Y. Liu, and Y. Lin. 2010. Analysis of 14 highly informative SNP markers on X chromosome by TaqMan SNP genotyping assay. *Forensic Sci. Int. Genet.* 4(5): e145–8.

Lim, E. J., H. Y. Lee, J. E. Sim, W. I. Yang, and K. J. Shin. 2009. Genetic polymorphism and haplotype analysis of 4 tightly linked X-STR duos in Koreans. *Croat. Med. J.* 50(3): 305–12.

Liu, Q. L., D. J. Lu, X. L. Wu, H. Y. Sun, X. Y. Wu, and H. L. Lu. 2008. Development of a five ChX STRs loci typing system. *Int. J. Legal Med.* 122(3): 261–5.

Liu, Q. L., D. J. Lu, X. G. Li et al. 2011. Development of the nine X-STR loci typing system and genetic analysis in three nationality populations from China. *Int. J. Legal Med.* 125(1): 51–8.

Liu, Q. L., D. J. Lu, L. Quan, Y. F. Chen, M. Shen, and H. Zhao. 2012. Development of multiplex PCR system with 15 X-STR loci and genetic analysis in three nationality populations from China. *Electrophoresis* 33(8): 1299–305.

Luczak, S., U. Rogalla, B. A. Malyarchuk, and T. Grzybowski. 2011. Diversity of 15 human X chromosome microsatellite loci in Polish population. *Forensic Sci. Int. Genet.* 5: e71–7.

Luo, H. B., Y. Ye, Y. Y. Wangat et al. 2011. Characteristics of eight X-STR loci for forensic purposes in the Chinese population. *Int. J. Legal Med.* 125(1): 127–31.

Machado, F. B., and E. Medina-Acosta. 2009. Genetic map of human X-linked microsatellites used in forensic practice. *Forensic Sci. Int. Genet.* 3(3): 202–4.

Martinez, R. E., M. L. Bravo, D. Aguirre et al. 2008. Genetic polymorphisms of four X-STR loci: DXS6797, DXS6800, HPRTB and GATA172D05 in a Peruvian population sample. *Forensic Sci. Int. Genet. Suppl. Ser.* 1: 153–4.

Martins, J. A., R. H. A. Silva, A. Freschi, G. G. Paneto, R. N. Oliveira, and R. M. B. Cicarelli. 2008. Population genetic data of five X-chromosomal loci in Bauru (Sao Paulo, Brazil). *Forensic Sci. Int. Genet. Suppl. Ser.* 1: 155–6.

Martins, J. A., J. C. Costa, G. G. Paneto et al. 2009. Genetic data of 10 X-chromosomal loci in Vitoria population (Espırito Santo state, Brazil). *Forensic Sci. Int. Genet. Suppl. Ser.* 2: 394–5.

Martins, J. A., R. H. Silva, A. Freschi, G. G. Paneto, R. N. Oliveira, and R. M. Cicarelli. 2010. X-chromosome genetic variation in Sao Paulo state (Brazil) population. *Ann. Hum. Biol.* 37(4): 598–603.

Massetti, S., E. Carnevali, M. Lancia et al. 2008. Analysis of 8 STR of the X-chromosome in two Italian regions (Umbria and Sardinia). *Forensic Sci. Int. Genet. Suppl. Ser.* 1: 157–9.

Matise, T. C., F. Chen, W. Chen et al. 2007. A second-generation combined linkage physical map of the human genome. *Genome Res.* 17(12): 1783–6.

Mukerjee, S., T. Ghosh, D. Kalpana, M. Mukherjee, and A. K. Sharma. 2010. Genetic variation of 10 X chromosomal STR loci in Indian population. *Int. J. Legal Med.* 124(4): 327–30.

Mulero, J. J., C. W. Chang, L. M. Calandro et al. 2006. Development and validation of the AmpFlSTR® Yfiler® PCR amplification kit: A male specific, single amplification 17 Y-STR multiplex system. *J. Forensic Sci.* 51(1): 64–75.

Mulero, J. J., C. W. Chang, R. E. Lagace et al. 2008. Development and validation of the AmpFlSTR® MiniFiler® PCR amplification kit: A miniSTR multiplex for the analysis of degraded and/or PCR inhibited DNA. *J. Forensic Sci.* 53(4): 838–52.

Nadeem, A., M. E. Babar, M. Hussain, and M. A. Tahir. 2009. Development of pentaplex PCR and genetic analysis of X chromosomal STRs in Punjabi population of Pakistan. *Mol. Biol. Rep.* 36(7): 1671–5.

Nakamura, Y., and K. Minaguchi. 2010. Sixteen X-chromosomal STRs in two octaplex PCRs in Japanese population and development of 15-locus multiplex PCR system. *Int. J. Legal Med.* 124(5): 405–14.

Nishi, T., A. Kurosu, Y. Sugano et al. 2013. Application of a novel multiplex polymerase chain reaction system for 12 X-chromosomal short tandem repeats to a Japanese population study. *Leg. Med. (Tokyo)* 15: 43–46.

Nothnagel, M., R. Szibor, O. Vollrath, C. Augustin, J. Edelman, M. Geppart et al. 2012. Collaborative genetic mapping of 12 forensic short tandem repeat (STR) loci on the human X chromosome. *Forensic Sci. Int. Genet.* 6: 778–84.

Oguzturun, C., C. R. Thacker, and D. Syndercombe Court. 2006. Population study of four X-chromosomal STR loci in the U.K. and Irish population. *Int. Congr. Ser.* 1288: 283–5.

Pasino, S., S. Caratti, M. Del Pero, A. Santovito, C. Torre, and C. Robino. 2011. Allele and haplotype diversity of X-chromosomal STRs in Ivory Coast. *Int. J. Legal Med.* 125(5): 749–52.

Pepinski, W., M. Skawronska, A. Niemcunowicz-Janica, E. Koc-Zorawska, J. Janica, and I. Soltyszewski. 2005. Polymorphism of four X-chromosomal STRs in a Polish population sample. *Forensic Sci. Int.* 151(1): 93–5.

Pepinski, W., A. Niemcunowicz-Janica, M. Skawronska et al. 2006. Polymorphism of four X-chromosomal STRs in a religious minority of old believers residing in northeastern Poland. *Int. Congr. Ser.* 1288: 307–9.

Pepinski, W., A. Niemcunowicz-Janica, M. Skawronska et al. 2007. X-chromosomal polymorphism data for the ethnic minority of polish Tatars and the religious minority of old believers residing in northeastern Poland. *Forensic Sci. Int. Genet.* 1(2): 212–4.

Pereira, R., I. Gomes, A. Amorim, and L. Gusmao. 2007. Genetic diversity of 10 X chromosome STRs in northern Portugal. *Int. J. Legal Med.* 121(3): 192–7.

Pereira, V., C. Tomas, A. Amorim, N. Morling, L. Gusmao, and M. J. Prata. 2010. Study of 25 X-chromosome SNPs in the Portuguese. *Forensic Sci. Int. Genet.* 5: 336–8.

Pico, A., A. Castillo, C. Vargas, A. Amorim, and L. Gusmao. 2008. Genetic profile characterization and segregation analysis of 10 X-STRs in a sample from Santander, Colombia. *Int. J. Legal Med.* 122(4): 347–51.

Pinto, N., L. Gusmao, and A. Amorim. 2010. Likelihood ratios in kinship analysis: Contrasting kinship classes, not genealogies. *Forensic Sci. Int. Genet.* 4(3): 218–9.

Pinto, N., L. Gusmao, and A. Amorim. 2011. X-chromosome markers in kinship testing: A generalisation of the IBD approach identifying situations where their contribution is crucial. *Forensic Sci. Int. Genet.* 5(1): 27–32.

Poetsch, M., H. Petersmann, A. Repenning, and E. Lignitz. 2005. Development of two pentaplex systems with X-chromosomal STR loci and their allele frequencies in a northeast German population. *Forensic Sci. Int.* 155(1): 71–6.

Poetsch, M., A. Sabule, H. Petersmann, V. Volksone, and E. Lignitz. 2006. Population data of 10 X-chromosomal loci in Latvia. *Forensic Sci. Int.* 157(2–3): 206–9.

Poetsch, M., D. El-Mostaqim, F. Tschentscher et al. 2009. Allele frequencies of 11 X-chromosomal loci in a population sample from Ghana. *Int. J. Legal Med.* 123(1): 81–3.

Poetsch, M., A. Knop, D. El-Mostaqim, N. Rakotomavo, and N. von Wurmb-Schwark. 2011. Allele frequencies of 11 X-chromosomal loci of two population samples from Africa. *Int. J. Legal Med.* 125(2): 307–14.

Qiagen. 2010. *Investigator Argus X-12 Handbook.* (April 2010).

Ribeiro Rodrigues, E. M., F. P. Leite, M. H. Hutz, J. Palha Tde, A. K. Ribeiro dos Santos, and S. E. dos Santos. 2008. A multiplex PCR for 11 X chromosome STR markers and population data from a Brazilian Amazon region. *Forensic Sci. Int. Genet.* 2(2): 154–8.

Ribeiro-Rodrigues, E. M., N. P. dos Santos, A. K. dos Santos et al. 2009. Assessing interethnic admixture using an X-linked insertion–deletion multiplex. *Am. J. Hum. Biol.* 21(5): 707–9.

Ribeiro-Rodrigues, E. M., T. D. Palha, E. A. Bittencourt, A. Ribeiro-Dos-Santos, and S. Santos. 2011. Extensive survey of 12 X-STRs reveals genetic heterogeneity among Brazilian populations. *Int. J. Legal Med.* 125(3): 445–52.

Robino, C., A. Giolitti, S. Gino, and C. Torre. 2006. Development of two multiplex PCR systems for the analysis of 12 X-chromosomal STR loci in a northwestern Italian population sample. *Int. J. Legal Med.* 120(5): 315–8.

Rodig, H., F. Kloep, L. Weissbach et al. 2010. Evaluation of seven X-chromosomal short tandem repeat loci located within the Xq26 region. *Forensic Sci. Int. Genet.* 4(3): 194–9.

Rodrigues, E. M., J. Palha Tde, A. K. Santos, and S. E. Dos Santos. 2010. Genetic data of twelve X-STRs in a Japanese immigrant population resident in Brazil. *Forensic Sci. Int. Genet.* 4(2): e57–8.

Samejima, M., Y. Nakamura, P. Nambiar, and K. Minaguchi. 2012. Genetic study of 12 X-STRs in Malay population living in and around Kuala Lumpur using Investigator™ Argus X-12 kit. *Int. J. Legal Med.* 126: 677–83.

Schmidtke, J., W. Kuhnau, D. Wand, J. Edelmann, R. Szibor, and M. Krawczak. 2004. Prenatal exclusion without involving the putative fathers of an incestuous father–daughter parenthood. *Prenatal Diagn.* 24(8): 662–4.

Schneider, P. M. 2007. Scientific standards for studies in forensic genetics. *Forensic Sci. Int.* 165(2–3): 238–43.

Schoske, R., P. M. Vallone, C. M. Ruitberg, and J. M. Butler. 2003. Multiplex PCR design strategy used for the simultaneous amplification of 10 Y chromosome short tandem repeat (STR) loci. *Anal. Bioanal. Chem.* 375(3): 333–43.

Serra, A., A. M. Bento, M. Carvalho et al. 2008. X-chromosome STR typing in deficiency paternity cases. *Forensic Sci. Int. Genet. Suppl. Ser.* 1: 162–3.

Shin, K. J., B. K. Kwon, S. S. Lee et al. 2004. Five highly informative X-chromosomal STRs in Koreans. *Int. J. Legal Med.* 118(1): 37–40.

Silva, D. A., F. S. N. Manta, M. Desiderio, C. Tavares, and E. F. de Carvalho. 2009. Paternity testing involving human remains identification and putative half sister: Usefulness of an X-hexaplex STR markers. *Forensic Sci. Int. Genet. Suppl. Ser.* 2: 230–1.

Silva, F., R. Pereira, L. Gusmao et al. 2010. Genetic profiling of the Azores islands (Portugal): Data from 10 X-chromosome STRs. *Am. J. Hum. Biol.* 22(2): 221–3.

Sun, R., Y. Zhu, F. Zhu et al. 2012. Genetic polymorphisms of 10 X-STR among four ethnic populations in northwest of China. *Mol. Biol. Rep.* 39(4): 4077–81.

Szibor, R. 2007. The X chromosome in forensic science: Past, present and future. In *Molecular Forensics*, ed. R. Rapley and D. Whitehouse, 103–126. West Sussex, England: John Wiley & Sons.

Szibor, R., S. Lautsch, I. Plate, and N. Beck. 2000. Population data on the X chromosome short tandem repeat locus HumHPRTB in two regions of Germany. *J. Forensic Sci.* 45(1): 231–3.

Szibor, R., J. Edelmann, M. T. Zarrabeitia, and J. A. Riancho. 2003a. Sequence structure and population data of the X-linked markers DXS7423 and DXS8377—Clarification of conflicting statements published by two working groups. *Forensic Sci. Int.* 134(1): 72–3.

Szibor, R., M. Krawczak, S. Hering, J. Edelmann, E. Kuhlisch, and D. Krause. 2003b. Use of X-linked markers for forensic purposes. *Int. J. Legal Med.* 117(2): 67–74.

Szibor, R., S. Hering, E. Kuhlisch et al. 2005. Haplotyping of STR cluster DXS6801–DXS6809–DXS6789 on Xq21 provides a powerful tool for kinship testing. *Int. J. Legal Med.* 119(6): 363–9.

Szibor, R., S. Hering, and J. Edelmann. 2006. A new website compiling forensic chromosome X research is now online. *Int. J. Legal Med.* 120(4): 252–4.

Szibor, R., J. Edelmann, S. Hering, I. Gomes, and L. Gusmao. 2009. Nomenclature discrepancies in the HPRTB short tandem repeat. *Int. J. Legal Med.* 123(2): 185–6.

Tabbada, K. A., M. C. De Ungria, L. P. Faustino, D. Athanasiadou, B. Stradmann-Bellinghausen, and P. M. Schneider. 2005. Development of a pentaplex X-chromosomal short tandem repeat typing system and population genetic studies. *Forensic Sci. Int.* 154(2–3): 173–80.

Tang, W. M., and K. Y. To. 2006. Four X-chromosomal STRs and their allele frequencies in a Chinese population. *Forensic Sci. Int.* 162(1–3): 64–5.

Tariq, M. A., O. Ullah, S. A. Riazuddin, and S. Riazuddin. 2008. Allele frequency distribution of 13 X-chromosomal STR loci in Pakistani population. *Int. J. Legal Med.* 122(6): 525–8.

Tariq, M. A., M. F. Sabir, S. A. Riazuddin, and S. Riazuddin. 2009. Haplotype analysis of two X-chromosome STR clusters in the Pakistani population. *Int. J. Legal Med.* 123(1): 85–7.

Tavares, C. C., L. Gusmao, C. Domingues et al. 2008. Population data for six X-chromosome STR loci in a Rio de Janeiro (Brazil) sample: Usefulness in forensic casework. *Forensic Sci. Int. Genet. Suppl. Ser.* 1: 164–6.

Tetzlaff, S., R. Wegener, and I. Lindner. 2012. Population genetic investigation of eight X-chromosomal short tandem repeat loci from a northeast German sample. *Forensic Sci. Int. Genet.* 6: e155–6.

Thiele, K., S. Loffler, J. Loffler et al. 2008. Population data of eight X-chromosomal STR markers in Ewe individuals from Ghana. *Forensic Sci. Int. Genet. Suppl. Ser.* 1: 167–9.

Tie, J., S. Uchigasaki, and S. Oshida. 2010. Genetic polymorphisms of eight X-chromosomal STR loci in the population of Japanese. *Forensic Sci. Int. Genet.* 4(4): e105–8.

Tillmar, A. O. 2012. Population genetic analysis of 12 X-STRs in Swedish population. *Forensic Sci. Int. Genet.* 6(2): e80–1.

Tillmar, A. O., P. Mostad, T. Egeland, B. Lindblom, G. Holmlund, and K. Montelius. 2008. Analysis of linkage and linkage disequilibrium for eight X-STR markers. *Forensic Sci. Int. Genet.* 3(1): 37–41.

Tomas, C., J. J. Sanchez, A. Barbaro et al. 2008. X-chromosome SNP analyses in 11 human Mediterranean populations show a high overall genetic homogeneity except in north-west Africans (Moroccans). *BMC Evol. Biol.* 8: 75.

Tomas, C., J. J. Sanchez, J. A. Castro, C. Borsting, and N. Morling. 2010. Forensic usefulness of a 25 X-chromosome single-nucleotide polymorphism marker set. *Transfusion* 50(10): 2258–65.

Tomas, C., V. Pereira, and N. Morling. 2012. Analysis of 12 X-STRs in Greenlanders, Danes and Somalis using Argus X-12. *Int. J. Legal Med.* 126(1): 121–8.

Toni, C., S. Presciuttini, I. Spinetti, and R. Domenici. 2003. Population data of four X-chromosome markers in Tuscany, and their use in a deficiency paternity case. *Forensic Sci. Int.* 137(2–3): 215–6.

Toni, C., S. Presciuttini, I. Spinetti, A. Rocchi, and R. Domenici. 2006. Usefulness of X-chromosome markers in resolving relationships: Report of a court case involving presumed half sisters. *Int. Congr. Ser.* 1288: 301–3.

Toscanini, U., G. Berardi, A. Gomez, and E. Raimondi. 2009a. X-STRs analysis in paternity testing when the alleged father is related to the biological father. *Forensic Sci. Int. Genet. Suppl. Ser.* 2: 234–5.

Toscanini, U., L. Gusmao, G. Berardi, and E. Raimondi. 2009b. Genetic data of 10 X-STR in two Native American populations of Argentina. *Forensic Sci. Int. Genet. Suppl. Ser.* 2: 405–6.

Tucker, V. C., A. J. Hopwood, C. J. Sprecher et al. 2011. Developmental validation of the PowerPlex® ESI 16 and PowerPlex® ESI 17 systems: STR multiplexes for the new European standard. *Forensic Sci. Int. Genet.* 5: 436–48.

Tucker, V. C., A. J. Hopwood, C. J. Sprecher et al. 2012. Developmental validation of the PowerPlex® ESX 16 and PowerPlex® ESX 17 systems. *Forensic Sci. Int. Genet.* 6: 124–31.

Turrina, S., and D. De Leo. 2004. Population genetic comparisons of three X-chromosomal STRs (DXS7132, DXS7133 and GATA172D05) in north and south Italy. *Int. Cong. Ser.* 1261: 302–4.

Turrina, S., R. Atzei, G. Filippini, and D. De Leo. 2007. Development and forensic validation of a new multiplex PCR assay with 12 X-chromosomal short tandem repeats. *Forensic Sci. Int. Genet.* 1(2): 201–4.

Turrina, S., G. Filippini, and D. De Leo. 2009. Genetic studies of eight X-STRs in a northeast Italian population. *Forensic Sci. Int. Genet. Suppl. Ser.* 2: 396–7.

Turrina, S., G. Filippini, and D. De Leo. 2011. Population genetic evaluation of 12 X-chromosomal short tandem repeats of Investigator™ Argus X-12 kit in north-east Italy. *Forensic Sci. Int. Genet. Suppl. Ser.* 3: e327–8.

Valente, C., I. Gomes, V. Pereira, A. Amorim, L. Gusmao, and M. J. Prata. 2009. Association between STRs from the X chromosome in a sample of Portuguese gypsies. *Forensic Sci. Int. Genet. Suppl. Ser.* 2: 391–3.

van Asch, B., R. Pinheiro, R. Pereira et al. 2010. A framework for the development of STR genotyping in domestic animal species: Characterization and population study of 12 canine X-chromosome loci. *Electrophoresis* 31(2): 303–8.

Weissenbach, J., G. Gyapay, C. Dib et al. 1992. A second-generation linkage map of the human genome. *Nature* 359(6398): 794–801.

Wiegand, P., and M. Kleiber. 2001. Less is more—length reduction of STR amplicons using redesigned primers. *Int. J. Legal Med.* 114(4–5): 285–7.

Wiegand, P., B. Berger, J. Edelmann, and W. Parson. 2003. Population genetic comparisons of three X-chromosomal STRs. *Int. J. Legal Med.* 117(1): 62–5.

Wu, W., H. Hao, Q. Liu, Y. Su, X. Zheng, and D. Lu. 2009. Allele frequencies of seven X-linked STR loci in Chinese Han population from Zhejiang province. *Forensic Sci. Int. Genet.* 4(1): e41–2.

Yoo, S. Y., N. S. Cho, S. W. Park et al. 2011. Genetic polymorphisms of eight X-STR loci of Mentype® Argus X-8 kit in Koreans. *Forensic Sci. Int. Genet. Suppl. Ser.* 3: e33–4.

Zalan, A., A. Volgyi, M. Jung, O. Peterman, and H. Pamjav. 2007. Hungarian population data of four X-linked markers: DXS8378, DXS7132, HPRTB, and DXS7423. *Int. J. Legal Med.* 121(1): 74–7.

Zalan, A., A. Volgyi, W. Brabetz, D. Schleinitz, and H. Pamjav. 2008. Hungarian population data of eight X-linked markers in four linkage groups. *Forensic Sci. Int.* 175(1): 73–8.

Zarrabeitia, M. T., T. Amigo, C. Sanudo, M. M. de Pancorbo, and J. A. Riancho. 2002a. Sequence structure and population data of two X-linked markers: DXS7423 and DXS8377. *Int. J. Legal Med.* 116(6): 368–71.

Zarrabeitia, M. T., T. Amigo, C. Sanudo, A. Zarrabeitia, D. Gonzalez-Lamuno, and J. A. Riancho. 2002b. A new pentaplex system to study short tandem repeat markers of forensic interest on X chromosome. *Forensic Sci. Int.* 129(2): 85–9.

Zarrabeitia, M. T., A. Alonso, A. Zarrabeitia, A. Castro, I. Fernandez, and M. M. de Pancorbo. 2004. X-linked microsatellites in two northern Spain populations. *Forensic Sci. Int.* 145(1): 57–9.

Zeng, X., A. Rakha, and S. Li. 2009. Genetic polymorphisms of 10 X-chromosome STR loci in Chinese Daur ethnic minority group. *Legal Med. (Tokyo, Japan)* 11(3): 152–4.

Zeng, X. P., Z. Ren, X. J. Wu et al. 2011a. Development of a 12-plex X chromosomal STR loci typing system. *Forensic Sci. Int. Genet. Suppl. Ser.* 3: e365–6.

Zeng, X. P., Z. Ren, J. D. Chen et al. 2011b. Genetic polymorphisms of twelve X-chromosomal STR loci in Chinese Han population from Guangdong province. *Forensic Sci. Int. Genet.* 5(4): e114–6.

Zhang, S. H., C. T. Li, S. M. Zhao, and L. Li. 2011. Genetic polymorphism of eight X-linked STRs of Mentype® Argus X-8 kit in Chinese population from Shanghai. *Forensic Sci. Int. Genet.* 5(1): e21–4.

Low Copy Number DNA Profiling

6

THERESA CARAGINE
KRISTA CURRIE
CRAIG O'CONNOR

Contents

6.1	Introduction	171
6.2	Forensic Applications of LT-DNA	173
6.3	Technical Considerations	174
6.4	LT-DNA Procedural Modifications	175
6.5	Statistical Considerations	181
6.6	Additional Considerations	183
References		184

6.1 Introduction

Low template (LT) DNA testing is a reliable state-of-the-art technology used to recover and detect small amounts of DNA. Given the wide variety of potential evidentiary samples, this analysis is a powerful tool that can enhance law enforcement's ability to identify or exclude individuals suspected of crimes. Samples that are suitable for LT-DNA testing are those that are of low quantity or quality. Accordingly, LT-DNA is also known as low level, low copy number, high sensitivity, trace, or "touch" DNA (Gill et al. 2000; Caddy et al. 2008; Gill et al. 2001; Balding and Buckleton 2009). The latter term refers to the predominant source of LT-DNA samples, handled items, on which only a few cells may be deposited. Sometimes categorized as a way to test the very smallest amounts of DNA (less than 100 pg, approximately less than 16 diploid cells), LT-DNA testing may exhibit increased stochastic effects.

LT-DNA analysis has been used for many decades and is certainly not a new or unique technique (Hahn et al. 2000; Handyside et al. 1989; Holzgreve 1997; Cornelison and Wold 1997). Different applications of LT-DNA have been used in many non-forensic disciplines such as clinical diagnostics, oncology, and immunology. For example, as early as 1989, a single-cell polymerase chain reaction (PCR) technique was employed to diagnose chromosomal abnormalities and determine the sex of the embryo (Sekizawa et al. 1996a,b). Preimplantation genetic diagnostics was a significant improvement to other detection methods available later in pregnancy such as amniocentesis and chorionic villus sampling (Cheung et al. 1996). In addition, single-cell reverse transcription PCR was used to check for other chromosomal abnormalities such as transcriptional failures and/or gene expression deficits that may lead to cancerous regions (Liss 2002).

Advancements in sequencing technologies have also benefitted the analysis of LT-DNA. Recently, researchers out of Austria isolated single tumor cells (lymphocytes, polar bodies, oocytes, colorectal, renal, breast, and blood cancer cells) with microdissection and micromanipulation techniques to perform whole genome amplification screening for chromosomal abnormalities (Geigl et al. 2007; Frumkin et al. 2008). Examining such low amounts of DNA allows the analyzer to pinpoint problematic areas that may be hidden when dealing with larger amounts of DNA.

LT-DNA testing was first used as a method for human identification by Taberlet et al. (1996) (Outbox 6.1) and by Findlay et al. (1997), who reported genotyping single cells using short tandem repeats (STR). Findlay amplified "usable" profiles (four or more STRs out of six) a vast majority of the time, showing that it was possible to analyze even the smallest amounts of DNA. In 1999, the Forensic Science Service (FSS) in Great Britain pioneered LT-DNA testing in forensic casework. This provided an alternative to mitochondrial DNA (mtDNA) analysis, which because of its matrilineal inheritance, lacks the individualization ability that nuclear DNA markers afford. Because LT-DNA testing is so sensitive, methods to assist in the interpretation of the results were integral to the system's framework (Gill et al. 2000; Gill and Buckleton 2010).

OUTBOX 6.1

In order to reliably genotype very small amounts of DNA, French researcher Pierre Taberlet and colleagues established two complementary methods. They developed a mathematical simulation of exactly how stochastic events would affect the amplification and eventual genotyping of very few picograms of DNA. The second step of their research used laboratory experiments to see if they correlated to the model and also to evaluate the frequency of any PCR-generated artifacts.

The mathematical model consisted of 100 samples with starting templates ranging from 0.01 to 10 diploid genomes and simulated PCR 1000 times. This examined the stochastic effects during PCR taking into account the random sampling of template amounts and assumed that even one molecule of DNA is amplifiable. In a similar manner, experimental procedures used diluted template DNA to see any concordance to the simulation. The results showed what would be expected intuitively; it is more probable that full genotypes will be generated with higher amounts of template DNA. In addition, as the amount of template DNA decreased, usable and even full profiles were still obtained.

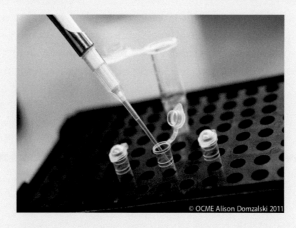

These studies demonstrated that low amounts of DNA, less than 100 pg, can produce reliable results with a 99% confidence level. This research was groundbreaking because it not only showed the ability to gain usable information from the smallest amounts of template DNA but also offered a number of guidelines to employ when performing these experiments such as multiple tube amplifications to account for sporadic events. Typically, DNA extracts are divided into two or three equal amounts for replicate amplifications.

6.2 Forensic Applications of LT-DNA

LT-DNA testing has the potential to help solve crimes and has been used in a wide variety of case types including homicides, assaults/sexual assaults, property crimes, arsons, hate crimes, missing persons investigations, and the identification of human remains for current casework, cold casework, and post-conviction cases (Outbox 6.2). By definition, LT-DNA analysis typically refers to the examination of low amounts of DNA and is a more sensitive way of doing traditional DNA analysis.

OUTBOX 6.2

POST CONVICTION EXONERATION

In November of 1968, the body of a 13-year-old girl was found near railroad tracks in a town in New Jersey. The girl had been raped and killed. She had gone missing the night before her body was discovered. There were several suspects in 1968; however, the case grew cold. The crime scene evidence, which included a white slip the young girl was wearing when she was killed, was put in storage.

Nearly four decades later, crime scene evidence, including the white slip, was submitted to the Department of Forensic Biology at the Office of Chief Medical Examiner of the City of New York for DNA testing. A DNA profile, generated with low template testing, was determined from a semen stain on the white slip, and it did not match an individual recently incarcerated for this crime. Rather, the DNA profile resulted in a database hit to an individual who had been one of the original suspects in 1968.

DATABASE HIT TO A SAMPLE FROM A TRIGGER OF A GUN

In 2006, a man at a nightclub in New York City pointed a gun at two other men and threatened to start shooting. Two on-duty police officers responded to the scene and were shot at by the suspect. The police officers returned fire hitting the suspect who survived. Although a loaded gun was recovered at the scene near the suspect, he claimed that he was an innocent bystander. A swab taken from the trigger and the grips of the gun was subjected to low template DNA testing and generated a male DNA profile. This DNA profile was uploaded to the New York State DNA database and matched the suspect who was found guilty at trial.

> **SAMPLE FROM AN AIRBAG**
>
> In 2008, two cars were racing through streets in New York City. One of the cars crashed into another car from which a 5-year-old boy was ejected and died as a result of his injuries. The person driving the vehicle responsible for the crash did not own the car and fled the scene on foot. The deployed air bag from the racing vehicle was submitted for DNA testing. A DNA profile, generated with low template testing, was determined and it matched the suspect who was identified and found guilty at trial.

This increased sensitivity is needed for items that are handled briefly and have lower amounts of DNA as opposed to biological fluids (blood, semen, saliva) that typically contain a much higher number of cells and therefore more DNA. These items include tools, clothing, jewelry, ligatures, cars, airbags, door knobs, windowsills, keys, lighters, matches, letters, envelopes, writing utensils, bottles and glasses, weapons, explosive devices, and fingerprints (fresh and archived) (Balogh et al. 2003; Bohnert et al. 2001; Horsman-Hall et al. 2009; Pizzamiglio et al. 2004; Polley et al. 2006; Rutty 2002; Sewell et al. 2008; van Hoofstat et al. 1999; van Oorschot and Jones 1987; van Oorschot et al. 1999; Van Renterghem et al. 2000)—basically anything that can be touched or handled for a length of time wherein only a limited number of cells are deposited. In addition, old, degraded, or compromised body fluid samples may also be candidates for LT-DNA testing. For example, charred and partial human remains of deceased individuals from fires and mass disasters may not yield sufficient DNA for routine testing (Whitaker et al. 1995; Budimlija et al. 2003).

A number of factors may contribute to the overall success of recovering DNA from a sample (Roeder et al. 2009). Some factors include the type of surface that was handled (porous material adhere cells better than nonporous material), the amount of time the object was handled, and the environmental conditions the item or the body fluid was subjected to before sampling (Kisilevsky and Wickenheiser 1999; Lowe et al. 2003; Oz et al. 1999). Biological fluids that contain higher amounts of DNA tend to be visible, and therefore more easily targeted for sampling. However, since skin cells and minute amounts of blood or saliva cells, for example, are invisible to the naked eye, it is helpful to know the exact areas for sampling in order to maximize the potential for DNA recovery.

6.3 Technical Considerations

Some laboratories tend to limit the term LT-DNA to any analysis that is below 200 or 100 pg, but in general most will agree that LT-DNA analysis can be applied to any technique that will exhibit increased stochastic effects (Caddy et al. 2008; Budowle et al. 2009). Difficulty lies in trying to agree on a general threshold for analysis since stochastic effects are viewed even with larger starting amounts of DNA and in mixtures. Nevertheless, these general thresholds serve as a starting point for analysis and can be used to triage samples.

Allelic drop-in and drop-out, heterozygous imbalance, and exaggerated stutter are representative random effects that must be accommodated in LT-DNA analysis (Whitaker et al. 2001; Lucy et al. 2007). For example, because of the sensitivity of the system, an allele, termed a drop-in allele, may be detected from DNA that was deposited after the sample was collected. Since only a small number of genomic templates are present in an LT-DNA sample tube, the primers for each allele at each locus may not collide and bind to their respective templates with the same efficiencies in the first few PCR cycles. This results in imbalance between heterozygous alleles or, in extreme cases, if this reaction never occurs, a total loss of one or both alleles, a phenomenon known as allelic drop-out (Budowle et al. 2001; Caragine et al. 2009).

Similarly, since there are so few copies of the original DNA template, if the processing enzyme slips off this template or the extending strand during an early cycle creating a stutter peak, the ratio of the height of the stutter peak to the true peak may be higher than that for a stutter product of high template (HT) DNA. Using traditional interpretation protocols, an allele from stutter or exogenous DNA may be perceived as a true allele; conversely, the absence of an allele, may lead to false homozygous determinations.

6.4 LT-DNA Procedural Modifications

LT-DNA testing is not a new technique as the instruments and general procedures used are exactly the same as those used in HT DNA testing with some modifications to account for the increased sensitivity. However, during the entire process, from evidence examination all the way through analysis, certain precautions should be implemented. Laboratory space is a major consideration, and efforts should be made to minimize potential areas of contamination (Caddy et al. 2008).

Most institutions that work with low levels of DNA tend to have a separate area to examine items and to perform DNA extractions. To ensure that no exogenous DNA taints the laboratory spaces, areas are sanitized on a regular basis and kept free of any amplified DNA. Personnel that work in these areas don proper protective attire such as laboratory coats, gloves, hair nets, face shields, and/or foot coverings to limit the chances of cross contamination. A common method of cleaning involves the use of 10% bleach followed by deionized water and 70% ethanol to sterilize the working areas and tools (Shaw et al. 2008; Gefridese et al. 2010). Exogenous DNA can be minimized on tubes and other plastic ware with UV irradiation or ethylene oxide fuming (Stratalinker® UV Crosslinker Instruction Manual; Tamariz et al. 2005).

Optimally, evidence should be collected and preserved using similar precautions (Outbox 6.3). If this does not occur, there is a risk that the evidence collector's DNA may mask that of the assailant if DNA is recovered. The ability to compare the DNA profiles of laboratory members and evidence collectors could help identify these scenarios. If, however, evidence is collected in a manner not consistent with current recommended practices, finding the assailant's DNA may still be significant depending on the context of the case.

OUTBOX 6.3   THE QUEEN VS. SEAN HOEY CASE (OMAGH BOMBING)

A notable case in the history of forensic LT-DNA testing is the case of the Queen vs. Sean Hoey, commonly referred to as the Omagh bombing case. Sean Hoey, an electrician, was arrested and charged with 58 counts associated with murder, causing explosions, and other related offences resulting from 13 bomb and mortar attacks. The attacks, which occurred between 1998 and 2001 in Northern Ireland and London, included a car bomb that exploded in a busy shopping area in the District of Omagh in Northern Ireland. The explosion killed 29 and injured hundreds of people, and was the deadliest event in 30 years of violence in Northern Ireland (Judgment: the *Queen v Sean Hoey*).

The prosecution built their case on evidence that they believed to show a connection between Sean Hoey and evidence recovered at the scenes of the attacks. The evidence was examined in two stages. The initial examination included reconstruction of the bombs and fiber comparisons. The evidence was then reexamined in 2000–2001 at the Forensic Science Service (FSS) using LT-DNA testing. DNA recovered from undetonated devices yielded the same DNA profile as that of Sean Hoey, and in September and November of 2003, more DNA matches were made to additional bombing devices. However, no association was made between Sean Hoey and the DNA from the detonated device used in the Omagh bombing.

In 1998 when the evidence was first collected, little precaution was taken to maintain the integrity of the evidence by minimizing the risk of cross-contamination among items. The presiding judge, Justice Weir, stated that the "freedom from possible contamination of each item throughout the entirety of the period between seizure and any examination" must be established (48). Regarding LT-DNA testing, as per Justice Weir, the LCN system had not been sufficiently validated by the international scientific community. Based on all the evidence presented at trial, Sean Hoey was found not guilty of all charges.

As a result of this case, the Crown Prosecution Service (CPS) temporarily suspended LT-DNA testing in the U.K. for three weeks during which current FSS LT-DNA cases were reviewed. The CPS concluded that there is "…no inherent unreliability in the LCN DNA analysis process *provided that it is carried out according to the prescribed processes, and that the results are properly interpreted*…" (CPS Press Release 2008).

Once properly equipped, forensic analysts then attempt to recover the maximum amount of DNA possible from the evidence using efficient sampling techniques. A swabbing or scraping technique (depending on the item), or tape lifts can be used as illustrated in Figure 6.1. Additional modifications include a specialized swab or swabbing solution, or type of tape (Caragine et al. 2009; Prinz et al. 2006; Hansson et al. 2009; Barash et al. 2010). After the sample is collected, more sensitive DNA extraction techniques featuring single tube extractions, automatic concentration of samples, and the use of commercial DNA extractions kits can be utilized (Greenspoon et al. 1998; Schiffner et al. 2005).

To account for the possibility of increased stochastic effects on analysis, modifications to the procedures are used throughout the testing process. The most critical of these modifications is viewed at the amplification step. Since PCR exponentially copies the DNA with each cycle, adding extra cycles will help to increase the overall amount of DNA amplified

Figure 6.1 The scraping technique is used to sample areas of material such as cotton using a sterile razor blade. The blade effectively removes skin cells from fabric, and the scrapings are directly digested.

(Rameckers et al. 1997). Post-PCR procedures are can also be modified to aid in the analysis of LT-DNA. Filtration of samples after the amplification step can be used to filter out unused primers, dye blobs, or other ions that can compete with the DNA during electrophorectic injection (Forster et al. 2008; Smith and Ballantyne 2007). After separation of the PCR products, careful interpretation of the results are performed to ensure that the overall conclusions from the sample are accurate. Replicate analysis, that is, redundant amplification, is used to verify that the alleles seen and subsequently genotyped are not stochastic artifacts (Outbox 6.4).

In order for forensic practitioners to ensure the accuracy of their results and to show that the results are reliable, robust, and reproducible, LT-DNA procedures must be extensively validated. An important element for validation is establishing thresholds. If quantitative thresholds are used, the accuracy of the quantitation system should be assessed by comparing the measured values of samples with their known concentrations. Samples with known genotypes should be amplified with a range of concentrations. The template amount for which allelic dropout and drop-in and intralocus peak imbalance are increased may define the threshold for LT-DNA.

Thresholds may also be applied to negative controls, and are based on the detection of spurious alleles. If no pattern is apparent among these alleles, then one can assume that they are random and likely originated from a single molecule. Therefore, these alleles will not affect other samples.

Criteria or thresholds for allelic assignment should also be determined and then verified through validation. Guidelines should consider intralocus peak imbalance, stutter peak heights, and the propensity for alleles to drop-out and drop-in as well as the size and the amplification efficiency of individual loci. To ensure reliability and support each component of the validation, experiments should be repeated. The efficacy of validations was evaluated in several court cases (Outboxes 6.5 and 6.6).

OUTBOX 6.4

Due to the small amount of DNA present and the enhanced sensitivity of the methods used to detect such DNA, stochastic effects may result. In order to accommodate these phenomena and to ensure accuracy and reliability, testing and interpretation protocols should be modified. This is commonly achieved through repeated amplifications. The extracted DNA is divided into equal amounts such that the same amount of DNA is amplified in each replicate. The number of repetitions may vary from 2 to 4 (Gill et al. 2000, 2010; Balding et al. 2009; Butler et al. 2010; Caragine et al. 2009; Benschop et al. 2011), with the salient feature being that the alleles that repeat are considered "confirmed." This ensures that the alleles seen and used for profile allele assignment are not spurious alleles.

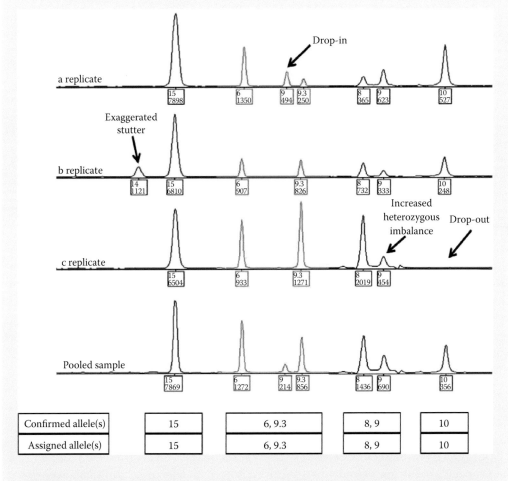

The first figure shows a single source (from one person) DNA profile. The confirmed alleles are shown at the bottom, as well as the alleles that were assigned to the profile based on validated interpretation protocols regarding peak imbalance and the amplification efficiency of the loci. For example, even though there is severe heterozygous

Low Copy Number DNA Profiling

imbalance in replicate c at the third locus, the 8 and 9 alleles may be assigned to the profile based on the data overall. Other stochastic effects such as drop-in, drop-out, and exaggerated stutter are apparent in this sample, although they are addressed by the interpretation protocols to ensure accurate allele assignment.

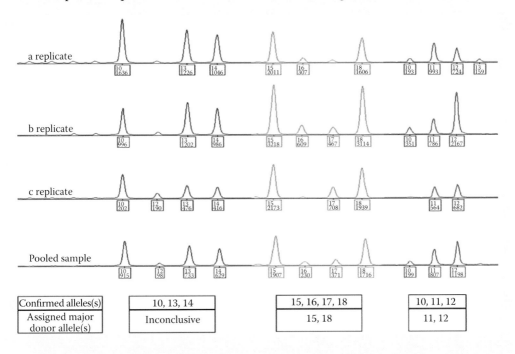

The second figure shows a mixture (two or more contributors) that can be partially deconvoluted. Notice that, at the first locus, the result is inconclusive or cannot be determined as the two tallest alleles are not apparent. However, alleles can be assigned at the other two loci. Guidelines to deduce contributor profiles must be verified extensively for accurate, robust, reproducible, and most importantly reliable results.

OUTBOX 6.5   QUEENS, NEW YORK FRYE HEARING

A seminal court case that highlights the use of LT-DNA testing in forensic casework is the case of the People of the State of New York vs. Hemant Megnath, presided by Justice Robert J. Hanophy of the Supreme Court, Queens County. This case is the first fully completed judgment regarding the use of LT-DNA autosomal testing in the United States.

The case involves a young woman who allegedly met a real estate agent named Hemant Megnath in 2005 when he was helping her find an apartment. In 2006, she came forward accusing Megnath of rape. He was arrested and released on bail, and was due back in court for a pretrial hearing at a later date. While out on bail in 2007, to prevent testimony at trial, Megnath allegedly slit her throat as she was leaving for work, fled in his automobile, and left her to die on the ground in the backyard of her home.

A search warrant was executed upon Megnath's automobile. Various samples of DNA evidence were recovered and tested by the New York City Office of Chief Medical Examiner using both high template (HT) and low template (LT) DNA testing as appropriate. Several samples linked Megnath to the murder including a car door swab (HT-DNA) that could be matched to the victim and a bloodstain on the seatbelt (LT-DNA) to which the victim could be a contributor.

In light of the DNA evidence as well as additional evidence, Megnath was charged with murder in the first degree and several other offenses. Prior to the murder trial, the defense requested a Frye hearing for the LT-DNA testing performed on evidence in the case. The *Frye* standard or *Frye* test evaluates whether a technique is generally accepted in the relevant scientific community and thus determines whether evidence tested with this procedure is admissible in a court of law. The court granted the defendant's request for a Frye hearing, which began in November of 2008. Following the testimony of eight expert witnesses called to testify by either the prosecution or the defense, and the review of written arguments, Judge Hanophy issued a final decision in February 2010.

Justice Robert J. Hanophy found that "the credible evidence presented at the hearing established that the OCME validation studies regarding LCN DNA typing yielded reliable and reproducible results in 100% of the samples tested. The court finds that the OCME has properly developed interpretation protocols for LCN DNA testing based upon its extensive validation studies and that when correctly performed these protocols consistently yield reliable and reproducible results… Therefore, based upon all of the evidence presented to the court during the hearing, the court finds that the People have demonstrated by ample credible evidence that LCN DNA testing with its increased amplification cycles as performed by the New York City OCME clearly passes the standard enunciated in *Frye* and is therefore admissible at trial…."

"Moreover, in addition to holding that the *Frye* standard of reliability has been met in this case, the court also finds that the standard enunciated in *Frye* pertains only to *novel* scientific techniques… the court finds that LCN DNA testing as performed by the OCME is not a *novel* scientific technique for the purposes of the *Frye* inquiry. …Indeed, both the LCN and HCN forms of DNA testing require the same steps to be taken. These steps, namely extraction, quantitation, amplification, and electrophoresis, are virtually identical in both HCN and LCN DNA testing. Similarly, the same issues such as stutter, allelic drop-out or drop-in occur in both forms of testing as well. In fact, the OCME has prepared and followed interpretation protocols for both HCN and LCN DNA testing to compensate for these scientific phenomena when and if they occur. These protocols were developed by the OCME based upon its validation studies and based upon similar protocols that have been used globally by other forensic scientists who perform HCN and LCN DNA testing…."

At trial, all evidence was presented, and Hemant Megnath was subsequently found guilty. He was sentenced to life in prison without parole. Although another Frye hearing was granted in New York County, the defendant pled guilty after four hours of Frye testimony from the prosecution's first witness. Three additional requests for Frye hearings were denied (*NY v Megnath* 2010).

OUTBOX 6.6   R V MICHAEL SCOTT WALLACE

On September 20, 2005, Michael Scott Wallace picked up a German tourist as she hitchhiked in New Zealand. He drove her to an area southwest of New Plymouth where he bludgeoned her with a metal bar, dragged her bleeding body into the bush, and stomped on her neck. As he was ripping off her shoes and socks, and unbuttoning her jeans, he was spooked by a passing car. He then picked up a knife and plunged it through her heart.

The woman's body was found later that day by a jogger who saw drag marks leading into the bush. Following an 18-day manhunt, Wallace was arrested for her murder. In 2007, he was convicted and sentenced to life in prison. He maintained his innocence.

The case was appealed on the grounds of inadequate representation. In addition, the strength of the fingerprint and DNA evidence, some of which were tested with LT-DNA techniques, was contested. The appeal was dismissed, and the court ruled that "LCN DNA evidence is not inherently unreliable" (*Wallace v Queen* 2010).

6.5   Statistical Considerations

The comparison of known DNA profiles to evidence samples containing small amounts of DNA or degraded DNA can be challenging, as many of the results produce mixtures and/or partial DNA profiles. Alleles from known contributors may be absent or, conversely, extraneous alleles that cannot be attributed to known contributors may be present. The biological model or the consensus approach described previously assigns repeating alleles from replicate amplifications. Other characteristics such as heterozygote imbalance, the amplification efficiency of the locus and kit, the propensity for stutter, and the amount of DNA amplified are also considered in order to determine or deduce a DNA profile from a mixture. If a DNA profile is so cautiously determined, statistical calculations frequently used in forensic laboratories such as the random match probability may be used.

Regarding mixtures or components of mixtures that cannot be deconvoluted, although criteria for qualitative comparisons may be delineated, modifications to common quantitative practices must be made. Specifically, the probability of drop-out and drop-in should be included in the statistical formulation. Peter Gill, John Buckleton, and colleagues at the FSS first suggested this statistical approach, which requires software using a likelihood ratio (LR) framework to incorporate these probabilities as well as the source genotypes to evaluate the strength of the results (Outbox 6.7).

OUTBOX 6.7   MODIFICATIONS TO LIKELIHOOD RATIOS FOR LT-DNA SAMPLES

In a conventional LR framework, the probability of the crime scene profile (G) is computed under two competing hypotheses: that of the prosecution (H_p) and that of the defense (H_d), which respectively state that the profile belongs to the suspect or to an unreleated individual. The LR is the ratio of these two probabilities:

$$LR = \frac{\Pr(G|H_p)}{\Pr(G|H_d)}$$

If the crime scene profile matches the suspect profile, then $\Pr(G|H_d)$ and $\Pr(G|H_d) = P$, where P is the estimated population frequency of the crime scene profile (i.e., the random match probability). Thus, in this scenario, if the evidence profile matches the suspect profile:

$$LR = \frac{1}{P}$$

When a crime scene profile reflects a mixture, the conventional LR can also be used by specifying H_p and H_d. In order to incorporate the probabilities of drop-out and drop-in, multiple replicate amplifications of an evidentiary sample can be considered. Let $\mathbf{R} = R_1, R_2, \ldots, R_n$ represent the alleles observed in amplification replicates 1 through n at a single locus in an evidentiary sample. The replicate data are used to compute

$$LR = \frac{\Pr(\mathbf{R}|H_p)}{\Pr(\mathbf{R}|H_d)}.$$

If H_p: S and H_d: U,

$$LR = \frac{\Pr(R|S)}{\sum_{j=1}^{m} \Pr(|U=G_j)\Pr(U=G_j)},$$

where S represents the suspect's alleles, U represents the alleles of an unknown contributor, and G_j represents the jth possibility for the genotype of the unknown contributor.

A critical step in incorporating allelic drop-out and drop-in to the modified LR framework will be to consider all possible genotypes for the unknown contributor in the denominator. If x distinct alleles are observed in the evidentiary profile at a locus, there will be m values comprising the set of possible genotypes of the unknown contributor, where:

$$m = \frac{x(x+1)}{2} + x + 1$$

This calculation treats all unobserved alleles as a single "other" allele. That is, the unknown contributor's genotype at the locus could include any pairwise combination of the observed alleles and the unobserved other allele. $\Pr(U = G_j)$ is the expected population frequency of G_j. Drop-out and drop-in rates can be incorporated into $\Pr(R_1,R_2,\ldots,R_n|S)$ and $\Pr(R_1,R_2,\ldots,R_n|U = G_j)$. The likelihood ratios for individual loci can be multiplied together to generate a single, composite LR for the profile comparison.

Such a method for LR analysis of mixed LT or degraded DNA profiles was first published in 2005 (Curran et al. 2005). Although a software implementation of this method, named LoComatioN, was published, to date, it has not been implemented in forensic casework (Gill et al. 2007). LoComatioN uses only two parameters to describe allelic drop-out and one parameter to describe drop-in. As these phenomena are a function of several factors (the starting quantity of DNA, the STR locus, the amplification kit, and the source genotype), this is limiting. In 2009, Tvedebrink et al. described an approach for estimating the probability of allelic dropout, and in 2010, David Balding published free software that computes single-locus likelihood ratios allowing for dropout of one or both alleles and drop-in of up to two alleles.

Similarly, the Office of the Chief Medical Examiner of the City of New York designed and validated the software termed the Forensic Statistical Tool (FST), which incorporates the probability of drop-out and drop-in in likelihood ratio calculations for samples containing one to four contributors. Like Balding's software, locus-specific partial drop-out of heterozygotes, complete drop-out of heterozygotes, and complete drop-out of homozygotes are considered. The drop-out rates used by FST, however, were estimated empirically as a function of the locus, the quantity of template DNA, the number of contributors to the sample, and the mixture ratio (i.e., deducible or nondeducible). Drop-in rates were estimated separately for LT and HT amplification (Mitchell et al. 2011). A θ correction for intraindividual (but not interindividual) population substructure, applied to homozygous genotypes, was included as recommended by the National Research Council (1996).

FST is the first such software that was validated for use in forensic casework. Nearly 500 LR values were generated for two- and three-person degraded and nondegraded purposeful mixtures and mixtures generated from touched items, the majority of which appeared degraded. The suspect pool tested for each sample included the DNA donors to the purposeful mixture or the individuals who handled the touched item, as well as over 1200 noncontributors. Overall, the LR results were consistent with manual comparisons, and good separation between the LRs of the true and the noncontributors was observed for all scenarios. The validation demonstrated that the FST program is a useful tool to apply a quantitative weight to otherwise qualitative assessments (Tvedebrink et al. 2009).

6.6 Additional Considerations

The weight of the association of any DNA profile recovered from an evidentiary item must be assessed in terms of the context of the case. The recovered DNA may have been deposited before or after the commission of the crime and have no relevance to the case. (If DNA is transferred during or after evidence collection, this is considered contamination.) How and when the DNA was deposited on an item cannot be determined. Moreover, with LT-DNA, it is often difficult to surmise the type of biological material that is the source of the DNA.

DNA may be transferred through direct or indirect contact. Casual contact such as hand shaking has been shown to be a vector for transfer of DNA (Farmen et al. 2008). If the source of the DNA is skin cells, the amount of transfer depends on the propensity of a person to shed, which is very variable. Hand washing also removes loose skin cells, thus minimizing transfer. The length of time of contact is also a factor (Lowe et al. 2003; van Oorschot and Jones 1987).

Once deposited, DNA from the first person who handled an item may not be detected if the object is subsequently handled by other individuals as the DNA of the original person

> **OUTBOX 6.8**   *R V REED AND ANOR* (2009)
>
> The Reed brothers were convicted of murder in August 2007. LT-DNA testing was performed on two pieces of plastic fragments found at the murder scene. They appealed their case in 2009. During the course of the appeal, the defense redirected their focus from the reliability of LT-DNA testing to the testimony of the forensic scientist regarding how the DNA was deposited on the plastic fragments.
>
> It was alleged that David and Terence Reed shook hands with persons who then transferred their DNA to the evidence. The scientist "regarded the possibility of secondary transfer as 'unrealistic'. There would have had to have been 'substantial contact' between David and Terence Reed and those other persons and those other persons would have had to have been in contact with the knives 'pretty quickly' for the cellular material from David and Terence Reed to have got onto the knives." Furthermore, as Terrance Reed was the only source of one of the samples, the scientist purported that in her experience, it was more likely that the DNA was deposited through primary transfer than secondary or tertiary transfer, which typically show mixtures. However, she clearly stated that the possibility of secondary transfer could not be eliminated.
>
> The court ruled the testimony was admissible and the conviction was upheld (*R v Reed and Anor* 2009).

may adhere to their hand(s) and be removed from the object. If these individuals then handle another item they may theoretically deposit small amounts of the first person's DNA as well as their own DNA. However, studies have shown that the original person could not be associated to the second object (Butler and Hill 2010; Tamariz 2009). This suggests that tertiary transfer is highly unlikely. The admissibility of an expert's opinion regarding modes of transfer was the subject of a notable appeals case in the United Kingdom (Outbox 6.8).

LT DNA analysis is a key component in the ever-improving landscape of sample individualization, and has proven to be a powerful tool in the criminal justice community when considered in the context of the case. Evidence that was formerly unusable may now yield informative results. To accommodate stochastic events and to ensure that the interpretation of the results is accurate and reliable, careful and thorough validation must be performed. Incorporation of the appropriate statistical model to this testing further enhances the capability of investigations.

References

Balding, D.J., and J. Buckleton. 2009. Interpreting low template DNA profiles. *Forensic Sci Int Genet* 4:1–10.

Balogh, M.K., J. Burger, and K. Bender. 2003. STR genotyping and mtDNA sequencing of latent fingerprint on paper. *Forensic Sci Int* 137:188–195.

Barash, M., A. Reshef, and P. Brauner. 2010. The use of adhesive tape for recovery of DNA from crime scene items. *J Forensic Sci* 55:1058–1064.

Benschop, C.C.G., C.P. van der Beek, and H.C. Meiland. 2011. Low template STR typing: Effect of replication number and consensus method on genotyping reliability and DNA database search results. *Forensic Sci Int Genet* 5:316–328.

Budimlija, Z.M., M.K. Prinz, A. Zelson-Mundorff et al. 2003. World Trade Center human identification project: Experiences with individual body identification cases. *Croat Med J* 44:259–263.

Budowle, B., D. Hobson, J. Smerick et al. 2001. Low copy number—Consideration and caution. Twelfth International Symposium on Human Identification.

Budowle, B., A.J. Eisenberg, and A. van Daal. 2009. Validity of low copy number typing and applications to forensic science. *Croat Med J* 50:207–217.

Butler, J.M., and C.R. Hill. 2010. Scientific issues with analysis of low amounts of DNA. *Profiles in DNA*. http://www.promega.com/profiles/1301/1301_02.html (accessed September 1, 2011).

Caddy, B., G.R. Taylor, and A.M.T. Linacre. 2008. A review of the science of low template DNA analysis. http://www.homeoffice.gov.uk/publications/police/790604/Review_of_Low_Template_DNA_1.pdf.

Caragine, T., R. Mikulasovich, J. Tamariz et al. 2009. Validation of testing and interpretation protocols for low template DNA samples using AmpFℓSTR® Identifiler®. *Croat Med J* 50:250–267.

Cheung, M.C., J.D. Goldberg, and Y.W. Kan. 1996. Prenatal diagnosis of sickle cell anaemia and thalassaemia by analysis of fetal cells in maternal blood. *Nat Genet* 14:264–268.

Cornelison, D.D.W., and B.J. Wold. 1997. Single-cell analysis of regulatory gene expression in quiescent and activated mouse skeletal muscle satellite cells. *Dev Biol* 191:270–283.

CPS Press Release. 2008. Review of the use of Low Copy Number DNA analysis in current cases: CPS statement. http://cps.gov.uk/news/latest_news/101_08/index.html (accessed June 15, 2011).

Curran, J.M., P. Gill, and M.R. Bill. 2005. Interpretation of repeat measurement DNA evidence allowing for multiple contributors and population substructure. *Forensic Sci Int* 148:47–55.

Farmen, R.K., R. Jaghø, P. Cortez et al. 2008. Assessment of individual shedder status and implication for secondary DNA transfer. *Forensic Sci Int Genet Suppl Ser* 1:415–417.

Findlay, I., T.P. Quirke, R. Frazier et al. 1997. DNA fingerprinting from single cells. *Nature* 389:555–556.

Forster, L., J. Thomson, and S. Kutranov. 2008. Direct comparison of post-28-cycle PCR purification and modified capillary electrophoresis methods with the 34-cycle 'low copy number' (LCN) method for analysis of trace forensic DNA samples. *Forensic Sci Int Genet* 2:318–328.

Frumkin, D., A. Wasserstrom, S. Itzkovitz, A. Harmelin, G. Rechavi, and E. Shapiro. 2008. Amplification of multiple genomic loci from single cells isolated by laser micro-dissection of tissues. *BMC Biotechnol* 8:17.

Gefridese, L.A., M.C. Powell, M.A. Donley et al. 2010. UV irradiation and autoclave treatment for elimination of contaminating DNA from laboratory consumables. *Forensic Sci Int Genet* 4:89–94.

Geigl, J.B., and M.R. Speicher. 2007. Single-cell isolation from cell suspensions and whole genome amplification from single cells to provide templates for CGH analysis. *Nat Protoc* 2:3173–3184.

Gill, P. 2001. Application of low copy number DNA profiling. *Croat Med J* 42:229–232.

Gill, P., and J. Buckleton. 2010. A universal strategy to interpret DNA profiles that does not require a definition of low-copy-number. *Forensic Sci Int Genet* 4:221–227.

Gill, P., J. Whitaker, C. Flaxman et al. 2000. An investigation of the rigor of interpretation rules for STRs derived from less than 100 pg of DNA. *Forensic Sci Int* 112:17–40.

Gill, P., A. Kirkham, and J. Curran. 2007. LoComatioN: A software tool for the analysis of low copy number DNA profiles. *Forensic Sci Int* 166:128–138.

Greenspoon, S.A., M.A. Scarpetta, M.L. Drayton et al. 1998. QIAamp Spin columns as a method of DNA isolation for forensic casework. *J Forensic Sci* 43:1024–1030.

Hahn, S., X.Y. Zhong, C. Troeger et al. 2000. Current applications of single-cell PCR. *Cell Mol Life Sci* 57:96–105.

Handyside, A.H., J.K. Pattinson, R.J. Penketh et al. 1989. Biopsy of human preimplantation embryos and sexing by DNA amplification. *Lancet* 1:347–349.

Hansson, O., M. Finnebraaten, I. Knutsen-Heitmann et al. 2009. Trace DNA collection—Performance of minitape and three different swabs. *Forensic Sci Int Genet Suppl Ser* 2:189–190.

Holzgreve, W. 1997. Will ultrasound-screening and ultra-sound-guided procedures be replaced by non-invasive techniques for the diagnosis of fetal chromosome anomalies? [editorial]. *Ultrasound Obstet Gynecol* 9:217–219.

Horsman-Hall, K.M., Y. Orihuela, and S.L. Karczynski. 2009. Development of STR profiles from firearms and fired cartridge cases. *Forensic Sci Int Genet* 3:242–250.

Judgment: the *Queen v Sean Hoey*. http://www.denverda.org/DNA_Documents/Hoey.pdf (accessed June 16, 2011).

Kisilevsky, A.E., and R.A. Wickenheiser, editors. DNA PCR profiling of skin cells transferred through handling. *Proceedings of the Annual Meeting of the Canadian Society of Forensic Science*; November 17–20, 1999; Edmonton, Alberta.

Liss, B. 2002. Improved quantitative real-time RT-PCR for expression profiling of individual cells. *Nucleic Acids Res* 30:e89.

Lowe, A., C. Murray, J. Whitaker et al. 2003. The propensity of individuals to deposit DNA and secondary transfer of low level DNA from individuals to inert surfaces. *Forensic Sci Int* 129:25–34.

Lucy, D., J.M. Curran, A.A. Pirie et al. 2007. The probability of achieving full allelic representation for LCN-STR profiling of haploid cells. *Sci Justice* 47:168–171.

Mitchell, A.A., J. Tamariz, K. O'Connell et al. 2011. Likelihood ratio statistics for DNA mixtures allowing for drop-out and drop-in. *Forensic Sci Int Genet Sup* 3:e240–e241.

National Research Council. 1996. *The Evaluation of Forensic DNA Evidence*. Washington, DC: National Academy Press.

NY v Megnath (2010). Available at http://www.denverda.org/DNA_Documents/Megnath.pdf.

Oz, C., J. Levi, and Y. Novoselski. 1999. Forensic identification of a rapist using unusual evidence. *J Forensic Sci* 44:860–862.

Pizzamiglio, M., A. Mameli, D. My et al. 2004. Forensic identification of a murderer by LCN DNA collected from the inside of the victim's car. *Int Congress Ser* 1261:437–439.

Polley, D., P. Mickiewicz, M. Vaughn et al. 2006. Investigation of DNA recovery from firearms and cartridge cases. *J Can Soc Forensic Sci* 39:217–228.

Prinz, M., L. Schiffner, J.A. Sebestyen et al. 2006. Maximization of STR DNA typing success for touched objects. *Int Congr Ser* 1288:651–653.

Rameckers, J., S. Hummel, and B. Herrmann. 1997. How many cycles does a PCR need? Determinations of cycle numbers depending on the number of targets and the reaction efficiency factor. *Naturwissenschaften* 84:259–262.

Roeder, A.D., P. Elsmore, and A. McDonald. 2009. Maximizing DNA profiling success form suboptimal quantities of DNA: A staged approach. *Forensic Sci Int Genet* 3:128–137.

Rutty, G.N. 2002. An investigation into the transference and survivability of human DNA following simulated manual strangulation with consideration of the problem of third party contamination. *Int J Leg Med* 116:170–173.

R v Reed and Anor (2009). Available at http://www.bailii.org/ew/cases/EWCA/Crim/2009/2698.html.

Schiffner, L., E. Bajda, M. Prinz et al. 2005. Optimization of a simple, automatable extraction method to recover sufficient DNA from low copy number DNA samples for generation of short tandem repeat profiles. *Croat Med J* 46:578–586.

Sekizawa, A., T. Kimura, M. Sasaki et al. 1996a. Prenatal diagnosis of Duchenne muscular dystrophy using a single fetal nucleated erythrocyte in maternal blood. *Neurology* 46:1350–1353.

Sekizawa, A., A. Watanabe, T. Kimura et al. 1996b. Prenatal diagnosis of the fetal RhD blood type using a single nucleated erythrocyte from maternal blood. *Obstet Gynecol* 87:501–505.

Sewell, J., I. Quinones, C. Ames et al. 2008. Recovery of DNA and fingerprints from touched documents. *Forensic Sci Int Genet* 2:281–285.

Shaw, K., I. Sesardic, N. Bristol et al. 2008. Comparison of the effects of sterilization techniques on subsequent DNA profiling. *Int J Legal Med* 122:29–33.

Smith, P.J., and J. Ballantyne. 2007. Simplified low-copy-number DNA analysis by post-PCR purification. *J Forensic Sci* 52:820–829.

Stratalinker® UV Crosslinker Instruction Manual, Model 2400 Catalog #400075 (120 V), #400076 (230 V) and #400676 (100 V). La Jolla, CA: Stratagene. Copyright 2002, Revision #122003 IN #70034-06.

Taberlet, P., S. Griffin, B. Goossens et al. 1996. Reliable genotyping of samples with very low DNA quantities using PCR. *Nucleic Acids Res* 24:3189–3194.

Tamariz, J. 2009. Investigation of the detection of DNA through secondary and tertiary transfer. Poster Session at the 20th International Symposium on Human Identification, Summerlin, Nevada, USA. October.

Tamariz, J., K. Voynarovska, M. Prinz et al. 2005. The application of ultraviolet irradiation to exogenous sources of DNA in plasticware and water for the amplification of Low Copy Number DNA. *J Forensic Sci* 51:790–4.

Tvedebrink, T., P.S. Eriksen, H.S. Mogensen et al. 2009. Estimating the probability of allelic drop-out of STR alleles in forensic genetics. *Forensic Sci Int Genet* 3:222–226.

van Hoofstat, D.E., D.L. Deforce, I.P. Hubert De Pauw et al. 1999. DNA typing of fingerprints using capillary electrophoresis: Effect of dactyloscopic powders. *Electrophoresis* 20:2870–2876.

van Oorschot, R.A.H., and M.K. Jones. 1987. DNA fingerprints from fingertips. *Nature* 387:767.

van Oorschot, R.A.H., I. Szepietowska, D.L. Scott et al. 1999. Retrieval of genetic profiles from touched objects. First International Conference for Human Identification in the Millennium; London.

van Renterghem, P., D. Leonard, and C. de Greef. 2000. Use of latent fingerprints as a source of DNA for genetic identification. *Prog Forensic Genet* 8:501–503.

Wallace v Queen (2010) (CA590/2007; judgment 3/3/10). Available at http://www.courtsofnz.govt.nz/from/decisions/judgments/html.

Whitaker, J.P., T.M. Clayton, A.J. Urquhart et al. 1995. Short tandem repeat typing of bodies from a mass disaster: High success rate and characteristic amplification patterns in highly degraded samples. *Biotechniques* 18:670–677.

Whitaker, J.P., E.A. Cotton, and P. Gill. 2001. A comparison of the characteristics of profiles produced with the AmpflSTR SGM Plus multiplex system for both standard and low copy number (LCN) STR DNA analysis. *Forensic Sci Int* 123:215–223.

Forensic DNA Mixtures, Approaches, and Analysis

THERESA CARAGINE
ADELE MITCHELL
CRAIG O'CONNOR

Contents

7.1	Types of Results	189
7.2	Number of Contributors	192
7.3	Mixture Ratios: Major/Minor Contributors	195
7.4	Statistics	196
References		202

7.1 Types of Results

DNA samples yielding no more than two alleles at each locus shall be deemed DNA from one person or single source. Since, as humans, we get half of our DNA from each parent, then each location shall have at most two alleles (e.g., 15, 16). If more than two alleles are viewed at any location, then the sample must have originated from more than one person, and it represents a mixture of DNA from those people. Laboratories have developed their own ways to deal with analysis of mixtures. Some spend copious amounts of time trying to deconvolute the mixture (or determine the profiles of one or more contributors to the mixture) and obtain individual profile(s) where others will deem the sample a mixture and calculate one of the statistics described below (Clayton et al. 1998). Some laboratories do not venture into interpretation of nondeducible mixtures and do nothing further if individual profiles cannot be determined.

Overall, the strategies for dealing with mixtures revolve around the attempt at deconvolution of the components and then applying some statistical weight to the findings. The DNA Commission of the International Society of Forensic Genetics has published recommendations on the interpretation of DNA mixtures and the eventual statistical analysis (Outbox 7.1). Among these recommendations are the use of likelihood ratios (LR) as the preferred statistical test as well as using extreme caution when attempting to deconvolute a mixture because of artifacts such as stutters and the effects of stochastic fluctuations. In kind, the Scientific Working Group for DNA Analysis Methods (SWGDAM) has also established a list of similar guidelines for the analysis of mixture samples (Outbox 7.2).

The interpretation of mixtures in forensic samples is challenging, and analysts must consider many factors. Among these are the number of mixture contributors, the relative amounts of each component, the determination of the major and the minor DNA profiles of the components if possible, and the eventual statistical analysis of the mixture. Laboratories and practitioners should develop and validate their own protocols and interpretation guidelines to define how mixtures should be treated. The overall weight of the evidence should accurately reflect the conclusions obtained from the mixture.

OUTBOX 7.1: ISFG RECOMMENDATIONS ON MIXTURE ANALYSIS

The DNA Commission of the International Society of Forensic Genetics has published two sets of recommendations on interpretation of forensic DNA mixtures. In the first (Gill et al. 2012), general guidance is given as follows:

1. LR is preferred, but RMNE can be acceptable in some circumstances. RMNE cannot be used for mixtures from which any of a suspect's alleles are missing.
2. The scientific community has the responsibility to use LRs in case notes and in court even if the results are more complex to explain and/or if the court does not want to hear results presented in this way.
3. LR may be calculated without considering peak heights.
4. Peak heights can be used to eliminate implausible allele pairings in the calculation of statistics or as a way to exclude a suspect.
5. The LR is composed of a prosecution hypothesis and a defense hypothesis, over which the respective parties should have some control.
6. If stutter peaks are similar in size to the alleles of a minor contributor to a mixture, alleles in stutter position should be included in statistical and qualitative assessments of the mixture.
7. If drop-out is required to explain the mixture given a particular set of contributors, the peak heights of surviving alleles should be low enough that drop-out of a partner allele is reasonably plausible.
8. If alleles at a particular locus are very close in height to the background noise, interpretation at that locus should not be attempted.
9. In any interpretation of low template mixtures, allelic drop-out and drop-in must be considered.

Low template mixtures are briefly addressed in the first publication, with a recommendation that heterozygous peak balance and allelic drop-out and drop-in should be considered in the assessment of a low template mixture. The second publication (Gill et al. 2006; Schneider et al. 2006) focuses exclusively on mixtures that may include drop-out and/or drop-in (i.e., stochastic effects). Detailed examples are given concerning estimation of drop-out and drop-in rates using probabilistic approaches. Recommendations of the Commission for mixtures with stochastic effects include

1. A probabilistic model should be used to evaluate the weight of evidence of a mixed sample considering drop-out and drop-in.
2. Estimates of drop-out and drop-in probabilities should be based on a laboratory's validation studies. These probabilities may depend on electropherogram peak height or on some other factor, such as DNA template amount.
3. Weight of evidence should be expressed as a likelihood ratio, not as RMNE.
4. Software should be used in order to avoid mistakes due to hand calculations.

OUTBOX 7.2: SWGDAM GUIDELINES ON DNA MIXTURE INTERPRETATION

The Scientific Working Group on DNA Analysis Methods (SWGDAM) is a group of approximately 50 scientists from DNA laboratories in the United States and Canada. In 2010, guidelines were published and are available on the FBI's website (SWGDAM 2010). It is first noted that STR mixture interpretation relies on analysts' expertise and professional judgment and that each laboratory's protocols must be validated and documented. The steps involved in mixture analysis are laid out as follows:

1. Identify and remove non-allelic peaks, such as stutter or dye blobs, if possible.
2. Apply a stochastic threshold, if one is used. If a stochastic threshold is not used, the laboratory should use some other, internally validated, method for identifying loci at which drop-out may have occurred.
3. Compute peak height ratios to determine which alleles may be paired together.
4. Estimate the number of contributors to the sample.
5. If a sample is determined to represent a mixture of DNA, it may contain one or more major contributors and one or more minor contributors. A laboratory should develop guidelines for mixture deconvolution, including when it is acceptable to assume that a known person, such as a victim, is a contributor to the mixture and to use that person's profile to help determine the profile(s) of other contributor(s).
6. The evidentiary results may be compared to the profile of a known individual, such as the suspect, if one is available. Laboratories are reminded that evidentiary samples should be interpreted as far as possible before a comparison to a known profile is made. Possible conclusions for such a comparison may include the following:
    a. The known individual cannot be excluded (i.e., is included) as a possible contributor to the mixture. A statistic is required whenever a positive association is made between a comparison person and a mixture.
    b. The individual is excluded. A statistic is not required for an exclusion.
    c. The DNA typing results are inconclusive.
    d. The DNA typing results from two or more items of evidence are consistent with originating from a common source.
7. Several different statistics may be calculated, depending on the sample characteristics, the availability of a comparison profile from a known person, and the procedures that have been validated by the individual laboratory. Formulae and sample calculations are presented in section 5 of the guidelines. As per NRC II recommendations (reference), a theta adjustment of 0.01 or 0.03 should be applied to homozygous genotypes. One or more of the following statistics may be calculated:
    a. RMP for single source samples and for single source profiles that have been deconvoluted from mixtures
    b. Unrestricted CPE/CPI/RMNE allowing contributors to have all possible combinations of alleles regardless of relative peak heights

c. Restricted CPE/CPI/RMNE incorporating information on the estimated number of contributors to the mixture and pairing peaks together only when their relative heights fall within a designated range
d. Unrestricted LR allowing contributors to have all possible combinations of alleles regardless of relative peak heights
e. Restricted LR taking relative peak heights into consideration and pairing peaks together only when their relative heights fall within a designated range

7.2 Number of Contributors

When estimating the number of contributors of any sample, the key is to look at the number and relative peak heights of the alleles (Cowell et al. 2007, 2011). Examinations should be done at each locus, looking for the loci with the maximum number of alleles observed (Butler et al. 2011). As with any drawn conclusions from a forensic sample, the entire profile needs to be evaluated. Because of sample complexities such as low template amounts, degradation, and inhibition, no conclusions should be made from just one locus.

SWGDAM defines a sample as a mixture if three or more alleles are present at one or more loci (SWGDAM 2010). For a two-person mixture, no more than four alleles should be seen at each locus. No more than six alleles at a locus will be seen for a three-person mixture, no more than eight alleles for a four-person mixture, and so on. The peak height ratios at a locus can also be used to infer that a mixture is present. Since the allelic content is composed of half maternal and paternal DNA, ideally, the two alleles should be equally represented in each cell, and therefore have peak heights that are approximately equal (given similar amplifications) (Curran et al. 2005).

For most situations, care should be taken in determining the number of contributors because of allele sharing. Each locus has a finite number of possible alleles and therefore, the more contributors in the mixture, the higher the chance of alleles being shared among them. Because of this, some laboratories will not attempt to deconvolute mixtures especially once it is determined that there are more than three contributors. Figure 7.1 shows results for a single source sample and for a two-person mixture. Once the number of contributors is calculated, analysis can be done at each locus to determine all of the possible genotypes for the given number of alleles.

For a two-person mixture with four alleles detected, there are six possible combinations of major and minor contributor genotypes. This number jumps considerably with a three-person mixture with six alleles detected. Figure 7.2 shows the number of possible combinations for a two-person mixture with four alleles detected and some of the possible combinations in a three-person mixture with six alleles detected. Notice though that if the maximum number of alleles is detected for the number of contributors, homozygous genotypes are no longer a possibility. If the minimum number of alleles is present, having a homozygous genotype is plausible and has to be taken into consideration.

Although locus-by-locus allele counting can provide an estimate of the minimum number of contributors to a mixture, it does not necessarily indicate the actual number of contributors to mixtures, particularly those with more than two contributors (Paoletti et al. 2005; Buckleton et al. 1998, 2007). That is, when only considering the maximum number of alleles seen at any locus, a three-person mixture could be classified as a mixture of at least two people and a

Forensic DNA Mixtures, Approaches, and Analysis

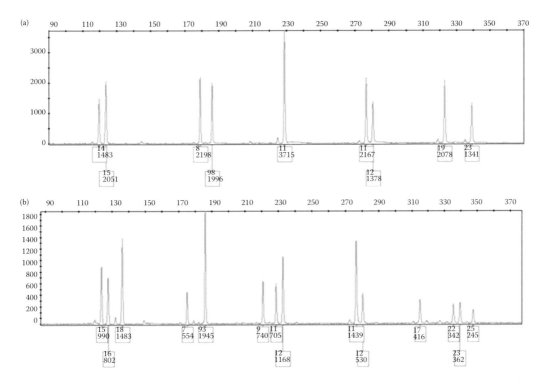

Figure 7.1 (a) A single-source sample (from one person) where no more than two alleles are present in each locus. (b) A mixture of at least two people where no more than four alleles are present in each locus. A three-person mixture would have no more than six alleles at each locus.

four-person mixture could be designated a mixture of at least three people and sometimes even as a mixture of at least two people. One published analysis of conceptual mixtures of individuals typed at 13 loci estimated that using the maximum number of alleles observed at any locus, 3.2% to 3.4% of all three-person mixtures would be categorized as mixtures of at least two people and approximately 76% of four-person mixtures would be classified as mixtures of at least two or at least three people (Paoletti et al. 2005). Analysis of conceptual mixtures of simulated individuals typed at SGM+ and Profiler Plus® loci gave similar results (Buckleton et al. 2007).

In order to calculate an LR, the number and identity of the contributors to the mixture must be specified under two mutually exclusive hypotheses: one held by the prosecution, typically including a suspect, and one held by the defense, typically replacing the suspect with an unknown, unrelated person. The LR is a measure of whether a mixture can be better explained with or without inclusion of the suspect (or another comparison person). An LR greater than 1 indicates more support for inclusion of the suspect than an unknown, unrelated person; an LR less than 1 indicates more support for inclusion of an unknown, related person, rather than the suspect. The effect of uncertainty in number of contributors on the LR has been explored in the forensics literature. It has been shown that the denominator of the LR (the likelihood of the defense hypothesis) is usually maximized by selecting the hypothesis with the minimum number of contributors required to explain the evidence (Brenner et al. 1996). In other words, for a given prosecution hypothesis, using the defense hypothesis with the minimum possible number of contributors will usually result in the lowest possible LR, that is, the LR that most favors the defendant. Based on results of the simulations performed by Buckleton et al. (2007), there is a "moderate risk" of obtaining a

(a)

Contributor 1	Contributor 2
11, 12	14, 15
11, 14	12, 15
11, 15	12, 14
14, 15	11, 12
12, 15	11, 14
12, 14	11, 15

Detected alleles: 11, 12, 14, 15

(b)

Contributor 1	Contributor 2	Contributor 3
11, 12	14, 15	17, 19
11, 14	12, 15	17, 19
11, 15	12, 14	17, 19
14, 15	11, 12	17, 19
12, 15	11, 14	17, 19
12, 14	11, 15	17, 19
11, 12	14, 17	15, 19
11, 12	14, 19	15, 17
11, 12	15, 19	14, 17
11, 12	15, 17	14, 19
11, 12	17, 19	14, 15
11, 17	14, 15	12, 19
11, 19	14, 15	12, 17
12, 19	14, 15	11, 17
12, 17	14, 15	11, 19
17, 19	14, 15	11, 12
11, 17	12, 14	15, 19
11, 19	12, 14	15, 17
15, 19	12, 14	11, 17
15, 17	12, 14	11, 19
	etc.	

Detected alleles: 11, 12, 14, 15, 17, 19

Figure 7.2 (a) Possible genotypes in a simple two-person mixture with four alleles detected with equal peak heights. (b) Some of the many different genotype combinations in a three-person mixture with six alleles detected with equal peak heights.

nonminimal LR when the defense hypothesis with the minimum number of contributors is used. Thus, underestimating the number of contributors may often be conservative.

However, one could argue that a better approach than opting for the minimum number of contributors to a mixture might be to determine the number of contributors best supported by the data. Several recent publications have explored this idea using maximum likelihood to estimate the most likely number of contributors to a mixture. Egeland et al. (2003) proposed a maximum likelihood estimator and presented a method for using diallelic markers, assuming Hardy–Weinberg equilibrium (HWE) in the population of origin. Haned et al. (2011a,b) extended the method to allow for multiallelic markers and for population substructure. The maximum likelihood method correctly estimated the number of contributors to two- and three-person mixtures more than 90% of the time, outperforming the maximum allele count method. For four- and five-person mixtures, the maximum likelihood method gave correct classifications of mixtures 64% to 79% of the time, which is a dramatic improvement over simple allele counting.

Another approach is to examine the characteristics of a mixture, such as the number of different alleles labeled at each locus or at all loci, or the total number of repeating alleles if multiple replicates are performed (Perez et al. 2011). Using this type of approach, a sample can be characterized as a three-person mixture without having any loci with five or more labeled alleles if, for example, there are many loci with four different alleles.

7.3 Mixture Ratios: Major/Minor Contributors

Once the number of contributors is established, deconvolution can be attempted. The major donor to a mixture can be defined as the person whose DNA is detected the most, whereas the minor donor is the person whose DNA is detected the least. For example, in a simple two-person mixture with four alleles present, the two highest peaks can be attributed to the major donor with the two lowest being those of the minor donor. Figure 7.3 shows a mixture of two people and the resulting deconvolution of the major and minor donors. This can be obvious in situations where the relative ratio of the major to minor is greater than 4:1 but can be ambiguous if the ratios are more closely associated. It should be noted that in order to avoid bias, mixture deconvolution should be done without reference to the comparison profile. However, the profile of a known contributor to the mixture, such as a victim with an intimate sample, can be used to help determine the profile of another contributor.

Estimating the ratio of the major to the minor component is useful in mixture deduction. For a two-person mixture: the sum of the peak heights of the major alleles divided by the sum of the peak heights for the minor component alleles at a locus with four peaks can give you a basal figure for the ratio of the contributors (Evett et al. 1991, 1998a,b; Gill et al. 1998). Once the ratio is established, it can then be used to infer the genotypes at the other loci with two or three alleles. Care should be taken if the number of contributors is greater than 2 or is ambiguous (relatives sharing alleles). For example, the major donor can be deconvoluted from a three-person mixture if the relative ratios of contributors are 8:1:1, but will be difficult in more complex mixtures where the ratios are closer, for example, 3:1:1.

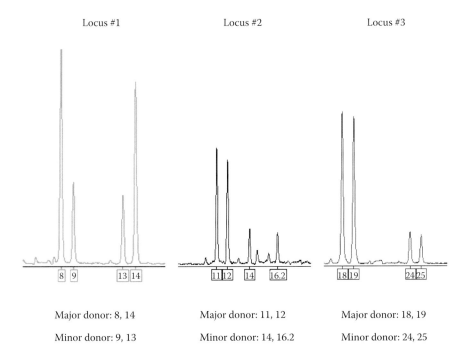

Figure 7.3 Deducible two person mixture with four apparent alleles. In this simplistic model, the two highest peaks in each locus can be attributed to the major donor, whereas the two shortest peaks can be attributed to the minor donor.

7.4 Statistics

Once a mixture is analyzed and the reference/known samples are compared, there are several possible conclusions that may be drawn. If a deconvolution is possible, then there can be a match between the comparison sample and the major/minor donor. If a deconvolution is not possible (or the known sample does not match the major/minor donor), then a direct comparison can be performed. If all of the alleles of the known contributor are detected in the mixture, that profile is said to be "included" as a possible contributor to the mixture. If most of the comparison profile is not detected in the mixture, then that person is said to be "excluded" as a contributor. If most of the known person's alleles are detected in the mixture and there is a reasonable scientific explanation for missing alleles, then a person could be categorized as "cannot be excluded/included." Sometimes, no conclusions can be made as to whether the person can be included or excluded as a contributor.

For any positive association between an individual and a mixture, a statistic should be computed. There are several different statistics that can be calculated for DNA mixtures. Each one evaluates a different aspect of the mixture and care should be taken when performing the calculations. If a profile can be deduced out of the mixture (i.e., the profile of the major and/or minor contributors), then the random match probability (RMP) can be used as if the profile originated from a single source sample. This statistic evaluates the rarity of the profile in that it represents the probability that a randomly selected person from the population would have a profile matching the deduced profile. RMP is calculated using genotype frequencies expected under HWE. Under HWE, the alleles carried on distinct chromosomes are independent of one another, and there is a predictable relationship between allele frequencies and genotype frequencies. That is, a person's two alleles at a particular locus can be represented as two random draws from the population, which can be visualized as a pool of independent chromosomes. The probability of being homozygous for Allele A is equal to the probability of drawing Allele A two times from the pool. If the frequency of Allele A is p_A, this probability is p_A times p_A or p_A^2. The probability of being heterozygous for Alleles A and B is equal to the probability of drawing Allele A followed by Allele B or drawing Allele B followed by Allele A from the pool. If the frequency of Allele A is p_A and the frequency of Allele B is p_B, this probability is two times p_A times p_B or $2p_A p_B$.

If, for a particular locus, one allele can be attributed to the major or minor contributor, but it is not clear whether the contributor is homozygous or heterozygous at that locus, RMP is calculated allowing the second allele to be anything. This is known as the "$2p$ rule." Instead of using $2p_A p_B$ for the estimated genotype frequency at the locus, only $2p_A$ is used. This is equivalent to assigning a frequency of 1.0 to the second allele, which covers all possibilities.

When independently inherited loci, such as the forensic autosomal STRs, are typed, RMP can be combined across loci using the product rule. RMP is calculated separately for each locus and then the individual values are multiplied together to obtain the RMP for the whole profile. RMP can be computed for a full profile or for a partial profile deduced from the mixture.

In cases where individual contributors' profiles cannot be deduced from a mixture, then Random Man Not Excluded (RMNE), also called combined probability of exclusion/inclusion (CPE/CPI), may be used (Ladd et al. 2001). This is the probability that a random person would be excluded/included as a contributor to the mixture. This takes into account all of the alleles detected but does not require an assumption of the number of contributors in the mixture. Although this approach does not consider the possible genotypes of the comparison profile, it is simple to explain conceptually to a jury. Consider a mixed locus with alleles 12, 13, 14, and 15; the relative peak heights of the alleles are irrelevant as all of the detected alleles, and consequently all possible genotypes, 12/12, 12/13, 12/14, 12/15, 13/13, 13/14, 13/15, 14/14, 14/15, and 15/15, are considered. Even if there are no indications that the mixture might contain more than two contributors, homozygous genotypes at loci with four alleles are allowed. That is, a suspect who is 12/12 at this hypothetical locus would be considered to be a possible contributor to the mixture, even though this would leave at least one unexplained allele at the locus if, in fact, it is a mixture of DNA from two people.

At each locus tested, the RMNE is calculated by adding up the frequencies of all alleles detected at that locus and squaring the sum. Using the example above, if the frequencies of the 12, 13, 14, and 15 alleles are 0.01, 0.13, 0.18, and 0.12, respectively, the RMNE for that locus is $(0.01 + 0.13 + 0.18 + 0.12)^2 = 0.44^2 = 0.194$. This is equivalent to the sum of all possible genotypes for a contributor to a mixture. The interpretation of the result is that the suspect could be a contributor to the mixture, and 19.4% of randomly selected individuals could also be contributors to the mixture. If multiple loci are tested, the overall RMNE is the product across all loci.

If one is willing to assume a number of contributors to a mixture, or if a mixture appears to contain some major contributors and some minor contributors, a "restricted RMNE" can be calculated. The restricted RMNE is essentially the same as the standard RMNE, but only a subset of detected alleles or possible genotypes are included in the calculation. The included alleles may consist of those that are above a certain height, representing only the major contributor(s) to the mixture. Or, if a mixture appears to consist of two contributors, RMNE can be calculated allowing potential contributors to have only the genotypes that would be consistent with two contributors. In the previous example, the locus with alleles 12, 13, 14, and 15, a restricted RMNE assuming two contributors would only allow potential contributors to be heterozygous, as the presence of a homozygote would require more than two contributors to account for all of the alleles labeled at this locus. The restricted RMNE in this situation would be $2(0.01)(0.13) + 2(0.01)(0.18) + 2(0.01)(0.12) + 2(0.13)(0.18) + 2(0.13)(0.12) + 2(0.18)(0.12) = 0.130$. Using the same interpretation of this number as above, 13% of randomly selected individuals could be contributors to this mixture. Thus, restricting the RMNE to two contributors makes it less likely that a randomly selected individual could be a contributor and increases the strength of the evidence for an individual that is included as a possible contributor.

Because RMNE is meant to describe all possible contributors to a mixture, one limitation is that it may not be used if drop-out is suspected. There is no generally accepted method for incorporating drop-out in the RMNE calculation. Alternatively, another statistic, the LR, may be used. The LR uses more available information and, therefore, gives a better estimate of the weight of the evidence than the RMNE.

The LR is a ratio of two probabilities. Traditionally, the numerator represents the probability of the genotype data under a scenario that is put forth by the prosecution (the "prosecution hypothesis"), and the denominator represents the probability of the same data under a scenario advocated by the defense (the "defense hypothesis"). Typically, the prosecution hypothesis includes the suspect or some other comparison person of interest and the defense hypothesis replaces that person with an unknown, unrelated person. For example, the prosecution hypothesis may be that the evidence sample represents a mixture of DNA from a suspect and victim and the defense hypothesis may be that it is a mixture of DNA from the victim and an unknown, unrelated person. The numerator will then be the probability of obtaining the alleles found in the mixture if the true contributors are the suspect and victim, and the denominator will be the probability of obtaining those alleles if the true contributors are the victim and an unknown, unrelated person.

If drop-out and drop-in are not modeled in the LR, the hypothesized contributors under each scenario must fully explain the mixture or the probability will be zero. That is, for example, in order for the prosecution to assert that the sources of the DNA in the mixture are the suspect and the victim, the evidence must represent a perfect combination of their alleles; none of either person's alleles can be missing from the mixture and no extra alleles that cannot be attributed to those two people can be labeled in the mixture.

Using the example from above, the locus with labeled alleles 12, 13, 14, and 15, an LR could be computed if the suspect and victim are each heterozygous such that together they have the 12, 13, 14, and 15 alleles. For example, if the suspect is 12/13 and victim is 14/15, the LR is computed as follows. The numerator represents the probability of obtaining the alleles labeled in the evidence, 12, 13, 14, 15, if the genotypes of the true contributors are 12/13 and 14/15. This probability is 1.0 because if the true contributors have alleles 12, 13, 14, and 15, and drop-out and/or drop-in are not modeled, the probability of obtaining 12, 13, 14, 15 from the evidence is 1.0. In the denominator, the victim explains the 14 and 15 alleles, so the unknown, unrelated contributor must explain the 12 and 13. The probability that a randomly selected person has genotype 12/13 is $2p_{12}p_{13} = 2(0.01)(0.13) = 0.0026$. Thus, the LR is $1/0.0026 = 385$. The interpretation is that the mixture is 385 times more probable if it originated from the suspect and victim than if it originated from the victim and an unknown, unrelated person.

For a two-person mixture with fewer than four labeled alleles at a locus, there are more options for the genotype of the unknown contributor. Consider, for example, the locus above with labeled alleles 12, 13, and 14, a suspect with genotype 12/13, and a victim with genotype 13/14. The numerator would again be 1.0, as the probability of obtaining 12, 13, and 14 from contributors with genotypes 12/13 and 13/14 is 1.0. In the denominator, the victim accounts for the 13 and 14 alleles, so an unknown person must carry a 12, but could be homozygous for the 12 allele or heterozygous 12/13 or 12/14. The denominator is then is $p_{12}^2 + 2p_{12}p_{13} + 2p_{12}p_{14} = 0.0063$ and the LR is 159. The denominator represents the probability that a randomly selected person could be the second contributor to the mixture, rather than the suspect.

In the examples above, the relatively rare 12 allele carried by the suspect and labeled in the mixture leads to a high LR for this locus. Because the 12 allele has a frequency of 1%, it would be unusual for a randomly selected person to be carrying that allele. A very different LR is obtained for the four-allele mixture if the victim genotype is 12/13 and the suspect genotype is 14/15. In this example, the victim accounts for the rare 12 allele and the unknown person is required to explain the more common 14 and 15 alleles. Here, the denominator would be $2p_{14}p_{15} = 2(0.18)(0.12) = 0.043$ and the LR would be 23. This time, the mixture is 23 times more probable if it originated from the suspect and victim than if it originated from the victim and an unknown, unrelated person. Thus, the LR is dependent on the frequencies of the alleles in the mixture, as well as the frequencies of the alleles carried by the comparison person. This is an element that is not considered by RMNE and is one of the strengths of the LR relative to RMNE.

Another strength of the LR over RMNE is that drop-out, drop-in, and other phenomena, such as probabilistic genotypes, as well as replicate amplifications of one sample can be modeled within the LR framework. To explore how drop-out and drop-in may be modeled using LR, consider the locus above with only the 13, 14, and 15 alleles labeled. If the victim's genotype is 14/15, a suspect with genotype 12/13 cannot be considered unless drop-out and drop-in are modeled. This is because, if they are not modeled, the probability of obtaining a mixture type that is not exactly equal to the alleles carried by the hypothesized contributors is zero. That is, without a model for drop-out and drop-in, the probability of obtaining 13, 14, and 15 from contributors with genotypes 12/13 and 14/15 is zero, making the entire LR equal to zero. The same result would be obtained if any of the victim's alleles were missing.

To model drop-out and drop-in, additional factors must be included for each contributor, namely, the probability of drop-out or no drop-out and the probability of drop-in or no drop-in. This must be done for each named contributor with a known profile (such as suspect, victim, or elimination sample) and for each possible genotype of each unknown, unrelated contributor. The value of the LR will depend on the allele frequencies, as well as the probabilities of drop-out and drop-in, and the number and frequency of the comparison person's alleles that do not appear in the mixture, that is, those that would have had to drop out if that person is a true contributor. Typically, partial heterozygous drop-out, total heterozygous drop-out, and total homozygous drop-out are parameterized separately. For heterozygotes, partial drop-out, D_1, total drop-out, D_2, and no drop-out, D_0, are mutually exclusive and represent all possible outcomes, so $\Pr(D_0) + \Pr(D_1) + \Pr(D_2) = 1.0$. Similarly, for homozygotes, total drop-out, D_{H1}, and no drop-out, D_{H0}, are mutually exclusive and represent all possible outcomes, so $\Pr(D_{H0}) + \Pr(D_{H1}) = 1.0$. There are many options for parameterization of drop-in. One option is to use separate values for drop-in of one allele (C_1) and for drop-in of two or more alleles (C_{2+}) at a single locus in a single amplification. Using this system, $\Pr(C_0) + \Pr(C_1) + \Pr(C_{2+}) = 1.0$.

When computing the LR for the locus with labeled alleles 13, 14, and 15, suspect genotype 12/13 and victim genotype 14/15, the numerator will not be 1.0 as in previous examples, as the probability of obtaining 13, 14, 15 from true contributors with genotypes 12/13 and 14/15 is not 1.0. Instead, the numerator will be equal to the probability of partial drop-out from a heterozygote (the suspect) times the probability of no drop-out from a heterozygote (the victim) times the probability of no drop-in. The probability of no drop-in is invoked because there are no labeled alleles that could not be attributed to the suspect and/or the victim.

In the denominator, the unknown contributor may have any of the genotypes listed in the first column of Table 7.1. The letter w denotes any allele not labeled in the mixture at that locus. For this locus, w includes any allele other than 13, 14, or 15, and $p_w = 1 - p_{13} - p_{14} - p_{15}$. The second column in Table 7.1 lists the estimated population frequency of the genotype. The third column shows the type of drop-out (or no drop-out) that would have had to occur if the unknown contributor had each particular genotype. The options are partial, total, or no drop-out from a heterozygote, and total or no drop-out from a homozygote, as described above. The fourth column lists the type of drop-in (or no drop-in) that would have had to occur if the sample is a mixture of DNA from the victim and an unknown person with each possible genotype. Drop-out or no drop-out of the known contributor (the victim) must also be considered in the denominator. Because the victim is heterozygous and both alleles are labeled in the mixture, the factor $\Pr(D_0)$ is used. The LR is then

Table 7.1 Drop-Out and Drop-In Rate Calculations Given Allele Frequency p

Unknown Genotype	Frequency	Drop-Out	Drop-In
13/13	p_{13}^2	D_{H0}	C_0
13/14	$2p_{13}p_{14}$	D_0	C_0
13/15	$2p_{13}p_{15}$	D_0	C_0
13/w	$2p_{13}p_w$	D_1	C_0
14/14	p_{14}^2	D_{H0}	C_1
14/15	$2p_{14}p_{15}$	D_0	C_1
14/w	$2p_{14}p_w$	D_1	C_1
15/15	p_{15}^2	D_{H0}	C_1
15/w	$2p_{15}p_w$	D_1	C_1
w/w	p_w^2	D_{H1}	C_1

$$\frac{\Pr(D_1)\Pr(D_0)\Pr(C_0)}{\Pr(D_0)\left[\left(p_{13}^2\right)\Pr(D_{H0})\Pr(C_1)+(2p_{13}p_{14})\Pr(D_0)\Pr(C_1)\ldots\left(p_w^2\right)\Pr(D_{H1})\Pr(C_1)\right]}$$

Central to the calculation of the LR incorporating drop-out and drop-in are the estimation of the probability of drop-out and drop-in. This can be done in a number of ways, including probabilistically or empirically, using electropherogram peak heights or DNA template quantity (Gill et al. 2012; Shrestha et al. 2006; Tvedebrink et al. 2009, 2012).

Calculation of LRs does not involve difficult math, but the process is tedious and should not be performed by hand, particularly for LR models that include drop-out and drop-in. With additional unknown contributors in the model, the number of combinations of genotypes that must be considered can become enormous. For example, in a model with five unknown contributors, such as suspect plus two unknowns in the numerator versus three unknowns in the denominator, at a locus with seven labeled alleles, there are 36 possible genotypes for each unknown contributor. All combinations of these genotypes for the unknown contributors in each scenario must be considered, totaling $36^2 + 36^3 = 47,952$ combinations. Several programs exist for performing some or all of these types of calculations.

One of the commonly used programs is Popstats, which is supplied by the FBI. In order to use popstats, the evidence must represent a perfect combination of DNA from all contributors to a mixture. If extra alleles are observed or alleles are not seen, other LR ratio programs that allow for drop-out and drop-in must be used. The programs include True Allele, LoComatioN, the LRMix module in Forensim, LikeLTD, Lab Retriever, and the NYC OCME Forensic Statistical Tool (FST).

In addition to the LR-based programs that exist for analysis of forensic DNA mixtures, other approaches are discussed in the literature. One such example is a Bayesian network approach; another is a regression-based approach that estimates the probability of drop-out using peak heights from electropherograms. The common thread for all of these approaches is that drop-out and drop-in are modeled in some way. This is an absolute requirement for analysis of complex DNA mixtures (Outbox 7.3).

OUTBOX 7.3: LIKELIHOOD RATIO PROGRAMS INCORPORATING DROP-OUT AND DROP-IN

True Allele (Perlin et al. 2001, 2009, 2011) assesses short tandem repeat (STR) DNA raw fluorescent sequencer files based on several key features such as peak height, shape, area, and position relative to a standard ladder as well as user-defined parameters. Using a statistical method called Markov chain Monte Carlo (MCMC), all reasonable DNA profiles for the contributors to the DNA mixture are determined. For some complex mixed samples, several potential genotypes for each contributor are generated. Once the potential profiles of individual contributors to a mixture are determined, the program computes a LR for any set of one to three contributors specified by the user. For every instance of drop-out that would be required to explain the mixture with a given set of hypothesized contributors, the resulting likelihood ratio is dropped by two orders of magnitude. That is, if a perfect match between hypothesized contributors and a mixture would produce a LR of 100,000, the LR for those same contributors and mixture in which one of them is missing an allele would be 1000.

LoComatioN, Forensim, likeLTD, and *FST* do not deconvolute DNA mixtures but simply compute a LR for scenarios specified by the user. By incorporating drop-out and drop-in probabilities in the LR calculation, a weight can be given to comparisons where a piece of DNA seen in a contributor profile is not seen in the mixture.

LoComatioN (Gill et al. 2007, 2008) is a software tool using likelihood ratios that was created by forensic scientists at the Forensic Science Services (FSS) in the United Kingdom in the mid-1990s. The probability of allele drop-out or the estimate of how likely a piece of DNA could not be amplified or be seen is incorporated into the calculation. Similarly, the probability of allele drop-in or how likely an exogenous piece of DNA could be observed is considered as well. The user defines these probabilities of drop-out and drop-in. Unlike the values used with FST, they are not determined based on empirical studies.

Forensim (Haned 2011, 2012) includes a module called "LRMix", which is another implementation of LoComatioN. LRMix does its initial calculations using user-specified drop-out and drop-in rates, but then performs a repeat analysis using drop-out rates ranging from 0.01 to 0.99 and produces a graph of the resulting LRs. In addition, the program performs a "Tippett test" (Gill et al. 2008) and produces a plot that shows expected LRs for the mixture being tested when simulated non-contributors are used as the comparison sample. This is similar to the non-contributor analysis that was performed during the validation of FST.

LikeLTD (Balding et al. 2009) was developed by David Balding at the University College in London. LikeLTD performs a maximum likelihood analysis. The LR is set up in the same way as LoComatioN, Forensim, and FST, but the drop-out and drop-in rates are not specified. Instead, the rates are chosen to separately maximize the likelihood in the numerator and in the denominator and then the ratio of the two likelihoods is computed. Thus, in this type of setup, the estimated drop-out and drop-in rates may be different under the prosecution hypothesis in the numerator and the defense hypothesis in the denominator.

Lab Retriever (Lohmueller and Rudin 2013) uses the LikeLTD software, coupled with estimation of drop-out probability using electropherogram peak heights.

This publication describes the use of Lab Retriever by defense experts in a case in California. The use of the software was challenged by the prosecution, a Frye hearing was conducted, and the results were accepted by the Court.

The Forensic Statistical Tool (FST) (Mitchell et al. 2011, 2012) is the LR-based method developed and validated by The Office of Chief Medical Examiner of the City of New York. FST uses the same LR setup as True Allele, LoComatioN, Forensim, and likeLTD. Drop-out probabilities are incorporated in the same manner as those in LoComatioN, Forensim, and likeLTD. FST, however, employs empirically-determined drop-out and drop-in probabilities that were estimated using over 2000 amplifications of purposeful DNA mixtures. Drop-out probabilities in FST depend on the amount of DNA in a sample, the number of contributors to the sample, their relative contributions, and the DNA locus tested. Another difference between FST and these other three programs is that FST does not model population substructure, but instead assumes that unknown contributors to a mixture are unrelated to one another and to any known or hypothesized contributors to the mixture.

References

Balding, D., and J. Buckleton. 2009. Interpreting low template DNA profiles. *Forensic Sci Int Genet* 4:1–10.

Brenner, C., R. Fimmers, and M.P. Baur. 1996. Likelihood ratios for mixed stains when the number of donors cannot be agreed. *Int J Legal Med* 109(4):218–219.

Buckleton, J.S., I.W. Evett, and B.S. Weir. 1998. Setting bounds for the likelihood ratio when multiple hypotheses are postulated. *Sci Justice* 38:23–26.

Buckleton, J.S., J.M. Curran, and P. Gill. 2007. Towards understanding the effect of uncertainty in the number of contributors to DNA stains. *Forensic Sci Int Genet* 1:20–28.

Butler, J., M. Coble, R. Cotton et al. 2011. Mixture Interpretation Using Scientific Analysis. http://www.cstl.nist.gov/strbase/training/MixtureWorkshop-ISHI2011-no-lit.pdf.

Clayton, T.M., J.P. Whitaker, R. Sparkes, and P. Gill. 1998. Analysis and interpretation of mixed forensic stains using DNA STR profiling. *Forensic Sci Int* 91:55–70.

Cowell, R.G., S.L. Lauritzen, and J. Mortera. 2007. Identification and separation of DNA mixtures using peak area information. *Forensic Sci Int* 166(1):28–34.

Cowell, R.G., S.L. Lauritzen, and J. Mortera. 2011. Probabilistic expert systems for handling artifacts in complex DNA mixtures. *Forensic Sci Int Genet* 5:202–209.

Curran, J.M., P. Gill, and M.R. Bill. 2005. Interpretation of repeat measurement DNA evidence allowing for multiple contributors and population substructure. *Forensic Sci Int* 148:47–53.

Egeland, T., I. Dalen, and P.F. Mostad. 2003. Estimating the number of contributors to a DNA profile. *Int J Legal Med* 117:271–275.

Evett, I.W., C. Buffery, G. Willott, and D. Stoney. 1991. A guide to interpreting single locus profiles of DNA mixtures in forensic cases. *J Forensic Sci Soc* 31:41–47.

Evett, I.W., L.A. Foreman, J.A. Lambert, and A. Emes. 1998a. Using a tree diagram to interpret a mixed DNA profile. *J Forensic Sci* 43(3):472–476.

Evett, I.W., P.D. Gill, and J.A. Lambert. 1998b. Taking account of peak areas when interpreting mixed DNA profiles. *J Forensic Sci* 43(1):62–69.

Gill, P., R.L. Sparkes, R. Pinchinet al. 1998. Interpreting simple STR mixtures using allelic peak areas. *Forensic Sci Int* 91:41–53.

Gill, P., C. Brenner, J. Buckleton et al. 2006. DNA commission of the International Society of Forensic Genetics: Recommendations on the interpretation of mixtures. *Forensic Sci Int* 160:90–101.

Gill, P., A. Kirkham, and J. Curran. 2007. *LoComatioN*: A software tool for the analysis of low copy number DNA profiles. *Forensic Sci Int* 166:128–138.

Gill, P., J. Curran, C. Neuman et al. 2008. Interpretation of complex DNA profiles using empirical models and a method to measure their robustness. *Forensic Sci Int Genet* 2:91–103.

Gill, P., L. Gusmão, H. Haned et al. 2012. DNA commission of the International Society of Forensic Genetics: Recommendations on the evaluation of STR typing results that may include drop-out and/or drop-in using probabilistic methods. *Forensic Sci Int Genet* 6:679–688.

Haned, H. 2011. *Forensim*: An open-source initiative for the evaluation of statistical methods in forensic genetics. *Forensic Sci Int* 5:265–268.

Haned, H., L. Pène, J.R. Lobry et al. 2011a. Estimating the number of contributors to forensic DNA mixtures: Does maximum likelihood perform better than maximum allele count? *J Forensic Sci* 56(1):23–28.

Haned, H., L. Pène, F. Sauvage et al. 2011b. The predictive value of the maximum likelihood estimator of the number of contributors to a DNA mixture. *Forensic Sci Int Genet* 5:281–284.

Haned, H., K. Slooten, and P. Gill. 2012. Exploratory data analysis for the interpretation of low template DNA mixtures. *Forensic Sci Int Genet* 6:762–774.

Ladd, C., H.C. Lee, N. Yang, and F.R. Bieber. 2001. Interpretation of complex forensic DNA mixtures. *Croat Med J* 42(3):244–246.

Lohmueller, K., and N. Rudin. 2013. Calculating the weight of evidence in low-template forensic DNA casework. *J Forensic Sci* 58:S243–S249.

Mitchell, A.A., J. Tamariz, K. O'Connell et al. 2011. Likelihood ratio statistics for DNA mixtures allowing for drop-out and drop-in. *Forensic Sci Int Genet Sup Ser* 3:e240–e241.

Mitchell, A.A., J. Tamariz, K. O'Connell et al. 2012. Validation of a DNA mixture statistics tool incorporating allelic drop-out and drop-in. *Forensic Sci Int Genet* 6:749–761.

Paoletti, D.R., T.E. Doom, C.M. Krane et al. 2005. Empirical analysis of the STR profiles resulting from conceptual mixtures. *J Forensic Sci* 50:1361–1366.

Perlin, M.W., and B. Szabady. 2001. Linear mixture analysis: A mathematical approach to resolving mixed DNA samples. *J Forensic Sci* 46:1372–1378.

Perlin, M.W., and A. Sinelnikov. 2009. An information gap in DNA evidence interpretation. *PLoS One* 4:e8327.

Perlin, M.W., M.M. Legler, C.E. Spencer et al. 2011. Validating TrueAllele® DNA mixture interpretation. *J Forensic Sci* 56:1430–1447.

Perez, J, A.A. Mitchell, N. Ducasse et al. 2011. Estimating the number of contributors to two-, three-, and four-person mixtures containing DNA in high template and low template amounts. *Croat Med J* 52(3):314–326.

Schneider, P.M., P. Gill, and A. Carracedo. 2006. Editorial on the recommendations of the DNA commission of the ISFG on the interpretation of mixtures. *Forensic Sci Int* 160:189.

Shrestha, S., S.A. Strathdee, K.W. Broman et al. 2006. Unknown biological mixtures evaluation using STR analytical quantification. *Electrophoresis* 27:409–415.

SWGDAM Interpretation Guidelines for Autosomal STR Typing by Forensic DNA Testing Laboratories. 2010. http://www.fbi.gov/about-us/lab/biometric-analysis/codis/swgdam.pdf.

Tvedebrink, T., P.S. Eriksen, H.S. Mogensen et al. 2009. Estimating the probability of allelic drop-out of STR alleles in forensic genetics. *Forensic Sci Int Genet* 3:222–226.

Tvedebrink, T., P.S. Eriksen, H.S. Mogensen et al. 2012. Statistical model for degraded DNA samples and adjusted probabilities for allelic drop-out. *Forensic Sci Int Genet* 6:97–101.

Forensic DNA Typing and Quality Assurance

DANIEL VANEK
KATJA DROBNIČ

Contents

8.1	Forensic DNA Typing and Quality Assurance	205
	8.1.1 QA in the United States	206
	8.1.2 European DNA Legislation	207
	8.1.3 QA in Europe: European Network of Forensic Science Institutes and International Organization for Standardization	207
	8.1.4 How to Establish an ISO 17025–Based Quality Management System?	214
	8.1.5 PT and Collaborative Exercises	217
	8.1.6 QA Outside United States and Europe	222
	8.1.7 Major Differences in Requirements between DAB versus ISO 17025	224
	8.1.8 Selection of PT Provider (Accredited versus Nonaccredited) from the View of Forensic Laboratories	224
8.2	Validation	226
	8.2.1 Definitions of Validation	227
	8.2.2 Level of Validation	228
	8.2.3 Procedure of Internal Validation	233
	8.2.3.1 Minimum Number of Samples for Internal Validation	234
	8.2.4 Documentation	234
	8.2.5 Reference Materials	235
	8.2.5.1 NIST SRMs	236
8.3	Costs and Benefits of the QA Implementation	238
8.4	Potential for Error in Forensic DNA Analysis	239
	8.4.1 Mishandling of Samples	240
	8.4.2 Contamination	241
	8.4.3 Error Rates in Forensic DNA Testing	246
	8.4.3.1 Should an Error Be Included in Calculations?	247
References		247

8.1 Forensic DNA Typing and Quality Assurance

The quality of DNA typing results is defined by the amount and condition of the sample being processed and strongly influenced by the practices taken in the forensic DNA laboratory. The recent years can be characterized by the movement toward developing and implementing quality assurance (QA) guidelines and quality control measures as a specific part of all forensic science disciplines. QA guidelines encompass everything from staff requirements to laboratory safety, control of document, collection of materials, validation of methods, reference material and controls, as well as criteria for interpretation and reporting of results. The American Bar Association has recommended that, "Crime laboratories and medical examiner officers should be accredited, examiners should be certified, and procedures should be standardized" (American Bar Association 2006). We can state that a QA system, if implemented properly, forces the laboratory to continuously improve its practices and procedures. Thus, the QA system improves the integrity of results released by the laboratory and provides evidence of its ability to perform tests reliably and competently.

8.1.1 QA in the United States

The individual identification, or (more properly) individualization, of a biological material has experienced rapid development since the employment of DNA analysis methods in the late 1980s. The QA process in the United States is a good example of a systematic evolution of a set of standards (Box 8.1). The group that played a key role in the preparation of guidelines for forensic DNA analysis was the Technical Working Group on DNA Analysis Methods (TWGDAM). This Federal Bureau of Investigation (FBI)–sponsored working group published the first set of QA guidelines for restriction fragment length polymorphism (RFLP) analysis in 1989. TWGDAM also drew up a series of recommendations and guidelines for the areas of QA, mitochondrial DNA (mtDNA) sequencing, proficiency tests, auditing, and statistical evaluation of the DNA typing results. The DNA Advisory Board (DAB) was established by the Director of the FBI under the DNA Identification Act of 1994. The DAB combined elements of the TWGDAM guidelines with emerging contemporary issues and submitted their recommendations to the Director of the FBI in 1997. The FBI Director subsequently issued the Quality Assurance Standards for Forensic DNA Testing Laboratories, which took effect on October 1, 1998, with an additional formal (overlapping) set of standards for Convicted Offender DNA Testing Laboratories (Federal Bureau of Investigation 2000). The revised Quality Assurance Standards for Databasing Laboratories took effect on July 1, 2009. In 1999, the name of TWGDAM was changed to the Scientific Working Group on DNA Analysis Methods (SWGDAM). The QA standards of DAB were preceded by the recommendations of the National Research Council (NRC) of the National Academy of Sciences. The main goal of the NRC Committee on DNA Technology in Forensic Sciences was to evaluate the reliability of DNA testing (National Research Council 1992). One of the NRC suggestions stated that there is an urgent need for an external monitoring of forensic laboratory QA and quality control procedures. The second commission of the National Academy of Sciences published *The Evaluation of Forensic DNA Evidence* in 1996. The NRC II report not only very clearly addressed the use of statistics but also suggested that laboratories should seek accreditation, and that proficiency testing (PT) should be performed on a regular basis. Its significant impact on the field of forensic genetics led to the Justice for All Act, which was signed by President George Bush in 2004. Some of the purposes of the

---

BOX 8.1

WHAT IS STANDARD?

A standard is an established norm or requirement, usually a formal document that establishes uniform technical criteria, methods, processes, and practices.

(From ISO. Available at http://www.iso.org. Accessed February 24, 2013.)

Act are to protect crime victims' rights, eliminate the substantial backlog of DNA samples collected from crime scenes and convicted offenders, and improve and expand the DNA testing capacity of federal, state, and local crime laboratories.

8.1.2 European DNA Legislation

The implementation of QA standards in European forensic DNA laboratories started significantly later compared to that in the United States. The recommendation no. 92 of the European Council in 1992 only defined basic rules on the use of DNA analysis within the criminal justice system. The EU Council recommendation from 1997 urged EU member states to establish their own national DNA databases, standardize the DNA technology used, implement national rules for personal data protection, define the rules for the exchange of DNA analysis results, and to create a network of compatible national DNA databases. The EU Council recommendation from 2001 defined the European Standard Set (ESS) of loci (D3S1358, D8S1179, D18S51, D21S11, FGA, TH01, and vWA) that ensured an overlap of the loci included in national databases across Europe. The Prüm Treaty in 2005 enabled European police forces to compare and exchange data more easily (Prüm Convention, 2005). The decision of the Justice and Home Affairs Council (Decision 2008/615/JHA) introduced a timeline for other EU nations to become compliant. The definition of the ESS of loci from 2001 (EU Council Resolution of 25 June 2001) was replaced by the EU Council recommendation from 2009 (EU Council Resolution of 30 November 2009), and now the set of loci comprises the following DNA markers: **D3S1358**, **VWA**, **D8S1179**, **D21S11**, **D18S51**, **HUMTH01**, **FGA**, D1S1656, D2S441, D10S1248, D12S391, and D22S1045 (loci overlapping with 13 CODIS [Combined DNA Index System] loci used in the United States are marked in bold).

8.1.3 QA in Europe: European Network of Forensic Science Institutes and International Organization for Standardization

Another European forensic science–related initiative dates back to 1993 when the directors of 11 Western Europe governmental forensic laboratories started to meet and the official founding meeting of the European Network of Forensic Science Institutes (ENFSI) took place in 1995. The aim of ENFSI is to ensure the quality of the development and delivery of forensic science throughout Europe. To achieve this goal, one of ENFSI's directions was the accreditation of all member European forensic laboratories, and thus some ENFSI members started with the implementation of the ISO (International Organization for Standardization) Guide 25:1990 (General requirements for the competence of calibration

BOX 8.2

WHAT IS ISO?

The *International Organization for Standardization* is a global federation of national standards bodies (ISO member bodies), e.g., the Czech Accreditation Institute (CAI) in Czech Republic, Slovenian Accreditation (SA), British Standard Institution (BSI) in the U.K. or the American National Standards Institute (ANSI) in the U.S.A.

(From ISO. Available at http://www.iso.org. Accessed February 24, 2013.)

Table 8.1 Documents of the ENFSI QCC

Title of Document	Year of Release
Management of Reference Materials, Reference Collections, and Databases	2002
Performance based standards for Forensic Science Practitioners	2004
Code of Conduct	2005
Guidance on the Conduct of Proficiency Tests and Collaborative Exercises within ENFSI	2005
Guidance on Best Practice for Management of Case Handling System in Forensic Laboratory	2005
Validation and Implementation of (New) Methods	2006
Guidance for Uncertainty of Measurement in Quantitative Analyses or Testing	2006
Template for Standard Operating Procedure	2006
Guidance for Best Practice Sampling in Forensic Science	2007
IT Software Validation	2007
Guidance on the Production of Best Practice Manuals within ENFSI	2008
Policy on Standards for Accreditation	2010
Policy on Proficiency Tests and Collaborative Exercises within ENFSI	2010

and testing laboratories) and ISO 45001:1989 (General criteria for the operation of testing laboratories) (Box 8.2). Both norms were replaced by the ISO 17025 in 1999. ISO 17025 was further revised in 2005. The ENFSI requires that all member laboratories be accredited, but so far almost one-half of all member laboratories remain nonaccredited—although the number of accredited laboratories within ENFSI has increased from 20 (2006) to 36 (2009). The current legislative act of the European Parliament (Council framework Decision 2009/905/JHA of November 30, 2009) requires that all forensic service providers carrying out laboratory activities in DNA testing are ISO 17025–accredited by November 2013. The ENFSI standing committee on Quality and Competence (QCC) have approved and released several documents to spur institutions toward achieving accreditation (Table 8.1). Besides QCC documents that apply to all types of forensic laboratories, the ENFSI Working Group also released more detailed recommendations that specifically apply to the field of forensic testing (Boxes 8.3 and 8.4). The recommendations of the ENFSI DNA Working Group are summarized in Table 8.2.

BOX 8.3

WHAT IS AN ISO STANDARD?

ISO standards are developed by technical committees comprising experts from the industrial, technical and business sectors which have asked for the standards. These experts may be joined by representatives of government agencies, testing laboratories, consumer associations, nongovernmental organizations and academic circles.

WHAT IS ISO 17025?

The ISO/IEC 17025 describes general requirements for the competence of testing and calibration laboratories.

(From ISO. Available at http://www.iso.org. Accessed February 24, 2013.)

BOX 8.4

WHAT IS THE ROLE OF INTERNATIONAL STANDARDS?

International Standards and their use in technical regulations on products, processes, and services play an important role in sustainable development and trade facilitation through the promotion of safety, quality, and technical compatibility.

(From ISO. Available at http://www.iso.org. Accessed February 24, 2013.)

The ENFSI (DNA WG) membership is quite restrictive, and normally only one representative per country is allowed. Exceptions from this rule are made especially for the ENFSI founding countries. ENFSI had 64 members in 36 countries by the end of 2012.

The number of forensic laboratories in Europe far exceeds the number of DNA laboratories organized within ENFSI, and therefore there exists a place for several scientific societies that include experts from DNA laboratories providing forensic testing services. A very significant role in the area of standardization and QA is played by the International Society for Forensic Genetics (ISFG). This scientific society not only organizes well-attended international conferences but also publishes a remarkable number of recommendations that cover crucial parts of forensic DNA testing. Table 8.3 lists the associations, working groups, and societies that are connected with the field of forensic DNA testing. Some of the bodies are active mainly in certain geographical regions (German DNA Profiling, Grupo Iberoamericano de Trabajo en Analisis de DNA, etc.), whereas other organizations work on an international basis (ISFG, International Association for Identification, etc).

Quality systems, either U.S.- or Europe-based, do not provide directions on how an organization should manage its affairs, but rather sets a series of requirements for an efficient system of management of forensic laboratories. The Quality Assurance Standards for Forensic DNA Testing Laboratories, as issued by the FBI Director and the European Standard ISO 17025 for Testing and calibration laboratories, have basically the same management and technical requirements. Most forensic laboratories are American Society of Crime Laboratory Directors/Laboratory Accreditation Board (ASCLD/LAB)- or 17025-accredited, but there also several other standardization and QA bodies (Table 8.4). The International Laboratory Accreditation Cooperation (ILAC) issued a series of recommendations that address specific areas of QA systems accredited according to ISO 17025:2005. Table 8.5 summarizes the ILAC guidelines and recommendations. The most important set for forensic laboratories is the ILAC G-19 guideline (ILAC G-19, 2002). The ISO/IEC 17025 describes the general requirements for the competence of testing and calibration laboratories, and ILAC G-19 is an interpretation of the ISO normative with respect to the type of testing used and the techniques involved. ILAC G-19 was meant for all forensic disciplines and thus contains little or no "DNA-specific" requirements such as how often should different types of laboratory instruments be calibrated, how often should the laboratory participate in external PT, and how are individuals within the laboratory involved in PT.

Table 8.2 Recommendations for Area of DNA Analysis

ISFG (International Society for Forensic Genetics)
Prinz, M., Carracedo, A., Mayr et al. 2007. DNA Commission of the International Society for Forensic Genetics (ISFG): Recommendations regarding the role of forensic genetics for disaster victim identification (DVI). *Forensic Sci. Int. Genet.* 1:3–12.
Gjertson, D.W., Brenner, C.H., Baur, M.P. et al. 2007. ISFG: Recommendations on biostatistics in paternity testing. *Forensic Sci. Int. Genet.* 1:223–231.
Gill, P., Brenner, C.H., Buckleton, J.S. et al. 2006. DNA commission of the International Society of Forensic Genetics: Recommendations on the interpretation of mixtures. *Forensic Sci Int.* 160:90–101.
Gusmao, L., Butler, J.M., Carracedo, A. et al. 2006. DNA Commission of the International Society of Forensic Genetics. DNA Commission of the International Society of Forensic Genetics (ISFG): An update of the recommendations on the use of Y-STRs in forensic analysis. *Forensic Sci. Int.* 157:187–197.
Morling, N., Allen, R.W., Carracedo, A. et al. 2002. Paternity Testing Commission of the International Society of Forensic Genetics: Recommendations on genetic investigations in paternity cases. *Forensic Sci. Int.* 129:148–157.
Gill, P., Brenner, C., Brinkmann, B. et al. 2001. DNA Commission of the International Society of Forensic Genetics: Recommendations on forensic analysis using Y-chromosome STRs. *Forensic Sci. Int.* 124:5–10.
Carracedo, A., Bär, W., Lincoln, P. et al. 2000. DNA Commission of the International Society for Forensic Genetics: Guidelines for mitochondrial DNA typing. *Forensic Sci. Int.* 110:79–85.
Bär, W., Brinkmann, B., Budowle, B. et al. 1997. DNA recommendations. Further report of the DNA Commission of the ISFG regarding the use of short tandem repeat systems. *Forensic Sci. Int.* 87:179–174.
Bär, W., Brinkmann, B., Lincoln, P. et al. 1993. Editorial: Statement by DNA Commission of the International Society for Forensic Haemogenetics concerning the National Academy of Sciences report on DNA Technology in Forensic Science in the USA. *Forensic Sci. Int.* 59:1–2.
Bär, W., Brinkmann, B., Lincoln P. et al. 1992. Editorial: Recommendations of the DNA Commission of the International Society for Forensic Haemogenetics relating to the use of PCR-based polymorphisms. *Forensic Sci. Int.* 55:1–3.
Brinkmann, B., Bütler, R., Lincoln, P. et al. 1992. Editorial: 1991 Report concerning recommendations of the DNA Commission of the International Society for Forensic Haemogenetics relating to the use of DNA polymorphisms. *Forensic Sci. Int.* 52:125–130.
Brinkmann, B., Bütler, R., Lincoln, P. et al. 1989. Editorial: Recommendations of the Society for Forensic Haemogenetics concerning DNA polymorphisms. *Forensic Sci. Int.* 43:109–111.
ENFSI DNA Working Group
DNA contamination prevention guidelines (version 2010)
Minimum validation guidelines in DNA profiling (version 2010)
Recommendations for the training of DNA staff (version 2010)
INTERPOL
Anticontamination guidelines
http://www.interpol.int/Public/Forensic/dna/dnahandbook.asp
OTHER
Bruce Budowle, Paolo Garofano, Andreas Hellman, Melba Ketchum, Sree Kanthaswamy, Walther Parson, Wim van Haeringen, Steve Fain and Tom Broad (2005) Recommendations for animal DNA forensic and identity testing, *International Journal of Legal Medicine*, Volume 119, Number 5, 295–30.

Forensic DNA Typing and Quality Assurance

Table 8.3 Associations, Working Groups, and Societies for the Field of Forensic DNA Testing

Organization	Website	Description
AAFS (American Academy of Forensic Sciences)	www.aafs.org	The American Academy of Forensic Sciences is a multidisciplinary professional organization that provides leadership to advance science and its application to the legal system. The objectives of the Academy are to promote integrity, competency, education, foster research, improve practice, and encourage collaboration in the forensic sciences.
ENFSI (European Network of Forensic Science Institutes)	www.ENFSI.org	ENFSI is the association of public forensic laboratory directors in Europe. ENFSI has been established with the purpose of sharing knowledge, exchanging experiences and coming to mutual agreements in the field of forensic science.
ISABS (International Society for Applied Biological Sciences)	www.ISABS.hr	ISABS is a nonprofit organization founded to promote, enhance and extend research, development and education in molecular biology as applied to clinical and molecular medicine, focusing on, but not limited to, molecular genetics, genomics, proteomics, forensic and anthropological genetics, and biotechnology.
DAB (DNA Advisory Board)	www.FBI.gov	DAB was established by the Director of the FBI under the DNA Identification Act of 1994. The Objectives of the DAB were essentially to develop quality assurance standards for forensic DNA testing.
SWGDAM (Scientific Working Group on DNA Analysis methods)	www.FBI.gov	Working group of forensic scientists that meets under the guidance of the Federal Bureau of Investigation (FBI)
TWGDAM (Technical Working Group on DNA Analysis Methods) changed to SWGDAM in 1999		
ISFG (International Society for Forensic Genetics)	www.ISFG.org	ISFG is an international association promoting scientific knowledge in the field of genetic markers analyzed for forensic purposes. ISFG was formerly known as the *International Society of Forensic Haemogenetics* (ISFH)
STADNAP	www.isfg.org/EDNAP/STADNAP	STADNAP was a network project funded by the EU and carried out by 20 partner laboratories.

(continued)

Table 8.3 Associations, Working Groups, and Societies for the Field of Forensic DNA Testing

Organization	Website	Description
EDNAP (European DNA Profiling Group)	www.isfg.org/EDNAP	EDNAP is a working group of ISFG.
GEDNAP (German DNA Profiling Group)	www.GEDNAP.org	Organizer of the GEDNAP Proficiency Tests
ABC (American Board of Criminalistics)	www.criminalistics.com	ABC provides professional competency certification in criminalistics.
FSS (Forensic Science Society)	www.forensic-science-society.org.uk	The main aim of the society is to encourage communication and collaboration by providing an arena in which forensic practitioners, researchers, academics and those working in related fields can congregate, communicate and invoke development of areas such as best practice, research, publication, quality, and ethics in forensic casework.
MAFS (Mediterranean Academy of Forensic Sciences)	www.mafs.info	The aim of the Academy is to realize an exchange of information and a collaboration among all the experts on the Forensic Sciences.
IAFS (International Association of Forensic Sciences)	www.iafs2011.mj.pt	IAFS is a worldwide association to bring together academics and practicing professionals of various disciplines in forensic science.
INTERPOL	www.interpol.int/public/forensic/dna/default.asp	Interpol European Working Party on DNA Profiling (IEWPDP) Interpol DNA Monitoring Expert Group (MEG)
IAI (International Association for Identification)	www.theiai.org	The IAI is the oldest and largest forensic professional association in the world
GITAD (Grupo Iberoamericano de Trabajo en Analisis de DNA)	www.gitad.org	
HITA (Human Identity Trade Association)	www.humanidentity.org	HITA is a nonprofit organization dedicated to advancing the human identity testing industry and increasing public awareness of forensic and parentage DNA testing products and services.
NFSTC (National Forensic Science Technology Center)	www.nfstc.org	NFSTC is dedicated to supporting the justice community through innovative research, programs, evaluation of the latest technologies, forensic science education, and laboratory quality reviews.
AFQAM (Association of Forensic Quality Assurance Managers)	www.afqam.org	AFQAM promotes standardized practices and professionalism in quality assurance management for the forensic community.

Table 8.4 Accreditation, Standardization, and Quality Assurance Bodies

Organization	Website	Description
ISO (International Organization for Standardization)	www.ISO.org	ISO is the world's largest developer and publisher of International Standards. ISO is a network of the national standards institutes of 159 countries (one member per country).
NIST (National Institute of Standards and Technology)	www.NIST.gov	NIST, an agency of the U.S. Department of Commerce, was founded in 1901 as the nation's first federal physical science research laboratory.
ILAC (International Laboratory Accreditation Cooperation)	www.ILAC.org	ILAC is the world's principal international forum for the development of laboratory accreditation practices and procedures, the promotion of laboratory accreditation as a trade facilitation tool, the assistance of developing accreditation systems, and the recognition of competent test facilities around the globe.
EA (European co-operation for Accreditation)	www.european-accreditation.org	Mission: defining, harmonizing, and building consistency in accreditation as a service in Europe, by ensuring common interpretation of the standards used by its members
IAF (International Accreditation Forum)	www.iaf.nu	The primary purpose of IAF is to ensure that its accreditation body members only accredit bodies that are competent to do the work they undertake and are not subject to conflicts of interest.
ASCLD (American Society of Crime Laboratory Directors)	www.ascld.org	ASCLD is a nonprofit professional society of crime laboratory directors and forensic science managers dedicated to promoting excellence in forensic science through leadership and innovation.
ASCLD/LAB (Laboratory accreditation Board of ASCLD)	www.ascld-lab.org	ASCLD/LAB offers voluntary accreditation to public and private crime laboratories in the United States and around the world. ASCLD/LAB was originally created as a committee of its mother organization, the American Society of Crime Laboratory Directors (ASCLD) in 1981. In 1984, ASCLD/LAB became a separate corporate entity with its own Board of Directors that is elected by a Delegate Assembly composed of the directors of accredited laboratories and laboratory systems.

(continued)

Table 8.4 Accreditation, Standardization, and Quality Assurance Bodies

Organization	Website	Description
ANSI (American National Standards Institute)	www.ansi.org	ANSI is a nonprofit organization that administers and coordinates the U.S. voluntary standardization and conformity assessment system.
FQS (Forensic Quality Services)	www.forquality.org	FQS is a not-for-profit membership organization providing the ISO/IEC 17025 accreditations to forensic testing agencies in the United States.
FSAB (Forensic Specialties Accreditation Board)	www.thefsab.org	The goal of this program is to establish a mechanism whereby the forensic community can assess, recognize, and monitor organizations or professional boards that certify individual forensic scientists or other forensic specialists.
IAAC (InterAmerican Accreditation Cooperation)	www.iaac.org.mx	IAAC is an association of accreditation bodies in the Americas.

Table 8.5 ILAG Guidelines and Procedures*

ILAC Number	Title/Subject
G3:1994	Guidelines for Training Courses for Assessors
G8:03/2009	Guidelines on the Reporting of Compliance with Specification
G9:2005	Guidelines for the Selection and Use of Reference Materials
G10:1996	Harmonized Procedures for Surveillance & Reassessment of Accredited Laboratories
G11:07/2006	ILAC Guidelines on Qualifications & Competence of Assessors and Technical Experts
G12:2000	Guidelines for the Requirements for the Competence of Reference Materials Producers
G13:08/2007	Guidelines for the Requirements for the Competence of Providers of Proficiency Testing Schemes
G17:2002	Introducing the Concept of Uncertainty of Measurement in Testing in Association with the Application of the Standard ISO/IEC17025
G18:04/2010	Guideline for the Formulation of Scopes of Accreditation for Laboratories
G19:2002	Guidelines for Forensic Science Laboratories
G20:2002	Guidelines on Grading of Non-conformities
G22:2004	Use of Proficiency Testing as a Tool for Accreditation in Testing
G24:2007	Guidelines for the Determination of Calibration Intervals of Measuring Instruments
P14:12/2010	ILAC Policy for Uncertainty in Calibration
P13:10/2010	Application of ISO/IEC 17011 for the Accreditation of Proficiency Testing Providers
P9:11/2010	ILAC Policy for Participation in Proficiency Testing Activities

* Guidelines and procedures available online at http://www.ilac.org/guidanceseries.html.

8.1.4 How to Establish an ISO 17025–Based Quality Management System?

The backbone of any quality management system (QMS) is the documentation. The number of instructions for the identification, layout, approval, and archiving of documents can be found on the websites of the different organizations involved in forensic testing.

The easiest (albeit underhanded) way is to obtain a copy of QMS documentation from an accredited body. This "template" can be relatively easily updated by implementing laboratory-specific solutions and procedures. The QA team of the forensic DNA laboratory wishing to be accredited should include an experienced QA manager who actually works in this institute or laboratory and the technical director of the laboratory. Another possibility is to use an external adviser; however, a lack of knowledge about that particular laboratory may lead to serious problems during an external audit performed by the national accreditation body.

The QMS documentation must cover all ISO 17025 requirements. The laboratory must thus describe in detail how particular requirements are complied with. We give an example on how to comply with the ISO 17025 requirement 5.3.3 (Accommodation and Environmental conditions), which requires that "There shall be effective separation between neighboring areas in which there are incompatible activities. Measures shall be taken to prevent cross-contamination." This ISO 17025 requirement is further interpreted by ILAC G-19 guideline that states that, "Special care is needed in forensic testing laboratories involved in the analysis or determination of trace levels of materials, including DNA. Physical separation of high-level and low-level work is required. Where special areas are set aside for this type of work, access to these areas should be restricted and the work undertaken carefully controlled. Appropriate records should be kept to demonstrate this control. It may also be necessary to carry out 'environmental monitoring' of equipment, work areas, clothing and consumables."

The section of the laboratory quality manual may describe the compliance with requirement 5.3.3 in the following way:

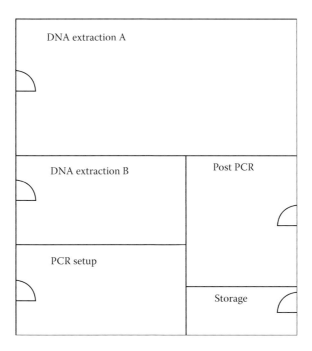

Figure 8.1 Sample layout of DNA laboratory.

The laboratory part of the DNA unit within the Forensic Institute is divided into five separate areas: DNA extraction A, DNA extraction B, polymerase chain reaction (PCR) setup, post PCR, and storage. The layout of the areas is shown in Figure 8.1; a description of the activities performed in different areas is given below.

Description of activities in the room *DNA extraction A*:
- Primary examination of submitted stains, artifacts and belongings, and photo documentation (described in SOP-01-01-99-Ac)
- Preliminary tests (blood, sperm, saliva, urine), hair analysis (described in SOP-01-02-98-DNA)
- DNA sampling (described in SOP-02-02-98-DNA)
- Manual DNA extraction from submitted samples stains, artifacts, and belongings (blood, saliva, sperm, touch samples, hair, bone) (described in SOP-03-03-98-DNA)
- Input of data to the Laboratory Information Management System (LIMS) (described in SOP-02-01-98-St)

Description of activities in the room *DNA extraction B*:
- Manual DNA extraction from reference samples (blood, mouth swab) (described in SOP-04-02-98-DNA)
- Automated DNA extraction from reference samples (blood, mouth swab) (described in SOP-05-02-98-DNA)
- Input of data to the LIMS (described in SOP-02-01-98-St)

Description of activities in the room *PCR setup*:
- Manual setup of PCR reactions (described in SOP-06-02-98-DNA)
- Manual setup of quantitative real-time PCR reactions (described in SOP-07-02-98-DNA)
- Preparation of DNA-free consumables and plastic by UV irradiation (described in SOP-08-03-98-DNA)
- Short-term storage of extracted DNA (described in SOP-09-01-98-DNA)
- Input of data to the LIMS (described in SOP-02-01-98-St)

Description of activities in the room *Post PCR*:
- PCR and quantitative real-time PCR (described in SOP-10-02-98-DNA)
- Setup of sequencing, single nucleotide polymorphism (SNP), and fragment (short tandem repeat [STR]) analysis (described in SOP-11-01-98-DNA)
- Sequencing, SNP, and STR analysis using the capillary electrophoresis (described in SOP-12-03-98-DNA)

Description of activities in the room *Storage*:
- Long-term storage of submitted artifacts (described in SOP-17-02-98-St)
- Long-term storage of the extracted DNA (described in SOP-09-01-98-DNA)

Access to all rooms is controlled, monitored, and restricted through the access control system (ACS) using magnetic cards. The use of ACS is described in the Standard Operating Procedure SOP-03-01-98-St.

Personnel who are allowed access to different areas within the DNA unit are required to wear a dedicated laboratory coat. The laboratory dressing code is described in the Standard Operating Procedure SOP-21-02-98-St.

The cleaning of all areas within the DNA laboratory unit is described in the Standard Operating Procedure SOP-09-02-98-St.

The monitoring of the background DNA contamination within the areas is controlled using the 4N6 XC test as described in the Standard Operating Procedure SOP-14-01-98-DNA.

8.1.5 PT and Collaborative Exercises

A serious problem, emerging with the continuously increasing sensitivity of the STR kits, is the cross-contamination of DNA samples during crime scene examination, transport, or processing in the laboratory. Another set of problems involves sample mix-up. Errors of this type are chronic and occur even at the best accredited DNA laboratories. Forensic DNA laboratories must ensure that their results are correct and that they meet the standards required by regulatory bodies. The performance of the particular DNA typing method used must be continuously controlled. Internal controls (PCR positive and negative controls) should be run with all batches of samples. Internal controls can reveal problems relating to assay sensitivity, laboratory contamination, and analysis of raw data from the sequencer using the specialized Expert System software line GeneScan, Genotyper, GeneMapper, i-Cubed, or TrueAllele (Box 8.5).

PT is a QA measure used to monitor performance and identify areas in which improvement may be needed. Proficiency tests may be classified as internal and external, and open and blind (Federal Bureau of Investigation 2000).

BOX 8.5

WHAT IS PT?

Proficiency testing (PT) is determination of laboratory performance by means of interlaboratory comparisons.

WHAT ARE INTERLABORATORY COMPARISONS?

Interlaboratory comparisons are organization, performance, and evaluation of tests on the same or similar items by two or more laboratories in accordance with predetermined conditions.

WHAT IS PT PROVIDER?

Proficiency testing (PT) provider is an organization that undertakes the design and conduct of proficiency testing scheme and takes responsibility for the evaluation.

WHAT DOES PROFICIENCY TESTING SCHEME MEAN?

Proficiency testing scheme is proficiency testing designed and operated in a specified area of testing, measurement, calibration, or inspection.

Source: ISO/IEC 17043:2010.

> **WHAT IS COLLABORATIVE EXERCISE?**
>
> A collaborative exercise does not require known excepted outcomes and is the most appropriate approach for troubleshooting, validation work, or the characterization of reference materials, although information for these purposes can also be generated incidentally from proficiency tests.
>
> (From ISO/IEC Guide 43-1:1997. Proficiency testing by interlaboratory comparisons—Part 1. Development and operation of proficiency testing schemes. International Organization for Standardization, Geneva. Available at http://www.iso.org. Accessed February 24, 2013.)

Internal proficiency tests are prepared and administered by the laboratory. External proficiency tests are conducted by an outside agency. An open proficiency test is a test that is identified to the participants as such. A blind proficiency test is a test that is submitted to the participants as a real case and the participant does not know it is a proficiency test. Blind PTs can be conducted in four different ways (Peterson et al. 2003a, 2003b):

1. The "case" is fully prepared by the law enforcement agency and sent to the target laboratory.
2. Specimen for the case is prepared by another laboratory.
3. The test is prepared within the laboratory and only the tested person (e.g., analyst) does not know he/she is participating in a test.
4. Random or purpose reanalysis of the case specimen by another analyst/laboratory/agency during the interrogation or appeals.

PT not only tests the suitability of the chosen method for the test conditions (e.g., the differential extraction is the sound procedure for the processing of postcoital vaginal swabs) but it also tests the ability of the laboratory to obtain the correct results. Table 8.6 summarizes some of the external proficiency test providers. Since testing the proficiency of the entire laboratory staff using commercially available PTs would be very costly, many accreditation bodies allow laboratories to pass one external PT per year, and the laboratory staff is tested using the internal tests.

Table 8.6 Proficiency Test Providers

GEDNAP Proficiency test	www.GEDNAP.org
IQAS Forensic Proficiency Testing Program	www.orchidcellmark.com/forensicdna/iqasproficiency.html
CTS (Collaborative Testing Services)	www.ctsforensics.com
LGC Standards Proficiency Testing	www.lgcpt.com
CAP (College of American Pathologists)	www.cap.org
Relationship Testing Workshop of the English Speaking Working Group of the ISFG	www.ISFG.org

Forensic DNA Typing and Quality Assurance

All PTs, external or internal, should test if the laboratory/individual is able to

1. Analyze the correct stain without sample mix-up
2. Determine the correct sample type (saliva, sperm, blood, etc.)
3. Avoid cross-contamination
4. Obtain the correct DNA profile or sequence
5. Come to the correct statistical interpretation
6. Report the results of the PT in writing in an error-free manner

It is mandatory that the design of the selected PT reflects the current technical level of the DNA typing and must be near practice-oriented as possible. The concept of GEDNAP (German DNA Profiling group) proficiency trials (Rand et al. 2002, 2004) meets all of the above set criteria, and participation in this PT is mandatory for many EU-based forensic

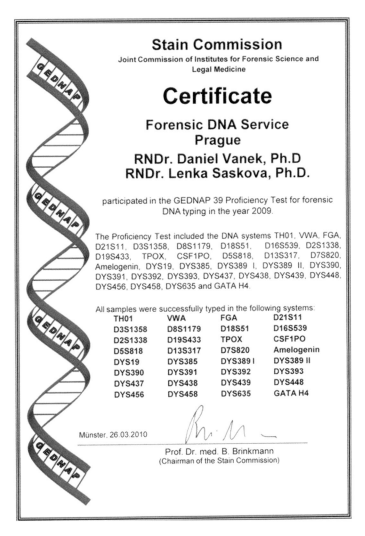

Figure 8.2 GEDNAP certificate—autosomal and Y-chromosome STR typing. GEDNAP, German DNA Profiling group.

DNA laboratories. The GEDNAP PTs are conducted once a year. The results of the analysis are evaluated by the organizer of the PT, and are presented in four categories:

Group I: Correct results obtained.
Group II: An allele in a mixture has not been detected.
Group III: This group has been removed and the option "reportable results" or "no reportable results" is no longer available.
Group IV: Incorrect results.

Certificates of participation in this PT are issued in the name of the institute that has actually undertaken the investigation, with the names of up to two staff members performing the tests (Figures 8.2, 8.3, and 8.4).

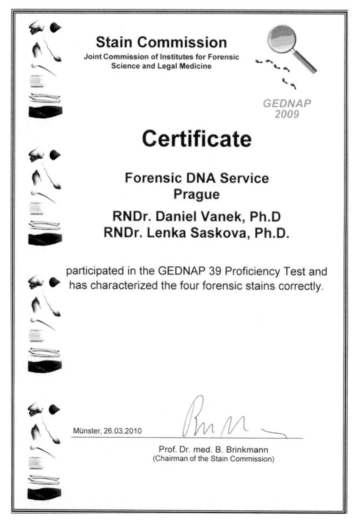

Figure 8.3 GEDNAP certificate—stain analysis.

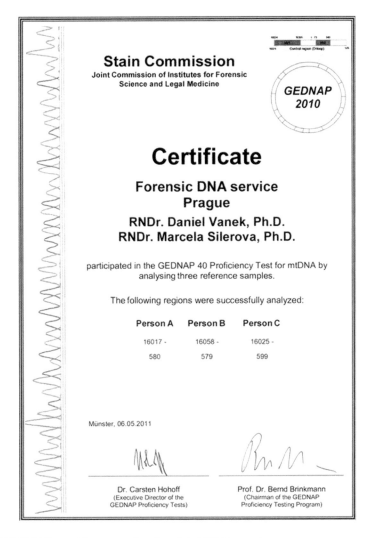

Figure 8.4 GEDNAP certificate—mitochondrial DNA testing.

Description of collaborative exercises (CE) is described in ENFSI document titled "Policy on Proficiency Tests and Collaborative Exercises within ENFSI" (Table 8.1): CEs do not require known expected outcomes and are the most appropriate approach for troubleshooting, validation work, or the characterization of reference materials, although information for these purposes can also be generated incidentally from proficiency tests (see Figure 8.5).

Differences between ISO 9001 and ISO 17025: Newcomers to the field of QA are not able to distinguish between certification and accreditation of the subject. Table 8.7 summarizes the major differences between those two standards (Box 8.6).

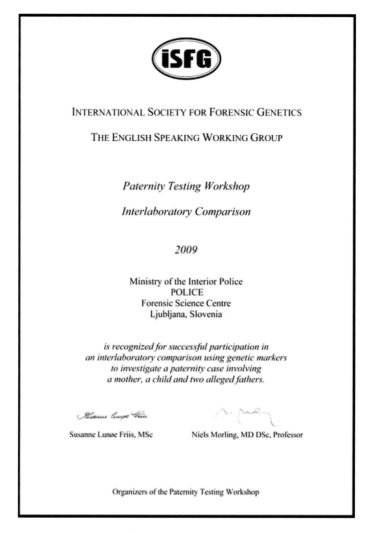

Figure 8.5 ISFG certificate—paternity testing interlaboratory comparison. ISFG, International Society for Forensic Genetics.

8.1.6 QA Outside United States and Europe

In Asia, among all laboratories working on DNA analysis, the following laboratories are accredited (some of them not on DNA analysis):

- South Korea—DNA analysis accredited by the Korea Laboratory Accreditation Scheme (KOLAS).
- Thailand—Central Institute for Forensic Science: ISO 17025 accreditation of DNA laboratory by the Department of Medical Sciences, Ministry of Health, Thailand, since January 2011.
- Singapore HSA—Biology/DNA: ASCLD/LAB Legacy since June 1996.

Table 8.7 Differences between ISO 9001 and ISO 17025

ISO 9001:2008	ISO 17025:2005
Certification not accreditation	Accreditation
Offered by the Certification Bodies (e.g., Lloyds Quality Assurance Register [LQAR], www.lrqa.com/)	Offered by the National Accreditation Bodies (e.g., ANSI)
Certification Bodies can be accredited to ISO 17021	National Accreditation Bodies must be members of ISO (International Standards Organization) and accredited to ISO 17011:2004
Applicable to all organizations irrespective of industry or product covers only management system	Applicable to the testing and calibration laboratories only; covers several technical competence requirements that are not covered in ISO 9001:2008
Certification against ISO 9001 does not in itself demonstrate the competence of the laboratory to produce technically valid data and results.	Accreditation is the procedure by which national body gives formal recognition that laboratory is competent to carry out technical valid forensic DNA results and work according to a management system. As well as that laboratories meet or exceed a list of standards depending on country they are from.
Certification can also be used as the measure of the competence of the staff	Accreditation applies only on the laboratory

BOX 8.6

WHAT IS QA?

Quality assurance (QA) comprises planned or systematic actions that are necessary to provide adequate confidence that a product or service will satisfy given requirements for quality.

WHAT IS QC?

Quality control (QC) comprises day-to-day operational techniques and activities used to fulfill requirements of quality.

WHAT IS VALIDATION?

Validation is the process of demonstrating that a laboratory procedure is robust, reliable, and reproducible in the hands of the personnel performing the test in that laboratory.

WHAT IS DEVELOPMENTAL VALIDATION?

Developmental validation (performed mainly by the manufacturers) is the acquisition of test data and determination of conditions and limitations of a new or novel DNA methodology for use on forensic samples.

WHAT IS INTERNAL VALIDATION?

Internal validation is an accumulation of test data within the laboratory to demonstrate that established methods and procedures perform as expected in the laboratory.

(From DNA Advisory Board Quality Assurance Standards. Available at http://www.fbi.gov/about-us/lab/biometric-analysis/codis/stds_testlabs. Accessed December 4, 2013.)

- Hong Kong Government Laboratory—Biology/DNA: ASCLD/LAB Legacy since June 1996. Likely seeking ISO 17025 accreditation in 2011 under the Hong Kong Laboratory Accreditation Scheme (HOKLAS).
- Indonesia—the Eijkman Institute for Molecular Biology is the foremost DNA laboratory in Indonesia; it is not accredited yet to our knowledge.
- Malaysia: KIMIA (Department of Chemistry)
  - Petaling Jaya (HQ): DNA laboratory has been accredited by the ASCLD/LAB since October 2005.
  - Sabah, Melaka, Perak, and Terengganu Branch laboratories—Biology (serology and hair screening only) ASCLD/LAB Legacy since January 2010.
  - Sarawak Branch laboratory—Biology/DNA: ASCLD/LAB Legacy since January 2010.

Other DNA laboratories in Taiwan, Macau, Brunei, Vietnam, and Philippines are not yet accredited. The Australian National Association of Testing Authorities (NATA) has similar requirements for DNA as those in the United States.

8.1.7 Major Differences in Requirements between DAB versus ISO 17025

In general, the requirements of ISO 17025 are less specific and stringent than the DAB Quality Assurance Standards for Forensic DNA Testing Laboratories (Table 8.8). At any rate, the laboratories accredited according to ISO 17025 must demonstrate compliance with the guidelines of the forensic professional community or with the requirements of any national accreditation body (like the interval of PT) that cover all requirements specifically noted in DAB.

There are two major differences between DAB and ISO 17025. The first one refers to requirements regarding personnel competence. ISO 17025 does not specifically define the criteria for personnel qualification and education, whereas DAB does. DAB defines the level of degree of education for the technical manager or leader, CODIS manager, examiner/analyst, and technician, and the program for their continuing education. The second difference refers to the requirement on participation in PT programs. ISO 17025 has no specific requirements regarding the participation in PT, whereas DAB explicitly defines that every staff involved in DNA analysis must participate in external PT on regular intervals (not exceeding 180 days) as well as criteria for evaluation of proficiency tests. In European countries, the requirements regarding the policy of participation in PTs and CEs that include, among others, the frequency of participation, identity of examiners involved, and establishment of criteria for evaluation of PTs and CEs results are detailed in documents published by national accreditation bodies as well as in ENFSI documents (Table 8.1).

8.1.8 Selection of PT Provider (Accredited versus Nonaccredited) from the View of Forensic Laboratories

Although the accreditation of testing laboratories has been in existence since the end of the 1980s, the accreditation of PT providers has only started few years ago (EUROLAB 2005; Schmidt et al. 2006). But as the demand for accreditation of forensic science providers has grown in past few years, it seems that the same trend will also affect PT providers. Demands

Table 8.8 Cross-Reference between DAB and ISO 17025

Standard	DAB[a] Requirements	Article	ISO 17025:2005 Requirements
1	Scope	1	Scope
2	Definitions	3	Terms and definitions
3	Quality assurance program	4.3	Document control
		4.2.2	Laboratory's management policies
4	Organization and management	4.1	Organization
		4.2	Management system
5	Personnel	5.2	Personnel
5.1.3	Specific requirements regarding continuing education for the technical manager or leader, examiner/analyst—referring on following current scientific literature and attendance at the relevant courses/seminars, etc., once per year	5.2	Personnel requirements very general and same as for other testing laboratories
5.2–5.4	Degree, experience and duty requirements for technical manager or leader, CODIS manager, examiner/analyst and technician. The qualification test must be successfully completed before starting the jobs		No specific requirements regarding the education degree, experience and duty
6	Facilities	5.3	Accommodation and environmental conditions
7	Evidence control	5.8	Handling of test and calibration items
8	Validation	5.4.5	Validation of method
9	Analytical procedures	5.4.1–5.4.4	Test and calibration methods
		5.9	Assuring the quality of test and calibration results
10	Equipment calibration and maintenance	5.5	Equipment
		5.6.1–5.6.2	Measurement traceability—general and calibration
11	Reports	5.10	Reporting the results
12	Review of case files and reports and annual monitoring of the testimony of each examiner	4.15	Management reviews
13	Proficiency testing—define the interval of external PT for the staff performing DNA analysis	5.9.1.b	Assuring the quality of test and calibration results—participation in interlaboratory comparison or proficiency-testing programs
14	Corrective action	4.11	Corrective action
15	Audits	4.14	Internal audits
16	Safety	5.3	Accommodation and environmental conditions
17	Subcontractor of analytical testing for which validated procedures exist	4.5	Subcontracting test and calibrations

[a] Quality Assurance Standards for Forensic DNA Testing Laboratories.

for a credible, independent process to assess and recognize the technical competence of PT providers that have come up in the mid-1990s have been increasing together with the number of PT providers. This process was certainly triggered by the publication of the international standard ISO/IEC 17011, which for the first time mentions the competence of PT providers and the appropriateness of their schemes. The need for harmonization and accreditation of PT providers has been recognized by PT providers and forensic laboratories alike. Both sides are eager for competent PT. For PT providers, this would be an effective way to convince customers of their technical competence, whereas for forensic laboratories this would mean a reduced risk of choosing an incompetent PT provider. The accreditation for PT providers is based on combinations of several normative documents, which illustrates the lack of harmony among the approaches adopted by European and U.S. accreditation bodies. Two major approaches can be observed: one that uses normative documents only, which are exclusively focused on PT (ISO/IEC Guide 43-1 1997; ISO/IEC Guide 43-2 1997; ILAC-G13 2007), or one that uses these documents in combination with international conformity assessment standards.

ENFSI and DAB have recognized very early on that participation in PT is a critical component of the QA programs of various laboratories, because ISO Guide 43 and ILAC G13 did not cover all issues regarding the criteria that PT providers are required to meet in order to be recognized as competent. ENFSI and DAB have published guidelines on the requirements regarding the competence of PT providers to assist forensic laboratories with the selection of PT providers (DNA Identification Act 1994; ENFSI Standing committee for quality and competence 2005). Similar actions were taken by different national accreditation bodies. ASCLD/LAB has a list of approved international providers of PTs and requires that every laboratory should undergo at least one test per discipline each year from one of the approved PT providers. An ENFSI guideline has had also an important role to help ENFSI DNA WG in terms of design, setting, and reporting of proficiency schemes. A step toward broader harmonization of accreditation of PT providers is a new ISO standard, ISO/IEC 17043:2010. The scope of this document is to specify general requirements for the competence of providers of PT schemes and for the development and operation of PT schemes. These requirements are intended to apply to all types of PT schemes, and they can be used as a basis for specific technical requirements for particular fields of application. However, the future will show if this will work according to plan. Currently, accreditation of PT providers is not regulated yet and is still done on a voluntary basis; however, it is expected that laboratories shall select those that follow the guidelines established by national accreditation bodies and ENFSI (in Europe).

Information on PT providers or a particular area of expertise is usually maintained on the websites of various accreditation bodies and ENFSI. One example is the European PT Information System (EPTIS) database, which has gone online since year 2000 (http://www.eptis.bam.de/en/index.htm). Today, 300 PT providers list about 1300 schemes on the database, but only four among them provide DNA forensic PT. Inclusion of a particular PT provider in a register does not necessarily imply that it had been granted accreditation or approval by accreditation bodies such as ASCLD/LAB. Currently, four of eight DNA PT providers on the ENFSI website are accredited. Three of them are also included in the ASCLD/LAB register.

8.2 Validation

Validation, together with training of personnel and calibration or checking of equipment, represents an essential pillar of the QA program in forensic laboratories. Even before a laboratory participates in a proficiency test, it should perform internal validation studies for any new method (equipment) by qualified and competent personnel from the laboratory.

8.2.1 Definitions of Validation

The most general international standard, ISO/IEC 17025 standard, which is recognized all over the world, defines validation as "confirmation by examination and provision of objective evidence that the particular requirements for a specified intended use are fulfilled"; this definition complied with ISO 8402 (lately replaced by ISO 9000) vocabulary on quality management and assurance from 1994. ISO/IEC 17025 does not require which method performance parameters need to be characterized in order to validate the method, but it does stipulate a general requirement that "validation shall be as extensive as is necessary to meet the needs of the given application." Further on, ISO/IEC 17025 makes clear that validation must provide information on accuracy, detection limit, precision, and range in the context of repeatability, reproducibility, and robustness. ISO/IEC 17025 also defines the list of methods that shall be validated (nonstandard methods, laboratory designed/developed methods, standard methods outside their intended scope, or their amplifications and modifications).

Despite its primary focus on chemical methods, Eurachem Guide (EURACHEM Guide 1998), the first extensive guide on method validation, is also useful for other methods; it much more precisely defines the process of validation as "the process of defining an analytical requirement and confirming that the method under consideration has performance capabilities consistent with what the application requires." The requirements that shall be fulfilled during validation studies include: evaluation of the method's performance capabilities; usage of equipment that is within specification, working correctly, and adequately calibrated; competence of the operator (having sufficient knowledge on the subject to make appropriate decisions from the observations). Eurachem contains also definitions of some important terms used in validation and some advice for determining each method performance parameter.

In Europe, the ENFSI QCC—realizing very early on that knowing how to design and properly maintain a validation process is a key to good QA—published the first general guidelines on validation and implementation of methods used in forensic laboratories. The QCC validation guidelines describe different approaches to validation based on the frequency of the test and are mainly focused on the strategy of validation and documentation. Today, the document has been withdrawn. Instead, individual working groups within the ENFSI prepared their own specific guidelines as well as DNA WG (ENFSI DNA Working Group 2010). The definition of validation in the ENFSI validation guidelines does not differ from that found in ISO documents. Within the United States, the first guidelines on validation forensic DNA analysis (only in criminal matters) were published in 1989 by the TWGDAM. Currently, recommendations and requirements on a validation forensic DNA analysis are defined in two documents: DAB standards and SWGDAM revised validation guidelines. SWGDAM defines validation as "a process by which a method is primarily assessed for: its reliability, whether it has suitable operational conditions for obtaining results and its limitations," whereas DAB's definition is simpler: "validation is a process by which a procedure is evaluated to determine its efficacy and reliability for forensic casework analysis." In comparison with the ENFSI validation guide, DAB standards, and especially

SWGDAM validation guidelines, defined which method performance parameters need to be characterized in order to validate that a forensic DNA analysis has complied with their requirements (see Section 8.2.2). The main reason why the validation guidelines on forensic DNA analysis were published in the United States almost 20 years earlier than the QCC ENFSI validation guidance in Europe, lies on the difference in their legal systems. In the United States—in contrast to Europe, where there are no standards for admissibility of scientific evidence—the federal courts only accept lines of scientific evidence that meet the federal rules of evidence. Two of them, Frye and Daubert rulings, require that the reliability of scientific evidence be determined.

If we summarize, the purpose of validation is to ensure and document that a particular method is "working properly" by demonstrating its robustness, reliability (accurateness), and limitations. A robust method is one that remains unaffected by small, but deliberate variations in method parameters and provides an indication of its reliability during normal usage (e.g., changing of annealing temperature during PCR will not influence the results). A reliable method refers to obtaining results that are *accurate* (correct and consistent), which is defined by precision (repeatability and reproducibility) and trueness (EURACHEM Guide 1998). Repeatable method means that the same results are obtained each time a sample is tested under repeatable conditions (same laboratory, same equipment, and same operator in short time intervals). For example, differences in size of particular alleles from STR profiling performed in the same laboratory, with the same operator, using the same equipment in a short period should be less than 0.5 bp using three times the standard deviation. Reproducible method means that the same results are obtained when a sample is tested in different laboratories or with a different operator or using different equipment. Trueness refers to the closeness of agreement between the average value obtained from a large set of test results and an accepted reference value. For example, the same STR profile is reported from a tested sample in different laboratories or with different operators or using different equipment. These interlaboratory tests are essential to demonstrate consistency in results from different laboratories especially in consideration of legislation of DNA databases and exchange of DNA profiles between countries.

8.2.2 Level of Validation

Validation is usually occurring at two different levels: development validation and internal validation. Neither ISO/IEC 17025 nor ENFSI validation guidelines explicitly distinguish between two stages of validation, whereas DAB standards and SWGDAM validation guidelines do. Pointing out the difference between both validations is crucial because it is directly related to the scope of validation. Both documents (DAB standards and SWGDAM validation guidelines) provide specific requirements that the organizer of the validation study needs to know in order to be in compliance with them. ISO/IEC 17025 or ENFSI validation guidelines lack a similar description of specific requirements.

These definitions (DNA Identification Act 1994; Federal Bureau of Investigation 2000) state that

- Developmental validation is the acquisition of test data and determination of conditions and limitations of a new or novel DNA methodology for use on forensic samples.
- Internal validation is an accumulation of test data within the laboratory to demonstrate that established methods and procedures perform as expected in the laboratory.

Forensic DNA Typing and Quality Assurance

Developmental validation presents the top level of validation and involves comprehensives and broad studies typically performed by manufacturers, technical organizations, academic institutions, and other laboratories (Collins et al. 2004; Swango et al. 2007; Mulero et al. 2008; Applied BioSystems 2009). Developmental validation shall precede the usage of a new method or technique for forensic DNA analysis. Typical development studies involves testing of new STR kits, new primers, new technique for quantification of DNA, new technology for separation and detection of STR alleles, and new methods for detection SNPs. Studies must demonstrate accuracy, precision, and reproducibility of a procedure. The standard development validation studies that may be performed according to SWGDAM validation guidelines are listed below. DAB requirements are shown in square brackets. Some of the studies are not necessary due to the method itself.

- Characterization of genetic markers (inheritance, mapping, detection, polymorphism) [8.1.2.1 DAB]
- Species specificity [8.1.2.2 DAB]
  - For techniques designed to type human DNA, the potential to detect DNA from forensically relevant nonhuman species should be evaluated. For techniques in which a species other than human is targeted for DNA analysis, the ability to detect DNA profiles from nontargeted species should be determined. The presence of an amplification product in the nontargeted species does not necessarily invalidate the use of the assay.
- Sensitivity studies: When appropriate, the range of DNA quantities able to produce reliable typing results should be determined [8.1.2.2 DAB].
- Stability studies: The ability to obtain results from DNA recovered from biological samples deposited on various substrates and subjected to various environmental and chemical insults has been extensively documented. In most instances, assessment of the effects of these factors on new forensic DNA procedures is not required. However, if substrates and/or environmental and/or chemical insults could potentially affect the analytical process, then the process should be evaluated using known samples to determine the effects of such factors [8.1.2.2 DAB].
- Reproducibility: The technique should be evaluated in the laboratory and among different laboratories to ensure the consistency of results. Specimens obtained from donors of known types should be evaluated.
- Case-type samples: The ability to obtain reliable results should be evaluated using samples that are representative of those typically encountered by the testing laboratory. When possible, consistency of typing results should be demonstrated by comparing results from the previous procedures to those obtained using the new procedure.
- Population studies: The distribution of genetic markers in populations should be determined in relevant population groups. When appropriate, databases should be tested for independence expectations [8.1.2.3 DAB].
- Mixture studies: The ability to obtain reliable results from mixed source samples should be determined [8.1.2.2 DAB].
- Precision and accuracy: The extent to which a given set of measurements of the same sample agree with their mean and the extent to which these measurements match the actual values being measured should be determined.

- PCR-based procedures: Publication of the sequence of individual primers is not required in order to appropriately demonstrate the accuracy, precision, reproducibility, and limitations of PCR-based technologies.
  - The reaction conditions needed to provide the required degree of specificity and robustness must be determined. These include thermocycling parameters, the concentration of primers, magnesium chloride, DNA polymerase, and other critical reagents.
  - The potential for differential amplification among loci, preferential amplification of alleles in a locus, and stochastic amplification must be assessed.
  - When more than one locus is coamplified, the effects of coamplification must be assessed (e.g., presence of artifacts).
  - Positive and negative controls must be validated for use.
  - Detection of PCR product
    - Characterization without hybridization.
      - When PCR product is characterized directly, appropriate measurement standards (qualitative and/or quantitative) for characterizing the alleles or resulting DNA product must be established.
      - When PCR product is characterized by DNA sequencing, appropriate standards for characterizing the sequence data must be established.
    - Characterization with hybridization.
    - Hybridization and wash conditions necessary to provide the required degree of specificity must be determined.
    - For assays in which the probe is bound to the matrix, a mechanism must be used to demonstrate whether adequate amplified DNA is present in the sample (e.g., a probe that reacts with an amplified allele(s) or a product yield gel).

Example: Development validation of particular STR kit according to DAB and SWGDAM requirements can be done in following manner:

- Characterization of genetic markers (inheritance, mapping, detection, polymorphism)—technique for mapping the position on chromosome (outside the scope of this chapter), a set of samples from related persons is analyzed to determine for relationship indices, polymorphism (rate of mutations), and detection (silent alleles)
- Species specificity—amplifications were performed on 1 to 10 ng genomic DNA from nonhuman samples (domestic and wild animal, bacteria and yeast) to evaluate a possibility to obtain reliable results (false positive) from other species.
- Sensitivity studies: dilution series of control DNA 007: 1 ng, 500 pg, 250 pg, 125 pg, 62.5 pg, 31.25 pg, and negative control (no range of DNA quantities able to produce reliable typing results should be determined in replicates of four.
- Stability studies: characterize amplification performance in the presence of either degraded or inhibited DNA or exposure to sunlight, humidity
  - Humic acid inhibition—series of 1 ng control DNA 007 with final concentration of 20, 50, and 80 ng/μL humic acid in replicates of three.
  - The efficiency of amplification in the presence of degraded DNA, deoxyribonuclease was used to digest DNA for different time intervals (30 s; for 1, 4, 8, 12, 20, and 30 min) and for 1, 2, 4, 8, and 24 h. One nanogram of undigested DNA or 4 ng of digested DNA from each time point was then added to individual reactions.

- Repeatability: 1 ng genomic control DNA (reference samples) was tested in replicates (four to eight) on the same instrument, by the same operator within a short time.
- Reproducibility: The same sets of samples are analyzed in different laboratories. Different samples (saliva, blood, semen) were tested by the same kit.
- Case-type samples: 20–30 casework types of DNA samples with final concentration of 0.1 ng/μL genomic DNA.
- Population studies: A set of unrelated samples from known populations (ethnic groups) is analyzed to determine the allele and genotype frequency (polymorphism) for calculation population statistics.
- Mixture studies: DNA from two samples of known profile DNA is mixed in various ratios ranging from, for example, 50:1 to 1:50. Total DNA input for the reaction was 1.0 ng.
- Precision studies: Size of individual STR allele is measured. All measured alleles should fall within the ±0.5-bp window around the measured size for the corresponding allele in the allelic ladder.
- PCR: degree of specificity and robustness
  - Annealing T studies: 2 degrees Celsius below or above optimal were tested.
  - Cycle studies: reduced and higher number of cycles than standard cycle were tested.
  - PCR reaction mix: concentration of primers, magnesium chloride, DNA polymerase, BSA were tested at ±10%.
    - Effect of coamplification: searching for artifacts.
    - Positive and negative controls must be validated for use: negative control water instead DNA, positive control—control DNA ad known references samples, allelic ladder and internal size standard.
    - Detection of PCR product from intralaboratory and interlaboratories studies:
      - Overall peak heights—minimum and maximum peak height alleles are measured from different studies (sensitivity, repeatability, inhibition)
      - Heterozygous studies (intralocus balance): the peak heights of the smaller and the larger alleles copa
      - Intercolor and interlocus balance
      - Stochastic amplification
- Stutter studies

Internal validation focuses on demonstration that general established procedures work as expected in one's own laboratory and is conducted to establish laboratory specific limits. Internal validation studies are much less extensive than development ones. Any new method or technique used in forensic DNA analysis before routine use needs to be validated. This is also done whenever conditions change (exchanging laser in capillary electrophoresis, new heat block in the PCR instrument, new chemicals with different characteristic for which a method was been validated, new instrument, modification of kit primers) or whenever a method is changed outside the original scope (new calling alleles at 50 RFU instead at 75 RFU). One of the DAB standards requires that any material modifications of analytical procedures shall be documented and subject to validation testing [8.1.3.4 DAB]. The standard internal validation studies that may be performed according to

SWGDAM validation guidelines are listed below. DAB requirements are shown in square brackets. Some of the studies are not necessary due to method itself.

- Known and nonprobative evidence samples (reference samples and case samples, authentic or from simulated case); if previous profiling is available, check the consistency of interpretation (exclusion/inclusion) [8.1.3.1 DAB].
- Reproducibility and precision: By using an appropriate control(s); human control DNA from the kit, internal standard, laboratories' own known reference samples, allelic ladder.
- Match criteria: For procedures that entail separation of DNA molecules based on size, precision of sizing must be determined by repetitive analyses of appropriate samples to establish criteria for matching or allele designation on empirical data [8.1.3.2 DAB].
- Sensitivity and stochastic studies: The laboratory must conduct studies that ensure the reliability and integrity of results. For PCR-based assays, studies must address stochastic effects and sensitivity levels.
- Mixture studies: When appropriate, forensic casework laboratories must define and mimic the range of detectable mixture ratios, including detection of major and minor components. Studies should be conducted using samples that mimic those typically encountered in casework (e.g., postcoital vaginal swabs).
- Contamination: The laboratory must demonstrate that its procedures minimize contamination that would compromise the integrity of the results. A laboratory should use appropriate controls and implement quality practices to assess contamination and demonstrate that its procedures minimize contamination.
- Qualifying test: The method must be tested using a qualifying test. This may be accomplished through the use of proficiency test samples or types of samples that mimic those that the laboratory routinely analyzes. This qualifying test may be administered internally, externally, or collaboratively [8.1.3.3 DAB].

Example: Typical internal validation of particular STR kit according to DAB and SWGDAM requirements can be done in the following manner:

- Population studies: A set of unrelated samples from known populations (ethnic groups) is analyzed to determine allele and genotype frequency (polymorphism) for calculation population statistics and relationship indices (if this has not done been before) (Drobnič et al. 2005).
- Repeatability and reproducibility may be concordant studies: Standard samples with known types: the kit positive control, DNA from staff members and NIST SRM (National Institute of Standards and Technology standard reference material) samples along with 5–10 nonprobative casework samples previously examined with other kits on same equipment by same analyst in short time, on different equipment or by different analyst.
- Precision studies: 5–10 injections of allelic ladders to define system precision across typical time and environmental conditions used in running a batch of samples.
- Sensitivity and stochastic studies: Samples covering the dynamic range of the STR kit: two sets of samples with the following dilution series: 5, 2, 1, 0.5, 0.2, 0.1, and 0.05 ng).

- Contamination studies: Negative control.
- Mixtures studies: Covering a range of ratios: two sets of samples with different allele combinations and the following ratios: 10:1, 3:1, 1:1, 1:3, and 1:10.
  - Detection of PCR product from different studies (population, sensitivity, repeatability, inhibition).
  - Overall peak heights—minimum and maximum peak height alleles are measured from different studies.
  - Heterozygous studies (intralocus balance): This was calculated by dividing the height of the lower allele of a heterozygous individual by the height of higher and then expressed as a percentage.
  - Intercolor and intracolor balance: Intercolor was assessed by first averaging heterozygous peaks and dividing homozygous peaks in half. Once normalized for diploidy, the lowest score for a locus labeled with a given dye was divided by the highest, and the result was reported as a percentage. Intercolor balance was calculated in a similar manner to intracolor, although the comparison was made across all loci regardless of dye label.
  - Stutter studies: the percentage of observed stutter at each locus is calculating by dividing the stutter peak height by the height of corresponding allele and then multiplying this value by 100.
- Stochastic amplification: determined by sensitivity studies.

8.2.3 Procedure of Internal Validation

Once the developer of a particular method or technique demonstrates its reliability and accuracy, the laboratory needs to confirm that the procedure is also valid when applied in-house (even if other laboratories have already conducted the internal validation). Because of their role in the criminal justice system, and specifically in the identification of people through DNA analysis, forensic laboratories in the United States are required to conduct internal validation according to DAB and SWGDAM rules, whereas in Europe or Australia there is no such legislation. It is entirely left to laboratories if they wish to follow professional standards. However, neither DAB standards nor SWGDAM guidelines have standardized a validation process as whole. The laboratory has to decide how it will conduct validation studies, and method performance parameters need to be characterized in order to validate the method. Still, ISO/IEC17025, DAB standards, and SWGDAM guidelines can be helpful in formulating an effective validation plan. There are already some publications on validation studies by reputable technical organizations (NIST), other scientists in relevant scientific journals, or validation guidelines from manufacturers (Butler et al. 2004; Butler 2005; Krenke 2005; Butler 2006; Promega 2006; Butler 2007; Loyo et al. 2009) that can be used to establish a good strategy. One very helpful database is the NIST STRBase website (http://www.cstl.nist.gov/strbase/), which provides a compilation of references of various validation studies plus ENFSI validation guidelines containing the plan and implementation of validation and responsibilities and authorities of individual personnel in the laboratory.

Internal validation must be performed whenever conditions change (new laser in capillary electrophoresis, new heat block in PCR instrument, new chemicals with different characteristic for which a method has been validated, new instruments, modification of kit primers), or whenever a method is changed outside the original scope (new calling alleles at 50 RFU instead at 75 RFU).

8.2.3.1 Minimum Number of Samples for Internal Validation

Forensic scientists occasionally misinterpret validation guidelines, leading to some incorrect views about how validations should be performed. One common misinterpretation of the SWGDAM guidelines for conducting an internal validation is the belief that it is necessary to have 50 samples per experiment or 50 samples of the same type per experiment. The 50 samples recommended in the guidelines actually refer to the minimum total sample size for an entire internal validation. In summary, the sample size has to be large enough to produce statistically valid results with methods such as the Student's t-test for assessing accuracy. NIST (Butler 2005) and ENFSI DNA WG (2010) consider between five and 10 samples to be sufficient for a validation experiment in forensics.

The following comprise the basic steps for an internal validation procedure:

- Prepare the plan
  - Define requirements (range and reliability; range—with the new kit, we would like to obtain a full STR profile from DNA smaller than 100 bp).
  - Quality standards depends on procedures (sensitivity—if we would like to determine of what level of input DNA with a new kit is expected to produce a full DNA profile, a dilution series of a control DNA from 1 ng to 3 pg must be evaluated).
  - Accepted criteria (sensitivity—at least 100 pg DNA must be used for DNA analysis to obtain reliable results—full profile STR with unknown samples; match criteria—deviation of allele size for calling allele—±0.5 bp window).
  - Design the test considering requirements.
- Conduct the test in controlled environment.
- Review the test results against the accepted criteria.
- Prepare validation report with statement of validation.
- Creation of the standard operating procedures with interpretation guidelines based on the validation studies.
- Prepare the implementation data.
  - Training plan of the personnel and the arrangements for competence assessment and PT.
- Each trained analyst passing a qualifying test for initial use in forensic casework (DAB and ISO/IEC 17205 requirement).

8.2.4 Documentation

Validation is, by definition, "confirmation by examination and provision of objective evidence that the particular requirements for a specified intended use are fulfilled." Through the entire chapter, the examination part of the definition was presented, but the second requirement—"provision of objective evidence"—is also very important. Because of this, documentation of validation results is required by the DAB standards and by ISO 17025 standards. Appropriate documentation will help ensure that application of the method from one occasion to the next is consistent. However, documentation does not have to be extensive. Laboratories shall record results and procedures used; all modifications according to ISO are also a statement of validity.

8.2.5 Reference Materials

Internal quality control measures are crucial for ensuring the QA of forensic DNA laboratories, and of course validation studies are part of the whole process. Most include the analysis of in-house control materials (extraction blank, reagent blank, amplification blank, positive control from samples with known profile) or uniform control materials from commercial kits (internal size standard, allelic ladder, and control DNA, etc.). QA can be enhanced by using reference materials, certified reference materials, or SRMs (ISO Guide 30 1992 and 2008) (see Box 8.7). The only worldwide provider of SRMs for forensic DNA analysis is the NIST (an agency of the U.S. Department of Commerce), which is responsible for developing national and international SRMs (http://www.nist.gov/srm/). Hundreds of SRMs are available from the NIST for three main purposes: (1) to help develop accurate methods of analysis; (2) to calibrate measurement systems used to facilitate exchange of goods, institute quality control, determine performance characteristics, or measure a property at the state-of-the-art limit; and (3) to ensure the long-term adequacy and integrity of measurement QA programs.

BOX 8.7

WHAT IS REFERENCE MATERIAL?

Reference Material (RM) is a material, sufficiently homogeneous and stable with respect to one or more specified properties, that has been established to be fit for its intended use in a measurement process.

WHAT IS CERTIFIED REFERENCE MATERIAL?

Certified Reference Material (CRM) is reference material characterized by a metrologically valid procedure for one or more specified properties, accompanied by a certificate that provides the value of the specified property, its associated uncertainty, and a statement of metrological traceability.

(From ISO Guide 30:1992 and 1:2008: Revision of definitions for reference material and certified reference material, International Organization for Standardization, Geneva, http://www.iso.org. Accessed December 4, 2013.)

WHAT IS NIST STANDARD REFERENCE MATERIAL (SRM)?

NIST Standard Reference Material® (SRM) is a CRM issued by NIST that also meets additional NIST-specific certification criteria and is issued with a certificate or certificate of analysis that reports the results of its characterizations and provides information regarding the appropriate use(s) of the material (NIST SP 260-136). The terms "Standard Reference Material" and the diamond-shaped logo that contains the term "SRM," are registered with the United States Patent and Trademark Office.

(From NIST. Available at http://www.nist.gov/srm/definitions.cfm. Accessed February 24, 2013.)

8.2.5.1 NIST SRMs

Eight NIST SRMs (SRM 2390, 2391a and b; 2392, 2391-I, 2394, 2395, and 2372) were developed according to DAB standards, and a laboratory performing forensic DNA analysis (paternity testing or crime case) is required to "check its DNA procedures annually or whenever substantial changes are made to protocol(s) against and appropriate and available NIST SRM or standard traceable to a NIST standard." All NIST SRMs for forensic DNA typing can also be used for QA when assigning values to in-house control materials. Short descriptions of SRMs for forensic purposes (available from NIST) are presented below, except for SRM2390, which is intended for use in RFLP testing (not presented). For more details, visit NIST's website (http://www.cstl.nist.gov/strbase/srm_tab.htm).

SRM 2391 to 2395 are used for control in typing and sequencing, whereas *SRM 2391b* is used in the standardization of forensic and paternity QA procedures and instructional law enforcement or nonclinical research purposes. SRM 2391b includes STR information for all genomic DNA samples in the SRM. The STR data include the FBI's CODIS 13 core STR loci, including loci that were commercially available at the time of renewal certification. Certified values for a total of 26 STR loci are included in this renewal issue. This SRM is composed of well-characterized human DNA in two forms: genomic DNA (10 components) and DNA to be extracted from cells that are spotted onto filter paper (components 11 and 12). A unit of SRM 2391b is composed of 12 frozen components packaged in one box.

SRM 2392 is intended to provide quality control when performing PCR and sequencing of human mtDNA for forensic identification, medical diagnosis, or mutation detection. It may also be used as a control when amplifying (PCR) and sequencing any DNA. SRM 2392 mtDNA Sequencing contains DNA extracted from two cell lines plus cloned DNA from a region that is difficult to sequence. This SRM is composed of well-characterized extracted human DNA from cell culture lines known as CHR and GM09947A as well as cloned DNA from the HV1 region of CHR. A unit of the SRM is composed of three frozen components packaged in one box. The certificate accompanying the SRM details the base pair sequences of the DNA, and the sequences of 58 unique primer sets that permit the amplification and sequencing of any specific area or the entire human mtDNA (strand).

SRM 2392-I Mitochondrial DNA Sequencing complements and adds another DNA template to SRM 2392 for the amplification and sequencing of human mtDNA. The selection of the HL-60 cell culture line for this additional DNA template was based on a suggestion from the FBI that this DNA would be particularly useful to the forensic community. It is certified for the sequences of the entire human mtDNA (16,569 bp) from a promyelocytic cell line (HL-60) prepared from the peripheral blood leukocytes from an individual with acute promyelocytic leukemia. A unit of SRM 2392-I consists of 65 µL of extracted DNA from cell culture line HL-60 at a nominal concentration of 1.4 ng/µL. This SRM can also be used for QA when assigning values to in-house control materials.

SRM 2394 Heteroplasmic Mitochondrial DNA Mutation Standard contains mixtures of a 285-bp PCR product from two different cell culture lines that differ by 1 bp. These mixtures contain varying ratios of the minor/major heteroplasmy including 1:99, 2.5:97.5, 5:95, 10:90, 20:80, 30:70, 40:60, and 50:50 (Figure 8.6). This SRM is intended to provide quality control in determining the sensitivity of heteroplasmic low-frequency single-nucleotide mutation detection techniques. SRM 2394 is intended to provide quality control benchmarks for forensic, medical, and DNA scientists who are trying to assess the detection

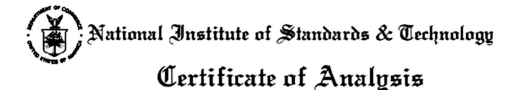

National Institute of Standards & Technology
Certificate of Analysis

Standard Reference Material® 2394

Heteroplasmic Mitochondrial DNA Mutation Detection Standard

This Standard Reference Material (SRM) is composed of human mitochondrial DNA mixtures which simulate different levels of heteroplasmy. SRM 2394 is intended to provide quality control benchmarks for forensic, medical, and DNA scientists to assess the detection sensitivity of low-frequency mutations, single nucleotide polymorphisms (SNPs) in either mitochondrial DNA (mtDNA) or in pooled nuclear DNA samples, or heteroplasmic sites in mtDNA. SRM 2394 is packaged in a single protective plastic box containing ten tubes: one tube containing the 100 % (by mass) polymorphic DNA, one tube containing the 100 % (by mass) CRS[1] DNA, and eight tubes containing different mass percentages of the polymorphic/CRS mtDNA mixtures (mass % polymorphic levels are 1 %, 2.5 %, 5 %, 10 %, 20 %, 30 %, 40 % and 50 %) (Table 1). Each vial contains 25 µL of DNA at a concentration of 8 ng/µL in 10 mM Tris-HCl, pH 8.5.

The DNA mixtures are constructed from the polymerase chain reaction (PCR) products from two different cell culture lines (CHR and GM09947A) that differ by one base pair (bp) at nucleotide position (np) 6371 in the 285 bp amplified region. The cell line (CHR) designated polymorphic has a T at np 6371 and the cell line (GM09947A) containing the CRS sequence has a C at that site.

Certified Values: Table 1 contains the certified values of polymorphic/CRS mtDNA mixtures. The uncertainty values calculated on the 50/50 mixture are reported in section, "Uncertainty Analysis". Table 2 contains the certified sequence information of the 285 bp amplicon from the two cell lines (CHR and GM09947A) and shows the one bp difference at np 6371. The entire 16,569 bp of mtDNA from both CHR and GM09947A was sequenced for SRM 2392, and that sequence can be found in reference 1, the Certificate of Analysis for SRM 2392 [2], and in NIST SP260-155 [3]. Table 3 contains the reference sequences of the forward and reverse primers used in the PCR amplifications of the 285 bp DNAs that were blended to construct these mixtures.

Expiration of Certification: The certification of SRM 2394 is valid until 31 July 2014, within the measurement uncertainties specified, provided the SRM is handled and stored in accordance with the instructions given in this certificate (see "Instructions for Use"). This certification is nullified if the SRM is damaged, contaminated, or modified.

Maintenance of SRM Certification: NIST will monitor this SRM over the period of its certification. If substantive technical changes occur that affect the certification before the expiration of this certificate, NIST will notify the purchaser. Registration (see attached sheet) will facilitate notification.

The overall direction and coordination of the technical measurements leading to the certification were performed by D.K. Hancock and B.C. Levin of the NIST Biotechnology Division.

The analytical determination, technical measurements, and analysis of data for the certification of this SRM were performed by D.K. Hancock, L.A. Tully, and B.C. Levin of the NIST Biotechnology Division. Complete details can be found in reference 4.

Vincent L. Vilker, Chief
Biotechnology Division

Gaithersburg, MD 20899
Certificate Issue Date: 09 June 2009
See Certificate Revision History on Last Page

Robert L. Watters, Jr., Chief
Measurement Services Division

---

[1] CRS refers to the Cambridge Reference Sequence [5] and the revised Cambridge Reference Sequence (rCRS) [6].

Figure 8.6 First page of SRM 2394 (Heteroplasmic Mitochondrial DNA Mutation Detection Standard).

sensitivity of low-frequency mutations, SNPs in either mtDNA or in pooled nuclear DNA samples, or heteroplasmic sites in mtDNA. SRM 2394 is packaged in a single protective plastic box containing 10 tubes: one tube containing the 100% (by mass) polymorphic DNA, one tube containing the 100% (by mass) CRS1 DNA, and eight tubes containing different mass percentages of the polymorphic/CRS mtDNA mixtures. Each vial contains 25 μL of DNA at a concentration of 8 ng/μL in 10 mM Tris–HCl (2-amino-2-(hydroxymethyl)-1,3-propanediol hydrochloride). The DNA mixtures are constructed from the PCR products from two different cell culture lines (CHR and GM09947A) that differ by 1 bp at nucleotide position (np) 6371 in the 285-bp amplified region.

SRM 2395 Human Y-Chromosome DNA Profiling Standard Certificate of Analysis for SRM 2395 has been updated with additional information for the five male components. This SRM is intended primarily for use in the standardization of forensic and paternity QA procedures for PCR-based genetic testing and for instructional law enforcement or nonclinical research purposes that involve the human Y-chromosome. These samples now have sequence information at more than 40 different Y-STR markers and typed at an additional three Y-STRs and 42 Y-SNPs. This SRM may be used for calibrating methods for Y-STR typing with any primer set or commercially available kit. In addition, SRM 2395 may be used to standardize the nomenclature for the field of genetic genealogy. This SRM can also be used for QA when assigning values to in-house control materials. It is not intended for any human or animal clinical diagnostic use. This SRM is composed of well-characterized human genomic DNA in liquid form. A unit of SRM 2395 consists of six frozen components (A through F) packaged in one box. Each component contains 50 μL of DNA at a concentration of approximately 1 ng/μL. There are five male samples and one female sample in this SRM.

SRM 2372 is intended primarily for use in the value assignment of human genomic DNA forensic quantitation materials. It is not intended for any human or animal clinical diagnostic use. SRM 2372 consists of three well-characterized human genomic DNA materials solubilized in 10 mmol/L Tris–HCl and 0.1 mmol/L ethylenediaminetetraacetic acid disodium salt (disodium EDTA) using deionized water adjusted to pH 8.0 (TE-4, pH 8.0 buffer). The three component genomic DNA materials, labeled A, B, and C, are respectively derived from a single male donor, multiple female donors, and multiple male and female donors. A unit of the SRM consists of one sterile 2-mL vial of each component, with each vial containing approximately 110 μL of DNA solution.

8.3 Costs and Benefits of the QA Implementation

Because of great evidentiary strength of DNA result, forensic DNA analysis has a great impact on decision of the court to convict or acquit a suspect. Consequently, it also has influence on society's confidence in the results of particular DNA laboratory and forensic science as whole. A false reported match between a crime sample and one's reference samples can have a dramatic consequence on a person's life—an innocent person can be convicted or a true perpetrator is acquitted. A key to minimizing the risk of error and generating reliable results is to ensure that laboratories have a stringent level of QA in the forensic DNA process. But implementation and maintenance of QA in forensic DNA laboratory is not free of charge. Practically every control component used to maintain high-level QA has an inherent cost, for example, calibration of equipment, validation of method, proficiency tests, reference material, and training of personnel.

However, when evaluating a QA program, it should be realized that the costs associated with the pursuit of quality are often offset by savings realized through improved processes. The benefits and add-on values of a good QA program, which are difficult to define in terms of costs, include:

- Compliance with regulation and professional standards
- Preparation for accreditation (accreditation is already mandatory in some legislation in the United States, the Netherlands, Australia, and in the future, in all European countries)
- Improved credibility in laboratory results in the court
- Reduced reanalysis, data correction, and unreliable results
- Improved service to customers

Furthermore, lack of quality may be costly. Low-quality data can undermine the integrity of DNA results and its acceptance as legal evidence. According to Westgard and Barry (1986), costs of quality can be grouped into prevention costs, appraisal costs, and failure costs. Prevention costs are the costs of carrying out laboratory operations according to good laboratory practices and proper QA guidelines to prevent unacceptable data being generated in the first place. Appraisal costs are connected with the monitoring of the functioning of the laboratory and quality system. They include, for example, evaluation of performance of analytical equipment and procedure, audits, proficiency program, and QA assessment. Correction costs are the costs of taking remedial measures if the quality system discovers deficits in the laboratory operations.

8.4 Potential for Error in Forensic DNA Analysis

Although methods used in forensic DNA analysis produce accurate and reliable results, and even when laboratories adhere to quality standards or are accredited for DNA analysis, errors still happen—however, the problem lies in recognizing these errors when they do occur.

Forensic DNA typing, as with any other analytical method, depends on human activity and as such is prone to error. The number of mistakes is directly related to complexity of the analytical process and the samples. Because forensic DNA typing is composed of many steps and variety of evidence, different types of errors can happen at every stage in the processing of DNA samples, from collection of samples at the field through laboratory work to interpretation of obtained results. There are errors that lead to false inclusion (false positive) or false exclusion (false negative). In the context of criminal legal proceedings, a false inclusion is the most serious, because it can lead to the conviction of an innocent person. By definition, a false inclusion or false match is a type of error in which a laboratory reported that two samples (probative and reference) came from the same source (person), when, in fact, they do not. In contrast, a false exclusion is a type of error in which a laboratory reported that two samples (probative and reference) came from different sources (persons) when, in fact, they came from the same source (person). Nowadays, the risk of false inclusion is increasing with the expansion of DNA databases and the exchange data between countries. As in criminal cases, errors also occur in relationship testing. In the case of paternity tests, a false inclusion means falsely including a man as the true father, whereas false exclusion implies falsely excluding the true father from paternity.

Errors can be grouped in the following ways:

- Sample mishandling
- Contamination
- Misinterpretation of results

8.4.1 Mishandling of Samples

Mishandling of samples can occur before samples arrive in the laboratory, that is, during collection of the samples or after that, during the laboratory process. In the laboratory, incorrect recording of results can also happen at any point when data are being recorded. Two types of mishandling errors are known as mix-ups or mislabeling of samples (see Box 8.8). Sample labeling problems are known to have caused false DNA incriminations in cases in United States, Australia, and New Zealand (Thompson et al. 2003; Thompson 2008). They can be avoided by means of comprehensive sample-handling and data entry policies, which include: proper training of crime scene officers and laboratory personnel,

---

BOX 8.8   MIXED-UP SAMPLES

The Philadelphia City Crime Laboratory reported that it had accidentally switched the reference samples of the defendant and victim in a rape case. The error led the laboratory to issue a report that mistakenly stated that the defendant was a potential contributor of what the analysts took to be "seminal stains" on the victim's clothing. The report also stated that the defendant's profile was "included" in a mixed sample taken from vaginal swabs. After the sample switch came to light, the laboratory reassessed the evidence and concluded that the "seminal stains" were actually bloodstains that matched the victim's DNA profile and that the defendant was excluded as a potential contributor to the vaginal sample.

Mislabelling of a blood sample in October 1991 in Kansas Bureau of Investigation contributed to a delay in identifying Douglas Belt as a suspect in several sex crimes. Belt went on to commit murder and rape. Had the sample not been mislabelled, Mr. Belt would have been arrested long before the other crimes were committed. D. Belt was being held in the Sedgwick County Jail on a charge of first-degree murder in the June 2002 killing and decapitation of Lucille Gallegos, 43, at the apartment complex where she worked as a maid. Months after Belt's arrest in November 2002, he was charged with seven rapes that took place between 1989 and 1994 in four Kansas counties. Belt also was charged Dec. 20, 2002, in Madison County, Ill., with three counts of aggravated criminal sexual assault stemming from a Nov. 22, 1992, attack on a 21-year-old mother of two near Granite City, Ill.

(From L. Brenner, B. Pfleeger. Amended report: investigation of the sexual assault of Danah H, Philadelphia, PA: Philadelphia Police Department DNA Identification Laboratory, 2000 Feb. 7, Lab No.: 97-70826; Tim Richardson. Kline orders FBI audit after error. *The Capital Journal*, June 5, 2003. Available at http://cjonline.com/stories/060503/bre_kbi.shtml. Accessed February 24, 2013.); L. Brenner, B. Pfleeger. Amended report: investigation of the sexual assault of Danah H. Philadelphia, PA: Philadelphia Police Department DNA Identification Laboratory; 2000 Feb. 7; Lab No.: 97-70826.

handling one item of evidence at a time, ensuring that reference samples and probative samples are analyzed separately (different times and places); ensuring that every transfer of DNA solution is witnessed by a second person. Also, redundancy in casework analysis by testing multiple evidentiary stains may serve as a check on the consistency of results. It reassures the forensic scientist when the same DNA profile is obtained from multiple samples in the same case.

The best way to detect mishandling errors in the laboratory is to obtain new samples from the original sources and retest them or to routinely perform duplicate testing. It is important to remember that this precaution or safeguarding cannot be done in all cases (minute samples) and cannot detect every error, such as mishandling of samples during sample collection or interpretation errors.

To avoid recording errors, it is important that all data records are reviewed by a second person. Retyping of data can be minimized using electronic transfer, for example, by introducing the LIMS in the laboratory process.

8.4.2 Contamination

Contamination means introducing foreign DNA in evidence sample by accident. Different kinds of contamination exist and can happen in different stages of the life of evidence samples (Thompson et al. 2003; Thompson 2008). We will focus only on contamination with biological materials because only these can produce errors in reporting DNA results (whereas nonbiological material contamination can result only from the inability to perform the test). Different biological materials have different consequence depending on the type of DNA analysis. For human DNA analysis, the laboratory shall focus only on contamination of human material, and animal or plant contamination does not matter, whereas for animal or plant DNA analysis the reverse holds true. Contamination implies the accidental transfer of DNA. Regarding when and how the contamination happened, National Research Council (NRC) distinguished three kinds of contamination (National Research Council 1996) inadvertent, mixed samples, and carryover. The first two occur in the pre-PCR process working area by genomic DNA. The last one occurs in the post-PCR working area by PCR products. The Forensic Science Service (FSS) was the first laboratory to report the detection of sporadic contamination derived from plastic tubes used in forensic DNA testing (Howitt 2003). The FSS identified 11 caseworks-contaminating DNA profiles in negative (blank) control samples in running more than 1 million samples during a 3-year period. Further investigation revealed that 10 out of 11 caseworks-contaminating DNA profiles came from the floor staff of the manufacturer of tubes. In 2004, Gill and Kirkham had already proposed the computer-based model for assessing contamination in the laboratory, and they have prepared the guidelines for the interpretation of DNA profiles. One of the well-known contamination cases is "The Phantom of Heilbronn" (see Box 8.9). This case caused a lot of trouble to the German police who, from 2007 to 2009, have been chasing a serial woman killer even across the borders of Germany. In the end, they concluded that a serial killer does not exist, but that the DNA profile recovered from 40 different crimes had been contaminated (Yeoman 2009) (see Box 8.9).

Inadvertent contamination can occur when evidence samples are deposited on a biological nonfree surface or during sample handling at the field or during a laboratory process. The first one, also called background contamination, is impossible to avoid. The latter one largely depends on a strict procedure for sample handling and behavior at every stage in the processing of DNA samples. Sample contamination can happen when handlers do not

BOX 8.9 THE PHANTOM OF HEILBRONN

They call her "the woman without a face," a serial killer who has stalked Western Europe for more than 15 years, murdering young and old while leaving no witness who could describe her. The Phantom of Heilbronn, as she has also become known—an allusion to the scene of her most notorious crime—has apparently killed, mugged and burgled across Germany, France, and Austria, leaving her DNA at 40 crime scenes.

Detectives have been on the trail of the Phantom since 2001, when a seven-year-old boy stepped on a heroin syringe in the German town of Gerolstein. DNA profile of trace was put into a central DNA databank and a match was found to two unsolved murders. The first was that of a 62-year-old woman who was strangled in 1993. The culprit had apparently left her DNA on a teacup. Fourteen years after that death, in 2007, the Phantom's appeared again when a 22-year-old policewoman was fatally shot from behind in a police car in Heilbronn.

Doubts about the Phantom's reality were raised when her DNA was found on the burnt body of an asylum-seeker in France. "Obviously that was impossible, as the asylum-seeker was a man and the Phantom's DNA belonged to a woman," Ernst Meiners, spokesman for the public prosecutor's office in Saarbrücken, West Germany, said. A second examination failed to find the Phantom's DNA on the body's fingerprints. "That aroused suspicions that the materials were contaminated."

In late March 2009, investigators came to the conclusion that the "Phantom" criminal did not exist, and the DNA recovered at the crime scenes had already been present on the cotton swabs used for collecting DNA samples.

All swabs came from the same factory, which employs several Eastern European women who fit the type the DNA was assumed to match. Although sterile, the swabs are not certified for human DNA collection or human DNA-free. Furthermore, Bavaria, a region central to these crimes, got their swabs from a different factory and subsequently had no reports of crimes committed by the Phantom.

Forensic science experts in Stuttgart have analysed hundreds of unused cotton swabs since fears were first raised in April last year, although no source of contamination has been found. Police in the Austrian city of Linz have also announced that they are investigating, after developing their own doubts about the existence of the Phantom.

(From Yeoman, F., *Times*, March 27, 2009.)

---

use DNA-free material (e.g., swab, reagent, tubes, tips) and can be introduced by personnel themselves (e.g., talking during handling of samples) or between samples, called cross-contamination (e.g., using unclean razor when sample is collected, transferring part of DNA or PCR solution from one tube to another) (see Box 8.10) (http://news.bbc.co.uk/2/hi/uk_news/england/3007854.stm).

Some precautions against inadvertent contamination include using personal protective equipment at all times (i.e., gloves, mask, cloth); purchasing human DNA-free, PCR inhibitor-free, and DNase-free or RNase-free reagent and material for the sample collection (see Box 8.9); no smoking or drinking at the crime scene; using fresh micropipette tip; careful cleaning of laboratory surfaces; keeping samples with high DNA amount (reference, blood, sperm) separated from samples with low DNA amount (touched evidence) during

BOX 8.10 CASE SAMPLES

CASE SAMPLE 1—CONTAMINATION BY PERSONNEL

City Crime Lab Director Fired

Baltimore crime analysts who had been contaminating evidence with their own DNA were discovered after updating their DNA elimination database. An initial review made in August 2008 showed that employees had contaminated DNA samples in two court cases, Robinson's and a handgun violation case that ended with a guilty plea in December 2007. This revelation led to the dismissal of the city Police Department's crime lab director and prompted questions yesterday from defense attorneys and forensic experts about the professionalism of the state's biggest and busiest crime lab. Mr. Koch, a former Anne Arundel County police officer who developed the forensics lab there, said supervisors had mistakenly believed since 2005 that the lab staff's DNA samples had been entered into the database.

CASE SAMPLE 2—CROSS-CONTAMINATION

In 2005, DNA sample in an Anoka County rape case accidentally got mixed into another sample during testing by the state Bureau of Criminal Apprehension lab. A 21-year-old Mohamed J. Abdullahi, a college student who is charged with raping a woman he and a friend met at a pool hall in Columbia Heights in April 2004. His DNA profile was also found in a sample from a separate case in Blue Earth County. The BCA scientist who was working on three separate cases somehow got some of Abdullahi's DNA sample into a sample container for the Blue Earth case. The wrong result was released to the county attorney's office before it could be corrected.

CASE SAMPLE 3—CROSS-CONTAMINATION

"Files from Orchid-Cellmark's Germantown, Maryland facility," according to William C. Thompson, "show dozens of instances in which samples were contaminated with foreign DNA or DNA was somehow transferred from one sample to another during testing. I recently reviewed the corrective action file for an accredited California laboratory operated by the District Attorney's Office of Kern County (Bakersfield). Although this is a relatively small laboratory that processes a low volume of samples (probably fewer than 1,000 per year), during an 18-month period, it documented multiple instances in which (blank) control samples were positive for DNA, an instance in which a mother's reference sample was contaminated with DNA from her child, several instances in which samples were accidentally switched or mislabelled, an instance in which an analyst's DNA contaminated samples, an instance in which DNA extracted from two different samples was accidentally combined into the same tube, falsely creating a mixed sample, and an instance in which a suspect tested twice did not match himself (probably due to another sample-labelling error)."

(From Thompson, W.C., "Tarnish on the 'gold standard:' Understanding recent problems in forensic DNA testing." *The Champion*, 30(1): 10–16 (January 2006). Bykowicz, J. and Fenton, J., "City crime lab director fired," *Baltimore Sun*, August 21, 2008. Chanen, D., "Defense attorneys raise concerns about DNA sample mixup," *Minneapolis Star Tribune*, May 20, 2005, Friday, Metro Edition, p. 1B.)

the whole laboratory process. The laboratory shall also keep the DNA profiles (preferably in DNA elimination database) of all personnel who might come into contact with the evidence (crime-scene officer, laboratory personnel, and others—prosecutors, manufacturing staff—depending on the judicial system). This procedure may help to detect unintentional contamination of DNA samples. The constant checking of the blank sample log information (extraction and PCR) shall be used for the detection of different, but not at all relevant contaminations.

Mixed samples are contaminated by their own nature (e.g., penile swab collected after intercourse is expected to contain a mixture of vaginal fluids and epithelial cells from penis, saliva swab collected from somebody's skin is expected to contain a mixture of the skin cells and epithelial cells from the mouth, saliva left on the end of a cigarette smoked by different persons is expected to contain a mixture of saliva fluids). Mixture samples are samples composed of materials from different individuals as a result of a criminal activity (rape or assault) or treatment (different persons smoke the same cigarette during the crime). Because the sensitivity of DNA tests have dramatically increased over the past two decades compared to the number of analyzed mixed samples, laboratories must have very stringent guidelines for the interpretation of mixed samples.

Carryover contamination appeared with PCR technology. It occurs when an amplified PCR product is transferred from the previous amplifications to the reaction mix of the succeeding amplification. The carryover product can then be amplified along with the target sample from evidence. There are a number of approaches to reduce the risk of carryover PCR contamination, such as separating the pre-PCR (sample preparation and PCR preparation) and post-PCR (PCR amplification and analysis) working areas, both with its own separate set of equipment, including pipettes, reagents, pipette tips, racks, use safety box, use positive displacement or barrier pipette tips, use of ultraviolet (UV) light, use of bleach for cleaning, and use of aliquots of PCR reagents. Detection of contaminations can be implemented through the mandatory use of positive and negative or blank controls in the process.

Errors in the interpretation of DNA results will not be discussed in detail in this chapter, because they are presented at individual forensic DNA analysis (STR, SNP, mtDNA). But we do need to remind the readers about these kinds of errors. Some of the most common misinterpretations of DNA results are as follows: wrong allele calling (reading; allele drop-out not detected, stutter or spike determined as allele or typing; allele is wrongly numbered) (Rand et al. 2004; Thompson et al. 2003; Thompson 2006, 2008); accuracy of frequency estimates of DNA profile (relatives or substructuring of population is not included in calculation); misinterpretation of coincidental match (frequency of profile is too high) (Džijan et al. 2009; BBC Panorama 2007; BBC News 2003); using the wrong equation for the calculation of relationship index (not included prior odds for victims in mass graves) (Gornik et al. 2002; Thompson et al. 2003; Thompson 2006; Karlsson et al. 2007; Thompson 2008; Thomsen 2009).

Another type of error can occur when the DNA analyst does not take into consideration the errors from cross-contamination of DNA samples and sample mix-ups (Thompson 2006, 2008; U.S. Department of Justice 2004; Sweeney and Main 2004). This is done on purpose mostly because analysts believe that their results are right anyway. W.C. Thompson classified these errors into a single group named *intentional planting or falsification of the results* (Thompson 2006). These errors occur when DNA analysts conceal the problems during DNA analysis, and instead continue with the analysis, and in order to do so they sometimes falsify their notes in such a way that the integrity of the DNA results became questionable and influence the scrutiny of the laboratory. For example, a negative (blank) extraction control should not contain any DNA, so when a positive result is produced, it indicates a problem. However, the DNA analysts concealed the contamination in the control samples because they did not run a negative control (when DNA appears in blank control during DNA quantification test) or excluded the blank from the analysis when it is actually positive in a particular run or used a negative control from another run (see Box 8.11).

BOX 8.11 THE FBI DNA LABORATORY: A REVIEW OF PROTOCOL AND PRACTICE VULNERABILITIES 2004

FBI analyst Jacqueline Blake prepared extraction blanks along with other samples and recorded the creation of these samples in her notes, but she dumped the portion of those samples that might have contained contaminated DNA before sending the samples through the computer-operated genetic analyzer program that typed the DNA. In addition to omitting the negative control tests, Blake falsified her laboratory documentation to conceal the shortcut she was taking to generate contamination-free testing results. Blake generated more than two years' worth of testing results before her omissions were finally caught, and even then her discovery was accidental. In April 2002, a colleague of Blake was working late one evening after Blake had left the Laboratory for the day, and noticed that the testing results displayed on Blake's computer were inconsistent with the proper processing of the control samples. Further inquiry by Laboratory personnel led to the discovery that Blake had failed to complete the negative control testing in the vast majority of her cases. The retesting of evidence in Blake's cases to date indicates that, while she did not properly conduct the contamination testing, the DNA profiles that she generated were accurate. But Blake's actions have caused many problems. Although the FBI Laboratory has not identified a case where Blake's misconduct interfered with the content of a DNA profile, Blake's failure to process the negative controls rendered all of her DNA analyses scientifically invalid. We found that her actions caused substantial adverse effects in at least five respects. First, it required the removal of 29 DNA profiles from the national registry of DNA profiles, known as NDIS, 20 of which have yet to be restored as of March 2004. Second, Blake's misconduct has delayed the delivery of reliable

DNA reports to contributors of DNA evidence. Retesting in many of Blake's cases has taken upward of two years to complete, leaving evidence contributors without information that they should have had long ago. Third, in a limited number of cases, Blake's faulty analysis is the only DNA information that is available. The previously submitted evidence was consumed in the testing process and new evidence samples cannot be obtained. Fourth, Blake's misconduct has adversely impacted the resources of the FBI and DOJ. The efforts that the FBI Laboratory and DOJ have had to expend on the corrective measures needed to address Blake's actions have been substantial. Both organizations have devoted thousands of hours of work to deal with the consequences of Blake's failure to comply with the FBI Laboratory's DNA protocols, a cost that does not include the funding expended for contractor support to retest evidence. State and local investigators and prosecutors who were notified of Blake's misconduct and instituted corrective measures in their cases also have had to expend additional resources. And lastly, we believe that Blake's misconduct, and the Laboratory's failure to detect it for a period exceeding two years, has damaged intangibly the credibility of the FBI Laboratory. The Blake controversy has fed into a perception that the Laboratory has unresolved management and employee oversight issues. Blake later resigned from the Laboratory and was investigated by the Department of Justice (DOJ or Department) for her misconduct. On May 18, 2004, Blake pled guilty in the United States District Court for the District of Columbia to a misdemeanor charge of providing false statements in her laboratory reports.

(From U.S. Department of Justice. Available at http://www.justice.gov/oig/special/0405/exec.htm. Accessed February 24, 2013.)

8.4.3 Error Rates in Forensic DNA Testing

The errors in forensic DNA analysis (criminal and civil) have been known for long time in many laboratories (Thompson et al. 2003), but very few systematic analyses were conducted (Thompson 2006, 2008). Most errors have come to light during PT (Rand et al. 2004; Thomsen 2009; Reich 2009) or during post-conviction DNA tests, like those advocated by the Innocence Project (US), when tests performed by other laboratories contradicted the original results. Some laboratories, in order to ensure the integrity of their results and maintain high QA, regularly monitor and note their errors but most of those results are confidential and as such rarely published.

Results of GEDNAP proficiency tests (PT) (autosomal STR analysis of stain) over the period from 1998 to 2002, which drew the participation of 79 laboratories in 1999 to 136 in 2002, showed that the error rate is relatively constant at 0.4% to 0.7% after using multiplex STR kits for analysis. The most common type of error has always been transcriptional errors followed by incorrect interpretation due to failure to recognize an error (Rand et al. 2004). Unfortunately, the newest results are not published outside the ENFSI group. The same level of discrepancies was reported for the results of the Paternity Testing Workshops (PTW), which are annually organized by the English Speaking Working Group (ESWG) of the ISFG for the period 2002 to 2008. The results showed a high degree of uniformity in typing results with observed errors for 0.1–0.3% of all PCR-based results (autosomal STR, X-STR, Y-STR). The numbers of participating laboratories increased from 46 in 2002 to 68 in 2008. Most are due to transcriptional errors. The highest degree of variation existed in complex scenarios with rare genetic constellations such as genetic inconsistencies/possible silent alleles, rare alleles, and haplotypes (Thomsen 2009). K.A. Reich and coauthors summarized the results of the PT conducted by the College of American Pathologists (CAP) for forensic DNA testing and identity testing from 1997 to 2003 and by the Collaborative Testing Service (CTS) from 2001 to 2003. They reported average discrepancies of 2.71% and 3.01% among all participating laboratories (Reich et al. 2009). The higher rate of errors in PT test reported by Reich et al. is mainly attributed to the higher percentage of laboratories that participated in CAP and CTS, using the RFLP technique and for separation of STR allele polyacrylamide gels than in PT test from GEDNAP and PTW ESWG (data not shown). The results reported by Reich are in concordance with GEDNAP PT results from 1995 to 1997, which reported observed errors between 1.2% and 2.1% (Rand et al. 2004).

In conclusion, as indicated by PT test results, the most common type of errors is transcriptional error, which clearly shows that human carelessness is the predominant source of error.

8.4.3.1 *Should an Error Be Included in Calculations?*

In NRC's opinion, error rates should not be combined with random match probability, because error rates change over time and are dependent on many factors (number of samples, redundancy in testing). NCR's recommendation is that the risk of errors should be properly considered in every case. In practice, most laboratories did not estimate error rate; neither did they combine it with random-match probability (National Research Council 1996). Some scientists have raised issues against this policy (Thompson et al. 2003; Thompson 2006). According to Thompson, "the problem with NCR argument (errors rates change) is that it equates the ability to appreciate the potential for a laboratory error with the ability to accurately estimate the probability of an error. The core of the fallacy is the erroneous assumption that the false positive probability, which is the probability that a match would be reported between two samples that do not match, is equal to the probability that a false match was reported in a particular case" (Thompson et al. 2003). Their approach relies on Bayes' theorem. Bayes' theorem indicates how a rational evaluator should adjust a probability assessment in light of new evidence. They presented how the probability of a false positive should influence a rational evaluator's belief in the proposition that a particular individual is the source of a biological specimen (Thompson 2006). For more details, we refer the reader to the practitioner guide number one from the Royal Statistical Society's Working Group on Statistics and the Law (Aitken et al. 2010).

References

Aitken, C., P. Roberts, and G. Jackson. 2010. 1—Fundamentals of Probability and Statistical Evidence in Criminal Proceedings—Guidance for Judges, Lawyers, Forensic Scientists and Expert Witnesses. Royal Statistical Society's Working Group on Statistics and the Law. http://www.maths.ed.ac.uk/~cgga/rss.pdf (accessed February 24, 2013).

American Bar Association. 2006. *Report of the ABA Criminal Justice Section's Ad Hoc Innocence Committee to Ensure the Integrity of the Criminal Process. Achieving Justice: Freeing the Innocent, Convicting the Guilty*, ed. P.C. Giannelli, and M. Raeder. Chicago: American Bar Association.

Applied BioSystems. 2009. The AmpFℓSTR® NGM™ PCR Amplification Kit: The Perfect Union of Data Quality and Data Sharing. Forensic News, September 2009. http://marketing.appliedbiosystems.com/images/Product_Microsites/NGM/downloads/NextGen_SS_v2.pdf (accessed February 24, 2013).

BBC News. 2003. Mystery Milly DNA link rejected.
BBC Panorama. 2007. Give Us Your DNA. http://news.bbc.co.uk/go/pr/fr/1/hi/programmes/panorama/7010687.stm (accessed February 24, 2013).
Butler, J.M. 2005. *Forensic DNA Typing: Biology, Technology, and Genetics of STR Markers*. London: Academic Press.
Butler, J.M. 2006. Debunking some urban legends surrounding validation within the forensic DNA community. *Profiles DNA* 9(2):3–6; available on-line at http://www.promega.com/profiles/902/ProfilesInDNA_902_03.pdf (accessed February 24, 2013).
Butler, J.M. 2007. Validation: What is it, why does it matter, and how should it be done? *Forensic News (Applied Biosystems)* January 2007.
Butler, J.M., C.S. Tomsey, and M.C. Kline. 2004. Can the validation process in forensic DNA typing be standardized? *Proceedings of the 15th International Symposium on Human Identification*. Madison, WI: Promega Corporation. http://www.promega.com/geneticidproc/ussymp15proc/oralpresentations/butler.pdf (accessed February 24, 2013).
Collins, P.J., L.K. Hennessy, C.S. Leibelt, R.K. Roby, D.J. Reeder, and P.A. Foxall. 2004. Developmental validation of a single-tube amplification of the 13 CODIS STR loci, D2S1338, D19S433, and amelogenin: The AmpFlSTR Identifiler PCR amplification kit. *J Forensic Sci* 49: 1265–1277.
Convention between the Kingdom of Belgium, the Federal Republic of Germany, the Kingdom of Spain, the French Republic, the Grand Duchy of Luxembourg, the Kingdom of the Netherlands and the Republic of Austria: May 27, 2005 on the stepping up of cross-border cooperation, particularly in combating terrorism, cross-border crime and illegal migration. Prüm convention. http://register.consilium.europa.eu/pdf/en/05/st10/st10900.en05.pdf (accessed February 24, 2013).
Council framework Decision 2009/905/JHA of November 30, 2009 on Accreditation of forensic service providers carrying out laboratory activities. www.Eur-lex.europa.eu/LexUriServ (accessed February 24, 2013).
Council of Europe. 1992. Recommendation No.R (92) 1 on the Use of Analysis of Deoxyribonucleic Acid (DNA) within the Framework of the Criminals Justice System, adopted by the Committee of Ministers on February 10, 1992 at the 470th meeting of the Miniásters´ Deputies.
DNA Identification Act of 1994. Public Law 103 322 (United States).
Drobnic, K., N. Pojskic, N. Bakal, and D. Marjanovic. 2005. Allele frequencies for 15 short tandem repeat loci in Slovenian population. *J Forensic Sci* 50:1505–1507.
Džijan, S., G. Ćurić, D. Pavlinić, M. Marcikić, D. Primorac, and G. Lauc. 2009. Evaluation of the reliability of DNA typing in the process of identification of war victims in Croatia. *J Forensic Sci* 54: 608–609.
ENFSI DNA Working Group. 2010. Recommended minimum criteria for the validation of various aspects of the DNA profiling process. http://www.enfsi.eu/sites/default/files/documents/minimum_validation_guidelines_in_dna_profiling_-_v2010_0.pdf.
ENFSI Standing committee for quality and competence. 2005. Guidance on the conduct of proficiency tests and collaborative exercises within ENFSI. http://www.enfsi.eu/sites/default/files/documents/bylaws/policy_on_ptce.pdf (accessed February 24, 2013).
EU Council Resolution of June 25, 2001 on the exchange of DNA analysis results. *Official J Eur Communities*, OJC 187:1–4.
EU Council Resolution of November 30, 2009 on the exchange of DNA analysis results. *Official J Eur Communities*, OJC 296:1–3.
EU Council Resolution of June 9, 1997 on the exchange of DNA analysis results. *Official J Eur Communities*, OJC 193: 2–3.
EURACHEM Guide. 1998. The fitness for purpose of analytical methods: a laboratory guide to method validation and related topics. http://www.eurachem.ul.pt/guides/valid.pdf (accessed February 24, 2013).
EUROLAB. 2005. Survey on the accreditation of proficiency test providers. *Technical Report* 1/2005. http://www.eurolab.org/documents/EL_11-01_05_145.PDF (accessed February 24, 2013).

Federal Bureau of Investigation. 2000. Quality assurance standards for forensic DNA testing laboratories and quality assurance standards for convicted offender DNA databasing laboratories. *Forensic Sci Commun* 2(3) (July).

Gill, P., and A. Kirkham. 2004. Development of a simulation model to assess the impact of contamination in casework using STRs. *J Forensic Sci* 49:485–491.

Gornik, I., M. Marcikić, M. Kubat, D. Primorac, and G. Lauc. 2002. The identification of war victims by reverse paternity is associated with significant risks of false inclusion. *Int J Legal Med* 116: 255–257.

Howitt, T. 2003. Ensuring the integrity of results: A continuing challenge in forensic DNA analysis. Paper presented at 14th International Symposium of Human Identification.

ILAC-G13:2007. Guidelines for the Requirements for the Competence of Providers of Proficiency Testing Schemes. https://www.ilac.org/documents/WhatsNew.G13_07_2007.pdf (accessed February 24, 2013).

ILAC-G19:2002. Guidelines for Forensic Science Laboratories. International Laboratory Accreditation Cooperation. https://www.ilac.org/documents/g19_2002.pdf (accessed February 24, 2013).

ISO 8402:1994. Quality management and quality assurance – Vocabulary. International Organisation for Standardization, Geneva. http://www.iso.org.

ISO Guide 30:1992 and 1:2008: Revision of definitions for reference material and certified reference material, International Organization for Standardization, Geneva, http://www.iso.org.

ISO/IEC 17011.2006. Conformity assessment—General requirements for accreditation bodies accrediting conformity assessment bodies. International Organisation for Standardization, Geneva. http://www.iso.org.

ISO/IEC 17025:2005. General requirements for the competence of testing and calibration laboratories. International Organisation for Standardization, Geneva. http://www.iso.org.

ISO/IEC 17043:2010. Conformity assessment—General requirements for proficiency testing. International Organisation for Standardization, Geneva. http://www.iso.org.

ISO/IEC Guide 43-1:1997. Proficiency testing by interlaboratory comparisons—Part 1. Development and operation of proficiency testing schemes. International Organisation for Standardization, Geneva. http://www.iso.org.

ISO/IEC Guide 43-1:1997. Proficiency testing by interlaboratory comparisons—Part 2. Selection and use of proficiency testing schemes by laboratory accreditation bodies. International Organisation for Standardization, Geneva. http://www.iso.org.

Justice and Home Affairs Council Decision 2008/615/JHA of June 23, 2008 on the stepping up of cross-border cooperation, particularly in combating terrorism and cross-border crime. http://eur-lex.europa.eu/LexUriServ/LexUriServ.do?uri=CELEX:32008D0615:EN:NOT (accessed February 24, 2013).

Karlsson, A.O., G. Holmlund, T. Egeland, and P. Mostad. 2007. DNA-testing for immigration cases: The risk of erroneous conclusions. *Forensic Sci Int* 2–3:144–149.

Krenke, B.E., L. Viculis, M.L. Richard et al. 2005. Validation of a male-specific, 12-locus fluorescent short tandem repeat (STR) multiplex. *Forensic Sci Int* 148:1–14.

Loyo, M.A., G. Caraballo, K. Sanchez, and H. Takiff. 2009. PowerPlex(R) 16 HS: Internal validation of a new tool for genetic analysis of forensic and parentage testing. *Forensic Sci Int Genet Suppl Ser* 2:33–35.

Mulero, J.J., N.J. Oldroyd, M.T. Malicdem, and J.K. Hennessy. 2008. Developmental validation of the AmpFℓSTR® SEfiler Plus™ PCR amplification kit: An improved multiplex with enhanced performance for inhibited samples. *Forensic Sci Int Genet Suppl Ser* 1(1):121–122.

National Research Council. 1992. *DNA Technology in Forensic Science.* Washington, D.C.: National Academy Press.

National Research Council. 1996. *The Evaluation of Forensic DNA Evidence.* Washington, D.C.: National Academy Press.

Peterson, J.L., G. Lin, M. Ho, Y. Chen, and R.E. Gaensslen. 2003a. The feasibility of external blind DNA proficiency testing: I. Background and findings. *J Forensic Sci* 48:21–31.

Peterson, J.L., G. Lin, M. Ho, Y. Chen, and R.E. Gaensslen. 2003b. The feasibility of external blind DNA proficiency testing: II. Experience with actual blind test. *J Forensic Sci* 48:32–40.

Promega Corporation. 2006. Internal validation of STR Systems. http://worldwide.promega.com/~/media/files/resources/validation%20guides/internal%20validation%20of%20str%20systems.pdf?la=en (accessed February 24, 2013).

Rand, S., M. Schurenkamp, and B. Brinkmann. 2002. The GEDNAP blind trial concept. *Int J Legal Med* 116:199–206.

Rand, S., M. Schurenkamp, C. Hohoff, B. Brinkmann. 2004. The GEDNAP blind trial concept: Part II. Trends and developments. *Int J Legal Med* 118:83–89.

Reich, K.A., L.A. Graffy, and P.W. Boonlayangoor. 2009. Utilizing proficiency testing survey results in forensic DNA laboratories. *Independent Forensics.* Lombard. http://www.promega.com/~/media/files/resources/conference%20proceedings/ishi%2015/poster%20abstracts/42graffy-boonlayangoor.pdf?la=en (accessed February 24, 2013).

Schmidt, A., Örnemark, M. Golze, and G.M. Henriksen. 2006. Surveys on the accreditation of providers of proficiency testing and external quality assessment schemes. *Accred Qual Assur* 11: 379–384.

Swango, K.L., W.R. Hudlow, M.D. Timken, and M.R. Buoncristiani. 2007. Developmental validation of a multiplex qPCR assay for assessing the quantity and quality of nuclear DNA in forensic samples. *Forensic Sci Int* 170:35–45.

Sweeney, A., and F. Main. 2004. Botched cases forces state to change DNA reports. *Chicago Sun Times*, Nov. 8, 2004.

Thompson, W.C. 2006. Tarnish on the 'gold standard:' Understanding recent problems in forensic DNA testing. *Champion* 30(1):10–16.

Thompson, W.C. 2008. The potential for error in forensic DNA testing (and how that complicates the use of DNA databases for criminal identification). Council for Responsible Genetics (CRG) and its national conference. Forensic DNA Databases and Race: Issues, Abuses and Actions held June 19–20, 2008, at New York University, www.gene-watch.org.

Thompson, W.C., F. Taroni, and C.G.G. Atiken. 2003. How the probability of a false positive affects the value of DNA evidence. *J Forensic Sci* 48:47–54.

Thomsen, A.T., C. Hallenberg, Bo Simonsen et al. 2009. A report of the 2002–2008 paternity testing workshops of the English speaking working group of the International Society for Forensic Genetics. *FSI Genet* 3:214–221.

U.S. Department of Justice, Office of the Inspector General. 2004. The FBI DNA laboratory: A review of protocol and practice vulnerabilities. May 2004. http://www.justice.gov/oig/special/0405/final.pdf (accessed February 24, 2013).

Westgard, J.O., and P.L. Barry. 1986. *Cost-Effective Quality Control: Managing the Quality and Productivity of Analytical Processes.* Washington, D.C.: AACC Press.

Yeoman, F. 2009. The phantom of Heilbronn, the tainted DNA and an eight-year goose chase. *Times*, March 27, 2009.

Uses and Applications II

Collection and Preservation of Physical Evidence

9

HENRY C. LEE
TIMOTHY M. PALMBACH
DRAGAN PRIMORAC
ŠIMUN ANĐELINOVIĆ

Contents

9.1	Introduction	253
	9.1.1 Sample Collection from Victim or Suspect	254
	9.1.1.1 Known Oral Swab Standards	254
	9.1.1.2 Liquid Urine and/or Fecal Material	254
	9.1.1.3 Vaginal Materials	255
	9.1.1.4 Nasal Mucous	255
	9.1.1.5 Bite Mark Evidence	255
	9.1.1.6 Skin Tissue	255
	9.1.1.7 Clothing or Personal Items	256
9.2	Recognition and Identification of Blood Evidence	256
	9.2.1 Presumptive Blood Tests	256
	9.2.2 Confirmatory Blood Tests	258
9.3	Collection Methods for Blood	260
	9.3.1 Dried Blood Stains	260
	9.3.2 Liquid Blood Samples	261
	9.3.3 Seminal Stains	262
	9.3.4 Stains from Other Physiological Fluids	263
9.4	Blood Stain Pattern Analysis	264
9.5	Crime Scene Reconstruction	265
9.6	Case Examples	266
	9.6.1 Concetta "Penney" Serra Homicide Case	266
	9.6.2 Identical Twins, New Orleans, LA	269
	9.6.3 Murder in a Bathroom, Middletown, CT	271
	9.6.4 Brown's Chicken Murders	273
References		276

9.1 Introduction

Despite significant advances in the ability to analyze blood and other forms of biological evidence, if the sample is not properly collected and preserved, no level of sophistication will correct for mishandling a biological sample (Marjanović and Primorac, 2013). In fact, most legal challenges regarding physical evidence such as blood evidence focus on the recognition, collection, and preservation matters rather than the scientific methodologies that were used to analyze the sample (Fisher and Fisher, 2012). The Scientific Working Group for DNA Analysis and Methods (SWGDAM) has suggested several best practices for the collection of blood found at crime scenes and is an excellent resource.

Not only does the collection and preservation have to be conducted in a manner that will preserve the evidentiary nature of the stain, but it has to be done in a process in which there is thorough documentation. The collection, packaging, and preservation of blood evidence at the crime scene should never take place until the crime scene analyst or forensic scientists has taken extra care to make sure that the bloodstain patterns have been extensively documented. This documentation may include written or audio notes, sketches, photographs, and/or video documentation. Generally, proper documentation will require several of these documentation methods. The particular method may have to be tailored to address a specific relevant fact. For example, the condition of a bloodstain—is it wet, dry, or coagulated—is an important factor for estimating the time frame since the blood was deposited (Lee et al., 2010). Recording this information will require notes as well as demonstrative methods such as photography. Furthermore, the general characteristics of the biological stain, its pattern, and the location of the pattern will be of great value in the reconstruction of events as well as a means for laboratory analysts to best determine which samples are most relevant (Lee and Harris, 2000).

Once a potential blood or biological stain is located, thorough documentation must be completed. Photographs must include overall views showing the object or pattern in relationship to the overall scene. Then there should be an intermediate photograph that shows the observed or potential pattern and its orientation to the object it is located on. Finally, there needs to be close-up photographs. Close-up photographs must be taken with the camera perpendicular to the stain or pattern area to prevent distortion of the pattern. In addition the photograph should be taken in a manner to ensure that the image is clear, properly exposed, and high contrast. Finally, the close-up photograph must be taken in a 1:1 format, which is rarely possible, or with a scale included. It is best to use a ruler with a circle and diagonal lines as this is a method of establishing that the photograph was taken properly. In addition to quality photos, it is very helpful to sketch the stain. Overall stain dimensions should be obtained. In addition, it is important to measure and record the precise location of that stain or pattern area on the object or within the crime scene (Figure 9.1).

9.1.1 Sample Collection from Victim or Suspect

The real value of unknown biological samples from the crime scene or evidence is to provide a link or association with the victim and/or suspect (Lee et al., 2001; Primorac and Marjanović, 2008). Proper known standards must be obtained.

9.1.1.1 Known Oral Swab Standards

Obtain oral swabs of known origin from rape victims and suspects to determine their DNA type. There are numerous swab units that can be obtained from commercial suppliers of forensic science laboratory and crime scene supplies. Be sure to properly dry and package the swabs so that there is no degradation of the DNA sample.

9.1.1.2 Liquid Urine and/or Fecal Material

These samples may be collected in a clean specimen jar. Label accordingly and refrigerate until submission to the laboratory for DNA and toxicological analysis.

Collection and Preservation of Physical Evidence

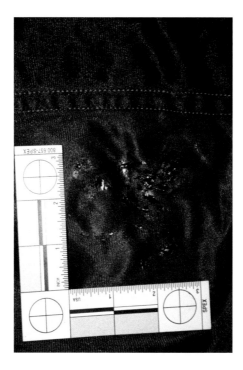

Figure 9.1 Examination quality, close-up view with scale of bloodstains on clothing.

9.1.1.3 Vaginal Materials

Vaginal secretions are usually encountered in connection with a sexual assault case. Most often, these samples are included in the sexual assault evidence kit and should be collected according to directions. Usually, these samples are obtained with a swab and may include a glass slide prepared for microscopic examination.

9.1.1.4 Nasal Mucous

Nasal mucous is occasionally found at crime scenes, on clothing, a handkerchief or tissue, or on a body. These materials should be air dried and packaged in a paper druggist fold. These samples can yield good DNA results.

9.1.1.5 Bite Mark Evidence

If evidence of a bite mark is found, collect a swabbing of the bite mark of the areas adjacent to and inside the observed bite mark. It is important to obtain examination quality photographs of the bite mark. Often the bite mark will visibly improve the following day as a result of the bruising and healing process. It may be beneficial to consult with a forensic odontologist during the photographic and documentation process.

9.1.1.6 Skin Tissue

Skin and tissue may be present under the fingernails in cases of sexual assault or violent confrontation. Package fingernail scrapings, which should be taken with a new orange stick, file, or any other device, in paper and then properly label. Alternatively, collect the fingernails individually by clipping and packaging them separately.

9.1.1.7 *Clothing or Personal Items*

All evidence potentially containing body fluid evidence should be handled with gloved hands to protect the collector and to reduce the possibility of contamination of the DNA evidence. In addition, care should be exercised to prevent cross-contamination between subsequently collected items. Change gloves often, and with instances involving small items the use of disposable tweezers is recommended. The proper packaging and storage of clothing is vital to the viability of any potential DNA evidence.

1. It is of primary importance that clothing that is wet be allowed to air dry at room temperature before packaging. The item is then packaged in paper bags or wrapped flat in paper. Portable drying units are available and can be of great assistance in the drying process.
2. Each item of clothing must be packaged separately.
3. Under no circumstances should this evidence be packaged in plastic bags or other airtight containers.
4. In order to preserve the trace evidence that may be present on the item, avoid excessive handling of clothing, and wrap in butcher paper before placing in the outer container, usually a paper bag.

9.2 Recognition and Identification of Blood Evidence

First, a particular stain or sample must be recognized as potential blood (James et al., 2005). Essential to this process is a preliminary evaluation of the scene or the object upon which the suspected stain is located. An understanding of basic bloodstain pattern analysis is of great value at this point in the process. It is critical to understand the basic principles in drop and pattern formation, how various forces or mechanisms that interact with the blood sample will effect the resulting pattern, and what—if any—interferences exist that may alter the resulting stain or pattern. Furthermore, an initial reconstruction of the crime scene, utilizing all available information—investigative, medical or injury patterns, physical evidence, and various forms of pattern evidence—can assist in searching for potential blood samples in appropriate locations.

Despite the ability to determine the potential for finding blood based on a careful analysis of the object or scene, it should not merely presumed that a stain is indeed blood. There are a variety of presumptive blood tests available that are very sensitive and can be used in laboratory and field, crime scene, applications.

9.2.1 Presumptive Blood Tests

Presumptive blood tests are designed to detect trace amounts of blood. They are very sensitive and detect minute traces of blood. They are, however, not specific and thus subject to the potential of false positives. False positives can occur when the sample tested contains chemical oxidants or substances with peroxidase activity. Since many common cleaning agents, such as bleach compounds, are strong chemical oxidants, it is likely that one will encounter this situation at crime scenes or on clothing that has been laundered. Moreover, these tests may yield false negatives because of chemical interferences. If one correctly understands that these are presumptive tests only and that additional confirmatory testing is required, the concerns related to false positives or negatives are properly addressed. As valuable as these tests and reagents may be to investigators and forensic scientists, there are certain precautions that must be heeded. Field tests are designed for screening purposes and should not be used in lieu of laboratory analysis and confirmation testing. As a general rule, if the amount of available sample for testing is so minute that there may not be sufficient material for a full array of testing, then, if at all possible, avoid field tests so as to preserve the sample for laboratory analysis.

Following a simple sequential procedure with prepared reagents is all that is required with most of these presumptive tests. The interpretation of testing results is often straightforward and simple, such as observing a color shift or color formation. Despite the simplicity of the basic procedure, it is helpful to have an understanding of the underlying chemical reaction or mechanism.

The tests are based on a chemical oxidation–reduction reaction in which blood, more specifically the heme portion of blood, acts as the catalyst. Heme, a principle component of hemoglobin, is a ferrous-bearing molecule located within a red blood cell. This hemoglobin molecule with its ferrous content is responsible for carrying oxygen throughout the body. The ferrous element is generally in a reduced state. The presumptive blood tests discussed hereafter are designed to detect the presence of the reduced ferrous molecule through an oxidation–reduction reaction that will convert a colorless reagent to a colored by-product (Figure 9.2).

Common presumptive blood test reagents include Phenolphthalein (Kastle–Meyer reagent), Leucocrystal Violet, and tetra-methylbenzidine. Since these different reagents have different sensitivities and may react slightly different in a more alkaline or acidic environment, the use of more than one reagent is recommended if questionable results are expected or obtained. These same reagents can be used as a bloody print enhancement reagent. Enhancement reagents are used to visualize and to increase the color contrast of transfer patterns such as bloodstains, fingerprints, footprints, shoe imprints, and other physical patterns. In some instances, enhancement reagents can be used for dual purposes: presumptive tests for biological substances and the development of physical patterns. To use reagents in this capacity, they must be mixed fresh and applied with a spray nebulizer

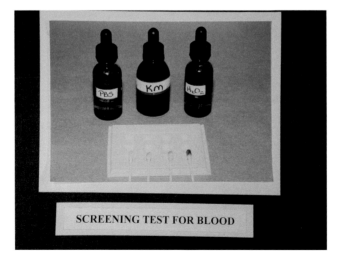

Figure 9.2 Phenolphthalein presumptive test for blood.

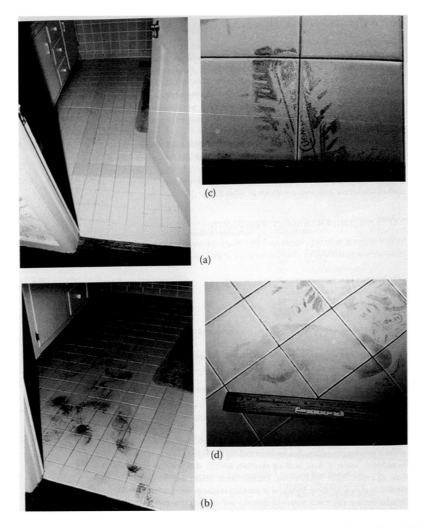

Figure 9.3 Enhancement of bloody foot impressions with tetra-methylbenzidine (TMB).

bottle. If the reagents are to be applied to a vertical surface great care must be exercised to avoid runs or drips through the pattern. Alternatively, these reagents can be mixed in a colloidal suspension with chloroform and ether so that there will not be any dripping on vertical surfaces (Figure 9.3).

9.2.2 Confirmatory Blood Tests

Additional samples must be taken and submitted to a forensic laboratory for confirmatory testing. There are a variety of confirmatory tests available in the laboratory. There is an immunoassay procedure, such as an ABA Hema Trace card, which is a simple and fast confirmatory test for human hemoglobin that is capable of being performed at the scene. This is an immunological (antigen/antibody) test that requires only the stain to be tested, a little pure water, and the commercial test kit. It has quickly become widely accepted and a good addition for modern crime scene processing kits (Figure 9.4).

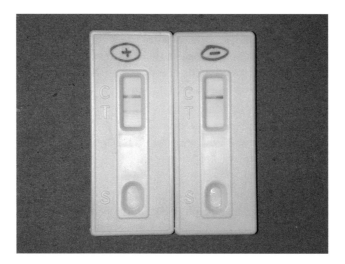

Figure 9.4 Hema trace card.

If there is insufficient sample available to collect for subsequent confirmatory testing, then do not conduct and field testing—rather, send the entire limited sample into the laboratory for analysis. There are minor differences in sensitivities of various reagents known; thus, controls should be conducted to understand the performance of a particular reagent.

Examples of presumptive reagents:

1. *Phenolphthalein (Kastle–Mayer reagent)*
   The test is conducted by rubbing a cotton swab that has been moistened in a distilled water on the suspected bloodstain. A drop of the phenolphthalein solution is added to the swab. After 30 seconds, a drop of hydrogen peroxide (3%) is added. A positive reaction is a pinkish color on the swab, which should occur within 15 seconds (Figure 9.5).

Figure 9.5 Luminol used to enhance blood stains covered over by a coat of latex paint.

2. *Leucomalachite green*

    This test is performed in the same way as the phenolphthalein test. A positive reaction is indicated by a greenish-blue color that will appear almost immediately on the area of the swab exposed to the suspected blood.

3. *Luminol, Bluestar, or Fluorescene*

    These reagents have fluorescent properties. These reagents are commonly utilized as bloody print enhancement reagents. They have great value on dark substrates, but there are added difficulties in photographing the enhanced patterns because of the need to work in low light conditions. With Luminol and Blue star, the reagent is simply sprayed onto the area to be checked for the presence of suspected bloodstains, but it must be viewed in total darkness. Bloodstains will yield luminescence or fluorescence within a few seconds. Over time the reaction will fade, but it can be refreshed with additional reagent. Luminol and Bluestar are very sensitive reagents, and work well with very small amounts of blood or older stains. However, that same level of sensitivity also makes them very susceptible to false positives. With Fluorescene, after the reagent is applied the sample must be viewed with ultraviolet light.

9.3 Collection Methods for Blood

After recognition and documentation of the blood evidence at the crime scene, the collection can begin. Generally, blood which is most susceptible to being destroyed or damaged. These are bloodstains that may be located in high volume traffic areas of the scene; bloodstains found in doorways, hallways, or roadsides; or blood that may be present at outdoor crime scenes. Bloodstains found on movable objects can be protected by temporarily moving the object to a safer location until the stain or object may be collected.

Blood at crime scenes will be found either as a dried stain or in a liquid state. If blood at the crime scene is liquid and not large enough to collect a liquid sample, and then let it air dry. If the object with the bloodstain is movable, then collect the entire object. To package blood and other biological fluids, there are several basic guidelines and precautions. Blood and bloodstained evidence should never be placed in airtight containers.

Bloodstained evidence frequently will have other trace evidence present. If the blood loosely holds the trace evidence, then it is proper to collect that trace evidence and place it in a druggist fold and then a sealed envelope. However, if a dried bloodstain holds the trace evidence tightly, do not attempt to collect the trace evidence. Carefully cover or protect the trace evidence while packaging the bloodstain. Bloodstained items should always be packaged separately to prevent cross-contamination. The packaging should be sealed at the scene before transportation. Do not allow bloodstained evidence to be exposed to excessive heat or humidity. If possible bloodstained evidence should be refrigerated, but the dried bloodstained evidence can be stored safely at room temperature and then submitted to forensic laboratories as soon as possible.

9.3.1 Dried Blood Stains

If possible, generally so with a moveable object, leave the blood on the object and package the entire object in a breathable container such as a paper bag. In most cases, wrapping the object in butcher paper first then sealing it in a paper bag is recommended. If the bloodstain is located

Collection and Preservation of Physical Evidence

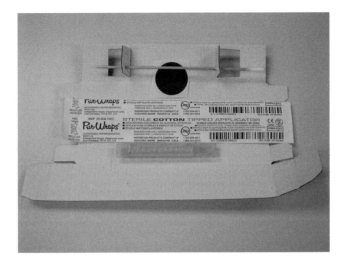

Figure 9.6 Swab safe box.

on a surface that can easily be removed, then cut an area around the bloodstain including the stain. If the surface material is fragile or easily broken, care should be taken to secure the cutout section to prevent breakage of the sample. Wrapping in a druggist's fold and then securing the object in a box works well. If the object cannot be moved, then the sample must be obtained by one of three methods: swabbing, scraping, or tape lift. In the vast majority of cases, swabbing is the preferred method. With scraping, use care to avoid contamination with the substrate from which you are scraping the sample. Tape lifting is rarely a good choice as subsequent extraction of the blood from the adhesive surface is complicated. The only rationale for that method is for preservation of the overall patter. However, good documentation methods are generally a more effective means of dealing with the pattern component. According to recognized guidelines such as the SWGDAM guidelines, dried bloodstains can be absorbed on to sterile cotton swabs. Moisten the swab with distilled water. Carefully swab the bloodstain with the swab. Absorb all the stain with a minimum of area consumed. Insert the swab into the swab-drying box. Allow the swab to dry once the stain has been collected. Seal the box and label all. Swab the bloodstain trying to avoid debris or other contaminants. Let the swab dry before placing in a swab box or container. There are commercially available swab boxes that allow for a rack to place the swab in while drying, and then the open box unit can be folded up and sealed (Figure 9.6).

9.3.2 Liquid Blood Samples

If the wet blood is small, then it should be collected with sterilized cotton swabs and then be allowed to air dry. The stain can be collected by the procedure discussed below for dried bloodstains. If the liquid bloodstain is large, then it may be collected by the following choices of collection procedures:

1. Absorb the liquid blood sample onto sterile cotton swabs. Allow the swatch to dry. The cotton swab must immediately be inserted into a swab box and sealed with evidence tape.
2. If the large wet bloodstain is located on a movable object (e.g., clothes or bed sheets), then wait until the bloodstain dries. Next, place paper in-between layers

of the clothing and collect the item carefully to avoid transfer or alteration of the bloodstain pattern. At the secure location the object should be unwrapped and laid out for continued air-drying. If new packaging is required because the blood soaked the original packaging, the original packaging should be maintained as trace evidence may be located on or within that original packaging. Sometimes the object, such as bloodstained clothing, must be cut to remove it from the scene; *do not* cut through the bloodstains as the pattern is often very important.

9.3.3 Seminal Stains

Semen stains found at the crime scene or on objects such as clothing are collected, packaged, and preserved in a similar way to bloodstains.

In sexual assault investigations, the crime scene investigation will also include the collection of evidence from the victim or suspect at the hospital. The victim of a sexual assault should be taken to the hospital for examination as soon as possible. A doctor assisted by a forensic nurse will usually conduct the examination and collect evidence. Forensic training for nurses is available nationwide and should be encouraged for hospital staff. Commercially available or forensic laboratory prepared sexual assault collection kits are used for the collection of the hospital specimens. These kits will include swabs, microscope slides, and various containers for the collection of a variety of evidence from the victim.

Recognition of a semen stain at a crime scene or on objects will often require examination with an Alternate Light Source (ALS units) or the use of a presumptive screening test. ALS units have proven to be an effective method for detection of dried semen stains. ALS does not work well on moist or wet semen stains. The most common ALS applications involve the use of a spectrum in the blue range, used in conjunction with an orange barrier filter. Alternatively, the use of an ALS unit with an ultraviolet wavelength, used with a yellow barrier filter, will also work to detect semen stains. Once the suspected stains are located, they should be thoroughly documented. With the use of ALS, specialized photographic methods are required. Once photographed, the entire object containing the stain should be seized and packed in a non-airtight container. If the stain is on a larger object that can be cut, such as a sofa cushion, then cut out the entire stain, making sure not to cut through any part of the pattern. As a last resort, as with dried bloodstains, absorb samples onto sterile swabs. However, because of the nature of seminal stains, minimize handling of suspected stains. The collected evidence should be placed in a primary container—swab box or druggist's fold—followed by an outer, secondary container that is not airtight. The container is sealed with evidence tape, marked appropriately, and preserved by refrigeration if possible. As with bloodstains, seminal stains may be stored at room temperature for a limited period until submission to the laboratory as soon as possible (Figure 9.7).

Use of an acid phosphatase presumptive test is fairly easy and applicable for both laboratory and crime scene purposes. The suspected stain is swabbed with a moistened swab, and a two-reagent process—alpha-naphthyl and Fast Blue B—is used. A positive presumptive test for semen yields a garnet red color formation. In addition, there is an ABA card immunoassay that will detect P30, a male-specific seminal enzyme.

Collection and Preservation of Physical Evidence

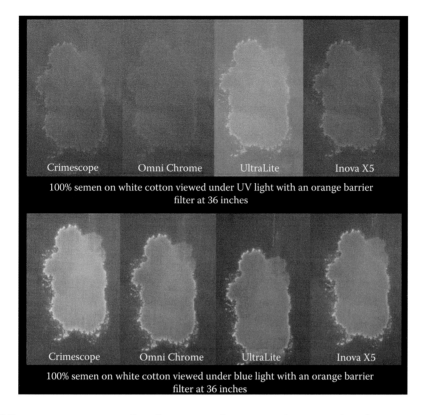

Figure 9.7 Semen stains viewed with various Alternate Light Source (ALS) units.

9.3.4 Stains from Other Physiological Fluids

Other physiological fluid stains are collected, packaged, and preserved in a similar way to the methods and techniques described for blood and seminal stains. These stains may be composed of saliva, urine, perspiration, vaginal excretions, or fecal material. Liquid samples should be collected with the sterile cotton swab method. Dried stains can be swabbed with a moist sterile swab. Saliva stains or bite marks can be collected with a moistened swab. Swab the area with a single swab and concentrate it in a limited area. The saliva standard is collected by use of sterile swabs, air drying, and the use of swab boxes. The saliva standard is best stored refrigerated but can be stored at room temperature for a limited period before submission to the forensic laboratory.

1. *Saliva*
   The presence of a starch-digesting enzyme called amylase is the basis of tests for saliva. Amylase is found in high concentrations in saliva, and the detection of this enzyme indicates the presence of saliva. An immunological test in the form of an ABA card test similar to the blood test is now available.
2. *Urine*
   The identification of urine is based on its characteristic color and odor, as well as the presence of characteristic chemical components, such as creatinine or urea.

3. *Fecal material*

    Fecal material usually has a characteristic color and odor. Chemical tests for the presence of urobilinogen are also conducted to identify feces.

4. *Gastric fluid*

    The identification of gastric fluid or stomach contents (vomit) occurs through chemical and microscopic analysis in addition to the detection of digestive enzyme activity.

5. *Perspiration*

    Perspiration is often present on clothing or other items that are submitted to the laboratory for examination. Its presence must be considered when analyzing evidentiary material for the presence of other body fluids. When testing articles of clothing, test a control sample to ensure that contamination from perspiration is not a factor. This is especially important if DNA analysis will be conducted.

6. *Epithelial cells and tissues*

    Microscopic examination is conducted of samples containing cellular materials. The morphological characteristics of epithelium and other tissues are used for their identification. Any nucleated cell has the potential for yielding DNA useful for analysis.

9.4 Blood Stain Pattern Analysis

Bloodstain patterns can reveal many facts about crimes—where, how, and when blood has been shed. Forensic scientists analyze the patterns of blood left behind at a crime scene in order to reconstruct the sequence of events; and the patterns of bloodstains can sometimes be as useful as the DNA results. Whereas DNA can tell us about the "who" of a crime, bloodstain patterns can tell us the "where, what, how, and when."

By carefully documenting and examining bloodstain patterns at a crime scene, investigators are able to reconstruct what happened during the crime. This must be based on detailed measurements and analysis of blood patterns. As with all pattern evidence, blood patterns must be interpreted with extreme care. Blood pattern examiners must also conduct controlled laboratory experiments to verify the interpretations they have made of the blood patterns. The following are some examples of what blood patterns at a crime scene can inform us:

1. The approximate distance between the blood source and the surface upon which the blood landed (also known as the target surface)
2. The approximate energy needed to create the resulting blood droplets at the impact site
3. The approximate direction the blood droplets were traveling in when they impacted the target surface
4. The position of the victim(s)/witness
5. The position of the suspect(s)
6. The direction in which a weapon or hand may have been swinging
7. The direction in which blood may have trailed after impact
8. Whether blood was then wiped or smeared by a person or object

9. The estimated amount of time that has elapsed since the blood was deposited on the target surface
10. The movement of the blood between focal points
11. What type of injury caused the blood deposits?
12. The approximate time lapse after blood deposited
13. The sequence of events during a crime such as shooting, dragging, bleeding
14. The sequence after the crime (such as cleaning, staging, alteration)

9.5 Crime Scene Reconstruction

Crime scene reconstruction is the process of determining the sequence of events, criminal activities, and logical predictions about what occurred during and after the crime. As such, reconstruction is a scientific fact-finding process. These five stages in the reconstruction process parallel the steps in scientific method. Like any scientific method, reconstruction must concentrate on the "testing stage." Only after exhaustive testing can one have confidence in the reconstruction. Each step in the process should follow a logical analysis model. It involves the scientific analysis of a crime scene, the interpretation of crime scene pattern evidence, the laboratory examination of forensic evidence, and also a systematic study of related information and the formulation of a logical hypothesis.

Reconstruction is a combination of the inductive and deductive aspects of logic and the combination of science and art. The steps and stages of reconstruction closely follow a basic scientific method approach. It involves consideration and incorporation of all investigative information along with results of examination of physical evidence and interpretation into a reasonable explanation of the crime and related events. Logic, careful observation, and considerable experience, both in crime scene investigation and forensic examination, are necessary for correct interpretation, objective analysis and ultimately, reconstruction. The following are the five separate steps commonly used in the process of reconstruction:

1. *Data collection*
   This step requires the accumulation of all information obtained at the scene, from physical evidence, and from the statements of witness/victim. This includes the condition of the evidence, patterns and impressions, injury of the victim, and the relative positions of victim and evidence. Investigators should review and organize all of these pieces of information and put the puzzle of the crime together.
2. *Conjecture*
   Before making any detailed analysis of the evidence, investigators may infer a possible explanation of the events involved in a crime. However, it is important to note at this stage that this possible explanation does not become the only explanation being considered at this time. In many cases, there may be several possible explanations.
3. *Hypothesis formation*
   Further accumulation of data is based on the detailed examination of the physical evidence, the continuing investigation, and additional reports. Scene examination includes interpretation of blood and impression patterns, gunshot patterns,

fingerprint evidence, and analysis of trace evidence. As these findings become clearer and their interrelationships emerge, it will lead to the formulation of an educated guess as to the probable course of events, that is, a hypothesis.

4. *Testing*

Once a hypothesis is formulated, further testing must be done to confirm or disprove the overall interpretation or specific aspects of it. This stage includes comparisons of samples collected at the scene with known standards and alibi samples, as well as chemical, microscopic and other analyses, and additional testing, as necessary. Some of this "testing" is the mental exercise of careful reexamination and evaluation of the evidence in terms of the hypothesis. At times testing will require an experimental design to address specific information associated with the case. It is important to understand and account for the potential variables. An example of this would be to construct and experiment to determine how long it would take for a given volume of blood to dry on a specific substrate.

5. *Theory formation*

Investigators may acquire additional information during the investigation about the condition of the victim or suspect, the activities of the individuals involved, accuracy of witness accounts, and other information about the circumstances surrounding the events. Testing and confirming the hypothesis involves integrating all the verifiable investigative information, physical evidence analysis and interpretation, and the results of experiments. When it has been thoroughly tested and verified by analysis, the hypothesis can be considered a plausible theory. Complete reconstructions are often not possible; however, partial reconstruction, or reconstructing certain aspects of the events without necessarily being able to reconstruct all of them, can be extremely valuable. Information developed through reconstruction can often lead to the successful solution of a case

It is important to understand that crime scene reconstruction is very different from "reenactment," "re-creation," or "criminal profiling." Reenactment generally refers to having the victim, suspect, witness, or other individual reenact events, based on their knowledge and recollection of the crime. Re-creation, on the other hand, is to replace the necessary items or actions back at a crime scene through original scene documentation. And criminal profiling is an analysis based on statistical and psychological factors of the criminal.

9.6 Case Examples

9.6.1 Concetta "Penney" Serra Homicide Case

On Friday, July 16, 1973, in New Haven, Connecticut, 21-year-old Penney Serra drove her father's blue Buick into a parking garage on Temple Street at 12:42 P.M. and parked the car on the ninth level. As Penney made her way through the garage to get down to Chapel Street, witnesses later said that Penney was chased on foot by a tall, skinny man with long dark hair. At 1:00 P.M., an employee spotted the body of someone lying in the fetal position at the base of stairwell, on the tenth level.

When the police arrived, they found Penney Serra dead. Her blue dress was covered in blood. Cuts and scrapes were found on Penney's wrist, finger, knee, and face, and an autopsy later revealed that Penney died of a small deep wound through her fifth and sixth ribs that penetrated her right ventricle of the heart. Penney's chest was full of blood. The medical examiner told police that it was impossible to determine the weapon, only that it was 3 inches in length. The medical examiner said that it took only a minute for Penney Serra to bleed to death. Police cornered off the crime scene and found her unlocked Buick parked at an erratic angle. Penney's car seat had blood on the outside door handle and door surfaces, on the steering shaft and driver's side floor. Police also found Penney's purse, wallet with $14.75, her shoes, a parking stamp with an entry time of 12:42 p.m., and unopened envelopes containing invoices for a dental patient. Behind the driver's seat, police found more blood stains on the floor and on a Rite Aid tissue box. Police found a trail of blood drops and splatters leading away from the Buick toward the stairwell up to the ninth and tenth levels. On the seventh level, police found a set of keys. Near the keys, was a man's handkerchief covered in blood. Bloodstains were found on levels 7, 6, and 5. Police were able to determine that the assailant's blood type was type O.

From the car's license plates, police traced the car to Penney's father John Serra and contacted him about his daughter. Police learned that Penney had an on–off relationship with a man named Phil DiLieto. After his alibi checked out, he was eliminated as a suspect. Police were then presented with another suspect. Anthony Golino's wife said that during one of their vicious arguments he wanted "to do her like Penney Serra." Golino's blood type was type O; however, after some speculation about whether Golino and Serra had a relationship, Golino was eliminated as a suspect. Police also interviewed Martin Cooratal. His dental bill was found on the dashboard of the Buick belonging to the Serra's. Cooratal was spotted in the parking garage and matched the description of the suspect. However, like Phil DiLieto, Cooratal's alibi was solid. For nearly two decades, Penney's murder had gone unsolved. John Serra was very critical about how the investigation was handled and put a lot of pressure on the state's attorney to find Penney's killer, so prosecutors looked to Dr. Henry Lee to investigate.

Dr. Lee and his staff began to reexamine all of the physical evidence left at the crime scene. On September 10, 1989, Dr. Lee reconstructed the crime scene at the Temple Street garage. With the help of his staff, Dr. Lee reviewed documents, photographs, and diagrams. According to a witness, the attacker ran back to the ninth floor and started Penney's Buick and drove down to the eighth floor and parked it at an extreme angle. The attacker then went down to the 7th floor, got into his own vehicle and drove it out of the garage and leaving a blood smeared ticket with the garage attendant at 1:01 p.m. Dr. Lee also examined latent fingerprints—prints not visible to the naked eye. Dr. Lee used chemicals to make these latent fingerprints visible. Still, it would take another 5 years to get a break in the case.

In 1994, Megan's Law in Connecticut was adapted. Once a person has committed a crime, it is likely they will repeat the same crime. Edward Grant of Waterbury beat his then fiancée so badly she was hospitalized. Grant was then booked and fingerprinted. Grant who had been badly wounded in Vietnam and had a metal plate in his head had been subject to violent mood swings. Grant's fingerprints were matched against latent fingerprints left on the bloody tissue box. Grant, who denied his involvement, offered no explanation as to how his prints were found at the crime scene. So detectives offered him a chance to exonerate himself by asking for a sample of his blood. The sample was sent to Dr. Lee's laboratory. Not only was Grant's blood type O, but it was a 300-million-to-one chance the blood belonged to Edward Grant. Sadly, by the time Edward Grant was arrested, John Serra had passed away.

At the trial, Dr. Lee testified about the crime scene reconstruction, latent fingerprints, and DNA blood evidence. Dr. Lee explained to the jury that the killer chased Penney Serra, who was running barefoot, up several levels of the garage into the stairwell where she ran into a dead end at the 10th floor. The killer stabbed Penney in the heart and she died on the steps. Dr. Lee explained the concept of a primary verses secondary crime scene. The murderer then ran down several levels, got into Penney's car, and left blood all over the car. The killer tried to stop some bleeding on his hand, a defensive wound probably inflicted by Penney, and grabbed the tissues in the back seat to soak up the blood. The killer then parked Penney's car on a sharp angle, got out, ran down another level and got into his own car and fled the scene leaving another bloody print on a parking ticket as he left. The fingerprints and DNA left at the scene matched Edward Grant. After a long trial, the jury found Edward Grant guilty of first-degree murder. Thanks to the advances in forensic science and Dr. Henry Lee, a pioneer in the field, finally Penney Serra's killer was brought to justice (Figures 9.8, 9.9, 9.10, and 9.11).

Figure 9.8 New clues point to new killer in Penney Serra case.

Figure 9.9 Grant's DNA obtained from remnants of handkerchief.

Collection and Preservation of Physical Evidence 269

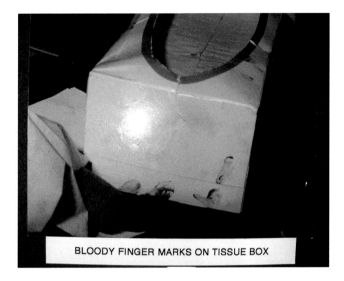

Figure 9.10 View of bloodstain on side of tissue box located in the victim's car.

Figure 9.11 View of the bloody stain on the tissue box after enhancement.

9.6.2 Identical Twins, New Orleans, LA

Damon Green, 25, decided to visit New Orleans with his girlfriend Scarlett Jamieson in the spring of 1988. Green and Jamieson checked into the Evergreen Plaza Inn in Metairie and spent time taking in the sites, going to jazz clubs, and gambling in local casinos. On April 4, the couple took a cab downtown, but Green returned to their hotel room alone. Around 12:30 A.M., the motel manager, Pat Watson, received a complaint about people arguing in room 202. No other guests had complained, so Watson dismissed the incident as a fight. Later that night, a motel guest walked by room 202 and noticed that the door was ajar. From the hallway, the guest saw Damon Green's body lying face down. Jefferson Parish Police quickly arrived at the scene. Damon Green was pronounced dead at the scene.

While crime scene investigators were collecting evidence, an informant contacted Jefferson Parish Police. The informant told detectives that he was at home when his roommate, James Ranna, appeared in the doorway with another individual named Neel. The informant told police that Neel had blood on his person and both Ranna and Neel were missing clothing. The informant knew where Neel lived and directed police to the home of Neel Thomas, 27, of New Orleans.

Police officers found Neel Thomas at his home. Detectives took Thomas to the police station for questioning. After being interrogated by police for several hours, Neel Thomas confessed to murdering Damon Green with James Ranna. However, when detectives questioned James Ranna, he refused to talk. Neel Thomas was arrested and charged with murder.

Two days later, Neel's identical twin brother, Arthur Thomas, left an audiotaped confession to the murder of Damon Green right before he committed suicide. Arthur Thomas said on the tape that he was the only one responsible for the murder and that his brother Neel had nothing to do with it. The tape gave very few details of the murder and left investigators with a lot of questions. Baffled, the district attorney contacted Dr. Henry Lee, the noted forensic investigator, to review the case.

After Dr. Lee arrived in New Orleans, he went to the crime scene at the Evergreen Plaza Inn to analyze the evidence. Dr. Lee reviewed the crime scene photographs as he moved through the motel room. Dr. Lee noticed instantly that there were signs of a struggle. The furniture has been tossed and the room appeared ransacked or searched, so robbery was a possible motive. A bloodstain on the floor appeared to be a footprint. Transfer smears along the lower part of the wall were also visible. Dr. Lee noted that a hairdryer cord had been wrapped around Damon Green's neck and he also sustained a single gunshot wound. However, a second 9-mm casing was found underneath Green's body.

The scientific facts in the case told Dr. Lee that at least two people attacked Damon Green. The attack started near the bed. Green was hit at least two times and caused medium velocity impact blood spatter on the mattress. Green then tried to get away by crawling down on the floor. This caused the transfer smears on the lower portion of the wall. A telephone cord was then wrapped around Green's neck. Someone attempted to strangle Green before shooting him once in the back of the head. Although the motel room had been processed with Ninhydrin, it only produced a few latent fingerprints from the bathroom area. Those prints matched Green's girlfriend, Scarlett Jamison. No other prints were recovered. Detectives seized cigarette butts from the scene to test for DNA, but Dr. Lee pointed out that the twins were identical. Therefore, the twins would share the same DNA.

After the district attorney reviewed Dr. Lee's analysis, he noted that the scientific facts in the case did not add up with the confession and Neel Thomas' account of what happened. Two bullet casings were recovered, but Neel said he fired his gun only once. No physical evidence linked Neel Thomas to the crime scene. Thomas later recanted his statement, so detectives went back to Neel Green to conduct a polygraph test.

When asked about his inconsistent statements, Neel Thomas said that his brother, Arthur Thomas, had given him some of the details. Because he did not actually commit the crime, the details were wrong. Prosecutors now believed that Neel Thomas tried to protect his twin brother by confessing to a crime he did not commit. "Because there is so much reasonable doubt as to Neel Thomas' guilt, justice dictates that we not go forward with the trial," Connick said. "Neel Thomas may have been involved in some way, but we have not proof. We're obligated to follow the evidence."

James Ranna later confessed to his part in the killing and said that Neel Thomas had nothing to do with Green's murder. Ranna was later sentenced to 5 years, as part of a plea deal, for accessory to manslaughter. The murder charges against Neel Thomas were later dropped.

9.6.3 Murder in a Bathroom, Middletown, CT

A week after a sociology professor at Connecticut's Middlesex Community College died of a heart attack; his widow went alone to clean out her husband's office. It was a quiet Sunday afternoon. Suddenly, gunshots were heard. A campus security guard found the widow, Susanna Srb, dead in a women's restroom. Police had a body, slumped, over a sink, but they had no witnesses and no leads. They called Dr. Henry Lee.

Dr. Lee drove quickly across Connecticut directly to the campus crime scene. He found Mrs. Srb's body fully clothed, including a jacket and scarf. The skirt and panties were partially pulled down. Although he saw no obvious signs of rape, Dr. Lee suspected an attempted rape, and he knew that in some uncompleted rapes the attacker has premature ejaculation. Dr. Lee searched the bathroom floor carefully and found three spots of semen. He estimated that the semen was probably only 3 or 4 hours old, roughly the same length of time that had elapsed since the death was reported by the security guard. Dr. Lee also found that the body had several bullet wounds in the upper back and two bullet wounds in the back of the head. But he found only one shell casing, covered by the victim's scarf. He suspected that the killer must have picked up the other shells, overlooking the one beneath the scarf.

Police gave Dr. Lee the statement the security guard had given them when they arrived on the scene. The guard, Todd McGrath, said he was on patrol outside the social sciences building when he heard several gunshots. He rushed into the building, looked on every floor, and found the body in the fourth floor women's restroom. He immediately called police, he said. The first officers on the scene asked McGrath if he carried a weapon. He said no, the campus guards were not armed. Police searched the building and the guard's office and found no guns.

Mr. McGrath was very cooperative, the police said, willingly submitting to a gunshot residue test on his hands. The test showed negative—no traces of gunpowder from having fired a gun. But Dr. Lee was intrigued by spots of water on the bathroom sink. When the bathroom was last cleaned? On Friday, the security guard said. Then Dr. Lee noticed paper towels in the wastebasket, as if someone had washed his or her hands recently. McGrath, who had been so cooperative, even turning over his shirt and a hair sample, suddenly balked at taking off his boots. Police immediately obtained a court order to allow them to seize McGrath's boots. When they removed them, Dr. Lee was amazed to find a 9-mm shell casings. McGrath, Dr. Lee surmised, had picked up the shells from the bathroom floor, hidden them in his boots and never had a chance to get rid of them. Police now had reason to arrest the security guard, and they did.

Dr. Lee continued to study the evidence. He found a spot of semen on Mr. McGrath's shirttail, and DNA tests showed that it matched the semen found on the floor near the body. Dr. Lee and police searched McGrath's car, but found only gun magazines in the trunk—no possible murder weapon. The car had heavy seat covers, however, so Dr. Lee had them removed. Under the right front seat cover, he found a 9-mm automatic pistol. Dr. Lee called his firearm examiner to come quickly to do tests that night. The test shots in the laboratory proved the weapon found in McGrath's car was the same one that had fired the bullets found in Mrs. Srb's body. The investigators' theory, based on Dr. Lee's accumulation of evidence, was that McGrath had happened upon Mrs. Srb in the deserted campus building and following her into the women's room planning to rape her. But she fought him, and, enraged by her resistance, he drew his gun, and started shooting, and kept shooting until he was certain she was dead.

At the trial, Dr. Lee took the jury through the evidence step by step—the DNA match in the semen on his shirt and the floor, the shell casing match with the ones in the boots, matching with the gun in the car. It was all circumstantial evidence. There were no witnesses. There could have been other possible explanations, according to the defense attorneys. But Dr. Lee uncovered so many pieces of evidence and had presented them so clearly in the courtroom, that the members of the jury were able to form an accurate and horrible picture of a despicable man killing a frightened, grieving woman. Todd McGrath was convicted of murder (Figures 9.12, 9.13, 9.14, and 9.15).

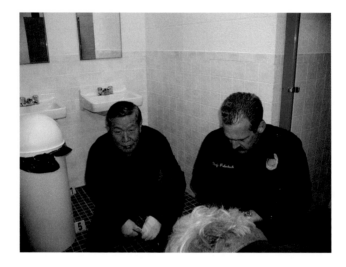

Figure 9.12 Dr. Lee and Prof. Tim Palmbach prepare to conduct a crime scene reconstruction in this case.

Figure 9.13 Relative positions of the body and key pieces of evidence are located to assist in the reconstruction.

Collection and Preservation of Physical Evidence 273

Figure 9.14 Suspected semen stain on victim's clothing, including a view with ALS.

Figure 9.15 Crucial role of DNA analysis in this case.

9.6.4 Brown's Chicken Murders

On January 9, 1993, seven bodies were found in the cooler areas of the Brown's Chicken Restaurant at 168 W. Northwest Highway, Palatine, Illinois. On January 28, 1994, Dr. Henry Lee of the Connecticut State Police Forensic Science Laboratory was contacted by Chief Jerry Bratcher of the Palatine Police Department and Attorney Patrick W. O'Brien, Special Prosecutor in Cook County, Illinois, to review the evidence and to assist in case investigation. From February 10 through 15, 1994, Dr. Lee met the investigative team, visited the crime scene, reviewed documents and photographs, and reexamined several pieces of physical evidence. After reviewing documents and photographs, and examining physical evidence, the following observations were made. The conclusions in the case read as follows:

1. The homicide scene was located at Brown's Chicken Restaurant, 168 W. Northwest Highway, Palatine, Illinois. All seven victims were shot and died as a result of multiple gunshot wounds. In addition, two of the victims also had knife wounds.
2. A large amount of cash receipts for the day were missing from the restaurant. The scene is consistent with a robbery–homicide type of case.
3. Firearm, footwear imprint, fingerprint, and a variety of other physical evidence was collected from this restaurant by detectives of Palatine Police Department and forensic scientists from Northern Illinois Crime Laboratory. This evidence was submitted to Northern Illinois Crime Laboratory for analysis.
4. Police evidence submission lists indicate that approximately (22) bullet/bullet fragments were submitted for examination. These bullet/bullet fragments were examined microscopically. Reports indicate that trace evidence was observed on some of the items.
5. The exact number of individual latent prints and palm prints developed during the examinations was not determined. All the latent prints developed at the scene and on items of evidence seized from the scene were examined by Latent Print Examiners of the Northern Illinois Crime Laboratory.
6. The exact number of individual footwear imprints developed during the examinations was not determined. All the footwear imprints where examined by examiners of the Northern Illinois Crime Laboratory.
7. One of the items of physical evidence was collected from a garbage can in the west side of the restaurant. This item consists of a chicken meal from the restaurant, labeled as "0093-157-9, 01/08/93–12/04/93." This item was stored in a freezer at Northern Illinois Crime Laboratory.
8. This item was taken out of the freezer and examined macroscopically and microscopically by Dr. Lee at the Serology Section of the Northern Illinois Crime Laboratory. The following items were found in the evidence:
    a. Thirty-seven strips of French fries
    b. Two intact biscuits
    c. One intact chicken leg
    d. One intact chicken breast
    e. One portion of chicken
    f. One portion of chicken wing
    g. Five pieces of bone remains
    h. Paper napkins, plates, box, stir, cups and verity of other materials

The five pieces of chicken bone appeared to be the remains of a chicken meal eaten by an individual or individuals. It is more likely that the DNA of the person or persons who ate the chicken will transfer and be deposited on those chicken bones. Subsequently, Dr. Lee recommended that those pieces of chicken bone be submitted to Cellmark Laboratory for DNA testing.

1. The crime scene documentation, location, and condition of the chicken meal, sales receipts of the last meal, and time that this last chicken meal was sold, all indicate

Collection and Preservation of Physical Evidence

Figure 9.16 Seven people massacred in Palatine.

Figure 9.17 Chicken parts and bones located during search.

Figure 9.18 Sales receipt showing date and time of purchase of meal consisting of fried chicken.

Figure 9.19 Chicken bones had DNA mixture.

that the physical evidence found on this chicken meal is a very important and leads toward linking a suspect to this case.
2. These observations and conclusions are based on a review of the submitted crime scene documentation, and the examination of the previously described items as physical evidence. This conclusion may be subject to change and/or modification based on submission of any new materials or information (Figures 9.16, 9.17, 9.18, and 9.19).

References

Fisher, A.B., and D. R. Fisher. 2012. *Techniques of Crime Scene Investigation*. Boca Raton, FL: Taylor and Francis.
James, S.H., P.E. Kish, and T.P. Sutton. 2005. *Principles of Bloodstain Pattern Analysis*. Boca Raton, FL: CRC Press.
Lee, H.C., E.M. Pagliaro, and K. Ramsland. 2010. *The Real World of Forensic Scientist*. Amherst, NY: Prometheus Books.
Lee, H.C., and H.A. Harris. 2000. *Physical Evidence in Forensic Science*. Tucson, AZ: Lawyers & Judges Publishing Company, Inc.
Lee, H.C., T.M. Palmbach, and M.T. Miller. 2001. *Henry Lee's Crime Scene Handbook*. San Diego: Elsevier Academic Press.
Marjanović, D., and D. Primorac. 2013. DNA Variability and molecular markers in forensic genetics. In *Forensic Genetics: Theory and Application*, ed. D. Marjanović, D. Primorac, L.L. Bilela et al., 75–98. Sarajevo: Lelo Publishing (Croatian edition).
Primorac, D., and D. Marjanović. 2008. DNA Analysis in forensic medicine and legal system. In: *DNA Analysis in Forensic Medicine and Legal System*, ed. D. Primorac, 1–59. Zagreb: Medicinska Naklada (Croatian edition).

Identification of Missing Persons and Mass Disaster Victim Identification by DNA

10

BARBARA A. BUTCHER
FREDERICK R. BIEBER
ZORAN M. BUDIMLIJA
SHEILA M. DENNIS
MARK A. DESIRE

Contents

10.1 Mass Fatality Incidents	277
10.2 Search and Recovery	278
10.3 Morgue Operations	279
10.4 Family Assistance Center	280
10.5 World Trade Center Remains Identification Project—New York City Office of Chief Medical Examiner Experience	280
References	289

10.1 Mass Fatality Incidents

Mass fatality events are not rare occurrences; in the first half of 2008 alone, 101 natural disasters were reported in seven countries, with approximately 229,043 fatalities. In addition, in the United States alone, each year approximately 60,000 people are reported missing, which has been termed "the largest silent mass catastrophe in North America." A mass fatality incident is defined as any situation where there are more fatalities than can be properly handled (recovered, examined, identified, and disposed of) using local resources only, with or without the presence of biological, chemical, and/or radiological contamination. By method, mass fatality incidents could be natural cataclysms, accidents, or deliberate acts such as terrorism.

Managing mass fatality incidents requires both a multidisciplinary scientific and logistic approach, with the ultimate goal of identifying as many victims as possible. The procedures involved are generally known as the disaster victim identification process. It has been shown that management of mass disasters must be multiaxial, with proper planning and preparedness essential for its success. It is necessary that the system be able to predict, cover, and demonstrate operational workflow. There are at least three operational areas that must be included in mass fatality incident management: Search and Recovery, Mortuary Operations and Family Assistance Operations (Butcher and DePaolo, 2011; Meyer, 2003).

10.2 Search and Recovery

Simply stated, search and recovery involves locating and removing human remains, including bodies and body parts as well as the personal effects of the victims. It is good policy to treat each and every mass disaster as a crime scene and to treat every single recovered fragment of biological or nonbiological origin as a separate evidence/exemplar item. The process has to be executed carefully, as it is easy to overlook fragmented remains when rescue and recovery operations occur simultaneously. Too often remains will be removed hurriedly by well-meaning rescue personnel in the search for living victims. We encourage responders to leave decedents in place unless they hamper rescue operations, and even then to involve mortuary personnel to assist in the removal of decedents. Transport to a body collection point or morgue from the crime/death scene(s) should follow the "chain of custody" procedures designated for morgue operations (Figures 10.1 and 10.2).

Figure 10.1 Exploration and recovery at World Trade Center (WTC) disaster site in New York City.

Figure 10.2 Morgue operations—triage area in New York City Office of Chief Medical Examiner (NYC OCME).

10.3 Morgue Operations

Depending on the circumstances of the case, magnitude, and fatality type, the mortuary operation should have as its goal the identification of victims and determination of the cause and manner of death. Physical organization of the morgue location should logically translate to the flow of the tasks themselves so as to maximize efficiency of movement. Stations should include, but not be limited to, the following: receiving, triage and cataloging, imaging, personal effects collection site with special imaging and cataloging station, fingerprinting, dental identification, anthropological examinations, pathology procedures, tissue sampling for toxicology, histology and forensic biology analyses, storage, transport, administrative and logistical support (such as information technologies) (Figure 10.3).

It is of utmost importance that every single examination be properly documented, and all samples taken from the body or body parts have to have the same unique identification code as the source case. Good practice requires that remains are escorted from station to station, together with the documentation (postmortem charts) necessary for further examination. Postmortem charts and case files will be finalized upon all scheduled and/or amended analyses completion. After the identification process is completed (which can take years in some cases) and each single case is technically and administratively reviewed, remains can be released to the next of kin.

Although remains recovery and rescue operations should be concurrent, staging of fatality management personnel and equipment should occur in such a way that guarantees full access for life safety operations, avoiding blocked roads and inaccessible equipment. Mass fatality incidents may include special conditions, such as contamination (chemical, biological, and/or radiological); therefore, it must not be forgotten that management of mass catastrophes does not only include the care of victims, but also the health and safety of workers. Personal protection and long-term health consequences have to be seriously considered for first responders, as well as personnel working off-site. If the remains are contaminated, they must undergo decontamination by trained and certified personnel before any further actions toward identification are taken.

Figure 10.3 Morgue operations—triage area in New York City Office of Chief Medical Examiner (NYC OCME).

10.4 Family Assistance Center

The Family Assistance Center (FAC) is dedicated to the reassociation of the deceased with the next of kin through the provision of identifying information and biological evidence from family members. The FAC should be established quickly after an incident, ensuring security and privacy for the families, and should include, among others, grief counselors and interpreters.

We also provide families with information that they may need in the days following the incident, ensure that family member data are recorded correctly, and allow investigators access to families so they can obtain information more efficiently. Data collection is crucial from the very first conversation at the FAC, as the opportunity might not present itself again. Antemortem data drive the identification process, and without proper information to be compared to recovered remains, the identification practice will fail. This information is compared to the postmortem findings collected during morgue operations.

It is reasonable to expect that mass catastrophe incidents will be followed by a level of confusion and chaos, especially in the hierarchy of control or leadership. Thus, it is necessary to state that the local authorities are always in charge of an incident happening within their jurisdiction. At some point, professional and general help from outside organizations may be needed including multiregional and international agencies, but the local authorities will remain in command.

10.5 World Trade Center Remains Identification Project—New York City Office of Chief Medical Examiner Experience

On September 11, 2001, two hijacked airplanes struck the Twin Towers at the World Trade Center (WTC) in New York City, killing 2753 people. This mass disaster was challenging on many levels to all of the organizations involved; the management of 21,906 recovered human remains fell to the New York City Office of Chief Medical Examiner (NYC OCME). Two major goals in mass fatality investigation are determining the cause of death and identifying the decedents. Identification is necessary for three reasons: to provide closure for the families, to issue a death certificate enabling people to conduct the "business of death" such as life insurance and wills, and to reconstruct the incident for criminal investigation purposes. The sheer magnitude of the disaster on 9/11 is demonstrated in the statistics: as of October 2013, there were 1635 identifications for the 2753 people reported missing. Of these, 1003 were identified by a single means, of which DNA analysis was responsible for 887 of the victims. DNA analysis, combined with anthropologic expertise, has become the standard method for identification in these types of disasters where fragmentation is overwhelming (Ballantyne, 1997; Bieber, 2002; Hirsch and Shaler, 2002; Holland et al., 2003; Bieber, 2004; Leclair et al., 2004; Biesecker et al., 2005).

The identification process has always been a team effort, one that consists of pathologists, anthropologists, forensic scientists, and law enforcement. During the WTC victim identification project, samples of tissue and bone were collected over the course of 7 years with the assistance of more than 20 different agencies. Because of the environmental conditions and the length of recovery time, DNA in the collected samples typically experienced degradation.

Two of the WTC remains identification project's greatest challenges was the number of fragmented remains, and the fact that this incident was an "open manifest" in nature. It occurred in areas accessible to the public, and so it was not possible to know who was present at the scene at the time of the attacks. In other words, in open manifest incidents it is theoretically possible that the last identified tissue fragment could be the only one belonging to a previously unreported victim. Conversely, a "closed manifest" incident is one in which there is a definitive list of victims (such as passenger lists in a plane crash), making the identification process less complex.

A decision to identify each victim versus each remain must be made early in the identification process of any mass fatality (Budimlija et al., 2003; Marchi and Chastain, 2002). For the WTC incident, the decision was to identify each human part. In such a high fragmentation event, this gives the best possibility of identifying all of the deceased. An open manifest could mean an unknown or varying number of victims as in the case of the September 11 tragedy, wherein the number has changed many times since 2001. During recent excavation and building at the "ground zero" site, additional human remains have been found years after the incident. The collapse of the towers produced 1.7 million tons of debris, including 0.5 million tons of steel. The excavation and debris cleaning strategy included transport of the debris to a field on Staten Island, where additional search operations and screening for remains were performed. A total of 16,000 individuals participated in this phase (Figures 10.4, 10.5, and 10.6).

Traditional forensic identification strategies may not work for some mass fatalities. During the WTC project, environmental conditions, an open manifest, and disintegration of the remains dictated the course of action. When high fragmentation is evident, dentures and fingerprinting will only identify remains containing those particular anatomical parts. This leads to the high importance of DNA analysis.

Figure 10.4 Ground zero.

Figure 10.5 Active search for possible human remains at Staten Island location.

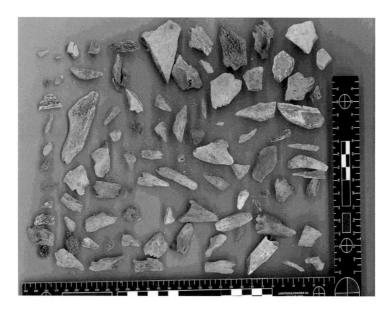

Figure 10.6 Fragments of human remains recovered from the World Trade Center (WTC) complex and buildings in its vicinity.

DNA procedures are not without concerns when used for identification in mass fatalities. Degradation, comingling, and the presence of unrelated biological material pose considerable challenges. Factors that break down DNA and interfere with analysis are the most powerful deterrents to a forensic molecular identification approach. DNA identification techniques are subject to the negative effects of inhibiting agents from material and chemicals that are commonly found in buildings and the environment. The environment of the location could bring heat, fire, water, bacteria, insects, and mold. All of them degrade DNA, breaking the complete molecule strand into smaller pieces and making the identification

Identification of Missing Persons and Mass Disaster Victim Identification by DNA 283

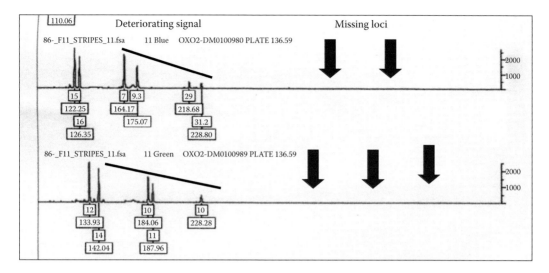

Figure 10.7 Typical partial short tandem repeat (STR) profile of the analyzed remain.

process more tedious. These factors were all present during the recovery phase of the WTC identification project (Figure 10.7).

Throughout the 11 years of the WTC identification project to date, all of the recovered samples have been tested multiple times. After new technologies were developed, validated, and approved for use in human identification, they too have been used in this operation. In addition to classic STR (short tandem repeats) panels used at the beginning, mini-STRs, single nucleotide polymorphisms, and mitochondrial DNA analysis were used, adding more combined results and making possible additional fragment to fragment links or new body identifications (Figures 10.8 and 10.9) (Bogenhagen and Clayton, 1974; Bing and Bieber, 1997; Bing et al., 2000; Mandrekar et al., 2002; Butler et al., 2003; Campobasso et al., 2003; Budowle, 2004; Budowle et al., 2004, 2005, 2006).

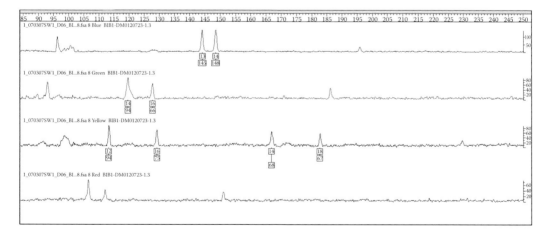

Figure 10.8 Partial profile of the sample using regular short tandem repeats (STRs).

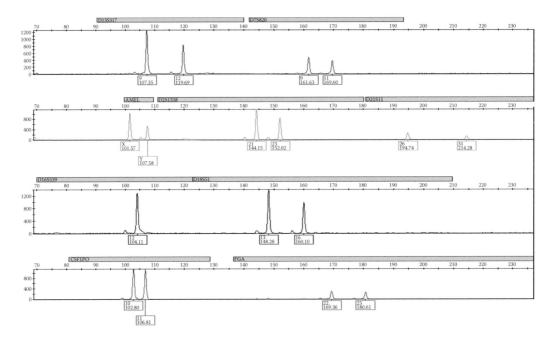

Figure 10.9 Additional results for the same sample using mini-STRs.

Antemortem data for DNA identification are samples generated while the victim was alive. These samples may be directly or indirectly compared to the victim. A direct reference sample includes such items as a toothbrush, razor, hairbrush, or clothing. Studies have shown that toothbrushes yield the highest abundance of DNA with the least chance of a mixture (if used by one person only). Razors are the second best source of DNA. Hairbrushes would be used only if other samples are not available because they have the highest probability of yielding more than one person's DNA. The ability to collect nonvictim hair lasts much longer than saliva cells on a toothbrush or skin cells on a razor, because these items are rinsed after each use. Other types of direct references include blood, tissue, and semen samples collected during previous medical procedures.

If the victim cannot be identified via direct match to a personal effect, formal kinship analysis is in order requiring statistical expertise and computer software designed to perform likelihood ratio analysis. Kinship analysis requires DNA samples from the victim's biological relatives, such as parents, siblings, and children. Family DNA samples are usually obtained from buccal swabs. The closer the biological relationship between the family member and the victim, the more useful the sample will be. For example, if one of the biological parents is missing, the key DNA sample to be collected is the child's, since the parents are not blood relatives to each other. More than 17,000 reference samples were collected following the WTC disaster. The success rates for reference samples are based on whether a DNA profile was generated with sufficient information to make comparisons to postmortem samples. Establishing which reference samples give the best results is useful for future mass fatalities. This will allow the efficient collection of samples from family members, leading to quicker identifications and return of remains to their family (Huffine et al., 2001; Wiegand and Kleiber, 2001).

Identification of Missing Persons and Mass Disaster Victim Identification by DNA 285

An additional problem in this incident was commingling of the remains. Because of the enormous forces created during the collapse of the 41,514-m-high towers (reduced to a 21-m-high hill dispersed over 6.47 ha [64,700 m^2]), many fragments were crushed together, causing anthropological–DNA conflict during the identification process (Figures 10.10, 10.11, 10.12, 10.13, 10.14, 10.15a,b). That is why the anthropologic verification and the DNA resampling project was initiated to verify the minimum number of individuals in each set of remains. As a result, 239 "new" cases were created.

Statistical calculations were performed on all DNA profiles to make valid comparisons. The chance of two random people having the same profile is calculated to give assurance that the right identification was made. Based on the number of victims, statistics

Figure 10.10 Hill of debris after WTC attack.

Figure 10.11 Site of the former World Trade Center (Copyright © 2011 NYC Port Authority).

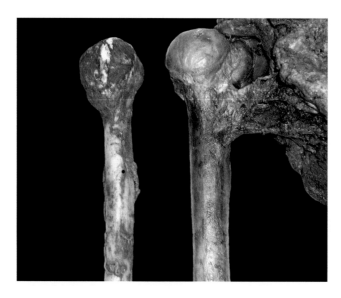

Figure 10.12 Two right humeri with same DNA profile (conflict).

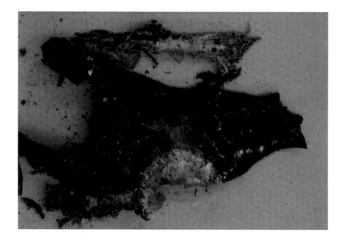

Figure 10.13 Possible morphological commingling of the remains.

for the threshold of direct matching was finalized at 1 in 200,000,000 for females and 1 in 2,000,000,000 for males (Biesecker et al., 2005). This number changes if the incident was a closed manifest. The difference in male and female identification was because there were more male victims. An extremely important role was that of the Kinship and Data Analysis Panel, composed of a group of DNA and statistical experts who made recommendations on testing performed and statistical thresholds for identification based on direct or kinship match (Brenner and Weir, 2003).

During the project, some true biological relationships were not determined until after DNA profiling was performed on multiple family members of the same victim. There were 2753 victims along with thousands of parental, children, and sibling profiles. A tremendous amount of information was collected: 6000 reports, which equaled 50,000 pages for

Figure 10.14 Set of remains with unknown number of the samples coming from different people.

17,000 reference samples. The final estimate was 1.2 million pieces of information from the FAC alone. Trying to organize 1.2 million pieces of information on paper or in a traditional database is nearly impossible, and managing this information is in itself a challenge. Agencies can quickly become overwhelmed with the amount of information absorbed during a mass fatality (Lorente et al., 2002; Leclair et al., 2004). Material and data flow between laboratories involved in processing WTC samples is shown in Figure 10.16, which illustrates (a) material and (b) data flow between laboratories involved in processing WTC samples. Laboratories included the Office of the Chief Medical Examiner (OCME, New York City), New York State Police (NYSP, Albany, NY), Myriad Genetics (Salt Lake City, UT), Orchid Cellmark (Dallas, TX), Celera Genomics (Rockville, MD), and Bode Technology Group (Springfield, VA). Physical materials shipped between laboratories included DNA extracts (solid red line), buccal swabs from biological relative reference samples (dashed red line), personal effects (dotted red line), recovered bones (solid blue line), and recovered tissue (dashed blue line). Note that most of the DNA samples were extracted at the NYSP and OCME laboratories although some tissue and bone were shipped directly to Myriad and Bode.

Delayed identification is in itself a source of heartache to families, and the inability to bury a loved one's remains compounds the pain. To delay the release because of data management issues is unacceptable. There are enough challenges to the identification process; removing this barrier allows better management of the identification team's resources. The Unified Victim Identification System is a product of the aftermath of WTC work. This software system tracks all antemortem and postmortem data collection, allows quick comparisons, and maintains chain of custody. It manages and coordinates all of the activities related to missing persons reporting, medical examiner operations, and victim identification, coordinating responses among city officials, medical examiner, law enforcement, FAC, hospitals, and other agencies. Similar systems have been used in management of different mass catastrophes around the world.

The multidisciplinary approach for mass fatality identification requires a centralized means for collecting antemortem data from the public. This must be accessible to a number

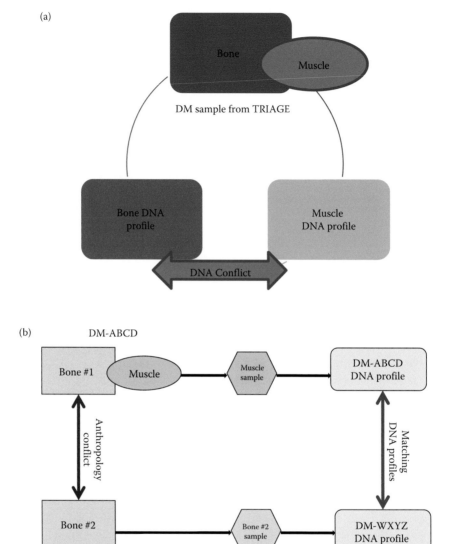

Figure 10.15 Types of anthropological–DNA conflicts.

of agencies for proper data input and review. We have developed a "call center" for use in emergencies, wherein trained operators take a brief missing persons report from any person believing their loved one was involved in the disaster. This enables us to quickly develop a list of the reported missing—in essence, a manifest that will require trimming down as people are found. The ability to communicate with citizens directly after a major incident and development of a valid manifest will prove much more efficient over traditional methods of collecting information. Thus, close communication between officials and citizens is more than an imperative; it is necessary for the sake of our public mandate: to serve citizens in their time of greatest need through the provision of technically excellent, compassionate service through efficient fatality management.

Identification of Missing Persons and Mass Disaster Victim Identification by DNA

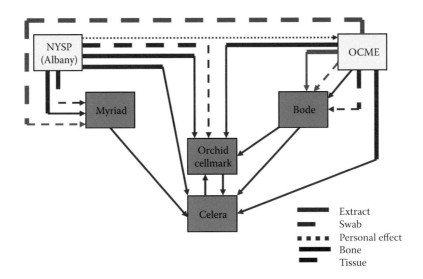

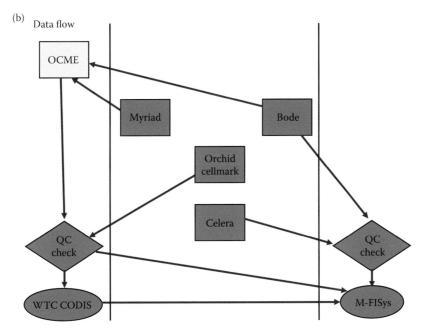

Figure 10.16 (a) Material flow between laboratories involved in processing World Trade Center samples. (b) Data flow between laboratories involved in processing World Trade Center samples. (Courtesy of National Institute of Justice World Trade Center Kinship and Data Analysis Panel and New York City Office of the Chief Medical Examiner.)

References

Ballantyne, J. Mass disaster genetics. *Nature Genetics* 15(4) (1997): 329–331.
Bieber, F.R. Reckoning with the dead. *Harvard Medical Alumni Bulletin* 76(2) (2002): 34–37.
Bieber, F.R. Science and technology of forensic DNA profiling. In Lazar, D. (ed.), *DNA and the Criminal Justice System*. Cambridge, MA: MIT Press, 2004: 23–62.

Biesecker, L.G., Bailey-Wilson, J.E., Ballantyne, J., Baum, H., Bieber, F.R., Brenner, C., Budowle, B., Butler, J.M., Carmody, G., Conneally, P.M., Duceman, B., Eisenberg, A., Forman, L., Kidd, K.K., Leclair, B., Niezgoda, S., Parsons, T.J., Pugh, E., Shaler, R., Sherry, S.T., Sozer, A., and Walsh, A. DNA Identifications after the 9/11 World Trade Center attack. *Science* 310 (2005): 1122–1123.

Bing, D.H. and Bieber, F.R. Collecting and handling samples for parentage and forensics DNA-based genetic testing. *Current Protocols in Human Genetics*. New York: John Wiley, 1997: 14.2.1–14.2.6.

Bing, D.H., Bieber, F.R., Holland, M.M., and Huffine, E.F. Isolation of DNA from forensic evidence. *Current Protocols in Human Genetics*. New York: John Wiley, 2000: 14.3.1–14.3.19 (Supplement 26).

Bogenhagen, D. and Clayton, D.A. The number of mitochondrial deoxyribonucleic acid genomes in mouse L and human HeLa cells. *Journal of Biological Chemistry* 249 (1974): 7791–7795.

Brenner, C.H. and Weir, B.S. Issues and strategies in the identification of World Trade Center victims. *Theoretical Population Biology* 63 (2003): 173–178.

Budimlija, Z.M., Prinz, M.K., Zelson-Mundorff, A., Wiersema, J., Bartelink, E., MacKinnon, G., Nazzaruolo, B.L., Estacio, S.M., Hennessey, M.J., and Shaler, R.C. World Trade Center human identification project: Experiences with individual body identification cases. *Croatian Medical Journal* 44(3) (2003): 259–263.

Budowle, B. SNP Typing strategies. *Forensic Science International* 146 (Suppl.) (2004): S139–142.

Budowle, B., Adamowicz, M., Aranda, X.G., Barna, C., Chakraborty, R., Cheswick, D., Dafoe, B., Eisenberg, A., Frappier, R., Gross, A.M., Ladd, C., Lee, H.S., Milne, S.C., Meyers, C., Prinz, M., Richard, M.L., Saldanha, G., Tierney, A.A., Viculis, L., and Krenke, B.E. Twelve short tandem repeat loci Y chromosome haplotypes: Genetic analysis on populations residing in North America. *Forensic Science International* 150(1) (2005): 1–15.

Budowle, B., Planz, J., Campbell, R., and Eisenberg, A. SNPs and microarray technology in forensic genetics: Development and application to mitochondrial DNA. *Forensic Science Review* 16 (2004): 22–36.

Budowle, B., Bieber, F.R., and Eisenberg, A.J. Forensic aspects of mass disasters: Strategic considerations for DNA based human identification. *Legal Medicine* 7(4) (2006): 230–243.

Butcher, B.A. and DePaolo, F. Mass fatality management. In: Reilly M.J. and Markenson, D.S., (ed.), *Health Care Emergency Management—Principles and Practice*. Jones & Bartlett Learning (2011): 423–445.

Butler, J.M., Shen, Y., and McCord, B.R. The development of reduced size STR amplicons as tools for analysis of degraded DNA. *Journal of Forensic Sciences* 48(5) (2003): 1054–1064.

Campobasso, C.P., Falamingo, R., and Vinci, F. Investigation of Italy's deadliest building collapse: Forensic aspects of a mass disaster. *Journal of Forensic Sciences* 48(3) (2003): 635–639.

Hirsch, C.S. and Shaler, R. 9/11 through the eyes of a medical examiner. *Journal of Investigative Medicine* 50 (2002): 1–3.

Holland, M.M., Cave, C.A., Holland, C.A., and Bille, T.W. Development of a quality, high throughput DNA analysis procedure for skeletal samples to assist with the identification of victims from the World Trade Center attacks. *Croatian Medical Journal* 44(3) (2003): 264–272.

Huffine, E., Crews, J., Kennedy, B., Bomberger, K., and Zinbo, A. Mass identification of persons missing from the break-up of the former Yugoslavia: Structure, function, and role of the International Commission on Missing Persons. *Croatian Medical Journal* 42(3) (2001): 271–275.

Leclair, B., Fregeau, C.J., Bowen, K.L., and Fourney, R.M. Enhanced kinship analysis and STR-based DNA typing for human identification in mass fatality incidents: The Swissair flight 111 disaster. *Journal of Forensic Sciences* 49(5) (2004): 939–953.

Leclair, B., Shaler, R., Carmody, G.R., Eliason, K., Hendrickson, B.C., Judkins, T., Norton, M.J., Sears, C., and Scholl, T. Bioinformatics and human identification in mass fatality incidents: The World Trade Center disaster. *Journal of Forensic Sciences* 52(4) (2006): 806–819.

Lorente, J.A., Entrala, C., Alvarez, J.C., Lorente, M., Carrasco, F., Arce, B., Heinrich, B., Budowle, B., and Villanueva, E. Social benefits of noncriminal genetic databases: Missing persons and human remains identification. *International Journal of Legal Medicine* 116(3) (2002): 187–190.

Mandrekar, P.V., Flanagan, L., and Tereba, A. Forensic extraction and Isolation of DNA from hair, tissue and bone. *Profiles in DNA* 5(2) (2002): 11–13.

Marchi, E. and Chastain, T. The sequence of structural events that challenged the forensic effort of the World Trade Center disaster. *American Laboratory* 34 (2002): 13–17.

Meyer, H.J. The Kaprun Cable Car fire disaster—Aspects of forensic organisation following a mass fatality with 155 victims. *Forensic Science International* 138(1–3) (2003): 1–7.

Wiegand, P. and Kleiber, M. Less is more—Length reduction of STR amplicons using redesigned primers. *International Journal of Legal Medicine* 114(4–5) (2001): 285–287.

11 Bioterrorism and Forensic Microbiology

ALEMKA MARKOTIĆ
JAMES W. LE DUC
JENNIFER SMITH

Contents

11.1 Definitions	293
11.2 History of Bioterrorism and Biological Warfare	293
11.3 Classification of Specific Bioterrorism Agents	294
11.4 Microbial Forensic Protocols and Practices	295
11.5 Criteria for Considering an Outbreak Unusual	306
11.6 Suspicious Infectious Diseases Outbreaks	307
11.7 Biosafety and Biosecurity	310
References	311

11.1 Definitions

Bioterrorism refers to release of microorganisms or their toxins with the intent to harm, sicken or kill humans, animals, or agricultural crops to further personal, political, or social objectives (CDC 2013).

Microbial forensics is a relatively new scientific discipline used to analyze possible bioterrorism attacks, or inadvertent microorganisms or toxins releases (Breeze et al. 2011).

11.2 History of Bioterrorism and Biological Warfare

Throughout history, there are examples of humans using biological organisms as weapons to sicken or kill perceived enemies either as acts of war by organized government states, or acts of terrorism by non-state–affiliated organizations or individuals. As early as in 6th century B.C., Assyrians poisoned the wells of their enemies with rye ergot, and Solon of Athens poisoned the water supply with hellebore (skunk cabbage), an herb purgative, during the siege of Krissa. In a naval battle (184 B.C.) against King Eumenes of Pergamon, Hannibal's forces hurled pots filled with snakes upon the enemy decks and won the battle (Philips 2005). In the fourteenth century, during the siege of Kaffa, the Tartar army catapulted plague-ridden corpses over the walls of the city and forced the defenders to surrender. During the French and Indian War (1767), the English general, Sir Jeffrey Amherst, gave blankets contaminated with smallpox to Indians loyal to the French. The resulting epidemic decimated the tribes, and the British successfully attacked Ft. Carillon (Breeze et al. 2011; Philips 2005). During WWI, the Germans attempted to spread cholera in Italy, plague in St. Petersburg, and dropped bombs with pathogenic microorganisms over Britain. A German-American, Dr. Anton Dilger, supported by the German government, had grown cultures of *Bacillus anthracis* and *Pseudomonas (Burkholderia) mallei* in his Washington, D.C. home. These agents were spread by dockworkers in Baltimore in an attempt to infect 3000 head of horses, mules, and cattle destined for the Allied troops in Europe. As a result, an estimated several hundred troops were affected (Philips 2005).

On June 17, 1925, the Geneva Protocol banned biological weapons. It was the first multilateral agreement that extended the prohibition of chemical agents to biological agents. In spite of the Geneva Protocol, during WW II, the Japanese released plague bacteria in Chuhsien, resulting in the death of 21 people (1940), and in 1941 the British experimented with anthrax at the Scottish coast (Philips 2005; Kasten 2002). More recently, in 1972 members of the right-wing "Order of the Rising Sun" possessed 30–40 kg of typhoid cultures, which they planned to use to poison the water supply in Chicago, St. Louis, and other midwest cities in the United States (Philips 2005). The same year, the Biological Weapons Convention (also known as Convention on the Prohibition of the Development, Production, and Stockpiling of Bacteriological and Toxin Weapons and on their Destruction) was signed by 103 nations. Like the Geneva Protocol, the Biological Weapons Convention was similarly ignored when in 1978 the Bulgarian dissident and writer Georgi Markov, then living in London, was injected with ricin by specially constructed umbrella and died within several days (Breeze et al. 2011; Philips 2005). In 1995, three briefcases designed to release botulinum were discovered in a Tokyo subway. In Zadar, Croatia in 1999, a perpetrator tried to rob a currency exchange office by using a syringe allegedly containing HIV as a weapon (Philips 2005). In 2001, anthrax letters were sent to several news media and U.S. senators (CDC 2001b; Jernigan et al. 2001; Atlas 2002; Greene et al. 2002; Tan et al. 2002; Hoffmaster et al. 2002). There are many other examples, and some of them will be described later in Section 11.6.

The threat of a bioterrist attack has been the topic of international concern. Numerous U.S. National Intelligence Estimates and reports as well as have examined the issues (Breeze et al. 2011).

In December 2008 in accordance with the Implementing Recommendations of the 9/11 Commission Act of 2007 (P.L. 110-53), the Commission on the Prevention of Weapons of Mass Destruction Proliferation and Terrorism submitted its report, "World at Risk" (2009). That report assessed the nation's activities, initiatives, and programs to prevent weapons of mass destruction proliferation and terrorism and provided concrete recommendations to address these threats. The report provided the following threat assessment that was unanimously expressed:

> Unless the world community acts decisively and with great urgency, it is more likely than not that a weapon of mass destruction (WMD) will be used in a terrorist attack somewhere in the world by the end of 2013. That weapon is more likely to be biological than nuclear.

11.3 Classification of Specific Bioterrorism Agents

The U.S. Centers for Disease Control and Prevention (CDC) has classified microbial agents that could be potentially used in bioterrorism attacks in three categories (2013):

A category—Anthrax (*B. anthracis*), botulism (*Clostridium botulinum* toxin), plague (*Yersinia pestis*), smallpox (variola major), tularemia (*Francisella tularensis*), viral hemorrhagic fevers (filoviruses [e.g., Ebola, Marburg] and arenaviruses [e.g., Lassa, Machupo])—biological agents that can be easily disseminated or transmitted from person to person, result in high mortality rates and have the potential for major public health impact, might cause public panic and social disruption, and require special action for public health preparedness.

B category—Brucellosis (*Brucella* species), epsilon toxin of *Clostridium perfringens*, food safety threats (e.g., *Salmonella* species, *Escherichia coli* O157:H7, *Shigella*), glanders (*Burkholderia mallei*), melioidosis (*Burkholderia pseudomallei*), psittacosis (*Chlamydia psittaci*), Q fever (*Coxiella burnetii*), ricin toxin from *Ricinus communis* (castor beans), staphylococcal enterotoxin B, typhus fever (*Rickettsia prowazekii*), viral encephalitis (alphaviruses, e.g., Venezuelan equine encephalitis, eastern equine encephalitis, western equine encephalitis), water safety threats (e.g., *Vibrio cholerae*, *Cryptosporidium parvum*)—biological agents, which are moderately easy to disseminate, result in moderate morbidity rates and low mortality rates, and require specific enhancements of diagnostic capacity and enhanced disease surveillance.

C category—Emerging infectious diseases (such as Nipah virus and hantavirus)—biological agents including emerging pathogens that could be engineered for mass dissemination in the future because of availability, ease of production and dissemination, and potential for high morbidity and mortality rates and major health impact.

Some basic facts about incubation periods, transmission modes, general signs and symptoms, treatment, and vaccines for specific bioterrorism agents are listed in Table 11.1.

11.4 Microbial Forensic Protocols and Practices

Thanks to modern technology and a tremendous increase in knowledge accumulated about molecular biology of microbial agents, scientists can more effectively and efficiently identify pathogens from a variety of matrices to include those from the environment. Progress in this area has been of interest to the traditional public health community for a very long time. It is also of interest to those communities that will be responsible for attribution following a criminal act or terrorist attack using a bioagent. To ultimately identify the individuals or organizations responsible for such an attack will very likely require rigorous use of modern microbiological analytic tools coupled with traditional forensic disciplines (Breeze et al. 2011). Recently, an expert government panel from the United States acknowledged that determination of a responsible party for a covertly planned or actual biological attack would be the culmination of a complex investigative process drawing on many different sources of information including technical forensic analysis of material evidence collected during the course of an investigation of a planned attack or material evidence resulting from an attack. Evidence collected as part of a biological attribution investigation will yield unique types of microbiological evidence that may be specific to the nature of the attack. As examples of such microbiological evidence, they cited the following: viable samples of the microbial agent, protein toxins, nucleic acids, clinical specimens from victims,

Table 11.1 Basic Facts about Specific Bioterrorism Agents

	A Category
Anthrax (*Bacillus anthracis*)	
Incubation	Few hours to 7 days.
Transmission mode	Zoonotic infection; Cattle or other livestock infected by spores from soil pass infection to humans; Person-to-person transmission very rare; Articles and soil contaminated with soil—remain infective for decades.
Basic signs/symptoms	Cutaneous—swelling, ulcer formation, black eschar—heals spontaneously in 1–2 weeks (80% cases).
	Pulmonary—edema factor (EF), protective antigen (PA) and lethal factor (LF) causes swelling and necrosis in mediastinum and lungs, septicemia and death in 2–5 days.
	Gastrointestinal—lethal swelling in the intestine.
Treatment	Penicillin or ciprofloxacin.
Vaccine	Vaccine is protective against invasive disease, recommended only for high-risk populations.
Botulism (*Clostridium botulinum toxin*)	
Incubation	12–36 hours.
Transmission mode	A spore-forming obligate anaerobic bacillus produce toxin in improperly processed, canned, low-acid or alkaline foods and in pasteurized and insufficiently boiled or cooked food; spores are common in the soil and can contaminate the food including honey.
Basic signs/symptoms	Toxin blocks acetylcholine release and interferes with the transmission of signals for muscle contraction (painful flaccid paralysis; swallowing, vision and speech are primarily affected, then breathing with case-fatality rate up to 10%).
Treatment	Amoxicillin + polyvalent antitoxin
Vaccine	Pentavalent (ABCDE) toxoid, 3 dose efficacy 100% against 25–250 LD_{50} in primates.
Plague (*Yersinia pestis*)	
Incubation	1–7 days.
Transmission mode	"Black death" of the Middle Ages; result of human contact with sylvatic rodents, but also infected domestic pets and their fleas (*Xenopsylla cheopis*); possible person-to-person transmission by *Pulex irritans* fleas; in secondary pneumonic plague respiratory droplets are the source of person-to-person transmission—devastating epidemics.
Basic signs/symptoms	Bubonic—inflamed, swollen and tender lymph nodes (inguinal area 90%), fever, may progress to septicemic plague with bloodstream dissemination with 50%–60% case-fatality in nontreated patients.
	Pneumonic—secondary involvement of lungs (due to bloodstream dissemination)—pneumonia with mediastinitis or pleural effusion—after interhuman transmission—primary pneumonic or pharyngeal plague.
Treatment	Gentamicin, Doxycycline.
Vaccine	Suspension of killed (formalin-inactivated) *Y. pestis*; efficacy has yet to be measured precisely in controlled studies; at least 2 vaccinated persons contracted pneumonic plague following *Y. pestis* exposure.
Smallpox (variola major)	
Incubation	7–17 days.
Transmission mode	Person-to-person transmission by infected saliva droplets that expose a susceptible person having face-to-face contact with the ill person; patients with smallpox are most infectious during the first week of illness; some risk of transmission lasts until all scabs have fallen off.

(continued)

Table 11.1 Basic Facts about Specific Bioterrorism Agents (Continued)

	A Category
Basic signs/symptoms	Initial symptoms—high fever, fatigue, head and backaches; characteristic rash follows in 2–3 days (most prominent on the face, arms, and legs), which starts with flat red lesions, become pus-filled and begin to crust early in the second week with scabs developing and then separate and falling off after about 3–4 weeks. Death occurs in up to 30% of cases.
Treatment	Supportive therapy.
Vaccine	Two calf-lymph–derived vaccines, Dryvax® (Wyeth Laboratories Inc., Marietta, PA); FDA licensed Smallpox (Vaccinia) Vaccine, Live-ACAM2000 (Sanofi Pastuer Biologics Co.), for active immunization against smallpox disease for persons determined to be at high risk for smallpox infection; Bavarian Nrodic's new vaccine for individuals where the traditional vaccine is contraindicated (for certain immune-compromised populations).

Tularemia (*Francisella tularensis*)

Incubation	1–14 days.
Transmission mode	Bite of arthropods—ticks, flies, and mosquitoes; handling or ingesting insufficiently cooked meat, drinking contaminated water, by inhalation of dust from contaminated soil, grain, or hay.
Basic signs/symptoms	Skin or mouth, swollen and painful lymph glands, swollen and painful eyes, and a sore throat with dry cough and progressive weakness.
Treatment	Streptomycin, Gentamicin, Doxycycline.
Vaccine	Live attenuated vaccine; 80% protection against 1–10 LD_{50}.

Viral hemorrhagic fevers filoviruses (e.g., Ebola, Marburg)

Incubation	Ebola virus: 2–21 days. Marburg virus: 3–9 days.
Transmission mode	Person-to-person transmission—direct contact with infected blood, secretions, organs, and semen; frequent nosocomial infections; reservoir—not known.
Basic signs/symptoms	Life-threatening multiorgan infection with fever, CNS involvement, bleeding from multiple sites, mostly mucous membranes with mortality of 50%–80%.
Treatment	Supportive therapy, patient isolation and strict nursing barrier is required.
Vaccine	Experimental vaccines were 100% effective in protecting a group of monkeys from the disease; based on either a recombinant Vesicular stomatitis virus or a recombinant Adenovirus carrying the Ebola spike protein on its surface; early human vaccine trials are underway.

Viral hemorrhagic fever arenaviruses (e.g., Lassa, Machupo, Junin virus)

Incubation	6–21 days.
Transmission mode	The reservoir/host of Lassa virus—rodent of the genus *Mastomys*; rodents shed the virus in urine and droppings and the virus can be transmitted through direct contact with these materials, through touching objects or eating food contaminated with these materials, or through cuts or sores; person-to-person transmission—direct contact with infected blood, secretions, organs, and semen; frequent nosocomial infections.
Basic signs/symptoms	Life-threatening multiorgan infection that fever, retrosternal pain (pain behind the chest wall), sore throat, back pain, cough, abdominal pain, vomiting, diarrhea, conjunctivitis, facial swelling, proteinuria, and mucosal bleeding; neurological problems (hearing loss, tremors, and encephalitis), with mortality of 50%–80%.
Treatment	Supportive therapy, i.v. ribavirin may tentatively reduce mortality if given before day 7 after the onset.
Vaccine	An experimental attenuated vaccine for Lassa fever made by reassortment of Lassa and Mopeia viruses; human vaccine efforts are under development; live, attenuated Candid 1 vaccine used in Argentina to protect against Argentine hemorrhagic fever due to Junin virus.

Table 11.1 Basic Facts about Specific Bioterrorism Agents (Continued)

	B Category
Brucellosis (*Brucella species*)	
Incubation	5–60 days.
Transmission mode	Contact with tissues, blood, urine, vaginal discharges, aborted fetuses, and especially placentas (through breaks in skins); ingestion of raw, contaminated unpasteurized milk or cheese from infected animals; no person-to-person transmission.
Basic signs/symptoms	The symptoms are non-specific and systemic, with fever, sweats, headache, anorexia, back pain, and weight loss being frequent; the chronic form of the disease can mimic miliary tuberculosis with suppurative lesions in the liver, spleen, and bone; the mortality is up to 5% in untreated individuals.
Treatment	Streptomycin + Gentamicin or Doxycycline; ciprofloxacin.
Vaccine	No human vaccine available; in 1996, RB51, animal live attenuated strain of *B. abortus* replaced the S19 strain, which was also a live attenuated vaccine.
Epsilon toxin of *Clostridium perfringens*	
Incubation	8–22 h.
Transmission mode	The epsilon toxin could probably be transmitted in contaminated food, water or by aerosol.
Basic signs/symptoms	The epsilon toxin—one of 12 protein toxins produced by *Clostridium perfringens*, a Gram positive, anaerobic spore-forming rod; toxin is a pore-forming protein; it causes potassium and fluid leakage from cells; little or no information about the effects of epsilon toxin on humans; extrapolation from studies with experimentally infected animals; neurologic signs or pulmonary edema may be possible.
Treatment	Supportive therapy, hyperimmune serum might be helpful if given soon after exposure.
Vaccine	No.
Food safety threats (e.g., *Salmonella* species, *Escherichia coli* O157:H7, *Shigella*)	
Incubation	*Salmonella*: 6–72 h. *E. coli* O157:H7: 3–8 days. *Shigella*: 1–3 days.
Transmission mode	Eating contaminated food that may look and smell normal; person-to-person transmission if hygiene or hand washing habits are inadequate. This is particularly likely among toddlers who are not toilet trained.
Basic signs/symptoms	With or without fever, abdominal cramps with often bloody diarrhea (dysentery).
Treatment	Supportive therapy; Trimethoprim/Sulfamethoxazole; Ceftriaxone; no treatment for *E. coli* O157:H7 (risk of enhancement of toxin release and hemolytic uremic syndrome).
Vaccine	DT TAB Diphtheria, tetanus, Salmonella typhi, Paratyphi A & B AVP (France) only available.
Glanders (*Burkholderia mallei*)	
Incubation	1–5 days.
Transmission mode	Animal exposure, cases of human-to-human transmission have been reported (two suggested cases of sexual transmission and several cases in family members who cared for the patients).
Basic signs/symptoms	Localized, pus-forming cutaneous infections, pulmonary infections, bloodstream infections, and chronic suppurative infections of the skin; generalized symptoms—fever, muscle aches, chest pain, muscle tightness, and headache; also excessive tearing of the eyes, light sensitivity, and diarrhea.

(continued)

Table 11.1 Basic Facts about Specific Bioterrorism Agents (Continued)

	B Category
Treatment	Tetracyclines, ciprofloxacin, streptomycin, novobiocin, gentamicin, imipenem, ceftrazidime, sulfonamides.
Vaccine	No.

Melioidosis (*Burkholderia pseudomallei*)

Incubation	Two days to many years.
Transmission mode	Inhalation of dust, ingestion of contaminated water, contact with contaminated soil especially through skin abrasions, and for military troops, by contamination of war wounds; person-to-person transmission can occur.
Basic signs/symptoms	Acute, localized infection: a nodule; can produce fever and general muscle aches. Pulmonary infection: mild bronchitis to severe pneumonia; chest pain is common; a nonproductive or productive cough with normal sputum is the hallmark of this form of melioidosis. Acute bloodstream infection: underlying illness such as HIV, renal failure, and diabetes; usually results in septic shock; respiratory distress, severe headache, fever, diarrhea, development of pus-filled lesions on the skin, muscle tenderness, and disorientation. Chronic suppurative infection: involves the organs of the body (the joints, viscera, lymph nodes, skin, brain, liver, lung, bones, and spleen).
Treatment	Imipenem, penicillin, doxycycline, amoxicillin–clavulanic acid, azlocillin, ceftazidime, ticarcillin–vulanic acid, ceftriaxone, and aztreonam.
Vaccine	No.

Psittacosis (*Chlamydia psittaci*)

Incubation	1–4 weeks.
Transmission mode	Inhalation of the agent from dessicated droppings, secretions and dust from feathers of infected birds, but also geese and pigeons may be reservoirs, rare person-to-person infections.
Basic signs/symptoms	Fatigue, fever, headache, rash, chills, and sometimes pneumonia.
Treatment	Azythromycin, doxicycline.
Vaccine	No.

Q fever (*Coxiella burnetii*)

Incubation	2–3 weeks.
Transmission mode	High numbers of *C. burnetii* are shed in the reproductive fluids and placentas of infected animals; infection of humans—inhalation of agents in small droplets or from inhalation of barnyard dust contaminated with *C. burnetii*. Infection via ingestion of contaminated dairy products, including tick bites and human-to-human transmission, are rare.
Basic signs/symptoms	Sudden onset of high fevers, severe headache, confusion, sore throat, nonproductive cough, nausea, vomiting, diarrhea, abdominal pain, and chest pain and pneumonia; weight loss can occur. Chronic Q fever, characterized by infection that persists for more than 6 months is uncommon but is a much more serious disease. Patients who have had acute Q fever may develop the chronic form as soon as 1 year or as long as 20 years after initial infection with serious complication—endocarditis and high fatality (65%).
Treatment	Azythromycin, doxicycline.
Vaccine	A human vaccine—developed and used successfully in Australia; before vaccination with whole cell vaccine, persons must have a skin test to verify they have not been previously exposed to *C. burnetii*.

Table 11.1 Basic Facts about Specific Bioterrorism Agents (Continued)

	B Category
Staphylococcal enterotoxin B	
Incubation	3–12 hours.
Transmission mode	Ingestion of food products containing toxins (contaminated milk and milk products); mostly "food handlers"—purulent discharges (infected finger, eye, abscess nasopharyngeal secretions).
Basic signs/symptoms	A sudden onset of high fever, chills, headache, myalgia, and nonproductive cough, with shortness of breath and retrosternal chest pain, also nausea, vomiting, and diarrhea (if person swallowed toxin).
Treatment	Supportive therapy.
Vaccine	No.
Typhus fever (*Rickettsia prowazekii*)	
Incubation	1–2 weeks.
Transmission mode	By *Pediculus humanus corporis* (body louse)—infected by feeding on the blood of patient with acute disease; lice may get infected by feeding on the blood of patients with Brill–Zinsser disease; people infected by rubbing crushed lice or their feces into the bite or abrasions.
Basic signs/symptoms	Headache, chills, prostration, fever, and generalized body aches; a macular eruption becomes petechial or hemorrhagic, then develops into brownish-pigmented areas; changes in mental status are common with delirium or coma; toxemia, myocardial and renal failure; lethality 10%–40%.
Treatment	Doxicycline.
Vaccine	No.
Viral encephalitis (alphaviruses [e.g., Venezuelan equine encephalitis, eastern equine encephalitis, western equine encephalitis])	
Incubation	Venezuelan equine encephalitis: 1–6 days. Eastern equine encephalitis: 5–15 days.
Transmission mode	Bite of infected mosquito; viruses are maintained in bird or rodent–mosquito cycle; cycle also involves horses.
Basic signs/symptoms	High fever, vomiting, stiff neck, and drowsiness, generalized, facial, or periorbital edema; paresis; major disturbances of autonomic function (impaired respiratory regulation or excess salivation) with up to 75% lethality; neurological sequelae in survivors; cognitive impairment.
Treatment	Supportive therapy.
Vaccine	Poorly immunogenic, investigational vaccines.
Water safety threats (e.g., *Vibrio cholerae, Cryptosporidium parvum*)	
Incubation	Cholera: 4 hours–5 days. Crptosporidosis: 1–12 days.
Transmission mode	Ingestion of fecally contaminated food or water, including water swallowed while swimming; by exposure to fecally contaminated environmental surfaces; and by the fecal–oral route from person to person.
Basic signs/symptoms	Sudden onset of vomiting, abdominal distension, headache, pain, little or no fever, followed by profuse watery diarrhea with a "rice water" appearance; dehydration, hypovolemia and shock. Symptoms may be less prominent in cryptosporidosis.
Treatment	Doxycyclin, Ciprofloxacin for cholera; Paromomycin for cryptosporidosis.
Vaccine	Cholera vaccine exists—not recommended because of its partial efficacy; no vaccine for cryptosporidosis.

(continued)

Table 11.1 Basic Facts about Specific Bioterrorism Agents (Continued)

	C Category
Emerging infectious diseases such as Nipah virus and hantavirus	
Incubation	Nipah virus: up to 1 month. Hantaviruses: 9–35 days
Transmission mode	Nipah virus was transmitted to humans, cats, and dogs through close contact with infected pigs; the natural reservoir is still under investigation; preliminary data suggest the bats of the genus *Pteropus*.
	The main reservoirs of hantaviruses are persistently infected small rodents (e.g., *Apodemus flavicollis, Apodemus agrarius, Myodes glareolus, Peromyscus maniculatus, Olygoryzomis longicaudatus*, etc.; virus shedding in rodent's saliva and excreta; contaminated dust inhalation; professional exposure; human-to-human transmission possible for Andes virus.
Basic signs/symptoms	Nipah virus: encephalitis (inflammation of the brain) characterized by fever and drowsiness and more serious central nervous system disease, such as coma, seizures, and inability to maintain breathing.
	Two clinical syndromes recognized in infections with hantaviruses: hemorrhagic fever with renal syndrome (HFRS) and hantavirus pulmonary syndrome (HPS); broad spectrum of clinical conditions has been recognized from unapparent or mild illness to a fulminant hemorrhagic disease with severe renal or cardiopulmonary failure and death.
Treatment	Supportive therapy, i.v. ribavirin treatment for hantaviruses may tentatively reduce mortality if given before day 7 after the onset.
Vaccine	No vaccine for Nipah virus. Hantavax—killed vaccine; DNA vaccines in investigation; no firm evidence for efficacy so far.
References	CDC 2013; Benenson 1995; Begovac 2006; Schönwald et al. 2000; Gilbert et al. 2006; Markotic et al. 1996; Meltzer et al. 2001; Le Duc et al. 2001; Jahrling et al. 2004; Engelthaler and Lewis 2004

laboratory equipment, dissemination devices and their contents, environmental samples, contaminated clothing, or trace evidence specific to the process that produced and/or weaponized the biological agent. This group acknowledged that there was a need for research and development efforts to enhance current capabilities in microbial forensics and that this would require effort among many elements of the national government (Breeze et al. 2011). The recommendations found in this national strategy may be relevant to many countries attempting to better prepare for the threat of bioterrorism.

Whether for a public health or a bioterrorism event, the identification, collection, and analysis of appropriate samples are critical prerequisites for successful detection of microbes. Not only can clinical material from infected patients be utilized for the detection, but very often animal or environmental samples will yield critical information. For successful identification of suspicious agents, it is important to use both classic microbiological techniques and modern molecular biology techniques. A list of samples that can be used for the detection of specific microbial agents is presented in Table 11.2. The DNA extraction protocols for some human samples do not necessarily correspond to the successful extraction of environmental samples: recovery may be low or PCR inhibitors may not be properly removed. Different DNA-based methods are necessary for identification of specific bioterrorism agents, and some of them are also listed in Table 11.2. In general,

Table 11.2 Molecular Biology Techniques for the Detection of Specific Bioterrorism Agents

Agent (References)	Specimens	Molecular Biology Diagnostic
A Category		
Anthrax (*Bacillus anthracis*) (Hoffmaster et al. 2002; NRC, NAS 2011; Melo et al. 2003)	Swabs, whole blood, fluids (pleural, bronchial, CSF), fresh tissue, blood clot, serum, citrated plasma, stool	Real-time polymerase chain reaction (PCR) assays for *B. anthracis* for the SmartCycler™, ABI/PE 7700 and 5700. MLVA (*multilocus variable-number of tandem [consecutive] repeat analysis*); examines a number of DNA segments within the chromosome or plasmids of *B. anthracis* that have specific repeat patterns of nucleotides (fundamental DNA units); the repeats may differ by sequence and length, and the number of times that they are repeated; specific pattern that identify different strains of the organism (more than 100 different strains of *B. anthracis* have been identified).
Botulism (*Clostridium botulinum* toxin) (McGrath et al. 2000; Kull et al. 2010; Fenicia et al. 2011)	Feces, enema, gastric aspirate or vomitus, serum, tissue or exudates, food specimens, swab samples (environmental or clinical), soil, water	Competitive reverse transcription-PCR. A real-time PCR method for detection and typing of BoNT-producing Clostridia types A, B, E, and F; a combination of a multiplex-immunoaffinity purification approach, followed by matrix-assisted laser desorption/ionization (MALDI)-based detection for the simultaneous identification of ricin, SEB, BoNT/A, and BoNT/B.
Plague (*Yersinia pestis*) (Lindler et al. 2001; Tomaso et al. 2003)	Lower respiratory tract (pneumonic): bronchial wash or transtracheal aspirate (≥1 ml); sputum; blood; aspirate of involved tissue (bubonic) or biopsied specimen; liver, spleen, bone marrow, lung	PCR methods for the amplification of *Y. pestis* DNA.
Smallpox (variola major) (Sulaiman et al. 2007; Le Duc et al. 2002; Jahrling et al. 2004; Eshoo et al. 2009; Fedele et al. 2006)	Vesicular material, scab specimens, biopsy lesions, serum	PCR identification of variola DNA in clinical specimens. GeneChip. A pan-Orthopoxvirus assay for identification of all members of the genus based on four PCR reactions targeting Orthopoxvirus DNA and RNA helicase and polymerase genes. The amplicons are detected using electrospray ionization-mass spectrometry (PCR/ESI-MS) on the Ibis T5000 system.
Tularemia (*Francisella tularensis*) (Versage et al. 2003; Barns et al. 2005; Whitehouse and Hottel 2007; Tomaso et al. 2007)	Blood samples, conjuctival swabs, environmental samples (water, mud, etc.)	PCR detection of *F. tularensis* DNA Multitarget real-time TaqMan PCR assay

(continued)

Table 11.2 Molecular Biology Techniques for the Detection of Specific Bioterrorism Agents (Continued)

Agent (References)	Specimens	Molecular Biology Diagnostic
Viral hemorrhagic fevers filoviruses (e.g., Ebola, Marburg) and arenaviruses (e.g., Lassa, Machupo) (Gunther et al. 2004; Hass et al. 2006; Zhai et al. 2007; Kurosaki et al. 2007)	Antemortem: sampling whole blood or serum. Postmortem: spleen, lung, kidney, liver, lymph nodes, heart, pancreas, pituitary, brain, or liver tissue, or heart blood	RT-PCR assays are available.
B Category		
Brucellosis (*Brucella* species) (Romero et al. 1995; Newby et al. 2003; Bogdanovich et al. 2004; Spicić et al. 2010; De Santis et al. 2011)	Blood, bone marrow, spleen, liver, joint fluid or abscesses, serum	PCR assay with primers derived from the 16S rRNA sequence of *Brucella abortus*. A real-time, genus-specific 5′ nuclease PCR assay for amplification of a 322-bp fragment of the *per* gene—developed for rapid (<2 h) identification of *Brucella* spp. from agar plates. Three approaches—SYBR Green I (a double-stranded DNA intercalating dye), 5′-exonuclease (enzymatically released fluors), and hybridization probes (fluorescence resonance energy transfer)—were evaluated for use in a real-time PCR assay to detect *Brucella abortus*. A high throughput system of MLVA-16 typing for *Brucella* spp. by using of the microfluidics technology.
Epsilon toxin of *Clostridium perfringens* (Al-Khaldi et al. 2004)	Stool, environmental samples	Polymerase chain reaction (PCR) assays can identify the epsilon toxin gene.
Food safety threats (e.g., *Salmonella* species, *Escherichia coli* O157:H7, *Shigella*) (Villalobo and Torres 1998; Hoorfar et al. 2000; Metzger-Boddien et al. 2004; Ibekwe et al. 2002; Sharma and Dean-Nystrom et al. 2003; Hauser et al. 2013; FDA 2013)	Blood, stool, food	A simple and ready-to-go test based on a 5′ nuclease (TaqMan) PCR technique was developed for identification of presumptive *Salmonella enterica* isolates. A commercially available PCR kit (AnDiaTec Salmonella sp. PCR-ELISA) was developed and evaluated for the detection of Salmonella sp. in food samples. Multiplex fluorogenic real-time PCR for detection and quantification of *Escherichia coli* O157:H7. Simultaneous identification of *Escherichia coli* of the O157:H7 serotype and their Shiga-like toxin type by MAMA/multiplex PCR. The use of PCR to amplify a specific *virA* gene fragment for detection of virulent bacteria of the genus *Shigella* and enteroinvasive *Escherichia coli*.

(continued)

Table 11.2 Molecular Biology Techniques for the Detection of Specific Bioterrorism Agents (Continued)

Agent (References)	Specimens	Molecular Biology Diagnostic
Glanders (*Burkholderia mallei*) (Pearson and Keim 2005; U'Ren et al. 2005; Tomaso et al. 2006; Antonov et al. 2008)	Blood, sputum, urine, or skin lesions	A 5′-nuclease real-time PCR assay targeting *fliP* for the Rapid Identification of *Burkholderia mallei* in clinical samples. Real-Time PCR TaqMan Assay for rapid identification and differentiation of *Burkholderia pseudomallei* and *Burkholderia mallei*. RAPD, variable amplicon typing scheme, Rep-PCR, BOX-PCR and multiple-locus variable-number tandem repeat analysis have been recommended for the rapid differentiation of *B. mallei* and *B. pseudomallei* strains.
Melioidosis (*Burkholderia pseudomallei*) (U'Ren et al. 2005; Meumann et al. 2006; Antonov et al. 2008)	Blood, sputum, urine, or skin lesions	Type III Secretion System Real-Time PCR assay for diagnosing melioidosis. Real-time PCR TaqMan assay for rapid identification and differentiation of *Burkholderia pseudomallei* and *Burkholderia mallei*.
Psittacosis (*Chlamydophila psittaci*) (Messmer et al. 1997; Heddema et al. 2006; Sachse et al. 2008)	Sputum, bronchoalveolar lavage, throat swab, serum	Nested, multiplex PCR for simultaneous detection of three species of chlamydiae in human and avian specimens. Genotyping of *Chlamydophila psittaci* DNA in human samples by ompA gene sequencing. The newly developed DNA microarray-based assay represents a promising diagnostic tool for tracing epidemiological chains, exploring the dissemination of genotypes and identifying non-typical representatives of *C. psittaci*.
Q fever (*Coxiella burnetii*) (Fournier and Raoult 2003; Hou et al. 2011; Frangoulidis et al. 2012)	Blood samples, serum, heart valve tissue	LightCycler Nested PCR (LCN-PCR), a rapid nested PCR assay that uses serum sampled early during the disease. PCR targeting single-copy genes such as *Com*1 and 16S rRNA genes. A simple method that combines the simplicity of conventional PCR with new technical and methodical enhancements, resulting in a fast, specific and easy method for the molecular detection of *C. burnetii*. Single-tube nested real-time PCR (STN-RT PCR) assay using the repetitive, transposon-like element IS1111 as the DNA target to facilitate early diagnosis of acute Q fever.
Staphylococcal enterotoxin B (Purschke et al. 2003; Fischer et al. 2007; Yang et al. 2010a,b)	Serum, culture isolates, food, environmental specimens, nasal swab, induced respiratory specimens, urine, stool/gastric aspirates	Biologically stable mirror-image oligonucleotide ligands—a DNA Spiegelmers to staphylococcal enterotoxin B. Quantitative real-time immuno-PCR approach for detection of staphylococcal enterotoxins. A combination of a multiplex-immunoaffinity purification approach, followed by matrix-assisted laser desorption/ionization (MALDI)–based detection for the simultaneous identification of ricin, SEB, BoNT/A, and BoNT/B. Electrical percolation-based biosensor for real-time direct detection of staphylococcal enterotoxin B. A lab-on-a-chip (LOC) that utilizes a biological semiconductor (BSC) transducer for label free analysis of Staphylococcal Enterotoxin B (SEB) (or other biological interactions) directly and electronically.

(continued)

Table 11.2 Molecular Biology Techniques for the Detection of Specific Bioterrorism Agents (Continued)

Agent (References)	Specimens	Molecular Biology Diagnostic
Typhus fever (*Rickettsia prowazekii*) (Jiang et al. 2003)	Blood samples, tissue samples, including skin biopsies	Real-time PCR duplex assay.
Viral encephalitis alphaviruses (e.g., Venezuelan equine encephalitis, eastern equine encephalitis, western equine encephalitis) (Vernet 2004; Wang et al. 2006)	Blood samples, serum, cerebrospinal fluid (CSF), human or animal tissues	Real-time PCR, DNA microarrays.
Water safety threats (e.g., *Vibrio cholerae*, *Cryptosporidium parvum*) (Chow et al. 2001; Tark et al. 2010, Balatbat et al. 1996; Chung et al. 1999; Gile et al. 2002; Tanriverdi et al. 2002)	Stool, vomitus, food and water samples	PCR method that selectively amplifies a DNA fragment within the *ctxAB* operon of *V. cholerae*. A PCR that amplifies *Vibrio cholerae* RTX (repeat in toxin) toxin gene. The label-free detection of cholera toxin is demonstrated using microcantilevers functionalized with ganglioside nanodisks. *Cryptosporidium parvum* mixed genotypes detection by PCR-restriction fragment length polymorphism analysis. Detection of *Cryptosporidium parvum* DNA in human feces by nested PCR. Detection and genotyping of oocysts of *Cryptosporidium parvum* by Real-Time PCR and Melting Curve Analysis. A PCR method for the quantitation of *Cryptosporidium parvum* oocysts in drinking water samples.
	C Category	
Emerging infectious diseases such as Nipah virus and hantavirus (Wacharapluesadee and Hemachudha 2007; Guillaume et al. 2004; Aitichou et al. 2005; Garin et al. 2001; Nordstrom et al. 2004)	Blood sample, serum, CSF, urine, human and animal tissues	Duplex nested RT-PCR (nRT-PCR) with internal control (IC) for the detection of Nipah virus RNA. A specific fluorogenic (5′ nuclease probe) Taqman real-time PCR assay for the diagnosis and detection of Nipah virus. Real-time RT-PCR for the detection of hantaviruses. Highly sensitive Taqman PCR detection of Puumala hantavirus. DNA microarrays for the detection of hantaviruses.

DNA-based analysis for the detection of specific bioterrorism agents includes: partial or whole genome sequencing, 16S rRNA sequencing, microarray analysis including pathogenicity array analyses, single-nucleotide polymorphism characterization, variable number tandem repeat analyses, and also a very important tool—antibiotic resistance gene characterization. Ideally, laboratories should be capable of identifying bioengineered microorganisms to distinguish a bioterrorist attack from a natural outbreak. For toxin detection, various methods are available today: different immunoasays produced by recombinant technology, biofunctional assays, protein or peptide-based assays, mass spectrometry, and DNA-based assays. Additionally, numerous non-DNA–based methods may be important for identification of microorganisms and to characterize preparation and process related signatures associated with the microbial forensic samples.

Microbial forensics, as with other traditional forensic disciplines, requires implementation and application of protocols and practices that will ultimately yield results that can be used either by decision makers in law enforcement and/or national/international security. Identification of appropriate samples, "chain of custody" records (documentation that tracks physical control of samples), the use of valid analytical protocols by trained personnel, adherence to quality assurance measures and ensuring the secure storage and preservation of samples—all are of critical importance for successful attribution investigations. Somewhat unique to this type of evidence is ensuring the safety of anyone involved with the collection and handling of microbial forensic evidence. When dealing with pathogenic microorganisms, there are special handling procedures that must be followed in order to adequately prevent additional harm to personnel and the surrounding community. All personnel involved in response to or handling of these bioagents must undergo specialized training to ensure the safety and security of the individuals involved and the surrounding environment. Manipulations of potentially infectious material should be conducted under the appropriate containment conditions, ideally using a biocontainment laboratory equipped with necessary safety equipment. Those individuals at risk of direct exposure to infectious material during collection or handling should wear appropriate personal protective equipment (CDC 2013; Breeze et al. 2011; Hoffmaster et al. 2002; Jahrling et al. 2004; U.S. Department of Health and Human Services 1999).

11.5 Criteria for Considering an Outbreak Unusual

The initial outbreak of an infectious disease may not immediately be recognized as that caused by a bioterrorism event because pathogenic microorganisms occur naturally all over the world. Early determination of an attack versus a naturally occurring outbreak is a goal of organizations responsible for attribution of bioterrorist events. It is important to identify unusual outbreaks in a timely fashion. The World Health Organization (WHO) Ad Hoc Group developed in 1998 criteria for considering some outbreaks unusual (Breeze et al. 2011; Dando et al. 1988).

These criteria include:

1. The first appearance of a disease outbreak in a nonendemic region.
2. An outbreak occurring outside of its normal season.
3. An uncommon route of transmission of pathogens.
4. The natural reservoir hosts or insect vectors do not occur in the region when the outbreak occurs.
5. The epidemiology of the diseases suggest an abnormal reduction in the incubation period of disease.
6. Characteristic of the pathogen differs from the known characteristics profile of the agent (strain type, sequence, antibiotic resistance, pattern, etc.).

7. An increased virulence of pathogens.
8. The pathogen is capable of establishing new natural reservoirs to facilitate continuous transmission.

These criteria are based on the knowledge that although microbial agents are ubiquitous, it is well known that many of them may survive in some climate conditions or biotopes and not in others, and that some infectious diseases are more common or exclusively appear in some seasons (Benenson 1995; Begovac et al. 2006). For example, detection of human influenza cases during the summer time would be an unusual outbreak worthy of further investigation. Thus, epidemiological investigations to differentiate naturally occurring outbreaks from those that are man-made are necessary. The microbial agents that can be transferred from animals to people (zoonoses) or by insects (vector-borne diseases) are closely linked to their animal reservoirs/vectors and to their specific biotopes. For example, the reservoirs for New World hantaviruses, which cause hantavirus pulmonary syndrome (HPS), are distinct small rodent species found only in North or South America. An outbreak of HPS in other parts of the world would be suspicious as a possible bioterrorism attack and worthy of careful investigations. Also, if an animal reservoir or vector appears to be transmitting a new microbial agent, this should be carefully analyzed as a possible deliberate distribution. It is also important to know the typical disease characteristics caused by potential threat pathogens such as the normal incubation time. If a disease appears with a short incubation period or enhanced or unusual clinical signs and symptoms, especially in a population experienced with that disease, this too could be considered as a suspicious disease outbreak. The transmission route for many pathogens is very well known, as well as consequent pathogenic mechanisms. Significant variance in the route of transmission or pathogenic mechanisms may be the result of an intentional spread of newly engineered microorganisms. This might include the appearance of an unexpectedly high rate of resistance to previously effective antibiotics, or enhanced virulence of a microorganism that previously caused only mild or moderate infections (CDC 2013; Breeze et al. 2011; Benenson 1995; Begovac et al. 2006). One of the very important tools to help differentiate natural epidemic and intentional attack is disease surveillance and global disease reporting system. There are several web-based programs, such as ProMed (http://www.promedmail.org) or GOARN (WHO's Global Outbreak Alert and Response Network, http://www.who.int/csr/outbreaknetwork), used as important tools in recognizing outbreaks and possible bioterrorist events. It is important to collect essential information on all significant factors of disease outbreaks to learn how and why the disease has occurred. For such purposes, routine monitoring of changes in disease patterns, documenting and understanding factors that have initiated changes, timely dissemination of information to decision makers, understanding the normal occurrence of diseases, and web-based programs for international disease surveillance for reporting of human and animal diseases can all contribute to a clearer understanding and better recognition of unusual events (Breeze et al. 2011).

11.6 Suspicious Infectious Diseases Outbreaks

As discussed, bioterrorism unfortunately has a rich, long history. However, today with modern microbial forensics methods and a tremendous increase in knowledge about microbial agents, including modern laboratory techniques and global disease reporting systems, it is much easier to distinguish naturally occurring infections from deliberately caused diseases. Several examples follow of outbreaks caused by accident or criminal acts.

- In April 1979 in the city of Yekaterinburg in the former Soviet Union an explosion occurred at a local military compound. Over the next several days in an area 4 km south and east of the compound, many residents began to develop high fevers and difficulty breathing. Approximately 200 patients died within a short period. Local doctors diagnosed it as an outbreak of inhalation anthrax; however, the government officially reported consumption of tainted meat from a cow suffering from anthrax as being responsible for the outbreak. Patients did not, however, display the symptoms of gastric or cutaneous anthrax, and autopsies of human victims indicated severe pulmonary edema and toxemia. Sheep and cattle in six different villages up to 50 km distant also died of anthrax. DNA analysis of tissue samples taken from victims indicated they had been exposed to several anthrax strains simultaneously, which is atypical in natural infection in cattle. It was subsequently concluded that the anthrax outbreak was the result of the accidental release from a bioweapons facility (Wade 1998; Walker et al. 1994; Sternbach 2003; Brookmeyer et al. 2001; Wilkening 2006).
- In the 1980s, a number of people became HIV-infected. None of these individuals had lifestyles that would put them a high risk for exposure to HIV. In 1986, when their dentist was found to be HIV-positive, it became clear that multiple patients who contracted HIV had visited the same dentist. DNA sequencing of samples from the patients, the dentist, a local control group, and a distinct control was performed. HIV nucleotide sequences from a number of patients were closely related (not exactly the same) to those from the dentist, and distinct from viruses obtained from control patients. The evidence strongly supported, but could not conclusively prove, HIV transmission from the dentist to patients during invasive dental care (Breeze et al. 2011). This was, however, nonintentional transmission of HIV from the dentist who was not aware that he was an HIV carrier.
- In September 1984, the Rajneeshee cult, an Indian religious cult, purposely contaminated salad bars in restaurants located in or near Dalles, Oregon, with *Salmonella typhimurium*. Over 750 persons were poisoned and about 40 hospitalized. The purpose of the attack was to influence the outcome of a local election as it was discovered a year later when members of the cult turned informants (Philips 2005; Carus 2001).
- In 1995, on more than 10 occasions Aum Shinrikyo attempted to disperse anthrax, botulinum toxin, Q fever, and Ebola virus against the general population and authority figures in Japan. Fortunately, no reported infections occurred, although they were successful in releasing sarin gas into the Tokyo subway system that year, resulting in 12 deaths and causing an estimated 6000 people to seek medical attention (Philips 2005).
- Among the last stories were the 2001 anthrax letter attacks that came in two waves. The first letters were sent to several U.S. news media offices in New York (ABC, CBC, NBC, and the *New York Post*) and Florida (American Media Inc.), and 3 weeks later the second set of letters were sent to two U.S. senators in Washington,

D.C., Tom Daschle of South Dakota and Patrick Leahy of Vermont. Five people were killed and 17 others were infected between October 4 and November 22. DNA sequencing of the *B. anthracis* isolated from the first victim who died from pulmonary anthrax was conducted, and it was identified as the Ames strain. This strain had only been used for research purposes. Since its original isolation, the Ames strain had been maintained at the U.S. Army Medical Research Institute of Infectious Diseases (USAMRIID) in Fort Detrick, Maryland. USAMRIID had shared the Ames strain with at least 15 bioresearch laboratories within the United States and six locations overseas (CDC 2001a; Jernigan et al. 2001; Atlas 2002; Greene et al. 2002; Tan et al. 2002; Hoffmaster et al. 2002; NRC, NAS 2011). The Federal Bureau of Investigation (FBI) was responsible for the investigation of the letter attacks. The FBI assigned squads of Special Agents who teamed with investigators from the U.S. Postal Service. The investigation, given the major case title of "Amerithrax," lasted nearly 9 years and involved almost 600,000 investigator work hours, 10,000 witness interviews, 80 searches, and 26,000 e-mail reviews including analysis of 4 million megabytes of computer memory in six continents. In addition, 29 government, university, and commercial laboratories were involved in scientific analyses, and close collaboration with other federal agencies was established. In June 2002, FBI scrutinized 20 to 30 scientists who might have had knowledge and opportunity to arrange the "letters attack." On August 6, 2002, a former USAMRIID scientist, Steven J. Hatfill, a biodefense expert, was indicated as a "person of interest" in the investigation. However, on August 8, 2008, Department of Justice (DOJ) cleared Steven Hatfill of involvement in the anthrax mailings and the federal government awarded Steven Hatfill $5.82 million to settle his violation of privacy lawsuit against DOJ. On July 29, 2008, another USAMRIID microbiologist Bruce E. Ivins committed suicide just as the FBI was about to file criminal charges against him. Following his suicide, the FBI released information that they believed linked him to the crime. Specifically, he has been responsible for the preparation and oversight of a flask containing a stock supply of *B. anthracis* that was identified as RMR-1029. The *B. anthracis* sent in the letters was believed to have originated from this stockpile. This link as well as other information led the FBI, the DOJ, and the U.S. Postal Inspection Service to officially close the case on February 19, 2010, concluding that Dr. Bruce Ivins alone mailed the *B. anthracis* laden letters (NRC, NAS 2011).

- Before the official closure of the case, the FBI asked the National Research Council (NRC) of the National Academy of Sciences (NAS) to conduct an independent review of the various scientific methods and procedures that had been used throughout the investigation. This was in response to issues that had been raised concerning the FBI's determination of RMR 1029 as the source of the *B. anthracis* used in the letters. For this purpose, a committee of esteemed scientist and experts was appointed. The committee was expected to determine whether microbial forensic techniques used during the FBI investigation that supported the final FBI conclusions met appropriate standards for scientific reliability. The review process focused on novel scientific methods that were developed for purposes of the FBI investigation.

After an extensive analysis of documents and facts provided by FBI the committee made the following conclusions:

1. The *B. anthracis* in the letters was the Ames strain and was not genetically engineered.
2. Multiple distinct colony morphological types or morphotypes of *B. anthracis* Ames were present in the letters.
3. The scientific link between the letter material and flask number RMR-1029 indicated by FBI as an origin source was not conclusive.
4. There was no evidence that silicon found in spores from the letter was intentionally added.
5. It was not possible to determine the amount of time needed to prepare the spore material; it was considered that the time might vary from 2 to 3 days or to several months.
6. The physicochemical and radiological analyses were of limited forensic value in this investigation.
7. There was no consistent evidence of *B. anthracis* Ames DNA in environmental samples collected from an overseas site.
8. During the time, new powerful scientific tools and methods (e.g., high-throughput next-generation DNA sequencing) were developed that might have given better insight if used during the investigation.

In general, the committee emphasized that should another biological attack occur, better communication between the public and policymakers should be established before, during, and after the investigation (NRC, NAS 2011).

11.7 Biosafety and Biosecurity

The field of forensic microbiology needs to harmonize the efforts of scientists in the field of infectious diseases, epidemiology, and microbiology with people who make political, administrative, and legal decisions in making biodefense policy. This is a prerequisite for building a national capacity to investigate bioterrorism. The research community in infectious diseases, microbiology, epidemiology, and immunology can contribute greatly through their skills and insights to the efforts to build national capacity for biodefense. Basic research, including genomics, vaccines, therapies, and a robust corps of talented and committed scientists who are trained and educated, are important to achieve national biosafety and biosecurity. Also, laboratory and public health experts should be aware of the responsibilities of law enforcement and national security organizations and programs that encourage joint investigations at the time of an initial outbreak should be established. A program to construct national biocontainment laboratories is essential to the success of biosecurity and biodefense, since it is important to ensure that pathogen collections are safely stored and access to pathogenic strains is appropriately limited. In the United States, the select agents rules are important aspects of each organization's biosecurity program.

Although biosecurity is inherently a government's responsibility, there are professional organizations (institutional, academic, and private sector) and individual experts that can help shape and develop guidelines on equitable and fair access and benefit-sharing of genetic resources, microbial agents and the ethical implications of biotechnology, international governance of biotechnology and biosafety, and vaccination programs including rule making (CDC 2013; Breeze et al. 2011; CDC 2001a; Atlas and Reppy 2005; Cook-Deegan et al. 2005; Atlas and Dando 2006; Bork et al. 2007).

References

Aitichou, M., Saleh, S.S., McElroy, A.K., Schmaljohn, C., and Ibrahim, M.S. 2005. Identification of Dobrava, Hantaan, Seoul, and Puumala viruses by one-step real-time RT-PCR. *J Virol Methods* 124:21–6.

Al-Khaldi, S.F., Myers, K.M., Rasooly, A. and Chizhikov, V. 2004. Genotyping of Clostridium perfringens toxins using multiple oligonucleotide microarray hybridization. *Mol Cell Probes* 18:359–67.

Antonov, V.A., Tkachenko, G.A., Altukhova, V.V. et al. 2008. Molecular identification and typing of *Burkholderia pseudomallei* and *Burkholderia mallei*: When is enough enough? *Trans R Soc Trop Med Hyg* 102(Suppl 1):S134–9.

Atlas, R.M. 2002. Bioterrorism: From threat to reality. *Annu Rev Microbiol* 56:167–85.

Atlas, R.M. and Reppy, J. 2005. Globalizing biosecurity. *Biosecur Bioterror* 3:51–60.

Atlas, R.M. and Dando, M. 2006. The dual-use dilemma for the life sciences: Perspectives, conundrums, and global solutions. *Biosecur Bioterror* 4:276–86.

Balatbat, A.B., Jordan, G.W., Tang, Y.J. and Silva Jr., J. 1996. Detection of *Cryptosporidium parvum* DNA in human feces by nested PCR. *J Clin Microbiol* 34:1769–72.

Barns, S.M., Grow, C.C., Okinaka, R.T., Keim, P. and Kuske, C.R. 2005. Detection of diverse new *Francisella*-like bacteria in environmental samples. *Appl Environ Microbiol* 71:5494–500.

Begovac, J., Božinović, D., Lisić, M., Baršić, B. and Schönwald, S., eds. 2006. *Infectology*. Zagreb: Profil.

Benenson, A.S., ed. 1995. *Control of Communicable Diseases Manual*. An official report of the American Public Health Association. Washington, DC: American Public health Association.

Bogdanovich, T., Skurnik, M., Lubeck, P.S., Ahren, P. and Hoorfar, J. 2004. Validated 5′ nuclease PCR assay for rapid identification of the genus *Brucella*. *J Clin Microbiol* 42:2261–3.

Bork, K.H., Halkjaer-Knudsen, V., Hansen, J.E. and Heegaard, E.D. 2007. Biosecurity in Scandinavia. *Biosecur Bioterror* 5:62–71.

Breeze, R.G., Budowle, B., Schutzer, S.E., Keim, P. and Morse, S. eds. 2011. *Microbial Forensics*, 2nd edition. Amsterdam: Elsevier Academic Press.

Brookmeyer, R., Blades, N., Hugh-Jones, M. and Henderson, D.A. 2001. The statistical analysis of truncated data: Application to the Sverdlovsk anthrax outbreak. *Biostatistics* 2:233–47.

Carus, W.S. 2001. *Bioterrorism and biocrimes*. The illicit use of biological agents since 1900 Center for Counterproliferation Research National Defense University, Washington, DC, 217 pp. http://www.ndu.edu/centercounter/Full_Doc.pdf.

Center for Disease Control and Prevention (CDC), Department of Health and Human Services (HHS). 2001a. Requirements for facilities transferring or receiving select agents. Final rule. *Fed Regist* 66:45944–5.

Centers for Disease Control and Prevention (CDC) 2001b. Update: Investigation of bioterrorism-related anthrax and adverse events from antimicrobial prophylaxis. *MMWR Morb Mortal Wkly Rep* 50:973–6.

Centers for Diseases Control and Prevention (CDC) 2013. Bioterrorism agents/diseases. http://www.bt.cdc.gov/Agent/agentlist.asp.

Chow, K.H., Ng, T.K., Yuen, K.Y. and Yam, W.C. 2001. Detection of RTX toxin gene in *Vibrio cholerae* by PCR. *J Clin Microbiol* 39:2594–7.

Chung, E., Aldom, J.E., Carreno, R.A. et al. 1999. PCR-based quantitation of *Cryptosporidium parvum* in municipal water samples. *J Microbiol Methods* 38:119–30.

Committee on review of the scientific approaches used during the FBI's investigation of the 2001 *Bacillus anthracis* mailings, Board on Life Sciences Division on Earth and Life Studies, Committee on Science, Technology, and Law Policy and Global Affairs Division, National

Research Council of the National Academies. 2011. *Review of the Scientific Approaches Used During the FBI's Investigation of the 2001 Anthrax Letters*. Washington, DC: National Academies Press.

Cook-Deegan, R.M., Berkelman, R., Davidson, E.M. et al. 2005. Issues in biosecurity and biosafety. *Science* 308:1867–8.

Dando, M., Pearson, G.S. and Kriz, B., eds. 1988. Scientific and technical means of distinguishing between natural and other outbreaks of disease. *Proceedings of the NATO Advanced Research Workshop*, Prague, Czech Republic, October 18–20.

De Santis, R., Ciammaruconi, A., Faggioni, G. et al. 2011. High throughput MLVA-16 typing for *Brucella* based on the microfluidics technology. *BMC Microbiol* 11:60.

Engelthaler, D.M., Lewis, K., eds. 2004. *Zebra Manual: A Reference Handbook for Bioterrorism Agents*. Phoenix, AZ: Arizona Department of Health Services, Division of Public Health Services.

Eshoo, M.W., Whitehouse, C.A., Nalca, A. et al. 2009. Rapid and high-throughput pan-Orthopoxvirus detection and identification using PCR and mass spectrometry. *PLoS One* 2009;e6342.

FDA, 2013. ORA Lab Manual, Volume IV, Section 2—Microbiology. http://www.cfsan.fda.gov/~ebam/bam-28.html.

Fedele, C.G., Negredo, A., Molero, F., Sanchez-Seco, M.P. and Tenorio, A. 2006. Use of internally controlled real-time genome amplification for detection of variola virus and other orthopoxviruses infecting humans. *J Clin Microbiol* 44:4464–70.

Fenicia, L., Fach, P., van Rotterdam, B.J. et al. 2011. Towards an international standard for detection and typing botulinum neurotoxin-producing Clostridia types A, B, E and F in food, feed and environmental samples: A European ring trial study to evaluate a real-time PCR assay. *Int J Food Microbiol* 145(Suppl 1):S152–7.

Fischer, A., von Eiff, C., Kuczius, T., Omoe, K., Peters, G. and Becker, K.A. 2007. quantitative real-time immuno-PCR approach for detection of staphylococcal enterotoxins. *J Mol Med* 85:461–9.

Fournier, P.E. and Raoult, D. 2003. Comparison of PCR and serology assays for early diagnosis of acute Q fever. *J Clin Microbiol* 41:5094–8.

Frangoulidis, D., Meyer, H., Kahlhofer, C. and Splettstoesser, W.D. 2012. 'Real-time' PCR-based detection of *Coxiella burnetii* using conventional techniques. *FEMS Immunol Med Microbiol* 64:134–6.

Garin, D., Peyrefitte, C., Crance, J.M., Le Faou, A., Jouan, A. and Bouloy, M. 2001. Highly sensitive Taqman PCR detection of *Puumala hantavirus*. *Microbes Infect* 3:739–45.

Gilbert, D.N., Moellering, R.C. Jr., Eliopoulos, G.M., Sande, M.A., eds. 2006. *The Sanford Guide for Antimicrobial Therapy 2006*. Sperryville, VA, USA: Antimicrobial Therapy Inc.

Gile, M., Warhurst, D.C., Webster, K.A., West, D.M. and Marshall, J.A. 2002. A multiplex allele specific polymerase chain reaction (MAS-PCR) on the dihydrofolate reductase gene for the detection of *Cryptosporidium parvum* genotypes 1 and 2. *Parasitology* 125:35–44.

Greene, C.M., Reefhuis, J., Tan, C. et al. 2002. CDC New Jersey Anthrax Investigation Team. Centers for Disease Control and Prevention. Epidemiologic investigations of bioterrorism-related anthrax, New Jersey, 2001. *Emerg Infect Dis* 8:1048–55.

Guillaume, V., Lefeuvre, A., Faure, C. et al. 2004. Specific detection of Nipah virus using real-time RT-PCR (TaqMan). *J Virol Methods* 120:229–37.

Gunther, S., Asper, M., Roser, C. et al. 2004. Application of real-time PCR for testing antiviral compounds against Lassa virus, SARS coronavirus and Ebola virus in vitro. *Antiviral Res* 63:209–15.

Hass, M., Westerkofsky, M., Muller, S., Becker-Ziaja, B., Busch, C. and Gunther, S. 2006. Mutational analysis of the Lassa virus promoter. *J Virol* 80:12414–9.

Hauser, E., Mellmann, A., Semmler, T. et al. 2013. Phylogenetic and molecular analysis of food-borne Shiga toxin-producing *Escherichia coli*. *Appl Environ Microbiol* February 15. [Epub ahead of print] PubMed PMID: 23417002.

Heddema, E.R., van Hannen, E.J., Duim, B. et al. 2006. An outbreak of psittacosis due to *Chlamydophila psittaci* genotype A in a veterinary teaching hospital. *J Med Microbiol* 55:1571–5.

Hoffmaster, A.R., Meyer, R.F., Bowen, M.D. et al. 2002. Evaluation and validation of a real-time polymerase chain reaction assay for rapid identification of *Bacillus anthracis*. *Emerg Infect Dis* 8:1178–82.

Hoorfar, J., Ahrens, P. and Radstrom, P. 2000. Automated 5′ nuclease PCR assay for identification of *Salmonella enterica*. *J Clin Microbiol* 38:3429–35.

Hou, M.Y., Hung, M.N., Lin, P.S. et al. 2011. Use of a single-tube nested real-time PCR assay to facilitate the early diagnosis of acute Q fever. *Jpn J Infect Dis* 64:161–2.

Ibekwe, A.M., Watt, P.M., Grieve, C.M., Sharma, V.K. and Lyons, S.R. 2002. Multiplex fluorogenic real-time PCR for detection and quantification of *Escherichia coli* O157:H7 in dairy wastewater wetlands. *Appl Environ Microbiol* 68:4853–62.

Jahrling, P.B., Hensley, L.E., Martinez, M.J. et al. 2004. Exploring the potential of variola virus infection of cynomolgus macaques as a model for human smallpox. *Proc Natl Acad Sci U S A* 101:15196–200.

Jernigan, J.A., Stephens, D.S., Ashford, D.A. et al. 2001. Anthrax Bioterrorism Investigation Team. Bioterrorism-related inhalational anthrax: The first 10 cases reported in the United States. *Emerg Infect Dis* 7:933–44.

Jiang, J., Temenak, J.J. and Richards, A.L. 2003. Real-time PCR duplex assay for *Rickettsia prowazekii* and *Borrelia recurrentis*. *Ann N Y Acad Sci* 990:302–10.

Kasten, F.H. 2002. Biological weapons, war crimes, and WWI. *Science* 296:1235–7.

Kull, S., Pauly, D., Störmann, B. et al. 2010. Multiplex detection of microbial and plant toxins by immunoaffinity enrichment and matrix-assisted laser desorption/ionization mass spectrometry. *Anal Chem* 82:2916–24.

Kurosaki, Y., Takada, A., Ebihara, H. et al. 2007. Rapid and simple detection of Ebola virus by reverse transcription-loop-mediated isothermal amplification. *J Virol Methods* 141:78–83.

Le Duc, J.W., Damon, I., Relman, D.A., Huggins, J. and Jahrling, P.B. 2001. Smallpox research activities: U.S. interagency collaboration. *Emerg Infect Dis* 8:743–5.

Lindler, L.E., Fan, W. and Jahan, N. 2001. Detection of ciprofloxacin-resistant *Yersinia pestis* by fluorogenic PCR using the LightCycler. *J Clin Microbiol* 39:3649–55.

Markotic, A., Le Duc, J.W., Hlaca, D. et al. 1996. Hantaviruses are likely threat to NATO forces in Bosnia and Herzegovina and Croatia. *Nat Med* 2:269–70.

McGrath, S., Dooley, J.S. and Haylock, R.W. 2000. Quantification of *Clostridium botulinum* toxin gene expression by competitive reverse transcription-PCR. *Appl Environ Microbiol* 66:1423–8.

Melo, A.C., Almeida, A.M. and Leal, N.C. 2003. Retrospective study of a plague outbreak by multiplex-PCR. *Lett Appl Microbiol* 37:361–4.

Meltzer, M.I., Damon, I., Le Duc, J.W. and Millar, J.D. 2001. Modeling potential responses to smallpox as a bioterrorist weapon. *Emerg Infect Dis* 7:959–69.

Messmer, T.O., Skelton, S.K., Moroney, J.F., Daugharty, H. and Fields, B.S. 1997. Application of a nested, multiplex PCR to psittacosis outbreaks. *J Clin Microbiol* 35:2043–6.

Metzger-Boddien, C., Bostel, A. and Kehle, J. 2004. AnDiaTec *Salmonella* sp. PCR-ELISA for analysis of food samples. *J Food Prot* 67:1585–90.

Meumann, E.M., Novak, R.T., Gal, D. et al. 2006. Clinical evaluation of a type III secretion system real-time PCR assay for diagnosing melioidosis. *J Clin Microbiol* 44:3028–30.

Newby, D.T., Hadfield, T.L. and Roberto, F.F. 2003. Real-time PCR detection of *Brucella abortus*: A comparative study of SYBR green I, 5′-exonuclease, and hybridization probe assays. *Appl Environ Microbiol* 69:4753–9.

Nordstrom, H., Johansson, P., Li, Q.G., Lundkvist, A., Nilsson, P. and Elgh, F. 2004. Microarray technology for identification and distinction of hantaviruses. *J Med Virol* 72:646–55.

Philips, M.B. 2005. Bioterrorism. A brief history. *North East Florida Med J* 56:32–35.

Purschke, W.G., Radtke, F., Kleinjung, F. and Klussmann, S. 2003. A DNA Spiegelmer to staphylococcal enterotoxin B. *Nucleic Acids Res* 31:3027–32.

Romero, C., Gamazo, C., Pardo, M. and Lopez-Goni, I. 1995. Specific detection of Brucella DNA by PCR. *J Clin Microbiol* 33:615–7.

Sachse, K., Laroucau, K., Hotzel, H., Schubert, E., Ehricht, R. and Slickers, P. 2008. Genotyping of *Chlamydophila psittaci* using a new DNA microarray assay based on sequence analysis of *ompA* genes. *BMC Microbiol* 8:63.

Schönwald, S., Baršić, B., Beus, A. et al. 2000. *Manual for the Therapy and Prevention of Infectious Diseases*. Zagreb: Birotisak.

Sharma, V.K. and Dean-Nystrom, E.A. 2003. Detection of enterohemorrhagic *Escherichia coli* O157:H7 by using a multiplex real-time PCR assay for genes encoding intimin and Shiga toxins. *Vet Microbiol* 93:247–60.

Spicić, S., Zdelar-Tuk, M., Racić, I., Duvnjak, S. and Cvetnić, Z. 2010. Serological, bacteriological, and molecular diagnosis of brucellosis in domestic animals in Croatia. *Croat Med J* 51:320–6.

Sternbach, G. 2003. The history of anthrax. *J Emerg Med* 24:463–7.

Sulaiman, I.M., Tang, K., Osborne, J., Sammons, S. and Wohlhueter, R.M. 2007. GeneChip resequencing of the smallpox virus genome can identify novel strains: A biodefense application. *J Clin Microbiol* 45:358–63.

Tan, C.G., Sandhu, H.S., Crawford, D.C. et al. 2002. Regional Anthrax Surveillance Team; Centers for Disease Control and Prevention New Jersey Anthrax Surveillance Team. Surveillance for anthrax cases associated with contaminated letters, New Jersey, Delaware, and Pennsylvania, 2001. *Emerg Infect Dis* 8:1073–7.

Tanriverdi, S., Tanyeli, A., Baslamisli, F. et al. 2002. Detection and genotyping of oocysts of *Cryptosporidium parvum* by real-time PCR and melting curve analysis. *J Clin Microbiol* 40:3237–44.

Tark, S.H., Das, A., Sligar, S. and Dravid, V.P. 2010. Nanomechanical detection of cholera toxin using microcantilevers functionalized with ganglioside nanodiscs. *Nanotechnology* 21:435502.

Tomaso, H., Reisinger, E.C., Al Dahouk, S. et al. 2003. Rapid detection of *Yersinia pestis* with multiplex real-time PCR assays using fluorescent hybridisation probes. *FEMS Immunol Med Microbiol* 38:117–26.

Tomaso, H., Scholz, H.C., Al Dahouk, S. et al. 2006. Development of a 5′-nuclease real-time PCR assay targeting fliP for the rapid identification of *Burkholderia mallei* in clinical samples. *Clin Chem* 52:307–10.

Tomaso, H., Scholz, H.C., Neubauer, H. et al. 2007. Real-time PCR using hybridization probes for the rapid and specific identification of *Francisella tularensis* subspecies *tularensis*. *Mol Cell Probes* 21:12–6.

U.S. Department of Health and Human Services. *Biosafety in Microbiological and Biomedical Laboratories*. Washington, DC: U.S. Government Printing Office; 1999.

U'Ren, J.M., Van Ert, M.N., Schupp, J.M. et al. 2005. Use of a real-time PCR TaqMan assay for rapid identification and differentiation of *Burkholderia pseudomallei* and *Burkholderia mallei*. *J Clin Microbiol* 43:5771–4.

Vernet, G. 2004. Diagnosis of zoonotic viral encephalitis. *Arch Virol Suppl* 18:231–44.

Versage, J.L., Severin, D.D., Chu, M.C. and Petersen, J.M. 2003. Development of a multitarget real-time TaqMan PCR assay for enhanced detection of *Francisella tularensis* in complex specimens. *J Clin Microbiol* 41:5492–9.

Villalobo, E. and Torres, A. 1998. PCR for detection of *Shigella* spp. in mayonnaise. *Appl Environ Microbiol* 64:1242–5.

Wacharapluesadee, S. and Hemachudha, T. 2007. Duplex nested RT-PCR for detection of Nipah virus RNA from urine specimens of bats. *J Virol Methods* 141:97–101.

Wade, N. 1908. Death at Sverdlovsk: A critical diagnosis. *Science* 209:1501–2.

Walker, D.H., Yampolska, O. and Grinberg, L.M. 1994. Death at Sverdlovsk: What have we learned? *Am J Pathol* 144:1135–41.

Wang, E., Paessler, S., Aguilar, P.V. et al. 2006. Reverse transcription-PCR-enzyme-linked immunosorbent assay for rapid detection and differentiation of alphavirus infections. *J Clin Microbiol* 44:4000–8.

Whitehouse, C.A. and Hottel, H.E. 2007. Comparison of five commercial DNA extraction kits for the recovery of *Francisella tularensis* DNA from spiked soil samples. *Mol Cell Probes* 21:92–6.

Wilkening, D.A. 2006. Sverdlovsk revisited: Modeling human inhalation anthrax. *Proc Natl Acad Sci U S A* 103:7589–94.

World at Risk Commission on the prevention of weapons of mass destruction proliferation and terrorism, 2009. Available from: www.preventwmd.gov/report/.

Yang, M., Sun, S., Bruck, H.A., Kostov, Y. and Rasooly, A. 2010a. Electrical percolation-based biosensor for real-time direct detection of staphylococcal enterotoxin B (SEB). *Biosens Bioelectron* 25:2573–8.

Yang, M., Sun, S., Bruck, H.A., Kostov, Y. and Rasooly, A. 2010b. Lab-on-a-chip for label free biological semiconductor analysis of staphylococcal enterotoxin B. *Lab Chip* 10:2534–40.

Zhai, J., Palacios, G., Towner, J.S. et al. 2007. Rapid molecular strategy for filovirus detection and characterization. *J Clin Microbiol* 45:224–6.

Forensic Animal DNA Analysis

12

MARILYN A. MENOTTI-RAYMOND
VICTOR A. DAVID
STEPHEN J. O'BRIEN
SREE KANTHASWAMY
PETAR PROJIĆ
VEDRANA ŠKARO
GORDAN LAUC
ADRIAN LINACRE

Contents

12.1 Introduction	317
12.2 Felid Forensic DNA Testing	321
12.2.1 Case Studies	322
12.2.2 Development of a Forensic Typing System for Genetic Individualization of Domestic Cat Samples	324
12.2.3 Validation Studies of Cat Multiplex	328
12.3 Canine Forensic DNA Testing	330
12.3.1 Case A Details—Fatal Dog Attack	332
12.3.2 Case B Details—Homicide	333
12.4 Bovine Forensic DNA Testing	335
12.5 Wildlife Forensic DNA Testing	337
12.5.1 mtDNA in Species Testing	339
12.5.2 Species Identification Using Loci on Mitochondrial Genome	340
12.5.3 Mitochondrial Sequence Analysis	341
12.5.4 SNP Analysis	342
12.5.5 Conclusions on Animal Testing	342
Acknowledgments	342
References	343

12.1 Introduction

The application of biological testing to allegations involving animals (other than human) and plants is relatively recent. This is a reflection of a very recent awakening to the scope of illegal trade in endangered species, a previous view that the crimes were of less significance than investigating crimes against people, and that few laboratories were willing and able to assist in such investigations. However, in the past decade there has been enormous increase in interest in the area, leading to a burgeoning number of research papers along with textbooks and recent reviews (Alacs et al. 2010; Linacre 2008, 2009; Linacre and Tobe 2011; Ogden et al. 2009; Tobe and Linacre 2010; Wilson-Wilde 2010a,b).

Since the mid-1990s, DNA testing in the field of forensic science has increased dramatically. Standardized and accredited crime scene processing and laboratory protocols, statistical models and analytical techniques for human DNA analysis has elevated this forensic science tool to the point of being the "gold standard" against which other forensic sciences are measured. As more research provides a deeper understanding of DNA analysis, applications have begun to span beyond the realm of human forensic science and into the area of nonhuman forensic DNA analysis.

Nonhuman DNA analysis has grown by leaps and bounds for many reasons both criminal and civil. For decades, laboratories have been testing horses and cattle for pedigree and other purposes. The study of cultivars to look at variation among plant species, the testing of sold fish and meat to verify that is correctly labeled, the Convention on International Trade in Endangered Species of Wild Flora and Fauna (CITES), as well as the ubiquitous presence of our pets and their fur winding up as trace evidence in crime scene material, have all lead to the expansion of the use of DNA technology to identify or characterize nonhuman samples.

Prior attempts at forensic analysis of animal hair relied on morphological criteria (Moore 1988; Peabody et al. 1983); identification could be made at a species, or at most, breed level. Isoelectric focusing of keratins were the first efforts used toward molecular characterization of animal specimens (Carracedo et al. 1987). The potential for genetic individualization of specimens was first demonstrated by Jeffreys et al. (1985a) using highly repetitive minisatellite loci to generate individual-specific "fingerprints" of human DNA. DNA fingerprinting methodology was used soon after in a highly publicized immigration test case (Jeffreys et al. 1985b). Jeffreys et al. were also the first to demonstrate the potential of DNA fingerprinting of companion animals—using human minisatellite DNA to generate multilocus DNA fingerprints of dog and cat DNA (Jeffreys and Morton 1987).

The Locard's Exchange Principle (James and Nordby 2003) allows forensic analysts to link original source and target surface and have a perspective on primary and secondary transfers. Animal biological evidence is not exempted from Locard's rule and has become very important for identification and individualization of trace or transfer biomaterial from crime scenes. In recent years, interest in animal forensic science has piqued, because of the abundance of animal evidence encountered in crime investigations. This is not surprising because more than 50% of households in the United States own at least (a) a cat (86.4 million cats in the United States) or a dog (78.2 million dogs in the United States, leading to high likelihoods of exchange of pet hair between a crime suspect and victim at the scene of the crime (Humane Society of the United States 2012). Because of the vast amounts of hairs shed by dogs and cats, D'Andrea et al. (1998) showed that animal hair in households are easily transferred and are very persistent; therefore, an intruder cannot exit a house a pet without coming into contact with and carrying the animal's hair away with him.

Most DNA analysis of animal biomaterial, however, only provide vital investigative leads and reconstruction of the crime; the analysis of animal DNA cannot contribute to the individualization of any human suspect, and usually this type of evidence needs to be supplemented by other forms of physical evidence. The earliest use of animal DNA analysis in a criminal case was the 1996 case of "Snowball" the cat, performed by Menotti-Raymond et al. (1997a) (see below). Since then, animal evidence has been used in myriad types of litigation including those involving traffic accidents (Schneider et al. 1999), murders (*State of California v. David Westerfield* 2002), bank robberies (Savolainen and Lunderberg 1999), and dog attack cases where there are human (Pádár et al. 2002) or nonhuman victims

(Pádár et al. 2001). Since 1996, canine DNA evidence has contributed to more than 20 criminal cases in Great Britain and the United States alone (Halverson and Basten 2005a). Because of the powerful links between a victim and suspect that can be made with animal evidence, the frequency of publications on nonhuman DNA has surpassed the number of publications on important aspects of human DNA analysis including single-nucleotide polymorphism (SNP), low copy number DNA analysis, and mixture analysis (Butler et al. 2010).

Animal forensic cases can be divided into three basic categories: (1) when the animal is a passive witness to a crime, such as when dog or cat hair and dander are used to link a suspect or perpetrator to the crime scene or victim; (2) when the animal is the victim such as abuse, theft, killing (poaching), violation of laws on endangered species and dog or cock fighting cases; (3) when the animal is a suspect, for example, when a dog or other animal attacks or mauls humans or other animals (Kanthaswamy 2009).

Cases of arson, homicide, and burglary are examples of human on human crimes that have been successfully investigated and resolved using DNA tests of material from animal "witnesses" found at the scene or on the suspect and/or victim. One of the earliest animal forensic cases was performed at the Veterinary Genetics Laboratory (VGL), University of California, Davis (UC Davis), that involved a homicide investigation. The blood trail found leading away from a murdered pub bouncer was identified by the U.K. Forensic Science Service (FSS) as canine in origin. The FSS requested the university's animal forensic laboratory to conduct DNA analysis of the blood samples from the crime scene. The analysis revealed that the DNA from the bloody trail matched a dog belonging to a man that had earlier been refused admittance into the pub. The disgruntled man had stabbed the bouncer to death when he was denied entry into the pub, and had wounded his own dog during the knife attack (Agronis 2002).

One of the most heinous acts against animals was by two teenagers in Largo, Florida, about 10 years ago when they decided to go on a 2-month-long animal cruelty spree (Inhumane.org). Samples of clothing, blood soaked leaves, metal rod, hockey stick, and llama blood samples were sent to the UC Davis animal forensic laboratory for DNA analysis. These evidence samples matched reference samples from three llamas. The youths had severely beaten the llamas, sodomized the nursing female llama with a broken golf club, and gouged out the eye of her Cria (baby llama), and they slashed the third llama in the face with a meat cleaver. The youths had also shot two bulls with crossbows, killed and decapitated turtles, and hurt several other livestock during their spree. Since then, one of them has been released after 3 years of probation, whereas the other is completing an 8-year jail sentence for grand larceny, burglary, and cocaine charges (Dreamin Demon; Tampa Bay Online 2010).

Animal suspects, victims, or witnesses cannot be interviewed or represent themselves or testify in court, but their biomaterial can independently and objectively link an animal suspect to an animal victim. Therefore, the analysis of animal DNA evidence is highly suited for nonhuman forensic science. Animal DNA testing that has standardized, validated, and accredited procedures facilitates greater reliance on objective scientific evidence when eyewitness accounts can be unreliable/biased (Himmelberger et al. 2008; Kanthaswamy et al. 2009; Smalling et al. 2010). Aside from providing important leads and links during an investigation, these testing confer a leverage to disprove, refute, or support alibis, leading to prompt confessions of human suspects and could deter habitual and or opportunistic human offenders from committing a crime in the first place.

Although there are many parallels between human and animal forensic DNA analysis, there are also some significant differences that impose some of the most formidable challenges on nonhuman forensic techniques. Like other fields of science, there is a need for a tactical and long-term strategy to face these challenges. Evidence recognition forms a highly pivotal step in sample processing and analysis. One of the more effective challenges to animal DNA techniques is attacking the integrity of the sample, including collection, preservation, and documentation at the crime scene, and analysis, interpretation, of the results and reporting at the forensic laboratory (Scharnhorst and Kanthaswamy 2011). The ability of an analyst to identify probative animal evidence from vast quantities of redundant, irrelevant, or unrelated items is a rare trait. Much of the evidence is not recognized as being probative or important by criminalists/forensic analysts (Scharnhorst and Kanthaswamy 2011).

Expertise and interest in government forensic laboratories in using animal DNA is still lacking. Therefore, the technical inability to obtain meaningful information about the source of animal biological samples without resorting to specialized laboratories has also contributed to why such evidence has not been utilized to its full potential in civil and criminal investigations (Scharnhorst and Kanthaswamy 2011). It is important to address issues and questions about animal forensic evidence before using this evidence in testing and in court so that this evidence can be appropriately represented in criminal cases. In almost every trial court system, nonhuman DNA evidence is not accorded the same weight as human DNA evidence and is frequently not inadmissible.

Compared to human forensic DNA analysis, animal forensic DNA analysis is still in its infancy, and since this is a relatively new field for a large fraction of the legal community, DNA results from animal evidence are more likely to experience admissibility issues. This type of situation prevails, for instance, in the United States, where the Frye and Daubert standards impart particular rules regarding the admissibility of evidence in the court of law. According to these admissibility standards, procedures involving animal evidence must gain general acceptance in the field (Cassidy and Gonzales 2005). At present, animal DNA methods and resources are not as developed as those in human forensic science. Of chief concern is the lack of an accredited and comprehensive quality assurance system (Budowle et al. 2005; Scharnhorst and Kanthaswamy 2011) for animal forensic DNA testing. Therefore, analysis of animal DNA is frequently not requested and as a result, is seldom performed in the majority of forensic laboratories. This results in a deficiency of animal forensic DNA testing experience, which in turn results in technical errors including poor interpretation of the electropherograms (van Asch et al. 2009) and other common mistakes (Scharnhorst and Kanthaswamy 2011) that can be corrected with sufficient training.

The lack of commercially available forensic DNA kits or comprehensive population genetic databases for most species that incorporate regional and breed specific information is another indication of how underdeveloped the field of animal forensic science is compared to its human counterpart. Most forensic canine and feline population databases do not contain sufficient sample sizes of outbred animals despite the fact that more than 50% of U.S. dog and approximately 98% of U.S. cat populations, respectively, are outbred animals, which is disconcerting. Comprehensive relational databases are needed for estimating the frequencies of common and rare alleles and haplotypes in a particular population and to give statistical weight or meaning to matches between genotype/haplotype from the questioned sample and a genotype/haplotype from a known reference sample. Exclusion probabilities, likelihood ratios, random match probabilities, and Bayesian methods are all examples of statistical assessments used to convey the significance of a match (Budowle et al. 2005).

Because the DNA statistical model relies heavily on the Hardy–Weinberg equilibrium, which requires random mating, the high occurrence of inbreeding and intensity of genetic subdivisions within and among domestic animal populations are major concerns especially for estimating animal profile frequencies and random match probability. Issues with inbreeding and genetic subdivision are even greater pitfalls for species that are threatened or endangered, because their numbers are much smaller (Ayres and Overall 1999; Ayres 2000; Kanthaswamy et al. 2009). Excessive inbreeding may require additional panels of markers for parentage and genetic identity analyses for endangered animal populations.

Participation in an interlaboratory proficiency testing program from an outside agency will ensure that the laboratory and its analysts are adhering to a successful quality assurance (QA)/quality control (QC) program that covers QA, QC, standard operating protocols pertaining to sample handling, storage and integrity, casework documentation, and whether the DNA analyses—including genotyping, statistical analysis, and interpretation—are consistent with scientific approaches (Scharnhorst and Kanthaswamy 2011; Budowle et al. 2005).

The sections that follow deal with Felid (cats) (Menotti-Raymond, David and O'Brien), Canids (dogs) (Kanthaswamy), Bovids (cattle) (Projić, Škaro and Lauc), and wildlife forensic DNA testing (Linacre).

12.2 Felid Forensic DNA Testing

Domesticated some 10,000 years ago (Vigne et al. 2004), *Felis sylvestris catus* has become one of the worlds' most popular household pets.

Gilbert et al. (1991) characterized the first cat multilocus minisatellite probes, which they used to assess genetic relatedness of individuals in lion prides and cheetah populations, and ultimately as a metric to infer the social dynamics of lion pride structure. Single locus minisatellite probes had not yet been well characterized in other species, other than human, before the report of simple tandem repeat (STR) markers (Tautz 1989; Weber and May 1989), which had clear advantages over minisatellites for forensic analysis.

Short in length and generally less than 100 base pairs (bp), and abundant and polymorphic in all eukaryotic genomes examined (Tautz 1989), several STRs could be amplified in a single reaction by the newly developed polymerase chain reaction (PCR), using nanogram amounts of DNA. Additionally, newly developing technologies provided a degree of sizing precision that had proved problematic with minisatellite analysis (Ziegle et al. 1992). STR markers ultimately revolutionized human forensic analysis (Fregeau and Fourney 1993).

With the routine use of STR loci applied to the genetic individualization of human samples, came the realization of the potential of samples of nonhuman origin. Genetic linkage maps incorporating STRs with coding loci were rapidly developed in companion and commercial animal species for the mapping and characterization of genes of health-related and commercial interest (Archibald et al. 1995; Barendse et al. 1994; Crawford et al. 1995; Dietrich et al. 1992; Ellegren et al. 1994; Jacob et al. 1995; Mellersh et al. 1997; Menotti-Raymond et al. 1999), thereby opening the potential of STR application to forensic analysis.

The National Cancer Institute's Laboratory of Genomic Diversity (LGD) has maintained an interest in the domestic cat as a model of human hereditary and infectious disease (O'Brien et al. 2002). Many hereditary disease pathologies reported in the cat bear homology to human hereditary disease (http://omia.angis.org.au/home/). In an effort to map and characterize feline disease-related genes, the LGD has developed genetic linkage and radiation hybrid maps, which incorporated genes and STR loci (O'Brien et al. 2002; Menotti-Raymond et al. 1999, 2003a, 2009; Murphy et al. 2000, 2007; Pontius et al. 2007). Domestic cat *STR* loci had additionally been used in conservation genetics applications in exotic felids (Culver et al. 2000, 2001; Eizirik et al. 2001; Johnson et al. 1999).

12.2.1 Case Studies

The first application of STR loci in genetic individualization of felid samples arose from a puma attack of a female jogger in California in 1994 (Culver, unpublished). Early on the morning of December 4, 1994, a woman hiking alone was attacked and killed by a puma in the Cuyamacha Rancho State Park near San Diego. Samples submitted by the California Department of Fish and Game from the mountain lion presumed responsible for the attack provided matching STR profiles for eight felid STR loci from DNA isolated from puma hairs found on the victim (Culver, unpublished). Culver, who was generating a phylogeny of puma subspecies, even had a database of pumas with which to calculate the match probability that another individual could have been the attacker (Culver et al. 2000).

The first application of domestic cat STR loci in genetic individualization of domestic cat samples came in response to an inquiry from the Royal Canadian Mounted Police (RCMP). The case involved a 32-year-old woman from Richmond, Prince Edward Island, Canada, who was reported missing on October 3, 1994. Within a few days, her car was found abandoned in a wooded setting within a few kilometers of her home and soon thereafter, within the same vicinity, a bag containing a man's leather jacket and tennis shoes, stained with blood determined to be that of the missing person. White hairs clinging to the jacket lining were identified by the RCMP's Halifax forensics laboratory (A.E. Evers, RCMP) to be cat hairs. These hairs were potential probative evidence in the case, as the only suspect, lived with his white cat, Snowball (Figure 12.1). The RCMP queried whether DNA fingerprint profiles could be generated and compared from DNA extracted from the hairs in the jacket and Snowball.

There had been multiple reports of DNA fingerprint profiles generated from hairs, including profiles from a human hair root (Higuchi et al. 1988) and from single hairs of free-ranging chimpanzees (Morin 1992). We were provided with 27 white hairs from the jacket lining and approximately 10 mL of blood drawn from the subpoenaed cat. Microscopic evaluation of the hairs identified one hair with a good root and possibly a small piece of attached tissue. We extracted approximately 17 ng of DNA from this hair root, and no DNA from three additional hair roots or any of the four hair shafts examined (Menotti-Raymond et al. 1997b).

Ten dinucleotide STR loci had been selected for the analysis (Menotti-Raymond et al. 1997b) based on robustness, absence of linkage with other proposed STRs, and low "stutter" (Hauge and Litt 1993). PCR products generated from DNA extracted from the one

Figure 12.1 Snowball, the suspect's pet cat. (From Menotti-Raymond, M. et al., *Nonhuman DNA Typing, Theory and Casework Applications*, ed. H.M. Coyle, CRC Press, Boca Raton, FL, 2008b. With permission from Taylor & Francis Group.)

hair root (Menotti-Raymond et al. 1997a) and subsequently, from the blood sample, were electrophoresed in the same gel.

A "match window" was established for each STR locus using guidelines developed by the RCMP to empirically determine a size difference threshold ("match window"), which would define acceptable variation in migration between any two measured alleles to conclude that they match (RCMP Biology Section Methods Guide 1996). The match window provided an empirical determination of precision for each STR locus from a sample of multiple ascertainments of identical alleles (Menotti-Raymond et al. 2005). Using match window criteria, *composite STR genotypes* amplified from hair root and blood DNA were determined to match at all 10 loci (seven heterozygous and three homozygous) STR loci (Figure 12.2) (Menotti-Raymond et al. 1997a).

A population genetic database was generated from cats in Prince Edward Island in order to calculate the likelihood of a random match between the hair genotype and a random individual in the population (Menotti-Raymond et al. 1997a). Additionally, a second database was generated from outbred cats from the Eastern United States (Menotti-Raymond et al. 1997a). Although the PEI database was small, the island sample was adequate (95% confidence interval) to detect any STR allele present at a frequency of 9.5% or higher (Menotti-Raymond et al. 1997a). The two populations showed appreciable allelic variation and remarkable population genetic similarity (as opposed to geographic population substructuring). The incidence of the composite hair genotype for the seven heterozygous loci, estimated using the product rule (Jones 1972) and minimum allele frequency estimates for rare alleles (Budowle et al. 1996), was 2.2×10^{-8} and 6.9×10^{-7} for the Prince Edward Island and the U.S. population databases, respectively (Menotti-Raymond et al. 1997a).

The cat evidence and additional human DNA evidence was presented to the Supreme Court of Prince Edward Island. On July 19, 1996, the jury convicted the defendant of second-degree murder.

Comparison of Snowball and Jacket Cat Hair STR Genotypes

STR Locus	DNA Source	Allele	Allele Size (bin)	Size Difference (bp) [Snowball Blood - Hair]	STR Locus Match Window (bp)		Conclusion
FCA 026	Snowball	1	147.83 (F)				
	Jacket cat hair	1	148.11 (F)	0.28	<	0.37	MATCH
	Snowball	2	143.73 (D)				
	Jacket cat hair	2	143.73 (D)	0	<	0.37	MATCH
FCA 043	Snowball	1	126.38 (C)				
	Jacket cat hair	1	126.29 (C)	0.09	<	0.59	MATCH
	Snowball	2	120.52 (B)				
	Jacket cat hair	2	120.50 (B)	0.02	<	0.59	MATCH
FCA 080	Snowball	1	259.27 (F)				
	Jacket cat hair	1	259.20 (F)	0.07	<	0.30	MATCH
	Snowball	2	253.39 (C)				
	Jacket cat hair	2	253.14 (C)	0.25	<	0.30	MATCH
FCA 088A	Snowball	1	121.91 (F)				
	Jacket cat hair	1	121.91 (F)	0	<	0.42	MATCH
	Snowball	2	110.50 (A)				
	Jacket cat hair	2	110.50 (A)	0	<	0.42	MATCH
FCA 126	Snowball	1	143.41 (D)				
	Jacket cat hair	1	143.41 (D)	0	<	0.53	MATCH
	Snowball	2	141.08 (C)				
	Jacket cat hair	2	141.08 (C)	0	<	0.53	MATCH
FCA 132	Snowball	1	152.73 (G)				
	Jacket cat hair	1	152.73 (G)	0	<	0.27	MATCH
	Snowball	2	150.69 (F)				
	Jacket cat hair	2	150.69 (F)	0	<	0.27	MATCH
FCA 149	Snowball	1	132.02 (E)				
	Jacket cat hair	1	131.80 (E)	0.22	<	0.29	MATCH
	Snowball	2	128.05 (C)				
	Jacket cat hair	2	128.07 (C)	0.02	<	0.29	MATCH
FCA 058	Snowball	1	229.03 (D)				
	Jacket cat hair	1	229.24 (D)	0.21	<	0.36	MATCH
FCA 090	Snowball	1	93.54 (A)				
	Jacket cat hair	1	93.54 (A)	0	<	0.46	MATCH
FCA 096	Snowball	1	210.95 (D)				
	Jacket cat hair	1	211.00 (D)	0.05	<	0.25	MATCH

Figure 12.2 Comparison of Snowball and Jacket Cat Hair STR Genotypes. Graphics presented to the Prince Edward Island jury presenting matching composite STR profiles generated from DNA extracted from an evidentiary cat hair and the blood of Snowball. (From Menotti-Raymond, M. et al., *Nonhuman DNA Typing, Theory and Casework Applications*, ed. H.M. Coyle, CRC Press, Boca Raton, FL, 2008b. With permission from Taylor & Francis Group.)

12.2.2 Development of a Forensic Typing System for Genetic Individualization of Domestic Cat Samples

This legal precedent introducing animal genetic individualization in a homicide trial stimulated interest in the forensic community for the potential use of animal DNA forensic analysis. Under support from the National Institute of Justice (NIJ), we have developed a peer-reviewed forensic typing system for genetic individualization of domestic cat samples that includes (1) a panel of 10 tetranucleotide loci, mapped in the cat genome, (2) a multiplex amplification protocol for the 10 loci and a gender identifying sequence tagged site (STS), and (3) a population genetic database that has been generated from a sample set of 1043 individuals representing 38 cat breeds and a small subset of outbred domestic cats. The panel has additionally undergone validation tests, as recommended by the DNA Advisory Board for Nonhuman DNA Testing (Coomber et al. 2007).

Forensic Animal DNA Analysis

In order to generate a forensic typing panel, 49 tri- and tetranucleotide STR loci were isolated from felid STR-enriched genomic libraries (Menotti-Raymond et al. 2005). Tetranucleotide STRs have been used for human profiling because they minimize the generation of "stutter band" products generated during PCR amplification (Hauge and Litt 1993), which can complicate the interpretation of genotypes from mixed DNA samples. The loci were incorporated into genetic maps of the domestic cat relative to 579 coding genes and 255 STR loci (Menotti-Raymond et al. 2003b) in order to select markers that were unlinked. Subsequently, the loci were screened in a small panel of outbred domestic cats and 28 cat breeds (3–10 animals/breed, $n = 213$), to identify a panel of markers with the highest discriminating power (Menotti-Raymond et al. 2005). A set of 11 highly polymorphic loci was initially selected for the typing panel. The loci were unlinked, demonstrated high heterozygosity across multiple cat breeds, and showed an absence of cross-species amplification (Menotti-Raymond et al. 2005).

A multiplex amplification protocol was developed so that the loci could be coamplified with as little as a nanogram of DNA, which included a gender identifying STS on the Y-chromosome, amplifying a fragment of the *SRY* gene (Menotti-Raymond et al. 2005). The PCR products of the 11 loci were designed in a size range from 100 to 415 bp, labeled with one of four fluorescent tags, with no allele overlap with adjacent loci and the SRY product detectable at 96 bp (Menotti-Raymond et al. 2005) (Figure 12.3). Validation studies of the multiplex demonstrated that complete product profiles could be generated with as little as 125 pg of genomic DNA, with an absence of "allele dropout" (Coomber et al. 2007).

A population genetic database of domestic cat breeds has been generated from the multiplex with which to compute composite match probabilities (Menotti-Raymond et al.

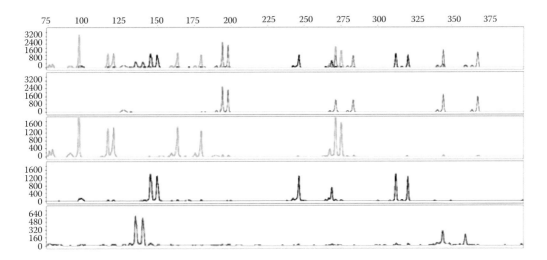

Figure 12.3 The PCR products of the 11 loci were designed in a size range from 100 to 415 bp, labeled with one of four fluorescent tags, with no allele overlap with adjacent loci and the SRY product detectable at 96 bp. Electropherogram of PCR products of 12-member multiplex amplified from 4 ng of male genomic DNA (upper panel). Lower panels demonstrate PCR products labeled with fluorescent tags FAM (3 STR), VIC (3 STR, SRY gene), NED (3 STR) and PET (2 STR). Note that one locus (FCA736) of the original proposed multiplex has since been dropped due to a high incidence of null alleles observed in cat breeds of Asian ancestry (53). (From Menotti-Raymond, M. et al., *Nonhuman DNA Typing, Theory and Casework Applications*, ed. H.M. Coyle, CRC Press, Boca Raton, FL, 2008b. With permission from Taylor & Francis Group.)

Table 12.1 Heterozygosities of Cat STR Multiplex Observed in Cat Breeds

Cat Breed	n	Origin[a]	Date(s) of Origin[a]	H_e STR
Abyssinian	29	Ethiopia	1860s	0.62
American Curl	9	Domestic mutation introduced into ASH	1980s	0.82
American Shorthair (ASH)	26	Domestic population—U.S.A.	1900	0.77
American Wirehair	9	Domestic mutation introduced into ASH	1966	0.75
Balinese	11	U.S.A.	1940s[b]	0.58
Bengal	13	U.S.A.	1963	0.80
Birman	43	Burma, with outcrosses	1930s[b]	0.60
Bombay	21	ASH and Burmese cross	1958	0.72
British Shorthair (BSH)	13	U.K.	1870s[b]	0.66
Burmese	50	Thailand	1350–1767	0.69
Chartreux	21	France, with outcrosses to BSH	14th century	0.74
Colorpoint Shorthair	14	U.S.A., U.K.	1947	0.70
Cornish Rex	41	Mutation in U.K. domestic population with outcrosses to SIA and other breeds	1950	0.71
Devon Rex	57	Mutation in U.K. domestic population with early outcrosses to ASH and other breeds	1960	0.66
Egyptian Mau	21	Egypt	early[c]	0.70
Exotic	18	U.S.A.	1966	0.75
Havana	49	Originated from crosses between OSH and other domestics	1951	0.55
Himalayan	19	U.S.A./U.K.	1950s/1920s	0.71
Japanese Bobtail	16	Japan	5th–10th century	0.76
Javanese	13	U.K./U.S.A.	1960s	0.68
Korat	11	Thailand	1350–1767	0.55
Maine Coon Cat	43	New England, U.S.A.	1860s	0.79

Breed	N	Origin	Date	H
Manx	29	U.K., Isle of Man	early	0.84
Norwegian Forest Cat	67	Norway	early	0.84
Ocicat	19	Cross between ABY and SIA	1964	0.65
Oriental Shorthair	33	U.K.	1950s	0.74
Persian	51	Iran	early	0.77
Ragdoll	43	U.S. domestic population with crosses to other breeds	1960s	0.76
Russian Blue	23	Russia	late 1800s	0.71
Scottish Fold	41	Mutation in British domestic population with crosses to BSH, ASH, and PER	1961	0.81
Selkirk Rex	28	Mutation in U.S. domestic population with crosses to PER, ASH, and BSH	1987	0.77
Siamese	34	Thailand	1350–1767	0.64
Singapura	14	Small founding population of cats of SE Asian origin	1971	0.57
Somali	24	U.S.A./Canada	1967	0.66
Sphynx	26	Domestic mutations with crosses to DRE and ASH	1966	0.78
Tonkinese	19	Cross between BUR and SIA	1950s	0.71
Turkish Angora	17	Turkey	early	0.81
Turkish Van	28	Turkey	early	0.76
Average				0.71
Complete breed set				0.87
Outbred cats				0.85

Note: ABY, Abyssinian; ASH, American Shorthair; BAL, Balinese; BSH, British Shorthair; BUR, Burmese; CSH, Colorpoint Shorthair; DRE, Devon Rex; EXO, Exotic; OSH, Oriental Shorthair; PER, Persian; SIA, Siamese; U.K., United Kingdom; ND, Not determined.

[a] From Vella and Robinson 1999 (53).
[b] Controversy over origin, some sources suggest an ancient origin.
[c] Early: prior to 1800.

2012) using a sample set of 1043 samples representing 38 cat breeds (Vella et al. 2003). A small sample set of outbred domestic cats ($n = 24$) was additionally genotyped to determine how well the multiplex would perform in the outbred cats, and demonstrated high heterozygosities for the multiplex (Menotti-Raymond et al. 2012).

Heterozygosities for the individual loci using all cats of recognized breed as a single dataset are quite high, ranging from 0.72 to 0.96, and are not significantly different than heterozygosities obtained for the outbred cat sample set (Table 12.1) (Menotti-Raymond et al. 2012). One locus, FCA736, was dropped from the multiplex because of an unusually high incidence of null alleles observed in breeds of Asian ancestry. We determined that the null allele phenomenon was due to a 12-bp deletion in one of the new primer binding sites for FCA736, which was designed to accommodate the multiplex, and was impossible to redesign because of size constraints with adjacent loci. The multiplex exhibits good potential for genetic individualization of domestic cat samples within cat breeds and in outbred domestic cats, with a probability of match (P_m) of 6.2×10^{-14}, using a conservative $\Theta = 0.05$ (Menotti-Raymond et al. 2012). The individual feline markers appear to be more discriminating than the CODIS markers used in human forensic applications (Menotti-Raymond et al. 2012).

12.2.3 Validation Studies of Cat Multiplex

We have followed quality assurance standards for DNA analysis advised by the DNA Advisory Board (DAB 1998) and recommendations for animal DNA and forensic testing (Budowle et al. 2005) before analysis of evidentiary samples, to ensure the accuracy, precision, and reproducibility of the system. We examined species specificity and reproducibility of the system in DNA extracted from multiple tissue types in addition to the ability of the typing system to identify DNA mixtures, and the mutation rate of the loci (Coomber et al. 2007).

Species specificity of the loci was examined by amplifying DNA extracted from a number of North American mammals, including badger, beaver, brown bear, chipmunk, cow, coyote, deer, dog, ferret, fox, goat, guinea pig, hamster, horse, human, mink, mole, mouse, ocelot, opossum, otter, pig, puma, rabbit, raccoon, sheep, skunk, and wolf, as well as two prokaryotes, *Sacchromyces cervesiae* and *Escherichia coli*. A high degree of specificity was observed, with PCR products of expected size obtained in members of the felid family (ocelot, puma, domestic cat) (Coomber et al. 2007). Amplification products were observed for two loci in the brown bear, another member of the Carnivore order. However, under the amplification conditions used in the multiplex, no other products were observed in any other mammalian species or in either of the prokaryotes.

Reproducibility of the typing system was examined by amplifying the multiplex in DNA extracted from blood and buccal samples from 13 unrelated domestic cats. Identical profiles were obtained from the two tissue types for each individual (Coomber et al. 2007).

Only one allele of a heterozygous individual may amplify if the quantity of DNA template is extremely low (Walsh et al. 1992). By amplifying the loci in a dilution series of domestic cat genomic DNA, we determined that 0.125 ng of DNA was required in the two sample DNAs to detect (with a minimum of threshold of 50 relative fluorescent units) both alleles of heterozygous loci. These results are similar to sensitivity studies observed with the CODIS loci, which require a minimal amount of 0.2 ng, and optimal amount of 1–2.5 ng of sample DNA (Micka et al. 1999).

The most common felid specimens that are likely to be associated with crime scenes are hair samples. The critical factor associated with STR testing of felid hair will be the quantity of genomic DNA available for analysis. Cat hair roots have proven to be roots a poor source of genomic DNA. We have obtained 30 ng of DNA from the very best fresh plucked guard hair roots, (Menotti-Raymond et al. 2000). This is on an order of 10 to 30 times less DNA than is available from a human hair root. Furthermore, the majority of cat hairs likely to be associated with cases are likely an even poorer source of DNA, if aged, shed, or originating from the undercoat (Menotti-Raymond et al. 2000).

We have designed a method to estimate feline genomic DNA yield through a quantitative, or "real-time" PCR-based assay in order to determine the likelihood of STR genotyping success without compromising product available for alternate analyses (i.e., mitochondrial DNA [mtDNA]) (Menotti-Raymond et al. 2003c). Quantitative PCR assays measure the amount of PCR product at the completion of each cycle. Thus, by comparing the PCR product profile generated from a DNA source of unknown quantity with the product profiles of a DNA dilution standard, it is possible to estimate the amount of DNA in the unknown (Menotti-Raymond et al. 2003c; Morrison et al. 1998). The assay targets small interspersed nuclear elements (SINEs), a highly abundant repetitive element composing approximately 10% of the cat nuclear genome (Pontius et al. 2007; Yuhki et al. 2003). SINEs are abundant in other mammalian genomes (Lander et al. 2001; Walker et al. 2002; Waterston et al. 2002) and have been used to quantify human genomic DNA (Mandrekar et al. 2001). The felid SINE assay is highly sensitive and can be performed rapidly using trace amounts of DNA and detects feline genomic DNA at a concentration of 10 femtograms (fg) in a 20-μL reaction, following 30 cycles of amplification (Menotti-Raymond et al. 2003c). The assay is not felid-specific, as primer pairs were designed in regions that demonstrate a high degree of sequence conservation across species (Menotti-Raymond et al. 2003c). However, we do not think that the lack of species specificity deters from the utility of the assay, as DNA mixtures are unlikely to be an issue with feline samples.

Samples, which do not yield sufficient nuclear DNA for felid STR testing, are potential candidates for mtDNA forensic analysis. mtDNA has become a valuable resource for forensic analysis of DNA samples that are either highly degraded or of insufficient quantity for STR profiling (Budowle et al. 1990; Holland and Parsons 1999). The cat mtDNA genome has been sequenced by Lopez et al. (1996), which demonstrated that the organization and gene content of the cat mtDNA genome largely resembles that of other placental mammals (Lopez et al. 1996). However, it is critical in characterizing polymorphisms in cat mtDNA that could be utilized for forensic analysis to avoid selecting nucleotides within the 50% of the mitochondrial genome that has been transposed into the nuclear genome. Translocated pieces of mtDNA into the nuclear genome have been reported in more than 64 different animal species (Bensasson et al. 2001). The cat transposed region, or "Numt," extends from the 3' end of the control region (CR) through 80% of the *COII* gene, and exhibits high sequence similarity with the homologous cytoplasmic mtDNA region, demonstrating 5.1% sequence divergence (Lopez et al. 1994).

Polymorphisms within the CR of human mtDNA have generally been used as a source of variation for forensic analysis (Gill et al. 1994; Holland et al. 1993; Holland and Parsons 1999). Fridez first characterized the discriminating power of the mitochondrial CR in outbred domestic cats using 21 polymorphic sites to identify 14 haplotypes in a survey of 50 cats (Fridez et al. 1999). Halverson and Basten (2005a) have since examined polymorphism within an 80-bp region of the CR with harbors a tandem repeat (Halverson and

Baster 2005a). Recently, an mtDNA typing system has been proposed in the domestic cat (Tarditi et al. 2011). A 402-bp region of the CR was first examined for informativeness in 174 random-bred cats representing four geographic regions in the United States. Thirty-two mtDNA haplotypes were observed, ranging in frequency from 0.6% to 27% (Tarditi et al. 2011). Grahn et al. (2011) have since reported on a population genetic database generated from 1394 cats, which includes individuals from 25 worldwide geographic locations and 26 cat breeds, for the 402-bp region. Twelve major haplotypes were observed and a random match probability of 11.8%.

A growing area of interest in human forensic analysis is the potential of forensic DNA phenotyping, which focuses on the prediction of the physical appearance of an individual from a DNA sample (Kayser and de Knijff 2011). Such phenotypes will be inferred largely by identifying SNPs/deletions/insertions that have been characterized as causative of distinctive morphological characteristics, such as eye or skin color or height. This is one area where animal genetics is perhaps advanced relative to human genetics! Multiple phenotypes in the domestic cat have been characterized on a molecular genetics level associated with coat color, pattern, hair length, and consistency. Mutations have been characterized in the domestic cat responsible for a range of coat color phenotypes, including melanism (black) (Eizirik et al. 2003), variants of the brown locus (brown and cinnamon) (Schmidt-Küntzel et al. 2005), dilute (Ishida et al. 2006), orange (unpublished), white (unpublished), as well as hair length (Drogemüller et al. 2007; Kehler et al. 2007), tabby pattern (Eizirik et al. 2010; Kaelin et al. 2012), hair consistency (Gandolfi et al. 2010), and three mutations at the albino locus responsible for progressive lack of melanin pigment deposition, generating albino (Imes et al. 2006), Siamese, and Burmese phenotypes (Lyons et al. 2005; Schmidt-Küntzel et al. 2005). Thus, it would be possible from a non-hair sample (i.e., blood, saliva, fecal or bone) to accurately predict the appearance of a cat, and even the likelihood of whether it is representative of a particular breed (Menotti-Raymond et al. 2008a). Mutations for all coat phenotypes can easily be characterized with site-directed PCR and sequence analysis (see referenced publications).

Animal forensic analysis is still in its infancy relative to human genetics. However, the genomic tools for STR profiling or mtDNA analysis have now been developed and peer reviewed, and population genetic databases have been generated with which to generate match probabilities. Biological evidence from animal specimens has the potential to play an important role in future investigations. What will facilitate the application of these genomic profiling systems is the development of commercial kits with the necessary QC-produced reagents, validation studies, and allelic ladders for the STR kit. An *ad hoc* survey of detectives across the country by the NIJ's Office of Science and Technology revealed a strong interest in using such evidence if it were available.

12.3 Canine Forensic DNA Testing

Unlike cats, which can be aggressive but rarely inflict significant injury, dogs can be perpetrators that produce significant injury. The occurrence of dog bites at the rate of 3.5 and 4.7 million dog bite injuries per year resulting in 238 mauling deaths annually and hundreds of millions of dollars in insurance claims (USA Today.com). Among the most vicious dog breeds, Rottweilers, American Pit Bulls, and their hybrids have been characterized as "dangerous" breeds by the U.S. Centers for Disease Control because they lead the statistics in dog bite–related fatalities (Kanthaswamy et al. 2009). The passage of "Cody's Law" in California to allow prosecutors to charge dog owners with a misdemeanor or a felony if their dogs attacked people was prompted by a vicious dog attack (Dog Bite Law). In this incident, a pack of 21 pit bull–like dogs attacked an 11-year-old California boy named Cody Fox. Analysis of canine hairs from the victim's clothing helped with the unambiguous identification of the two biting dogs. Both dogs were euthanized but since the case preceded "Cody's Law" (Governor Gray Davis of California signed the legislation in August 1999), the dog's owner was only found guilty of misdemeanor for training dogs to attack and causing injury despite the intensity of the attack. Besides attacking people, usually children, animals tend to attack other animals. In one case investigation in which the author was involved, the killing of a cat by neighbor's dog resulted in the victim's owner paying about $500 to prove that the DNA from canine hairs found in the cat's mouth came from the suspected dog. As a result, the owners of the dog had to spend $1000 on a dog fence to prevent further cat kills (Tresniowski et al. 2006).

Animal forensic science is still a much younger science relative to human forensics. Budowle et al. (2005) proposed a set of recommendations for improving the quality standards of this field. In accordance to these guidelines to ease the exchange of information between laboratories, Kanthaswamy et al. (2009) and Tom et al. (2010) developed a DNA identification reagent kit consisting of standardized genetic markers, a common nomenclature and a publicly accessible database for canine forensic DNA analysis. The reagent kit, which is now commercially available along with the published nomenclature system and the public database have already promoted collaboration among different laboratories in a murder investigation (Canine Genotypes™ 1.1, Finnzyme Diagnostics; Thermo Scientific, Vantaa, Finland). Similarly, to improve nonhuman forensic genetic testing and promote the use of this forensic evidence in civil and criminal investigations, genetic identification resources akin to the ones developed by Kanthaswamy et al. (2009) and Tom et al. (2010) have to be developed for other species as well. The loci in Canine Genotypes™ are presented in Table 12.2.

Recent court challenges to canine DNA analysis have demonstrated a need for animal forensic DNA testing that has been validated according to human forensic guidelines. The potential for widespread acceptance of animal DNA evidence was somewhat diminished by the September 2003 decision by the Washington Court of Appeals that excluded canine DNA evidence in the 1998 case of the *State of Washington v. Kenneth Leuluaialii*. The appellate court ruled that the trial court judge had erred by failing to establish "whether the scientific community generally accepted that the specific loci used by PE Zoogen in the present case were highly polymorphic and appropriate for forensic use." The court also stated that "current canine DNA testing and mapping focuses on the goals of paternity testing, breed testing and cancer and research studies. There is little indication that polymorphic loci and alleles in canine DNA have been sufficiently studied such that probability estimates are appropriate for the forensic use applied in this case" (Docket Number 43507-8-I).

Two additional examples of the use of animal forensic DNA typing in casework are described below. The DNA analysis for Case A was almost not performed because of the numerous media releases on the dismissal of canine DNA evidence in a 2003 appeals court hearing in the State of Washington. The author had to convince the prosecutor in Case A to allow the laboratory to continue with the DNA testing because the controversy with the DNA results in Washington trial did not impact his case. The second case investigation

Table 12.2 Locus Descriptions for Canine Genotypes™ Panel 1.1 Kit Microsatellites and Amelogenin Marker

Locus Name	Chromosome	Repeat Motif	Size Range (bp)[a]	Dye Color[b]
AHTk211	26	di	79–101	Blue
CXX279	22	di	109–133	Blue
REN169O18	29	di	150–170	Blue
INU055	10	di	190–216	Blue
REN54P11	18	di	222–244	Blue
INRA21	21	di	87–111	Green
AHT137	11	di	126–156	Green
REN169D01	14	di	199–221	Green
AHTh260	16	di	230–254	Green
AHTk253	23	di	277–297	Green
INU005	33	di	102–136	Black
INU030	12	di	139–157	Black
Amelogenin	X	–	174–218	Black
FH2848	2	di	222–244	Black
AHT121	13	di	68–118	Red
FH2054	12	tetra	135–179	Red
REN162C04	7	di	192–212	Red
AHTh171	6	di	215–239	Red
REN247M23	15	di	258–282	Red

Note: Finnzymes Canine Genotypes™ Panel 1.1 Manual; http://diagnostics.finnzymes.fi/pdf/canine_genotypes_manual_1_1_low.pdf.

[a] Size ranges are based on information provided by ISAG and data generated by Finnzymes Diagnostics. The data represents a large selection of dog breeds. However, some breeds may have alleles outside the ranges provided.

[b] Dye colors are listed as they appear following electrophoresis with Filter Set G5.

was successfully performed because most of the issues raised by the court ruling had been put to rest by developments in the animal forensic DNA analysis community including the study by Halverson and Basten (2005b).

12.3.1 Case A Details—Fatal Dog Attack

In 2003, Vivian Anthony, a mother of three and a grandmother of 17 from Columbus, Ohio was severely mauled by two large Rottweilers. She suffered massive trauma and blood loss and was on life support for 2 days before she died when her heart, lungs, and kidneys failed. A former physician was charged with manslaughter and/or murder because his dogs were implicated in the brutal attack. He denied it was his dogs, but DNA from dog saliva on victim's clothing (see Figure 12.4) matched his dogs. Faced with this irrefutable evidence the dogs' owner pled guilty and was incarcerated for 6 months for one count of reckless homicide and one count of assault involuntary manslaughter. Both his dogs were subsequently destroyed. This was the first case in Ohio where dog DNA was used in a trial.

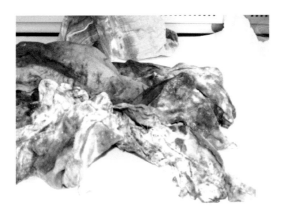

Figure 12.4 DNA from dog saliva on victim's clothing. Victim's clothing from the fatal dog attack case. (File photo courtesy of VGL.)

12.3.2 Case B Details—Homicide

In 2009, two southwest London youths were attacked by rival gang members in a park. First, the boys were chased down and attacked with dogs, and then the dogs' owners kicked and punched the victims, and stabbed them with knives. One youth was stabbed 16 times and died, whereas another survived nine stab wounds, also suffered from knife wounds. There were no eyewitnesses. Available CCTV footage, which recorded the attack, could not help identify the assailants with certainty because they had on hooded sweatshirts. In this case, the canine blood and saliva contributed to the physical evidence used by the British Crown Court to convict the defendants. The genetic identity testing kit (Finnzyme Canine Genotypes™ 1.1 Multiplex STR Reagent kit; see Kanthaswamy et al. 2009) used to analyze the canine evidence was developed by the author and his colleagues at the Department of Anthropology, UC Davis, and the population genetic database used by the Crown's experts from LGC Forensics, U.K., composed of a sample set of U.K. dogs that represented 14 distinct dog breeds (R. Ogden, personal communication). An animal forensic expert from Questgen Forensics in Davis, California, and the author reviewed the results of LGC Forensics' DNA analysis and verified the veracity of the DNA analysis (J. Halverson, personal communication).

As a consequence of the Washington ruling, animal DNA evidence (canine in particular) was excluded from pending court cases in California and threatened the admissibility of this type of evidence in other states in the United States, including the previously stated case from Ohio (Case A). This ruling clearly underscored the need for validated nonhuman DNA analysis procedures that meets the rules of scientific acceptance and reliability. Recent developments of more informative genetic markers have demonstrated the polymorphic variability in nonhuman DNA analysis, leading to the same confidence of high discrimination as its human counterpart (Butler et al. 2010; Kanthaswamy et al. 2009). The second case example presented above used a canine forensic genetic testing kit, that is, the Finnzyme Canine Genotypes™ 1.1 Multiplex STR Reagent kit (Figure 12.5), which was developed and validated using a panel of canine-specific STRs using

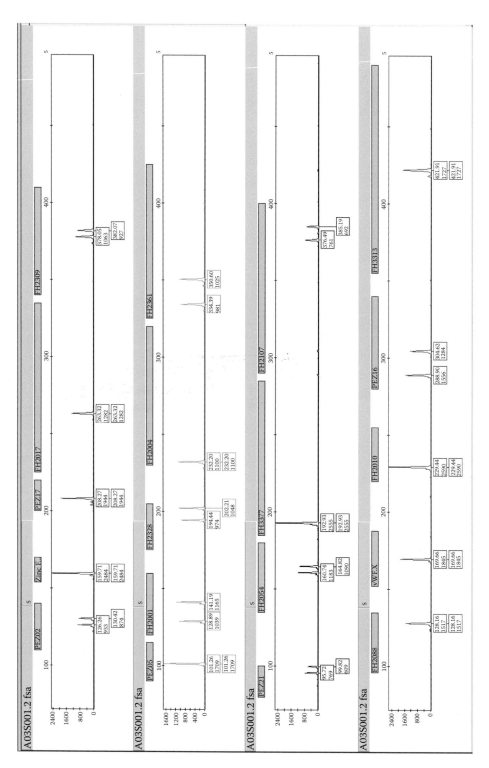

Figure 12.5 Canine forensic genetic testing kit, i.e., the Finnzyme Canine Genotypes™ 1.1 Multiplex STR Reagent kit. GeneMapper electropherograms of amplified positive control dog (F-863) by the Finnzymes Canine 1.1 Multiplex STR Reagent kit and the allelic distribution at each of the 19 loci. (From Kanthaswamy, S. et al., *J Forensic Sci*, 54, 829–840, 2009. With permission.)

funding from the U.S. NIJ. Therefore, current nonhuman genetic testing techniques have now begun to address issues raised by the courts and are at the verge of adding another powerful tool of DNA typing methods to the forensic science arsenal.

12.4 Bovine Forensic DNA Testing

The recently sequenced bovine (*Bos taurus*) genome suggests a continued genetic exchange between wild cattle populations during their coexistence over a wide geographical range (Adelson 2008). At present, since the domestication event of cattle 8000 years ago, more than 50 distinct breeds are recognized. As a consequence of the close integration of cattle into the food chain of humans, forensically relevant cases involving cattle, such as identity forgery or cattle theft, are relatively common. Some of these examples have been published recently, demonstrating the relevance of this issue in forensic casework (van de Goor and van Haeringen 2007).

Microsatellites, also known as simple-sequence repeats (SSRs) or short-tandem repeats (STRs), represent specific sequences of DNA consisting of tandemly repeated units of one to six nucleotides, present as mono-, di-, tri-, tetra-, penta-, and hexanucleotide repeats, respectively. These sequences are abundant in prokaryotic and eukaryotic genomes, occurring in both coding and noncoding regions. Compared to other molecular markers, SSRs are uniquely characterized by their simplicity, abundance, ubiquity, variation, codominance, and multialleles among genome (Powell et al. 1996). So, they have become a common tool broadly used in aspects of genetic mapping, molecular evolution and systematic taxonomy in most genomes.

The conventional methods of generating microsatellite markers from genomic libraries (Weising et al. 2005) are being replaced rapidly by *in silico* mining of microsatellite sequences from DNA sequence databases (Jayashree et al. 2006; Korpelainen et al. 2007). Several search tools are available for mining microsatellite repeats in assembled genome sequences: BLASTN (Basic Local Alignment Search Tool; Temnykh et al. 2001), MISA (MIcroSAtellite; Thiel et al. 2003), Tandem-Repeats Finder (Benson 1999), TROLL (tandem repeat occurrence locator; Castelo et al. 2002), etc.

This approach has been used to identify microsatellite markers and investigate the type and distribution of repeat motifs in the expressed sequence tags of cattle. UniGene sequences of cattle were systematically searched for microsatellite markers using the "SSRFinder" Perl program. The dinucleotide repeat motifs were the most abundant SSRs in cattle, accounting for 54%, followed by 22%, 13%, 7%, and 4% for tri-, hexa-, penta-, and tetranucleotide repeats, respectively. Depending on the length of the repeat unit, the length of microsatellites varied from 14 to 86 bp. Among the di- and trinucleotide repeats, AC/TG (57%) and AGC (12%) were the most abundant type (Yan et al. 2008). These results indicated that the abundance of the different repeat motifs in the SSRs, as detected in UniGene sequences, was variable and thus unevenly distributed. These results are in agreement with previous reports on several animal species (Gupta et al. 2002; Ju et al. 2005; Perez et al. 2005; Rohrer et al. 2002; Serapion et al. 2004), which showed that the abundance of different repeats varied extensively depending on the species examined (Toth et al. 2000).

Table 12.3 30 Most Common Cattle STRs Recommended by International Society for Animal Genetics (ISAG)

No	Markers	Chr	Primer Sequences (5′–3′) F & R		Reference
1	ETH225(D9S1)	9	GATCACCTTGCCACTATTTCCT	ACATGACAGCCAGCTGCTACT	Steffen et al. (1993)
2	ETH152(D5S1)	5	TACTCGTAGGGCAGGCTGCCTG	GAGACCTCAGGGTTGGTGATCAG	Steffen et al. (1993)
3	HEL1(D15S10)	15	CAACAGCTATTTAACAAGAA	AGGCTACAGTCCATGGGATT	Kaukinen and Varvio (1993)
4	ILSTS005(D10S25)	10	GGAAGCAATGAAATCTATAGCC	TGTTCTGTGAGTTTGTAAGC	Brezinsky et al. (1993a)
5	HEL51(D21S15)	21	GCAGGATCACTTGTTAGGGA	AGACGTTAGTGTACATTAAC	Kaukinen and Varvio (1993)
6	INRA0052(D12S4)	12	CAATCTGCATGAAGTATAAATAT	CTTCAGGCATACCCTACACC	Vaiman et al. (1992)
7	INRA035(D16S11)	16	ATCCTTTGCAGCCTCCACATTG	TTGTGCTTTATGACACTATCCG	Vaiman et al. (1994)
8	INRA063(D1855)	18	ATTTGCACAAGCTAAATCTAACC	AAACCACAGAAATGCTTGGAAG	Vaiman et al. (1994)
9	MM8(D2S29)	2	CCCAAGGACAGAAAAGACT	CTCAAGATAAGACCACAC	Mommens et al. (1994)
10	MM12(D9S20)	9	CAAGACAGGTGTTTCAATCT	ATCGACTCTGGGGATGATGT	Mommens et al. (1994)
11	HEL9(D8S4)	8	CCCATTCAGTCTTCAGAGGT	CACATCCATGTTCTCACCAC	Kaukinen and Varvio (1993)
12	CSRM60(D10S5)	10	AAGATGTGATCCAAGAGAGAGGCA	AGGACCAGATCGTGAAAGGCATAG	Moore et al. (1994)
13	CSSM663(D14S31)	14	ACACAAATCCTTTCTGCCAGCTGA	AATTTAATGCACTGAGGAGCTTGG	Barendse et al. (1994)
14	ETH185(D17S1)	17	TGCATGGACAGAGCAGCCTGGC	GCACCCCAACGAAAGCTCCCAG	Steffen et al. (1993)
15	HAUT24(D22S26)	22	CTCTCTGCCTTTGTCCCTGT	AATACACTTTAGGAGAAAAATA	Harlizius (comm.pers.)
16	HAUT27(D26S21)	26	TTTTATGTTCATTTTTTGACTGG	AACTGCTGAAATCTCCATCTTA	Harlizius (comm.pers.)
17	ETH3(D19S2)	19	GAACCTGCCTCTCCTGCATTGG	ACTCTGCCTGTGGCCAAGTAGG	Solinas Toldo et al. (1993)
18	ETH104(D5S3)	5	GTTCAGGACTGGCCCTGCTAACA	CCTCCAGCCCACTTTCTCTTCTC	Solinas Toldo et al. (1993)
19	INRA0325(D11S9)	11	AAACTGTATTCTCTAATAGCAC	GCAAGACATATCTCCATTCCTT	Vaiman et al. (1994)
20	INRA023(D3S10)	3	GAGTAGAGCTACAAGATAAACTTC	TAACTACAGGGTGTTAGATGAACTCA	Vaiman et al. (1994)
21	BM2113(D2S26)	2	GCTGCCTTCTACCAAATACCC	CTTAGACAACAGGGGTTTGG	Bishop et al. (1994)
22	BM1818(D23S21)	23	AGCTGGGAATATAACCAAAGG	AGTGCTTTCAAGGTCCATGC	Bishop et al. (1994)
23	BM1824(D1S34)	1	GAGCAAGGTGTTTTTCCAATC	CATTCTCCAACTGCTTCCTTG	Bishop et al. (1994)
24	HEL135(D11S15)	11	TAAGGACTTGAGATAAGGAG	CCATCTACCTCCCATCTTAAC	Kaukinen and Varvio (1993)
25	ILSTS006(D7S8)	7	TGTCTGTATTTCTGCTGTGG	ACACGGAAGGCGATCTAAACG	Brezinsky et al. (1993b)
26	ILSTS030(D2S44)	2	CTGCAGTTCTGCATATGTGG	CTTAGACAACAGGGGTTTGG	Kemp et al. (1995)
27	ILSTS0344(D5S54)	5	AAGGGTCTAAGTCCACTGGC	GACCTGGTTTAGCAGAGAGC	Kemp et al. (1995)
28	ILSTS0332(D12S31)	12	TATTTAGAGTGGCTCAGTGCC	ATGCAGACAGTTTTAGAGGG	Kemp et al. (1995)
29	ILSTS0113(D14S16)	14	GCTTGCTACATGGAAAGTGC	CTAAAATGCAGAGCCCTACC	Brezinsky et al. (1993c)
30	ILSTS0541(D21S44)	21	GAGGATCTTGATTTTGATGTCC	AGGGCCACTATGGTACTTCC	Kemp et al. (1995)

In 2006, the International Society for Animal Genetics (ISAG) recommended nine microsatellite loci (TGLA227, BM2113, ETH10, SPS115, TGLA126, TGLA122, INRA023, ETH225, BM1824) as the International Panel of Microsatellites for Cattle Parentage Testing (ISAG Panel) with a suggestion that other markers should be added to this panel to increase the efficacy in parentage testing (ISAG 2006). Recently, additional three markers (BM1818, ETH3, and TGLA53) were suggested as candidate loci in cattle parentage analysis (ISAG 2008). At the moment, there are two firms, Finnzymes and Applied Biosystems, offering reagent kits for meat traceability and cattle parentage testing. Finnzymes, as part of Thermo Fisher Scientific, has three combinations of genotyping kit in the product portfolio: Bovine Genotypes™ Panels 1.2 (12 STR loci), 2.2 (6 STR loci), and 3.1 (18 STR loci). Applied Biosystems offers the StockMarks® for Cattle Bovine Parentage Typing Kit encompassing 11 STR loci. All herein mentioned reagent kits contain primers that amplify microsatellite markers recommended by ISAG for routine use in parentage testing and identification (Table 12.3). The alleles of all loci displayed the dinucleotide repeat structure.

In some developed countries, paternity identification using microsatellite markers is established for their cattle populations (Carolino et al. 2009; Cervini et al. 2006; Curi and Lopes 2002; Ozkan et al. 2009; Radko et al. 2005; Radko 2008; Řehout et al. 2006; Tian et al. 2008). In addition, microsatellites were used for the analyses of genetic diversity in cattle (Čítek and Řehout 2001; Čítek et al. 2006; Czerneková et al. 2006; Grzybowski and Prusak 2004; Zaton-Dobrowolska et al. 2007; Zhou et al. 2005).

12.5 Wildlife Forensic DNA Testing

Interpol recognized wildlife crime as the second most prevalent crime worldwide after drug trafficking (Wyler and Sheikh 2008). The value of the trade is often quoted as $20 billion per year (Alacs et al. 2010) although the exact figure is unknown as very little illegal trade is ever detected (Alacs and Georges 2008). The only figures are based on cross-border security surveys (Blundell and Mascia 2005, 2006). Furthermore, the low penalties and potential large financial gain lead to extensive and lucrative trade in endangered species (Alacs and Georges 2008; Linacre and Tobe 2011; Tobe and Linacre 2010). The scope and range of wildlife crime is wide and encompasses a variety of criminal activities such as poaching and illegal hunting of mammals (Caniglia et al. 2010; Ferreira et al. 2011; McGuire et al. 2009; Sanches et al. 2011), reptiles (Dubey et al. 2011; Meganathan et al. 2010), and fish (Schwenke et al. 2006; Shivji et al. 2002); instances of animal cruelty (Merck 2007); and the use of animal derivatives in traditional medicines (Hsieh et al. 2006, 2011; Lo et al. 2006; Peppin et al. 2008).

It is the illegal trade in endangered species either as whole organisms, body parts, or as ornaments and medicines where international legislation can assist (Figures 12.6, 12.7, and 12.8). Forensic investigations can only be undertaken if legislation has been breached. The international trade in endangered species is monitored and regulated through recommendations made by CITES. These recommendations are enforced at a national level by legislation. Normally, this legislation stipulates the names of the species that are protected. It is for these reasons that much interest at a research level has focused on methods of species identification.

Figure 12.6 Illegal trade in endangered species as whole organisms, body parts, or as ornaments and medicines. Effect of poaching on a species where there may be less than 1000 Bengali tigers in the wild. (Image courtesy of the Wildlife Institute of India.)

Figure 12.7 Illegal trade in endangered species as whole organisms, body parts, or as ornaments and medicines. An ivory sculpture from which DNA can be isolated to determine if it is a CITES listed species.

Genetic testing to enforce national legislation is rarely performed by the same operational laboratories that undertake human identification. This is primarily attributable to time and cost constraints, and the fact that most operational laboratories are accredited to undertake standard operating procedures relevant to human identification only. Most wildlife forensic investigations are undertaken by universities on an *ad hoc* basis and performed by university academics who may not routinely undertake work for the criminal justice system. Good laboratory practice has been encouraged, and guidelines have been suggested for QC and QA within a wildlife forensic laboratory as much of the work is undertaken by academic institutions (Budowle et al. 2005; Linacre et al. 2011; Ogden 2010).

Figure 12.8 Illegal trade in endangered species as whole organisms, body parts, or as ornaments and medicines. A seized traditional East Asian medicine that was found to contain CITES listed species.

12.5.1 mtDNA in Species Testing

It may be that the seized material has been processed to create a sculpture, ornament, item of clothing, food, or supposed medicinal product. DNA can be obtained at trace levels from ivory (Gupta et al. 2011; Lee et al. 2009b; Wasser et al. 2008, 2009) to determine if the material is from a species listed on CITES. Ornaments might include the handles of knives actually made from rhino (Hsieh et al. 2003), where again the genetic material will be at trace levels. Food products include soups (Shivji et al. 2002; Hsieh et al. 2006, 2011; Lo et al. 2006; Holmes et al. 2009) and potions. In all such cases, the DNA isolated may be at levels too small for the detection of nuclear DNA; hence, mitochondrial loci are used predominantly in species testing (Tobe and Linacre 2009).

mtDNA testing has been used in human DNA testing as there are well-described benefits (Dissing et al. 2007; Gill et al. 1994; Hopwood et al. 1996; Ivanov et al. 1996; Nelson and Melton 2007). Human identification has centered on a small section of mtDNA that does not encode the hypervariable regions constituting less than 1000 of the 16,569 bases that make up the human mitochondrial genome. The mitochondrial genome in eukaryotes encodes a total of 37 genes, 22 of which encode tRNA molecules, two encode rRNA molecules, and the other 13 encode proteins involved primarily with the process of oxidative respiration (Tobe and Linacre 2009). The number of genes on the mitochondrial genome is largely invariant for all vertebrate mitochondrial genomes, but the order of the genes may alter (Pereira 2000). The order of the loci on the mitochondrial genome is the same within mammalian species, but can differ between taxonomic classes; for instance, the order is different between avian and mammalian mitochondrial genomes. Vertebrate mtDNA has two strands of different densities: the heavy (or H-strand) and the light (or L-strand). The H-strand is the sense strand for one protein-coding gene (*ND6*) and eight tRNA genes, and the L-strand is the sense strand for the other 12 protein-coding genes, two rRNA genes, and 12 tRNA genes (Percira 2000).

Although the use of mitochondrial loci in species testing has many benefits, there are a few disadvantages as well. Part of the benefit of mitochondrial loci is that they are inherited maternally and as a haplotype with no recombination. Maternal inheritance can be a problem if hybrid species occur, particularly if the mother is from a nonprotected species and the father is from a species that is protected. Another issue to consider when working with mtDNA is the possible amplification of nuclear mitochondrial insertions (NUMTs); these occur when mitochondrial sequences transpose to the nuclear genome (Lopez et al. 1994; Ricchetti et al. 2004; Richly and Leister 2004). As a result, the sequence data obtained may be from nuclear DNA and not mitochondrial, leading to erroneous species identification. Many NUMTs are no longer active in the nuclear DNA, leading to loci termed pseudogenes, and these loci soon develop mutations such that they can be identified.

12.5.2 Species Identification Using Loci on Mitochondrial Genome

A classic characteristic for a locus used in species identification is that it should exhibit very little variation for any member of the same species such that all members of that species have the same, or very nearly the same, sequence; this is referred to as intraspecies variation. The second characteristic is that this same locus should exhibit sufficient differences from any member of the next closest species, being interspecies DNA sequence variation. The main locus used in taxonomic and phylogenetic studies until recently was cytochrome *b* (cyt *b*) (Irwin et al. 1991; Kocher et al. 1989; Kuwayama and Ozawa 2000; Linacre 2009; Parson et al. 2000). This occurs between bases 14,747 and 15,887 in human mtDNA (Anderson et al. 1981) and encodes a protein 380 amino acids in length. The cyt *b* locus has been used extensively in taxonomic and forensic studies (Alacs et al. 2010; Tobe and Linacre 2010; Wilson-Wilde 2010a), including turtle eggs and shells (Hsieh et al. 2006; Moore et al. 2003), peafowl (Gupta et al. 2005), crocodile parts (Meganathan et al. 2009, 2010), Tibetan antelope (Lee et al. 2006), pangolins (Hsieh et al. 2011), rhino (Hsieh et al. 2003), and tiger body parts (Kitpipit et al. 2009; Wan and Fang 2003).

The cytochrome *c* oxidase I (COI) mitochondrial gene locus was adopted by the Barcode for Life Consortium (BOLD) (www.boldsystems.org) (Hajibabaei et al. 2006a; Hebert et al. 2003; Ivanova et al. 2007; Lambert et al. 2005; Ratnasingham and Herbert 2007). COI is found between bases 5904 and 7445 in human mtDNA (Anderson et al. 1981). COI was used initially in the identification of invertebrate species (Ball et al. 2005; Hajibabaei et al. 2006a,b; Janzen et al. 2005; Smith et al. 2006) and became the locus of choice in forensic entomology (Cywinska et al. 2006; Nelson and Melton 2007; Nelson et al. 2007) before being adopted by BOLD.

The two mitochondrial loci, cyt *b* and COI, are used predominantly in forensic wildlife although other loci such as the two ribosomal loci, the 12 and 16 S genes, have been used (Balitzki-Korte et al. 2005; Melton and Holland 2007; Mitani et al. 2009). These ribosomal genes are considered to be evolving slowly compared to the protein encoding loci and therefore exhibit fewer differences between recently diverged species, the exception being the use to identify the domestic dog from gray wolves (Melton and Holland 2007). The ND2 and ND5 loci can be used in avian species because of their low intraspecies variation and relatively high interspecies variation (Boonseub et al. 2009).

The recent Commission for the International Society for Forensic Genetics (ISFG) recommended that there should be some comment as to which locus was used in any analysis (Linacre et al. 2011). Currently, there is no standardized locus in species testing. It is most likely that different loci will exhibit varying inter- and intraspecies similarity depending on whether examining insects, other invertebrates, fish, reptiles, birds, or mammals. Mammals are the only taxonomic group where a detailed studied has been performed to determine whether cyt *b* or COI have fewer false positive and false negative identifications

(Tobe et al. 2009, 2010). In this study, cyt *b* was found to outperform COI (Tobe et al. 2009, 2010).

Both COI and cyt *b* are large genes and currently not all of the loci are used in species identification—rather, sections of the loci are used; for the cyt *b* gene this is typically the first 400 bases (Hsieh et al. 2001) and for COI the first 645 bp (Hebert et al. 2003). The use of either section of the cyt *b* or COI loci is aided by the design of universal primers; these are primers that can be used in PCR to amplify a section of the gene from all the species from which the primers are designed.

The standard approach to species testing is to use either universal or species-specific primers and amplify a section of the mitochondrial genome. It may be that even after the standard amplification process there is still insufficient DNA present to analyze further (Tobe and Linacre 2009). In such instances, internal nested primers may be used with the initial amplification product as a template for further amplifications. This method has been used previously in instances of tortoise shell (Hsieh et al. 2006; Lee et al. 2009a), Tibetan antelope (Lee et al. 2006), and rhino horn (Hsieh et al. 2003) analyses.

12.5.3 Mitochondrial Sequence Analysis

Regardless of which primer pair has been used, amplification followed by sequencing is performed to obtain data for analysis. These sequences are then compared to a reference DNA sequence. Reference DNA sequences need to be obtained from known sources, and in this regard voucher specimens are ideal and were recommended by the ISFG Commission (Linacre et al. 2011). DNA sequence data can be otherwise found at repositories such as the National Center for Biotechnology Information (NCBI) database of genetic information called GenBank (www.ncbi.nlm.nih.gov/genbank/). In August 2009, the database reached 108 million sequences (Linacre and Tobe 2011) with more than 3800 taxa being added every month (Benson et al. 2011). Software including the Basic Local Alignment Search Tool (BLAST) compares the target sequence fragment to the data deposited on GenBank and produces a similarity score of the 100 most closely matched sequences. These are listed with the closest similarity at the top of the list. A similarity score of 100% to a known species, with the next closest being less than a similarity of 95%, would be ideal. This situation rarely occurs with a similarity score of 99% between the sequence tested and the nearest sequence being more typical. There is no consensus on how many differences and over what number of bases is either due to intra- or interspecies variation. It was reported that one base variation per 400 bp sequence is acceptable intraspecies variation (Linacre 2009), although the only detailed examination was in mammalian species (Tobe et al. 2010).

This method of species identification works well when the sample contains DNA from only one species, as would be expected if a sample is a shell, a single hair, or taken from muscle tissue. When the sample comes from a mixture, such as in traditional medicines, then the universal primers will amplify all the species leading to a mixed DNA sequence. However, methods of detection can be used to identify the DNA of a particular species within a mixture by techniques requiring species specific primers (Tobe and Linacre 2007, 2008a,b). These detection methods can be used to identify species based on banding patterns, but additionally the species-specific bands can be excised from the gel and sequenced. The use of species-specific loci requires some knowledge of the probable species present because it is not possible to design a species-specific assay for all known species.

12.5.4 SNP Analysis

Performing a complete DNA sequence analysis from a section of DNA provides the maximum amount of genetic information. It may be that there are only a few DNA bases that differ (are polymorphic), and that much of the DNA sequence is of little value as the species being considered share much of their DNA. In such instances, if the polymorphic DNA bases are known, these bases can be targeted specifically. A recent example of this is the identification of SNPs within the mitochondrial genome of tigers (Kitpipit et al. 2011). In this instance, small parts of the tiger mitochondrial genome are amplified in one reaction using primers specific for the genus *Panthera*. These small sections are used as a template for SNP testing, detecting up to 11 SNPs in one reaction. Because the sections amplified in the first instance are short, no more than 300 bases, this type of DNA testing works very effectively on highly degraded DNA and was developed for the detection of these highly endangered mammalian species that may be found in traditional East Asian medicine. Additionally, because the SNP detection can be species-specific, the test works well even when there is a complex mixture of other species.

12.5.5 Conclusions on Animal Testing

Species testing using loci on the mitochondrial genome has become a standard method in conservation ecology and phylogenetic studies. These same processes were transferred to a forensic science arena after minimal validation. There remains no standardization as to which locus to use, nor which primer sets have been validated. Furthermore, many of the studies used in taxonomic research do not consider the opportunity for false inclusions or exclusions, although these factors are crucial and should be known before use of the test in the criminal justice system. It should be noted that many who undertake species testing in a forensic context come from a background where this type of testing is familiar, from taxonomic or evolutionary studies. Moreover, operational forensic science laboratories typically work to maximum capacity with examination of human DNA, and have little spare capacity to consider developing methods for species testing that may only be undertaken on occasion. The net result would be that many requests would not be undertaken. Wildlife forensic science in many countries therefore remains the domain of university academics, where analysis can be performed to appropriate standards provided that recommendations, such as those published by the ISFG, are adhered to.

Acknowledgments

Felid section: The authors gratefully acknowledge the National Institute of Justice for funding of this research through interagency agreements to the National Cancer Institute's Laboratory of Genomic Diversity. We gratefully acknowledge the Cat Fanciers' Association and The International Cat association for their considerable help in facilitating sample collection from cat breeds. We also thank the hundreds of independent cat breeders who provided us with blood and buccal swab samples of their cats and pedigrees. Without the help of these individuals and organizations, this research would not have been possible. Amy Snyder, Leslie Wachter, and Nikia Coomber provided excellent technical assistance in this project. Grant Number 1999-IJ-R-A079 awarded by the National Institute of Justice, Office of Justice Programs, and U.S. Department of Justice supported this project. Points of view in this document are those of the author and do not necessarily represent the official position or policies of the U.S. Department of Justice. *Canine section*: The author thanks the VGL for allowing the review of the closed case forensic files for use in this research. Case A and the dog/cat case are from the period when the author headed the laboratory's forensic unit. The author was assisted by the staff of the forensic unit in the completion of work related to the cases, and had taken their contributions into account in the preparation of their case reports. This project was partially funded by the National Institute of Justice (NIJ Grant No. 2004-DN-BX-K007) awarded to the author. Wildlife section: The author wishes to acknowledge the Ministry of Justice, South Australia, for the ongoing support of forensic science at Flinders University.

References

Adelson, D.L. 2008. Insights and applications from sequencing the bovine genome. *Reprod Fertil Dev* 20:54–60.
Agronis, A. 2002. *UC Davis Mag* 19:3 (UC Davis).
Alacs, E. and A. Georges. 2008. Wildlife across our borders: A review of the illegal trade in Australia. *Aust J Forensic Sci* 40(2):147–60.
Alacs, E., A. Georges, N. FitzSimmons, and J. Robertson. 2010. DNA detective: A review of molecular approaches to wildlife forensics. *Forensic Sci Med Pathol* 6(3):180–84.
Anderson, S., A.T. Bankier, B.G. Barrell et al. 1981. Sequence and organization of the human mitochondrial genome. *Nature* 290(5806):457–65.
Archibald, A.L., C.S. Haley, J.F. Brown et al. 1995. The PIGMap consortium linkage map of the pig (*Sus scrofa*). *Mamm Genome* 6:157–75.
Ayres, K.L. and A.D.J. Overall. 1999. Allowing for within-subpopulation inbreeding in forensic match probabilities. *Forensic Sci Int* 103:207–16.
Ayres, K.L. 2000. Relatedness testing in subdivided populations. *Forensic Sci Int* 114:107–15.
Balitzki-Korte, B., K. Anslinger, C. Bartsch, and B. Rolf. 2005. Species identification by means of pyrosequencing the mitochondrial 12S rRNA gene. *Intern J Leg Med* 119(5):291–4.
Ball, S.L., P.D.N. Hebert, S.K. Burian, and J.M. Webb. 2005. Biological identifications of mayflies (Ephemeroptera) using DNA barcodes. *J North Am Benth Soc* 24(3):508–24.
Barendse, W., S.M. Armitage, L.M. Kossarek et al. 1994. A genetic linkage map of the bovine genome. *Nat Genet* 6:227–35.
Bensasson, D., D. Zhang, D.L. Hartl, and G.M. Hewitt. 2001. Mitochondrial pseudogenes: Evolution's misplaced witnesses. *Trends Ecol Evol* 16:314–21.
Benson, D.A., I. Karsch-Mizrachi, D.J. Lipman, J. Ostell, and E.W. Sayers. 2011. GenBank. *Nucleic Acids Res* 39:D32–7.
Benson, G. 1999. Tandem repeats finder: A program to analyze DNA sequences. *Nucleic Acids Res* 27:573–580.
Blundell, A.G. and M.B. Mascia. 2005. Discrepancies in reported levels of international wildlife trade. *Conserv Biol* 19(6):2020–5.
Blundell, A.G. and M.B. Mascia. 2006. Data on wildlife trade. *Conserv Biol* 20(3):598–9.
Boonseub, S., S.S. Tobe, and A.M.T. Linacre. 2009. The use of mitochondrial DNA genes to identify closely related avian species. *Forensic Sci Int Genet* 2(1):275–7.
Budowle, B., D.E. Adams, C.T. Comey et al. 1990. Mitochondrial DNA: A possible genetic material suitable for forensic analysis. In: *Advances in Forensic Sciences*, ed. H.C. Lee and R.E. Gaensslen. Chicago: Medical Publishers.
Budowle, B., K.L. Monson, and R. Chakraborty. 1996. Estimating minimum allele frequencies for DNA profile frequency estimates for PCR-based loci. *Int J Legal Med* 108:173–6.
Budowle, B., P. Garofano, A. Hellmann et al. 2005. Recommendations for animal DNA forensic identity testing. *Int J Legal Med* 119(5):295–302.

Butler, J.M., P.M. Scheneider, and A. Carracedo. 2010. Journal update. *Forensic Sci Int* 4:143–4.

Caniglia, R., E. Fabbri, C. Greco, M. Galaverni, and E. Randi. 2010. Forensic DNA against wildlife poaching: Identification of a serial wolf killing in Italy. *Forensic Sci Int Genet* 4(5):334–8.

Carolino, I., C.O. Sousa, S. Ferreira, N. Carolino, F.S. Silva, and L.T. Gama. 2009. Implementation of a parentage control system in Portuguese beef-cattle with a panel of microsatellite markers. *Genet Mol Biol* 32:306–11.

Carracedo, A., J.M. Prieto, L. Concheiro et al. 1987. Isoelectric focusing patterns of some mammalian keratins. *J Forensic Sci* 32:93–9.

Cassidy, B.G. and R.A. Gonzales. 2005. DNA testing in animal forensics. *J Wildl Manage* 69:1454–62.

Castelo, A.T., W. Martins, and G.R. Gao. 2002. TROLL—Tandem repeat occurrence locator. *Bioinformatics* 18:634–6.

Cervini, M., F. Henrique-Silva, N. Mortari, and E. Matheucci Jr. 2006. Genetic variability of 10 microsatellite markers in the characterization of Brazilian Nellore cattle (*Bos indicus*). *Genet Mol Biol* 29:486–90.

Čítek, J. and V. Řehout. 2001. Evaluation of the genetic diversity in cattle using microsatellites and protein markers. *Czech J Anim Sci* 46:393–400.

Čítek, J., L. Panicke, V. Řehout, and H. Procházková. 2006. Study of genetic distances between cattle breeds of Central Europe. *Czech J Anim Sci* 51:429–36.

Coomber, N., V.A. David, S.J. O'Brien et al. 2007. Validation of a short tandem repeat multiplex typing system for genetic individualization of domestic cat samples. *Croat Med J* 48:547–55.

Crawford, A.M., K.G. Dodds, A.J. Ede et al. 1995. An autosomal genetic linkage map of the sheep genome. *Genetics* 140:703–24.

Culver, M., W.E. Johnson, J. Pecon-Slattery et al. 2000. Genomic ancestry of the American puma (*Puma concolor*). *J Hered* 91:186–97.

Culver, M., M.A. Menotti-Raymond, and S.J. O'Brien. 2001. Patterns of size homoplasy at 10 microsatellite loci in pumas (*Puma concolor*). *Mol Biol Evol* 18:1151–6.

Curi, R.A. and C.R. Lopes. 2002. Evaluation of nine microsatellite loci and misidentification paternity frequency in a population of Gyr breed bovines. *Braz J Vet Res Anim Sci* 39:129–35.

Cywinska, A., F.F. Hunter, and P.D.N. Hebert. 2006. Identifying Canadian mosquito species through DNA barcodes. *Med Vet Entomol* 20(4):413–24.

Czerneková, V., T. Kott, Z. Dudková, A. Sztankoová, and J. Soldat. 2006. Genetic diversity between seven Central European cattle breeds as revealed by microsatellite analysis. *Czech J Anim Sci* 51:1–7.

D'Andrea, E., F. Fridez, and R. Conquoz. 1998. Preliminary experiments on the transfer of animal hair during simulated criminal behavior. *J Forensic Sci* 43:1257–8.

Dietrich, W., H. Katz, S.E. Lincoln et al. 1992. A genetic map of the mouse suitable for typing intraspecific crosses. *Genetics* 131:423–47.

Dissing, J., J. Binladen, A. Hansen, B. Sejrsen, E. Willerslev, and N. Lynnerup. 2007. The last Viking king: A royal maternity case solved by ancient DNA analysis. *Forensic Sci Int* 166(1):21–7.

DNA Advisory Board (DAB). 1998. Quality assurance standards for forensic DNA testing laboratories. *Forensic Sci Commun* 2000:2.

Dog Bite Law, 16. http://www.dogbitelaw.com/plain-english-overview-of-dog-bite-law/when-dogs-bite-people.html

Dreamin Demon, 14. http://www.dreamindemon.com/forums/showthread.php?38050-Robert-Pettyjohn-animal-abusing-scum-faces-40-years-in-prison

Drogemüller, C., S. Rufenacht, B. Wichert et al. 2007. Mutations within the FGF5 gene are associated with hair length in cats. *Anim Genet* 38:218–21.

Dubey, B., P.R. Meganathan, and I. Haque. 2011. DNA mini-barcoding: An approach for forensic identification of some endangered Indian snake species. *For Sci Int Genet* 5(3):181–4.

Eizirik, E., J.H. Kim, M. Menotti-Raymond et al. 2001. Phylogeography, population history and conservation genetics of jaguars (*Panthera onca*). *Mol Ecol* 10:65–79.

Eizirik, E., N. Yuhki, W.E. Johnson et al. 2003. Molecular genetics and evolution of melanism in the cat family. *Curr Biol* 13:448–53.

Eizirik, E., V.A. David, Buckley-Beason et al. 2010. Defining and mapping mammalian coat pattern genes: Multiple genomic regions implicated in domestic cat stripes and spots. *Genetics* 184:267–75.

Ellegren, H., B.P. Chowdhary, M. Johansson et al. 1994. A primary linkage map of the porcine genome reveals a low rate of genetic recombination. *Genetics* 137:1089–100.

Ferreira P.B., R.A. Torres, and J.E. Garcia. 2011. Single nucleotide polymorphisms from cytochrome b gene as a useful protocol in forensic genetics against the illegal hunting of manatees: *Trichechus manatus*, *Trichechus inunguis*, *Trichechus senegalensis*, and *Dugong dugon* (Eutheria: Sirenia). *Zoologia* 28(1):133–8.

Finnzymes Canine Genotypes™ Panel 1.1 Manual. 2012. http://diagnostics.finnzymes.fi/pdf/canine_genotypes_manual_1_1_low.pdf.

Fregeau, C.J. and R.M. Fourney. 1993. DNA typing with fluorescently tagged short tandem repeats: A sensitive and accurate approach to human identification. *BioTechniques* 15:100–19.

Fridez, F., S. Rochat, and R. Coquoz. 1999. Individual identification of cats and dogs using mitochondrial DNA tandem repeats. *Sci Justice* 39:167–71.

Gandolfi, B., C.A. Outerbridge, L.G. Beresford et al. 2010. The naked truth: Sphynx and devon rex cat breed mutations in KRT71. *Mamm Genome* 21:509–15.

Gilbert, D.A., C. Packer, A.E. Pusey et al. 1991. Analytical DNA fingerprinting in lions: Parentage, genetic diversity, and kinship. *J Hered* 82:378–86.

Gill, P., P.L. Ivanov, C. Kimpton, R. Piercy, N. Benson, G. Tully et al. 1994. Identification of the remains of the Romanov family by DNA analysis. *Nat Genet* 6(2):130–5.

Grahn, R.A., J.D. Kurushima, N.C. Billings et al. 2011. Feline non-repetitive mitochondrial DNA control region database for forensic evidence. *Forensic Sci Int Genet* 5:33–42.

Grzybowski, G. and B. Prusak. 2004. Genetic variation in nine European cattle breeds as determined on the basis of microsatellite markers: III. Genetic integrity of the Polish Red cattle included in the breeds preservation programme. *Anim Sci Pap Rep* 22:45–56.

Gupta, P.K., S. Rustgi, S. Sharma, R. Singh, N. Kumar, and H.S. Balyan. 2002. Transferable EST-SSR markers for the study of polymorphism and genetic diversity in bread wheat. *Mol Genet Genomics* 270:315–23.

Gupta, S.K., S.K. Verma, and L. Singh. 2005. Molecular insight into a wildlife crime: The case of a peafowl slaughter. *For Sci Internat* 154(2–3):214–7.

Gupta, S.K., K. Thangaraj, and L. Singh. 2011. Identification of the source of ivory idol by DNA analysis. *J Forensci Sci* 56(5):1343–5.

Hajibabaei, M., M.A. Smith, D.H. Janzen, J.J. Rodriguez, J.B. Whitfield, and P.D.N. Hebert. 2006a. A minimalist barcode can identify a specimen whose DNA is degraded. *Mol Ecol Notes* 6(4):959–64.

Hajibabaei, M., D.H. Janzen, J.M. Burns, W. Hallwachs, and P.D.N. Hebert 2006b. DNA barcodes distinguish species of tropical Lepidoptera. *Proc Natl Acad Sci U S A* 103(4):968–71.

Halverson, J.L. and C. Basten 2005a. DNA identification of animal-derived trace evidence: Tools for linking victims and suspects. *Croat Med J* 46:598–605.

Halverson, J.A. and C. Basten 2005b. A PCR multiplex and database for forensic DNA identification of dogs. *J Forensic Sci* 50:352–63.

Hauge, X.Y. and M. Litt. 1993. A study of the origin of 'shadow bands' seen when typing dinucleotide repeat polymorphisms by the PCR. *Hum Mol Genet* 2:411–5.

Hebert, P.D.N., S. Ratnasingham, and J.R. de Waard. 2003. Barcoding animal life: Cytochrome *c* oxidase subunit 1 divergences among closely related species. *Proc R Soc Lond Series B Biol Sci* 270(Suppl 1):S96–9.

Higuchi, R., C.H. von Beroldingen, G.F. Sensabaugh et al. 1988. DNA typing from single hairs. *Nature* 332:543–6.

Himmelberger, A.L., T.F. Spear, J.A. Satkoski, D.A. George, W.T. Garnica, V.S. Malladi et al. 2008. Forensic utility of the mitochondrial hypervariable 1 region of domestic dogs, in conjunction with breed and geographic information. *J Forensic Sci* 53: 81–9.

Holland, M.M., D.L. Fisher, L.G. Mitchell et al. 1993. Mitochondrial DNA sequence analysis of human skeletal remains: Identification of remains from the Vietnam War. *J Forensic Sci* 38:542–53.

Holland, M.M. and T.J. Parsons. 1999. Mitochondrial DNA sequence analysis—Validation and use for forensic casework. *Forensic Sci Rev* 11:1–49.

Holmes, B.H., D. Steinke, and R.D. Ward. 2009. Identification of shark and ray fins using DNA barcoding. *Fish Res* 95(2–3):280–8.

Hopwood, A.J., A. Mannucci, and K.M. Sullivan. 1996. DNA typing from human faeces. *Int J Legal Med* 108(5):237–43.

Hsieh, H.M., H.L. Chiang, L.C. Tsai et al. 2001. Cytochrome *b* gene for species identification of the conservation animals. *Forensic Sci Int* 122(1):7–18.

Hsieh, H.M., L.H. Huang, L.C. Tsai et al. 2003. Species identification of rhinoceros horns using the cytochrome b gene. *Forensic Sci Int* 136(1–3):1–11.

Hsieh, H.M., L.H. Huang, L.C. Tsai et al. 2006. Species identification of *Kachuga tecta* using the cytochrome b gene. *J Forensic Sci* 51(1):52–6.

Hsieh H.M., J.C.I. Lee, J.H. Wu et al. 2011. Establishing the pangolin mitochondrial D-loop sequences from the confiscated scales. *Forensic Sci Int Genet* 5(4):303–7.

Humane Society of the United States. U.S. Pet Ownership Statistics 2011–2012. http://www.humanesociety.org/issues/pet_overpopulation/facts/pet_ownership_statistics.html

Imes, D.L., L.A. Geary, R.A. Grahn et al. 2006. Albinism in the domestic cat (*Felis catus*) is associated with a tyrosinase (TYR) mutation. *Anim Genet* 37:175–8.

Inhuman.Org http://www.inhumane.org/data/BEldred.html.

Irwin, D., T. Kocher, and A. Wilson. 1991. Evolution of the cytochrome b gene of mammals. *J Mol Evol* 32(2):128–44.

ISAG Conference. 2006. Porto Seguro, Brazil. Cattle Molecular Markers and Parentage Testing Workshop. Available at: http://www.isag.org.uk/ISAG/all ISAG2006_CMMPT.pdf.

ISAG Conference. 2008. Amsterdam, the Netherlands. Cattle Molecular Markers and Parentage Testing Workshop. Available at: http://www.isag.org.uk/ISAG/all/ISAG2008_CattleParentage.pdf.

Ishida, Y., V.A. David, E. Eizirik et al. 2006. A homozygous single-base deletion in MLPH causes the dilute coat color phenotype in the domestic cat. *Genomics* 88:698–705.

Ivanov, P.L., M.J. Wadhams, R.K. Roby, M.M. Holland, V.W. Weedn, and T.J. Parsons. 1996. Mitochondrial DNA sequence heteroplasmy in the Grand Duke of Russia Georgij Romanov establishes the authenticity of the remains of TsarNicholas II. *Nat Genet* 12(4):417–20.

Ivanova, N.V., T.S. Zemlak, R.H. Hanner, and P.D.N. Hebert. 2007. Universal primer cocktails for fish DNA barcoding. *Mol Ecol Notes* 7(4):544–8.

Jacob, H.J., D.M. Brown, R.K. Bunker et al. 1995. A genetic linkage map of laboratory rat, *Rattus norvegicus*. *Nat Genet* 9:63–9.

James, S.H. and J.J. Nordby (eds). 2003. *Forensic Science: An Introduction to Scientific and Investigative Techniques*. 2nd ed. Boca Raton, FL: CRC Press.

Janzen, D.H., M. Hajibabaei, J.M. Burns, W. Hallwachs, E. Remigio, and P.D.N. Hebert. 2005. Wedding biodiversity inventory of a large and complex *Lepidoptera fauna* with DNA barcoding. *Philos Trans R Soci B Biol Sci* 360(1462):1835–45.

Jayashree, B., R. Punna, P. Prasad et al. 2006. A database of simple sequence repeats from cereal and legume expressed sequence tags mined in silico: Survey and evaluation. *In Silico Biol* 6:607–20.

Jeffreys, A.J., V. Wilson, and S.L. Thein 1985a. Individual-specific 'fingerprints' of human DNA. *Nature* 316:76–9.

Jeffreys, A.J., J.F. Brookfield, and R. Semeonoff 1985b. Positive identification of an immigration test-case using human DNA fingerprints. *Nature* 317:818–9.

Jeffreys, A.J. and D.B. Morton. 1987. DNA fingerprints of dogs and cats. *Anim Genet* 18:1–15.

Johnson, W.E., F. Shinyashiku, M. Menotti-Raymond et al. 1999. Molecular genetic characterization of two insular Asian cat speices, Bornean Bay Cat and Iriomote Cat. In: *Evolutionary Theory and Processes: Modern Perspectives*, ed. S.P. Wasser, 223–48. Netherlands: Kluwer Academic Publishers.

Jones, D.A. 1972. Blood samples: Probability of discrimination. *J Forensic Sci Soc* 12:355–9.
Ju, Z., M.C. Wells, A. Martinez, L. Hazlewood, and R.B. Walter. 2005. An in silico mining for simple sequence repeats from expressed sequence tags of zebrafish, medak, Fundulus and Xiphophorus. *In Silico Biol* 5:439–63.
Kaelin, C., X. Xiao, L.Z. Hong et al. 2012. Specifying and sustaining pigmentation patterns in domestic and wild cats. *Science* 337:1536–1541.
Kanthaswamy, S. 2009. The development and validation of a standardized canine STR panel for use in forensic casework (NIJ Grant No. 2004-DN-BX-K007 Final Report). National Institute of Justice (NIJ) National Criminal Justice Reference Service (http://www.ncjrs.gov/pdffiles1/nij/grants/226639.pdf).
Kanthaswamy, S., B.K. Tom, A.M. Mattila et al. 2009. Canine population data generated from a Multi-Plex STR Kit for use in forensic casework. *J Forensic Sci* 54:829–40.
Kayser, M. and P. de Knijff. 2011. Improving human forensics through advances in genetics, genomics and molecular biology. *Nat Rev Genet* 12:179–92.
Kehler, J.S., V.A. David, A.A. Schäffer et al. 2007. Four independent mutations in the feline fibroblast growth factor 5 gene determine the long-haired phenotype in domestic cats. *J Hered* 98:555–66.
Kitpipit, T., A. Linacre, and S.S. Tobe. 2009. Tiger species identification based on molecular approach. *Forensic Sci Int Genet* 2(1):310–2.
Kitpipit, T., S.S. Tobe, A.C. Kitchener, P. Gill, and A. Linacre. 2011. The development and validation of a single SNaPshot multiplex for tiger species and subspecies identification—Implications for forensic purposes. *For Sci Int Genet* 10.1016/j.fsigen.2011.06.001.
Kocher, T.D., W.K. Thomas, A. Meyer, S.V. Edwards, S. Paabo, F.X. Villablanca et al. 1989. Dynamics of mitochondrial DNA evolution in animals—Amplification and sequencing with conserved primers. *Proc Natl Acad Sci U S A* 86(16):6196–200.
Korpelainen, H., K. Kostamo, and V. Virtanen. 2007. Microsatellite marker identification using genome screening and restriction–ligation. *Biotechniques* 42:479–86.
Kuwayama, R. and T. Ozawa. 2000. Phylogenetic relationships among European red deer, wapiti, and sika deer inferred from mitochondrial DNA sequences. *Mol Phylogenet Evol* 15(1):115–23.
Lambert, D.M., A. Baker, L. Huynen, O. Haddrath, P.D.N. Hebert, and C.D. Millar. 2005. Is a large-scale DNA-based inventory of ancient life possible? *J Hered* 96(3):279–84.
Lander, E.S., L.M. Linton, B. Birren et al. 2001. Initial sequencing and analysis of the human genome. *Nature* 409:860–921.
Lee, J.C.I., L.C. Tsai, C.Y. Yang et al. 2006. DNA profiling of shahtoosh. *Electrophoresis* 27(17):3359–62.
Lee, J.C.I., L.C. Tsai, S.P. Liao, A. Linacre, and H.M. Hsieh 2009a. Species identification using the cytochrome b gene of commercial turtle shells. *Forensic Sci Int Genet* 3(2):67–73.
Lee, J., H.M. Hsieh, and L.H. Huang 2009b. Ivory identification by DNA profiling of cytochrome *b* gene. *Int J Legal Med* 123(2):117–21.
Linacre, A. 2008. The use of DNA from non-human sources. *Forensic Sci Int Genet Suppl Ser* 1(1):605–6.
Linacre, A. (ed) 2009. *Forensic Science in Wildlife Investigations*. Boca Raton, FL: CRC Press, 178 pp.
Linacre, A. and S.S. Tobe. 2011. An overview to the investigative approach to species testing in wildlife forensic science. *Invest Genet* 2(2).
Linacre, A., L. Gusmao, W. Hecht et al. 2011. ISFG: Recommendations regarding the use of non-human (animal) DNA in forensic genetic investigations. *Forensic Sci Int Genet* 5(5):501–5.
Lo, C.F., Y.R. Lin, H.C. Chang, and J.H. Lin. 2006. Identification of turtle shell and its preparations by PCR-DNA sequencing method. *J Food Drug Anal* 14(2):153–8.
Lopez, J.V., N. Yuhki, R. Masuda et al. 1994. Numt, a recent transfer and tandem amplification of mitochondrial DNA to the nuclear genome of the domestic cat. *J Mol Evol* 39(5):174–90.
Lopez, J.V., S. Cevario, and S.J. O'Brien. 1996. Complete nucleotide sequence of the domestic cat (*Felis catus*) mitochondrial genome and a transposed mtDNA tandem repeat (*Numt*) in the nuclear genome. *Genomics* 33:229–46.

Lyons, L.A., D.L. Imes, H.C. Rah et al. 2005. Tyrosinase mutations associated with Siamese and Burmese patterns in the domestic cat (*Felis catus*). *Anim Genet* 36:119–26.

Mandrekar, M.N., A.M. Erickson, K. Kopp et al. 2001. Development of a human DNA quantitation system. *Croat Med J* 42:336–9.

McGuire, S.M., G.P. Emodi, G.D. Shore, R.A. Brenneman, and E.E. Louis Jr. 2009. Characterization of 21 microsatellite marker loci in the silky sifaka (*Propithecus candidus*). *Conserv Genet* 10(4):985–8.

Meganathan, P.R., B. Dubey, and I. Haque. 2009. Molecular identification of crocodile species using novel primers for forensic analysis. *Conserv Genet* 10(3):767–70.

Meganathan, P.R., B. Dubey, K.N. Jogayya, N. Whitaker, and I. Haque. 2010. A novel multiplex PCR assay for the identification of Indian crocodiles. *Mol Ecol Resour* 10(4):744–7.

Mellersh, C.S., A.A. Langston, G.M. Acland, M.A. Fleming et al. 1997. A linkage map of the canine genome. *Genomics* 46:326–36.

Melton, T. and C. Holland. 2007. Routine forensic use of the mitochondrial 12S ribosomal RNA gene for species identification. *J Forensic Sci* 52(6):1305–7.

Menotti-Raymond, M.A., V.A. David, and S.J. O'Brien 1997a. Pet cat hair implicates murder suspect. *Nature* 386:774.

Menotti-Raymond, M.A., V.A. David, J.C. Stephens et al. 1997b. Genetic individualization of domestic cats using feline STR loci for forensic applications. *J Forensic Sci* 42:1039–51.

Menotti-Raymond, M.A., V.A. David, L.A. Lyons et al. 1999. A genetic linkage map of microsatellites in the domestic cat (*Felis catus*). *Genomics* 57:9–23.

Menotti-Raymond, M.A., V.A. David, and S.J. O'Brien. 2000. DNA yield from single hairs (wool and guard/shed and plucked), success rate in amplifying STR and mtDNA targets, estimating DNA yield using multicopy target. Eleventh International Symposium on Human Identification; Biloxi, MS. http://www.promega.com/geneticidproc/ussymp11proc/abstracts/menotti_raymond.pdf.

Menotti-Raymond, M.A., V.A. David, Z.Q. Chen et al. 2003a. Second generation integrated genetic linkage/radiation hybrid maps of the domestic cat. *J Hered* 94:95–106.

Menotti-Raymond, M.A., V.A. David, R. Agarwala et al. 2003b. Radiation hybrid mapping of 304 novel microsatellites in the domestic cat genome. *Cytogenet Genome Res* 102:272–6.

Menotti-Raymond, M.A., V.A. David, L. Wachter et al. 2003c. Quantitative polymerase chain reaction-based assay for estimating DNA yield extracted from domestic cat specimens. *Croat Med J* 44:327–31.

Menotti-Raymond, M.A., V.A. David, L.L. Wachter et al. 2005. An STR forensic typing system for the genetic individualization of domestic cat (*Felis catus*) samples. *J Forensic Sci* 50:1061–70.

Menotti-Raymond, M.A., V.A. David, S.M. Pflueger et al. 2008a. Patterns of molecular genetic variation among cat breeds. *Genomics* 91:1–11.

Menotti-Raymond, M. et al. 2008b. STR-based forensic analysis of felid samples from domestic and exotic cats. In: *Nonhuman DNA Typing, Theory and Casework Applications*, ed. H.M. Coyle. Boca Raton, FL: CRC Press.

Menotti-Raymond, M.A., V.A. David, A.A. Schäffer et al. 2009. An autosomal genetic linkage map of the domestic cat, *Felis silvestris catus*. *Genomics* 93:305–13.

Menotti-Raymond, M.A., V.A. David, B.S. Weir et al. 2012. A population genetic database of cat breeds developed in coordination with a domestic cat STR multiplex. *J Forensic Sci* 57(3):596–601.

Merck, M.D. 2007. *Veterinary Forensics: Animal Cruelty Investigations*. Oxford: Blackwell Publishing.

Micka, K.A., E.A. Amiott, T.L. Hockenberry et al. 1999. TWGDAM validation of a nine-locus and a four-locus fluorescent STR multiplex system. *J Forensic Sci* 44:1243–57.

Mitani, T., A. Akane, T. Tokiyasu, S. Yoshimura, Y. Okii, and M. Yoshida. 2009. Identification of animal species using the partial sequences in the mitochondrial 16S rRNA gene. *Legal Med (Tokyo)* 11(Suppl 1):S449–50.

Moore, J.E. 1988. A key for the identification of animal hairs. *J Forensic Sci Soc* 28:335–59.

Moore, M.K., J.A. Bemiss, S.M. Rice, J.M. Quattro, and C.M. Woodley. 2003. Use of restriction fragment length polymorphisms to identify sea turtle eggs and cooked meats to species. *Conserv Genet* 4(1):95–103.

Morin, P.A. 1992. Paternity exclusion using multiple hypervariable micorsatellite loci amplified from nuclear DNA of hair cells. In: *Paternity in Primates: Genetic Tests and Theories*, ed. R.D. Martin, A.F. Dixson, and E.J. Wickings, 63–81. Basel: Karger.

Morrison, T.B., J.J. Weis, and C.T. Wittwer. 1998. Quantification of low-copy transcripts by continuous SYBR green I monitoring during amplification. *BioTechniques* 24:954–8.

Murphy, W.J., B. Davis, V.A. David et al. 2007. A 1.5-Mb-resolution radiation hybrid map of the cat genome and comparative analysis with the canine and human genomes. *Genomics* 89:189–96.

Murphy, W.J., S. Sun, Z. Chen et al. 2000. A radiation hybrid map of the cat genome: Implications for comparative mapping. *Genome Res* 10:691–702.

Nelson, K. and T. Melton. 2007. Forensic mitochondrial DNA analysis of 116 casework skeletal samples. *J Forensic Sci* 52(3):557–61.

Nelson, L.A., J.F. Wallman, and M. Dowton. 2007. Using COI barcodes to identify forensically and medically important blowflies. *Med Vet Entomol* 21(1):44–52.

O'Brien, S.J., M. Menotti-Raymond, W.J. Murphy et al. 2002. The feline genome project. *Annu Rev Genet* 36:657–86.

Ogden, R., N. Dawnay, and R. McEwing. 2009. Wildlife DNA forensics—Bridging the gap between conservation genetics and law enforcement. *Endangered Species Res* 9(3):179–95.

Ogden, R. 2010. Forensic science, genetics and wildlife biology: Getting the right mix for a wildlife DNA forensics lab. *Forensic Sci Med Pathol* 6(3):172–9.

Ogden, R., R. J. Mellanby, D. Clements, A. G. Gow, R. Powell, and R. McEwing. 2012. Genetic data from 15 STR loci for forensic individual identification and parentage analyses in U.K. domestic dogs (*Canis lupus familiaris*). *Forensic Sci Int Genet* 6(2):e63–5. doi: 10.1016/j.fsigen.2011.04.015.

Ozkan, E., M.I. Soysal, M. Ozder, E. Koban, O. Sahin, and I. Togan. 2009. Evaluation of parentage testing in the Turkish Holstein population based on 12 microsatellite loci. *Livest Sci* 124:101–6.

Pádár, Z., M. Angyal, B. Egyed et al. 2001. Canine microsatellite polymorphisms as the resolution of an illegal animal death case in a Hungarian Zoological Gardens. *Int J Legal Med* 115:79–81.

Pádár, Z., B. Egyed, K. Kontadakis et al. 2002. Canine STR analyses in forensic practice. *Int J Legal Med* 116:286–8.

Parson, W., K. Pegoraro, H. Niederstatter, M. Foger, and M. Steinlechner. 2000. Species identification by means of the cytochrome *b* gene. *Int J Legal Med* 114(1):23–8.

Peabody, A.J., R.J. Oxborough, P.E. Cage, and I.W. Evett. 1983. The discrimination of cat and dog hairs. *J Forensic Sci Soc* 23:121–9.

Peppin, L., R. McEwing, G.R. Carvalho, and R. Ogden. 2008. A DNA-based approach for the forensic identification of Asiatic black bear (*Ursus thibetanus*) in a traditional Asian medicine. *J Forensic Sci* 53(6):1358–62.

Pereira, S.L. 2000. Mitochondrial genome organization and vertebrate phylogenetics. *Genet Mol Biol* 23:745–52.

Perez, F., J. Ortiz, M. Zhinaula, C. Gonzabay, J. Calderon, and F.A. Volckaert. 2005. Development of EST-SSR markers by data mining in three species of shrimp, *Litopenaeus vannamei*, *Litopenaeus stylirostris* and *Trachypenaeus birdy*. *Mar Biotechnol (N.Y.)* 7:554–69.

Pontius, J.U., J.C. Mullikin, D.R. Smith et al. 2007. Initial sequence and comparative analysis of the cat genome. *Genome Res* 17:1675–89.

Powell, W., G.C. Machray, and J. Provan. 1996. Polymorphism revealed by simple sequence repeats. *Trends Plant Sci* 1:215–22.

Radko, A., A. Zyga, T. Zabek, and E. Slota. 2005. Genetic variability among Polish Red, Hereford and Holstein–Friesian cattle raised in Poland based on analysis of microsatellite DNA sequences. *J Appl Genet* 46:89–91.

Radko, A. 2008. Microsatellite DNA polymorphism and its usefulness for pedigree verification of cattle raised in Poland. *Ann Anim Sci* 8:311–21.

Ratnasingham, S. and P.D.N. Hebert. 2007. BOLD: The Barcode of Life Data System (www.barcodinglife.org). *Mol Ecol Notes* 7(3):355–64.

Řehout, V., E. Hradecká, and J. Čítek. 2006. Evaluation of parentage testing in the Czech population of Holstein cattle. *Czech J Anim Sci* 51:503–9.

Ricchetti, M., F. Tekaia, and B. Dujon. 2004. Continued colonization of the human genome by mitochondrial DNA. *PLoS Biol* 2(9):1313–24.

Richly, E. and D. Leister. 2004. NUMTs in sequenced eukaryotic genomes. *Mol Biol Evol* 21(6):1081–4.

Rohrer, G.A., S.C. Fahrenkrug, D. Nonneman, N. Tao, and W.C. Warren. 2002. Mapping microsatellite markers identified in porcine EST sequences. *Anim Genet* 33:372–6.

Royal Canadian Mounted Police. 1996. Interpretation of STR Profiles. *Royal Canadian Mounted Police; Biology Section Methods Guide.*

Sanches, A., W.A.M. Perez, M.G. Figueiredo et al. 2011. Wildlife forensic DNA and lowland tapir (*Tapirus terrestris*) poaching. *Conserv Genet Resourc* 3(1):189–93.

Savolainen, P. and J. Lunderberg. 1999. Forensic evidence based on mtDNA from dog and wolf hairs. *J Forensic Sci* 44: 77–81.

Scharnhorst, G. and S. Kanthaswamy. 2011. An assessment of scientific and technical aspects of closed canine forensics DNA investigations—Case series from the University of California, Davis. *Croat Med J* 52(3):280–92.

Schmidt-Küntzel, A., E. Eizirik, S.J. O'Brien et al. 2005. Tyrosinase and tyrosinase related protein 1 alleles specify domestic cat coat color phenotypes of the albino and brown loci. *J Hered* 96:289–301.

Schneider, P.M., Y. Seo, and C. Rittner. 1999. Forensic mtDNA hair analysis excludes a dog from having caused a traffic accident. *Int J Legal Med* 112:315–16.

Schwenke, P.L., J.G. Rhydderch, M.J. Ford, A.R. Marshall, and L.K. Park. 2006. Forensic identification of endangered Chinook Salmon (*Oncorhynchus tshawytscha*) using a multilocus SNP assay. *Conservat Genet* 7(6):983–9.

Serapion, J., H. Kucuktas, J. Feng, and Z. Liu. 2004. Bioinformatics mining of type I microsatellites from expressed sequence tags of channel catfish (*Ictalurus punctatus*). *J Mar Biotechnol* 6:364–77.

Shivji, M., S. Clarke, M. Pank, L. Natanson, N. Kohler, and M. Stanhope. 2002. Genetic identification of pelagic shark body parts for conservation and trade monitoring. *Conserv Biol* 16(4):1036–47.

Smalling, B.B., J.A. Satkoski, B.K. Tom et al. 2010. The significance of regional and mixed breed canine mtDNA databases in forensic science. *Bentham Open-Open Forensic Sci J (Special Issue)* 3:22–32.

Smith, M.A., N.E. Woodley, D.H. Janzen, W. Hallwachs, and P.D.N. Hebert. 2006. DNA barcodes reveal cryptic host-specificity within the presumed polyphagous members of a genus of parasitoid flies (Diptera: Tachinidae). *Proc Natl Acad Sci U S A* 103(10):3657–62.

Tampa Bay Online. 2010. http://www2.tbo.com/content/2010/jul/27/man-convicted-notorious-animal-cruelty-case-back-j/.

Tarditi, C.R., R.A. Grahn, J.J. Evans, J.D. Kurushima, and L.A. Lyons. 2011. Mitochondrial DNA sequencing of cat hair: An informative forensic tool. *J Forensic Sci* 56(Suppl 1):S36–46.

Tautz, D. 1989. Hypervariability of simple sequences as a general source for polymorphic DNA markers. *Nucleic Acids Res* 17:6463–71.

Temnykh, S., G. DeClerck, A. Lukashova, L. Lipovich, S. Cartinhour, and S. McCouch. 2001. Computational and experimental analysis of microsatellites in rice (*Oryza sativa* R.): Frequency, length variation, transposon associations, and genetic marker potential. *Genome Res* 11:1441–52.

Thiel, T., W. Michalek, R.K. Varshney, and A. Graner. 2003. Exploiting EST databases for the development and characterization of gene-derived SSR-markers in barley (*Hordeum vulgare* L.). *Theor Appl Genet* 106:411–22.

Tian, F., D. Sun, and Y. Zhang. 2008. Establishment of paternity testing system using microsatellite markers in Chinese Holstein. *J Genet Genom* 35:279–84.

Tobe, S.S. and A. Linacre. 2007. Species identification of human and deer from mixed biological material. *Forensic Sci Int* 4;169(2–3):278–9.

Tobe, S.S. and A. Linacre. 2008a. A multiplex assay to identify 18 European mammal species from mixtures using the mitochondrial cytochrome *b* gene. *Electrophoresis* 29(2):340–7.

Tobe, S.S. and A. Linacre. 2008b. A method to identify a large number of mammalian species in the U.K. from trace samples and mixtures without the use of sequencing. *Forensic Sci Int Genet* 1(1):625–7.

Tobe, S.S. and A. Linacre. 2009. Identifying endangered species from degraded mixtures at low levels. *Forensic Sci Int Genet* 2(1):304–5.

Tobe, S.S., A. Kitchener, and A. Linacre. 2009. Cytochrome b or cytochrome c oxidase subunit I for mammalian species identification—An answer to the debate. *Forensic Sci Int Genet* 2(1):306–7.

Tobe, S.S., A. Kitchener, and A. Linacre. 2010. Reconstructing mammalian phylogenies: A detailed comparison of the cytochrome *b* and cytochrome oxidase subunit *i* mitochondrial genes. *PLoS ONE* 5(11):e14156.

Tobe, S.S. and A. Linacre. 2010. DNA typing in wildlife crime: Recent developments in species identification. *Forensic Sci Med Pathol* 6(3):195–206.

Tom, B.K., M.T. Koskinen, M.R. Dayton, A.M. Mattila, E. Johnston, and D. Fantin. 2010. Development of a nomenclature system for a canine STR multiplex reagent kit. *J Forensic Sci* 55:597–604.

Toth, G., Z. Gaspari, and J. Jurka. 2000. Microsatellite in different eukaryotic genome, survey and analysis. *Genome Res* 10:1967–81.

Tresniowski, A., J.S. Podesta, and H. Breur. 2006. *People Mag*, January 23rd issue.

van Asch, B., C. Albarran, A. Alonso, R. Angulo, C. Alves, and E. Betancor. 2009. Forensic analysis of dog (*Canis lupus familiaris*) mitochondrial DNA sequences: An inter-laboratory study of the GEP-ISFG working group. *Forensic Sci Int Genet* 4:49–54.

Van de Goor, L.H.P. and W.A. van Haeringen. 2007. Identification of stolen cattle using 22 microsatellite markers. *Nonhuman DNA Typing: Theory and Casework Applications*, 122–3. Boca Raton, FL: CRC.

Vella, C.M., L.M. Shelton, J.J. Mcgonagle, and T.W. Stanglein. 2003. *Robinson's Genetics for Cat Breeders and Veterinarians.* 4th ed. Edinburgh: Butterworth Heinemann.

Vigne, J.D., J. Guilaine, K. Debue et al. 2004. Early taming of the cat in Cyprus. *Science* 304:259.

Walker, J.A., D. Hughs, and M. Batzer. 2002. SINE based PCR for the identification of species-specific DNA. Thirteenth International Symposium on Human Identification, Phoenix, Arizona.

Walsh, P.S., H.A. Erlich, and R. Higuchi. 1992. Preferential PCR amplification of alleles: Mechanisms and solutions. *PCR Methods Appl* 1:241–50.

Wan, Q.H. and S.G. Fang. 2003. Application of species-specific polymerase chain reaction in the forensic identification of tiger species. *Forensic Sci Int* 131(1):75–8.

Wasser, S.K., W.J. Clark, O. Drori et al. 2008. Combating the illegal trade in African elephant ivory with DNA forensics. *Conserv Biol* 22(4):1065–71.

Wasser, S.K., B. Clark, and C. Laurie. 2009. The ivory trail. *Sci Am* 301(1):68–74.

Waterston, R.H., K. Lindblad-Toh, E. Birney et al. 2002. Initial sequencing and comparative analysis of the mouse genome. *Nature* 420:520–62.

Weber, J.L. and P. E. May. 1989. Abundant class of human DNA polymorphisms which can be typed using the polymerase chain reaction. *Am J Hum Genet* 44:388–96.

Weising, K., H. Nybom, K. Wolff, and G. Kahl. 2005. *DNA Fingerprinting in Plants. Principles, Methods and Applications.* 2nd ed. Boca Raton, FL: CRC Press, Taylor & Francis Group.

Wilson-Wilde, L. 2010a. Combating wildlife crime. *Forensic Sci Med Pathol* 6(3):149–50.

Wilson-Wilde, L. 2010b. Wildlife crime: A global problem. *Forensic Sci Med Pathol* 6(3):221–2.

Wyler, L.S. and P.A. Sheikh. 2008. International Illegal Trade in Wildlife: Threats and U.S. Policy. *Congressional Research Service Report for Congress, RL34395*, August 22, 2008, Washington, DC: The Library Congress.

Yan, Q., Y. Zhang, H. Li et al. 2008. Identification of microsatellites in cattle unigenes. *J Genet Genomics* 35:261–6.

Yuhki, N., T. Beck, R.M. Stephens et al. 2003. Comparative genome organization of human, murine and feline MHC class II region. *Genome Res* 13:169–79.

Zaton-Dobrowolska, M., J. Čítek, A. Filistowicz, V. Řehout, and T. Szulc. 2007. Genetic distance between the Polish Red, Czech Red, and German Red cattle estimated based on selected loci of protein coding genes and DNA microsatellite sequences. *Anim Sci Pap Rep* 25:45–54.

Zhou, G.L., H.G. Jin, Q. Zhu, S.L. Guo, and Y.H. Wu. 2005. Genetic diversity analysis of five cattle breeds native to China using microsatellites. *J Genet* 84:77–80.

Ziegle, J.S., Y. Su, K.P. Corcoran et al. 1992. Application of automated DNA sizing technology for genotyping microsatellite loci. *Genomics* 14:1026–31.

Application of DNA-Based Methods in Forensic Entomology

13

JEFFREY D. WELLS
VEDRANA ŠKARO

Contents

13.1 Introduction	353
13.2 Methods of Insect DNA Analysis	356
13.3 Analysis of Human DNA Extracted from Insects	359
13.3.1 Case Study 1: A Caddis-Fly Casing in Service of Criminalistics	360
13.3.2 Case Study 2: Identity of Maggots Found on Outside and Inside of Body Bag	360
13.3.3 Case Study 3: Human and Insect mtDNA Analysis from Maggots	361
13.3.3.1 Insect mtDNA	361
13.3.3.2 Human mtDNA	362
13.3.4 Case Study 4: Genotyping of Human DNA Recovered from Mosquitoes Found at a Crime Scene	362
References	363

13.1 Introduction

Nonhuman DNA analysis plays an important role in forensic investigations. It includes invertebrates and vertebrates. Whereas the DNA analysis of vertebrates has equal application in both human and animal forensics, DNA analysis of invertebrates has applications almost exclusively as an auxiliary tool in human forensics. Among the invertebrates, the most common sources of forensic evidence are arthropods. Hundreds of arthropod species, in particular, the insects, and primarily flies (Diptera) and beetles (Coleoptera), are attracted to corpses. Forensic entomology uses insects to help law enforcement determine the cause, location, and time of death of a human corpse as well as to detect poisons, drugs, and physical neglect and abuse. With information gathered at a crime scene and the study of insects found there (i.e., distribution, biology, and behavior), it may be possible to accurately determine how long a body has been at a specific location or to connect the suspect to the crime scene (Anderson and Cervenka 2002; Catts and Goff 1992; Catts and Haskell 1990; Greenberg and Kunich 2002). However, the great majority of forensic entomological analyses are aimed at estimating the time of death. By conventional signs of postmortem decay (paleness or *pallor mortis*, hypostasis or *livor mortis*, coldness or *algor mortis*, stiffness or *rigor mortis*, etc.), time of death (postmortem interval [PMI]) can be accurately estimated for the first 2–3 days. However, by forensic entomology (the study of present necrophage species and developmental stages of insects that feed on the corpse), the time elapsed from the moment of death to the preservation of insect specimens from the corpse can be accurately calculated for the interval of 1 day to several weeks. Insect specimens should be considered physical evidence at the crime scene and processed as any other biological material, following the recommended procedures for quality assurance for collection, preservation, packaging, and transport in order to prevent contamination or destruction of evidence and to guarantee the chain of custody (Amendt et al. 2007; Amendt and Hall 2007; Budowle et al. 2005; Lord and Burger 1983; Randall et al. 1998; Ritz-Timme et al. 2000). The groups of invertebrates and their role in forensics are listed in Table 13.1, and the most important insects are listed in Table 13.2 (Gunn 2009).

Table 13.1 Groups of Invertebrates in Forensics

Invertebrates' Role in Forensics	Invertebrate Grouped According to Their Role in Forensics	Invertebrate Representatives
Forensic indicators in cases of murder or suspicious death	Those that are attracted to dead bodies	Detritivores (many different flies, dermestid beetles, pyralid and tineid moths, mites, earthworms, and miscellaneous invertebrates such as nematodes, slugs, snails, Collembola, Diplura, and millipedes)
		Carnivores and parasitoids (carabid and staphylinid beetles, burying beetles, histerid beetles, ants and wasps, centipedes and spiders)
	Those that leave dead bodies	Fleas and lice
	Those that become accidentally associated with the dead body and/or the crime scene	Miscellaneous insects
Cause of death	Those with direct response (anaphylactic reaction caused by stinging a person in the mouth or throat)	Honeybee, wasp, hornets
	Those with indirect response (causing distraction or panic attacks that results in a fatal accident—unexplained vehicle accidents)	Spiders, wasps and bees, bumblebees
Forensic indicators in cases of neglect and animal welfare	Those that infest wounds or which live on the body surface (possible consequences: pain and distress, fatal infection of the blood with bacteria)	Fleas, lice, and mites larvae of Diptera (the maggots of blowflies and flesh flies)
Food spoilage and hygiene litigation	Those that make a foodstuff unsuitable for human consumption (feces, cast skins, or bite marks) and transmit diseases by transferring pathogens such as bacteria, viruses, protozoa, and nematodes	Blowflies and other diptera beetles, ants and wasps, butterflies and moths, cockroaches, mites, and many other insects
Subject of the illegal trade	Those that are captured above the allowed quota (results in population decline, habitat destruction, and possible extinction and therefore some species become more valuable for illegal trade)	Mole cricket *Gryllotalpa gryllotalpa*, the large blue butterfly, *Maculinea arion*, the swallowtail butterfly, *Papilio machaon*, the medicinal leech, *Hirudo medicinalis*, and apus, *Triops cancriformis* (e.g., lobsters, oysters)

Table 13.2 Most Important Insects in Forensics[a]

Flies (order Diptera)	Blowflies (family Calliphoridae)
	Flesh flies (family Sarcophagidae)
	House flies (family Muscidae)
	Cheese flies (family Piophilidae)
	Scuttle flies (family Phoridae)
	Lesser corpse flies (family Sphaeroceridae)
	Lesser house flies (family Fanniidae)
	Black scavenger flies (family Sepsidae)
	Sun flies (family Heleomyzidae)
	Black soldier fly (family Stratiomyidae)
Beetles (order Coleoptera)	Rove beetles (family Staphylinidae)
	Hister beetles (family Histeridae)
	Carrion beetles (family Silphidae)
	Ham beetles (family Cleridae)
	Carcass beetles (family Trogidae)
	Leather/hide beetles (family Dermestidae)
	Scarab beetles (family Scarabaeidae)
	Sap beetles (family Nitidulidae)
Mites (Class Acari)	family Tyroglyphidae
	family Oribatidae
Moths (order Lepidoptera)	Clothing moths (family Tineidae)
	Grease moth (family Pyralinae)
Hymenoptera (order Hymenoptera)	Wasps (family Vespidae)
	Ants (family Formicidae)
	Bees (superfamily Apoidea)

[a] The list is incomplete because many other species can be found on corpse and there may be significant differences with regard to climatic conditions.

Case study 1 (see below) describes a forensic case from 1948 in which the use of caddisfly casings was introduced as a tool for forensic investigation to estimate the time of death and link the forensic sample to the victim (Caspers 1952).

In forensic entomology, not only is the insect's life cycle important, but species identification important as well. Recognition by morphological characteristics is often difficult and requires comprehensive knowledge of a taxonomic group. One reason for the complexity of this kind of recognition is that species of the class Insecta constitute more than 80% of the entire animal kingdom. Also, there are a limited number of experts able to identify forensically important insect larvae at the species level. Since some closely related species grow at very different rates, precise identification of species can be crucial to the analysis. Because of these difficulties, taxonomic misidentification of a specimen can lead to corresponding uncertainty in an estimation of PMI which, depending on the species and temperature, could be miscalculated by more than a week. For some groups of insects, differentiating the morphological larvae stage is not possible. In these cases, an alternative would be rearing the larvae to adult. However, this method is slow, because of the duration of the process, and not reliable since rearing sometimes fails. An alternative approach is DNA analysis, not only to identify forensically important species in any life stage, but also to possibly identify human DNA that may be present in the insects' guts. In situations in which larvae are found in the absence of a corpse, it would be useful to know on what (and if it was a human, on whom) they were feeding. More recently proposed forensic insect molecular genetics techniques include population genetic methods for inferring the postmortem relocation of a corpse (Picard and Wells 2012) and the use of gene expression levels to estimate specimen age (Tarone et al. 2007; Boehme et al. 2013).

13.2 Methods of Insect DNA Analysis

The first step in the typical analysis of forensic insect material is DNA extraction followed by amplification by polymerase chain reaction (PCR), cycle sequencing, and comparison of obtained results with reference samples. An example of sequencing results is shown in Figure 13.1a. DNA extraction can be performed by one of the commonly used methods such as phenol/chloroform, cetyltrimethylammonium bromide, Chelex extraction, or by commercially available extraction kits (e.g., QIAamp® Tissue Kit or DNAzol®). DNA analysis (mitochondrial or nuclear origin) enables insect species identification in each stage of its life cycle, and also in cases where only parts of the insect are available. To remain suitable for DNA extraction, entomological samples collected on and around the corpse are usually stored in 95%–100% alcohol and/or by freezing. Adult insects preserved by traditional pinning may be suitable for DNA extraction, but this is not optimal for genetic analysis. Contamination with foreign DNA is minimized by washing the surface of the specimen with ethanol or a weak bleach solution (Linville and Wells 2002), and by avoiding the gut when removing tissue for DNA extraction. Amplification and sequence analysis of DNA follows the usual protocols.

DNA sequencing is certainly the most common method used in forensic entomology, primarily because it is the most informative among the available methods of DNA analysis and because the availability of a large number of PCR primers makes it easy to analyze small overlapping DNA regions such as is necessary for degraded tissue. A variety of commercial kits, computer-based genetic analyzers, and software-aided sequence analysis made sequencing widely available and relatively easy to perform. Furthermore, many companies now offer sequencing at a relatively affordable price, although using such a service is not appropriate for casework. The most common locus for specimen identification is some portion of the mitochondrial DNA (mtDNA) genes for cytochrome *c* oxidase subunits one and two (COI and COII) (e.g., Wells and Williams 2007). Because mtDNA is abundant in the cell and haploid, it is easier to sequence compared to nuclear DNA, and its use in forensic entomology preceded the similar DNA barcoding (Hebert et al. 2003). Proposed alternatives to COI + II include a different mtDNA gene (e.g., Zehner et al. 2004b) or the nuclear internal transcribed spacer regions (e.g., Nelson et al. 2008). A few publications have included more than one locus in a single analysis based on sequence data (Stevens 2003; Stevens et al. 2002; Wallman et al. 2005). Other authors have also suggested identification using the DNA fragment size genotypes produced by restriction digestion of PCR product (PCR-RFLP; e.g., Schroeder et al. 2003), randomly amplified polymorphic DNA (RAPD; e.g., Benecke 1998), or amplified fragment length polymorphism (AFLP; e.g., Picard and Wells 2012). An example of the AFLP profile is shown in Figure 13.1b.

Whatever the genotype used, certain steps are required to design a reliable species-diagnostic test (see discussion in Wells and Stevens 2008). One is the assembly of a trusted and relevant database of reference genotypes. Many published genotypes are erroneous (Brunak et al. 1990), so an investigator must be confident that the reference specimens were

Application of DNA-Based Methods in Forensic Entomology 357

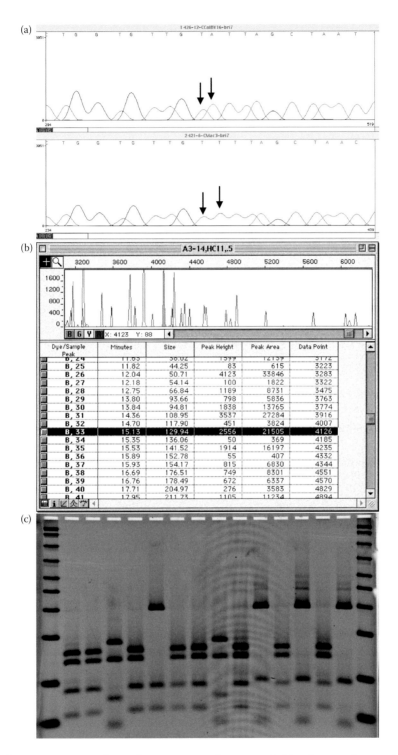

Figure 13.1 Two sequence electopherograms with arrows indicating a polymorphic site distinguishing the blow flies Compsomyiops callipes (a) and Cochliomyia macellaria. (b) An AFLP profile for the blow fly Phormia regina. (c) An acrylamide gel image showing heteroduplex profiles used to distinguish species of Sarcophagidae.

correctly identified and that the record contains no artifact. Another common problem has been the use of an incomplete taxon sample for the reference database. A specimen could be misidentified by a given procedure if it is not a species represented in the database. As with any forensic analytical procedure, a species-diagnostic test should be empirically validated.

The interpretation of a species-diagnostic test can vary according to the technology. One can expect there to be intraspecific genetic variation, especially in the loci useful for distinguishing close relatives. Phylogenetic analysis is a convenient way to handle large and complex genetic datasets with intraspecific variation, and identification is straightforward in taxa that show reciprocal monophyly for the locus (Wells and Williams 2007), as is not always the case (Will and Rubinoff 2004; Wells et al. 2007). Polymerase chain reaction-restriction fragment length polymorphism techniques (PCR-RFLP) depends on an exact match between members of a species. Therefore, PCR-RFLP would be unreliable to the extent that there was intraspecific variation in the relevant restriction enzyme sites. However, as with any method, the measure of reliability is empirical validation rather than theoretical concerns. Random amplified polymorphic DNA (RAPD), is problematic in that the results have been shown to be extremely sensitive to experimental conditions (MacPherson et al. 1993; Perez et al. 1998).

The difference between intraspecific sequence variation and the genetic distance between close relatives has been shown for some comparisons. For *Chrysomya rufifacies* and *Chrysomya albiceps*, it was detected that there was less than 1% mismatch within a species and about 3% mismatch between species. Similar levels of variation was found for flesh flies (family Sarcophagidae) *Sarcophaga argyrostoma* and *Sarcophaga crassipalpis* (within-species variation, 1%; between-species variation, about 3%) and blow flies *C. vicina* and *C. vomitoria* (within-species variation, <1%; between-species variation, about 5%). An example of heteroduplex profiles used to distinguish species of Sacrophagidae is shown in Figure 13.1c. Although variation within species is generally less than the variation between species, a careful interpretation is necessary because data on sequences and variations are still not complete. An additional problem is the possible breeding between different species and major differences in sequences of the same species from geographically distant locations (Wells and Sperling 2001; Wells et al. 2001; Zehner et al. 2004b). More research needs to be published in order to make clear whether all populations of a given species can be identified by a particular gene sequence, and whether there are problems caused by subspecies or cryptic species (Gunn 2009; Wells and Stevens 2008).

Although RAPD has been used for species determination (see Case study 2), a result may be difficult to reproduce. However, RAPD profiles can be generated from any well-preserved tissue sample, and the method does produce complex genotypes that are more suitable for population genetic analysis than are the data from a typical sequencing experiment (Stevens and Wall 2001).

For animal studies, the role of RAPD profiles has been replaced by the more repeatable genotypes produced by AFLP (Vos et al. 1995). This included recent investigations of calliphords (Baudry et al. 2003; Alamalakala et al. 2009; Picard and Wells 2009). Although AFLP has been proposed for forensic insect specimen identification (Alamalakala et al. 2009; Picard and Wells 2012). The method requires a large amount of nondegraded specimen DNA, and the protocol is more complicated than sequencing. Therefore, it is not likely to be used for this purpose except for taxa that cannot be distinguished by standard sequence-based methods. The extreme polymorphism of AFLP genotypes makes them useful for applications such as population assignment, which in theory could be used to reconstruct the postmortem transportation of a corpse (Picard and Wells 2010).

In many scientific disciplines, the use of microsatellite loci has now become established as the gold standard for revealing population-level genetic variation. One of the key advantages of using microsatellite loci is the ability to reliably recognize alleles and to interpret the resulting genetic data using powerful conventional Mendelian/Hardy–Weinberg-based analyses. Microsatellite typing methods have been developed for some carrion-feeding Calliphoridae (Florin and Gyllenstrand 2002; Torres and Azeredo-Espin 2008; Rodrigues et al. 2009.). However, we are not aware of them having been applied in a forensic context.

Recently developed species-diagnostic methods, such as pyrosequencing (Balitzki-Korte et al. 2005), heteroduplex analysis (Boakye et al. 1999), real-time PCR (Wetton et al. 2004), or microarrays (Belosludtsev et al. 2004), are used for non-insect taxa, and it can be concluded that such methods will play a greater role in forensic entomology. Because a variety of oligonucleotide probes can be placed on a single microarray chip, a single microarray could be produced that is capable of performing virtually every genotyping task needed for a forensic insect specimen (Wells and Stevens 2008).

13.3 Analysis of Human DNA Extracted from Insects

An insect can also be a useful source of vertebrate DNA. It has been shown that it is possible to obtain mtDNA-, AFLP-, and STR data from adult insect blood meals that can lead to individualization of the ingested host blood (Lord et al. 1998; Vos et al. 1995; Mukabana et al. 2002; Spitaleri et al. 2006). Forensic analysis of DNA recovered from a larvae's digestive tract can be used to identify what the larvae was feeding on (Wells et al. 2001; Zehner et al. 2004a). Case study 3 describes a laboratory experiment demonstrating that larval crops may be a suitable source of DNA for identification of both the insect and its gut contents, and therefore mtDNA analysis may potentially be used to associate a larva with a human corpse, even if physical contact between the two is not observed (Wells et al. 2001). Potential practical applications of such gut content analysis include: (1) resolving a chain-of-evidence dispute because the source of the larvae's food is disputed, (2) determining if the human corpse was present at a site at which only larvae are found, (3) detecting that a larva on a corpse had moved to the victim from another food source, and (4) identifying sperm DNA that was otherwise unretrievable because of larval feeding (Wells et al. 2001; Clery 2001).

Primer specificity is very important for determining the host DNA in the insect gut. Identification of vertebrates is typically based on the mtDNA cytochrome *b* (cyt *b*) gene, and less often on the noncoding D-loop. An investigator can select PCR primers that will amplify only vertebrate or only insect mtDNA, thus avoiding sequencing problems that result from a mixture. Because of the importance of genetic analysis of the contents of an insect's gut in crime cases, it is important to follow the recommendations for the collection, storage, and further handling of the specimen. If the larvae are too small for dissection or the larva's crop is no longer visible, DNA extraction may be less successful. Although some limitations of the technique still have to be explored, the results of studies so far suggest that the larvae's foregut, the crop, is the best source of larval gut contents, and that the mature larvae (third-instar larvae: 15–20 mm long, 4–6 days old) actively feeding on the corpse can be considered the best source of human DNA (Campobasso et al. 2005; Zehner et al. 2004a). Live larvae no longer on the corpse must be quickly located and preserved before the gut contents are digested (Linville and Wells 2002; Campobasso et al. 2005).

Case study 4 describes a forensic case that showed that it is possible to successfully obtain a complete human genetic profile even if the DNA is recovered from small and biologically contaminated traces such as mosquito blood trace on the wall (Spitaleri et al. 2006). Other studies have also shown that the contents of a larvae crop can be used to obtain a human or nonhuman mtDNA haplotype (Wells et al. 2001; Linville and Wells 2002) or a human Y-STR profile (Clery 2001). It is even possible to successfully analyze human hypervariable regions using beetle larvae that fed on human bone (DiZinno et al. 2002). DiZinno and colleagues presented arguments for the potential forensic utility of carrion insect gut analysis. Nevertheless, for a successful analysis, carefully controlled conditions are required as well as using relatively sensitive mtDNA techniques and/or undegraded tissue. It is still not certain if a less sensitive but more informative STR analysis is likely to work given the typically degraded state of human corpses during actual death investigations.

13.3.1 Case Study 1: A Caddis-Fly Casing in Service of Criminalistics

Caddisflies are moth-like insects that have aquatic larvae that make protective cases of silk decorated with gravel, sand, twigs, or other debris (Caspers 1952; in English described by Benecke 2001). In a case from 1948, a caddisfly (most likely *Limnephilus flavicornis L.*) casing was used to suggest a possible PMI. Also, because of the fibers it was built out of, it could also be connected to the socks of a dead woman. A few years later, Hubert Caspers described the case and since then the use of caddisfly casings was introduced as a tool for forensic investigation.

The naked body of a dead woman had been found in a moat of a windmill. She wore only a pair of red socks and she was wrapped in a sack. On one sock a caddisfly casing was found and it contained red fibers on its top and bottom. Sock fibers had obviously been used to build the casing. The conclusion was that the larva had mostly built its casing before entering the sack in which the deceased was wrapped in. The larva then finished the casing (fibers on top) and attached it to the red sock (fibers on bottom). According to the duration of the attachment procedure, which takes at least a few days, it was estimated that the body had been in the water for at least 1 week.

13.3.2 Case Study 2: Identity of Maggots Found on Outside and Inside of Body Bag

In October 1997, a severely decomposed body was analyzed for insect evidence in order to determine the PMI (Benecke 1998). The average size of maggots found on the corpse and on the outside of the closed body bag was 9 mm. The question was whether the maggots were of the same or different species. In one case, they might have squeezed themselves through tiny holes and came from outside the bag (e.g., to find a place for pupation). The other option suggested that a second oviposition of another species might have taken place after the body was put into the bag. Moreover, pupae were found on the floor beneath the corpse. They could not be directly related to any specific body in the morgue, that is, they may have fallen down from several unknown corpses. Because of the different developing time of fly species, PMI estimation is only possible if the insect species is known and can thus be used as a time since death indicator. It is difficult to identify the species of maggots, particularly in the early stage of their life cycle. Therefore, RAPD, a quick and inexpensive

DNA test known to help in cases dealing with a variety of species, was used. The method was used to determine if the maggots that were found on the inside of a body bag were identical (a) with maggots found on the outside of the bag and (b) with pupae found on the floor under the corpse. Four single maggots found at the body bag were compared to a pupa found near the corpse, a blow fly from another case, and an adult silphid carrion beetle (negative control). Comparison of overall, the RAPD DNA profile showed that the profiles of maggots collected at the body bag were of the same genus. It was also found that the pupa was not a different (later) stage of one of the maggots from the actual corpse. There was no similarity between the DNA profiles of the two fly species and between the profiles produced by the different primers. The DNA profile of the beetle (negative control) did not show any similarities to the blow fly profiles.

Although this was not a serious evaluation of the utility of a method for specimen identification, it has been shown that RAPD results could be transferred from research to practical applications such as some urgent forensic investigations. In this example, RAPD profiles were used to distinguish *Lucilia sericata*, *Calliphora vicina* (=*erythrocephala*), and a beetle.

13.3.3 Case Study 3: Human and Insect mtDNA Analysis from Maggots

Usually, the estimate of the PMI is based on the age of a fly larva (maggot), if all of their development and feeding occurred on the victim (Wells et al. 2001). This is the only case when the age of the larva is relevant to PMI. However, there is a variety of situations where an alternative method for associating a maggot with a corpse would have been useful for forensic investigation: (1) the investigators discover maggots but no corpse, (2) the maggots are not found directly on a corpse, and an alternative food source is nearby, (3) the maggots are found on a corpse, but may have come from somewhere else. In this study, it was demonstrated that mtDNA sequence data can be obtained from both the gut contents and the gut itself from maggots that had fed on human tissue. These data can be used to identify both the human corpse upon which the maggot was feeding and the species of the maggot itself.

Eggs were obtained from wild *Cynomya cadaverina* collected near San Antonio, Texas, and the resulting larvae were raised on human liver tissue that had been removed during an organ transplant procedure. Third larval instars were preserved in 70% ethanol at about 12 mm in length and stored at −20°C until used DNA extraction. Other larvae were allowed to complete development to the adult stage. In order to produce reference haplotypes to be compared to human and fly DNA from the maggot crops, liver donor donated blood as well and the thoracic muscle from one of the emerged adult flies was used. The DNA of each sample type was extracted following the protocols from Qiagen DNA extraction kits. PCR analysis was done for a region of the fly cytochrome oxidase subunit one (COI) and a region of the human hypervariable region II (HVII).

13.3.3.1 Insect mtDNA

Successful amplification of a fragment of COI was obtained in all tested larval crops and the adult fly. Three of the larvae differed from the adult due to the silent T to A substitution, whereas two produced COI haplotypes identical to the adult fly. The COI primers failed to amplify DNA from the reference blood sample, as expected.

13.3.3.2 Human mtDNA

The HV2 primers failed to amplify DNA from the reference adult fly, as expected. A fragment of HV2 from the blood reference sample and from three of five crop extracts was successfully amplified. All HV2 sequences were identical and differed from the standard human sequence by the presence of a guanine at position 263 and by a cytosine at position 315.1.

Evidently, maggot crops can be a suitable source of DNA for identification of both the insect and its gut contents showing that mtDNA analysis may potentially be used to associate a maggot with a human corpse, even if physical contact between the two is not observed.

13.3.4 Case Study 4: Genotyping of Human DNA Recovered from Mosquitoes Found at a Crime Scene

Often, a mosquito will rest very close to the location of its recent blood meal, thereby potentially being a source of blood from a person who is no longer present (Spilateri et al. 2006). The possibility of amplifying human DNA from blood meals of Diptera species was demonstrated in a casework that occurred in Sicily. The corpse of a transsexual prostitute was discovered in a vegetated area near a beach. There were marks on the body suggesting strangulation as the cause of death. A suspect, a distinguished businessman, was identified because his car had been seen in the area around the time of the murder. A search of the suspect's home, which was located inland and far from the beach, did not yield any conventional biological evidence. There were, however, the remains of a blood-fed mosquito that had apparently been swatted against a wall, leaving both a bloodstain and the damaged insect specimen. Both the stain and insect specimen were collected, along with sample of the suspect's clothing that contained traces of sand grains and vegetation fragments. An optimized DNA extraction was carefully carried out in order to yield DNA. PCR amplification and STRs profiling at 15 human genetic loci was then performed on the extracted DNA, using AmpFLSTR Identifiler (Applied Biosystems). A nearly complete 15 loci profile was obtained either from the blood on filter paper or from the blood debris obtained by scraping the stain on the wall. So, it was confirmed that each signal corresponded to a given human STR amplicon and not to unspecific amplification of unknown DNA. The obtained profile fully matched the victim's profile, confirming her presence at least in the neighborhood of the suspect's home. Entomological examination attributed the mosquito to the species whose members, under normal conditions, do not travel the distance between the beach and the suspect's home. Electron microscopy and chemical x-ray microanalysis of the sand collected from the soles of the sneakers revealed that the grains found in the suspect's sneakers were consistent with the ones sampled at the beach. Leaf fragments found on suspect's clothing were from the same plant species whose bushes were hiding the victim's body. In addition, this plant was not reported to flourish in any medium other than sandy soil. If taken alone, all of the aforementioned occurrences could not have served as an unquestionable link between the suspect and the murder. However, when taken into consideration together with police observation and significant scientific data, the extreme rarity of such joined events can lead the jury toward a conviction of second-degree murder.

Results of this case report showed that it is possible to successfully obtain a complete genetic profile even if DNA is recovered from poor and biologically contaminated traces.

It is crucial to treat the evidence properly and to evaluate it with respect to other factors revealed by a thorough approach to the crime scene. The applied analytical strategy has proved to be a powerful tool for the investigation, which enabled the authorities to obtain the DNA profile of the major suspect of the murder.

References

Alamalakala, L., S.R. Skoda, and J.E. Foster. 2009. Amplified fragment length polymorphism used for inter- and intraspecific differentiation of screwworms (Diptera: Calliphoridae). *Bull Entomol Res* 99:139–149.

Amendt, J.R., C.P. Campobasso, E. Gaundry, C. Reiter, H. LeBlance, and M. Hall. 2007. Best practice in forensic entomology: Standards and guidelines. *Int J Legal Med* 121:90–104.

Amendt, J.R. and M. Hall. 2007. Forensic entomology—Standards and guidelines. *Proceedings of the 6th International Congress of the Baltic Medico-Legal Association—New Technologies in Forensic Medicine. Forensic Sci Int* 169(1):S27.

Anderson, G.S. and V.J. Cervenka. 2002. Insects associated with the body: Their use and analyses. In: *Advances in Forensic Taphonomy—Method, Theory and Archaeological Perspectives*, ed. W.D. Haglund and M.H. Sorg, 173–200. Boca Raton, FL: CRC Press.

Balitzki-Korte, B., K. Anslinger, C. Bartsch, and B. Rolf. 2005. Species identification by means of pyrosequencing the mitochondrial 12S rRNA gene. *Int J Leg Med* 119:291–294.

Baudry, E., J. Bartos, K. Emerson, T. Whitworth and J.H. Werren. 2003. Wolbachia and genetic variability in the birdnest blowfly *Protocalliphora sialia*. *Mol Ecol* 12:1843–1854.

Belosludtsev, Y.Y., D. Bowerman, R. Weil, N. Marthandan, R. Balog, K. Luebke et al. 2004. *Biotechniques* 37:654–658, 660.

Benecke, M. 1998. Random amplified polymorphic DNA (RAPD) typing of necrophageous insects (Diptera, Coleoptera) in criminal forensic studies: Validation and use in praxi. *Forensic Sci Int* 98:157–168.

Benecke, M. 2001. A brief history of forensic entomology. *Forensic Sci Int* 120(1–2):2–14.

Boakye, D.A., J. Tang, P. Truc, A. Merriweather, and T.R. Unnasch. 1999. Identification of bloodmeals in haematophagous Diptera by cytochrome B heteroduplex analysis. *Med Vet Entomol* 13:282–287.

Boehme, P., P. Spahn, J.R. Amendt, and R. Zehner. 2013. Differential gene expression during metamorphosis: A promising approach for age estimation of forensically important *Calliphora vicina* pupae (Diptera: Calliphoridae). *Int J Legal Med* 127:243–249.

Brunak, S., J. Englebrecht, and S. Knudsen. 1990. Cleaning up gene databases. *Nature* 343:123.

Budowle, B., P. Garofano, A. Hellman, M. Ketchum, S. Kanthaswamy, W. Parson et al. 2005. Recommendations for animal DNA forensic and identity testing. *Int J Legal Med* 119:295–302.

Campobasso, C.P., J.G. Linville, J.D. Wells, and F. Introna. 2005. Forensic genetic analysis of insect gut contents. *Am J Forensic Med Pathol* 26(2):161–165.

Caspers, H. 1952. Ein Köcherfliegen-Gehäuse im Dienste der Kriminalistik [A caddis-fly casing in the service of criminalistics]. *Arch Hydrobiol* 46:125–127, 155.

Catts, E.P. and M.L. Goff. 1992. Forensic entomology in criminal investigations. *Annu Rev Entomol* 37:253–272.

Catts, E.P. and N.H. Haskell. 1990. *Entomology and Death—A Procedural Guide*. Clemson: Joyce's Print shop.

Clery, J.M. 2001. Stability of prostate specific antigen (PSA) and subsequent Y-STR typing of *Lucilia (Phaenicia) sericata* (Meigen) (Diptera: Calliphoridae) maggots reared from a simulated postmortem sexual assault. *Forensic Sci Int* 120:72–76.

DiZinno, J.A., W.D. Lord, M.B. Collins-Morton, M.R. Wilson and M.L. Goff. 2002. Mitochondrial DNA sequencing of beetle larvae (Nitidulidae: *Omosita*) recovered from human bone. *J Forensic Sci* 47:1337–1339.

Florin, A.B. and N. Gyllenstrand. 2002. Isolation and characterization of polymorphic microsatellite markers in the blowflies *Lucilia illustris* and *Lucilia sericata*. *Mol Ecol Notes* 2:113–116.

Greenberg, B. and J.C. Kunich. 2002. *Entomology and the Law—Flies as Forensic Indicators*. Cambridge: Cambridge Univ. Press.

Gunn, A. 2009. *Essential Forensic Biology*. Hong Kong: Wiley-Blackwell.

Hebert, P.D., A. Cywinska, S.L. Ball, and J.R. deWaard. 2003. Biological identifications through DNA barcodes. *Proc Biol Sci* 270(1512):313–321.

Linville, J.G. and J.D. Wells. 2002. Surface sterilization of a maggot using bleach does not interfere with mitochondrial DNA analysis of crop contents. *J Forensic Sci* 47:1055–1059.

Lord, W.D. and J.F. Burger. 1983. Collection and preservation of forensically important entomological materials. *J Forensic Sci* 28:936–944.

Lord, W.D., J.A. DiZinno, M.R. Wilson, B. Budowle, D. Taplin, and T.L. Meinking. 1998. Isolation, amplification, and sequencing of human mitochondrial DNA obtained from human crab louse, *Pthirus pubis* (L.), blood meals. *J Forensic Sci* 43:1097–1100.

MacPherson, J.M., P.E. Eckstein, G.J. Scoles, and A.A. Gajadhar. 1993. Variability of the random amplified polymorphic DNA assay among thermal cyclers, and effects of primer and DNA concentration. *Mol Cell Probes* 7:293–299.

Mukabana, W.R., W. Takken, P. Seda, G.F. Killeen, W.A. Hawley, and B.G. Knols. 2002. Extent of digestion affects the success of amplifying human DNA from blood meals of *Anopheles gambiae* (Diptera: Culicidae). *Bull Entomol Res* 92:233–239.

Nelson, L.A., J.F. Wallman, and M. Dowton. 2008. Identification of forensically important *Chrysomya* (Diptera: Calliphoridae) species using the second ribosomal internal transcribed spacer (ITS2). *Forensic Sci Int* 177:238–247.

Perez, T., J. Albornozm and A. Dominguez. 1998. An evaluation of RAPD fragment reproducibility and nature. *Mol Ecol* 7:1347–1357.

Picard, C.P., M.H. Villet, and J.D. Wells 2012. Amplified fragment length polymorphism confirms reciprocal monophyly in *Chrysomya putoria* and *Chrysomya chloropyga*: A correction of reported shared mtDNA haplotype. *Med Vet Entomol* 26:116–119.

Picard, C.J. and J.D. Wells. 2009. Survey of the genetic diversity of *Phormia regina* (Diptera: Calliphoridae) using amplified fragment length polymorphisms. *J Med Entomol* 46:664–670.

Picard, C.J. and J.D. Wells. 2010. The population genetic structure of North American *Lucilia sericata* (Diptera: Calliphoridae), and the utility of genetic assignment methods for reconstruction of postmortem corpse relocation. *Forensic Sci Int* 195:63–67.

Picard, C.J. and J.D. Wells. 2012. A test for carrion fly full siblings: A tool for detecting postmortem relocation of a corpse. *J Forensic Sci* 57:535–538.

Randall, B.B., M.F. Fierro, and R.C. Froede. 1998. Practice guideline for forensic pathology. *Arch Pathol Lab Med* 122:1056–1064.

Ritz-Timme, S., G. Rochholz, H.W. Schutz, M.J. Collins, E.R. Waite, C. Cattaneo et al. 2000. Quality assurance in age estimation based on aspartic acid racemisation. *Int J Legal Med* 114:83–86.

Rodrigues, R.A., A.M.L. Azeredo-Espin, and T.T. Torres. 2009. Microsatellite markers for population genetic studies of the blowfly *Chrysomya putoria* (Diptera: Calliphoridae). *Mem Inst Oswaldo Cruz* 104:1047–1050.

Schroeder, H., H. Klotzbach, S. Elias, C. Augustin, and K. Pueschel. 2003. Use of PCR-RFLP for differentiation of calliphorid larvae (Diptera, Calliphoridae) on human corpses. *Forensic Sci Int* 132:76–81.

Spitaleri, S., C. Romano, E. Di Luise, E. Ginestra and L. Saravo. 2006. Genotyping of human DNA recovered from mosquitoes found on a crime scene. *International Society for Forensic Genetics; Progress in Forensic Genetics 11: Proceedings of the 21st International ISFG Congress, International Congress Series* 1288:574–557.

Stevens, J.R. 2003. The evolution of myiasis in blowflies (Calliphoridae). *Int J Parasitol* 33:1105–1113.

Stevens, J. and R. Wall. 2001. Genetic relationships between blowflies (Calliphoridae) of forensic importance. *Forensic Sci Int* 120:116–123.

Stevens, J.R., R. Wall and J.D. Wells. 2002. Paraphyly in Hawaiian hybrid blowfly populations and the evolutionary history of anthropophilic species. *Insect Mol Biol* 11:141–148.

Tarone, A.M., K.C. Jennings, and D.R. Foran. 2007. Aging blow fly eggs using gene expression: A feasibility study. *J Forensic Sci* 52:1350–1354.

Torres, T.T. and A.M.L. Azeredo-Espin. 2008. Characterization of polymorphic microsatellite markers for the blowfly *Chrysomya albiceps* (Diptera: Calliphoridae). *Mol Ecol Resour* 8:208–210.

Vos, P., R. Hogers, M. Bleeker, M. Reijans, T. van de Lee, M. Hornes et al. 1995. AFLP: A new technique for DNA fingerprinting. *Nucleic Acids Res* 23:4407–4414.

Wallman, J.F., R. Leys, and K. Hogendoom. 2005. Molecular systematics of Australian carrion breeding blowflies (Diptera: Calliphoridae) based on mitochondrial DNA. *Invertebr Syst* 19:1–15.

Wells, J.D., T. Pape, and F.A.H. Sperling. 2001. DNA-based identification and molecular systematics of forensically important Sarcophagidae. *J Forensic Sci* 46(5):1098–1102.

Wells, J.D., F. Introna Jr., G. DiVella, C.P. Campobasso, J. Hayes, and F.A. Sperling. 2001. Human and insect mitochondrial DNA analysis from maggots. *J Forensic Sci* 46:685–687.

Wells, J.D. and F.A.H. Sperling. 2001. DNA-based identification of forensically important Chrysomyinae (Diptera: Calliphoridae). *Forensic Sci Int* 120(1):110–115.

Wells, J.D. and J.R. Stevens. 2008. Application of DNA-based methods in forensic entomology. *Annu Rev Entomol* 53:103–120.

Wells, J.D., R. Wall, and J.R. Stevens. 2007. Phylogenetic analysis of forensically important Lucilia flies based on cytochrome oxidase I: A cautionary tale for forensic species determination. *Int J Legal Med* 121:229–233.

Wells, J.D. and D.W. Williams. 2007. Validation of a DNA-based method for identifying Chrysomyinae (Diptera: Calliphoridae) used in a death investigation. *Int J Legal Med* 121:1–8.

Wetton, J.H., C.S. Tsang, C.A. Roney, and A.C. Spriggs. 2004. An extremely sensitive species-specific ARMs PCR test for the presence of tiger bone DNA. *Forensic Sci Int* 140:139–135.

Will, K.W. and D. Rubinoff. 2004. Myth of the molecule: DNA barcodes for species cannot replace morphology for identification and classification. *Cladistics* 20:47–55.

Zehner, R., J. Amendtand, and R. Krettek. 2004a. STR typing of human DNA from fly larvae fed on decomposing bodies. *J Forensic Sci* 49(2):337–340.

Zehner, R., J. Amendt, S. Schütt, J. Sauer, R. Krettek, and D. Povolný. 2004b. Genetic identification of forensically important flesh flies (Diptera: Sarcophagidae). *Int J Legal Med* 118(4):245–247.

14
Forensic Botany
Plants as Evidence in Criminal Cases and as Agents of Bioterrorism

HEATHER MILLER COYLE
HENRY C. LEE
TIMOTHY M. PALMBACH

Contents

14.1 Introduction	367
14.2 Evidence Collection	368
14.3 Overview of Techniques	370
14.3.1 Microscopy	370
14.3.2 Species Identification	371
14.3.3 DNA Individualization	372
14.3.3.1 Amplified Fragment Length Polymorphisms	372
14.3.3.2 Short Tandem Repeats	373
14.4 Palynology and Mycology	374
14.5 Missing Persons	375
14.5.1 Location of the Body	375
14.6 Toxicology	376
14.6.1 Bioterrorism	377
14.6.2 Drug Enforcement	378
14.7 Summary	379
References	379

14.1 Introduction

Forensic botany is the combination of many plant biology disciplines and their application to matters of law (Coyle 2005). Encompassed within forensic botany are aspects of plant anatomy, plant growth and behavior, plant genetics and population studies, and most recently, botanical classification based on morphology combined with plant molecular biology (DNA). The forensic applications are analogous to those in human identification (HID) and include recognition of plant evidence at the crime scene, collection and preservation of plant evidence, maintenance of a chain of custody, an understanding of scientific test methods, validation of novel forensic tests, construction of reference population databases, and testimony for the admissibility of this evidence in a court of law (Coyle 2005; De Forest et al. 1983; Miller Coyle et al. 2001; Lee and Ladd 2001; Lee et al. 2001).

This chapter is written to provide a general overview of forensic botany and highlights some of the specialty areas such as palynology (pollen identification), drug enforcement, and bioterrorism aspects in criminal casework. Plants as criminal evidence are highly varied in species but are prevalent in homicide as well as physical and sexual assaults as trace evidence. Almost every forensic science laboratory has examples of plant evidence, either intact or as fragments in their trace analysis section (e.g., grass species, tree species, algae, diatoms). Typically, a species identification is made first, followed by further individualizing tests by chemistry or DNA. In addition, drug and toxicology laboratories house many different examples of plant-derived evidence. Although this chapter focuses on criminal case applications, there is almost an equal area of civil case application that includes plants and plant products. Those areas in the United States are under the jurisdiction of the Food and Drug Administration and commercial and private plant breeders who seek to protect the quality of their plant-based products or new genetic varieties under the Plant Variety Protection Act and the Plant Patent Act (Coyle 2005).

The number of ways that plants can present as biological evidence is innumerable but a few examples include:

- Grass stains on a dress after sexual assault or *Cannabis* samples at a homicide scene
- Plant leaves and stems caught in the undercarriage of a vehicle at a secondary crime scene where a victim's body is dumped
- Seeds caught in the pants cuffs of a burglar who is fleeing the scene of a home invasion
- Stomach contents with vegetable matter from a victim's last meal to support time of death or an alibi to a location
- Pollen to date the season of burial of skeletal remains in a mass grave
- Geographic profiling to determine possible regions where a plant sample could have originated either via the plant itself or its packaging

All of these examples serve to assist the forensic community in establishing a linkage between a person and an object, a person and a scene, or a suspect to a victim. When there is little other biological evidence, plants can provide valuable evidence to a case but are often overlooked because of the microscopic size of pollen grains or small seeds. The first and most crucial step in using plants as evidence is the recognition of the evidentiary value of this form of biological evidence. This is not always straightforward since many investigators lack the specialized training, and tremendous numbers of different plant species may be encountered that vary considerably with geography (Coyle 2005; Miller Coyle et al. 2001; Lee et al. 2001). Only the drug forms of plants (e.g., marijuana, opium, cocaine, hashish) appear to be consistent in species across all types of crime scenes.

14.2 Evidence Collection

Crime scene investigators need to consider the potential for locating botanical evidence and modify their search methods and search locations to maximize the likelihood of locating this type of evidence. Before releasing a crime scene report, a review of the entire crime scene should be made to determine if any botanical evidence might have been overlooked for collection (Coyle 2005; Lee et al. 2001). In particular, one should search for the presence of small plant leaves, seeds, pollen grains, algae, grass, plant-based drugs, wood chips, and stems in or on a body, clothing, and at the scene. Known reference samples and control samples also need to be collected for later comparison and with plant evidence, one sample is not typically sufficient. For example, if a holly leaf is found clutched in a victim's hand at a secondary scene, and if three holly bushes are present at the suspect's residence, then at least three samples (one from each bush) must be collected for later comparison. There are many different genetic varieties of the same plant species; therefore, for later individualization (either by distinct morphological features or by genetic analysis), a known reference sample from each must be collected for comparison to one at the scene. If plant samples are small, investigators may have to obtain reference samples from a wide variety of potential plant donors located in or adjacent to the scene (Coyle 2005). If no reference samples are found at the scene, then herbarium or collection plant reference samples can be used as comparative references for species identification (Coyle 2005).

One of the most critical and time-consuming factors in forensic botany is performing the appropriate comparisons. From a single piece of evidence, it is often possible to make an identification of the plant species; however, to perform individualization back to a single source, often a comparative reference database must be created. For HID by DNA methods, forensic scientists use a preexisting population reference database to assess the individuality of the sample (Lee et al. 1998). With most plant cases, a reference population database will need to be constructed from (a) the same geographic region as for the crime scene or (b) a representative sampling of the plant species if it is very mobile and widely transportable (e.g., crops or drugs). Unfortunately, with very few exceptions, the comparative databases do not exist, and part of the overall examination process may require the creation of a specific database. Some places to look for reference databases are academic research institutions already performing population genetic studies on the plant of interest or private seed and plant technology companies that are involved in breeding genetic varieties of plant species for profit (Coyle 2005). Herbariums and botanical gardens may also allow for their private collections to be used as references.

The actual process of collecting plant evidence follows traditional forensic collection methods including crime scene documentation, photography, individual packaging of samples in paper envelopes, or pollen collection on swabs for sterile containers (Coyle 2005; De Forest et al. 1983; Miller Coyle et al. 2001; Lee et al. 2001). Each item of evidence must be wrapped separately. and a chain of custody and evidence seals must be maintained. For any unusual forms of plant evidence (e.g., stomach contents, vomit, fecal matter), preservation may be a bit different, such as collection in a plastic cup and storage at 4°C since drying may destroy the features of the plant cells such that species identification will be difficult. For plant DNA testing and appropriate archival of the sample for long term storage, fast technology for analysis of nucleic acids (FTA) cards can be used in a similar fashion as for HID testing (Allgeier et al. 2011). When in doubt, a consultation with the forensic botanist or forensic science laboratory would be recommended.

Crime scene documentation should include intermediate and close-up photos of the different plant sources in the crime scene (Lee et al. 2001). Investigators should carefully evaluate the entire scene such as to collect reference samples from any plants that appear to be of the same species as the questioned botanical material (Coyle 2005). Before proceeding to court with the plant evidence after it has been compared and tested, one should consider the following issues that may affect admissibility in court (Coyle 2005):

- An appropriate review of the evidence and testing procedures has been made from the crime scene through forensic laboratory testing.
- A chain of custody has been documented and maintained.
- Has plant evidence ever been admitted in this particular court system?

- What are the qualifications of the examiners performing the tests and writing the reports?
- Was it possible to perform a blind comparison for identification purposes?
- Is a representative population database of samples or DNA profiles available for comparative purposes? Was it created specifically for this case?
- Is the genetic reproductive strategy of the plant species known? For example, is it clonally propagated and likely to have more than one match genetically? Or is it seed propagated and therefore, individual in its genetic profile?
- Are all of the participants in the case apprised of the benefits and limitations of the botanical evidence to avoid discrepancies in testimony?
- What is the known and potential error rate related to this type of analysis?
- Are the procedures based on "generally accepted" scientific principles?

Botanical evidence is inherently variable because of the great number of plant species found on land and in water. A comprehensive understanding between crime scene personnel, forensic scientists, academic practitioners, and attorneys can lend clarity to a judge and jury when plant evidence is presented in a court of law.

14.3 Overview of Techniques

14.3.1 Microscopy

Simple, cost-effective procedures such as light microscopy and use of a comparison microscope are time-honored methods in forensic science. Most forensic science laboratories have a trace evidence section that collects fibers (many of which are of plant origin: cotton, jute), hairs, paint chips, soil, and pollen grains off of evidence using a vacuuming or scrape-down procedure. The vacuuming process uses a modified handheld vacuum with a specialized capture filter to collect microscopic pollen (Lee et al. 2001). Larger fragments of plant evidence are collected with forceps and placed in individual paper envelopes to preserve the color, shape, and surface features such as trichomes, glands, stomata, and other important features useful for species identification. The scrape-down procedure uses a spatula to physically scrape items such as clothing and carpets to loosen small fragments of plant material for collection and storage in paper envelopes (Lee et al. 2001). Resins such as hashish can be scraped into a druggist fold and placed in a secondary envelope for transport. FTA collection cards are usually used at the scene and are sealed in a secondary envelope before being sent to the laboratory for analysis (Allgeier et al. 2011).

Microscopy magnifies the physical features of a plant fragment so that vein patterns, stomata patterns, trichome shapes, and other features that lend uniqueness to a plant can be more readily identified. Typically, a botanical key is used to search a reference database for the plant species that possess those characteristics and enables species identification (Coyle 2005). A comparison microscope is useful if one is trying to determine a physical match between two plant fragments. Often, plant evidence is small and to the unaided eye, not very unique. Under magnification, however, if both fragments are placed under the same magnification on a comparative microscope, the images can be overlaid and matched analogous to two corresponding puzzle pieces. The leaf surface patterns of stomata and trichomes can provide additional information to the comparison as well as having a direct physical match between two edges. Be wary of using microscopy as a formal means of identification on some plant species because many appear morphologically similar and are not easy to distinguish with this method. For those species or for fragmented plant material, DNA identification is the optimal technique (Kress et al. 2009; Wiltshire 2006; Kress et al. 2005). These methods for species identification parallel HID testing on human hairs.

14.3.2 Species Identification

The proper identification of a plant to the species level is useful if one is trying to establish a connection to a primary or secondary crime scene (Coyle 2005; Lee et al. 1998). Frequently, victims are killed at one location and relocated to a secondary location in an attempt to hide the crime. Many cases include plant evidence found clutched in the victim's hand, attached to the victim's clothing, on a blanket, or in a vehicle that was used to transport the victim. In other cases, ask whether wood fragments collected from a vehicle's undercarriage can be matched back to a tree at the scene of the crime; or if pine needles found inside the barrel of a gun can be linked to the tree where the attacker was crouched and waiting for the victim; if leaf samples from a serial killer's body dumping site match to each other; if drug samples from a suitcase match those recovered from a suspected thief. These may sound like odd forms of associative evidence, but these are based on real cases where plant individualization by DNA was requested. Plant species identification answers the question of "What type of plant is this and what type of region does it grow in?," but cannot provide individualization to answer the question "Did this fragment come from this tree and only this tree?" Plant species identification is a necessary first classification step in trying to establish a forensic linkage. and then individualization is performed if sufficient sample quantity and quality is available for subsequent DNA testing (amplified fragment length polymorphisms [AFLP]; Figure 14.1).

Figure 14.1 Pollen from drug samples can be used to try and determine a geographic origin if mixed with pollen of indigenous species grown in the same location. The presence of cystolith hairs on *Cannabis* leaves is considered one form of botanical identification by microscopy. All botanical evidence should be properly photographed, documented, and packaged in clean paper envelopes with a chain of custody maintained.

Species identification can be made by assessing the characteristic physical features of the plant evidence, however, in many instances, the fragment is too small or lacks sufficient detail to make a conclusive identification microscopically. When this situation occurs, DNA can also be used to make a species identification (Kress et al. 2009; Wiltshire 2006; Kress et al. 2005; Bruni et al. 2010). Using commercially available plant DNA extraction kits, plant DNA can be obtained in less than 2 hours. A portion of the recovered plant DNA is then amplified using specific polymerase chain reaction (PCR) primers to selectively target a region that is specific to that plant species. There are hundreds of plant sequences that can be used for identification (nuclear genes: *18S, ITS1, ITS2*; chloroplast genes: *rbcL, atpB, ndhF*), and many searchable databases (e.g., http://www.ncbi.gov) are available for confirmation of sequence information (Coyle 2005; Kress et al. 2009; Wiltshire 2006; Kress et al. 2005; Bruni et al. 2010). A recent publication by Bruni et al. (2010) established that the matK plastid DNA marker was the most useful for rapid identification of poisonous plant species by DNA barcoding. This is especially important for use at poison control centers where rapid identification is a must to help ascertain best treatment for ingested poison victims. This particular approach can also be used for the species identification of trace plant particulates found on outer packaging for drugs as one example or on outer layers of clothing collected from a suspect as another example. The ability to effectively perform this analysis using DNA from trace particulates may be invaluable for geosourcing samples that are being trafficked across borders and can then be regulated by federal and international restrictions. In addition, many grass species are challenging to identify microscopically because of the extreme similarity of their morphological features. In Australia, a DNA identification system was developed as a tool for progressive identification of grass samples to different taxonomic levels (Ward et al. 2005, 2009).

14.3.3 DNA Individualization

14.3.3.1 Amplified Fragment Length Polymorphisms

The AFLP methodology was first described by Vos et al. (1995) and represents a method for genotyping single source plant samples for a large number of markers to generate a highly discriminating DNA band pattern. The AFLP technique is based on the detection of genomic restriction fragments after PCR amplification. Band patterns are produced without prior sequence knowledge, using a limited set of amplification primers. The AFLP method permits the inspection of polymorphisms at a large number of loci within a short period of time and requires only a small amount of DNA (about 10 ng). Although other DNA typing techniques exist and have been developed for specific applications, AFLP analysis is the method of choice since it has general applicability to any single source plant sample regardless of the species (Coyle 2005; Miller Coyle et al. 2001; Vos et al. 1995). It is similar to mitochondrial DNA sequencing, however, in that if two or more samples of the same species or if two different species are present, then the AFLP profile will appear as a mixture and will be difficult to interpret because of its complexity even with known reference samples for comparisons (Coyle 2005).

Several commercial kits for AFLP analysis are available and all follow the general procedure of

- One-step digestion–ligation reaction where short DNA adaptors are added to each end of a DNA fragment
- PCR preselective amplification of a subset of DNA fragments with one primer set to generate a pool of fragments

Forensic Botany

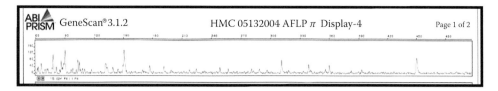

Figure 14.2 An example of an AFLP profile from *Sutera* leaves. Relative fluorescence units (RFU) are represented on the *y*-axis (0–1000); size (0–460 nucleotide bases) is represented by the *x*-axis.

- PCR selective amplification and fluorescent labeling of the subset of DNA fragments to generate an individualizing "barcode"
- Detection of the DNA fragments with a DNA sequencer and genetic analysis with computerized software to convert the "barcode" to a numeric code for comparisons

Figure 14.2 illustrates a typical AFLP profile that differs from short tandem repeats (STR) analysis in that discrete alleles are not typed. AFLP is most analogous to the original HID method developed by Alec Jeffreys known as multilocus typing by RFLP (restriction fragment length polymorphisms) and used bins to score the DNA fragments even if the fragments showed slight differences in mobility during gel electrophoresis (De Forest et al. 1983; Miller Coyle et al. 2001; Lee and Ladd 2001; Lee et al. 2001, 1998; Vos et al. 1995). The binning process involves creating "software bins" based on the size of DNA fragments and are arbitrary in the sense that the size accommodation of the bins can be anywhere from 30 to 100 bases, and as long as the scoring process is consistent, then the presence or absence of a peak is either 1 or 0, respectively. The greater the number of DNA fragments scored, the higher the degree of certainty of a match between the evidentiary sample and the known reference comparative sample (Coyle 2005).

14.3.3.2 Short Tandem Repeats

STRs, also known as simple sequence repeats (SSRs) or minisatellites, are highly individualizing and excellent for detection of mixed samples once they have been developed and validated for the plant species of interest. If a panel of STR markers is already in existence or the time and resources required are not a limitation, then they are a good method to use (Newton and Vo 2003; Craft et al. 2007; Howard et al. 2009; Štambuk et al. 2007; Howard et al. 2008). Often, however, a case may occur where an unusual plant species is recovered but there is no interest in developing a panel of STR markers for a single case since that may take approximately 2 years to complete and validate. In those cases, AFLP is a good choice since the technology is applicable to any plant species as long as it is a single-source sample. As technology advances, automation and chip-based technology for STR markers or single nucleotide polymorphisms may become a rapid and viable alternative for genetic individualization of plant samples and rapid construction of plant databases.

STR markers are usually multiplexed into a single reaction tube for PCR amplification and then detected by a DNA sequencer as discrete DNA fragments of a specific size when compared to an allelic ladder. Unlike human identity testing in forensics, each plant species may not have an allelic ladder available and therefore, sizing of the DNA fragment would be based on the software analysis and fragment mobility without a commercial reference. Computer software can generate fixed bins for fragment scoring and allele assignment. Nevertheless, many different animal and plant species have had STR marker panels developed, and some kits are even commercially available (Coyle 2005; Ward et al. 2005; Newton and Vo 2003; Craft et al. 2007). For plant species that are highly inbred, AFLPs have been suggested when STR markers may fail to distinguish between individual plants because of high levels of genetic similarity (Coyle 2005; Miller Coyle et al. 2001; Vos et al. 1995).

14.4 Palynology and Mycology

Pollen analysis, or palynology, is useful for providing forensic linkages as well as establishing time of death (Mildenhall et al. 2006; Mathewes 2006; Brown 2006; Bryant and Jones 2006; Montali et al. 2006; Mildenhall 2006a, b). An excellent review of the current state of forensic palynology was recently published by Walsh and Horrocks (2008) that discusses the need for greater utilization of this form of trace evidence as well as the development of rigorous standards. Pollen and spores can be obtained as microscopic trace material from a wide range of items including bodies. The usefulness of this form of evidence lies in the combination of species information that gleaned the uniqueness of the geographic distribution of the species collected, and the resistance of pollen and spores to mechanical and chemical destruction, all of which make it an ideal form of plant evidence. One of the best known cases utilizing palynology is that of a mass grave site in Magdeburg, Germany. In February of 1994, a mass grave was discovered that contained 32 male skeletons. Both the identities of the victims and that of the murderers were unknown. Because of the all-male population and close range of ages, it was presumed that these young men may have been part of a military troop. Two hypotheses were proposed: the victims were killed and disposed of in the spring of 1945 by the Gestapo at the end of World War II (Coyle 2005) or the victims were Soviet soldiers killed by the secret police in June of 1953 after the German Democratic revolt (De Forest et al. 1983). Twenty-one skulls were analyzed and seven showed high levels of pollen from plantain, lime tree, and rye. All three of these plant species release pollen during the months of June and July, which supports the second hypothesis (Miller Coyle et al. 2005). A similar application of palynology has been used for the first time to investigate war crimes in Bosnia (Brown 2006).

Pollen may also be useful in cases such as homicide, sexual and physical assault, forgery, armed robbery, or drug trafficking (Coyle 2005; Brown 2006; Montali et al. 2006; Mildenhall 2006a, b). Pollen analysis can help convict criminals, but its greatest application is for providing associative evidence for investigative leads. To identify the source of a pollen grain, its physical features are evaluated under the microscope (Coyle 2005). Those physical features include size, shape, aperture type, ornamentation or surface pattern, wall structure and composition, and mode of pollen grain dispersal. Control samples are typically samples of surface soil, mud, or water from a scene with the evidentiary samples being pollen grains recovered from almost any other material (Mildenhall et al. 2006). Examples of types of materials that have been examined for pollen in forensic cases include air filters, fruit, honey, heroin, leather, leaves, opium, packing materials, soil, wood, weapons, carpet, coffee, paper money, clothing, and shoes (Coyle 2005).

A case example from New Zealand includes the recovery and identification of *Cannabis* pollen from farmland that had been purchased and resold by a family that was suspected of growing and distributing marijuana. Since pollen remains in soil samples for an extended period, a sample was collected from the floorboards of the farm building owned by the family over the previous 5 years. Authorities were then able to seize the family's assets based on an estimate of their total income (Coyle 2005; Mildenhall et al. 2006). Another interesting application of pollen identification is to try and track heroin and cocaine back to its geographic origin where it was grown and packaged. If pollen types are rare and specific to a region, they can be informative for determining a geographic point of origin, which is useful when tracking illegal airport shipments for example (Coyle 2005; Mildenhall et al. 2006; Mathewes 2006; Bryant and Jones 2006).

Mycology is the use of fungi in criminal investigations. Fungi have been used in at least 13 criminal cases, and the potential uses are providing trace evidence, estimating postmortem intervals, determining time of deposit, cause of death, hallucinations, poisonings, and aid in locating burial sites. An excellent review has recently been published by Hawksworth and Wiltshire (2011), and additional case examples are in Forensic Botany (Coyle 2005). Even bryophytes have received some good press for crime-solving; studies have been performed to establish their forensic usefulness on footwear (Virtanen et al. 2007). Bryophytes remain attached after drying to shoes for many, hours and bryophyte DNA is recoverable and typable in most scenarios tested, making them a useful plant species for casework.

14.5 Missing Persons

14.5.1 Location of the Body

The uses of botany in forensic investigations include the identification of plant fragments, age dating of roots and stems, and using vegetation as an indication of surface disturbance. Many agencies such as Necrosearch International (http://www.necrosearch.org/) provide specialized training in forensic botany as a tool for identification of clandestine graves and for dating of time of burial. Plants grow back at predictable rates after an area of land has been disturbed such as in a burial. If vegetation patterns are different from the surrounding areas, then based on the type of vegetation and the estimated rate of plant growth, one can predict how long the body has been at that site.

Leaf litter has been used successfully to locate a missing person in a murder–suicide case where little other evidence was present. This case illustrates how geographic profiling of plant species and their growth constraints can limit the large geographic areas that need to be searched to find a girl's body. On June 4, 2002, two forensic scientists from the California Department of Justice called the Biological Sciences Department at California State University (Chico, CA) seeking a trained forensic botanist to assist them in "profiling" the location of a missing girl from botanical evidence recovered from her father's vehicle (Coyle 2005; Schierenbeck 2003). The father of the child had died of a self-inflicted gunshot wound in his truck along a mountainous highway in Butte County, CA. A jacket with the girl's blood was found on the seat next to the father along with some plant leaves but extensive searches of the area did not lead to recovery of her body. Based on the plant species present in the sample, it was determined that the botanical sample came from the top centimeter of leaf litter in the area. The plant species included greenleaf manzanita (*Arctostaphylos patula*), canyon live oak (*Quercus chrysolepsis*), interior live oak (*Quercus wislizenii*), white fir (*Abies concolor*), ponderosa pine (*Pinus ponderosa*), and black oak (*Quercus kelloggii*). Identification of plant species was made by comparison to known reference samples available in the California State University Herbarium by morphological characteristics. The habitat and growth conditions were then considered in the area where the vehicle was found and certain regions of the mountains could be eliminated based on the fact that these plant species did not coincide together in growth habitat. The composition and dark color of the leaf litter suggested high organic matter content and indicated the body was likely in a fairly dense forest canopy. After considering light conditions for growth and elevation requirements, five possible areas were designated on maps, and teams of search personnel and dogs were brought in to attempt to recover the body of the missing girl. Eventually, the body of the girl was located at the fifth area within a few miles of where the truck was found and the girls' body was retrieved from between two fallen trees and was wrapped carefully in a blanket by the father after her homicide. On retracing the possible routes of the father during the homicide, all of the plant species were encountered and this case illustrates how with the proper training, experience, and cooperation of personnel, case closure can be achieved (Coyle 2005; Schierenbeck 2003).

14.6 Toxicology

Plants can be identified using both biological and chemical methods, and once characterized, those methods can often be used in a forensic assessment to differentiate between a properly consumed food item and a poisonous plant. This becomes relevant in forensic science to aid in establishing the cause and manner of death and for determining if legal action is appropriate for a wrongful death (Coyle 2005; Moffat 1980).

One of the key aspects to a wrongful death lawsuit is establishing if the decedent intentionally ingested a substance that was harmful or if they knowingly exceeded the recommended dosage. In addition, establishing if death or severe illness was due to a food contaminant introduced during manufacturing is important in civil lawsuits. Standard toxicology assays have been developed to measure toxic substances in blood or to assess changes in tissue that can aid in the identification of many common plant poisons. Regardless of whether the decedent was poisoned, or the death was due to accident or suicide, the forensic scientist still must link information from the autopsy back to a source and time of ingestion (De Forest et al. 1983; Lee et al. 2001). The first step in this process is the proper identification of the plant species (Bruni et al. 2010).

In a survey of forensic literature, the most common sources of plant-related deaths have been identified. Poison Control Centers across the United States received more than 57,000 inquiries related to plant exposures in 2003 (Nelson et al. 2007). Eighty-five percent of these inquiries involved children under the age of 6 years. For small children, in particular, most of these plant exposures result from accidental ingestion. However, in some cases, poisoning is a possibility. Typically, plant samples are recovered at autopsy, in the mouth, digestive tract, or stomach. Trained forensic botanists can recover the undigested cellulose material from the acidic stomach environment, rinse them, and rehydrate the cells for microscopic analyses. The indigestible plant cells have a characteristic shape that can be used to identify contents of a last meal, for example (e.g., tomato, potato, spinach). This technique has been used successfully for the identification of a body found on a remote road. The victim's last meal was consistent with that from a Mexican restaurant.

Area restaurants were surveyed and witnesses who recognized her led law enforcement to a suspect whom was later convicted (Coyle 2005). It is even possible to obtain DNA from plant seeds that have been recovered at autopsy from the stomach (Coyle 2005).

14.6.1 Bioterrorism

Even more of an increasing concern in the forensic community, is the potential for easily accessible plants and their poisons to be used as bioterrorist agents to instill mass hysteria, serious health hazards, and panic in the general population (Schier et al. 2007; He et al. 2007).

Some plant poisons for concern include the following:

- Abrin—Obtained from *Abrus precatorius* L., a weed common to the tropics and subtropics. The seeds are decorative and used for jewelry, handicrafts, and good luck charms, and are imported into the United States by tourists. The toxin is a plant lectin that inhibits cellular protein synthesis and is extremely toxic. Abrin is contained in the hard seed coat and is not released until chewed and digested in the gastrointestinal tract where they can cause severe distress. Abrin ingestion may not result in death, however, especially if the seed remained intact during digestion. If injected or inhaled, abrin can cause multisystem organ failure even in small doses (Nelson et al. 2007; Schier et al. 2007; He et al. 2007; Ler et al. 2006).
- Aconitine—Obtained from a perennial plant (*Aconitum* species) with a geographic distribution of North America, Eurasia. The entire plant is poisonous including the seeds; aconitine acts as a sodium channel activator. These plants are used in some herbal products and are the likely source of exposure. Symptoms are predominantly cardiological and neurological in nature and lead to convulsions and coma before death (Nelson et al. 2007).
- Ricin—Obtained from an annual (*Ricinus communis* L.) plant that grows to 15 feet in the tropics. It grows naturally in the West Indies, also as a weed in the United States (Florida, Texas, California, and Hawaii). It is commercially available as well and sold for its foliage. The toxin is contained in the seeds and not released until the seed coat is broken. The symptoms are similar to that of abrin. There are several forensic cases where ricin has been used to successfully poison others, including a famous case involving a spy, Georgi Markov (Moffat 1980; Nelson et al. 2007; Schier et al. 2007; He et al. 2007; Ler et al. 2006).
- Strychnine—Obtained from *Strychnos nux-vomica* L., which is a small tree native to Asia and cultivated in Hawaii to produce strychnine. The whole plant, including the seeds, is poisonous. The seeds are used in some countries as a medicinal herbal product. Seizures and hyperthermia may result from large doses (Nelson et al. 2007).

Although these are just four major poisons obtained from plants, there are an infinite variety of less well-known and much less traceable poisonous plants, many of which have poisonous seeds. Before death, the species identification of a plant may aid in treatment of the patient, put their level of risk in context, and add a time frame for the development of clinical findings (Nelson et al. 2007). After death, the microscopic or DNA-based identification of ingested seeds can aid in establishing cause of death (Nelson et al. 2007).

14.6.2 Drug Enforcement

The introduction of a highly invasive plant species is very heavily illustrated by the illegal cultivation of marijuana on public federal land in the United States. To avoid detection and protect forfeiture assets from being seized during drug confiscation operations, various violent criminal organizations are using U.S. federal parks for the cultivation of their drug crops (Office of National Drug Control Policy 2007). Bioterrorism is usually defined as "any biological agent that is used to instill fear and widespread panic among the general public." The use of federal land to grow marijuana has resulted in the public encountering dangerous booby traps designed to protect the illegal crops, environmental contamination of water and soil by pesticides, and herbicides used in the illegal growing process, and ultimately, an increased health hazard to the consumer of the illegal drug as the THC content in these plants has risen sharply in the past 10 years to almost 20% (Office of National Drug Control Policy 2007). It is estimated that for every acre of forest planted with marijuana, 10 acres are damaged by toxic chemicals, and to restore the areas that have been damaged by chemicals and clean up the refuse left by illegal camp sites for cultivation, it costs U.S. taxpayer $11,000 per acre (Office of National Drug Control Policy 2007). In response to this ever-growing threat, the Office of National Drug Control Policy (ONDCP) established the National Marijuana Initiative to coordinate Federal, State, and local law enforcement to target, disrupt, and dismantle marijuana-growing operations in seven states (California, Kentucky, Hawaii, Oregon, Tennessee, Washington, and West Virginia). In 2007, these agencies eradicated more than 6.8 million marijuana plants, clearly indicating that this is not a small problem (Office of National Drug Control Policy 2007). In addition, more than $18 million of funding and 8000 hours of National Guard support has been provided toward this public safety and national security problem in 2007 alone. As long as there is increased drug demand for high THC content marijuana and it is difficult to purchase outside the U.S. border, then the illegal cultivation will continue within the borders. According to recent ONDCP data, almost 17 million Americans age 12 or older have used marijuana within the past month. Since 1998, the average potency for marijuana has doubled, which means that the toxic and neurological effects from long-term smoking are also likely to be significantly more severe (Office of National Drug Control Policy 2010). The typical health effects for this Schedule I drug (high level of abuse; no scientifically proven medical value) include distorted perceptions, impaired coordination, difficulty in thinking and problem solving, significant memory loss, and increased rates of anxiety, depression, suicide, and schizophrenia (Office of National Drug Control Policy 2010). These medical issues are above and beyond the smoking aspect of marijuana, which is similar in health effects to that described by the U.S. Surgeon General for tobacco products and includes emphysema and lung cancer (U.S. Department of Health and Human Services 2010). The intentional illicit growth on federal land and intentional distribution of this drug to Americans by Mexican drug cartels, which have infiltrated many states and cities in the United States. and involve violent gang activity, can be viewed as a larger act of bioterrorism against Americans and reinforces the need for this plant substance to be controlled and maintained as a Schedule I drug at this time.

The Commerce Clause refers to Article 1, Section 8, and Clause 3 of the U.S. Constitution, which gives Congress the power "to regulate commerce with foreign nations, and among the several states, and with the Indian tribes" (Gerace 1999). This can be invoked to assist in the regulation of dangerous goods (i.e., bioterrorist threats, drug enforcement) that are sold between different countries, states, and territories. Once we view items that cross large geographic areas and affect populations on a global scale, we move into the arena of public health and foreign policy requiring high levels of interagency cooperation. In forensic science, we have seen this effectively utilized for human, animal, and drug trafficking interdiction.

Cannabis can be collected from crime scenes (Coyle 2005; De Forest et al. 1983; Miller Coyle et al. 2001; Lee et al. 2001) and identified by either chemical or DNA methods (De Forest et al. 1983; Miller Coyle et al. 2001; Vos et al. 1995). The ability to sort *Cannabis* varieties or uniquely individualize samples is still an open-ended question. Certainly, AFLP analysis on fresh plant material can generate highly complex DNA profiles for matching to evidentiary samples (Coyle 2005; Miller Coyle et al. 2001). Other types of markers are under scrutiny including intergenic spacer (IGS) sequences and STR markers. There are multiple *Cannabis*-specific PCR primers available (Coyle 2005; Hsieh et al. 2003), and recently the rDNA IGS structure of *Cannabis sativa* showed that of 77 different marijuana samples, the DNA products grouped these samples into four different categories (Hsieh et al. 2005). In addition, STR panels for *Cannabis* are under development and assessment (Allgeier et al. 2011; Howard et al. 2008, 2009) for general implementation in forensic laboratories in the near future. The genetics of *C. sativa* and its breeding history are not fully elucidated as yet but as more DNA markers are developed and further surveys of different types of marijuana populations continue, the ability to use DNA to both identify and further individualize *Cannabis* varieties will soon be at hand.

14.7 Summary

Perhaps inherent to our human nature is a need to organize and classify objects and organisms that we encounter. Many items and organisms are easy to identify and characterize; however, others are more difficult because of a lack of physical features, or conflicting traits such as morphology and molecular markers that do not agree. Botanical evidence is useful in criminal and civil cases but is still a much underutilized resource. Plant evidence is commonly associated with crime scenes and victims and suspects; typically, stems, leaves, seeds, and flowers are useful as trace botanicals. Not just intact plant materials, but also microscopic pollen and plant resins may be used to demonstrate an associative link between scenes and evidentiary items. Pollen is also useful, if unique to a season or location, for tracing geographic origin or for estimating a season of death and burial. As more is learned about forensic botany and this type of evidence is used more often in court, plant associative evidence will be increasingly viewed as crucial for resolving the case both by species identification and subsequent individualization of the sample. In addition, with a focus on bioterrorism and drug enforcement, the global aspects of plant evidence and drug trafficking affect foreign and national security and health policies as monitoring of border control becomes essential for detection and elimination of these threats.

References

Allgeier, L., J. Hemenway, N. Shirley, T. LaNier, and H. Miller Coyle. 2011. Field testing of DNA collection cards for *Cannabis sativa* with a single hexanucleotide marker. *J. Forensic Sci.* 56(5): 1245–1249.

Brown, A.G. 2006. The use of forensic botany and geology in war crimes investigations in NE Bosnia. *Forensic Sci. Int.* 163(3): 204–210.

Bruni, I., F. De Mattia, and A. Galimberti. 2010. Identification of poisonous plants by DNA barcoding approach. *Int. J. Legal Med.* 124(6): 595–603.

Bryant, V.M., and G.D. Jones. 2006. Forensic palynology: Current status of a rarely used technique in the United States of America. *Forensic Sci. Int.* 163(3): 183–197.

Coyle, H.M. 2005. *Forensic Botany: Principles and Applications to Criminal Casework*. Boca Raton, FL: CRC Press.

Craft, K.J., J.D. Owens, and M.V. Ashley. 2007. Application of plant DNA markers in forensic botany: Genetic comparison of *Quercus* evidence leaves to crime scene trees using microsatellites. *Forensic Sci. Int.* 165(1): 64–70.

De Forest, P.R., R.E. Gaensslen, and H.C. Lee. 1983. *Forensic Science: An Introduction to Criminalistics*. New York: McGraw-Hill.

Gerace, T.A. 1999. The toxic-tobacco law: "Appropriate remedial action." *J. Public Health Policy* 20(4): 394–407.

Hawksworth, D.L., and P.E. Wiltshire. 2011. Forensic mycology: The use of fungi in criminal investigations. *Forensic Sci. Int.* 206(1–3): 1–11.

He, X., D.L. Brandon, G.Q. Chen, T.A. McKeon, and J.M. Carter. 2007. Detection of castor contamination by real-time polymerase chain reaction. *J. Agric. Food Chem.* 55(2): 545–550.

Howard, C., S. Gilmore, J. Robertson, and R. Peakall. 2008. Developmental validation of a *Cannabis sativa* STR multiplex system for forensic analysis. *J. Forensic Sci.* 53(5): 1061–1067.

Howard, C., S. Gilmore, J. Robertson, and R. Peakall. 2009. A *Cannabis sativa* STR genotype database for Australian seizures: Forensic applications and limitations. *J. Forensic Sci.* 54(3): 556–563.

Hsieh, H.M., C.L. Liu, L.C. Tsai et al. 2005. Characterization of the polymorphic repeat sequence within the rDNA IGS of *Cannabis sativa*. *Forensic Sci. Int.* 152(1): 23–28.

Hsieh, H.M., R.J. Hou, L.C. Tsai et al. 2003. A highly polymorphic STR locus in *Cannabis sativa*. *Forensic Sci. Int.* 131(1): 53–58.

Kress, W.J., D.L. Erickson, F.A. Jones et al. 2009. Plant DNA barcodes and a community phylogeny of a tropical forest dynamics plot in Panama. *Proc. Natl. Acad. Sci. U. S. A.* 106(44): 18621–18626.

Kress, W.J., K.J. Wurdack, E.A. Zimmer, L.A. Weigt, and D.H. Janzen. 2005. Use of DNA barcodes to identify flowering plants. *Proc. Natl. Acad. Sci. U. S. A.* 102(23): 8369–8374.

Lee, H.C., and C. Ladd. 2001. Preservation and collection of DNA evidence. *Croat Med J.* 42: 225.

Lee, H.C., C. Ladd, C.A. Scherczinger, and M.T. Bourke. 1998. Forensic applications of DNA typing, collection and preservation of DNA evidence. *Am. J. Forensic Med. Pathol.* 19: 10.

Lee, H.C., T.M. Palmbach, and M.T. Miller. 2001. *Henry Lee's Crime Scene Handbook*. Boston: Academic Press.

Ler, S.G., F.K. Lee, and P. Gopalakrishnakone. 2006. Trends in detection of warfare agents: Detection methods for ricin, staphylococcal enterotoxin B and T-2 toxin. *J. Chromatogr. A* 1133(1–2): 1–12.

Mathewes, R.W. 2006. Forensic palynology in Canada: An overview with emphasis on archaeology and anthropology. *Forensic Sci. Int.* 163(3): 198–203.

Mildenhall, D.C. 2006a. An unusual appearance of a common pollen type indicates the scene of the crime. *Forensic Sci. Int.* 163(3): 236–240.

Mildenhall, D.C. 2006b. *Hypericum* pollen determines the presence of burglars at the scene of a crime: an example of forensic palynology. *Forensic Sci. Int.* 163(3): 231–225.

Mildenhall, D.C., P.E. Wiltshire, and V.M. Bryant. 2006. Forensic palynology: Why do it and how it works. *Forensic Sci Int.* 163(3): 163–172.

Miller Coyle, H., C. Ladd, T. Palmbach, and H.C. Lee. 2001. The Green Revolution: Botanical contributions to forensics and drug enforcement. *Croat. Med. J.* 42(3): 340–345.

Miller Coyle, H., C.L. Lee, W.Y. Lin, H.C. Lee, and T.M. Palmbach. 2005. Forensic botany: Using plant evidence to aid in forensic death investigation. *Croat. Med. J.* 46(4): 606–612.

Moffat, A.C. 1980. Forensic pharmacognosy—Poisoning with plants. *J. Forensic Sci. Soc.* 20(2): 103–109.

Montali, E., A.M. Mercuri, G. Trevisan Grandi, and C.A. Accorsi. 2006. Towards a "crime pollen calendar"—Pollen analysis on corpses throughout one year. *Forensic Sci. Int.* 163(3): 211–223.

Nelson, L.S., R.D. Shih, and M.J. Balick. 2007. *Handbook of Poisonous and Injurious Plants*, New York: Springer.

Newton, C., and T. Vo. 2003. Application of microsatellite markers in plant forensics. *Can. Soc. Forensic Sci. J*. 36: 57.

Office of National Drug Control Policy. 2007. Marijuana on public lands. www.WhiteHouseDrugPolicy.gov. Accessed January 23, 2011.

Office of National Drug Control Policy. 2010. Marijuana: Know the facts. www.WhiteHouseDrugPolicy.gov. Accessed January 23, 2011.

Schier, J.G., M.M. Patel, M.G. Belson et al. 2007. Public health investigation after the discovery of ricin in a South Carolina postal facility. *Am. J. Public Health*. 97(Suppl. 1): S152–S157.

Schierenbeck, K.A. 2003. Forensic biology. *J. Forensic Sci*. 48(3): 696.

Štambuk, S., D. Sutlović, P. Bakarić, Š. Petričević, and Š. Andjelinović. 2007. Forensic botany: Potential usefulness of microsatellite-based genotyping of Croatian olive (*Olea europaea* L.) in forensic casework. *Croat. Med. J*. 48(4): 556–562.

U.S. Department of Health and Human Services. 2010. How tobacco smoke causes disease: The biology and behavioral basis for smoking-attributable disease. www.surgeongeneral.gov. Accessed January 23, 2011.

Virtanen, V., H. Korpelainen, and K. Kostamo. 2007. Forensic botany: Usability of bryophyte material in forensic studies. *Forensic Sci. Int*. 172(2–3): 161–163.

Vos, P., R. Hogers, M. Bleeker, M. Reijans, T. van de Lee, M. Hornes, A. Frijters, J. Pot, J. Peleman, M. Kuiper, and M. Zabeau. 1995. AFLP: A new technique for DNA fingerprinting. *Nucleic Acids Res*. 23: 4407.

Walsh, K.A., and M. Horrocks. 2008. Palynology: Its position in the field of forensic science. *J Forensic Sci*. 53(5): 1053–1060.

Ward, J., R. Peakall, S.R. Gilmore, and J. Robertson. 2005. A molecular identification system for grasses: A novel technology for forensic botany. *Forensic Sci. Int*. 152(2–3): 121–131.

Ward, J., S.R. Gilmore, J. Robertson, and R. Peakall. 2009. A grass molecular identification system for forensic botany: A critical evaluation of the strengths and limitations. *J. Forensic Sci*. 54(6): 1254–1260.

Wiltshire, P.E. 2006. Hair as a source of forensic evidence in murder investigations. *Forensic Sci. Int*. 163(3): 241–248.

Recent Developments and Future Directions in Human Forensic Molecular Biology

III

Forensic Tissue Identification with Nucleic Acids

15

DMITRY ZUBAKOV
MANFRED KAYSER

Contents

15.1 Introduction	386
15.2 Classical Tests for Forensic Tissue Identification	386
15.2.1 Peripheral Blood	386
15.2.2 Menstrual Blood	387
15.2.3 Parturient Blood	388
15.2.4 Fetal and Neonatal Blood	388
15.2.5 Nasal Blood	388
15.2.6 Saliva	388
15.2.7 Semen	389
15.2.8 Vaginal Secretion	389
15.2.9 Skin	390
15.3 RNA-Based Approaches to Forensic Tissue Identification	391
15.3.1 Stability of RNA	391
15.3.2 mRNA Markers for Forensic Tissue Identification	392
15.3.3 Multiplex mRNA Systems for Parallel Determination of Various Cell Types	394
15.3.4 Stand-Alone mRNA Assays for Identifying Particular Body Fluids	399
15.3.4.1 Nasal Blood	399
15.3.4.2 Parturient Blood	400
15.3.4.3 Neonatal Blood	400
15.3.4.4 Sweat	400
15.3.4.5 Vaginal Secretion versus Skin	400
15.3.5 miRNA Markers for Forensic Tissue Identification	401
15.3.6 Unsolved Issues in RNA-Based Forensic Tissue Identification	404
15.4 DNA Methylation Approach to Forensic Tissue Identification	405
15.5 Nonhuman DNA and RNA for Forensic Tissue Identification	407
15.5.1 Saliva	407
15.5.2 Mouth-Expirated Blood	407
15.5.3 Vaginal Secretion	408
15.5.4 Feces	409
15.6 Outlook	410
Acknowledgments	411
References	411

15.1 Introduction

Biological samples from human body fluids such as saliva and vaginal secretion, liquid tissues such as blood and semen, and skin epithelium recovered from crime scenes often provide DNA useful for individual identification of the sample donor. After DNA-based donor identification, establishing the cellular origin of biological traces used for DNA profiling represents the second most crucial tasks in forensic biology casework because it allows reconstructing the sequence of crime events and evaluating the crime relevance of the forensic evidences collected. There are obvious scenarios where the forensic relevance of the DNA profile obtained at a crime scene, and thus for the crime involvement of the DNA-identified sample donor, can be supported or not by the determination of the cellular type of the biological material used for DNA profiling. For instance, if a DNA profile was obtained from a determined semen stain, the DNA-identified sample donor (or his defense attorney) would have more problems explaining how his DNA ended up at the scene of crime and in denying his direct involvement in the crime, relative to a determined saliva sample that could have been transmitted via wet speech independent of the crime event. Even if the cell type of the stain appears obvious, such as from visual inspection (e.g., the red to dark color of blood), it needs to be proved by an objective and accurate method (e.g., obviously not all red to dark stains are human blood). Therefore, forensic investigators require tools for unequivocal, confirmatory proof of the cellular origin of biological crime scene stains, which can be defended in court, most often in combination with identity information of the stain donor obtained from DNA. Conventional methods used for the determination of the cellular type of biological stains that mostly rely on chemical, enzymatic, or immunological tests are mostly presumptive and prone to false positive or false negative interpretation outcomes. Only very few classical and nonmolecular methods, such as the microscopic identification of sperm cells, can be considered conclusive. Another disadvantage of traditional tests is their destructive influence to the samples, which often makes the subsequent extraction of DNA for individual identification purposes impossible. Furthermore, given that most of the classical tests are only capable of identifying one body fluid/tissue, the amount of biological material required to perform the number of tests needed to obtain a positive result may become a limiting factor in many cases. Many currently used methods for biological stain determination were developed a long time ago—for example, the Kastle–Meyer phenolphthalein test (Cox 1991) was first proposed for blood identification in 1903—and thus ignore the wealth of scientific knowledge achieved since. In recent years, a progress in forensic science lead to the development of novel physical, biochemical, and biological methods that allow sensitive, specific, nondestructive, and parallel establishing of the cellular origin of biological traces found at crime scenes. In the current chapter, after briefly summarizing the currently applied presumptive tests, we focus on the use of nucleic acids, that is, RNA as well as DNA, for forensic tissue identification as the most promising and extensively developed approach in the past decade.

15.2 Classical Tests for Forensic Tissue Identification

15.2.1 Peripheral Blood

Peripheral or venous blood is the most common body fluid encountered at crime scenes and its determination allows concluding the occurrence of violent crime. Bloodstains are not always obvious because of the manner in which they are formed, the time since their placement, or because of cleaning attempts. Numerous presumptive and several confirmatory tests for blood identification were proposed. Fast and simple exploratory detection of bloodstains directly at the crime scene is achieved by using alternative light sources, such as Polilight (Vandenberg and vanOorschot 2006), which contains a range of wavelengths and can even reveal stains covered by paint. These light sources, however, must be used with caution regarding subsequent DNA analysis for identification purposes, since certain ultraviolet wavelengths can damage the DNA evidence in a sample (Virkler and Lednev 2009). The Luminol test was one of the first presumptive blood tests that investigators still often use at a crime scene, and it has been around for more than 40 years (Barni et al. 2007; Virkler and Lednev 2009). The test is based on the ability of iron contained in blood hemoglobin to react with the chemical substance luminol in the presence of an oxidant and alkaline solution. In practice, the luminol solution is sprayed over the area where blood is expected, and in case of positive results emits a blue glow that can be seen in a darkened room. This test is known to be the most sensitive and the least prone to false positive or false negative results of all the presumptive blood tests currently in use (Webb et al. 2006). It is considered especially useful in cases when blood was attempted to be cleaned away (Kent et al. 2003). Furthermore, there are several different catalytic tests commonly used to presumptively identify blood based on the peroxidase-like activity of the heme group (Gaensslen and National Institute of Justice (U.S.) 1983). They include the historically first but still popular Kastle–Meyer test, which relies on the heme-induced oxidation of colorless phenolphthalin into bright pink phenolphthalein. Another simple blood test uses benzidine, which yields blue color when oxidized by blood hemoglobin in the presence of ethanol/acetic acid solution (Cox 1991). Catalytic tests are very sensitive; therefore, they are not prone to false negative outcomes, but can produce false positive identifications in cases of contamination of bloodstains with chemical oxidants and plant peroxidases (Virkler and Lednev 2009). Classical confirmatory blood tests include microscopic identification of blood cells and fibrin threads, either directly in a liquid blood or after specific staining in dried blood spots (Gaensslen and National Institute of Justice (U.S.) 1983). Commonly used chemical confirmatory blood tests include two so-called crystal tests because of the ability of hemoglobin to form colored feathery crystals after reaction with pyridine (Takayama method) or halides (Teichman test) (Shaler 2002). Modern immunological methods for definitive identification of human blood are based on antibody against human hemoglobin (Kashyap 1989). Of these methods, the ABAcard® HemaTrace strip test is the most extensively validated, sensitive, specific, and rapid (Hochmeister et al. 1999b).

15.2.2 Menstrual Blood

Distinguishing between peripheral and menstrual blood is one of the most demanded and difficult tasks in forensic analysis and interpretation of evidence in sexual offence investigations. Obviously, it is crucial to differentiate between peripheral blood indicating violent crime and menstrual blood not indicating any crime at all. Several presumptive methods proposed for menstrual blood are available based on the detection of fibrinogen and fibrin degradation products, or based on the electrophoretic separation pattern of lactate dehydrogenase isozymes (Asano et al. 1972; Whitehead and Divall 1974; Baker et al. 2011). More recently, immunochemical detection of three protein targets was evaluated: matrix metalloproteinase 14 (MMP14), estrogen receptor α (ERα), and fibrinogen (Gray et al. 2012). The two former immunomarkers appeared to be specific to menstrual blood; however, they need to be confirmed in a larger sample set, including specimens from different days of menstrual cycle, before this test can be considered reliable.

15.2.3 Parturient Blood

The diagnosis of pregnancy from forensic bloodstains can be useful in cases of infanticide, criminal abortions, and for missing person identification. Pregnancy diagnostics from bloodstains has been demonstrated via radioimmunoassay detection of pregnancy-specific proteins such as human chorionic gonadotrophin, human placental lactogen, total estriol, and progesterone (Vergote et al. 1991). However, the sensitivity achievable by this approach limits its application in a forensic context, where often minute amounts of dried blood have to be investigated (Vallejo 1990).

15.2.4 Fetal and Neonatal Blood

Establishing that the blood found at a crime scene originated from a fetus or a neonate may contribute to the resolution of some cases of illegal interruption of pregnancy or murder of a neonate. Serological methods are used to distinguish between human fetal hemoglobin (HbF) and human α-fetoprotein (AFP) for fetal and neonatal blood identification, respectively (Wraxall 1972; Baxter and Rees 1974). Another method for the identification of HbF in bloodstains by reverse-phase high-performance liquid chromatography was described (Inoue et al. 1989). Despite the high specificity to fetal blood of the proposed immunomarkers in most of the cases, certain medical conditions, such as hereditary persistence of fetal hemoglobin (Jacob and Raper 1958) or hepatic carcinomas with elevated AFP level (Wee et al. 2003), may lead to false positive identifications, which should be investigated carefully.

15.2.5 Nasal Blood

Forensic confirmation of epistaxis or nose bleeding might provide important clues in legal investigation of violent crimes. Although nose bleeding can be a result of violence, it does not have to be. Consequently, the determination of nasal blood would allow less strong conclusion about violent crime compared with peripheral blood. However, no classical test is available for specific nasal blood identification so that nasal blood in its differentiation from any other human blood sources is regarded as one of the most difficult bloodstains to identify (Sakurada et al. 2012).

15.2.6 Saliva

Saliva stains are often present at violent and sexual assault crime scenes. Forensic specimens requiring testing for saliva may include swabs (sampled from different body areas such as the neck, breasts, or genitals), clothing, and linen. Compared to other body fluids, saliva is rich in amylase enzyme, which digests polysaccharides coming with food (Whitehead and Kipps 1975). German investigator B. Mueller conceived the use of alpha-amylase as a forensic target to validate the presence of saliva on a given surface as early as in 1928 (Gaensslen and National Institute of Justice [U.S.] 1983). Later, starch-iodine (Schill and Schumacher 1972), Phadebas® (Myers and Adkins 2008), and Abacus SALIgAE-saliva (Pang and Cheung 2008) presumptive tests were developed for the identification of saliva by colorimetric detection of the enzymatic activity of alpha-amylase. These tests are prone to false positive results with urine, breast milk, and feces, as well as with some hygiene products (Pang and Cheung 2008; Myers and Adkins 2008). More sensitive and specific identification of saliva can be achieved with immunological methods. Amylase-based enzyme-linked immunosorbent assay (ELISA) tests (Quarino et al. 2005) or strip immunochromatographic Rapid Stain IDentification (RSID-Saliva) test (Old et al. 2009) can distinguish salivary amylase from other amylases, such as pancreatic and bacterial amylases. Another ELISA test described (Akutsu et al. 2010) uses antibodies against statherin, an enzyme that is supposedly more specific to saliva than alpha-amylase is.

15.2.7 Semen

Semen/ejaculate determination reflects one of the most crucial evidence to conclude (male-induced) sexual assault. Semen is a liquid tissue consisting of sperm cells (spermatozoids) and seminal fluid secreted by seminal vesicles, seminiferous tubules, prostate gland, and bulbourethral gland. A typical human ejaculate consists of 1.5–5 mL of semen and contains 40–250 million spermatozoa; however, male individuals may have different sperm counts: abnormally low (oligospermia) or complete absence of spermatozoids (azoospermia) due to pathological conditions or vasectomy (sterilization) procedure. In 1.9% of semen stains in sexual assault cases, no sperm cells can be detected for various reasons (Willott 1982). Therefore, highly specific microscopic identification of semen, such as Christmas tree staining of sperm cells (Allery et al. 2001), can produce false negative results in cases of oligo- or azoospermia, as well as in a situation where the analyzed area of semen stain is too small and by chance appears to be free from spermatozoids. Detection of semen-specific acid phosphatase (SAP) in the seminal fluid allows for more universal semen identification as it does not rely on the presence of sperm cells. There are different variations of the SAP test; most of them are based on the ability of SAP to catalyze the hydrolysis of phosphates, which results in the formation of a product that can be visualized with chromogenic reactions (Schiff 1978; Kaye 1949). However, SAP is also naturally present in other body fluids such as vaginal secretions, urine, and sweat (Gaensslen and National Institute of Justice (U.S.) 1983), making SAP tests presumptive but not conclusive. More reliable semen identification can be achieved by immunological detection of other semen-specific proteins. The most popular of these tests are RSID semen strip test (Pang and Cheung 2007) and variants of prostate-specific antigen (PSA) test (Hochmeister et al. 1999a; Masibay and Lappas 1984). However, in the latter case false positive results cannot be excluded because PSA can also be found in detectable levels in other body fluids, such as female urine and breast milk (Yokota et al. 2001; Schmidt et al. 2001).

15.2.8 Vaginal Secretion

The determination of vaginal secretion may have an important evidential value in assumed sexual assault cases. Currently, a test utilizing acid-Schiff reagent for the detection of glycogen-rich cells is used for presumptive identification of vaginal fluid (Randall 1988). However, glycogenation level in vaginal epithelial cells varies depending on the menstrual cycle and reproductive age, which could lead to false negative results; false positive identification is also possible because of occasional staining of male buccal and urogenital tract epithelial cells with acid-Schiff reagent (Virkler and Lednev 2009). Although other enzymatic, histological, and chemical tests were proposed, none of them is sufficiently sensitive and specific for confirmatory identification of vaginal secretion (Virkler and Lednev 2009; French et al. 2008).

15.2.9 Skin

Given that the so-called "touch DNA" (van Oorschot and Jones 1997) does not necessarily come from shed or printed skin cells, but may also originate from speech-induced saliva transfer (Warshauer et al. 2012), the necessity for identification of skin epithelial cells (and their differentiation from saliva cells) became obvious. Currently, there is no widely used

Table 15.1 Classical Methods for Forensic Tissue Identification Described in Current Review

Body Fluid/Tissue	Method	Presumptive	Confirmatory
Peripheral blood	Polilight	×	
	Microscopic identification		×
	Kastle–Meyer	×	
	Benzidin	×	
	Luminol	×	
	Crystal tests (Teichman and Takayama)		×
	ABAcard HemaTrace		×
Menstrual blood	Immunodetection of fibrinogen and fibrin degradation products	×	
	Immunodetection of matrix metalloproteinase 14 (MMP14) and estrogen receptor α (ERα)	×	
	Lactate dehydrogenase (LDH) isozymes	×	
Parturient blood	Immunoassay pregnancy hormones		×
Fetal/neonatal blood	Immunodetection of fetal hemoglobin (HbF) and α-fetoprotein (AFP)		×
Nasal blood	None		
Saliva	Polilight	×	
	Starch-iodine amylase test	×	
	Phadebas amylase test	×	
	Abacus SALIgAE-saliva	×	
	Amylase ELISA		×
	RSID-saliva		×
	Statherin ELISA		×
Semen	Microscopic identification		×
	Semen-specific acid phosphatase (SAP) chemical test	×	
	RSID-semen	×	
	Prostate-specific antigen (PSA) immunoassay	×	
Vaginal secretion	Histological staining	×	
Skin	Polilight	×	
	Histological staining	×	

presumptive test for the positive identification of skin material, although alternative light sources, such as the Polilight, can indicate the presence of epithelial cells (Vandenberg and van Oorschot 2006), and a histological approach to distinguishing between skin, buccal, and vaginal epithelium was proposed (French et al. 2008). Hence, the ability to conclusively identify and confirm the presence of skin epithelial cells with classical tests is currently lacking (Table 15.1).

15.3 RNA-Based Approaches to Forensic Tissue Identification

The use of RNA for forensic tissue identification is based on the knowledge that various genes are expressed at different levels in different cell types, a phenomenon referred to as differential gene expression. Methods for very sensitive quantification of RNA molecules are well established, although were initially developed outside the forensic field. Unlike most of the methods mentioned in the previous section, RNA-based forensic tissue identification is not prone to false positive identification of biological stains due to contamination of the samples with environmental agents and biological materials derived from animals, since most RNA-based tests are human-specific. Of practical forensic relevance is that RNA isolation can be performed in parallel with DNA extraction from the same sample, so that no precious biological material needed for DNA profiling is wasted. One of the major concerns about forensic application of RNA analysis, including—but not only for—forensic tissue identification, is that in contrast to DNA, RNA is generally considered more prone to *in vitro* degradation effects (Alberts et al. 2008). The idea of using RNA markers in forensic applications was considered unrealistic until several studies had demonstrated long-term persistency of particular RNA markers in a variety of postmortem scenarios and *in vitro* collected samples of various tissue types (Marchuk et al. 1998; Inoue et al. 2002; Bahar et al. 2007; Johnson et al. 1986; Zubakov et al. 2009; Koppelkamm et al. 2011).

15.3.1 Stability of RNA

The phosphodiester bonds in an RNA molecule are 200 times less stable than those in DNA (at neutral pH and in the presence of physiological concentrations of Mg^{2+}) (Lindahl 1996). Because of the presence of 2'-OH groups, RNA molecules are in general more prone to hydrolysis, especially in the presence of bivalent cations such as Ca^{2+} or Mg^{2+} and at alkali pH, that is, physiological conditions normally occurring in living cells (Adams et al. 1986). However, compared to DNA, RNA has stronger *N*-glycosidic links; therefore, the rate of depurination of RNA is considerably slower than those in DNA (Lindahl and Nyberg 1972). Notably, hydrolytic depurination, along with interstrand cross-linking, is one of the major factors causing spontaneous fragmentation of ancient DNA (Paabo and Wilson 1991; Hansen et al. 2006). RNA molecules are less prone to depurination and more resistant to oxidation (Thorp 2000); therefore, there can be cases where in dry postmortem conditions RNA is more stable than DNA. For instance, a next-generation sequencing (NGS) study of 725-year-old desiccated maize seeds (Fordyce et al. 2013) showed that RNA was better preserved than DNA, as confirmed by longer sequencing reads obtained in RNA sequencing versus DNA sequencing of the same samples. RNA degradation has been explored in a variety of postmortem tissues and body fluids (Marchuk et al. 1998;

Inoue et al. 2002; Bahar et al. 2007; Zubakov et al. 2009; King et al. 2001; Koppelkamm et al. 2011). It has been found that in certain tissues such as brain (Johnson et al. 1986) or skeletal muscles (Bahar et al. 2007), high-quality RNA may be extracted even after very long postmortem intervals (up to several days). Several studies showed that particular messenger RNA (mRNA) markers can be detected in extremely old (up to 23 years) dried blood or saliva stains (Kohlmeier and Schneider 2012; Zubakov et al. 2009). Other types of RNA molecules with forensic relevance, such as microRNA (miRNA) and ribosomal RNA (rRNA), demonstrated even higher postmortem stability in tissues or dried body fluid stains compared to mRNA (Jung et al. 2010; Zubakov et al. 2010; Bahar et al. 2007), which is explained by the structural and biological peculiarities of these RNA molecules. Both rRNA and miRNAs are protected from RNases by binding proteins and subcellular compartmentalization; in addition, the small size of miRNA and secondary structures of rRNA make them less prone to physical and chemical, as well as enzymatic fragmentation *ex vivo*. However, despite the large number of evidences of exceptional stability of RNA molecules in a variety of postmortem scenarios, in many forensically relevant situations RNA degrades rapidly. Humidity and UV radiation appear to be the most detrimental environmental factors affecting RNA stability in dried body fluid stains (Setzer et al. 2008).

15.3.2 mRNA Markers for Forensic Tissue Identification

The first forensic paper describing postmortem RNA synthesis was published in 1984 (Oehmichen and Zilles 1984); however, until the end of the twentieth century, RNA forensic reports were scarce. In 2002, Martin Bauer was the first forensic researcher who proposed to exploit the potential of mRNA markers in forensic tissue identification by using gene expression differences between peripheral and menstrual blood (Bauer and Patzelt 2002). One year later, Juusola and Ballantyne (2003) presented a prototype method of mRNA-based identification of blood, semen, and saliva. These and other studies (Haas et al. 2009; Fleming and Harbison 2010a; Bauer and Patzelt 2003) developed forensically applicable mRNA-based tools from marker knowledge previously published in studies that investigate the expression of genes among various body tissues and cell types, which also are summarized in databases such as BioGPS (Wu et al. 2009). Although such an approach has the advantage that the information about tissue specificity of RNA markers usually comes from multiple datasets, thereby increasing the reliability of such knowledge, the disadvantage is that the *in vitro* degradation behavior of the mRNA markers can hardly be predicted by *in silico* analysis. Consequently, mRNA markers with strong differential expression between forensically relevant tissues as ascertained from such resources are not necessarily useful in forensic application, because they may be prone to strong *in vitro* degradation. As a result, extensive subsequent studies to prove the *in vitro* stability of the ascertained mRNA markers need to be carried out before the markers can be proposed for forensic applications. In 2008, the first systematic study was published that took the RNA degradation issue into account already in the mRNA marker selection by carrying out a genome-wide expression analysis in blood and saliva stains that were stored at various time intervals (Zubakov et al. 2008) (Figure 15.1). This study could not only highlight mRNA markers with strong expression differences between forensically relevant body fluids, but from the study design it is expected that the identified markers are stable over time *in vitro* as subsequent work indeed had demonstrated (Zubakov et al. 2009).

Forensic Tissue Identification with Nucleic Acids

Figure 15.1 Unsupervised hierarchical cluster analysis of genome-wide microarray gene expression data based on differential mRNAs of saliva (Sa) and peripheral blood (B) from several individuals identified from Affymetrix U133 plus 2.0 dataset of Zubakov et al. (2008). Down-regulated genes are colored green; red corresponds to up-regulated transcripts.

The European DNA Profiling Group (EDNAP) of the International Society of Forensic Genetics has so far published three collaborative exercises on forensic tissue identified, namely, studies to validate sets of mRNA markers for the identification of blood (Haas et al. 2011, 2012), saliva (Haas et al. 2013), and semen (Haas et al. 2013) (Table 15.2). The organizing laboratories (Institute of Legal Medicine, University of Zürich and National Center for Forensic Science, University of Central Florida) preevaluated published and unpublished mRNA markers for specificity and sensitivity and selected the best candidate mRNA markers for collaborative analysis in several forensic genetic laboratories. Most of the participating laboratories had no prior experience with RNA, so that these studies allow users to evaluate the performance of the mRNA markers and analysis methods in a situation close to real forensic casework. EDNAP exercises on mRNA marker validation for vaginal secretion, menstrual blood, and skin are under way.

Table 15.2 EDNAP-Validated mRNA Markers for Identification of Blood, Saliva, and Semen

Targeted Tissue/Body Fluid	mRNA Markers	Suitability of Markers
Peripheral blood	HBB (Hemoblobin beta)	Good candidates: high sensitivity
	HBA (Hemoglobin alpha)	
	ALAS2 (Aminolevulinic acid synthase)	Good candidates: medium sensitivity
	CD3G (T-cell surface glycoprotein CD3, gamma chain)	
	ANK1 (Ankyrin 1, erythrocytic)	
	PBGD (Porphobilinogen deaminase)	
	SPTB (β-Spectrin)	
	AQP9 (Aquaporin 9)	Not specific
Saliva	HTN3 (Histatin 3)	Good candidates
	STATH (Statherin)	
	MUC7 (Mucin 7)	
	PRB1-3 (Proline-rich protein BstNI subfamily 1)	Not sensitive
	PRB4 (Proline-rich protein BstNI subfamily 4)	
	SPRR2A (Small proline rich protein 2A)	Not specific
	KRT13 (Keratin 13)	
Semen	PRM1 (Protamin 1)	Good candidates
	PRM2 (Protamin 2)	
	TGM4 (Transglutaminase 4)	
	KLK3 (Kalikrein 3)	
	SEMG1 (Semenogelin 1)	
	SPANXB [SPANX family member B]	Inconsistent results during singleplex testing
	HSFY [Heat shock transcription factor, Y-linked]	
	SZPBP [Zona pellucida binding protein]	
	ODF1 [Outer dense fiber of sperm tails 1]	
	BPY2 [Basic charged, Y-linked 2]	

15.3.3 Multiplex mRNA Systems for Parallel Determination of Various Cell Types

One of the advantages of mRNA-based approach over the majority of conventional methods for forensic tissue identification is the possibility to simultaneously analyze several markers that target several forensically relevant body fluids and other cell types in parallel. This is achieved by using multiplex reverse transcription-polymerase chain reaction (RT-PCR) technique, which allows single-tube amplification of several RNA markers from the same RNA-derived cDNA sample. In 2005, Juusola et al. proposed the first multiplex PCR assay for parallel mRNA-based identification of peripheral and menstrual blood, saliva, semen, and vaginal secretion (Juusola and Ballantyne 2005) as a follow-up of their earlier study (Juusola and Ballantyne 2003), where they identified tissue-specific markers and tested previously proposed mRNA markers for menstrual blood (Bauer and Patzelt 2002) with singleplex PCR assays. An octaplex system comprising two markers per each forensically relevant body fluid was designed, and the products of end-point PCR were detected by capillary electrophoresis. This method allowed discrimination between all targeted body fluids with a sensitivity of <200 pg–12 ng of total RNA input, depending on the tissue, which in principle makes it suitable for forensic casework use. In 2007, Juusola and Ballantyne published quantitative RT-PCR assays using the same set of body fluid markers, except vaginal secretion, and additionally including a reference gene *GAPDH* for normalization purposes (Juusola and Ballantyne 2007). This system consisted of triplex amplification of two tissue-specific makers per each body fluid together with *GAPDH* reference gene, and allowed relative quantification of the amount of RT-PCR product per each mRNA marker directly from the PCR and without additional electrophoresis step. The sensitivity of quantitative PCR (qPCR) detection was estimated at 1 pg–2 ng of input total RNA per reaction, depending on the tissue, exceeding the limit of end-point PCR assays (such as CE-based) by almost 1 order of magnitude. In 2009, Haas et al. combined the mRNA markers for peripheral and menstrual blood, semen, saliva, and vaginal secretion (Haas et al. 2009) proposed by Juusola and Ballantyne (2003, 2005, 2007). Multiplex end-point PCR assays were developed and tested with a panel of samples from different body fluids, including mock forensic samples and aged stains. In parallel with mRNA profiling, immunological and enzymatic methods of body fluid identification were applied, including benzidine and Hexagon OBTI tests for blood, alpha-amylase test for saliva, and acid phosphatase test for semen. This study demonstrated that the sensitivity of the mRNA assay is comparable to that of classical tests for body fluids identification. Moreover, Haas et al. expressed the sensitivity of the PCR assays in volume of body fluids used for RNA extraction and subsequent PCR analysis, which clearly showed that mRNA profiling works successfully with extremely small amounts of biological material (from 10^{-5} μL for blood to 1 μL for saliva) and thus should satisfy real forensic casework requirements. Another end-point multiplex PCR assay was presented by Fleming and Harbison in 2010 including an alternative set of mRNA markers for identification of peripheral and menstrual blood, saliva, and semen (Fleming and Harbison 2010a) (Table 15.3). One notable difference of this system compared to previously proposed ones was the use of an mRNA marker targeting the transglutaminase 4 gene (*TGM4*) for the identification of seminal fluid not containing sperm cells, which are particularly useful for the identification of semen containing no or a small number of sperm cells such as in ejaculates of azoospermic or oligospermic males, respectively. The previously suggested semen mRNA markers PRM1 and PRM2 (Haas et al. 2009; Juusola and Ballantyne 2005) are expressed in sperm cells only and cannot be used for identification of sperm-free ejaculate samples. By that time, several reports (Nussbaumer et al. 2006; Zubakov et al. 2008) had indicated that MUC4 and HBD1, two previously suggested vaginal markers, cross-react with saliva, which could lead to false-positive identification, as later confirmed by the dedicated study of Jakubowska et al. (2013). Therefore, later Fleming and Harbison extended their system by adding bacterial rRNA markers for identification of vaginal secretion (Fleming and Harbison 2010b). In 2012, Lindenbergh et al. proposed an additional end-point PCR 19-plex system based on a new set of mRNAs. This system integrated the mRNA markers used in previously published multiplex assays and also included

Table 15.3 mRNA Markers Implemented in Multitissue Multiplex PCR Assays for Forensic Tissue Identification

Targeted Tissue/Body Fluid	Marker (Gene Name)	Multiplex PCR Assays				
		Juusola and Ballantyne (2005)	Haas et al. (2009)	Fleming and Harbison (2010)	Lindenbergh et al. (2012)	Roeder and Haas (2012)
Semen	KLK3 (kallikrein-related peptidase 3)					×
	PRM1 (protamine 1)	×	×			×
	PRM2 (protamine 2)	×	×			×
	SEMG1 (semenogelin 1)			×		×
	TGM4 (transglutaminase 4)			×	×	×
Saliva	STATH (statherin)	×	×	×	×	×
	HTN3 (histatin 3)	×	×	×	×	×
	SMR3B (submaxillary gland androgen regulated protein 3B)					×
	PRB4 (proline-rich protein BstN1 subfamily 4)					×
	MUC7 (mucin 7)					×
Vaginal secretion	HBD1 (beta defensin 1)	×	×		×	×
	MUC4 (mucin 4)	×	×		×	×
	Ljen (16s ribosomal RNA)					×
	Lgas (16S–23S intergenic spacer region)			×		
	Lcris (16S–23S intergenic spacer region)			×		
Peripheral blood	HBB (hemoglobin subunit beta)		×		×	×
	CD93 (cluster of differentiation 93)				×	
	AMICA1 (adhesion molecule, interacts with CXADR antigen 1)				×	

	Gene						
	ALAS2 (delta-aminolevulinate synthase)					×	×
	PRF1 (perforin)					×	×
	GlycoA (glycophorin A)			×		×	×
	PF4 (platelet factor 4)					×	×
	SPTB (erythrocytic spectrin beta chain)	×				×	×
	PBGD (hydroxymethylbilane synthase)	×				×	×
Menstrual blood	MMP11 (metallopeptidase 11)	×		×		×	×
	MSX1 (msh homeobox 1)					×	×
	SFRP4 (secreted frizzled-related protein 4)					×	×
	LEFTY2 (left–right determination factor 2)					×	×
	MMP10 (metallopeptidase 10)					×	×
	MMP7 (metallopeptidase 7)	×			×	×	×
	Hs202072 (uncharacterized LOC100505776)					×	
Skin	CDSN (corneodesmosin)				×		
	LOR (loricrin)				×		
Mucosa	KRT4 (keratin 4)				×		
	KRT13 (keratin 13)				×		
	SPRR2A (small proline-rich protein 2A)				×		
Housekeeping (all tissues)	GAPDH (glyceraldehyde-3-phosphate dehydrogenase)	×			×	×	
	18S rRNA (small subunit ribosomal rRNA)	×			×		
	TEF (transcription elongation factor SII)		×			×	
	UCE (E2 ubiquitin conjugating enzyme UbcH5B)		×			×	
	ACTB (actin beta)		×				

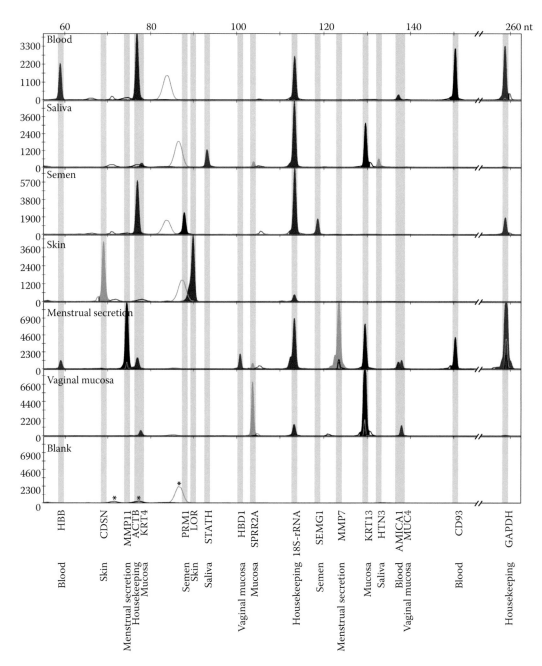

Figure 15.2 Typical overlay electropherograms of single source blood, saliva, semen, skin, menstrual secretion and vaginal mucosa as produced by the 19-plex assay reported by Lindenbergh et al. (2012). The blank corresponds to an RT-PCR with no RNA input (no RNA RT-PCR). Peaks representing marker signals are filled. Empty peaks represent bleed-through signals or by-products (e.g., red peak at 72 and 105 nt) that are both due to overamplification. All signals are cDNA-derived, as no signals were observed in the minus RT controls. (Reprinted with permission from Lindenbergh, A. et al., *Forensic Sci Int Genet*, 6, 5, 565–577, 2012.)

new mRNA markers identified by Zubakov et al. (2008), and most importantly the mRNA markers for skin samples discovered by Visser et al. (2011) (Table 15.3), which for the first time allowed handlers to detect skin epithelial cells, that is, touched objects using a multiplex mRNA system that in parallel also identifies all forensically relevant body fluids. Up to date, this multiplex assay (Figure 15.2) provides the most complete coverage of forensically relevant body fluids and tissues, and is already in use by the Netherlands Forensic Institute for practical forensic casework (Lindenbergh et al. 2013).

The development of multiplex PCR mRNA assays for forensic tissue identification is still in progress as more specific and more sensitive markers are being discovered, for instance, by the application of genome-wide screening approaches using expression microarrays or next-generation RNA sequencing techniques (Park et al. 2013; Benschop et al. 2012; Hanson et al. 2012; Hanson and Ballantyne 2013). For instance, Park et al. (2013) recently performed a systematic microarray gene expression profiling study of different body fluids and developed a multiplex RT-qPCR system using a completely new set of mRNA markers. Very recently, Roeder and Haas published a new multiplex system comprising five mRNA markers per each body fluid (some of them were newly identified) and proposed a novel scoring method for forensic tissue identification (Roeder and Haas 2012). Other investigators conducted validation studies on mRNA markers previously used in multiplex PCR assays (Richard et al. 2012; Jakubowska et al. 2013).

15.3.4 Stand-Alone mRNA Assays for Identifying Particular Body Fluids

Some cell types are less often found at crime scenes, and therefore are not targeted (yet) with the multitissue mRNA systems developed thus far, as discussed in the previous section. However, stand-alone mRNA assays have been developed as described in the following subsections.

15.3.4.1 Nasal Blood

In 2012, Sakurada et al. demonstrated that distinguishing nasal blood from other body fluids, including other blood sources such as peripheral and menstrual blood, by means of mRNA markers, including hemoglobin (*HBB*), statherin (*STATH*), and histatin 3 (*HTN3*), is feasible. Nasal blood comprises peripheral blood and nasal secretions, and the assay uses expression markers specific to both fluid types. The general blood component was identified by high expression of the *HBB* gene. *STATH* is overexpressed in nasal secretion; however, this gene is also present in saliva. Therefore, the third marker, *HTN3*, which is present in saliva, but not in nasal samples, was used to distinguish between nasal blood and saliva. Menstrual blood showed the same expression patterns of *HBB*, *STATH*, and *HTN3* genes as nasal blood; to discriminate between nasal and menstrual blood, the menstrual blood–specific protein matrix metallopeptidase 7 gene (*MMP7*) can be used. Notably, the results of Sakurada et al. were obtained for a very limited number of samples; therefore, further validation of the mRNA markers is required. However, as a proof of principle, this study demonstrated that mRNA markers may provide a unique tool for identification of nasal blood, which cannot be determined with any other existing method thus far, but, as outlined above, can be of high forensic relevance.

15.3.4.2 Parturient Blood

In 2010, Gauvin et al. introduced a pregnancy diagnostics method from blood based on mRNA markers. The authors analyzed 11 genes with previously reported pregnancy-specific expression pattern and found that two of them, human placental lactogen (*hPL*) and beta subunit of human chorionic gonadotropin (β*hCG*) genes, appeared to be most useful for distinguishing parturient blood from the blood of nonpregnant women. This study demonstrated, using blood samples collected at different stages of pregnancy, that *hPL* transcripts were present in all samples tested throughout the pregnancy, whereas β*hCG* transcripts were detectable until half of the second trimester but not at later gestation stages (whereas expression of both genes was absent in blood samples from nonpregnant women). Therefore, this approach not only allows differentiating blood of pregnant women from blood of nonpregnant ones but, especially if further validated, may also allow estimating the age of gestation from bloodstains of pregnant women. The sensitivity of TaqMan® PCR assay used by the authors was sufficient to detect *hPL* mRNA from 0.25 cm^2 dried bloodstain, which in principle makes it suitable for forensic applications.

15.3.4.3 Neonatal Blood

In 2006, Alvarez and Ballantyne proposed a method for identifying fetal blood using mRNA markers targeting the developmentally regulated gamma hemoglobin (*HBG*) gene. They discovered two novel gene isoforms, designated *HBG1n* and *HBG2n*, that exhibit an extremely restricted pattern of gene expression being confined to newborns. Multiplex quantitative reverse transcription PCR (qRT-PCR) assays incorporating these novel mRNAs have been designed, tested, and evaluated for their potential forensic use. Additional evaluation of this effect shall be performed.

15.3.4.4 Sweat

Sakurada et al. developed a real-time RT-PCR assay for the single mRNA marker dermicidine (*DCD*) and demonstrated that it allows highly sensitive identification of sweat relative to other body-fluid stains. *DCD* transcripts were detected in different mock forensic sweat samples, which demonstrated the high potential of the marker for forensic casework analysis (Sakurada et al. 2010). More evaluation work of this marker is needed.

15.3.4.5 Vaginal Secretion versus Skin

Distinguishing between epithelial cells originated from vaginal secretion, saliva, and skin appeared to be one of the most difficult tasks in RNA-based forensic tissue identification. All previously proposed mRNA markers (but also miRNA and bacterial DNA markers as described below) fail to clearly and definitively differentiate these forensically relevant epithelial cell types. Recently, two studies aiming to identify new epithelial markers were undertaken using whole transcriptome profiling by means of next-generation RNA sequencing techniques (RNA-Seq). In 2012, Hanson et al. reported the discovery of new specific and sensitive mRNA markers for forensic identification of skin. One of the markers, LCE1C, is particularly highly sensitive (5–25 pg of input total RNA) and was detected in the majority of skin samples tested including touched objects, but absent in other tissues and body fluids. In 2013, Hanson and Ballantyne performed another RNA-Seq study in order to identify new vaginal secretion mRNA markers. After a detailed evaluation of >200 candidates from the tens of thousands of mRNA species found in vaginal samples, six new and highly promising candidates mRNA markers were identified. *CYP2B7P1* and *MYOZ1* consistently demonstrated high specificity and sensitivity for vaginal samples when used in a qualitative capillary electrophoresis–based assay and allowed differentiating between vaginal secretion and other body fluids containing significant numbers of epithelia, particularly saliva and skin. *CYP2B7P1* was shown to be exceedingly specific with no detectable cross-reactivity with other forensically relevant body fluids/tissues noted to date. These newly identified mRNA markers for skin and vaginal secretion appear very promising and—if positively validated further—may eventually be included in multiplex RNA systems for parallel identification of various forensically relevant tissues. Furthermore, these two studies convincingly demonstrated that modern genome-wide transcriptome screening technologies such as RNA-seq can be very helpful in the search for new mRNA markers including cases where previously applied approaches failed in marker delivery.

15.3.5 miRNA Markers for Forensic Tissue Identification

miRNAs belong to a class of small nonprotein coding RNA molecules of 18 to 22 nucleotides in length that function as negative regulators of gene expression. Associated with a multiprotein complex, the RNA-induced silencing complex miRNAs hybridize to the 3′ UTR of specific mRNA targets causing inhibition of mRNA translation and/or mRNA degradation (Bartel 2004). A number of miRNAs have been implicated in the processes of embryonic development, cell proliferation and differentiation, as well as in the pathogenesis of many human diseases such as cancer, neurodegenerative diseases, and metabolic disorders (Bartel 2004; Sood et al. 2006). Taking into account the involvement of miRNAs in developmental processes, it is not surprising that several recent studies revealed tissue-specific expression patterns of many miRNA markers (Liang et al. 2007; Sood et al. 2006). Physical fragmentation of RNA molecules due to degradation dictates the use of very short RT-PCR amplicons; therefore, intrinsically short miRNAs hold great promise for representing ideal forensic biomarkers. Also, the high abundance of certain miRNAs in living cells should provide very sensitive detection of tissue-specific miRNA markers. In 2009, Hanson et al. first introduced miRNA profiling for the purposes of forensic tissue identification. They surveyed the expression of 452 human miRNAs via quantitative RT-PCR analysis in several body fluids, including peripheral and menstrual blood, saliva, semen, and vaginal secretions. In this study, miRNA expression was shown to be generally, but not exclusively, specific to particular tissues or body fluids; however, the partly strong differences in expression levels of selected miRNA tissue markers allowed quantitative distinction between the tissues studied (Table 15.4). Zubakov et al. (2010) profiled the expression levels of 718 miRNAs in menstrual and venous blood, saliva, semen, and vaginal secretion by means of microarrays using LNA™-modified oligonucleotides (Exiqon, Vedbæk, Denmark) as capture probes. They observed that samples from different individuals belonging to the same body fluid tend to cluster together and that different body fluids display distinct miRNA expression signatures (Figure 15.3).

After qPCR validation of 14 mRNA candidate markers that showed the most differential expression in the microarray dataset, Zubakov et al. found good concordance between expression levels measured by microarray and qPCR only for blood and semen miRNA markers. For saliva and vaginal secretion miRNA candidate markers however, the differential expression obtained with microarrays was not confirmed using qPCR (Table 15.4). When they tested the expression of seven of the miRNAs proposed by Hanson et al. for

Table 15.4 miRNA Markers Suggested for Forensic Tissue Identification

Targeted Tissue/ Body Fluid	Hanson et al. (2009) SYBR Green qPCR	Zubakov et al. (2010) Microarray with LNA Capture Probes	Zubakov et al. (2010) TaqMan qPCR	Courts and Madea (2011) Geniom Biochips	Courts and Madea (2011) SYBR Green qPCR	Wang et al. (2013) TaqMan Array Human MicroRNA Cards	Wang et al. (2013) TaqMan qPCR
Peripheral blood	miR451, miR16	miR20a, miR106a, miR185	miR20a, miR106a, miR185, miR144	miR126, miR150, miR451	miR126, miR150, miR451	miR486, miR16	miR486, miR16
Menstrual blood	miR451, miR412	miR185*, miR144				miR214	miR214
Semen	miR135b, miR10b	miR943, miR135a, miR10a, miR507	miR943, miR135a, iR10a, miR507, miR891a			miR888, miR891a	miR888, miR891a
Saliva	miR658, miR205	miR583, miR518c*, miR208b		miR200c, miR203, miR205	miR200c, miR203, miR205	miR138-2	
Vaginal secretion	miR124a, miR372	miR617, miR891a				miR124a	

Source: Wang, Z. et al., *Forensic Sci Int Genet*, 7, 2, 230–239, 2013. With permission.

body-fluid identification, Zubakov et al. were able to partially replicate the results for semen- and blood-specific miRNAs, but not the miRNA markers proposed for saliva and vaginal secretion. A possible explanation for the discrepancies observed between both studies could be the different qPCR technologies used in these two studies. In particular, Hanson et al. used Qiagen miScript primer assays that utilize simple oligonucleotide primers and SYBR green chemistry for miRNA detection and quantification, whereas Zubakov et al. used Applied Biosystem miRNA TaqMan assays for qPCR validation that make use of special stem-loop primer design. In principle, stem-loop primers outperform conventional PCR primers in terms of RT efficiency and specificity; they are capable of discriminating related miRNAs that differ by as little as one nucleotide only, and they are not affected by genomic DNA contamination (Chen et al. 2005). Also, unlike LNA capture probes used in microarrays by Zubakov et al. or simple oligonucleotide primers used by Hanson et al. in qPCR assays, TaqMan miRNA assays with stem-loop primers are specific only to mature miRNAs and not to their unprocessed precursors (Chen et al. 2005). The latter factor may be especially important as some particular miRNAs may have different tissue localization of their precursor and mature forms, as was exemplified for miR-891a miRNA by Zubakov et al. Northern blot hybridization with specific LNA probe performed by Zubakov et al. revealed that the miR-891a precursor form is abundant in saliva, whereas the mature miRNA was detected exclusively in semen samples. Initially, miR-891a was picked up from microarray analysis as saliva-specific marker, but TaqMan qPCR validation showed its high specificity to semen. Later, other researches, also using the stem-loop

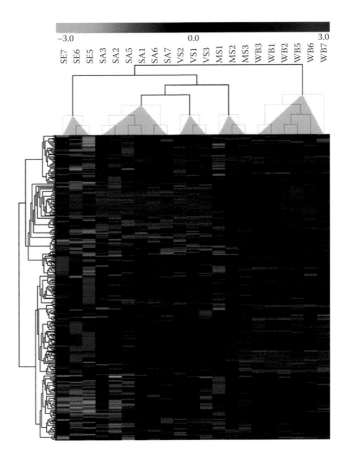

Figure 15.3 Unsupervised hierarchical cluster analysis of microarray expression data based on 458 miRNAs of five forensically relevant body fluids, i.e., semen (*SE*), saliva (*SA*), vaginal secretion (*VS*), menstrual blood (*MS*), and whole venous blood (*WB*) from several individuals as reported by Zubakov et al. (2010). Down- and upregulated miRNAs are highlighted with green and red, respectively. (Reprinted with permission from Zubakov, D. et al., *Int J Legal Med*, 123, 1, 71–74, 2010.)

primers technology, confirmed that miR-891a is indeed expressed exclusively in semen (Wang et al. 2013). In a more recent study in search for miRNA markers by Courts and Madea (2011), miScript assays with SYBR green primers were used to compare miRNA expression signatures in blood and saliva. Furthermore, Wang et al. applied TaqMan array human miRNA cards that utilize stem-loop primers for screening and confirmation of miRNA markers across venous and menstrual blood, saliva, semen, and vaginal secretions (Wang et al. 2013). Both groups obtained contradictory results regarding the confirmation (or not) of previously proposed tissue-specific miRNA markers (Table 15.4). However, given that the miRNA analysis technology seriously influences the outcome of miRNA profiling, as well as the fact that the sample size was relatively small in all of studies conducted so far, it is not surprising that different research groups produced different sets of miRNA tissue markers. Currently, it can be stated that at least for venous blood and semen, reliable tissue-specific miRNA markers exist, whereas for other body fluids such as vaginal secretion, saliva, and menstrual blood, additional exploratory and confirmation studies may be required. In general, the forensic relevance of miRNA markers for forensic tissue identification is justified by their high stability and high sensitivity level of analysis in principle providing strong advantages when dealing with small amounts of aged samples. For instance, Zubakov et al. (2010) found no decrease in miRNA abundance after 1 year of bloodstain storage, and showed that individual miRNA markers can be detected from as little as 0.1 pg of total RNA, which is more sensitive than all forensically used mRNA quantification methods available thus far.

15.3.6 Unsolved Issues in RNA-Based Forensic Tissue Identification

The concept of the tissue specificity of m/miRNA markers is a matter of discussion. Gene expression is hardly ever totally restricted to one particular tissue. Albeit strong expression differences are observed for several genes between cell types, including those of forensic relevance, background transcription is ubiquitous. Practically any m/miRNA species can be detected at some level in any tissue provided that the detection method applied is sensitive enough. Therefore, the specificity of m/miRNA markers aimed to be applied to forensic tissue identification need to be defined in relative quantitative terms but not in an absolute way. For instance, epithelial cells of skin, vaginal or saliva origin are biochemically and physiologically very close to each other and express similar sets of genes. Reliable discrimination between those epithelial cells from different body locations using mRNA markers can only be performed in a quantitative way. This task, however, requires accurate normalization of transcription levels, which in most gene expression studies is achieved by applying reference genes (sometimes called housekeeping genes) that are equally expressed in all samples studied. However, most of the RNA assays, including multiplex assays targeting multiple tissues, developed for forensic tissue identification thus far are based on the end-point PCR method, which does not allow normalization and quantification by use of reference genes. In this situation, the only available option is the normalization of the input RNA, that is, equalizing RNA content in the samples before cDNA synthesis and PCR amplification. This approach requires accurate RNA quantification of human RNA in forensic samples, which is not trivial. For instance, certain relevant body fluids such as saliva and vaginal secretion naturally contain bacterial RNA, which cannot be distinguished from human RNA using currently available quantification methods, including ultraviolet absorbance, microcapillary electrophoresis, or fluorescence-based measurements. Consequently, no adequate method for normalization of end-point PCR assays currently exists. Quantitative PCR assays for forensic tissue identification may allow overcoming this caveat by using reference genes. However, although many studies had suggested and had used reference genes, including for forensic tissue identification applications, until recently (Moreno et al. 2012) no systematic attempts to identify reliable reference genes for the quantification of mRNA for forensic applications were undertaken. Moreover, some authors reported that classically used reference genes such as 18S rRNA or *GAPDH* are not equally expressed across forensic samples (Juusola and Ballantyne 2007; Haas et al. 2009; Fleming and Harbison 2010a). Currently, the lack of proper normalization methods does not allow reliable distinguishing of saliva, vaginal secretion, and does not allow conclusive identification of mixed forensic stains. On the other hand, although qRT-PCR does allow accurate quantification of gene expression by using reference genes, currently available qRT-PCR technologies are not suitable for the simultaneous analysis of several m/miRNA markers as necessary in parallel determination of several forensic tissue types. This dilemma of having advantages and disadvantages at both currently used RNA approaches to forensic tissue identification, and thus not yet having available the ideal method, will need to be solved by technological advances allowing fully quantitative, accurate, and reliable multiplex RNA analysis in the future.

15.4 DNA Methylation Approach to Forensic Tissue Identification

Differential DNA methylation, currently the most intensively studied epigenetic mechanism, has essential roles in cellular processes including gene regulation, development, and genomic imprinting (Holliday and Pugh 1975; Bird 1986). DNA methylation occurs at the 5′ position of the pyrimidine ring of cytosine in CpG dinucleotides and influences DNA stability and transcriptional silencing without changing the primary DNA sequence. Different cell types have different methylation patterns (Byun et al. 2009), and chromosome segments called tissue-specific differentially methylated regions (tDMRs) are known to show different DNA methylation profiles according to cell or tissue type (Song et al. 2005; Ohgane et al. 2008; Illingworth et al. 2008). The first study demonstrating that a DNA methylation approach to forensic tissue identification is feasible was published in 2011. In this study, Frumkin et al. affiliated with the Nucleix Ltd. company, identified 15 DNA loci with differential methylation patterns across forensically relevant cell types using methylation-sensitive restriction endonuclease PCR (MSRE-PCR) technology (Frumkin et al. 2011). A single-tube multiplex end-point PCR assay was designed, separation of PCR products was performed with capillary electrophoresis, and the assay was tested on 50 DNA samples from blood, saliva, semen, and skin epidermis. In all cases analyzed, the source tissue was successfully identified (Figure 15.4). This method also allowed correct identification of semen and saliva in mixed stains; moreover, the approximate quantification of each component was possible (LaRue et al. 2013). In principle, the identification of other mixtures should be feasible as well. About 1 ng of DNA was required for the analysis; due to the use of short amplicons, the assay has great potential for successful analysis of degraded DNA samples. Importantly, the proposed method provides the possibility to combine tissue identification with STR profiling in one PCR reaction.

Recently, Nucleix Ltd. proposed the first commercial product for methylated DNA tissue identification, the so called DSI-Semen™ kit, for distinguishing semen from non-semen samples (Wasserstrom et al. 2013) using the MSRE approach. A forensic validation study confirmed that the DSI-Semen kit provides a reliable and sensitive tool for the determination of DNA derived from semen (LaRue et al. 2013). Besides the MSRE approach, which requires having available restriction enzymes that can recognize the methylation status of the particular DNA loci, the methylation status of DNA can be evaluated with alternative methods based on sodium bisulfite treatment of DNA samples, a process that converts unmethylated cytosines to uracil. For instance, the bisulfite DNA sequencing approach was used by Lee et al. (2012) to identify tDMRs specific to peripheral and menstrual blood, saliva, semen, and vaginal secretion. A similar method that utilized pyrosequencing of bisulfite modified DNA was used by Madi et al. (2012) to define epigenetic markers that display differential methylation patterns between blood, saliva, semen, and epithelial tissue. Both studies demonstrated high specificity and robustness of the identified tDMR markers. However, bisulfite treatment–based methods require relatively large amounts of DNA compared to MSRE; usually more than 100 ng of DNA is used for successful bisulfite treatment. Madi et al. (2012), however, report the use of 1.5–30 ng for bisulfite modification of epithelial samples. Moreover, conventional

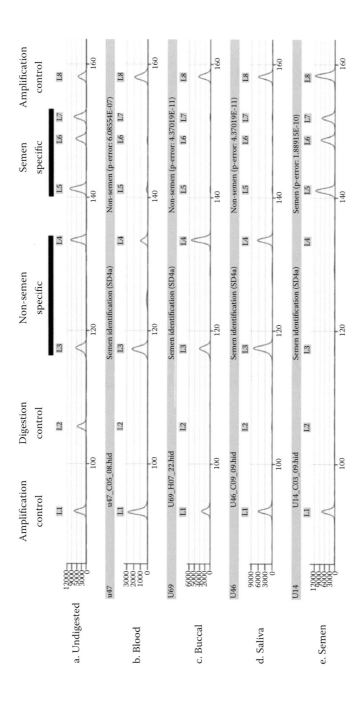

Figure 15.4 Profiles of multiple tissue types using the DSI Semen assay as reported by LaRue et al. (2013). Eight amplification peaks exist in the assay and can be observed in an undigested sample. Numbered from left to right, there are two peaks that serve as positive amplification controls (peaks L1 and L8), one peak as a control for restriction digestion (peak L2), two peaks to indicate methylation patterns specific for nonsemen tissues (peaks L3 and L4), and three peaks to indicate methylation patterns present in semen (peaks L5, L6, and L8). (Reprinted with permission from LaRue, B.L. et al., *Int J Legal Med*, 127, 2, 299–308, 2013.)

DNA sequencing and especially pyrosequencing are expensive and technically challenging methods of DNA methylation analysis. In general, although it just started to be developed, the DNA methylation approach has high potential for forensic tissue identification. Its major advantage over m/miRNA markers is that the information on cell type and individual identity is obtained from the same molecule, namely, DNA. Such approach in principle reduces the chance of false matching a cell type with an individual. However, the factors causing variation in DNA methylation patterns, including the influence of environmental factors or disease status, as well as restriction-site polymorphism in different populations that could affect MSRE markers detection, need to be studied more comprehensively for the identified candidate tDMR markers before the DNA methylation approach to forensic tissue identification can be applied in routine forensic practice.

15.5 Nonhuman DNA and RNA for Forensic Tissue Identification

The human body hosts numerous and diverse bacterial species that can be different at different localities within/on the body and can play important roles in maintaining healthy homeostasis. Recent studies of human microbiota revealed that despite high variability across individuals, certain tissues and anatomical locations of the body possess characteristic sets of bacterial species (Human Microbiome Project 2012; Turnbaugh et al. 2007), which in principle provide possibilities for forensic tissue identification. Bacterial species traditionally identified by PCR using *16S* and *23S* rRNA genes, either from RNA or DNA, which display enormous interspecies diversity. With respect to forensics, bacterial DNA analysis was first evaluated for the identification of persons by matching the bacteria on an object to the skin-associated bacteria of an individual who touched the object (Fierer et al. 2010) and for the verification of human bite marks (Brown et al. 1984). Recently, several articles appeared on the use of bacterial DNA/RNA for the identification of those forensically relevant body fluids that naturally have high bacterial loads, that is, vaginal secretion, saliva, and feces.

15.5.1 Saliva

Saliva contains 10^7 to 10^9 bacteria per mL (Lazarevic et al. 2012); of these, streptococci are the most abundant (Socransky and Manganiello 1971). In 2009, Nakanishi et al. demonstrated that saliva can be reliably distinguished from semen, urine, vaginal fluid, and skin via PCR detection of two *Streptococcus* species (*S. salivarius* and *S. mutans*). *S. salivarius* was detected in all individual saliva samples tested, whereas *S. mutans* was only detected in 60% of individual saliva stains. The reported detection limit for *S. salivarius* was 10 pg (corresponding to the amount of DNA in approximately 5.0×10^3 bacteria). This study demonstrated the forensic potential of bacterial DNA analysis for saliva identification.

15.5.2 Mouth-Expirated Blood

Another application of bacterial DNA analysis is in distinguishing bloodstains that were mouth-expirated, and thus are expected to be mixed with saliva, from the blood spatter of peripheral blood, for example, due to gunshots, stabbing, or trauma (and thus not mixed with saliva). So far, this task, which may have serious evidential value, especially in the investigation of murder cases, could only be accomplished by bloodstain pattern analysis.

However, in many cases, especially when the bloodstains were deposited on absorbing surfaces, blood pattern analysis appears to be uninformative. Recent publications on expirated blood identifications by Power et al. (2010) and Donaldson et al. (2010) make use of *Streptococcus* DNA analysis. Both studies use end-point PCR with streptococci-specific primers thus allowing very sensitive identification of saliva present in bloodstains (down to 0.01 μL), indicating the presence of mouth-expirated blood, even after several months upon deposition of the stains.

15.5.3 Vaginal Secretion

Another body fluid, which is naturally inhabited by high numbers of bacteria from different species, is vaginal secretion. Among them, *Lactobacillus* species are predominant in normal vaginal samples, and high *L. crispatus* and *L. jensenii* abundance was specific to normal vaginal secretion samples, whereas *L. iners* was abundant in all categories including bacterial vaginosis (Zozaya-Hinchliffe et al. 2010; Witkin et al. 2007). In 2010, Fleming and Harbison proposed bacterial RNA markers based on 16S–23S rRNA intergenic spacer region for identifying vaginal-specific *L. crispatus* and *L. gasseri*. They incorporated the *Lactobacilli* RNA vaginal markers into multiplex PCR assay that can detect human mRNA markers for peripheral and menstrual blood, saliva, and semen (Fleming and Harbison 2010b). In 2012, Giampaoli et al. proposed the use of DNA qPCR assays for detection of *L. crispatus* and *L. gasseri* in forensic samples. The results of vaginal identification were comparable to those obtained by Fleming and Harbison; however, the authors claimed that the DNA-based assay should be less sensitive to sample degradation, although no dedicated experiments on this were performed. In 2012, Akutsu et al. evaluated 16S rRNA primers specific to several *Lactobacillus* species as well as *Gardnerella vaginalis* and *Atopobium vaginae* for vaginal secret identification (Akutsu et al. 2012). In general, the results were compatible with the previous findings, except that *L. gasseri* was detected not only in vaginal samples but also in male and female gastric mucosa. Jakubowska et al. (2013) developed multiplex PCR assays for the detection *Lactobacillus crispatus* and *L. gasseri/L. johnsonii* rRNA along with mRNA markers for the identification of vaginal secretion. In this study, *Lactobacilli* markers appeared to be sensitive and specific to vaginal secretion; however, the authors were concerned upon the observed correlation of the health status and the detection rate of bacterial amplicons, as well as the possible influence of antibiotic treatment on the variation of vaginal microbiota particularly the presence of *Lactobacilli* in vaginal secretion. A more systematic study of vaginal microflora aimed to the *de novo* identification of vaginal microbial markers was undertaken by Benschop et al. (2012) using DNA NGS technology and microarray analysis. First, the authors performed comparative DNA NGS analysis of large number of samples of different tissues that, unlike previous studies, also included skin from different anatomical locations, and selected characteristic bacterial species of vaginal origin. Corroborating the previous finding, they found that most of the vaginal sequence reads (59%) corresponded to species within the genus *Lactobacillus*, followed by the genus *Gardnerella* (21%). Based on NGS data, a total of 389 oligonucleotide 16S rDNA probes (covering 101 genera) were spotted on microarrays to analyze the microbial species in DNA extracts from forensic samples. The result of microarray profiling indicated that the previously suggested marker species *L. crispatus* and *L. gasseri/L. johnsonii* were indeed highly abundant in vaginal samples and were not detected in saliva, semen, or blood. However, when compared to skin samples, none of the identified bacterial species appeared to be strictly specific to vaginal secretion. The authors concluded that reliable establishing of vaginal origin of biological samples would require the analysis of multiple bacterial species within various genera, and should be performed using a probabilistic approach.

15.5.4 Feces

The identification of feces may provide an important evidence in particular crimes, including illegal waste dumping and sexual assaults. Previously, a method of feces detection by the analysis of steroids using gas chromatography was proposed; however, it requires considerable amounts of material and is technically challenging (Nakanishi et al. 2013). The microbial flora of the human intestinal tract comprises more than 10^{11} bacterial cells per gram of colonic content (Stephen and Cummings 1980). *Bacteroides* species account for 20% to 52% of fecal bacteria (Dick et al. 2005) and are used for feces identification in environmental studies. Nakanishi et al. investigated the specificity of three *Bactroides* species (*B. uniformus*, *B. vulgatus*, and *B. thetaiotaomicron*) using various body specimens (feces,

Table 15.5 Bacterial Species Proposed for Forensic Tissue Identification

Targeted Tissue/Body Fluid	Genus Name	Species Name	DNA/RNA	Marker (Gene Name)	Reference
Saliva	Streptococcus	S. salivalius	DNA	Glucosyltransferase (*GTF*)	Nakanishi et al. (2009)
	Streptococcus	S. mutans	DNA		
Mouth-expired blood	Streptococcus	S. mutans	DNA	Glucosyltransferase (*GTF*)	Donaldson et al. (2010)
	Streptococcus	S. sanguinis	DNA		
	Streptococcus	S. gordonii	DNA		
	Streptococcus	S. salivarius	DNA		
	Streptococcus	S. sp.	DNA	16S rRNA	Power et al. (2010)
Vaginal secretion	Lactobacillus	L. crispatus	RNA	16S–23S rRNA intergenic spacer	Fleming and Harbison (2010), Jakubowska et al. (2013)
	Lactobacillus	L. gasseri/L. johnsonii	RNA		
	Lactobacillus	L. crispatus	DNA	16S rRNA	Giampaoli et al. (2012)
	Lactobacillus	L. gasseri	DNA		
	Lactobacillus	L. iners	DNA	16S rRNA	Akutsu et al. (2012)
	Lactobacillus	L. crispatus	DNA		
	Lactobacillus	L. jensenii	DNA		
	Lactobacillus	L. gasseri	DNA		
	Gardnerella	G. vaginalis	DNA		
	Atopobium	A. vaginae	DNA		
	Lactobacillus	L. sp.	DNA	Whole genome sequencing	Benschop et al. (2012)
	Gardnerella	G. vaginalis	DNA		
Feces	Bacteroides	B. uniformis	DNA	RNA polymerase β-subunit (*rpoB*)	Nakanishi et al. (2013)
	Bacteroides	B. vulgatus	DNA		
	Bacteroides	B. thetaiotaomicron	DNA	α-1-6 Mannanase (*MANBA*)	

blood, saliva, semen, urine, vaginal fluids, and skin surfaces), and evaluated this method of feces detection using forensic samples (Nakanishi et al. 2013). A TaqMan quantitative PCR approach showed that DNA specific to these species was present in fecal samples at high frequency; either *B. uniformis* or *B. vulgatus* were detected in all of the fecal samples tested, whereas *B. thetaiotaomicron* was present only in about half of fecal samples. *Bacteroides uniformis* and *B. thetaiotaomicron* were not found in other body samples, but *B. vulgatus* was detected in one of the six vaginal fluid samples. These results suggest that the use of several *Bacteroides* species is crucial; however, in general, the identification of fecal samples with bacterial DNA analysis appears to be a promising approach.

As expected from the relatively limited studies available thus far, the identification of particular body fluids using bacterial D/RNA markers has high forensic potential (Table 15.5). This approach provides the means for discriminating between body fluids, such as saliva and vaginal secretion, which are hardly distinguishable by means of human m/miRNA markers. However, further development of this approach including careful evaluation of the candidate D/RNA markers is required. Besides the evaluation of tissue specificity, the influence of pathological condition such as vaginosis or candidiasis and their antibiotic treatment on microbial D/RNA marker abundance, as well as human specificity of particular species, need to be tested carefully. Finally, the most suitable bacterial D/RNA markers identified may be integrated into multiplex PCR assays in combination with human D/RNA markers, which would allow parallel identification of different tissues. This way, for instance, the issue of the presence of *Lactobacillus* species in skin samples may be overcome by using human-specific skin markers. So far, only vaginal bacterial markers have been implemented into multiplex assay together with human mRNA markers for the identification of multiple forensically relevant cell types (Fleming and Harbison 2010b; Jakubowska et al. 2013).

15.6 Outlook

During the past decade, the development of nucleic acid markers and analysis methods for forensic tissue identification advanced considerably. Currently, nucleic acid–based methods are widely recognized by the forensic community and in certain countries, such as the Netherlands and New Zealand, RNA profiling for forensic tissue identification is already implemented into operational casework. It is expected that nucleic acid–based forensic tissue identification will become more widespread in practical casework if more commercial assays would become available. Future research development of R/DNA-based tissue identification methods is likely to be targeted toward increasing the reliability of distinguishing those forensically relevant tissue types—which are biologically similar (e.g., saliva, skin, and vaginal secretion)—from each other. Promising approaches toward solving this issue include comprehensive biomarker screening via whole transcriptome and epigenome analyses using NGS technologies. To avoid possible misassignments of individuals and cell types in case of mixed biological sample, it is preferable to use the same biomolecule for individual identification and for tissue identification purposes, which is now achievable by using DNA methylation tissue markers. For RNA tissue identification markers, this potential caveat should be eliminated with the development of reliable RNA quantification and gene expression normalization methods.

Acknowledgments

We thank Cordula Haas for useful comments on the manuscript. We are grateful to the numerous colleagues who published on forensic tissue identification, whose work we had the privilege to partly summarize here. The authors in their original work on forensic tissue identification were supported by funding from the Netherlands Forensic Institute (NFI) and by a grant from the Netherlands Genomics Initiative (NGI)/Netherlands Organization for Scientific Research (NWO) within the framework of the Forensic Genomics Consortium Netherlands (FGCN).

References

Adams, R. L. P., Knowler, J. T. and Leader, D. P. 1986. *The biochemistry of the nucleic acids.* 10th ed. London: Chapman and Hall.
Akutsu, T. et al. 2012. Detection of bacterial 16S ribosomal RNA genes for forensic identification of vaginal fluid. *Leg Med (Tokyo),* 14(3): 160–2.
Akutsu, T. et al. 2010. Applicability of ELISA detection of statherin for forensic identification of saliva. *Int J Legal Med,* 124(5): 493–8.
Alberts, B., Wilson, J. H. and Hunt, T. 2008. *Molecular biology of the cell.* 5th ed. New York: Garland Science.
Allery, J. P. et al. 2001. Cytological detection of spermatozoa: comparison of three staining methods. *J Forensic Sci,* 46(2): 349–51.
Alvarez, M. and Ballantyne, J. 2006. The identification of newborns using messenger RNA profiling analysis. *Anal Biochem,* 357(1): 21–34.
Asano, M., Oya, M. and Hayakawa, M. 1972. Identification of menstrual blood stains by the electrophoretic pattern of lactate dehydrogenase isozymes. *Forensic Sci,* 1(3): 327–32.
Bahar, B. et al. 2007. Long-term stability of RNA in post-mortem bovine skeletal muscle, liver and subcutaneous adipose tissues. *BMC Mol Biol,* 8: 108.
Baker, D. J., Grimes, E. A. and Hopwood, A. J. 2011. D-dimer assays for the identification of menstrual blood. *Forensic Sci Int,* 212(1–3): 210–4.
Barni, F. et al. 2007. Forensic application of the luminol reaction as a presumptive test for latent blood detection. *Talanta,* 72(3): 896–913.
Bartel, D. P. 2004. MicroRNAs: genomics, biogenesis, mechanism, and function. *Cell,* 116(2): 281–97.
Bauer, M. and Patzelt, D. 2002. Evaluation of mRNA markers for the identification of menstrual blood. *J Forensic Sci,* 47(6): 1278–82.
Bauer, M. and Patzelt, D. 2003. Protamine mRNA as molecular marker for spermatozoa in semen stains. *Int J Legal Med,* 117(3): 175–9.
Baxter, S. J. and Rees, B. 1974. The immunological identification of foetal haemoglobin in bloodstains in infanticide and associated crimes. *Med Sci Law,* 14(3): 163–7.
Benschop, C. C. et al. 2012. Vaginal microbial flora analysis by next generation sequencing and microarrays; can microbes indicate vaginal origin in a forensic context? *Int J Legal Med,* 126(2): 303–10.
Bird, A. P. 1986. CpG-rich islands and the function of DNA methylation. *Nature,* 321(6067): 209–13.
Brown, K. A. et al. 1984. The survival of oral streptococci on human skin and its implication in bite-mark investigation. *Forensic Sci Int,* 26(3): 193–7.
Byun, H. M. et al. 2009. Epigenetic profiling of somatic tissues from human autopsy specimens identifies tissue- and individual-specific DNA methylation patterns. *Hum Mol Genet,* 18(24): 4808–17.
Chen, C. et al. 2005. Real-time quantification of microRNAs by stem-loop RT-PCR. *Nucleic Acids Res,* 33(20): e179.
Courts, C. and Madea, B. 2011. Specific micro-RNA signatures for the detection of saliva and blood in forensic body-fluid identification. *J Forensic Sci,* 56(6): 1464–70.

Cox, M. 1991. A study of the sensitivity and specificity of four presumptive tests for blood. *J Forensic Sci*, 36(5): 1503–11.

Dick, L. K. et al. 2005. Host distributions of uncultivated fecal Bacteroidales bacteria reveal genetic markers for fecal source identification. *Appl Environ Microbiol*, 71(6): 3184–91.

Donaldson, A. E. et al. 2010. Using oral microbial DNA analysis to identify expirated bloodspatter. *Int J Legal Med*, 124(6): 569–76.

Fierer, N. et al. 2010. Forensic identification using skin bacterial communities. *Proc Natl Acad Sci U S A*, 107(14): 6477–81.

Fleming, R. I. and Harbison, S. 2010a. The development of a mRNA multiplex RT-PCR assay for the definitive identification of body fluids. *Forensic Sci Int Genet*, 4(4): 244–56.

Fleming, R. I. and Harbison, S. 2010b. The use of bacteria for the identification of vaginal secretions. *Forensic Sci Int Genet*, 4(5): 311–5.

Fordyce, S. L. et al. 2013. Deep sequencing of RNA from ancient maize kernels. *PLoS One*, 8(1): e50961.

French, C. E. et al. 2008. A novel histological technique for distinguishing between epithelial cells in forensic casework. *Forensic Sci Int*, 178(1): 1–6.

Frumkin, D. et al. 2011. DNA methylation-based forensic tissue identification. *Forensic Sci Int Genet*, 5(5): 517–24.

Gaensslen, R. E. and National Institute of Justice (U.S.), 1983. *Sourcebook in forensic serology, immunology, and biochemistry*. Washington, D.C.: U.S. Dept. of Justice For sale by the Supt. of Docs., U.S. G.P.O.

Gauvin, J. et al. 2010. Forensic pregnancy diagnostics with placental mRNA markers. *Int J Legal Med*, 124(1): 13–7.

Giampaoli, S. et al. 2012. Molecular identification of vaginal fluid by microbial signature. *Forensic Sci Int Genet*, 6(5): 559–64.

Gray, D., Frascione, N. and Daniel, B. 2012. Development of an immunoassay for the differentiation of menstrual blood from peripheral blood. *Forensic Sci Int*, 220(1–3): 12–8.

Haas, C. et al. 2013. RNA/DNA co-analysis from human saliva and semen stains—results of a third collaborative EDNAP exercise. *Forensic Sci Int Genet*, 7(2): 230–9.

Haas, C. et al. 2012. RNA/DNA co-analysis from blood stains—results of a second collaborative EDNAP exercise. *Forensic Sci Int Genet*, 6(1): 70–80.

Haas, C. et al. 2011. mRNA profiling for the identification of blood—results of a collaborative EDNAP exercise. *Forensic Sci Int Genet*, 5(1): 21–6.

Haas, C. et al. 2009. mRNA profiling for body fluid identification by reverse transcription endpoint PCR and realtime PCR. *Forensic Sci Int Genet*, 3(2): 80–8.

Hansen, A. J. et al. 2006. Crosslinks rather than strand breaks determine access to ancient DNA sequences from frozen sediments. *Genetics*, 173(2): 1175–9.

Hanson, E. et al. 2012. Specific and sensitive mRNA biomarkers for the identification of skin in 'touch DNA' evidence. *Forensic Sci Int Genet*, 6(5): 548–58.

Hanson, E. K. and Ballantyne, J. 2013. Highly specific mRNA biomarkers for the identification of vaginal secretions in sexual assault investigations. *Sci Justice*, 53(1): 14–22.

Hanson, E. K., Lubenow, H. and Ballantyne, J. 2009. Identification of forensically relevant body fluids using a panel of differentially expressed microRNAs. *Anal Biochem*, 387(2): 303–14.

Hochmeister, M. N. et al. 1999a. Evaluation of prostate-specific antigen (PSA) membrane test assays for the forensic identification of seminal fluid. *J Forensic Sci*, 44(5): 1057–60.

Hochmeister, M. N. et al. 1999b. Validation studies of an immunochromatographic 1-step test for the forensic identification of human blood. *J Forensic Sci*, 44(3): 597–602.

Holliday, R. and Pugh, J. E. 1975. DNA modification mechanisms and gene activity during development. *Science*, 187(4173): 226–32.

Human Microbiome Project, C. 2012. Structure, function and diversity of the healthy human microbiome. *Nature*, 486(7402): 207–14.

Illingworth, R. et al. 2008. A novel CpG island set identifies tissue-specific methylation at developmental gene loci. *PLoS Biol*, 6(1): e22.

Inoue, H., Kimura, A. and Tuji, T. 2002. Degradation profile of mRNA in a dead rat body: basic semi-quantification study. *Forensic Sci Int,* 130(2–3): 127–32.

Inoue, H. et al. 1989. Identification of fetal hemoglobin in blood stains by high performance liquid chromatography. *Z Rechtsmed,* 102(7): 437–44.

Jacob, G. F. and Raper, A. B. 1958. Hereditary persistence of foetal haemoglobin production, and its interaction with the sickle-cell trait. *Br J Haematol,* 4(2): 138–49.

Jakubowska, J. et al. 2013. mRNA profiling for vaginal fluid and menstrual blood identification. *Forensic Sci Int Genet,* 7(2): 272–8.

Johnson, S. A., Morgan, D. G. and Finch, C. E. 1986. Extensive postmortem stability of RNA from rat and human brain. *J Neurosci Res,* 16(1): 267–80.

Jung, M. et al. 2010. Robust microRNA stability in degraded RNA preparations from human tissue and cell samples. *Clin Chem,* 56(6): 998–1006.

Juusola, J. and Ballantyne, J. 2003. Messenger RNA profiling: a prototype method to supplant conventional methods for body fluid identification. *Forensic Sci Int,* 135(2): 85–96.

Juusola, J. and Ballantyne, J. 2005. Multiplex mRNA profiling for the identification of body fluids. *Forensic Sci Int,* 152(1): 1–12.

Juusola, J. and Ballantyne, J. 2007. mRNA profiling for body fluid identification by multiplex quantitative RT-PCR. *J Forensic Sci,* 52(6): 1252–62.

Kashyap, V. K. 1989. A simple immunosorbent assay for detection of human blood. *J Immunoassay,* 10(4): 315–24.

Kaye, S. 1949. Acid phosphatase test for identification of seminal stains. *J Lab Clin Med,* 34(5): 728–32.

Kent, E. J., Elliot, D. A. and Miskelly, G. M. 2003. Inhibition of bleach-induced luminol chemiluminescence. *J Forensic Sci,* 48(1): 64–7.

King, K., Flinter, F. A. and Green, P. M. 2001. Hair roots as the ideal source of mRNA for genetic testing. *J Med Genet,* 38(6): E20.

Kohlmeier, F. and Schneider, P. M. 2012. Successful mRNA profiling of 23 years old blood stains. *Forensic Sci Int Genet,* 6(2): 274–6.

Koppelkamm, A. et al. 2011. RNA integrity in post-mortem samples: influencing parameters and implications on RT-qPCR assays. *Int J Legal Med,* 125(4): 573–80.

LaRue, B. L., King, J. L. and Budowle, B. 2013. A validation study of the Nucleix DSI-Semen kit—a methylation-based assay for semen identification. *Int J Legal Med,* 127(2): 299–308.

Lazarevic, V. et al. 2012. Analysis of the salivary microbiome using culture-independent techniques. *J Clin Bioinforma,* 2: 4.

Lee, H. Y. et al. 2012. Potential forensic application of DNA methylation profiling to body fluid identification. *Int J Legal Med,* 126(1): 55–62.

Liang, Y. et al. 2007. Characterization of microRNA expression profiles in normal human tissues. *BMC Genomics,* 8: 166.

Lindahl, T. 1996. The Croonian Lecture, 1996: endogenous damage to DNA. *Philos Trans R Soc Lond B Biol Sci,* 351(1347): 1529–38.

Lindahl, T. and Nyberg, B. 1972. Rate of depurination of native deoxyribonucleic acid. *Biochemistry,* 11(19): 3610–8.

Lindenbergh, A. et al. 2012. A multiplex (m)RNA-profiling system for the forensic identification of body fluids and contact traces. *Forensic Sci Int Genet,* 6(5): 565–77.

Lindenbergh, A., Maaskant, P. and Sijen, T. 2013. Implementation of RNA profiling in forensic casework. *Forensic Sci Int Genet,* 7(1): 159–66.

Madi, T. et al. 2012. The determination of tissue-specific DNA methylation patterns in forensic biofluids using bisulfite modification and pyrosequencing. *Electrophoresis,* 33(12): 1736–45.

Marchuk, L. et al. 1998. Postmortem stability of total RNA isolated from rabbit ligament, tendon and cartilage. *Biochim Biophys Acta,* 1379(2): 171–7.

Masibay, A. S. and Lappas, N. T. 1984. The detection of protein p30 in seminal stains by means of thin-layer immunoassay. *J Forensic Sci,* 29(4): 1173–7.

Moreno, L. I. et al. 2012. Determination of an effective housekeeping gene for the quantification of mRNA for forensic applications. *J Forensic Sci,* 57(4): 1051–8.

Myers, J. R. and Adkins, W. K. 2008. Comparison of modern techniques for saliva screening. *J Forensic Sci,* 53(4): 862–7.

Nakanishi, H. et al. 2009. A novel method for the identification of saliva by detecting oral streptococci using PCR. *Forensic Sci Int,* 183(1–3): 20–3.

Nakanishi, H. et al. 2013. Identification of feces by detection of Bacteroides genes. *Forensic Sci Int Genet,* 7(1): 176–9.

Nussbaumer, C., Gharehbaghi-Schnell, E. and Korschineck, I. 2006. Messenger RNA profiling: a novel method for body fluid identification by real-time PCR. *Forensic Sci Int,* 157(2–3): 181–6.

Oehmichen, M. and Zilles, K. 1984. Postmortale DNS- und RNS-Synthese: Erste Untersuchungen an menschlichen Leichen. *Z Rechtsmed,* 91(4): 287–94

Postmortale DNS- und RNS-Synthese: Erste Untersuchungen an menschlichen Leichen. *Z Rechtsmed,* 91(4): 287–94.

Ohgane, J., Yagi, S. and Shiota, K. 2008. Epigenetics: the DNA methylation profile of tissue-dependent and differentially methylated regions in cells. *Placenta,* 29 Suppl A: S29–35.

Old, J. B. et al. 2009. Developmental validation of RSID-saliva: a lateral flow immunochromatographic strip test for the forensic detection of saliva. *J Forensic Sci,* 54(4): 866–73.

Paabo, S. and Wilson, A. C. 1991. Miocene DNA sequences—a dream come true? *Curr Biol,* 1(1): 45–6.

Pang, B. C. and Cheung, B. K. 2007. Identification of human semenogelin in membrane strip test as an alternative method for the detection of semen. *Forensic Sci Int,* 169(1): 27–31.

Pang, B. C. and Cheung, B. K. 2008. Applicability of two commercially available kits for forensic identification of saliva stains. *J Forensic Sci,* 53(5): 1117–22.

Park, S. M. et al. 2013. Genome-wide mRNA profiling and multiplex quantitative RT-PCR for forensic body fluid identification. *Forensic Sci Int Genet,* 7(1): 143–50.

Power, D. A. et al. 2010. PCR-based detection of salivary bacteria as a marker of expirated blood. *Sci Justice,* 50(2): 59–63.

Quarino, L. et al. 2005. An ELISA method for the identification of salivary amylase. *J Forensic Sci,* 50(4): 873–6.

Randall, B. 1988. Glycogenated squamous epithelial cells as a marker of foreign body penetration in sexual assault. *J Forensic Sci,* 33(2): 511–4.

Richard, M. L. et al. 2012. Evaluation of mRNA marker specificity for the identification of five human body fluids by capillary electrophoresis. *Forensic Sci Int Genet,* 6(4): 452–60.

Roeder, A. D. and Haas, C. 2012. mRNA profiling using a minimum of five mRNA markers per body fluid and a novel scoring method for body fluid identification. *Int J Legal Med.*

Sakurada, K. et al. 2010. Detection of dermcidin for sweat identification by real-time RT-PCR and ELISA. *Forensic Sci Int,* 194(1–3): 80–4.

Sakurada, K. et al. 2012. Identification of nasal blood by real-time RT-PCR. *Leg Med (Tokyo),* 14(4): 201–4.

Schiff, A. F. 1978. Reliability of the acid phosphatase test for the identification of seminal fluid. *J Forensic Sci,* 23(4): 833–44.

Schill, W. B. and Schumacher, G. F. 1972. Radial diffusion in gel for micro determination of enzymes: I. Muramidase, alpha-amylase, DNase 1, RNase A, acid phosphatase, and alkaline phosphatase. *Anal Biochem,* 46(2): 502–33.

Schmidt, S. et al. 2001. Prostate-specific antigen in female urine: a prospective study involving 217 women. *Urology,* 57(4): 717–20.

Setzer, M., Juusola, J. and Ballantyne, J. 2008. Recovery and stability of RNA in vaginal swabs and blood, semen, and saliva stains. *J Forensic Sci,* 53(2): 296–305.

Shaler, R. C. 2002. *Modern forensic biology.* Upper Saddle River, NJ: Prentice Hall.

Socransky, S. S. and Manganiello, S. D. 1971. The oral microbiota of man from birth to senility. *J Periodontol,* 42(8): 485–96.

Song, F. et al. 2005. Association of tissue-specific differentially methylated regions (TDMs) with differential gene expression. *Proc Natl Acad Sci U S A,* 102(9): 3336–41.

Sood, P. et al. 2006. Cell-type-specific signatures of microRNAs on target mRNA expression. *Proc Natl Acad Sci U S A,* 103(8): 2746–51.

Stephen, A. M. and Cummings, J. H. 1980. The microbial contribution to human faecal mass. *J Med Microbiol,* 13(1): 45–56.

Thorp, H. H. 2000. The importance of being r: greater oxidative stability of RNA compared with DNA. *Chem Biol,* 7(2): R33–6.

Turnbaugh, P. J. et al. 2007. The human microbiome project. *Nature,* 449(7164): 804–10.

Vallejo, G. 1990. Human chorionic gonadotropin detection by means of enzyme immunoassay: a useful method in forensic pregnancy diagnosis in bloodstains. *J Forensic Sci,* 35(2): 293–300.

van Oorschot, R. A. and Jones, M. K. 1997. DNA fingerprints from fingerprints. *Nature,* 387(6635): 767.

Vandenberg, N. and van Oorschot, R. A. 2006. The use of Polilight in the detection of seminal fluid, saliva, and bloodstains and comparison with conventional chemical-based screening tests. *J Forensic Sci,* 51(2): 361–70.

Vergote, G. et al. 1991. Forensic determination of pregnancy hormones in human bloodstains. *J Forensic Sci Soc,* 31(4): 409–19.

Virkler, K. and Lednev, I. K. 2009. Analysis of body fluids for forensic purposes: from laboratory testing to non-destructive rapid confirmatory identification at a crime scene. *Forensic Sci Int,* 188(1–3): 1–17.

Visser, M. et al. 2011. mRNA-based skin identification for forensic applications. *Int J Legal Med,* 125(2): 253–63.

Wang, Z. et al. 2013. Screening and confirmation of microRNA markers for forensic body fluid identification. *Forensic Sci Int Genet,* 7(1): 116–23.

Warshauer, D. H. et al. 2012. An evaluation of the transfer of saliva-derived DNA. *Int J Legal Med,* 126(6): 851–61.

Wasserstrom, A. et al. 2013. Demonstration of DSI-semen—A novel DNA methylation-based forensic semen identification assay. *Forensic Sci Int Genet,* 7(1): 136–42.

Webb, J. L., Creamer, J. I. and Quickenden, T. I. 2006. A comparison of the presumptive luminol test for blood with four non-chemiluminescent forensic techniques. *Luminescence,* 21(4): 214–20.

Wee, A., Thamboo, T. P. and Thomas, A. 2003. alpha-Fetoprotein-producing liver carcinomas of primary extrahepatic origin. *Acta Cytol,* 47(5): 799–808.

Whitehead, P. H. and Divall, G. B. 1974. The identification of menstrual blood—the immunoelectrophoretic characterisation of soluble fibrinogen from menstrual bloodstain extracts. *Forensic Sci,* 4(1): 53–62.

Whitehead, P. H. and Kipps, A. E. 1975. The significance of amylase in forensic investigations of body fluids. *Forensic Sci,* 6(3): 137–44.

Willott, G. M. 1982. Frequency of azoospermia. *Forensic Sci Int,* 20(1): 9–10.

Witkin, S. S., Linhares, I. M. and Giraldo, P. 2007. Bacterial flora of the female genital tract: function and immune regulation. *Best Pract Res Clin Obstet Gynaecol,* 21(3): 347–54.

Wraxall, B. G. 1972. The identification of foetal haemoglobin in bloodstains. *J Forensic Sci Soc,* 12(3): 457–8.

Wu, C. et al. 2009. BioGPS: an extensible and customizable portal for querying and organizing gene annotation resources. *Genome Biol,* 10(11): R130.

Yokota, M. et al. 2001. Evaluation of prostate-specific antigen (PSA) membrane test for forensic examination of semen. *Leg Med (Tokyo),* 3(3): 171–6.

Zozaya-Hinchliffe, M. et al. 2010. Quantitative PCR assessments of bacterial species in women with and without bacterial vaginosis. *J Clin Microbiol,* 48(5): 1812–9.

Zubakov, D. et al. 2008. Stable RNA markers for identification of blood and saliva stains revealed from whole genome expression analysis of time-wise degraded samples. *Int J Legal Med,* 122(2): 135–42.

Zubakov, D. et al. 2009. New markers for old stains: stable mRNA markers for blood and saliva identification from up to 16-year-old stains. *Int J Legal Med,* 123(1): 71–4.

Zubakov, D. et al. 2010. MicroRNA markers for forensic body fluid identification obtained from microarray screening and quantitative RT-PCR confirmation. *Int J Legal Med,* 124(3): 217–26.

Evolving Technologies in Forensic DNA Analysis

16

CASSANDRA D. CALLOWAY
HENRY ERLICH

Contents

16.1	Introduction	417
16.2	mtDNA and HV+ HaploArray	418
16.3	HV+ HaploArray	418
16.4	Distribution of HV+ Mitotypes	420
16.5	mtDNA and Forensics Mixtures	421
16.6	Potential Forensics Applications of NGS	421
16.7	Forensics Mixtures and NGS	422
16.8	Overview of 454 Genome Sequencing Technology	422
16.9	HLA 454 Assay for Mixture Analysis	424
16.10	mtDNA 454 GS Assay for Mixture Analysis	424
16.11	NGS for Forensic DNA Analysis	426
16.12	Conclusion	426
References		427

16.1 Introduction

Polymerase chain reaction (PCR) based genetic analysis has revolutionized the practice of forensics since the first application of the HLA-DQA1 test in the *Pennsylvania vs Pestinikis* case in 1986 (Blake et al. 1992). In recent years, the focus of forensics genetic analysis has been primarily on length polymorphisms, panels of short tandem repeat (STR) markers and, to some extent, on sequence polymorphism of the mitochondrial (mt) DNA hypervariable regions (HVI and II) (Butler 2004; Holland and Parsons 1999; Sullivan, Hopgood, and Gill 1992; Just et al. 2004). The STR markers have been analyzed by capillary electrophoresis and the mtDNA HVI and HVII sequence markers by dideoxy chain termination Sanger sequencing or by various probe-based methods, such as the Linear Array (Melton et al. 2005; Gabriel et al. 2003; Divne et al. 2005; Roberts and Calloway 2007; Chong et al. 2005). Recently, novel technologies and new genetic markers have been developed that promise to have a significant impact on the future of forensic genetic analysis. In this chapter, we discuss two of these new developments: (1) the expansion of the mtDNA linear array technology to include informative polymorphisms throughout the whole ~16-kb mitochondrial genome and (2) next-generation sequencing (NGS) technologies that can provide a single high-throughput platform for sequence and tandem repeat polymorphisms and, because of the clonal sequencing property of NGS, a digital readout for analyzing forensics mixtures.

16.2 mtDNA and HV+ HaploArray

mtDNA is an informative and useful target for the forensic genetic analysis of limited and/or degraded samples. However, there are several inherent limitations to targeting only the HVI/II, independent of the method of analysis. The power of discrimination is limited for all population groups because of the presence of a few common HVI/II sequences in all populations tested (Melton et al. 2005; Date Chong et al. 2005; Gabriel et al. 2003). Therefore, additional sequence polymorphisms outside the HVI/II regions need to be targeted to increase the power of discrimination of mtDNA analysis in addition to the HV regions.

The HV regions of the human mitochondrial genome have a high sequence diversity and, accordingly, have been the primary area of study for identification purposes. Several systems targeting the HVI/II regions of the mitochondrial genome are currently being used by forensic laboratories for the analysis of highly degraded or limited DNA. It has been widely reported that some HVI/HVII types are common among all populations, even though the overall distribution of mtDNA HVI/II sequences is highly skewed toward rare types. Approximately 7% of Caucasians share the most common HVI/II sequence differing from the rCRS at 263G, and 13 additional HV sequences are shared among >0.5% of the population (Parsons and Coble 2001). A higher proportion share the same mtDNA linear array types (mitotypes) because not all polymorphic sites within the HVI/II regions are targeted with the HVI/HVII probe panel. Approximately 10% of Caucasians share the most common HVI/HVII mitotype (1111111111). Sequencing the entire HVI/II region can discriminate somewhat between these mitotypes but the problem posed by relatively common sequences remains. The primary limitation for current mtDNA testing, based on HVI/II, reflects the small number of common mtDNA types.

To increase the discrimination power of mtDNA testing, additional sequence variation outside of the commonly targeted HVI/HVII regions needs to be analyzed. Although the noncoding region has the highest degree of sequence diversity, this region only accounts for ~10% of the mitochondrial genome. Variation in the rest of the noncoding region, referred to as variable regions (VR) I/II, as well as the coding region (CR) can be targeted to increase the discrimination of current assays that only target the HVI/HVII regions. Several assays targeting CR sequence have been developed and include SNaPshot (Life Technologies™, Foster City, CA, USA) and GeneChip® Human Mitochondrial Resequencing Array (Affymetrix®, Santa Clara, CA, USA) (Brandstatter, Parsons, and Parson 2003; Quintans et al. 2004; Kline et al. 2005). These assays can be used following HVI/HVII sequencing to further distinguish common HV haplogroups. Alternatively, assays that target polymorphic sites distributed throughout the mitochondrial genome can be used. We discuss here in detail the HV+ HaploArray that allows for simultaneous analysis of hypervariable region *and* CR sites in a single assay.

16.3 HV+ HaploArray

The HV+ HaploArray uses a multiplex PCR to amplify 15 regions and a panel of 105 immobilized probes to target 61 polymorphic sites distributed across the entire mitochondrial genome (Calloway et al. 2009). Polymorphic sites to subdivide common HVI/HVII types as well as some rapidly evolving polymorphic sites were identified and targeted to help improve the discrimination power. Also, several HVI and HVII sites were added to improve the robustness of the HVI/II Linear Array Assay by reducing the number of "0" and "weak" signal types caused by destabilizing probe mismatches. Primers and probes targeting these sites were designed and optimized to work under the existing amplification and hybridization conditions of the HVI/HVII linear array assay.

A 5-plex and a 10-plex PCR were developed targeting regions ranging in size from 314 to 444 and from 103 to 183 bp, respectively. Primers were designed to generate laddered product sizes in order to visualize the amplicons by gel electrophoresis. Two multiplex PCRs were required in order to target both of the hypervariable and variable regions because the primer sets overlap. The probe panel corresponding to the 5-plex amplification consists of 59 probes and targets variation at 14 HVI sites, 11 HVII sites, eight CR sites, and two VRI sites (Calloway et al. 2009). The 10-plex PCR probe panel consists of 46 probes and targets 17 CR sites, four VRI sites, and five VRII sites (Calloway et al. 2009). Representative typing results from the 5-plex and 10-plex panels are shown in Figure 16.1.

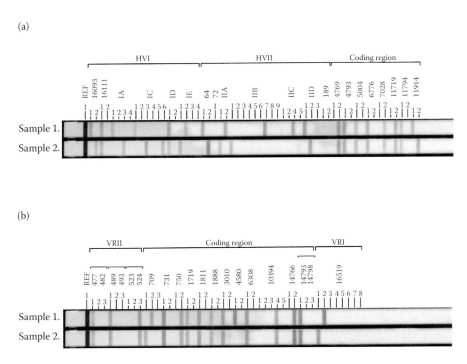

Figure 16.1 Representative typing results from HV+ HaploArray 5-plex and 10-plex probe panels. Representative typing results from five individuals are shown for the (a) 5-plex and (b) 10-plex HV+ HaploArray probe panels. The HV+ HaploArray 5-plex probe panel consists of 59 probes and targets 35 polymorphic sites (14 HVI, 11 HVII, eight coding region [CR], and two variable region I [VRI] sites). The HV+ HaploArray 10-plex PCR probe panel consists of 46 probes and targets 26 polymorphic sites (17 CR, four VRI, and five VRII sites). Four categories of probe signals can be observed: a single positive signal, weak probe signal resulting from a destabilizing mismatch within the probe binding region (Sample 2 IIC w5), no signal from a highly destabilizing mismatch (Samples 1 and 2 189 0), or 2 signals arising from a mixture of two sequences or heteroplasmy.

16.4 Distribution of HV+ Mitotypes

A population database was generated by amplifying and typing 674 samples from four population groups (194 African American, 197 U.S. Caucasian, 197 U.S. Hispanic, and 86 Japanese) with the 5-plex and 10-plex PCR and probe panels. The genetic diversity values ($h = (1 - \sum \text{freq}^2) * (n/1 - n)$) were calculated for each of the population groups and were compared to the values obtained with the HVI/HVII linear array assay. The h values are reported in Table 16.1 for each population group for both the HVI/HVII assay and the HV+ assay. The discrimination power was greatly improved for the Japanese, U.S. Hispanic, and U.S. Caucasian populations with a slight increase for the African American population for the new HV+ assay compared to the HVI/HVII assay. The discrimination power for the U.S. Caucasian population was increased from 0.9768 with the HVI/HVII assay to 0.9946 with the HV+ assay. The discrimination power was increased from 0.9449 to 0.9893 for the U.S. Hispanic population using the HV+ assay. The number of observed and unique mitotypes increased for all populations as well, with the greatest increases for the U.S. Hispanic and U.S. Caucasian populations; there was an ~35% increase in the number of observed types and an ~43% increase in the number of unique types for both population groups.

Table 16.1 Power of Discrimination and Number of Unique Types

Population Group (n)	No. of Observed Mitotypes			No. of Unique Mitotypes			h Value	
	HVI/HVII	HV+	% Increase	HVI/HVII	HV+	% Increase	HVI/HVII	HV+
African American (194)	137	153	10.5%	111	136	18.4%	0.993	0.9938
U.S. Caucasian (197)	99	149	33.6%	73	128	43.0%	0.9768	0.9946
U.S. Hispanic (197)	91	141	35.5%	67	116	42.2%	0.9449	0.9893
Japanese (86)	58	73	20.5%	48	64	25.0%	0.9806	0.9948

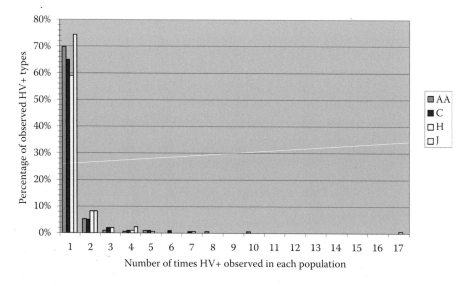

Figure 16.2 Distribution of HV+ types in 674 individuals from four population groups.

The distribution of HV+ mitotypes was also determined for each population group (Figure 16.2). Approximately 60%–75% of the HV+ types were unique (occurring only one time in the sampled population), ~5%–8% of the HV+ types occurred twice, and less than 2% of the HV+ types occurred three or more times.

A sensitivity study demonstrated that both the 5-plex and 10-plex assays are highly sensitive assays, with the 10-plex being more sensitive. Typable results were observed with ~5 pg of DNA (~730 mtDNA copies) for the 5-plex at 34 cycles and ~1 pg (~75 mtDNA copies) for the 10-plex PCR. At 38 cycles, ~0.5 pg of DNA (~75 mtDNA copies) input yielded typable results for both the 5-plex and 10-plex assays. Results from a mixture study show that a minor component present at ~10% is reliably detected with the 5-plex assay and at ~5% with the 10-plex assay. The 5-plex and 10-plex probe panels were shown to be primate-specific, and all primers were shown to be primate-specific except the pair targeting the 16s rRNA region. Up to 48 samples can be typed manually or automated in less than 2 hours, and the data can now be quickly analyzed using the Strip Scan Mitotyper software.

16.5 mtDNA and Forensics Mixtures

mtDNA is a valuable forensics genetic target because of the high copy number per cell, allowing the analysis of minute amounts of specimen DNA that could not be analyzed with chromosomal genetic markers (Higuchi et al. 1988; Holland and Parsons 1999; Sullivan, Hopgood, and Gill 1992). The haploid nature of the mtDNA genome (excluding the issue of heteroplasmy) is also valuable, particularly in the analysis of forensic mixtures, a specimen category notably difficult to analyze and interpret with chromosomal markers. If each individual contributor to a mixture is represented by one mtDNA sequence, a forensics mixture with n different sequences would be expected to contain at least n different contributors. Although less informative than a panel of chromosomal markers, mtDNA analysis can thus provide a useful estimate of the number of different contributors and, thus, facilitate and simplify the analysis of chromosomal marker data, if available, from the same mixture. Resolving the number of different mtDNA sequences in a mixture can be challenging with Sanger sequencing or with Linear Array genotyping data. However, the clonal sequencing aspect of NGS platforms addresses this issue and allows a digital readout based on the number of sequence reads for each individual mtDNA sequence. The long reads provided by the Roche 454 sequencing system (see below) facilitate this approach to the analysis of forensics mixtures. In the next section, we discuss this technology and its application to forensic analysis.

16.6 Potential Forensics Applications of NGS

NGS systems, such as the GS FLX and GS Junior (Roche Diagnostics/454), Genome Analyzer HiSeq and MiSeq (Illumina®/Solexa), SOLiD™ and Ion Torrent™ (Life Technologies™), and PacBio Rs (Pacific Biosciences®) are characterized by massively parallel, clonal sequencing (Mardis 2008). These systems have the potential to address several challenging issues in forensics analysis (Berglund, Kiialainen, and Syvanen 2011). The analysis of forensics specimens that are mixtures (>1 contributor) remains one of the most problematic issues in forensics from both a technical and statistical perspective. The clonal sequencing aspect of

NGS provides the opportunity to recover different sequence reads for every genetic variant (alleles) present in the mixture. Thus, the number of sequence reads recovered for a particular variant provides a digital readout and allows a quantitative analysis of the contributors.

16.7 Forensics Mixtures and NGS

Degraded and mixed DNA samples are often encountered in forensic cases and pose interpretation challenges. Genetic markers such as nuclear biallelic SNPs (Budowle 2004; Budowle and van Daal 2008; Sobrino, Brion, and Carracedo 2005) and mtDNA (Wilson et al. 1995), alternatives to the standard panels of STR markers, are often used to analyze limited and/or degraded DNA. However, there are some limitations to these approaches. Nuclear bi-allelic SNP markers do not allow for efficient detection and resolution of mixtures and the discrimination power of mtDNA markers is relatively low. Although STR and mtDNA markers allow for *detection* of mixtures, they do not allow for the physical separation of the components (DNA defined alleles) within a mixture. In general, mixtures pose significant problems for analysis with nuclear genetic markers because, in principle, a number of different genotypes can be consistent with the presence of more than two detected alleles. One statistical approach to the interpretation of multiple alleles in a mixed sample is to consider all possible genotypes and then sum their frequencies in the relevant population. The analysis of mtDNA markers in a suspected mixture can avoid some of these problems associated with nuclear markers since each contributor is expected to have one primary mtDNA sequence, although the issue of potential heteroplasmy will have to be addressed.

16.8 Overview of 454 Genome Sequencing Technology

As noted above, NGS systems, based on clonal sequencing, promise to facilitate the analysis of such forensic mixtures. The 454 DNA sequencing technology is a scalable, highly parallel pyrosequencing system that can be used for *shotgun* sequencing of whole genomes, exomes, or sequencing of DNA products generated by PCR (Goldberg et al. 2006; Meyer et al. 2007; Meyer, Stenzel, and Hofreiter 2008). The 454 system, like other NGS systems, requires the preparation of a "library" of DNA fragments that contain system-specific adaptor sequences (see below) and multiplex ID (MID) tags that identify the individual sample from which the amplicon sequence was generated (Bentley et al. 2009). The power of the MID tags is that the same genomic region from many different individuals can be prepared and then pooled before the sequencing run (Bentley et al. 2009). The genotyping software can then, based on the MID tag, assign each clonally derived sequence read to the appropriate individual and, based on the genomic primer sequence, to the appropriate locus. For the 454 system with "fusion" primers (primers containing the A and B adaptor sequences and MID tags at the 5' end), the adaptor sequences are incorporated into the library fragment to be sequenced; these sequences are complementary to the capture oligonucleotides coupled to "A" and "B" beads. Following a limiting dilution and mixture of the fragments with beads, single strands of the individual DNA molecules are captured by the beads and amplified to millions of copies within a microdroplet in an emulsion PCR (emPCR). After the DNA containing beads are recovered from the emulsion, the individual beads are placed in wells of a picotiter plate, and a pyrosequencing reaction, in which the individual nucleotide triphosphates are flowed in a defined sequence across the plate, is carried out. The incorporation of bases is detected by a CCD (charge-couple device) camera as a flash of light, generated by a coupled reaction in which the release of a pyrophosphate during base incorporation activates luciferin. Bi-direction sequencing for the fragments can be achieved because the single strands captured by the A and B beads allow both forward and reverse reads of the library fragment to be identified.

Next generation sequence chemistries and software, as well as the strategies for incorporating the adaptor sequences and MIDs, can be different for genome shotgun versus amplicon sequencing. In general, genome shotgun sequencing involves fragmenting the genomic DNA and then ligating on the MID tags and the A and B adaptors. For amplicon sequencing, this sequence information can be present in the "fusion" primers, which contain the genomic primer as well as these additional sequences. The benefit of this approach is the simplicity of library preparation; the relevant library is ready as soon as the genomic PCR is completed. However, for a given amplicon, n different primer pairs are required for n MID tags. Consequently, the fusion primer approach is attractive when the number of MID tags (and samples) is not very large (e.g., <20) but can become less practical with larger numbers. In principle, an amplicon library can also be prepared by long-range genomic PCR followed by cleavage and ligation of the MIDs and adaptor sequences to the fragmented DNA. However, this approach may not be successful for forensic samples because the DNA is often degraded and limited in quantity.

An approach that combines the generation of a library with a single PCR and the convenience of combinatorial MID tags is the "4 primer method," in which an "inner" primer

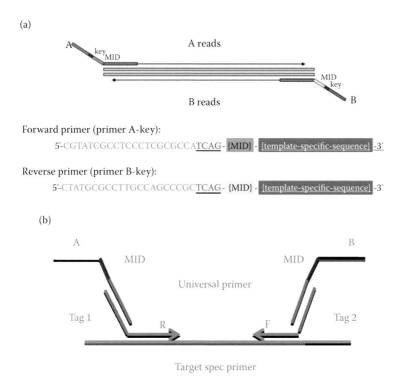

Figure 16.3 Two approaches for direct amplicon sequencing. (a) Fusion primer. (b) Four primer approach.

pair with genomic target sequence as well as a "universal" tag at the 5′ end is combined with an "outer" primer set with the universal tag at the 3′ end and the MID tags and A and B adaptors at the 5′ end of the primers (Figure 16.3). An elegant microfluidics device has been designed by Fluidigm, the 48 × 48 Access Array system, to automate the PCRs for up to 48 different primer pairs and 48 genomic DNA samples, each with a different MID tag.

16.9 HLA 454 Assay for Mixture Analysis

We have used both fusion primers as well as the Fluidigm 4 primer system to achieve high-resolution and high-throughput HLA class I and class II typing (Holcomb et al. 2011; Moonsamy et al. 2013; Bentley et al. 2009). Using the fusion primer approach, 14 different primer pairs were used to amplify exon 2, 3, and 4 for class I, exon 2 for DPB1, DQA1, and the DRB loci (DRB1, DRB3, DRB4, and DRB5) and exons 2 and 3 for DQB1 with up to 12 MID tags. As with other PCR-based HLA typing systems, the challenge is to design primers that are specific for a given gene and amplify all alleles of that gene with relatively equal efficiency. Although HLA locus-specific primers are generally desirable, the clonal sequencing property of 454 GSFLX sequencing in conjunction with the Assign™ATF 454 software (Conexio Genomics, Ltd.) allows the use of generic primer pairs, such as the DRB primers that amplify exon 2 of DRB1, DRB3, DRB4, and DRB5 to determine the genotypes at all of these loci. In addition, the clonal sequencing allows the separation of coamplified genomic sequences (pseudo-genes, related genes, etc.) that would contribute "noise" or "background" under conventional sequencing of the target gene. The robustness of this system was evaluated in an alpha study trial of eight laboratories, several of which had no previous experience with 454 sequencing or the Assign™ATF 454 software (Conexio Genomics, Ltd), and high concordance was achieved at all loci (Holcomb et al. 2011).

In general, a high-resolution HLA typing for all loci for seven samples can be completed on the GS Junior, the small desktop instrument, in 3–4 days and for about 40–50 samples on the GS FLX in about 5–6 days. The use of automation, such as liquid handling robots or the Fluidigm Access Array system, can significantly reduce hands on time. The throughput of these systems is limited by the need to recover at least 50 sequence reads per amplicon for reliable genotype assignments.

The 454 DNA sequencing technology has been successfully used to analyze mixtures using the highly polymorphic HLA system in mixtures of cell line DNAs as well as in microchimeric blood samples (Bentley et al. 2009). A 1% minority component could be readily detected and typed. These initial studies used the 454 "standard" chemistry for amplicon sequencing that provides sequence reads of about 250 bp in length, but with the current Titanium sequencing chemistry and signal processing software, the average read length is about 450 bp for amplicon sequencing and, with the GS FLX+ about 800 bp for shotgun sequencing. These long reads allow both forward and reverse sequencing (bi-directional coverage) as well as setting phase for linked polymorphisms within the amplicon.

16.10 mtDNA 454 GS Assay for Mixture Analysis

Recently, we have developed a 454 GS Junior assay targeting the HVI/HVII regions in a duplex PCR using MID fusion primers. A 10 bp MID tag is used to barcode samples for parallel sequencing as described above. A combinatorial approach using eight unique forward and reverse MID tags allow for barcoding and sequencing 64 samples in a single 454 GS Junior run with ~500 reads per amplicon. This strategy greatly reduces the number of MID fusion primers needed (16 vs. 128 per amplicon). Additional MID tags could be added to increase the number of samples sequenced in parallel. However, the sensitivity for detecting mixtures is dependent on the depth of coverage (the number of observed sequence reads per amplicon). Thus, both the throughput and sensitivity should be considered when developing an assay and the experimental strategy and may differ depending on the application.

A mixture study was conducted to determine the sensitivity of the mtDNA HVI/HVII duplex 454 sequencing assay for detecting minor components in a mixture. Two DNA samples were mixed together at various ratios based on mtDNA copy number determined by quantitative PCR. Results for two of the mixed samples (10% and 1%) are shown in Figure 16.4 with ~600–1000 reads per amplicon. The minor base was detected at each of the mixed base positions, and the observed frequencies were similar to the expected frequencies using 454 sequencing but not by Sanger sequencing.

Not only is NGS more sensitive for detecting minor components in a mixture, it is also more quantitative because it provides a digital readout counting the number of sequence reads corresponding to the individual components. Linear Array, Sanger sequencing, and 454 NGS results from a buccal and blood sample from a heteroplasmic individual are shown in Figure 16.5. Heteroplasmy was detected at position 16093 in only the buccal sample

(a)

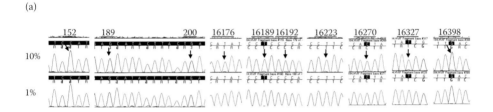

(b)

	HVII				HVI							
	454	Minor Base Frequency at Mixed Base Position			454	Minor Base Frequency at Mixed Base Position						
	Total reads	152	189	200	Total reads	16176	16189	16192	16223	16270	16327	16398
10%	800	6.88%	7.50%	7.62%	998	8.02%	6.21%	6.21%	8.42%	6.51%	8.12%	7.01%
1%	623	0.96%	0.96%	1.12%	641	1.40%	0.62%	0.62%	2.03%	1.09%	1.25%	1.56%

Figure 16.4 (a) Sanger sequencing results from 1% and 10% mixtures. (b) 454 next-generation sequencing results from 1% and 10% mixtures.

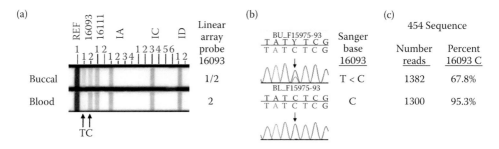

Figure 16.5 (a) Linear Array, (b) Sanger sequencing, and (c) 454 NGS results from a buccal and blood sample from a heteroplasmic individual.

with the Linear Array and Sanger sequencing, whereas heteroplasmy was detected in both the buccal and blood sample using the more sensitive 454 NGS assay. Approximately 67% and 95% of the 454 sequence reads showed a C at position 16093 in the buccal and blood sample, respectively. This comparison of methods for detection of heteroplasmy shows the higher sensitivity of the 454 NGS technology compared to more conventional methods for analysis of mtDNA sequence mixtures.

The HVI/HVII 454 sequencing assay was also shown to be highly sensitive. A sensitivity study was conducted by amplifying for 34 cycles various dilutions of a DNA sample that was quantified to determine the mtDNA copy number. Amplification and sequencing was successful for samples with DNA amounts of ~1 pg and 100–500 mtDNA copies (data not shown). This study was successful in showing that NGS results can be obtained from significantly lower amounts of DNA than previously reported or recommended by the manufacturer (pg amounts compared to ng or µg amounts). This study also demonstrates the feasibility of using NGS for the analysis of forensically relevant samples that are often limited in amount of DNA.

16.11 NGS for Forensic DNA Analysis

As noted above, NGS technologies offer a high-throughput solution for massively parallel sequencing that can be applied to the analysis of forensic markers. Additionally, the clonal sequencing aspect of the NGS technologies allow for sensitive and quantitative mixture detection and analysis. Another potential benefit of the NGS platforms is that this technology could allow the analysis of STR polymorphism as well as sequence polymorphisms such as mtDNA or HLA in a single run (Erlich et al. 2011; Bentley et al. 2009). Recently, the currently widely used markers for forensic analysis, the STR markers and the mtDNA HVI, were analyzed on the 454 platform. Others are exploring the potential to sequence a panel of autosomal SNP and STR markers in parallel. We and others are developing assays to allow for full mitochondrial genome sequencing. These studies have demonstrated the potential utility of using NGS for analysis of forensic samples, including the challenging category of mixed specimens.

16.12 Conclusion

The field of forensic DNA analysis continues to advance as new assays and technologies are developed. Addition of new informative markers to existing assays can improve discrimination power, including CR SNPs of the mitochondrial genome and additional STRs to expand the CODIS (Combined DNA Index System) panel. New technologies can offer higher throughput solutions, which could help in reducing the significant DNA backlog experienced in most forensics laboratories. Parallel analysis of multiple forensically relevant markers made possible with NGS can greatly improve the throughput (one platform for all markers) as well as discrimination power. The clonal aspect of NGS platforms can also offer significant improvements such as greater sensitivity for mixture analysis, allowing for analysis and interpretation of samples that previously could not be analyzed. Addition of new markers and implementation of new technologies, such as the current platforms for NGS, have the potential to revolutionize the field of forensic DNA analysis.

References

Bentley, G., R. Higuchi, B. Hoglund et al. 2009. High-resolution, high-throughput HLA genotyping by next-generation sequencing. *Tissue Antigens* 74(5):393–403.

Berglund, E. C., A. Kiialainen, and A. C. Syvanen. 2011. Next-generation sequencing technologies and applications for human genetic history and forensics. *Investig Genet* 2:23.

Blake, E., J. Mihalovich, R. Higuchi, P. S. Walsh, and H. Erlich. 1992. Polymerase chain reaction (PCR) amplification and human leukocyte antigen (HLA)-DQ alpha oligonucleotide typing on biological evidence samples: casework experience. *J Forensic Sci* 37(3):700–26.

Brandstatter, A., T. J. Parsons, and W. Parson. 2003. Rapid screening of mtDNA coding region SNPs for the identification of west European Caucasian haplogroups. *Int J Legal Med* 117(5):291–8.

Budowle, B. 2004. SNP typing strategies. *Forensic Sci Int* 146(Suppl):S139–42.

Budowle, B., and A. van Daal. 2008. Forensically relevant SNP classes. *Biotechniques* 44(5):603–8, 610.

Butler, J. M. 2004. Short tandem repeat analysis for human identity testing. *Curr Protoc Hum Genet* Chapter 14:Unit 14 8.

Calloway, C. D., S. M. Stuart, and H. A. Erlich. 2009. Development of a multiplex PCR and linear array probe assay targeting informative polymorphisms within the entire mitochondrial genome. https://www.ncjrs.gov/App/Publications/abstract.aspx?ID=250297.

Date Chong, M., C. D. Calloway, S. B. Klein, C. Orrego, and M. R. Buoncristiani. 2005. Optimization of a duplex amplification and sequencing strategy for the HVI/HVII regions of human mitochondrial DNA for forensic casework. *Forensic Sci Int* 154(2):137–48.

Divne, A. M., M. Nilsson, C. Calloway, R. Reynolds, H. Erlich, and M. Allen. 2005. Forensic casework analysis using the HVI/HVII mtDNA linear array assay. *J Forensic Sci* 50(3):548–54.

Erlich, R. L., X. Jia, S. Anderson et al. 2011. Next-generation sequencing for HLA typing of class I loci. *BMC Genomics* 12:42.

Gabriel, M. N., C. D. Calloway, R. L. Reynolds, and D. Primorac. 2003. Identification of human remains by immobilized sequence-specific oligonucleotide probe analysis of mtDNA hypervariable regions I and II. *Croat Med J* 44(3):293–8.

Goldberg, S. M., J. Johnson, D. Busam et al. 2006. A Sanger/pyrosequencing hybrid approach for the generation of high-quality draft assemblies of marine microbial genomes. *Proc Natl Acad Sci U S A* 103(30):11240–5.

Helmuth, R., N. Fildes, E. Blake et al. 1990. HLA-DQ alpha allele and genotype frequencies in various human populations, determined by using enzymatic amplification and oligonucleotide probes. *Am J Hum Genet* 47(3):515–23.

Higuchi, R., C. H. von Beroldingen, G. F. Sensabaugh, and H. A. Erlich. 1988. DNA typing from single hairs. *Nature* 332(6164):543–6.

Holcomb, C. L., B. Hoglund, M. W. Anderson et al. 2011. A multi-site study using high-resolution HLA genotyping by next generation sequencing. *Tissue Antigens* 77(3):206–17.

Holland, M., and T. Parsons. 1999. Mitochondrial DNA sequence analysis—validation and use for forensic casework. *Forensic Sci Rev* 11:21–50.

Just, R. S., J. A. Irwin, J. E. O'Callaghan et al. 2004. Toward increased utility of mtDNA in forensic identifications. *Forensic Sci Int* 146(Suppl):S147–9.

Kline, M. C., P. M. Vallone, J. W. Redman, D. L. Duewer, C. D. Calloway, and J. M. Butler. 2005. Mitochondrial DNA typing screens with control region and coding region SNPs. *J Forensic Sci* 50(2):377–85.

Mardis, E. R. 2008. Next-generation DNA sequencing methods. *Annu Rev Genomics Hum Genet* 9:387–402.

Melton, T., G. Dimick, B. Higgins, L. Lindstrom, and K. Nelson. 2005. Forensic mitochondrial DNA analysis of 691 casework hairs. *J Forensic Sci* 50(1):73–80.

Meyer, M., U. Stenzel, and M. Hofreiter. 2008. Parallel tagged sequencing on the 454 platform. *Nat Protoc* 3(2):267–78.

Meyer, M., U. Stenzel, S. Myles, K. Prufer, and M. Hofreiter. 2007. Targeted high-throughput sequencing of tagged nucleic acid samples. *Nucleic Acids Res* 35 (15):e97.

Moonsamy, P. V., T. Williams, P. Bonella et al. 2013. High throughput HLA genotyping using 454 sequencing and the Fluidigm Access Array System for simplified amplicon library preparation. *Tissue Antigens* 81(3):141–9.

Parsons, T. J., and M. D. Coble. 2001. Increasing the forensic discrimination of mitochondrial DNA testing through analysis of the entire mitochondrial DNA genome. *Croat Med J* 42(3):304–9.

Quintans, B., V. Alvarez-Iglesias, A. Salas, C. Phillips, M. V. Lareu, and A. Carracedo. 2004. Typing of mitochondrial DNA coding region SNPs of forensic and anthropological interest using SNaPshot minisequencing. *Forensic Sci Int* 140(2–3):251–7.

Roberts, K. A., and C. Calloway. 2007. Mitochondrial DNA amplification success rate as a function of hair morphology. *J Forensic Sci* 52(1):40–7.

Sobrino, B., M. Brion, and A. Carracedo. 2005. SNPs in forensic genetics: a review on SNP typing methodologies. *Forensic Sci Int* 154(2–3):181–94.

Sullivan, K. M., R. Hopgood, and P. Gill. 1992. Identification of human remains by amplification and automated sequencing of mitochondrial DNA. *Int J Legal Med* 105(2):83–6.

Wilson, M. R., J. A. DiZinno, D. Polanskey, J. Replogle, and B. Budowle. 1995. Validation of mitochondrial DNA sequencing for forensic casework analysis. *Int J Legal Med* 108(2):68–74.

17

Prediction of Physical Characteristics, such as Eye, Hair, and Skin Color Based Solely on DNA

ELISA WURMBACH

Contents

17.1	Summary	429
17.2	Introduction	430
17.3	Single Nucleotide Polymorphisms	431
17.4	Complex Traits	434
17.5	Pigmentation	435
17.6	Development of Forensic Tests to Predict Eye Color From DNA	437
17.7	Other Visible Traits: Hair and Skin Color and Height	440
17.8	Validation of Forensic Predictors Based Solely on DNA	442
17.9	Validation of Forensic DNA Assays	445
17.10	Future Directions	447
17.11	Conclusions	448
	Acknowledgments	448
	References	448

17.1 Summary

For more than 2.5 decades, DNA has been used in the forensic field for identifications. Usually, DNA is used as a template to generate profiles, which are then compared to stored profiles in databanks. A match might lead to the identification of a person, or link a person to a location; key is the comparison of profiles. On the other hand, recent scientific findings allow the development of novel DNA tests, which utilize the inherent information of the DNA, including the sequence at certain locations. Variations at distinct loci of the genome can be used alongside other applications to predict visible features. This approach could be useful in forensics, especially when a profile does not lead to a match in databases. An application would be, for example, the assistance in identification of human remains or missing persons.

It would be desirable to describe many visible characteristics accurately and in detail by utilizing DNA tests. However, most traits, such as eye, hair, skin color, and height are considered complex, because they depend on several genes as well as environmental factors.

This chapter summarizes the most recent advances in molecular genetics, which have led to the development of some predictive DNA tests of physical features. As most information is known about eye color, this review will focus on pigmentation. Pigmentation is elucidated, which involves synthesis of melanin and its regulation and packaging. Variations in genes involved in this process, as well as in their regulatory regions, are thought to be responsible for the many shades of eye, hair, and skin color.

The identification of variants that correlate significantly with particular traits is the first step toward a predictive DNA test. Usually, in pigmentation, several of these variants describe a phenotype more accurately than a single one does, reflecting its complexity. A predictive test further includes the information of how to translate the genotype into physical description. Validation of a predictive test then adds confidence to the prediction, including error rate and/or applicability to certain populations.

17.2 Introduction

Beginning in the late 1980s/early 1990s, DNA analysis has been routinely used in forensic casework (Jobling and Gill 2004). Usually, DNA analysis implies generation of characteristic DNA profiles of individuals, which currently include amplification of several hypervariable short tandem repeats (STRs) by polymerase chain reaction (PCR). The profiles are then compared to DNA databases of offender profiles. A match would lead to an identification of an individual, allowing the suspect to be investigated. A match could also link persons to certain locations, by retrieving some cells (shed skin cells, or saliva, for instance) from the place of interest. Such forensic tests provide strong evidence in criminal cases. To avoid false convictions, these tests are standardized and validated. The utilization of several loci (13 in the United States [Budowle and Moretti 1999], and most likely the number will eventually rise [Hill, Butler, and Vallone 2009]) decreases a random match probability (to approximately 1 in 10^{15}; Butler 2006), which makes it virtually certain that any two samples with the same profile would come from the same individual, or from monozygotic twins (Jobling and Gill 2004).

Phenotypic DNA tests are different types of DNA tests. These are novel tests that use the inherent information of the DNA, its sequence, to determine variable characteristics of individuals from whom the DNA was taken. One piece of phenotypic information, the gender, is already included in conventional DNA profiles (Butler 2006). Phenotypic DNA tests rely on genetically inherited characteristics, such as facial features, height, eye, hair, and skin color (Figure 17.1). For forensic scientists, these tests are important, because they can lead to identifications of individuals.

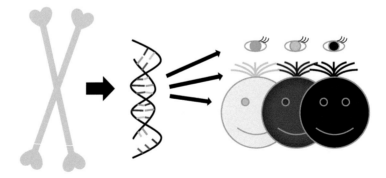

Figure 17.1 When only human remains (bones, body parts, blood) are found, visible characteristics are lost. From these remains, however, DNA can be extracted, and the inherent information can be used to predict eye, hair, and skin color of an individual. This prediction can assist in making identifications.

Forensic anthropologists routinely determine some physical characteristics such as age, gender, height, and ethnicity (European, Asian, or African) based on available skeletonized human remains. Although these metrics are extremely useful, they are no longer able to discern them if only partial remains are available. In addition, it is not possible to reveal eye, hair, and skin color from skeletal remains.

Table 17.1 gives an overview of recent approaches to describe visible features. The focus is on pigmentation, eye, hair, and skin coloration, and includes height. It shows the selected variants used in association studies and predictive tests, and their validations. Table 17.2 lists the variants of Table 17.1, their location in association to the genes, as well as the function of the gene products. Many association studies and predictive tests are based on the same variants.

Most is known about human eye color (Liu et al. 2009; Mengel-From et al. 2010; Spichenok et al. 2011; Valenzuela et al. 2010; Walsh, Liu et al. 2011). Thus, the first predictor published described this trait (Walsh, Liu et al. 2011). Another predictor was developed independently, which can be used to describe eye and skin color of individuals (Spichenok et al. 2011). These predictors and their validations are reviewed.

17.3 Single Nucleotide Polymorphisms

The comparison of two randomly selected human genomes reveals that they are 99.9% identical. The remaining 0.1% contains variations between individuals, including copy number variations (e.g., insertions and deletions of nucleotides) and single nucleotide polymorphisms (SNPs). SNPs are abundant; they occur at a frequency of approximately 1 in 1000 base pairs (bp) throughout the genome and are believed to be stably inherited (Brookes 1999). Based on sequencing results, the human genome contains at least 11 million SNPs, and approximately 7 million of these occur with a minor allele frequency (MAF) of at least 5% across the entire human population (Kruglyak and Nickerson 2001). An additional 4 million SNPs exist with a MAF between 1% and 5%, and there are innumerable very rare single-base variants, most of which exist only in single individuals (Hinds et al. 2005). Current information about SNPs is available at the Single Nucleotide Polymorphism database (Frazer et al. 2007).

Most of the SNPs are silent and have no impact on gene function or phenotype. However, SNPs may change encoded amino acids, when located in exons. They can also have other effects on proteins, including truncations (premature stop codon), and cause missense mutations, frame-shifts, or incorrect splicings. Furthermore, they can influence gene expression if they are located in a promoter or an enhancer region, affect mRNA stability, or change subcellular localization of the mRNA (Shastry 2009). This shows clearly that SNPs can have functional consequences and that they can contribute to phenotypes, suggesting associations with human traits, including height, curly hair, build, and facial structures, or susceptibility to many common diseases, such as diabetes or schizophrenia (Martin, Boomsma, and Machin 1997).

The identification of SNPs associated with a phenotype requires the comparison of genotypes with the corresponding phenotypes, followed by statistical analysis to determine the significance of the association.

Table 17.1 Overview of Recent Publications in Predicting Eye, Hair, Skin Color, and Height

Phenotype	Reference	Selected SNPs/Markers	Association with Phenotype	Predictor	Validation of Predictor	Various Populations
Eye color	Valenzuela et al. (2010)	rs12913832, rs16891982, rs1426654	Probability of eye color variance: 76%			Yes
	Mengel-From et al. (2010)	rs12913832, rs1129038, rs11636232, rs1800407	light (blue, green, and gray), dark (brown and hazel)			No
	Pospiech (2011)	rs12913832, rs12896399, rs1408799, rs1800407	blue, green, hazel, brown			No
	Liu et al. (2009)	rs12913832, rs1800407, rs12896399, rs16891982, rs1393350, rs12203592	Prediction accuracy: brown: 93%, blue: 91%, intermediate: 73%			No
	Walsh et al. (2011)	rs12913832, rs1800407, rs12896399, rs16891982, rs1393350, rs12203592		Eye-color-predictor-1: brown, intermediate, blue	Error rate[a]: 6.1%,[b] 31%[c] Call rate: 77.5%,[a] 74%[b]	Yes[d]
	Spichenok et al. (2011)	rs12913832, rs16891982, rs1545397, rs885479, rs6119471, rs12203592		Eye-color-predictor-2: brown, green, not brown (green, blue), not blue (brown green)	Error rate[e]: 3% Call rate[e]: 100%	Yes
Hair color	Valenzuela et al. (2010)	rs16891982, rs1426654, rs12913832	Probability of total melanin content: 76%			Yes
		rs1426654, rs16891982, rs1805007	Probability of phenotypic variance (eu- to pheomelanin ratio): 43%			Yes

Prediction of Physical Characteristics Based Solely on DNA

	Branicki et al. (2011) Walsh et al. (2013)	Y152OCH, N29insA, rs1805006, rs11547464, rs1805007, rs1805008, rs1805009, rs1805005, rs2228479, rs1110400, rs885479, rs1042602, rs4959270, rs28777, rs683, rs2402130, rs12821256, rs2378249, rs12913832, rs1800407, rs16891982, rs12203592, rs12896399, rs1393350	Prediction accuracy[d]: red: 80%, blond (blond and light brown): 69.5%, brown (light brown and dark brown): 78.5%, black (dark brown and black): 87.5%	**Hair-color-predictor:** red, blond, brown, black	Call rate[f]: 100%	Yes[g]
Skin color	Valenzuela et al. (2010)	rs1426654, rs16891982, rs2424984	Probability of total skin reflectance: 45.7%			
	Spichenok et al. (2011)	rs12913832, rs16891982, rs1545397, rs885479, rs6119471, rs1426654, rs12203592		**Skin-color-predictor:** not white, not dark	Error rate[e]: 1% Call rate[e,h]: 81%	Yes
Height	Weedon and Frayling (2008)	44 loci				
	Lango Allen et al. (2010)	More than 600 variants				
	Lanktree et al. (2011)	64 loci				Yes

[a] Error rate depends on the distribution of the eye colors brown, intermediate/green, and blue (Pneuman et al. 2011).
[b] Tested on 3840 individuals of several populations across Europe, with a distribution of eye color: brown 48.4%, intermediate 7%, blue 44.6% (Walsh et al. 2011).
[c] Tested 803 samples of various populations with an eye color distribution of the European population of brown 34.1%, green/intermediate 31.5%, blue 34.4% (Pneuman et al. 2011).
[d] Tested on 803 samples of various populations, including African-American, South Asian, East Asian, European, and mixed populations (Pneuman et al. 2011).
[e] Verification of additional 251 samples confirmed earlier testing of 554 samples (Pneuman et al. 2011).
[f] Based on 308 European test set (Walsh et al. 2013).
[g] Tested on various European populations and on HGDP-CEPH cell lines from 51 populations (Walsh et al. 2013).
[h] The call rate varies broadly between populations, ranging from 15% to 95% (Pneuman et al. 2011).

Table 17.2 Variation Types of Selected SNPs in Pigmentation

Protein Function	Gene	SNP	Variation Type/Location
Possible involved in protein ubiquitination	HERC2	rs12913832	Located in predicted transcription factor binding site of OCA2
		rs1129038	3′ UTR
		rs11636232	Synonymous
Integral membrane protein involved in transport of small molecules	OCA2	rs1800407	Missense
		rs1545397	Intron
Transporter protein that mediates melanin synthesis	SLC45A2	rs16891982	Missense
		rs28777	Intron
Potassium-dependent sodium/calcium exchanger protein	SLC24A5	rs1426654	Missense
Potassium-dependent sodium/calcium exchanger protein	SLC24A4	rs12896399	5′ region
		rs2402130	Intron
Melanosomal enzyme involved in melanin biosynthetic pathway	TYRP1	rs1408799	5′ region
		rs683	3′ UTR
Involved in conversion of tyrosine to melanin	TYR	rs1393350	Intron
		rs1042602	Missense
Regulation of interferons	IRF4	rs12203592	Intron
Exocyst complex	EXOC2	rs4959270	3′ region (between IRF4 and EXOC2)
Receptor protein that controls melanogenesis	MC1R	rs201326893	Nonsense
		N29insA	INDEL
		rs1805006	Missense
		rs11547464	Missense
		rs1805007	Missense
		rs1805008	Missense
		rs1805009	Missense
		rs1805005	Missense
		rs2228479	Missense
		rs1110400	Missense
		rs885479	Missense
Signaling protein in pigmentation	ASIP	rs6119471	5′ region, predicted transcription factor binding site
		rs2424984	Intron
Ligand of tyrosine kinase receptor	KITLG	rs12821256	5′ region
Membrane protein with GPI anchor activity	PIGU	rs2378249	Intron

17.4 Complex Traits

Many visible characteristics, such as pigmentation, height, and build are influenced by more than one gene and by other factors, including environmental aspects, age, or diet. These are therefore considered complex traits (Lettre 2011; Sturm and Larsson 2009). Despite the progress in human genome sequencing, still many genetic markers responsible for variations of these traits are not yet unambiguously identified (Lango Allen et al. 2010; Yamaguchi and Hearing 2009).

Prediction of Physical Charactcristics Based Solely on DNA

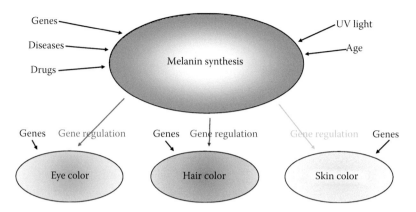

Figure 17.2 Pigmentation is a complex trait. Melanin pigments cause the various shades of eye, hair, and skin coloration. Its synthesis depends on several genes, UV light, diseases, drugs, and age. Gene expression may be differentially regulated.

The only trait that has undergone serious investigation is the pigmentation of irises, hair, and skin. However, predictions with 100% accuracy are not possible yet. It is estimated that more than 100 genes are involved in melanin synthesis (Yamaguchi and Hearing 2009). Melanin pigments are found in the iris, hair, and skin, but the synthesis could be regulated independently, in a tissue-dependent manner, and may involve different genes. Diseases, drugs, age, and UV light may also affect melanin synthesis, positively or negatively (Figure 17.2).

Height is another visible trait, and 80%–90% of its variation relies on heritable factors (Lanktree et al. 2011; Weedon and Frayling 2008). Height, also a complex trait, can be affected by diet, nutrition, and morbidity—particularly infections (Cole 2003). Just recently, it was pointed out that height is an indicator of childhood nutrition, disease, and poverty (Subramanian, Ozaltin, and Finlay 2011). However, in some populations over the past 150 years physiological and environmental influences have contributed to an increase in human height with grown-up children being taller on average than their same sex parent (Cole 2003). Nevertheless, tall parents usually have tall children.

It was rather difficult to identify genetic variants that influence human height, especially since it is very likely that numerous variants contribute in small portions to its heritability (Weedon and Frayling 2008). Recently, genome-wide association (GWA) studies identified hundreds of loci associated with height (Lango Allen et al. 2010; Lanktree et al. 2011; Weedon and Frayling 2008; Yang et al. 2010), most of which are connected with pathways involved in skeletal growth (Lango Allen et al. 2010). However, the main outcome of these GWA studies provides more biological insight than predictive power (Lango Allen et al. 2010).

17.5 Pigmentation

Melanin pigments are packaged into melanosomes; these are specialized subcellular compartments, which are exported to adjacent keratinocytes, where most of the melanin is found (Barsh 2003). Differences in pigmentation arise from variation in number, size, composition, and distribution of melanosomes (Lin and Fisher 2007). During melanogenesis, two forms of melanin pigment particles are produced: the brown-black eumelanin and the red-yellow pheomelanin (Simon 2009). Both melanins are found in the iris, hair, and skin. However, the ratio of eumelanin to pheomelanin varies between individuals, as can be seen in the many shades of eye, hair, and skin color of different people (Sturm and Frudakis 2004).

Within the European population, blue- and brown-eyed individuals can have all shades of hair colors, but in other geographical regions, populations with darker skin tones tend to have darker eye and hair colors (Frost 2006). Skin coloration correlates strongly with environmental ultraviolet (UV) radiation levels and can be explained by varying physiological requirements of photoprotection on one hand and vitamin D synthesis on the other (Jablonski and Chaplin 2000; Norton et al. 2007). This suggests that the genetic determinants for pigmentation in various populations are distinct, although they have been subject to a common set of evolutionary forces (Chaplin 2004; Frudakis et al. 2003).

To reduce complexity, previous studies focused either on a single trait in one population, or on a few genes. The gene products of *MC1R*, *SCL45A2*, *OCA2*, *HERC2*, *ASIP*, and *SLC24A5* are good candidates for pigmentation, since they are either associated with melanosomes, involved in melanosome transport, the uptake by keratinocytes, or are part of the melanin biogenesis cascade itself (Branicki et al. 2008; Branicki, Brudnik, and Wojas-Pelc 2009; Shekar et al. 2008; Stokowski et al. 2007; Valverde et al. 1995). In addition, GWA studies identified new alleles associated with distinct human pigmentation traits (Han et al. 2008; Sulem et al. 2007, 2008). Much progress was made with association studies performed in European populations. The *OCA2* locus was identified as a major contributor for eye color determination (Duffy et al. 2007; Frudakis, Terravainen, and Thomas 2007; Frudakis et al. 2003; Sturm and Frudakis 2004). *OCA2* codes for a 12 pass-transmembrane protein that is localized on melanosomes and is involved in melanin synthesis (Sitaram et al. 2009).

Another GWA study focusing on eye color identified *HERC2*, a gene directly adjacent to *OCA2*, implying that variants within this region contribute to the human iris color (Kayser et al. 2008). Fine mapping of the *OCA2-HERC2* loci performed in European populations identified two SNPs (rs12913832 and rs1129038; see Tables 17.1 and 17.2), located in the 86th intron and the 3′ UTR of *HERC2* that showed strong correlation with blue and brown eye color, respectively (Eiberg et al. 2008; Sturm et al. 2008). Intron 86 of the *HERC2* gene contains a highly conserved enhancer region of *OCA2*, and the SNP rs12913832 is directly located in a predicted transcription factor binding site (Sturm et al. 2008). Studies in human melanocytes detected binding of transcription factors to this side. The A allele of rs12913832 led to elevated OCA2 expression, which was reduced in lighter pigmented cells carrying the G allele (Visser, Kayser, and Palstra 2012).

Lighter eye colors (blue, gray, and amber) as well as lighter hair colors (blonde and red) are more common in the European population, in particular in the Northeastern part and the East Baltic (Frost 2006). It was shown that light eye and light hair color as well as brown eye and dark hair color are significantly correlated in Scottish and Danish populations (Mengel-From et al. 2009). Variations in the *MC1R* gene have been associated with red hair (Branicki et al. 2007; Mengel-From et al. 2009; Valverde et al. 1995). This gene codes for the melanocortin 1 receptor, a G-protein–coupled receptor controlling melanogenesis (Abdel-Malek et al. 1999). Additional variations in *SLC45A2*, *KITLG*, *TYR*, *SLC24A5*, *IFR4*, *EXOC2*, *TYRP1*, *ASIP*, and in *HERC2/OCA2* (see Table 17.2) were associated with lighter or darker hair colors (Branicki et al. 2011; Mengel-From et al. 2009; Valenzuela et al. 2010). However, further analysis is required to make useful predictions (Bouakaze et al. 2009).

Skin coloration is an obvious feature in which humans differ. It is associated with various populations based on geographic regions, including, among others, Africans, Europeans, and East Asians (Westerhof 2007). Examinations of SNP variations in six genes (*SLC45A2*, *SLC24A5*, *OCA2*, *TYR*, *ASIP*, and *MC1R*) found evidence that the light skin color of Europeans and East Asians arose independently of each other (McEvoy, Beleza, and Shriver 2006; Norton et al. 2007; Tishkoff and Verrelli 2003). This indicates that several markers may be necessary to make predictions of skin coloration (Spichenok et al. 2011; Valenzuela et al. 2010), demonstrating an even increased complexity of this feature.

17.6 Development of Forensic Tests to Predict Eye Color From DNA

Forensic tests that are able to predict visible features solely from the inherent information of DNA are valuable tools in identifying human remains by gathering information that can serve as legitimate investigative leads (Figure 17.1).

Among the SNPs located near or within pigmentation genes, some may correlate strongly with eye, hair, or skin coloration. Because of the complexity of these pigmentation traits (e.g., being dependent on several genes and other factors), the identified SNPs may not be sufficient to describe these traits in detail, even though their correlation may be statistically significant.

As described above, fine mapping of the *HERC2-OCA2* region pointed to the SNP rs12913832, which is strongly associated with the blue and brown eye color (Eiberg et al. 2008; Sturm et al. 2008). The correlation is statistically highly significant: 150 blue-eyed Danes were homozygote for GG on this position and 46-brown eyed Danes were heterozygote (GA), resulting in a *p* value of 6.12^{-46} (Eiberg et al. 2008). Other studies, performed on European descendants, corroborated the statistically significant correlation (Pneuman et al. 2012; Sturm et al. 2008). It was stated that individuals carrying the genotype AA have an 80% probability of having brown eyes, whereas individuals with the GG genotype have a 1% probability of having brown eyes (Sturm et al. 2008).

However, one SNP is not sufficient to describe eye color in full detail. More SNPs were added to improve the predictive value (Table 17.1) (Liu et al. 2009; Mengel-From et al. 2010; Valenzuela et al. 2010): One study found that the probability of the eye color variation depended up to 76% on three SNPs from three genes (Valenzuela et al. 2010). Four SNPs from two genes were used in a second study to show high correlation with light (blue, blue/green, and green) and dark (hazel and brown) eye colors (Mengel-From et al. 2010). A third study demonstrated that four SNPs from four genes could affect the determination of the eye colors blue, green, hazel, and brown (Pospiech et al. 2011); and a fourth study developed a model using six SNPs from six genes to discriminate between the eye colors blue, brown, and intermediate (neither blue nor brown) (Liu et al. 2009). All these correlations had at least one SNP in common: rs12913832 (Tables 17.1 and 17.2), because of its highly significant correlation with the blue and brown eye color (Eiberg et al. 2008; Sturm et al. 2008).

Whereas the first three studies selected certain SNPs and investigated their correlation to various eye colors, the fourth study added instructions, describing how to interpret the genotype in order to predict the eye color. This makes the predictor independent of the performing analyst or laboratory: the genotypes for the selected SNPs were determined and following the instructions resulted in an output, the predicted eye color. Since the output depends on both, the selected SNPs and the instructions, the output could vary when using different instructions along with the same SNPs. Hence, an eye color predictor includes both, a selection of SNPs and instructions describing how to interpret the genotypes (Figure 17.3).

Two independent eye color predictors were published so far (Table 17.1). Each used six SNPs from six genes, of which three were identical: rs12913832 (*HERC2*), rs16891982 (*SLC45A2*), and rs12203592 (*IRF4*) (Spichenok et al. 2011; Walsh, Liu et al. 2011). For two of the three common SNPs, rs12913832 and rs16891982, it was shown that they are associated with reduced melanin content in cultured human melanocytes, therefore supporting their role in pigmentation (Cook et al. 2009). Still, both predictors rely profoundly on rs12913832. This SNP carries approximately 75% of the predictive information (Valenzuela et al. 2010), whereas additional SNPs improve the accuracy of the prediction.

Both predictors grouped the many shades into three eye color bins—blue, green/intermediate, and brown—according to their melanin content (Figure 17.4). This procedure has

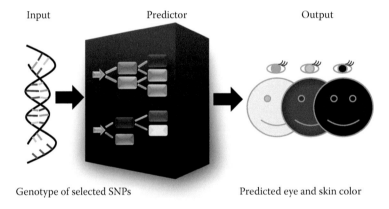

Figure 17.3 A phenotypic predictor uses an input [selected short-tandem repeats (SNPs)] and instructions resulting in an output (predicted eye and skin color).

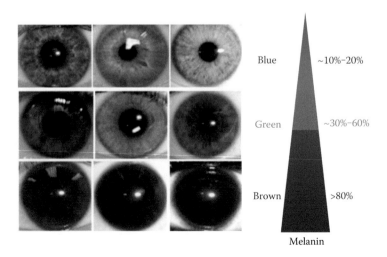

Figure 17.4 Grouping of skin colors into three bins according to their melanin content.

several advantages: it can reduce the noise of data, lead to higher numbers of samples per bin and to fewer groups, which makes it easier to work with.

Eye-color-predictor-1 (Walsh, Liu, et al. 2011). The genotypes of six SNPs (rs12913832 [*HERC2*], rs1800407 [*OCA2*], rs12896399 [*SLC24A4*], rs16891982 [*SLC45A2*], rs1393350 [*TYR*], and rs12203592 [*IRF4*]; see Tables 17.1 and 17.2) were used to calculate probabilities of having blue, brown, and intermediate (neither blue nor brown) eye color. For each individual (sample), three probabilities were calculated, one for every eye color, which add up to 1.0. The eye color for which the probability is ≥0.7 determined the outcome, the predicted eye color (lowering the probability may increase the error rate; Walsh, Wollstein et al. 2011). The eye-color-predictor-1 was named "IrisPlex" (Walsh et al. 2010; Walsh, Liu et al. 2011; Walsh, Wollstein et al. 2011).

Eye-color-predictor-2 (Spichenok et al. 2011). The genotypes of six SNPs (rs12913832 [*HERC2*], rs16891982 [*SLC45A2*], rs1545397 [*OCA2*], rs885479 [*MC1R*], rs6119471 [*ASIP*], rs12203592 [*IRF4*]; Tables 17.1 and 17.2) were used to predict the eye colors brown, green, not brown (green or blue), or not blue (brown or green). The prediction relied mainly on rs12913832. In a conservative approach, G/G homozygotes at this SNP were predicted to have non-brown (green or blue) eyes, and G/A heterozygotes and A/A homozygotes of having non-blue (brown or green) eyes. For a positive description, additional SNPs were necessary. In an exclusive approach, only homozygote genotypes were considered (Table 17.3). The brown eye color was predicted by the following genotypes: A/G or A/A at rs12913832, plus C/C at rs16891982, and/or T/T at rs16891982, and/or A/A at rs885479, and/or G/G at rs6119471. The green eye color was predicted by the following genotypes: A/G at rs12913832, plus T/T at rs12203592, or G/G at rs12913832, plus C/C at rs16891982.

Both predictors are more accurate than association studies using three or four SNPs to describe the eye color (Mengel-From et al. 2010; Pospiech et al. 2011; Valenzuela et al. 2010). In the following step, these predictors were validated, assessing their accuracy. This can be done by applying the predictor on numerous additional samples, ideally from various populations and comparing the outcome (prediction) with the corresponding eye color.

Table 17.3 Eye-Color-Predictor-2

Gene	SNP ID	Genotype	Eye Color (Predicted)
HERC2	rs12913832	G/G	Not brown
		G/A	Not blue
		A/A	
HERC2	rs12913832	A/A	Not blue
IRF4	rs12203592	T/T	
HERC2	rs12913832	G/A	Green
IRF4	rs12203592	T/T	
HERC2	rs12913832	G/G	Green
SLC45A2	rs16891982	C/C	
HERC2	rs12913832	A/A or G/A	Brown
SLC45A2	rs16891982	C/C	
HERC2	rs12913832	A/A or G/A	Brown
OCA2	rs1545397	T/T	
HERC2	rs12913832	A/A or G/A	Brown
MC1R	rs885479	A/A	
HERC2	rs12913832	A/A or G/A	Brown
ASIP	rs6119471	G/G	

17.7 Other Visible Traits: Hair and Skin Color and Height

Of all visible traits studied so far, the eye color is best predicted (Kayser and de Knijff 2011). In addition to eye color, hair color also varies widely in European populations (Mengel-From et al. 2009). In these populations, it can be observed, and was corroborated using statistical tests, that the light eye color is found significantly more often with fair or blond hair, and the brown eye color with dark hair, suggesting that these traits are not independent (Mengel-From et al. 2009).

Hair color is thought to be the result of at least two parameters, the total melanin (combined amounts of eu- and pheomelanin) and the ratio of eu- to pheomelanin (Table 17.1) (Valenzuela et al. 2010). Certain variations of the *MC1R* gene have been associated with red hair because of elevated levels of pheomelanin (Branicki et al. 2007; Mengel-From et al. 2009; Valverde et al. 1995). A model for hair color prediction, based on 385 individuals, included for this reason two compound markers located in the *MC1R* gene ("R" and "r", which refer to seven and four variations, respectively; see Table 17.2 for more details on these compound markers), as well as 11 SNPs associated with 10 additional genes, distinguishing hair colors of blond, brown, red, and black hair (Branicki et al. 2011).

The eye-color-predictor-1, "IrisPlex," was recently expanded with 18 new SNPs used for hair color prediction. The system is now called the "HIrisPlex," and it uses 24 SNPs from 11 genes. The prediction guide approach was based on a 308-sample testing set and showed that the hair color prediction had an average of 69.5% correct calls for blonde (blonde and light brown), 78.5% for brown (light brown and dark brown), 87.5% for black (dark brown and black), and 80% for red (Walsh et al. 2013).

Besides the European population, individuals from the Solomon Islands and Equatorial Oceania show the highest prevalence of blond hair; they also have the darkest skin pigmentation outside Africa (Norton et al. 2006). A GWA study pointed to a gene involved in melanogenesis, tyrosine-related protein 1 (*TYRP1*). Resequencing revealed an unknown polymorphism, which causes a missense mutation in exon two of *TYRP1*, which correlates very well with the blond hair phenotype (Kenny et al. 2012). Further evidence suggests that the blond hair color arose independently and is not associated with the European blond hair (Kenny et al. 2012).

Skin color seems to be a more complex trait because of rising evidence that the fair skin color of European and East Asian populations arose independently (Edwards et al. 2010; McEvoy, Beleza, and Shriver 2006; Norton et al. 2007; Tishkoff and Verrelli 2003). A predictor for light skin color would therefore include SNPs that correlate with the light skin color of Europeans as well as SNPs that correlate with the light skin color of East Asians. The increased complexity and the limited access to samples of various populations make it difficult to develop and create a skin color predictor based on DNA. Nevertheless, one skin color predictor was published, which utilized seven SNPs to distinguish between being not light and being not dark (Table 17.1) (Spichenok et al. 2011).

Skin-color-predictor (Spichenok et al. 2011). The genotypes of seven SNPs that are located in or near pigmentation genes (rs12913832 [*HERC2*], rs16891982 [*SLC45A2*], rs1545397 [*OCA2*], rs1426654 [*SLC24A5*], rs885479 [*MC1R*], rs6119471 [*ASIP*], rs12203592 [*IRF4*]; see Table 17.2) were used to predict the skin color as being not light or not dark. The different shades of skin colors were binned into three groups: light, medium, and dark (Figure 17.5). The predicted skin color of being not light included medium and dark; similarly, the predicted skin color of being not dark included medium and light. The selected seven SNPs

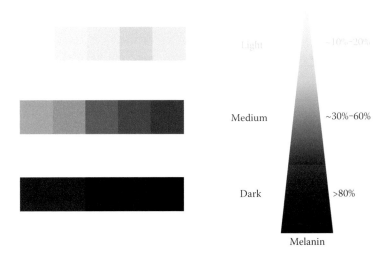

Figure 17.5 Grouping of eye colors into three bins according to their melanin content.

could be linked with light or dark skin coloration; six of them (rs12913832, rs16891982, rs1545397, rs1426654, rs885479, and rs12203592) are variable in populations with light skin colors, whereas rs6119471 is variable in populations with dark skin color. Only homozygote genotypes of at least two SNPs, because of the complexity of this trait, were used to predict the lighter skin color. The non-dark skin color (light or medium) was predicted at the presence of any two of the following genotypes: G/G (rs12913832), G/G (rs16891982), T/T (rs1545397), A/A (rs1426654), A/A (rs885479), and/or T/T (rs12203592). Non-light skin color (medium or dark) was predicted by G/G at rs6119471.

Currently, phenotypic tests that describe eye, hair, and skin color are almost established. Further research will identify additional SNPs, which after verification could improve these tests, leading to more accurate predictions. Modern methods such as next generation sequencing (Metzker 2010) allow sequencing of nuclear DNA of many individuals expressing various shades of eye, hair and skin color to find novel markers. The 1000 Genomes Project is an international effort to produce an extensive catalog of human genetic variations, including SNPs from individuals of several populations (Durbin et al. 2010). Results from the pilot phase revealed approximately 15 million gene variants, of which more than half had never been observed before (Katsnelson 2010). Data from this project will be highly valuable to achieve a better understanding of the genetic differences between individuals and populations, including visible characteristics.

Height is another characteristic that is subject to many research investigations. However, less is known about it compared to pigmentation. It is estimated that 80%–90% of its variation depends on heritable factors (Lanktree et al. 2011; Weedon and Frayling 2008). To date, many SNPs—40 to several hundred (Table 17.1)—were associated with height (Lango Allen et al. 2010; Lanktree et al. 2011; Weedon and Frayling 2008). The individual effect of each identified SNP seemed to be small and considering all SNPs simultaneously would explain 45% of its variance (Yang et al. 2010). Moreover, the average human height appears to vary with certain geographic populations, with the Northern Europeans being currently the tallest, followed by Southern Europeans and North Americans (Cole 2003).

17.8 Validation of Forensic Predictors Based Solely on DNA

The validation of predictive forensic DNA tests can define error rates and rates for inconclusive outcomes, thereby increasing the level of confidence. Interpretations of these results therefore appear more reliable. Both eye color predictors and the skin color predictor were validated on numerous samples by comparing the prediction (outcome) with the corresponding phenotype.

Validation of eye-color-predictor-1 (Walsh, Wollstein et al. 2011). The predictor was applied on DNA samples of 3840 individuals from seven countries across Europe. Overall, 48.4% of them had brown eyes, 44.6% had blue eyes, and 7.0% had an undefined eye color (non-blue, non-brown, green, or papillary rings of alternative color). The threshold was set to ≥0.7, as recommended (Walsh et al. 2010; Walsh, Wollstein et al. 2011). The outcome for 2975 samples was conclusive, leading to a call rate of 77.5% (the rate of inconclusive outcomes is therefore 22.5%). A total of 182 errors occurred in the predictions, causing an error rate of 6.1%.

In addition, the eye-color-predictor-1 was validated on 803 samples from various geographic populations, as shown in Table 17.4 (Pneuman et al. 2012). In three of the tested populations (African-American, South Asian, and East Asian) no error occurred. The call

Table 17.4 Validation of Eye-Color-Predictor-1 in 803 Samples of Various Populations

Population	Eye Color	Eye Color Predicted	$n \geq 0.7$	Error	Inconclusive
AA (43)	Brown (43)	Brown	38	0	5
SA (27)	Brown (27)	Brown	27	0	0
EA (35)	Brown (35)	Brown	34	0	1
E (555)	Blue (191)	Blue	149	3	39
		Intermediate/green	0		
		Brown	3		
	Green (175)	Blue	71	116	59
		Intermediate/green	0		
		Brown	45		
	Brown (189)	Blue	1	1	72
		Intermediate/Green	0		
		Brown	116		
Mixed (143)	Blue (11)	Blue	8	0	3
		Intermediate/Green	0		
		Brown	0		
	Green (20)	Blue	9	14	6
		Intermediate/Green	0		
		Brown	5		
	Brown (112)	Blue	1	1	24
		Intermediate/green	0		
		Brown	87		
Total		803	594	135	209
%				31%[a]	26%

Note: AA, African-American; SA, South Asian; EA, East Asian; E, European; Mixed, mixed populations.

[a] Calculated for European population.

rate was 74%, which was similar to the previous validation. Of 594 predictions out of 803 samples tested, 135 errors were found, which occurred in European and mixed populations. Mixed populations include samples from individuals whose parents were not associated with the same population, as well as Hispanics, since they include individuals of European, Native American, and African ancestry in all possible combinations (Tishkoff and Kidd 2004). The error rate for the eye color predictors was calculated considering only the European population, which shows a wide eye color variation. Including other populations into the error rate calculation would increase the sample size, but not the probability of having blue, green or brown eye colors and therefore could lead to lower error rates. The error rate for the eye-color-predictor-1 was calculated to 31%. This outcome differs a lot from the validation performed earlier and might be explained by the distribution of the samples into the three eye color bins: 34.1% had brown eyes, 34.4% blue, and 31.5% had green (non-blue, non-brown) eyes (see above). Further analysis revealed that the majority of errors occurred by predicting either the blue or the brown eye color for green-eyed individuals (see Table 17.4). When disregarding the errors related to the green eye color, the error rate dropped to 1.5%, showing that the error rate depends on the sample composition (Pneuman et al. 2012).

Validation of eye-color-predictor-2 (Spichenok et al. 2011). The predictor was applied on 554 individuals from various populations, including African-American ($n = 33$), South Asians ($n = 25$), East Asians ($n = 22$), European or European descendants ($n = 379$), and mixed populations ($n = 95$). The predicted eye color (brown, green, not brown, and not blue) was then compared to eye color bins: blue, green, and brown, into which the self-reported eye colors were grouped (Figure 17.4 and Table 17.5). The outcome showed that the predictor can be used in all populations tested, has a call rate of 100% and an error rate of approximately 1%.

Further testing of additional 251 samples led to similar results (Table 17.5), with a slightly higher error rate of 3% (Pneuman et al. 2012). Further error assessment revealed that 10 of 11 errors that were found out of 805 eye color predictions arose from rs12913832, the primary SNP on which the eye color prediction relies.

Validation of skin-color-predictor (Spichenok et al. 2011). The predictor was applied on 554 individuals from various populations, including African-American ($n = 33$), South Asians ($n = 25$), East Asians ($n = 22$), European or European descendants ($n = 379$), and mixed populations ($n = 95$). The predicted skin color was then compared to self-reported skin color bins (Table 17.6). A consequence of utilizing only homozygote genotypes is that for several samples the data were insufficient to predict skin coloration: from 554 samples tested, 398 (72%) prediction could be made (Spichenok et al. 2011). No error occurred in the prediction for most populations, including African-American ($n = 12$), South Asian ($n = 2$), East Asian ($n = 7$) and European ($n = 345$), revealing that these SNPs are good markers for pigment-related features. However, overall, two errors were found in the mixed populations. Further testing of additional 251 samples led to 203 predictions of which two were erroneous (Table 17.6), leading to an error rate of 1% (Pneuman et al. 2012). Future research may identify further markers, which could improve the call rate.

Evidently, all three predictors can be applied on various populations. This point is especially critical when developing forensic DNA tests. If human remains are found, it is possible that visible characteristics are lost. The biological evidence may not allow an association to a specific population but could serve as source for DNA. People travel or migrate even in areas where one population is more prominent. Therefore, DNA tests that utilize genetic information should be validated using various populations.

Table 17.5 Validation of Eye-Color-Predictor-2 in 805 Samples of Various Populations

Population	Eye Color Bin	Eye Color Predicted	Errors in Prediction
AA (33[a] + 10[a])	Brown (33 + 10)	Brown (21 + 8)	0
		Not blue (12 + 2)	
SA (25 + 2)	Brown (25 + 2)	Brown (22 + 1)	0
		Not blue (3 + 1)	
EA (22 + 13)	Brown (22 + 13)	Brown (21 + 12)	0
		Not blue (1 + 1)	
E (379 + 178)	Blue (121 + 71)	Not brown (118 + 66)	3 + 5
		Not blue (3 + 5)	
	Green (119 + 57)	Green (1 + 0)	0
		Not blue (66 + 26)	
		Not brown (52 + 23)	
	Brown (139 + 50)	Brown (5 + 1)	1 + 0
		Not blue (133 + 49)	
		Not brown (1 + 0)	
Mixed (95 + 48)	Blue (1 + 10)	Not brown (1 + 9)	0 + 1
		Not blue (0 + 1)	
	Green (12 + 8)	Green (1 + 1)	0 + 1
		Not blue (7 + 3)	
		Not brown (4 + 1)	
		Brown (0 + 1)	
	Brown (82 + 30)	Brown (35 + 17)	0
		Not blue (47 + 13)	
Total	554 + 251	554 + 251	11
%			3%[b]

Note: AA, African-American; SA, South Asian; EA, East Asian; E, European; Mixed, mixed populations.

[a] Nonrelated individuals, the first number indicates the number of samples when introducing the eye-color-predictor-2 (Spichenok et al. 2011) and the second number indicates the number of samples used for the verification (Pneuman et al. 2012).

[b] Calculated for European population, by using the data of the verification (Pneuman et al. 2012), which resulted in the most conservative error rate.

Corroboration of predictive DNA tests by other laboratories can assess the robustness of predictors, because sample sets and binning procedures may vary and lead to higher error rates.

Nevertheless, validations of these predictors are important, since most of the SNPs were selected based on correlation studies rather than functional studies (Branicki et al. 2008, 2011; Eiberg et al. 2008; Han et al. 2008; Liu et al. 2009; Mengel-From et al. 2010; Pospiech et al. 2011; Spichenok et al. 2011; Sturm et al. 2008; Valenzuela et al. 2010). Only for three of the selected SNPs (rs12913832, rs16891982, and rs1426654) was it shown that they lead to reduced melanin content in cultured human melanocytes, supporting their functional role in pigmentation (Cook et al. 2009). Thus, it is unclear how much the SNPs listed in Table 17.2 functionally contribute to the eye, hair, and skin color. Therefore, the group of mixed populations, where parents were associated with different populations, represents the ultimate test group.

Table 17.6 Validation of Skin-Color-Predictor in 805 Samples of Various Populations

Population	Skin Color (Self-Reported)	Skin Color Predicted	Errors in Prediction	Inconclusive
AA (33[b] + 10[b])	Medium (4 + 0)	Not white (1 + 0)	0	3 + 0
	Dark (29 + 10)	Not white (11 + 4)	0	18 + 6
SA (25 + 2)	Medium (25 + 0)	Not dark (2 + 0)	0	23 + 0
	Dark (0 + 2)	Not dark (0 + 1)	0 + 1	0 + 1
EA (22 + 13)	Light (20 + 11)	Not dark (7 + 2)	0	13 + 9
	Medium (2 + 2)	Not dark (0 + 1)	0	2 + 1
E (379 + 178)	Light (379 + 178)	Not dark (345 + 169)	0	34 + 9
Mixed (95 + 48)	Light (44 + 29)	Not dark (18 + 21)	0	26 + 8
	Medium (35 + 13)	Not dark (6 + 1)/not white (3 + 1)	0	26 + 11
	Dark (16 + 6)	Not white (3 + 2)/not dark (2 + 1)	2 + 1	11 + 3
Total		554 + 251	4	156 + 48
%			1%[b]	Range 28% to 19%

Note: AA, African-American; SA, South Asian; EA, East Asian; E, European; Mixed, mixed populations.

[a] Nonrelated individuals, the first number indicates the number of samples when introducing the skin color predictor (Spichenok et al. 2011), and the second number indicates the number of samples used for the verification (Pneuman et al. 2012).

[b] Calculated by using only data of the verification (Pneuman et al. 2012), which resulted in the most conservative error rate.

17.9 Validation of Forensic DNA Assays

The assays used in research to identify SNPs for their correlations with visible features, as well as for the validations, can differ from assays that are used in forensic routine procedures. Often, single assays are used in research, which provide more flexibility. A forensic test, however, would rather utilize a multiplex assay, which uses the available template DNA more efficiently. Several methods are available to conduct multiplex SNP assays, including mass spectrometry (Ross et al. 1998), multiplex SNaPshot assay (Sanchez and Endicott 2006), and high-throughput SNP genotyping platforms (Mead et al. 2008).

Here, the multiplex SNaPshot assay is used as an example, because it utilizes the same equipment as STR testing and therefore could easily be incorporated into the forensic routine of most laboratories. Key elements of this assay include a multiplex PCR followed by a single-base primer extension that creates fluorescently labeled oligonucleotides of distinct lengths. The fluorescent signal intensity is separated by length and detected by multicolor capillary electrophoresis (Figure 17.6). The eye-color-predictor-1, which includes six SNPs, was combined into such an assay (Walsh, Liu et al. 2011). Before the assay is incorporated into forensic routine, it should be validated. This validation verifies the technical procedure and differs substantially from the validations of predictors in pigmentation, which define the error rate, the call rate, and the applicability of various populations.

The Scientific Working Group on DNA Analysis Methods (SWGDAM) provided guidelines for the validation of DNA-based assays (SWGDAM 2004). Important elements for the validation are reproducibility, reliability, robustness, sensitivity, and different sources of

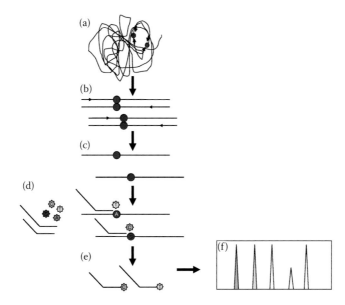

Figure 17.6 Outline of SNaPhot assay (multiplex SNP assay), shown for two SNPs: (a) Multiplex PCR performed on genomic DNA with several primer pairs (two SNPs are shown, in blue and red circles). (b) PCR amplicons and primers (arrows). (c) Purified and denatured PCR products. (d) Annealing of unlabeled primers, followed by primer extension with fluorescently labeled ddNTPs. (e) Purification and (f) detection of fluorescently labeled oligonucleotides by multicolor capillary electrophoresis.

DNA (see below). The multiplex SNaPshot assay of the eye-color-predictor-1 was validated following these guidelines and is used as an example (Walsh et al. 2010).

Reproducibility. The same samples will always lead to the same results under the same working conditions. **Example eye-color-predictor-1**: Forty samples were tested by two external laboratories. All results and interpretation from one laboratory were correct, whereas the other laboratory had two incorrect genotypes due to missing alleles (Walsh et al. 2010).

Reliability. The DNA assay can be corroborated by other methods. **Example eye-color-predictor-1**: The six SNPs were identified by a different genotyping method than the multiplex SNaPshot assay (Liu et al. 2009; Walsh et al. 2010).

Robustness. Robustness describes the conditions under which the DNA test must be performed to lead to reliable results. The wider the range of these conditions, the more robust is the DNA test. **Example eye-color-predictor-1**: The optimum range of template DNA for the multiplex SNaPshot assay was 250–500 pg (Walsh et al. 2010).

Sensitivity. Sensitivity refers to the range (lowest and highest) of (nuclear) DNA concentrations required to ensure consistent results. **Example eye-color-predictor-1**: The optimum range of template DNA for the assay was 250–500 pg, but it also worked with samples as low as 31 pg (Walsh et al. 2010).

Different sources of DNA. Different DNA extraction procedures may influence the multiplex assay and thus the forensic DNA test. It is important to verify that DNA extracted from bones, soft tissues, and blood, as well as DNA samples that mimic case-work samples, lead to reliable results. In addition, it may be interesting for some forensic tests, to assess whether degraded tissues and aged DNA could be used as input. **Example eye-color-predictor-1**: DNA extracted from buccal cells, hair, semen, and from touched items led to a full genotype profile (Walsh et al. 2010).

17.10 Future Directions

At present, descriptive DNA tests are developed, and probably soon they will be routinely used in the forensic field. In the near future, these predictors will be improved and will be able to describe features in more detail.

However, more information could be drawn from DNA. A recent discovery found that one SNP determines whether the *ear wax* is wet or dry. This was the first time that a visible genetic trait could be described by an SNP (Dean 2006). Ear wax is secreted from apocrine glands in the ear canal of humans and other mammals. Its function is not clear, but the existence of the two types was noted. The SNP, rs17822931 causes a missense mutation in a transporter protein encoded by the gene *ABCC11*. G/G homozygotes and heterozygotes have wet ear wax, and A/A-homozygotes have dry ear wax (Yoshiura et al. 2006).

Adding these kinds of genetic markers to the descriptive DNA tests will increase the information that can be obtained from DNA. Further examples include hair thickness, curly hair, baldness, and freckles.

Hair thickness, another visible trait, is determined by an SNP that causes a missense mutation in the *EDAR* gene. G/G homozygotes at rs3827760 have thick hair, and heterozygotes and A/A homozygotes have thinner hair (Fujimoto et al. 2008).

Curly hair comes from the shape of the bulb at the base of the hair follicle (Thibaut et al. 2005). Four SNPs are associated with curly or straight hair (Eriksson et al. 2010): rs17646946 (near the gene *TCHH*), rs499697 (430kb away from *LCE3E*), rs7349332 (in an intron of *WNT10A*), and rs1556547 (near *OFCC1*).

Four SNPs were associated with *baldness*: rs1385699 (a nonsynonymous SNP in *EDA2R*) (Prodi et al. 2008), rs6152 (located within the *AR* gene) (Ellis et al. 2007), and two SNPs, located on chromosome 20p11: rs2180439 (Hillmer et al. 2008) and rs1160312 (Richards et al. 2008).

Three SNPs were associated with freckles, small flat areas of skin with increased amounts of melanin (Eriksson et al. 2010): rs2153271 (*BNC2*), rs12203592 (*IRF4*), and rs1540771 (located between *IRF4* and *EXOC2* on chromosome 6).

In addition to these phenotypic markers describing visible characteristics, there are SNPs that correlate very well with particular populations and show substantial differences in their allele frequencies across populations. These SNPs could be used to associate DNA samples from individuals with specific populations, for example, European or Asian or African (Bamshad et al. 2004). A genome-wide approach, for example, could separate 51 populations into five known continental groups: African, European, Asian, American, and Oceanian (Li et al. 2008). In addition, there were attempts to reduce the number of SNPs required for this separation (Kersbergen et al. 2009; Lee et al. 2010; Phillips et al. 2007). However, for historically admixed populations or for individuals whose ancestors were from different continental regions, or for individuals from separated geographical regions, assumptions are still difficult to make (Bamshad et al. 2004; Jobling and Gill 2004).

Finally, one example shows how these markers could be used in combination for detailed descriptions of individuals. A set of SNPs, as described above, which included a selection of pigmentation, phenotypic, and population markers, was used to make a description of an individual, and even allowed publishers to create a drawing based solely on the information gained from DNA (Rasmussen et al. 2010). An estimated 4000-year-old preserved hair sample from an ancient human Eskimo culture of Greenland was used to extract genomic DNA. The DNA was sequenced using new technologies, which are reviewed elsewhere

(Metzker 2010), and more than 350,000 SNPs were identified (Rasmussen et al. 2010). The individual was male. Twenty-three SNPs were used for functional assessment and were compared to published HapMap frequencies. The HapMap consortium is a multicountry effort to identify and catalog genetic variants, including SNPs and their frequencies in various populations (Consortium 2003; Sachidanandam et al. 2001). For most of the 23 tested variants, the Eskimo DNA was closest to Asian DNA. In agreement with phenotypic markers, it was concluded that the Eskimo had brown eyes, did not have an European light skin coloration, had dark and thick hair (confirmed with morphological examinations), an increased risk of baldness, had shovel-graded front teeth, had the dry-type of ear wax, and was adapted to cold climate (Rasmussen et al. 2010). This information was used to draw a rough approximation of his face, which was printed on the cover of *Nature* in February 2010 (Vol. 463, February 11, 2010).

17.11 Conclusions

Inherited information from DNA can be used to make descriptions of visible features of individuals. Advances in molecular biology brought attention to genetic variations, in particular SNPs, which contribute to various phenotypes. This has important implications in the forensic field, especially for the identification of individuals from human remains without having a match of a DNA profile to databases or known samples. Although making identifications based on STR DNA profiles could be considered as the classical use of DNA in forensics, novel DNA techniques that lead to the description of eye, hair, and skin color are currently in development.

These DNA tests are based on statistically significant associations of genotypes with phenotypes. Depending on the complexity of these features, multiple SNPs are used in combination. For a standardized test, a predictor consists of a selection of SNPs plus instructions on how to interpret the genotypes. Validations can add confidence to the predictions, by defining error and call rates. They also can assess the applicability to various populations, which is an important concern in forensics.

Further research will improve these tests in precision and will eventually be complemented by other markers for additional characteristics.

Acknowledgments

I am grateful for the support of New York City Office of Chief Medical Examiner. For helpful discussions and careful reading of this manuscript, I thank Drs. Andreas Jenny and Veronique Bourdon, as well as Amanda Pneuman, Leigh Anne Sharek, Samantha (Xiu Hui) Pook, Patrick Carney, Vladimir Mushailov, Katie Hart, and Shey Kimura.

References

SWGDAM (Scientific Working Group on DNA Analysis Methods). 2004. Revised Validation Guidelines. *Forensic Sci Commun* 6(3).

Abdel-Malek, Z., I. Suzuki, A. Tada, S. Im, and C. Akcali. 1999. The melanocortin-1 receptor and human pigmentation. *Ann N Y Acad Sci* 885:117–33.

Bamshad, M., S. Wooding, B. A. Salisbury, and J. C. Stephens. 2004. Deconstructing the relationship between genetics and race. *Nat Rev Genet* 5(8):598–609.

Barsh, G. S. 2003. What controls variation in human skin color? *PLoS Biol* 1(1):E27.

Bouakaze, C., C. Keyser, E. Crubezy, D. Montagnon, and B. Ludes. 2009. Pigment phenotype and biogeographical ancestry from ancient skeletal remains: inferences from multiplexed autosomal SNP analysis. *Int J Legal Med* 123(4):315–25.

Branicki, W., U. Brudnik, T. Kupiec, P. Wolanska-Nowak, A. Szczerbinska, and A. Wojas-Pelc. 2008. Association of polymorphic sites in the *OCA2* gene with eye colour using the tree scanning method. *Ann Hum Genet* 72(Pt 2):184–92.

Branicki, W., U. Brudnik, T. Kupiec, P. Wolanska-Nowak, and A. Wojas-Pelc. 2007. Determination of phenotype associated SNPs in the *MC1R* gene. *J Forensic Sci* 52(2):349–54.

Branicki, W., U. Brudnik, and A. Wojas-Pelc. 2009. Interactions between HERC2, OCA2 and MC1R may influence human pigmentation phenotype. *Ann Hum Genet* 73(2):160–70.

Branicki, W., F. Liu, K. van Duijn, et al. 2011. Model-based prediction of human hair color using DNA variants. *Hum Genet* 129(4):443–54.

Brookes, A. J. 1999. The essence of SNPs. *Gene* 234(2):177–86.

Budowle, B., and T. R. Moretti. 1999. Genotype profiles for six population groups at the 13 CODIS short tandem repeat core loci and other PCRB-based loci. In *Forensic Sci Commun*.

Butler, J. M. 2006. Genetics and genomics of core short tandem repeat loci used in human identity testing. *J Forensic Sci* 51(2):253–65.

Chaplin, G. 2004. Geographic distribution of environmental factors influencing human skin coloration. *Am J Phys Anthropol* 125(3):292–302.

Cole, T. J. 2003. The secular trend in human physical growth: a biological view. *Econ Hum Biol* 1(2):161–8.

Consortium, International HapMap. 2003. The International HapMap Project. *Nature* 426(6968):789–96.

Cook, A. L., W. Chen, A. E. Thurber et al. 2009. Analysis of cultured human melanocytes based on polymorphisms within the SLC45A2/MATP, SLC24A5/NCKX5, and OCA2/P loci. *J Invest Dermatol* 129(2):392–405.

Dean, L. 2006. Don't put anything smaller than your elbow in your ear. In *Coffee Break: Tutorials for NCBI Tools*, eds. L. Dean and J. McEntyre. Bethesda, MD: National Center for Biotechnology Information (U.S.). http://www.ncbi.nlm.nih.gov/books/NBK2333/.

Duffy, D. L., G. W. Montgomery, W. Chen et al. 2007. A three-single-nucleotide polymorphism haplotype in intron 1 of OCA2 explains most human eye-color variation. *Am J Hum Genet* 80(2):241–52.

Durbin, R. M., G. R. Abecasis, D. L. Altshuler et al. 2010. A map of human genome variation from population-scale sequencing. *Nature* 467(7319):1061–73.

Edwards, M., A. Bigham, J. Tan et al. 2010. Association of the OCA2 polymorphism His615Arg with melanin content in east Asian populations: further evidence of convergent evolution of skin pigmentation. *PLoS Genet* 6(3):e1000867.

Eiberg, H., J. Troelsen, M. Nielsen, A. Mikkelsen, J. Mengel-From, K. W. Kjaer, and L. Hansen. 2008. Blue eye color in humans may be caused by a perfectly associated founder mutation in a regulatory element located within the HERC2 gene inhibiting OCA2 expression. *Hum Genet* 123(2):177–87.

Ellis, J. A., K. J. Scurrah, J. E. Cobb, S. G. Zaloumis, A. E. Duncan, and S. B. Harrap. 2007. Baldness and the androgen receptor: the AR polyglycine repeat polymorphism does not confer susceptibility to androgenetic alopecia. *Hum Genet* 121(3–4):451–7.

Eriksson, N., J. M. Macpherson, J. Y. Tung et al. 2010. Web-based, participant-driven studies yield novel genetic associations for common traits. *PLoS Genet* 6(6):e1000993.

Frazer, K. A., D. G. Ballinger, D. R. Cox et al. 2007. A second generation human haplotype map of over 3.1 million SNPs. *Nature* 449(7164):851–61.

Frost, P. 2006. European hair and eye color. A case of frequency-dependent sexual selection? *Evol Hum Behav* 27:85–103.

Frudakis, T., T. Terravainen, and M. Thomas. 2007. Multilocus OCA2 genotypes specify human iris colors. *Hum Genet* 122(3–4):311–26.

Frudakis, T., M. Thomas, Z. Gaskin et al. 2003. Sequences associated with human iris pigmentation. *Genetics* 165(4):2071–83.

Fujimoto, A., R. Kimura, J. Ohashi et al. 2008. A scan for genetic determinants of human hair morphology: EDAR is associated with Asian hair thickness. *Hum Mol Genet* 17(6):835–43.

Han, J., P. Kraft, H. Nan et al. 2008. A genome-wide association study identifies novel alleles associated with hair color and skin pigmentation. *PLoS Genet* 4(5):e1000074.

Hill, C. R., J. M. Butler, and P. M. Vallone. 2009. A 26plex autosomal STR assay to aid human identity testing. *J Forensic Sci* 54(5):1008–15.

Hillmer, A. M., F. F. Brockschmidt, S. Hanneken et al. 2008. Susceptibility variants for male-pattern baldness on chromosome 20p11. *Nat Genet* 40(11):1279–81.

Hinds, D. A., L. L. Stuve, G. B. Nilsen et al. 2005. Whole-genome patterns of common DNA variation in three human populations. *Science* 307(5712):1072–9.

Jablonski, N. G., and G. Chaplin. 2000. The evolution of human skin coloration. *J Hum Evol* 39(1):57–106.

Jobling, M. A., and P. Gill. 2004. Encoded evidence: DNA in forensic analysis. *Nat Rev Genet* 5(10):739–51.

Katsnelson, A. 2010. 1000 Genomes Project reveals human variation. *Nature News:* doi:10.1038/news.2010.567.

Kayser, M., and P. de Knijff. 2011. Improving human forensics through advances in genetics, genomics and molecular biology. *Nat Rev Genet* 12(3):179–92.

Kayser, M., F. Liu, A. C. Janssens et al. 2008. Three genome-wide association studies and a linkage analysis identify HERC2 as a human iris color gene. *Am J Hum Genet* 82(2):411–23.

Kenny, E. E., N. J. Timpson, M. Sikora et al. 2012. Melanesian blond hair is caused by an amino acid change in TYRP1. *Science* 33(6081):554.

Kersbergen, P., K. van Duijn, A. D. Kloosterman, J. T. den Dunnen, M. Kayser, and P. de Knijff. 2009. Developing a set of ancestry-sensitive DNA markers reflecting continental origins of humans. *BMC Genet* 10:69.

Kruglyak, L., and D. A. Nickerson. 2001. Variation is the spice of life. *Nat Genet* 27(3):234–6.

Lango Allen, H., K. Estrada, G. Lettre et al. 2010. Hundreds of variants clustered in genomic loci and biological pathways affect human height. *Nature* 467(7317):832–8.

Lanktree, M. B., Y. Guo, M. Murtaza et al. 2011. Meta-analysis of dense gene-centric association studies reveals common and uncommon variants associated with height. *Am J Hum Genet* 88(1):6–18.

Lee, Y. L., S. Teitelbaum, M. S. Wolff, J. G. Wetmur, and J. Chen. 2010. Comparing genetic ancestry and self-reported race/ethnicity in a multiethnic population in New York City. *J Genet* 89(4):417–23.

Lettre, G. 2011. Recent progress in the study of the genetics of height. *Hum Genet* 129(5):465–72.

Li, J. Z., D. M. Absher, H. Tang et al. 2008. Worldwide human relationships inferred from genome-wide patterns of variation. *Science* 319(5866):1100–4.

Lin, J. Y., and D. E. Fisher. 2007. Melanocyte biology and skin pigmentation. *Nature* 445(7130):843–50.

Liu, F., K. van Duijn, J. R. Vingerling et al. 2009. Eye color and the prediction of complex phenotypes from genotypes. *Curr Biol* 19(5):R192–3.

Martin, N., D. Boomsma, and G. Machin. 1997. A twin-pronged attack on complex traits. *Nat Genet* 17(4):387–92.

McEvoy, B., S. Beleza, and M. D. Shriver. 2006. The genetic architecture of normal variation in human pigmentation: an evolutionary perspective and model. *Hum Mol Genet* 15(Spec No 2):R176–81.

Mead, S., M. Poulter, J. Beck et al. 2008. Successful amplification of degraded DNA for use with high-throughput SNP genotyping platforms. *Hum Mutat* 29(12):1452–8.

Mengel-From, J., C. Borsting, J. J. Sanchez, H. Eiberg, and N. Morling. 2010. Human eye colour and HERC2, OCA2 and MATP. *Forensic Sci Int Genet* 4(5):323–8.

Mengel-From, J., T. H. Wong, N. Morling, J. L. Rees, and I. J. Jackson. 2009. Genetic determinants of hair and eye colours in the Scottish and Danish populations. *BMC Genet* 10:88.

Metzker, M. L. 2010. Sequencing technologies—the next generation. *Nat Rev Genet* 11(1):31–46.

Norton, H. L., J. S. Friedlaender, D. A. Merriwether, G. Koki, C. S. Mgone, and M. D. Shriver. 2006. Skin and hair pigmentation variation in Island Melanesia. *Am J Phys Anthropol* 130(2):254–68.

Norton, H. L., R. A. Kittles, E. Parra et al. 2007. Genetic evidence for the convergent evolution of light skin in Europeans and East Asians. *Mol Biol Evol* 24(3):710–22.

Phillips, C., A. Salas, J. J. Sanchez et al. 2007. Inferring ancestral origin using a single multiplex assay of ancestry-informative marker SNPs. *Forensic Sci Int Genet* 1(3–4):273–80.

Pneuman, A., Z. M. Budimlija, T. Caragine, M. Prinz, and E. Wurmbach. 2012. Verification of eye and skin color predictors in various populations. *Leg Med (Tokyo)* 14(2):78–83.

Pospiech, E., J. Draus-Barini, T. Kupiec, A. Wojas-Pelc, and W. Branicki. 2011. Gene–gene interactions contribute to eye colour variation in humans. *J Hum Genet* 56(6):447–55.

Prodi, D. A., N. Pirastu, G. Maninchedda et al. 2008. EDA2R is associated with androgenetic alopecia. *J Invest Dermatol* 128(9):2268–70.

Rasmussen, M., Y. Li, S. Lindgreen et al. 2010. Ancient human genome sequence of an extinct Palaeo-Eskimo. *Nature* 463(7282):757–62.

Richards, J. B., X. Yuan, F. Geller et al. 2008. Male-pattern baldness susceptibility locus at 20p11. *Nat Genet* 40(11):1282–4.

Ross, P., L. Hall, I. Smirnov, and L. Haff. 1998. High level multiplex genotyping by MALDI-TOF mass spectrometry. *Nat Biotechnol* 16(13):1347–51.

Sachidanandam, R., D. Weissman, S. C. Schmidt et al. 2001. A map of human genome sequence variation containing 1.42 million single nucleotide polymorphisms. *Nature* 409(6822):928–33.

Sanchez, J. J., and P. Endicott. 2006. Developing multiplexed SNP assays with special reference to degraded DNA templates. *Nat Protoc* 1(3):1370–8.

Shastry, B. S. 2009. SNPs: impact on gene function and phenotype. *Methods Mol Biol* 578:3–22.

Shekar, S. N., D. L. Duffy, T. Frudakis et al. 2008. Linkage and association analysis of spectrophotometrically quantified hair color in Australian adolescents: the effect of OCA2 and HERC2. *J Invest Dermatol* 128(12):2807–14.

Simon, J. D. 2009. Seeing red: pheomelanin synthesis uncovered. *Pigment Cell Melanoma Res* 22(4):382–3.

Sitaram, A., R. Piccirillo, I. Palmisano et al. 2009. Localization to mature melanosomes by virtue of cytoplasmic dileucine motifs is required for human OCA2 function. *Mol Biol Cell* 20(5):1464–77.

Spichenok, O., Z. M. Budimlija, A. A. Mitchell et al. 2011. Prediction of eye and skin color in diverse populations using seven SNPs. *Forensic Sci Int Genet* 5(5):472–478.

Stokowski, R. P., P. V. Pant, T. Dadd et al. 2007. A genomewide association study of skin pigmentation in a South Asian population. *Am J Hum Genet* 81(6):1119–32.

Sturm, R. A., D. L. Duffy, Z. Z. Zhao et al. 2008. A single SNP in an evolutionary conserved region within intron 86 of the HERC2 gene determines human blue-brown eye color. *Am J Hum Genet* 82(2):424–31.

Sturm, R. A., and T. N. Frudakis. 2004. Eye colour: portals into pigmentation genes and ancestry. *Trends Genet* 20(8):327–32.

Sturm, R. A., and M. Larsson. 2009. Genetics of human iris colour and patterns. *Pigment Cell Melanoma Res* 22(5):544–62.

Subramanian, S. V., E. Ozaltin, and J. E. Finlay. 2011. Height of nations: a socioeconomic analysis of cohort differences and patterns among women in 54 low- to middle-income countries. *PLoS One* 6(4):e18962.

Sulem, P., D. F. Gudbjartsson, S. N. Stacey et al. 2008. Two newly identified genetic determinants of pigmentation in Europeans. *Nat Genet* 40(7):835–7.

Sulem, P., D. F. Gudbjartsson, S. N. Stacey et al. 2007. Genetic determinants of hair, eye and skin pigmentation in Europeans. *Nat Genet* 39(12):1443–52.

Thibaut, S., O. Gaillard, P. Bouhanna, D. W. Cannell, and B. A. Bernard. 2005. Human hair shape is programmed from the bulb. *Br J Dermatol* 152(4):632–8.

Tishkoff, S. A., and K. K. Kidd. 2004. Implications of biogeography of human populations for 'race' and medicine. *Nat Genet* 36(11 Suppl):S21–7.

Tishkoff, S. A., and B. C. Verrelli. 2003. Patterns of human genetic diversity: implications for human evolutionary history and disease. *Annu Rev Genomics Hum Genet* 4:293–340.

Valenzuela, R. K., M. S. Henderson, M. H. Walsh et al. 2010. Predicting phenotype from genotype: normal pigmentation. *J Forensic Sci* 55(2):315–22.

Valverde, P., E. Healy, I. Jackson, J. L. Rees, and A. J. Thody. 1995. Variants of the melanocyte-stimulating hormone receptor gene are associated with red hair and fair skin in humans. *Nat Genet* 11(3):328–30.

Visser, M., M. Kayser, and R. J. Palstra. 2012. HERC2 rs12913832 modulates human pigmentation by attenuating chromatin-loop formation between a long-range enhancer and the OCA2 promoter. *Genome Res* 22(3):446–55.

Walsh, S., A. Lindenbergh, S. B. Zuniga et al. 2010. Developmental validation of the IrisPlex system: Determination of blue and brown iris colour for forensic intelligence. *Forensic Sci Int Genet* 5(5):464–71.

Walsh, S., F. Liu, K. N. Ballantyne, M. van Oven, O. Lao, and M. Kayser. 2011. IrisPlex: a sensitive DNA tool for accurate prediction of blue and brown eye colour in the absence of ancestry information. *Forensic Sci Int Genet* 5(3):170–80.

Walsh, S., F. Liu, A. Wollstein et al. 2013. The HIrisPlex system for simultaneous prediction of hair and eye colour from DNA. *Forensic Sci Int Genet* 7(1):98–115.

Walsh, S., A. Wollstein, F. Liu et al. 2011. DNA-based eye colour prediction across Europe with the IrisPlex system. *Forensic Sci Int Genet* 6(3):330–40.

Weedon, M. N., and T. M. Frayling. 2008. Reaching new heights: insights into the genetics of human stature. *Trends Genet* 24(12):595–603.

Westerhof, W. 2007. Evolutionary, biologic, and social aspects of skin color. *Dermatol Clin* 25(3):293–302, vii.

Yamaguchi, Y., and V. J. Hearing. 2009. Physiological factors that regulate skin pigmentation. *Biofactors* 35(2):193–9.

Yang, J., B. Benyamin, B. P. McEvoy et al. 2010. Common SNPs explain a large proportion of the heritability for human height. *Nat Genet* 42(7):565–9.

Yoshiura, K., A. Kinoshita, T. Ishida et al. 2006. A SNP in the ABCC11 gene is the determinant of human earwax type. *Nat Genet* 38(3):324–30.

Molecular Autopsy

18

GRACE AXLER-DIPERTE
FREDERICK R. BIEBER
ZORAN M. BUDIMLIJA
ANTTI SAJANTILA
DONALD SIEGEL
YINGYING TANG

Contents

18.1 Molecular Autopsy	453
18.1.1 Molecular Autopsy—Definition(s)	454
18.2 Molecular Genetics	454
18.2.1 Genetics and Genomics in Sudden Natural Death	454
18.2.1.1 Introduction	454
18.2.1.2 Positive Autopsy	455
18.2.1.3 Negative Autopsy	462
18.3 Postmortem Toxicogenetics	463
18.3.1 Toxicogenetics and Medico-Legal Death Investigation	463
18.3.2 Development of Pharmacogenetic Concept to Pharmacogenomics	463
18.3.3 Adverse Drug Reactions	465
18.3.4 Medico-Legal Investigation of Death	465
18.3.4.1 Postmortem Toxicogenetics and CoD Investigation	466
18.4 Microgenomics	469
18.5 Conclusion	471
References	472

18.1 Molecular Autopsy

Multidisciplinary investigations involving forensic pathology, forensic toxicology, and forensic genetics are crucial for public safety and legal protection. Autopsy (necropsy, postmortem examination), according to *Merriam-Webster's Dictionary*, is "… an examination of a body after death to determine the cause of death (CoD) or the character and extent of changes produced by disease …" The word has origin in Greek αυτοπσια, New Latin autopsia—act of seeing with one's own eyes [from auto- (self) + opsis (sight), appearance].

The forensic autopsy refers to the postmortem examination of a body performed with the intent of determining the specific cause and legal manner of death (MoD) (e.g., homicide, suicide, accidental, natural, or undetermined) in question; a complete forensic autopsy may require evaluation of evidence attached to the body and/or found at the scene of death/crime, and reconstruction of the scene itself and may even include a psychological component based on historical data. It is performed by designated professionals—medical examiners or qualified coroners—forensic pathologists in both cases, to determine if the death was consequence of natural causes or violence. It is necessary to determine whether the death was a result of criminal act per se, or if the foul play was involved in the cases of alleged suicides, accidents, or at the first sight, "natural deaths." Specifically, in the case of accidental deaths, it is important to determine the mechanism of occurrence, to ensure that the accident itself caused the death and to help to prevent future accidents of the same nature. A forensic autopsy is often necessary for medical reasons, too, for example, to confirm if appropriate medical care was given, and to determine if underlined conditions leading to the death in question could cause additional fatalities or affect other family members in any way. In addition, forensic autopsies have to be performed to check if the CoD in a single case could influence public health, as well as to serve as a first step in the identification of unidentified human remains.

Generally, medico-legal autopsies are unique and reliable sources of valuable data to investigate unexpected deaths. They also serve as data source for characterization of health and way of life at a population level. The information obtained from postmortem investigations, when properly archived and analyzed, can reveal patterns and trends that, ultimately, can be used by decision-makers to implement preventive actions and to develop health policies. More sophisticated information can be obtained from molecular autopsies, toxicology and drug interaction studies, postmortem toxicogenetics, and population genetics.

18.1.1 Molecular Autopsy—Definition(s)

Similar to the plasticity of biological systems, their flexibility to adapt, and even to succumb to the rules of "determined chaos" (idiom considered as oxymoron for many years), proposing the firm definition of the term "molecular autopsy" at the present stage of the applied sciences' progress is not an easy task. Since the definition(s) have to show many facets of the purpose of molecular autopsy, the authors of this chapter have decided to describe the term in more agile way, especially because this is still relatively new pathology/forensic sciences concept, an idea that is yet to be widely endorsed within professional communities. Known as "genetic," "silent," even "liquid" autopsy, it could be considered as a tool emerging out of molecular pathology, which is the study of biochemical and biophysical cellular mechanisms as the basic factors in diseases. So far, the molecular autopsy has been described by many as a "… number of genetic tests performed on the tissues taken during the autopsy to confirm the presence of suspect gene mutations linked to the sudden, unexpected death without visible morphological substrate…"

Being in agreement with all of the aforementioned thoughts and definitions regarding this relatively recently accepted neologism, we include in this chapter several additional procedures and techniques that have been used in forensic sciences and/or forensic pathology, being under strict meaning of the term "molecular autopsy" or not. In general, this chapter deals with principles and standards of molecular genetics and its application in the cases of sudden deaths, postmortem toxicogenetics, and microgenomics.

18.2 Molecular Genetics

18.2.1 Genetics and Genomics in Sudden Natural Death

18.2.1.1 Introduction
In addition to investigating death as a result of violence, accident, or suicide, forensic pathologists (medical examiners) also determine the immediate and underlying CoD as a

consequence of natural disease processes, often occurring outside hospitals and not under the care of a medical professional. Among natural deaths that medical examiners investigate, the majority are sudden and unexpected (Gill and Scordi-Bello 2010). The presence of natural disease is frequently discovered only at autopsy. Such deaths in adults are often associated with diseases affecting the cardiovascular system, such as coronary artery atherosclerosis, hypertensive heart disease, and pulmonary thromboembolism (Barsheshet et al. 2011). In infants and children, in addition to cardiac arrhythmias associated with structurally normal or abnormal hearts (Ilina et al. 2011), developmentally related diseases such as metabolic disorders and sudden unexplained death in epilepsy (SUDEP), may also manifest themselves. In the United States, the incidence of sudden cardiac death (SCD) at all ages is greater than 350,000 persons per year (Anderson et al. 1994).

The ability to identify underlying causes of death, especially genetic disorders, not only allows for the determination of CoD in a decedent, but may also help avoid a recurring tragedy within a family. Recently, molecular studies of the common disorders have shifted from examining a single or a few candidate genes in a few families (genetics) to the whole human genome level in a population (genomics), where the involvement and interaction of many genes and pathways are looked at simultaneously (Podgoreanu Schwinn 2005). Understanding the genetic components of the common diseases, coupled with the use of molecular genetic testing, has become an increasingly important tool in medical examiners' armamentarium in determining CoD or contributing factors to death.

A positive autopsy finding or the knowledge of an immediate CoD usually gives medical examiners clues as to which diseases should be tested for. A negative autopsy finding, on the other hand, can be a vexing challenge for medical examiners. Recent advances in "molecular autopsies" have begun to uncover the mystery surrounding sudden unexplained natural deaths by identifying mutations that can result in or predispose an apparently healthy individual to sudden death (Di Paolo et al. 2004). This article reviews the current knowledge of genetics and genomics in the diseases that are commonly seen in sudden natural deaths.

18.2.1.2 Positive Autopsy

In this section, the pathological conditions described are listed largely based on the frequency with which they are seen at the autopsy. More extensive reviews of a specific disease and the genes and loci linked to that disease is referenced and summarized in Table 18.1.

1. *Cardiovascular System*
   - *Coronary artery atherosclerosis.* Cardiovascular disease is the most common (Barsheshet et al. 2011) cause of natural unexpected death seen by medical examiners. The stenosis or occlusion of a particular coronary artery results in ischemia of the supplied region of myocardium, which may result in acute myocardial infarction and/or a lethal cardiac arrhythmia. Fatal coronary artery atherosclerosis tends to occur in the proximal aspects of the left main coronary artery, right main coronary artery, at branching points and the proximal aspects of the left anterior artery and left circumflex arteries. Acute critical occlusion of a vessel may result in the sudden onset of symptoms. The acute change may result from complications within an atheromatous plaque including plaque rupture with subsequent thrombosis of the vessel and sudden hemorrhage into a plaque resulting in loss of luminal area and subsequent

Table 18.1 Genetics and Genomics in Sudden Natural Deaths

Condition	Genes or Locus
Coronary artery atherosclerosis (familial)	
Coronary artery disease, autosomal dominant	MEF2A (Wang et al. 2003) and LRP6 (Mani et al. 2007)
Coronary artery atherosclerosis (sporadic)	
Ventricular fibrillation in acute myocardial infarction	21q21(CXADR) (Bezzina et al. 2010)
Coronary artery disease	9p21.3 (CDKN2A, CDKN2B) (Preuss et al. 2010)
Myocardial infarction, premature, susceptibility to	1p34–36 (LRP8) (Wang et al. 2004), 1q25.1(TNFSF4) (Wang et al. 2005; Ria et al. 2011), 6p12.1 (GCLC) (Koide 2003) 6q25.1(ESR1) (Aouizerat et al. 2011) 13q12 (ALOX5AP) (Helgadottir et al. 2004)
Hypertension	
Monogenic forms of hypertension	AGTR1 (Gribouval et al. 2005), PTGIS (Nakayama et al. 2003)
	ECE1 (Funke-Kaiser et al. 2004), RGS5 (Chang et al. 2007), ATP1B1 (Chang et al. 2007), SELE (Chang et al. 2007), AGT (Markovic et al. 2005), HYT3 (Angius et al. 2007), AGTR1 (Gribouval et al. 2005), ADD1 (Manunta et al. 1999), HYT6 (Joe et al. 2009), CYP3A5 (Thompson et al. 2004), NOS3 (Wang et al. 1996), HYT4 (Gong et al. 2003), HYT2 (Xu et al. 1999), HYT1 (Hilbert et al. 1991), NOS2A (Rutherford et al. 2001), HYT5 (Wallace et al. 2006), PTGIS (Nakayama et al. 2003)
Hypertension, genome-wide	1p36 (Newton-Cheh et al. 2009), 3p22.1 (Padmanabhan et al. 2010), 4q21 (Newton-Cheh et al. 2009), 10q21 (Newton-Cheh et al. 2009), 10q24 (Newton-Cheh et al. 2009; Padmanabhan et al. 2010), 10p12 (Padmanabhan et al. 2010), 11p15 (Padmanabhan et al. 2010), 12q21 (Padmanabhan et al. 2010), 12q24 (Newton-Cheh et al. 2009) (Padmanabhan et al. 2010), 12q24.21 (Padmanabhan et al. 2010), 15q24 (Newton-Cheh et al. 2009; Padmanabhan et al. 2010), 17q21 (Newton-Cheh et al. 2009), 17q21 (Newton-Cheh et al. 2009)
Pulmonary thromboembolism	Factor V Leiden (Tang et al. 2011), Prothrombin G20210A (Tang et al. 2011), protein S (Ikejiri et al. 2010), protein C (Vossen et al. 2005), antithrombin (Vossen et al. 2005)
Aortic aneurysm rupture	
Marfan syndrome	FBN1 (Dietz et al. 1991)
Ehlers-Danlos syndrome type IV	COL3A1 (Schwarze et al. 1997)
Genomic risks	9p21.3 (rs7025486 in DAB2IP) (Gretarsdottir et al. 2010)
Intracranial Berry aneurysm	2q33.1 (Bilguvar et al. 2008), 9p21 (Yasuno et al. 2010), 7q11.2 (Akagawa et al. 2006), 19q13 (van der Voet et al. 2004), 1p36.13–p34.3 (Nahed et al. 2005), 5p15.2–p14.3 (Verlaan et al. 2006), 14q23 (Mineharu et al. 2008), 2p15–q14 (Roos et al. 2004), 11q24–25 (Ozturk et al. 2006)
Cardiomyopathy	
Arrhythmogenic right ventricular cardiomyopathy (ARVC)	TGFB3 (Rampazzo et al. 2003), PKP2 (Gerull et al. 2004), DSC2 (Syrris et al. 2006), DSG2 (Syrris et al. 2007), JUP (Asimaki et al. 2007), DSP (Yang et al. 2006), TMEM43 (Merner et al. 2008), RyR 2 (Tiso et al. 2001)

(continued)

Table 18.1 Genetics and Genomics in Sudden Natural Deaths (Continued)

Condition	Genes or Locus
Hypertrophic cardiomyopathy (HCM)	MYH7 (Geisterfer-Lowrance et al. 1990), MYBPC3 (Watkins et al. 1995), TNNC1 (Hoffmann et al. 2001), TNNT2 (Thierfelder et al. 1994), TNNI3 (Kimura et al. 1997), TPM1 (Richard et al. 2003), ACTC (Richard et al. 2003), MYL2 (Richard et al. 2003), MYL3 (Richard et al. 2003), PLN (Landstrom et al. 2011), LAMP2 (Charron et al. 2004), PRKAG2 (Kelly et al. 2009), NEXN (Wang et al. 2010)
Dilated cardiomyopathy (DCM) reviewed by Dellefave and McNally (2010)	**Sarcomere genes:** *MYH7, MYH6, MYL3, MYL2, TPM1, TNNT2, TNNI3, TNNC1, MYBPC3, ACTC* **Z-disk genes:** *TTN, LDB3, VCL, CSRP3, TCAP, TTID* **Nuclear genes:** *TAZ, LMNA, EMD, EYA4, SYNE1, TMPO, RBM20* **Mitochondrial genes:** *TTR, FRA, Cox15, HFE* **Cytoskeletal-membrane linkers genes:** *DMS, SGCG, SGCD, SGCB, SBCA, FKRP* **Others genes:** *ANKRD1, LAMP2, SCN5A, DES, ZASP, NEBL, DNM1L*
Left ventricular noncompaction cardiomyopathy (LVNC)	DTNA (Ichida et al. 2001), LDB3 (Vatta et al. 2003), TAZ (Bleyl et al. 1997), ACTC (Klaassen et al. 2008), MYH7 (Klaassen et al. 2008), TNNT2 (Klaassen et al. 2008), MYBPC3 (Vatta et al. 2003)
Sudden cardiac death susceptible loci	2q24.2 (Arking et al. 2011)
Monogenic forms of diabetes	
Maturity-onset diabetes of the young (MODY)	HNF4A (Yamagata et al. 1996), GCK (Vionnet et al. 1992), HNF1A (Vaxillaire et al. 1997), IPF1 (Stoffers et al. 1997), HNF1B (Horikawa et al. 1997), NeuroD1 (Malecki et al. 1999), KLF11 (Neve et al. 2005), CEL (Raeder et al. 2006), PAX4 (Plengvidhya et al. 2007), BLK (Borowiec et al. 2009)
Neonatal diabetes mellitus (NDM)	GCK (Njolstad et al. 2001), INS (Edghill et al. 2008), KCNJ11 (Gloyn et al. 2004), SUR1 (Ellard et al. 2007)
Cardiac conduction system	
Long QT Syndromes (George et al. 1995; Berthet et al. 1999; Priori et al. 1999; Splawski et al. 2000; Plaster et al. 2001; Mohler et al. 2003; Splawski et al. 2005; Cronk et al. 2007; Medeiros-Domingo et al. 2008; Chen et al. 2007; Ueda et al. 2008; Yang et al. 2010)	KCNQ1 (LQT1); KCNH2 (LQT2); SCN5A (LQT3); ANK2 (LQT4); KCNE1 (LQT5); KCNE2 (LQT6); KCNJ2 (LQT7); CACNA1C (LQT8); CAV3 (LQT9); SCN4B (LQT10); AKAP9 (LQT11); SNTA1 (LQT12); KCNJ5 (LQT13)
Brugada Syndrome (Chen et al. 1998; London et al. 2007; Antzelevitch et al. 2007; Watanabe et al. 2008; Delpon et al. 2008; Hu et al. 2009; Ueda et al. 2009)	SCN5A; GPD1L; SCN1B; SCN2B; KCNE3; CACNA1C; SCN3B; CACNB2
Catecholaminergic polymorphic ventricular tachycardia (CPVT) (Priori et al. 2001; di Barletta et al. 2006)	RYR2 (Priori et al. 2001); CASQ2 (Lahat et al. 2004)
Short QT syndrome	KCNH2 (Hong et al. 2005); KCNQ1 (Bellocq et al. 2004); KCNJ2 (Priori et al. 2005)
Ventricular fibrillation	KCNJ8/KATP (Haissaguerre et al. 2009); DPP6 (Alders et al. 2009); CACNA1C (Aouizerat et al. 2011)
Wolff–Parkinson–White syndrome	PRKAG2 (Zhang et al. 2011)

(continued)

Table 18.1 Genetics and Genomics in Sudden Natural Deaths (Continued)

Condition	Genes or Locus
Sick sinus syndrome	HCN4 (Ueda et al. 2009), SCN5A (Gui et al. 2010)
Cardiac conduction defect	AKAP10 (Patel et al. 2010)
Progressive familial heart block, Type IB	TRPM4 (Kruse et al. 2009)
Atrial fibrillation and sudden death	KCNA5 (Yang et al. 2009)
Progressive cardiac conduction disease, atrial arrhythmias, and sudden death	LMNA (Marsman et al. 2011)
Central neural systems	
Sudden explained death in epilepsy	KCNA1 (Tu et al. 2011), HCN1 –4 (Tu et al. 2011), SCN5A (Aurlien et al. 2009; Tu et al. 2011), KCNH2 (Tu et al. 2011)

myocardial ischemia. As most cases of sudden natural death due to coronary artery atherosclerosis occur within hours of the onset of symptoms, the medical examiner does not usually find acute myocardial infarction upon macroscopic and microscopic examination of the heart (Little and Applegate 1996). In addition, there are gender differences in the severity and extent of coronary artery atherosclerosis in relationship to the fatal outcome. Women who died of fatal ischemic heart disease tend to have less severe and extensive coronary atherosclerosis (Smilowitz et al. 2011) at autopsy. Therefore, coronary artery atherosclerosis should be considered even when there is no typical autopsy finding of coronary arteries occlusion, especially in women.

Coronary artery atherosclerosis is often the result of multiple factors, include lifestyle (e.g., smoking, high-fat diet, lack of exercise), compounding disorders (e.g., hypertension, diabetes mellitus, hyperlipidemias), and genetic or genomic risks (Sivapalaratnam et al. 2011). A few studies of families where coronary artery atherosclerosis is clearly dominantly inherited with multiple affected individuals and often early-onset have allowed the identification of the candidate genes, such as *MEF2A* (Wang et al. 2003) and LRP6 (Mani et al. 2007), which are responsible for the autosomal dominant form of the coronary artery disease. Testing for these genes might be helpful for identifying the underlying CoD in young victims with family histories.

In the general population, however, coronary artery atherosclerosis usually follows a multifactorial disease model where multiple genetic loci (genomic risks) are involved and may have synergistic effects on the development of disease. For instance, genomic risks for myocardial infarct have been mapped to several loci, including 1p34-36 (LRP8) (Wang et al. 2004), 1q25.1 (TNFSF4) (Wang et al. 2005; Ria et al. 2011), 6p12.1 (GCLC) (Koide et al. 2003), 6q25.1 (ESR1) (Aouizerat et al. 2011), 13q12 (ALOX5AP) (Helgadottir et al. 2004); genomic risk for ventricular fibrillation in acute myocardial infarction to 21q21(CXADR) (Bezzina et al. 2010), and risks of coronary artery diseases to 9p21, where one single nucleotide polymorphism at position rs1333049 confers a 29% increase in risk for myocardial infarct per copy (Preuss et al. 2010; Table 18.1). Further research will provide insight into the roles of these loci in sudden unexpected death (SUD).

Unlike in familial studies, genome wide search for risks of coronary artery atherosclerosis usually involves large number of clinically well-characterized individuals with (cohorts) or without (controls) the disease. Furthermore, it is likely that genomic risks can be skewed by sampling differences such as ethnicity or medical care conditions; therefore, it is not surprising to see some genomic risks previously identified that cannot be confirmed in different studies. However, knowing the genomic risks of a disease may allow personalized medical care or preventive management before a disease manifests itself.

- *Hypertensive cardiovascular disease.* Hypertensive cardiovascular disease may be reliably diagnosed from the pathological features of an enlarged heart showing concentric left ventricular hypertrophy in the absence of valve disease or cardiomyopathy. Characteristic hypertensive changes may be observed in the kidney and in older individuals, the lenticulostriate vessels within the basal ganglia of the brain. Hypertensive cardiovascular disease can lead to sudden death through several mechanisms, including invoking a cardiac arrhythmia, precipitating rupture of an abdominal aortic aneurysm, dissection of the thoracic aorta, or causing an intracerebral hemorrhage (Burke et al. 1996).

  Hypertension affects 24% of the U.S. adult population (Burt et al. 1995), with a disproportionate burden in minorities, particularly African-Americans (Hajjar and Kotchen 2003). Hypertension is another multifactorial disease with lifestyle (e.g., high-salt diet, obesity), compounding common diseases or conditions (e.g., diabetes mellitus or aging) and multiple genes involved in regulating and maintaining the blood pressure (Franceschini et al. 2011).

  Numerous genes associated with essential hypertension (e.g., *ECE1* [Funke-Kaiser et al. 2003], *RGS5* [Chang et al. 2007], *ATP1B1* [Chang et al. 2007], *SELE* [Chang et al. 2007], *AGT* [Markovic et al. 2005], *HYT3* [Angius et al. 2002], *AGTR1* [Gribouval et al. 2005], *ADD1* [Manunta et al. 1999], *HYT6* [Joe et al. 2009], *CYP3A5* [Thompson et al. 2004], *NOS3* [Wang et al. 1996], *HYT4* [Gong et al. 2003], *HYT2* [Xu et al. 1999], *HYT1* [Hilbert et al. 1991], *NOS2A* [Rutherford et al. 2001], *HYT5* [Wallace et al. 2006], *PTGIS* [Nakayama et al. 2003]) have been identified, and genome-wide hypertension associated loci mapped (reviewed by Franceschini et al. 2011; Table 18.1). These genomic risks are important in clinical diagnosis and management.

  The monogenic forms of hypertension (autosomal dominant or recessive inheritance) usually manifested at an early age (children or young adult) in families with a history of moderate to severe hypertension, and may be associated with underlying clinical syndromes such as Bilginturan's disease (autosomal-dominant hypertension with brachydactyly) (Bahring et al. 2008). Hypertension can also be exacerbated during pregnancy (Rafestin-Oblin et al. 2003; Geller et al. 2000). A suppressed rennin–angiotensin–aldosterone system is believed to the common pathophysiological basis for all hereditary forms of hypertension (reviewed by Simonetti et al. 2011). Genes involved in the renin–angiontensin system or regulatory pathways, such as angiotensin receptor 1 (*AGTR1*) (Gribouval et al. 2005) and prostaglandin 12 synthase (*PTGIS*) (Nakayama et al. 2003), can be the candidate genes for the monogenic form of hypertension (Simonetti et al. 2011). Molecular testing of the monogenic forms of hypertension might be particular useful in young victims or death during pregnancy when related to hypertension.

- *Deep venous thrombosis and pulmonary embolism (PE).* Fatal PE is typically seen as a massive and acute embolus blocking central vascular zones and results in hemodynamic compromise. Fatal PE often is associated with deep venous thromboembolism (VTE). VTE is also a multifactorial disease and accounts for approximately 100,000 deaths annually in the United States (Heit et al. 1999).

  Most deaths are associated with the three predisposing factors described by Virchow: stasis, hypercoagulability, and vascular injury (Kumar et al. 2010). Hypercoagulability risk factors include the thrombophilias, that is, diseases or conditions associated with an increased thrombotic risk. The heritable thrombophilias include mutations in factor V Leiden and prothrombin, deficiencies of antithrombin, proteins C and S, among other rare factors in the coagulation cascade (reviewed by Favaloro and Lippi 2011).

  Ethnicity also plays a significant role in fatal PE. The study of a large number of PE incidents in New York City revealed that PE victims are predominately Blacks and died at an earlier age when compared to Whites. Consistent with other studies, Blacks were found less likely to be carriers of the common thrombophilia variants found in Whites (Tang et al. 2011). Future research is warranted to understand these unique ethnic aspects of fatal PE.

- *Aortic aneurysm rupture.* Typically an elderly individual, often with common risk factors such as coronary artery atherosclerosis or hypertension, complains of severe abdominal pain culminating in collapse from catastrophic hemorrhage, which is easily observable at autopsy.

  Aortic dissection may also occur in individuals with genetic disorders of collagen formation such as Marfan syndrome (McKusick 1991) or Ehlers–Danlos syndrome type IV (vascular type; Beighton et al. 1997) (Table 18.1). Postmortem molecular testing for the genes causing these syndromes (*FBN1* [Dietz et al. 1991] and *COL3A1* [Schwarze et al. 1997]) can be useful in determining the underlying CoD.

  Recently, a genome-wide risk for abdominal aortic aneurysm has been mapped to 9p21.3, where an "A" allele of rs7025486 (within the *DAB2IP* gene) was also found to be associated with, in addition to abdominal aortic aneurysm, myocardial infarction, peripheral arterial disease, and PE (Gretarsdottir et al. 2010).

- *Cardiomyopathy.* Common types of the cardiomyopathies seen at autopsy, including dilated, hypertrophic, left ventricular noncompaction, as well as arrhythmogenic right ventricular cardiomyopathy, often have gross and/or microscopically distinct features at autopsy in adults (reviewed by Grant and Evans 2009). Cardiomyopathy is the well-known substrate of cardiac arrhythmia, which when is not reversed immediately may result in SUD. Death can occur in an individual who has no previous clinical symptom or complain and the diagnosis of a specific type of cardiomyopathy is made only at autopsy. In addition, deaths may occur as a result of cardiac arrhythmias before the cardiac myocardium structural changes become apparent at autopsy. This is especially true in infants and children when the heart is still under development, and structural changes are less apparent (Tester and Ackerman 2009). Therefore, cardiomyopathy should not be excluded in autopsy negative infant or childhood deaths.

  The etiologies for dilated cardiomyopathy can be quite diverse: they may occur as a consequence of remote viral myocarditis, chronic exposure to cardiac toxins (derived from medications or from a metabolic disease), or defects of one of the sarcomeric, dystrophin complex or others genes. Hypertrophic cardiomyopathy, on the other hand, is more a genetic disorder due to defects in any of the genes encoding or processing sarcomere proteins. It is characterized by disarray of myocardial fibers on microscopic examination of heart muscle. Pathologically, arrythmogenic right ventricular cardiomyopathy (ARVC) shows inflammatory and fibrofatty infiltration of the right ventricle. A few genes encoding the desmosomal proteins are responsible for ARVC. The pathogenesis of the common cardiomyopathies has been reviewed in detail by Watkins et al. (2011) (Table 18.1).

At the molecular level, all types of cardiomyopathies are genetically heterogeneous: multiple genes (Table 18.1) are responsible for each type, and there are many different rare variants within each gene. Furthermore, different types of cardiomyopathies may have similar molecular bases. For instance, the *MYH7* gene is responsible for all three types of cardiomyopathies: hypertrophic, dilated, and left ventricular noncompaction (Klaassen et al. 2008). Cardiomyopathy can be a disease alone or part of a genetic syndrome, such as Emery Dreifuss dystrophy (Bonne et al. 2000) and LEOPARD syndrome (multiple lentigines, electrocardiographic conduction abnormalities, ocular hypertelorism, pulmonic stenosis, abnormal genitalia, retardation of growth, and sensorineural deafness) (Carvajal-Vergara et al. 2010). In addition, genotype–phenotype correlation in cardiomyopathies is complicated by reduced penetrance, variable expressivity, genetic and allelic heterogeneity, as well as genetic and environmental modifier effects. Nevertheless, molecular testing of candidate genes responsible for cardiomyopathies are especially useful when the structure changes are subtle and do not meet the classic morphological diagnosis.
- *Genome-wide risk of sudden cardiac death.* A meta-analysis of five genome-wide association studies, which involved 1283 SCDs, identified the BAZ2B locus as associated with SCD (Arking et al. 2011). Additional validation is needed to confirm the role of this locus in SCD in individuals other than European decedents.

2. *Other Systems*
- *Intracranial hemorrhage.* Sudden and unexpected death attributable to the central nervous system most commonly occurs as a consequence of intracranial hemorrhage, often in individuals with hypertension. Hemorrhage into the subarachnoid space most commonly occurs from rupture of an aneurysm within the circle of Willis at the base of the brain. Genetic and genomic risks for hypertension described above, therefore, can contribute to intracranial hemorrhage. In addition, genomic risks responsible for intracranial berry aneurysm were mapped to a handful of loci (Table 18.1). The significance of these loci in intracranial hemorrhage is to be determined.
- *Diabetes mellitus.* Diabetes and mellitus can lead to SUD likely to be the consequence of uncontrolled blood sugar level. Often, in retrospect, a young individual may have had a history of polydipsia and polyuria and been vaguely unwell before death ensued (Secrest et al. 2011). Postmortem toxicological examination can reveal above normal glucose concentration within the vitreous humor of the eye. Microscopic examination of the kidney may show vacuolization of the collecting ducts.

Several loci are responsible for the monogenic forms of the diabetes (reviewed by Steck and Winter 2011; Table 18.1). These are the genes that control either the production or release of insulin. There are two major classifications of monogenic diabetes: maturity-onset diabetes of the young and neonatal diabetes mellitus (NDM) (Steck and Winter 2011). Testing for these loci can be particular helpful in determining the status of the disease in decedents who were not previously clinically diagnosed.

- *Others.* Other less frequent yet broad group of diseases can also be the causes of SUD. These diseases may include metabolic diseases that are not on newborn screening panels (e.g., mitochondrial diseases) or chromosomal abnormalities that not detected by the prenatal karyotyping. Diagnosis of these diseases can be very challenging during postmortem investigation because the autopsy findings are often subtle or nonspecific. Consider consulting a clinical geneticist, molecular or cytogenetic specialist when needed.

18.2.1.3 Negative Autopsy

Sudden unexplained death is sudden death in apparently healthy individuals (newborn through adult), which remains unexplained after a thorough investigation including a complete autopsy, examination of the death scene, and review of clinical history. The genetic risks of SUDs have been studied from several different perspectives, with the majority of these studies focused on cardiac conduction and the central nervous system, which are discussed in this section.

1. *Cardiac conduction system.* Irregular heart rhythms that degenerate into ventricular fibrillation can be fatal if not reversed immediately. It is estimated that the 10%–15% of sudden unexplained deaths in infants (Arnestad et al. 2007; Ackerman et al. 2001; Millat et al. 2009; Chung et al. 2007) and up to 35% in children and adults (Tester and Ackerman 2009) can be attributable to a group of heritable cardiac arrhythmia syndromes, collectively known as channelopathies (i.e., cardiac ion channel diseases), including long QT syndrome, Brugada syndrome, short QT syndrome, catecholaminergic polymorphic ventricular tachycardia, and familiar ventricular fibrillation (reviewed by Cerrone and Priori 2011; Table 18.1). Each of the syndromes is typically associated with a characteristic change in electrocardiograms (reviewed by Cerrone and Priori 2011). At the molecular level, channelopathies are caused either by abnormal functions of the cardiac ion channels (Na^+, K^+, and Ca^{2+}), or protein processing. Genes encoding for most of the cardiac ion channels have been mapped and cloned (Cerrone and Priori 2011). Postmortem molecular testing for these genes can be the only way of making the diagnosis because electrocardiogram or functional cardiac conduction system evaluation is not feasible at autopsy. Although syncope is a primary symptom in some heritable channelopathies, only about half of patients exhibit symptoms. Therefore, channelopathies are often undetected before sudden death in apparently healthy individuals who have structurally normal hearts.
2. *Central nervous system.* Medical examiners are aware of deaths in individuals with epilepsy in the absence of morphological abnormalities. SUDEP has been extensively studied, yet the causes remained largely unknown. It is thought that various and overlapping pathophysiologic events may contribute to sudden death in epilepsy.

Respiratory events (e.g., airway obstruction) or cardiac arrhythmia leading to cardiac arrest may play an important role (Tu et al. 2011). Recently, molecular links to several genes in SUDEP have been suggested, including hyperpolarization-activated cyclic-nucleotide gated cation (HCN1–4) channels (Tu et al. 2011), Kv1.1 Shaker-like potassium channels gene *Kcna1* (Tu et al. 2011), cardiac ion channel genes *SCN5A* (Aurlien et al. 2009; Tu et al. 2011) and *KCNH2* (Tu et al. 2011). Continued research efforts are expected to lead additional SUDEP gene identification.

18.3 Postmortem Toxicogenetics

18.3.1 Toxicogenetics and Medico-Legal Death Investigation

With new research tools such as next generation sequencing and bioinformatics, the amount of information available from forensic sample types has multiplied. New directions in forensic research promises near-future use of previously unavailable biological data from crime scene and mortuary samples. The new biological analysis matrices include relative age of the sample donor (Zubakov et al. 2010; Ou et al. 2012), geographical origin of the sample donor (Kersbergen et al. 2009), cellular origin of the sample (Hanson et al. 2009), external visible traits of the sample donor (Liu et al. 2010), tissue type determination by expression analysis (Visser et al. 2011), as well as data relating to microbial forensics (Budowle et al. 2008). Similarly, the same genomic advances are of utmost importance in the future for medico-legal autopsy practice, where these new directions lead to molecular autopsies (Tester et al. 2012) in a multidisciplinary CoD investigation utilizing knowledge from pathology, toxicology and genetics. This section describes how forensic genetics can be integrated to forensic toxicology and forensic pathology to form postmortem toxicogenetics, and thus aid the medico-legal CoD investigations in the future.

18.3.2 Development of Pharmacogenetic Concept to Pharmacogenomics

The concepts of pharmacogenetics are generally considered to be the observations in the 1950s, although earlier documents of the idea of the individual differences in the response to ingested substances can be traced at least to the time of Pythagoras (ca. 510 B.C.). First, Hockwald et al. (1952) found that the antimalarial drug primaquine causes intravascular hemolysis, specifically among the African-Americans, but rarely among Caucasians. At about same time, the interindividuals differences response to tuberculosis drug, isoniazid, was studied in humans. After Carson et al. (1956) described the basis for the above-mentioned primaquine-induced intravascular hemolysis (deficiency of the red cell enzyme glucose-6-phosphate dehydrogenase), and Lehman and Ryan (1956) and Kalow and Staron (1957) described inherited variation in the neuromuscular paralysis time after succinylcholine administration (deficiency in the metabolizing enzyme pseudocholinesterase), Motulsky (1957) summarized the concept of a genetic basis the phenomenon of inherited variation in response to drug treatment. The term "pharmacogenetics" was then coined, and referred at that time to monogenic traits. In the 1970s, the variability in the drug response was first time uncovered using molecular genetics methods. The disadvantageous effects of debrisoquine (Mahgoub et al. 1977) and spartaine (Eichelbaum et al. 1979) to the patient were shown to be due to cytochrome P450 (CYP) monooxygenase variation, and CYP2D6 was characterized in 1988 as the first polymorphic gene affecting the drug response (Gonzalez et al. 1988).

It is now known that the drug metabolizing enzymes are part of the xenobiotic metabolism system, which protects us against the harmful effects of foreign compounds introduced to our body. The metabolism of xenobiotics can be divided into three concerted phases with the aim of detoxifying and removing foreign substances from our cells: modification (phase I), conjugation (phase II), and excretion (phase III). In phase I, the functional groups of the foreign compounds are modified by variety of enzymes. The most common, and largely studied is the CYP family enzymes (such as CYP2D6, CYP2C9, CYP2C19). In the subsequent phase II reaction enzymes conjugate the modified xonobiotics with enzymes (e.g., acetate, sulfate, glutathione, and glucuronic acid) in a reaction that is catalyzed by transferases (e.g., glutathione *S*-transferases, *N*-acetyltransferases [NAT1 and NAT2], thiopurine methyltransferase). After phase II reactions, the xenobiotic conjugates and their metabolites are excreted from cells. This phase III metabolism includes the multidrug resistance protein (MPR) family, which is part of the ATP-binding cassette transporters, which can catalyze the ATP-dependent transporters and thus remove phase II products to the extracellular matrix to be excreted from the body (Figure 18.1).

The development of genomics and bioinformatics has progressed the straightforward ideas of PGt to a new complex level that is currently better described as pharmacogenomics (Wang et al. 2011).

Because of the great interest and importance of the emerging field(s), the U.S. Department of Health and Human Services, Food and Drug Administration have released the Definitions for Genomic Biomarkers, Pharmacogenomics, Pharmacogenetics, Genomic Data and Sample Coding Categories, defining *pharmacogenomics* (PGx) as the study of variations of DNA and RNA characteristics as related to drug response, and *pharmacogenetics* (PGt) as a subset of PGx defined as the study of variations in DNA sequence as related to drug response.

Essentially, PGt and PGx research focuses on understanding the underlying inherited molecular factors in individual variation in drug, and offer a molecular-level explanation to medical practitioners' notion of "responders" and "nonresponders." Juxtaposing pharmacology with genetics, PGt and PGx now form—together with pharmacokinetics and pharmacodynamics—the basis of clinical pharmacology. The ultimate objective of PGt and PGx is to aid clinicians to prescribe the most effective and safe, individually tailored drug treatment for the patient, and to minimize the adverse effects of those drugs.

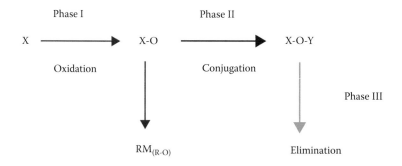

Figure 18.1 Xenobiotic metabolisms. → Phase I, → phase II, → phase III. $RM_{(R-O)}$ is a usually minor reactive metabolite of the oxidation in the phase I reaction, see the text for details.

18.3.3 Adverse Drug Reactions

According to the World Health Organization (WHO) adverse drug reaction (ADR) is defined as an unintended response to a drug occurring at a conventional dose and used for disease in prophylaxis, diagnosis, therapy, or for modification of physiological functions (WHO, factsheet No. 293). ADRs are divided into type A and B reactions (Pirmohamed et al. 1998), and further to type C–F reactions (Edwards and Aronson 2000). These definitions enable medical and pharmaceutical societies a) to compare the ADRs of different drugs and to collect them into a large database (see, e.g., www.who-umc.org/).

Although the efficacy of a drug is scrutinized in a clinical trial, more than 50% of ADRs are detected after marketing approval (Moore et al. 1998). In the clinical setting, ADRs are difficult to identify and report. The number of reported ADRs depends on several factors, such as type of ADRs (mild, severe, fatal) and reporting institution (general practice, general hospital, university hospital), which can affect the data generated, and thus the data can remain underreported. Studies on pharmacovigilance, defined by WHO as a science and activities relating to the detection, assessment, understanding and prevention of adverse effects or any other drug-related problem, and pharmacoepidemiology, defined as nonexperimental study of the use and effects of drugs in large number of people (International Society of Pharmacoepidemiology, www.pharmacoepi.org; Launiainen et al. 2011), offer a means of detecting ARDs. Lazarou et al. (1998) found that about 100,000 deaths of U.S. hospital inpatients occur per year because of ADRs, and calculated an overall incidence of fatal ADRs to be 0.32% among U.S. hospital inpatients. This figure would make ADRs one of the leading CoD in the United States. In a prospective Scandinavian study (Ebbesen et al. 2001) with 13,992 patients of internal medicine, the incidence of lethal ADR was estimated to be 0.95%, and Pirmohamed et al. (2004) observed a 6.5% ADR rate for hospital admissions in 18,820 patients, and 0.15%–2.3% ADR lethality, both of the figures being in gross agreement with the estimates from United States. As a conclusion, it is clear that ADR-related deaths and the possibilities of using PGt and PGx methods in investigation of those are of great medico-legal interest (Musshoff et al. 2010; Sajantila et al. 2010).

18.3.4 Medico-Legal Investigation of Death

In a case of a (suspected) unnatural death (i.e., death not due to a disease) a medico-legal autopsy is often the method of choice in determining the CoD and MoD. The WHO has defined CoD as the (a) disease or trauma that initiated the train of morbid events leading directly to death or (b) the circumstances of the accident of violence that produced the fatal injury, and MoD as the way or circumstances that led to the CoD.

Most countries classify and tabulate CoD data according to the WHO's standard procedure (International Statistical Classification of Diseases and Related Health Problems [ICD] 1993). CoD is classified using disease codes and external (E codes) and Injury (I codes), whereas MoD is discretely classified as natural, occupational, accidental, suicide, homicide, or war. Standard procedures for CoDs and MoDs are the basis for international comparative statistics (Lahti 2005).

"Autopsy negative" cases (Cohle and Sampson 2001; Corrado et al. 2001) where traditional autopsy is not sufficient to determine the CoD and/or MoD, are considered of special nature in the medico-legal field. In addition to a thorough description of macroscopical findings, further auxiliary tests (histology, toxicology, biochemistry, imaging, etc.)

are requested. Examples of autopsy negative cases are the different types of SUDs such as SCDs, SUDEP, sudden infant death syndrome, or untypical drug-related deaths. Series of negative cases are abundantly reported, and the incidence varies depending on the institution, population, and particularly among age groups.

One of the most widely used and important auxiliary analyses is forensic toxicology, which is derived from the toxicological study of the adverse effects of drugs and chemicals on humans, and bases its practice in published and widely accepted scientific methods. The objective of forensic toxicology is to analyze drugs, poisons, and any toxic agents in living and deceased individuals for legal purposes. An important part of forensic toxicology, and its most essential duty in CoD investigation, is interpreting the results in a scientific manner.

Academic research is an inherent part of forensic toxicology whose aims are to develop and evaluate new analytical methods and instrumentation (Ojanperä et al. 2012), to study postmortem effects (e.g., postmortem redistribution of drugs) (Pounder and Jones 1990) and postmortem storage and handling of samples, and to interpret the results in the medico-legal context (Koski 2005).

As in other forensic disciplines, forensic toxicology uses methods derived from innovations in clinical medicine and academic laboratories (e.g., analytical chemistry, pharmacology, and clinical chemistry). Forensic toxicology can be further divided into subdisciplines, such as human performance toxicology (e.g., effects of alcohol and drugs in traffic, alcohol- and drug-facilitated crimes), doping control, workplace testing, and postmortem toxicology (i.e., death investigation toxicology), each having its own specific topics and questions.

Numerous factors contribute to a successful drug therapy, and the same issues are also involved when considering interpretation of forensic toxicology results. These include patient's compliance, adherence, external factors, diseases, and genetics. Patients' behavioral habits may be of surprising medico-legal importance; for example, borrowing and sharing of medication (Petersen et al. 2008), co-use of prescription drugs with over-the-counter drugs, abuse of prescribed drugs (Zaracostas 2007), and enhancement of the effect of illicit drugs with prescription drugs (Vuori et al. 2003) are common knowledge for experienced forensic pathologists and forensic toxicologists.

Forensic toxicology results are particularly important in the investigation of CoD and MoD in cases of intoxication, particularly in the evaluation of suicides by alcohol, drugs, or their combination (Koski et al. 2003). The data from drug screening is also important in assessment of clinical maltreatment and medical negligence (Madea et al. 2009).

It is important to note that polypharmacy is a common phenomenon in modern society, and this is also reflected in forensic autopsy material. Therefore, it is important that data on drug–drug interactions from clinical data are to be translated into the forensic toxicology case-work interpretation (Launiainen et al. 2011) and can be mirrored against the data obtained from large forensic toxicology databases (Launiainen et al. 2009). Conversely, new insights can be revealed from forensic data (Launiainen et al. 2010), which can be used as an information basis for drug safety.

18.3.4.1 *Postmortem Toxicogenetics and CoD Investigation*
Although CoD and MoD of some ambiguous deaths cannot be unequivocally established via traditional autopsy, there are newly developed molecular approaches that may solve some of the SUDs. For application of genetic testing the mortuary work, the term "molecular autopsy" has been coined (Tester et al. 2012). The use of molecular autopsies for autopsy negative cases are obvious: first, genetic analyses are likely to reduce the number

of undetermined CoDs and MoDs, and second, the mechanisms underlying such deaths can be better understood, and third, genetic information on risks for sudden, unexpected death or for fatal ADR are of utmost importance to the relatives.

The data from clinical PGt (i.e., individualize drug treatment by facilitating the choice of the proper drug and dose for each individual based on genetic information) can be expected to be translated to the postmortem setting to aid in the determination of CoD and MoD. Indeed, demonstrative cases have been published, some promising research projects have been published, and few academic dissertations have been defended in this field. The promise of PGx in the medico-legal context has also stimulated discussion of individualized rights (Wong et al. 2010).

The cost-effective use of toxicogenetics in the medico-legal investigation of death requires integrating research in forensic pathology, toxicology, and genetics. The data points to be collected in a medico-legal autopsy for systematic toxicogenetic studies should include relevant medical history and other background information combined with detailed knowledge on the pathophysiological of the diseased, concentrations of all drugs and their relevant metabolites in the body at the time of death, and genotype data from the gene pathway related to the drugs found. Attempts for such studies have been published, particularly drugs or case-control studies (cf. Levo et al. 2003; Koski et al. 2006).

All the data collected in this manner should be considered along with the fact that an individual's pathophysiological phenotype affecting drug efficacy (the ability of a drug to produce the desired therapeutic effect) depends to some degree on the genetic constitution of an individual and several other factors. These additional factors include developmental stage, physiological and environmental factors, and diseases or specific conditions (Table 18.2).

Deducing phenotype from the genotype is crucial for clinical PGx, and some guidelines already exist for genotype–drug dosage relationship (Kirchheiner et al. 2004; Kirchheiner 2008). Before large-scale genotyping possibilities, phenotype from an individual was

Table 18.2 Factors Affecting an Individual's Pathophysiological Phenotype Related to Drug Efficacy

Genetic variation
  Gene deletions, gene duplications
  Sequence variation
Developmental stage
  Age
  Gender
Physiological factors
  Mental and physical stress
  Hormonal changes
  Periodical changes (seasonal and circadian factors)
Environmental factors
  Personal environmental history
  Lifestyle
  Diet
  Exposure to environmental toxins
  Concomitant use of alcohol(s) and drug(s)
Potential, specific associations
  Diabetes, obesity, gut microbiology

obtained by urine analysis one drug at a time, and calculating the "metabolic ratio" equaling to the metabolic capacity for the drugs, one by one (Dahl et al. 1995). The impact of genetic variation among patients to drug efficacy is now better known, and prediction of the phenotype from genotype is of a significant potential for individualized drug therapy. However, even with the new analytical methods and easily accessible genotype data, the probability of prediction of the phenotype from the genotype has challenges. The *CYP2D6* gene is the best characterized example, with four phenotypic classes, which can be deduced from genotypes: individuals, who lack the functional enzyme are poor metabolizers (PMs); carriers of two decreased-function variants or a combination of one decreased-function variant and one nonfunctional variant are intermediate metabolizers; individuals, who have at least one fully functional variant are extensive metabolizers; and finally, the carriers of active gene duplication or multiplication or another mutation increasing the enzyme activity inherited together with a functional variant are ultrarapid metabolizers (UMs).

Postmortem studies likely will prove useful in understanding the genotype–phenotype relationship, for determining CoD and MoD in ambiguous medico-legal cases, and for contributing to applications in the clinical or personalized medicine setting.

Case examples. Medico-legal cases are often a source for problems and thus for new research. The following are shortly described few selected cases, which are of considerable medico-legal interest. The examples are related to psychiatric disorders, use of analgesics, and suicides, which constitute a large proportion of medico-legal autopsies cases in the western world.

18.3.4.1.1 Poor Metabolizing Genotypes and Intoxication

Sallee et al. (2000) showed in their report how a genetically determined poor drug metabolism led to fatal drug intoxication in the case a 9-year-old boy, who had behavioral problems and died of fluoxetine intoxication. A high concentration of fluoxetine and its major active metabolite norfluoxetine was found in forensic toxicological analysis, which led to a further investigation of the case. Toxicogenetic analysis of the case revealed that the child had a completely defective *CYP2D6* gene resulting in a poor ability to metabolize fluoxetine.

Druid et al. (1999) and Levo et al. (2003) showed that *CYP* genotyping was feasible in postmortem sample material, and that genetic variation correlated with the observed phenotype, that is, parent drug/metabolite ratio. Interestingly, in two studies on fatal drug intoxications, the PM subjects were found to be underrepresented when compared to the general population, but no clear explanation has been offered for this observation. *CYP* genotyping and PM has been used to aid interpretation of postmortem toxicology results in oxycodone (Jannetto et al. 2002), methadone (Wong et al. 2003), and fentanyl-related deaths (Jin et al. 2005).

18.3.4.1.2 Ultrarapid Metabolizing Genotypes and Intoxication

UM drug metabolism can be associated also with severe or fatal ADRs, if the enzyme catalyzes the conversion of a prodrug into an active compound. Three case reports involving CYP2D6 and codeine have recently been described. Koren et al. (2006) reported a case, where a neonate was found dead at the age of 13 days. Postmortem analysis revealed that the CoD of the child was morphine intoxication. The newborn received the morphine via breast milk from the mother who had been prescribed codeine for episiotomy pain. Codeine is O-demethylated to morphine in a reaction catalyzed by CYP2D6. After the child's death, it was discovered that the mother carried an active *CYP2D6* gene duplication associated with increased metabolism of codeine to morphine, which was lethal to the neonate. This case illustrates well that without pharmacogenetic testing and careful medico-legal consideration, the case could have been interpreted as an infanticide or medical misconduct/negligence. Whereas the CoD was morphine poisoning, the MoD was demonstrably accidental.

The CYP2D6 genotyping has been used interestingly to analyze cases with MoD classified as suicide. From a previous study (Zackrisson et al. 2010), it was known that the frequency of *CYP2D6* gene duplication was higher than in normal populations, indicating UM metabolism, and Ahlner et al. (2010) analyzed two medico-legal autopsy groups: (violent) suicides and natural deaths. They found that the *CYP2D6* gene with more than two functional alleles occurred more than 10-fold greater in the suicide group, when compared with the group of natural deaths.

In addition to CYP genes, cases possibly explained by transporter genes (Neuvonen et al. 2011) may prove to be interesting for medico-legal purposes.

The above-described cases demonstrate that applying pharmacogenetic concepts can have tremendous clinical and medico-legal interest. Toxicogenetic approach should also ultimately add important data for drug safety and translation of this knowledge to bedside medicine.

18.4 Microgenomics

Microgenomics is considered to be the field of quantitative molecular analysis of macromolecules obtained from a single or small amount of isolated cell, which were isolated, collected, and examined according to precise micromanipulating techniques, such as laser microdissection techniques (LMTs). LMT refers to a marriage of existing light microscopic instrumentation to newer technology utilizing pulsed laser beams.

Two procedures are the essential parts of LMT: laser cutting and capture. In 1996, Emmert-Buck's group (Simone et al. 1998) developed the first laser microdissection device, described by its inventors as a "reliable method to procure pure populations of cells from specific microscopic regions of tissue sections, in one step, under direct visualization." The device has since then found extensive applications in both basic and applied sciences. The general principle of LMT is that, by assistance of regular or inverted microscope, finely focused laser beam (up to one micrometer) of certain wavelength literally cuts and assists in the collection of a cluster or single cell(s) from routinely and/or especially prepared histology slides. Constructional use of different microscope systems as a part of LMTs depends on the material collection method. An orthodox upright microscope is used for gravity-based material collection, as opposed to an inverted microscope, which is used for material pickup by an adhesive polymer or collection based on the force of laser-induced energetic pulse (catapulting). Laser systems used for LMTs are infrared (IR; for laser capture microdissection) or ultraviolet (UV; for laser cutting). A UV beam destroys material where it cuts, whereas an IR laser is not harmful to nearby tissue components (Figures 18.2, 18.3, and 18.4). Based on combination of the tissue collection principle and the type of laser used, LMTs include microisolation devices based on the principles of gravity, adhesion, or catapulting. Adhesive polymer, ethylene vinyl acetate thermoplastic film is used in precise apposition to the tissue specimen, capturing cellular material after controlled cutting by contact between the specimen and the clear plastic cap where it is located. The weakness of this system is that, because of the presence of physical contact between tissues and the capturing device, contamination is not absolutely avoidable. In the case of propelling (laser microdissection pressure catapulting) system—P.A.L.M., there is no contact

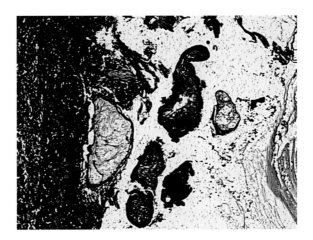

Figure 18.2 FFPET (formalin fixed paraffin embedded tissues) slide of POC before tissue cutting and capturing. H&E staining; magnification 10×; Arcturus-Veritas Laser Capture Microscope (IR laser).

between specimen and collection device. The sample is literally forced up for a couple of micrometers into the collecting device. This principle is based on the energy of the virtual cloud formed by UV laser of slightly changed focus distance (Micke et al. 2005).

The manner of specimen preparation has to be contingent with the nature of the biological material to be analyzed (Burgemeister 2005; Di Francesco et al. 2000; Ehrig et al. 2001; Espina et al. 2006). If it is possible (which in the case of applied forensics is very often not the case), the material of interest should be collected on a chemically inert supporting synthetic polymeric membrane—polyethylene terephthalate (PET) or polyethylene naphthalate (PEN)—that may be mounted onto a glass slide or a metallic frame. In this way, during the laser cutting of an area containing a discrete cell type, a portion of supporting membrane is collected together with the sample.

An adequate slide preparation and cellular morphology preservation is critical to correctly distinguish different cell types (Pinzani et al. 2006). To be able to visualize elements

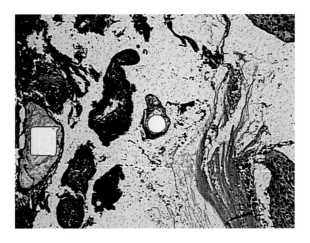

Figure 18.3 FFPET (formalin fixed paraffin embedded tissues) slide of POC after tissue cutting and capturing. H&E staining; magnification 10×; Arcturus-Veritas® Laser Capture Microscope (IR laser).

Molecular Autopsy

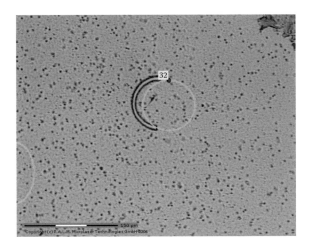

Figure 18.4 Sperm cell mounted on PEN (polyethylene naphthalate) membrane slide to be isolated by UV laser. NFR staining; magnification 20; PALM-Zeiss®.

of interest, different staining techniques are recommended, including immunofluorescence. Staining procedures are to be chosen from those which would not interfere with laboratory procedures that are to follow physical cell isolation (DNA extraction, quantitation, amplification, etc.) (Titford 2005).

Substrates for forensic LMT separation have included the following: products of conception from formalin fixed paraffin embedded tissues for paternity testing in criminal cases (Bauer et al. 2002; Budimlija et al. 2005; Robino et al. 2006), trace cellular material from debris, individual telogen hairs, as well as various cellular mixtures, especially epithelial versus sperm cells (or male vs. female cells), obtained after sexual assaults (Allery et al. 2001; Anslinger et al. 2005; Bienvenue et al. 2006; Collins et al. 1994; Di Martino et al. 2004; Elliott et al. 2003; Jobling et al. 1997; Johnson et al. 2005; Murray et al. 2006; Seidl et al. 2005; Soares-Viera et al. 2007; Vandewoestyne et al. 2009). In addition, forensic LMT has been successfully used in quality control issues in clinical pathology laboratories (cases of possible tissue contamination and mix-ups, malignant tissue introduction by a transplanted organ[s], etc.) (Popiolek et al. 2003, 2006).

18.5 Conclusion

Although molecular testing is not yet commonly performed at autopsy, it is expected to become an integrated component in forensic investigations in the near future. A multidisciplinary team approach, involving medical examiners, scene investigators, epidemiologists, molecular geneticists, research scientists, and clinicians, has allowed us to gain an understanding of the genetic and genomic causes of sudden natural death. Such collaborative work will continue to be the theme of sudden natural death investigations and research.

Important issues need to be considered when performing genetic testing, including family consent, privacy, emotional stress, uncertainty of the clinical effects of a unreported or undercharacterized genetic variant, as well as the possibility of uncovering a previously unrecognized nonpaternity. Recognizing some of these important issues, the

Genetic Information Nondiscrimination Act was signed into law in the United States in 2009 (Asmonga 2008), which protects individuals from the misuse of genetic information in health insurance and employment and protects the individual's privacy and rights. Furthermore, genetic counseling is expected to play an important role in postmortem molecular testing and family consultation.

References

Ackerman, M.J., B.L. Siu, W.Q. Sturner et al. 2001. Postmortem molecular analysis of SCN5A defects in sudden infant death syndrome. *JAMA* 286(18):2264–9.

Ahlner, J., A.L. Zackrisson, B. Lindblom, and L. Bertilsson. 2010. CYP2D6, serotonin and suicide. *Pharmacogenomics* 11(7):903–5.

Akagawa, H., A. Tajima, Y. Sakamoto et al. 2006. A haplotype spanning two genes, ELN and LIMK1, decreases their transcripts and confers susceptibility to intracranial aneurysms. *Hum Mol Genet* 15(10):1722–34.

Alders, M., T.T. Koopmann, I. Christiaans et al. 2009. Haplotype-sharing analysis implicates chromosome 7q36 harboring DPP6 in familial idiopathic ventricular fibrillation. *Am J Hum Genet* 84(4):468–76.

Allery, J.P., N. Telmon, R. Mieusset, A. Blanc, and D. Rougé. 2001. Cytological detection of spermatozoa: Comparison of three staining methods. *J Forensic Sci* 46(2):349–51.

Anderson, R.E., R.B. Hill, D.W. Broudy, C.R. Key, and D. Pathak. 1994. A population-based autopsy study of sudden, unexpected deaths from natural causes among persons 5 to 39 years old during a 12-year period. *Hum Pathol* 25(12):1332–40.

Angius, A., E. Petretto, G.B. Maestrale et al. 2002. A new essential hypertension susceptibility locus on chromosome 2p24–p25, detected by genomewide search. *Am J Hum Genet* 71(4):893–905.

Anslinger, K., B. Mack, B. Bayer, B. Rolf, and W. Eisenmenger. 2005. Digoxigenin labelling and laser capture microdissection of male cells. *Int J Legal Med* 119(6):374–7.

Antzelevitch, C., G.D. Pollevick, J.M. Cordeiro et al. 2007. Loss-of-function mutations in the cardiac calcium channel underlie a new clinical entity characterized by ST-segment elevation, short QT intervals, and sudden cardiac death. *Circulation* 115(4):442–9.

Aouizerat, B.E., E. Vittinghoff, S.L. Musone et al. 2011. GWAS for discovery and replication of genetic loci associated with sudden cardiac arrest in patients with coronary artery disease. *BMC Cardiovasc Disord* 11:29.

Arking, D.E., M.J. Junttila, P. Goyette et al. 2011. Identification of a sudden cardiac death susceptibility locus at 2q24.2 through genome-wide association in European ancestry individuals. *PLoS Genet* 7(6):e1002158.

Arnestad, M., L. Crotti, T.O. Rognum et al. 2007. Prevalence of long-QT syndrome gene variants in sudden infant death syndrome. *Circulation* 115(3):361–7.

Asimaki, A., P. Syrris, T. Wichter, P. Matthias, J.E. Saffitz, and W.J. McKenna. 2007. A novel dominant mutation in plakoglobin causes arrhythmogenic right ventricular cardiomyopathy. *Am J Hum Genet* 81(5):964–73.

Asmonga, D. 2008. Getting to know GINA. An overview of the Genetic Information Nondiscrimination Act. *JAHIMA* 79(7):18, 20, 2.

Aurlien, D., T.P. Leren, E. Tauboll, and L. Gjerstad. 2009. New SCN5A mutation in a SUDEP victim with idiopathic epilepsy. *Seizure* 18(2):158–60.

Bahring, S., M. Kann, Y. Neuenfeld et al. 2008. Inversion region for hypertension and brachydactyly on chromosome 12p features multiple splicing and noncoding RNA. *Hypertension* 51(2):426–31.

Barsheshet, A., A. Brenyo, A.J. Moss, and I. Goldenberg. 2011. Genetics of sudden cardiac death. *Curr Cardiol Rep* Jul 26.

Bauer, M., A. Thalheimer, and D. Patzelt. 2002. Paternity testing after pregnancy termination using laser microdissection of chorionic villi. *Int J Legal Med* 116:39–42.

Beighton, P., A. De Paepe, B. Steinmann, P. Tsipouras, and R.J. Wenstrup. 1997. Ehlers–Danlos syndromes: Revised nosology, Villefranche, Ehlers–Danlos National Foundation (USA) and Ehlers–Danlos Support Group (U.K.). *Am J Med Genet* 28;77(1):31–7.

Bellocq, C., A.C. van Ginneken, C.R. Bezzina et al. 2004. Mutation in the KCNQ1 gene leading to the short QT-interval syndrome. *Circulation* 109(20):2394–7.

Berthet, M., I. Denjoy, C. Donger et al. 1999. C-terminal HERG mutations: The role of hypokalemia and a KCNQ1-associated mutation in cardiac event occurrence. *Circulation* 99(11):1464–70.

Bezzina, C.R., R. Pazoki, A. Bardai et al. 2010. Genome-wide association study identifies a susceptibility locus at 21q21 for ventricular fibrillation in acute myocardial infarction. *Nat Genet* 42(8):688–91.

Bienvenue, J.M., N. Duncalf, D. Marchiarullo, J.P. Ferrance, and J.P. Landers. 2006. Microchip-based cell lysis and DNA extraction from sperm cells for application to forensic analysis. *J Forensic Sci* 51(2):266–73.

Bilguvar, K., K. Yasuno, M. Niemela et al. 2008. Susceptibility loci for intracranial aneurysm in European and Japanese populations. *Nat Genet* 40(12):1472–7.

Bleyl, S.B., B.R. Mumford, M.C. Brown-Harrison et al. 1997. Xq28-linked noncompaction of the left ventricular myocardium: Prenatal diagnosis and pathologic analysis of affected individuals. *Am J Med Genet* 31;72(3):257–65.

Bonne, G., E. Mercuri, A. Muchir et al. 2000. Clinical and molecular genetic spectrum of autosomal dominant Emery–Dreifuss muscular dystrophy due to mutations of the lamin A/C gene. *Ann Neurol* 48(2):170–80.

Borowiec, M., C.W. Liew, R. Thompson et al. 2009. Mutations at the BLK locus linked to maturity onset diabetes of the young and beta-cell dysfunction. *Proc Natl Acad Sci U S A* 25;106(34):14460–5.

Budimlija, Z.M., M. Lechpammer, D. Popiolek, F. Fogt, M. Prinz, and F.R. Bieber. 2005. Forensic applications of laser capture microdissection: Use in DNA-based parentage testing and platform validation. *Croat Med J* 46(4):549–55.

Budowle, B. et al. 2008. Criteria for validation of methods in microbial forensics. *Appl Environ Microbiol* 74(18):5599–607.

Burgemeister, R. 2005. New aspects of laser microdissection in research and routine. *J Histochem Cytochem* 53(3):409–12.

Burke, A.P., A. Farb, Y.H. Liang, J. Smialek, and R. Virmani. 1996. Effect of hypertension and cardiac hypertrophy on coronary artery morphology in sudden cardiac death. *Circulation* 94(12):3138–45.

Burt, V.L., P. Whelton, E.J. Roccella et al. 1995. Prevalence of hypertension in the U.S. adult population. Results from the Third National Health and Nutrition Examination Survey, 1988–1991. *Hypertension* 25(3):305–13.

Carson, P.E., C.L. Flanagan, C.E. Ickes, and A.S. Alving. 1956. Enzymatic deficiency in primaquine-sensitive erythrocytes. *Science* 124:484–5.

Carvajal-Vergara, X., A. Sevilla, S.L. D'Souza et al. 2010. Patient-specific induced pluripotent stem-cell-derived models of LEOPARD syndrome. *Nature* 465(7299):808–12.

Cerrone, M., and S.G. Priori. 2011. Genetics of sudden death: Focus on inherited channelopathies. *Eur Heart J* 32:2109–18.

Chang, Y.P., X. Liu, J.D. Kim et al. 2007. Multiple genes for essential-hypertension susceptibility on chromosome 1q. *Am J Hum Genet* 80(2):253–64.

Charron, P., E. Villard, P. Sebillon et al. 2004. Danon's disease as a cause of hypertrophic cardiomyopathy: A systematic survey. *Heart* 90(8):842–6.

Chen, L., M.L. Marquardt, D.J. Tester, K.J. Sampson, M.J. Ackerman, and R.S. Kass. 2007. Mutation of an A-kinase-anchoring protein causes long-QT syndrome. *Proc Natl Acad Sci U S A* 104(52):20990–5.

Chen, Q., G.E. Kirsch, D. Zhang et al. 1998. Genetic basis and molecular mechanism for idiopathic ventricular fibrillation. *Nature* 392(6673):293–6.

Chung, S.K., J.M. MacCormick, C.H. McCulley et al. 2007. Long QT and Brugada syndrome gene mutations in New Zealand. *Heart Rhythm* 4(10):1306–14.

Cohle, S.D., and B.A. Sampson. 2001. The negative autopsy: Sudden cardiac death or other? *Cardiovasc Pathol* 10:219–22.

Collins, K.A., P.N. Rao, R. Hayworth et al. 1994. Identification of sperm and non-sperm male cells in cervicovaginal smears using fluorescence in situ hybridization: Applications in alleged sexual assault cases. *J Forensic Sci* 39(6):1347–55.

Corrado, D., C. Basso, and G. Thiene. 2001. Sudden cardiac death in young people with apparently normal heart. *Cardiovasc Res* 50:399–408.

Cronk, L.B., B. Ye, T. Kaku et al. 2007. Novel mechanism for sudden infant death syndrome: Persistent late sodium current secondary to mutations in caveolin-3. *Heart Rhythm* 4(2):161–6.

Dahl, M.L., I. Johansson, L. Bertilsson, M. Ingelman-Sundberg, and F. Sjöqvist. 1995. Ultrarapid hydroxylation of debrisoquine in a Swedish population. Analysis of the molecular genetic basis. *J Pharmacol Exp Ther* 274(1):516–20.

Dellefave, L., and E.M. McNally. 2010. The genetics of dilated cardiomyopathy. *Curr Opin Cardiol.*

Delpon, E., J.M. Cordeiro, L. Nunez et al. 2008. Functional effects of KCNE3 mutation and its role in the development of Brugada syndrome. *Circ Arrhythm Electrophysiol* 1(3):209–18.

di Barletta, M.R., S. Viatchenko-Karpinski, A. Nori et al. 2006. Clinical phenotype and functional characterization of CASQ2 mutations associated with catecholaminergic polymorphic ventricular tachycardia. *Circulation* 114(10):1012–9.

Di Francesco, L.M., S.K. Murthy, J. Luider, and D.J. Demetrick. 2000. Laser capture microdissection-guided fluorescence in situ hybridization and flow cytometric cell cycle analysis of purified nuclei from paraffin sections. *Mod Pathol* 13(6):705–11.

Di Martino, D., G. Giuffré, N. Staiti, A. Simone, M. Le Donne, and L. Saravo. 2004. Single sperm cell isolation by laser microdissection. *Forensic Sci Int* 146S:S151–3.

Di Paolo, M., D. Luchini, R. Bloise, and S.G. Priori. 2004. Postmortem molecular analysis in victims of sudden unexplained death. *Am J Forensic Med Pathol* 25(2):182–4.

Dietz, H.C., G.R. Cutting, R.E. Pyeritz et al. 1991. Marfan syndrome caused by a recurrent de novo missense mutation in the fibrillin gene. *Nature* 352(6333):337–9.

Druid, H., P. Holmgren, B. Carlsson, and J. Ahlner. 1999. Cytochrome P450 2D6 (CYP2D6) genotyping on postmortem blood as a supplementary tool for interpretation of forensic toxicological results. *Forensic Sci Int* 99(1):25–34.

Ebbesen, J., I. Buajordet, J. Eriksen et al. 2001. Drug-related deaths in a department of internal medicine. *Arch Intern Med* 161:2317–23.

Edghill, E.L., S.E. Flanagan, A.M. Patch et al. 2008. Insulin mutation screening in 1,044 patients with diabetes: Mutations in the INS gene are a common cause of neonatal diabetes but a rare cause of diabetes diagnosed in childhood or adulthood. *Diabetes* 57(4):1034–42.

Edwards, I.R., and J.K. Aronson. 2000. Adverse drug reactions: Definitions, diagnosis, and management. *Lancet* 356(9237):1255–9.

Ehrig, T., S.A. Abdulkadir, S.M. Dintzis, J. Milbrandt, and M.A. Watson. 2001. Quantitative amplification of genomic DNA from histological tissue sections after staining with nuclear dyes and laser capture microdissection. *J Mol Diagn* 3(1):22–5.

Eichelbaum, M., N. Spannbrucker, and B. Steincke, and H.J. Dengler. 1979. Defective N-oxidation of sparteine in man: A new pharmacogenetic defect. *Eur J Clin Pharmacol* 16:183–7.

Ellard, S., S.E. Flanagan, C.A. Girard et al. 2007. Permanent neonatal diabetes caused by dominant, recessive, or compound heterozygous SUR1 mutations with opposite functional effects. *Am J Hum Genet* 81(2):375–82.

Elliott, K., D.S. Hill, C. Lambert, T.R. Burroughes, and P. Gill. 2003. Use of laser microdissection greatly improves the recovery of DNA from sperm on microscope slides. *Forensic Sci Int* 137(1):28–36.

Espina V., J.D. Wulfkuhle, V.S. Calvert et al. 2006. Laser-capture microdissection. *Nature Protoc* 1:586–603.

Favaloro, E.J., and G. Lipp. 2011. Coagulation update: What's new in hemostasis testing? *Thromb Res* 127(Suppl 2):S13–6.

Franceschini, N., A.P. Reiner, and G. Heiss. 2011. Recent findings in the genetics of blood pressure and hypertension traits. *Am J Hypertens* 24(4):392–400.

Funke-Kaiser, H., F. Reichenberger, K. Kopke et al. 2003. Differential binding of transcription factor E2F-2 to the endothelin-converting enzyme-1b promoter affects blood pressure regulation. *Hum Mol Genet* 12(4):423–33.

Geisterfer-Lowrance, A.A., S. Kass, G. Tanigawa et al. 1990. A molecular basis for familial hypertrophic cardiomyopathy: A beta cardiac myosin heavy chain gene missense mutation. *Cell* 62(5):999–1006.

Geller, D.S., A. Farhi, N. Pinkerton et al. 2000. Activating mineralocorticoid receptor mutation in hypertension exacerbated by pregnancy. *Science* 289(5476):119–23.

George Jr., A.L., T.A. Varkony, H.A. Drabkin et al. 1995. Assignment of the human heart tetrodotoxin-resistant voltage-gated Na+ channel alpha-subunit gene (SCN5A) to band 3p21. *Cytogenet Cell Genet* 68(1–2):67–70.

Gerull, B., A. Heuser, T. Wichter et al. 2004. Mutations in the desmosomal protein plakophilin-2 are common in arrhythmogenic right ventricular cardiomyopathy. *Nat Genet* 36(11):1162–4.

Gill, J.R., and I.A. Scordi-Bello. 2010. Natural, unexpected deaths: Reliability of a presumptive diagnosis. *J Forensic Sci* 55(1):77–81.

Gloyn, A.L., E.R. Pearson, J.F. Antcliff et al. 2004. Activating mutations in the gene encoding the ATP-sensitive potassium-channel subunit Kir6.2 and permanent neonatal diabetes. *N Engl J Med* 350(18):1838–49.

Gong, M., H. Zhang, H. Schulz et al. 2003. Genome-wide linkage reveals a locus for human essential (primary) hypertension on chromosome 12p. *Hum Mol Genet* 12(11):1273–7.

Gonzalez, F.J., R.C. Skoda, S. Kimura et al. 1988. Characterization of the common genetic defect in humans deficient in debrisoquine metabolism. *Nature* 331:442–6.

Grant, E.K., and M.J. Evans. 2009. Cardiac findings in fetal and pediatric autopsies: A five-year retrospective review. *Pediatr Dev Pathol* 12(2):103–10.

Gretarsdottir, S., A.F. Baas, G. Thorleifsson et al. 2010. Genome-wide association study identifies a sequence variant within the DAB2IP gene conferring susceptibility to abdominal aortic aneurysm. *Nat Genet* 42(8):692–7.

Gribouval, O., M. Gonzales, T. Neuhaus et al. 2005. Mutations in genes in the renin–angiotensin system are associated with autosomal recessive renal tubular dysgenesis. *Nat Genet* 37(9):964–8.

Gui, J., T. Wang, R.P. Jones, D. Trump, T. Zimmer, and M. Lei. 2010. Multiple loss-of-function mechanisms contribute to SCN5A-related familial sick sinus syndrome. *PLoS ONE* 5(6):e10985.

Haissaguerre, M., S. Chatel, F. Sacher et al. 2009. Ventricular fibrillation with prominent early repolarization associated with a rare variant of KCNJ8/KATP channel. *J Cardiovasc Electrophysiol* 20(1):93–8.

Hajjar, I., and T.A. Kotchen. 2003 Trends in prevalence, awareness, treatment, and control of hypertension in the United States, 1988–2000. *JAMA* 290(2):199–206.

Hanson, E.K. et al. 2009. Identification of forensically relevant body fluids using a panel of differentially expressed microRNAs. *Anal Biochem* 387(2):303–14.

Heit, J.A., M.D. Silverstein, D.N. Mohr, T.M. Petterson, W.M. O'Fallon, and L.J. Melton 3rd. 1999. Predictors of survival after deep vein thrombosis and pulmonary embolism: A population-based, cohort study. *Arch Intern Med* 159(5):445–53.

Helgadottir, A., A. Manolescu, G. Thorleifsson et al. 2004. The gene encoding 5-lipoxygenase activating protein confers risk of myocardial infarction and stroke. *Nat Genet* 36(3):233–9.

Hilbert, P., K. Lindpaintner, J.S. Beckmann et al. 1991. Chromosomal mapping of two genetic loci associated with blood-pressure regulation in hereditary hypertensive rats. *Nature* 353(6344):521–9.

Hockwald, R.S., J. Arnold, C.B. Clayman, and A.S. Alving. 1952. Toxicity of primaquine in Negroes. *J Am Med Assoc* 149:1568–70.

Hoffmann, B., H. Schmidt-Traub, A. Perrot, K.J. Osterziel, and R. Gessner. 2001. First mutation in cardiac troponin C, L29Q, in a patient with hypertrophic cardiomyopathy. *Hum Mutat* 17(6):524.

Hong, K., P. Bjerregaard, I. Gussak, and R. Brugada. 2005. Short QT syndrome and atrial fibrillation caused by mutation in KCNH2. *J Cardiovasc Electrophysiol* 16(4):394–6.

Horikawa, Y., N. Iwasaki, M. Hara et al. 1997. Mutation in hepatocyte nuclear factor-1 beta gene (TCF2) associated with MODY. *Nat Genet* 17(4):384–5.

Hu, D., H. Barajas-Martinez, E. Burashnikov et al. 2009. A mutation in the beta 3 subunit of the cardiac sodium channel associated with Brugada ECG phenotype. *Circ Cardiovasc Genet* 2(3):270–8.

Ichida, F., S. Tsubata, K.R. Bowles et al. 2001. Novel gene mutations in patients with left ventricular noncompaction or Barth syndrome. *Circulation* 103(9):1256–63.

Ikejiri, M., A. Tsuji, H. Wada et al. 2010. Analysis three abnormal Protein S genes in a patient with pulmonary embolism. *Thromb Res* 125(6):529–32.

Ilina, M.V., C.A. Kepron, G.P. Taylor, D.G. Perrin, P.F. Kantor, and G.R. Somers. 2011. Undiagnosed heart disease leading to sudden unexpected death in childhood: A retrospective study. *Pediatrics* 2011.

International Statistical Classification of Diseases and Related Health Problems. 10th Revision (ICD-10). 1993. Vol 2: Instruction Manual. World Health Organisation, Geneva.

Jannetto, P.J., S.H. Wong, S.B. Gock, E. Laleli-Sahin, B.C. Schur, and J.M. Jentzen. 2002. Pharmacogenomics as molecular autopsy for postmortem forensic toxicology: Genotyping cytochrome P450 2D6 for oxycodone cases. *J Anal Toxicol* 26:438–47.

Jin, M., S.B. Gock, P.J. Jannetto, J.M. Jentzen, and S.H. Wong. 2005. Pharmacogenomics as molecular autopsy for forensic toxicology: Genotyping cytochrome P4503A4*1B and 3A5*3 for 25 fentanylcases. *J Anal Toxicol* 29(7):590–8.

Jobling, M.A., A. Pandya, and C. Tyler-Smith. 1997. The Y chromosome in forensic analysis and paternity testing. *Int J Legal Med* 110:118–24.

Joe, B., Y. Saad, N.H. Lee et al. 2009. Positional identification of variants of Adamts16 linked to inherited hypertension. *Hum Mol Genet* 18(15):2825–38.

Johnson, E.D., R.C. Giles, J.H. Warren, J.I. Floyd, and R.W. Staub. 2005. Analysis of non-suspect samples lacking visually identifiable sperm using a Y-STR 10-plex. *J Forensic Sci* 50(5):1–3.

Kalow, W., and N. Staron. 1957. On distribution and inheritance of atypical forms of human serum cholinesterase, as indicated by dibucaine numbers. *Can J Biochem Physiol* 35(12):1305–20.

Kelly, B.P., M.W. Russell, J.R. and G.J. Hennessy. 2009. Severe hypertrophic cardiomyopathy in an infant with a novel PRKAG2 gene mutation: Potential differences between infantile and adult onset presentation. *Pediatr Cardiol* 30(8):1176–9.

Kersbergen, P., K. van Duijn, A.D. Kloosterman, J.T. den Dunnen, M. Kayser, and P. Knijff. 2009. Developing a set of ancestry-sensitive DNA markers reflecting continental origins of humans. *BMC Genet* 10:69 doi:10.1186/1471-2156-10-69.

Kimura, A., H. Harada, J.E. Park et al. 1997. Mutations in the cardiac troponin I gene associated with hypertrophic cardiomyopathy. *Nat Genet* 16(4):379–82.

Kirchheiner, J. 2008. CYP2D6 phenotype prediction from genotype: Which system is the best? *Clin Pharmacol Ther* 83:225–27.

Kirchheiner, J., K. Nickchen, M. Bauer et al. 2004. Pharmacogenetics of antidepressants and antipsychotics: The contribution of allelic variations to the phenotype of drug response. *Mol Psychiatry* 9:442–73.

Klaassen, S., S. Probst, E. Oechslin et al. 2008. Mutations in sarcomere protein genes in left ventricular noncompaction. *Circulation* 117(22):2893–901.

Koide, S., K. Kugiyama, S. Sugiyama et al. 2003. Association of polymorphism in glutamate–cysteine ligase catalytic subunit gene with coronary vasomotor dysfunction and myocardial infarction. *J Am Coll Cardiol* 41(4):539–45.

Koren, G., J. Cairns, D. Chitayat, A. Gaedigk, and S.J. Leeder. 2006. Pharmacogenetics of morphine poisoning in a breastfed neonate of a codeine-prescribed mother. *Lancet* 368:704.

Koski, A. 2005. Interpretation of postmortem toxicology results. Academic dissertation. Helsinki University Printing House, Helsinki. http://e-thesis.helsinki.fi.

Koski, A., I. Ojanperä, and E. Vuori. 2003. Interaction of alcohol and drugs in fatal poisonings. *Hum Exp Toxicol* 22(5):281–7.

Koski, A., J. Sistonen, I. Ojanpera, M. Gergov, E. Vuori, and A. Sajantila. 2006. CYP2D6 and CYP2C19 genotypes and amitriptyline metabolite ratios in a series of medicolegal autopsies. *Forensic Sci Int* 158(2–3):177–83.

Kruse, M., E. Schulze-Bahr, V. Corfield et al. 2009. Impaired endocytosis of the ion channel TRPM4 is associated with human progressive familial heart block type I. *J Clin Invest* 119(9):2737–44.

Kumar, D.R., E. Hanlin, I. Glurich, J.J. Mazza, and S.H. Yale 2010. Virchow's contribution to the understanding of thrombosis and cellular biology. *Clin Med Res* 8(3–4):168–72.

Lahat, H., E. Pras, and M. Eldar. 2004. A missense mutation in CASQ2 is associated with autosomal recessive catecholamine-induced polymorphic ventricular tachycardia in Bedouin families from Israel. *Ann Med* 36(Suppl 1):87–91.

Lahti, R. 2005. From findings to statistics: An assessment of Finnish medical cause-of-death information in relation to underlying-cause coding. Academic dissertation, Helsinki University Printing House, Helsinki. http//e-thesis.helsinki.fi.

Landstrom, A.P., B.A. Adekola, J.M. Bos, S.R. Ommen, and M.J. Ackerman. 2011. PLN-encoded phospholamban mutation in a large cohort of hypertrophic cardiomyopathy cases: Summary of the literature and implications for genetic testing. *Am Heart J* 161(1):165–71.

Launiainen, T., A. Sajantila, I. Rasanen, E. Vuori, and I. Ojanperä. 2010. Adverse interaction of warfarin and paracetamol: Evidence from a post-mortem study. *Eur J Clin Pharmacol* 66(1):97–103.

Launiainen, T., E. Vuori, and I. Ojanperä. 2009. Prevalence of adverse drug combinations in a large post-mortem toxicology database. *Int J Legal Med* 123(2):109–15.

Launiainen, T., I. Rasanen, E. Vuori, and I. Ojanperä. 2011. Fatal venlafaxine poisonings are associated with a high prevalence of drug interactions. *Int J Legal Med* 125(3):349–58.

Lazarou, J., B.H. Pomeranz, and P.N. Corey. 1998. Incidence of adverse drug reactions in hospitalized patients: A meta-analysis of prospective studies. *JAMA* 279:1200–05.

Lehmann, H., and E. Ryan 1956. The familial incidence of low pseudocholinesterase level. *Lancet* 271:124.

Levo, A., A. Koski, I. Ojanperä, E. Vuori, and A. Sajantila. 2003. Post-mortem SNP analysis of *CYP2D6* gene reveals correlation between genotype and opioid drug (tramadol) metabolite ratios in blood. *Forensic Sci Int* 135:9–15.

Little, W.C., and R.J. Applegate. 1996. Role of plaque size and degree of stenosis in acute myocardial infarction. *Cardiol Clin* 14(2):221–8.

Liu, F., A. Wollstein, P.G. Hysi et al. 2010. Digital quantification of human eye color highlights genetic association of three new loci. *PLoS Genet* 6:e1000934.

London, B., M. Michalec, H. Mehdi et al. 2007. Mutation in glycerol-3-phosphate dehydrogenase 1 like gene (*GPD1-L*) decreases cardiac Na$^+$ current and causes inherited arrhythmias. *Circulation* 13;116(20):2260–8.

Madea, B., F. Musshoff, and J. Preuss. 2009. Medical negligence in drug associated deaths. *Forensic Sci Int* 190(1–3):67–73.

Mahgoub, A., J.R. Idle, L.G. Dring, R. Lancaster, and R.L. Smith. 1977. Polymorphic hydroxylation of Debrisoquine in man. *Lancet* 2:584–86.

Malecki, M.T., U.S. Jhala, A. Antonellis et al. 1999. Mutations in NEUROD1 are associated with the development of type 2 diabetes mellitus. *Nat Genet* 23(3):323–8.

Mani, A., J. Radhakrishnan, H. Wang et al. 2007. LRP6 mutation in a family with early coronary disease and metabolic risk factors. *Science* 315(5816):1278–82.

Manunta, P., M. Burnier, M. D'Amico et al. 1999. Adducin polymorphism affects renal proximal tubule reabsorption in hypertension. *Hypertension* 33(2):694–7.

Markovic, D., X. Tang, M. Guruju, M.A. Levenstien, J. Hoh, A. Kumar, and J. Ott. 2005. Association of angiotensinogen gene polymorphisms with essential hypertension in African-Americans and Caucasians. *Hum Hered* 60(2):89–96.

Marsman, R.F., A. Bardai, A.V. Postma et al. 2011. A complex double deletion in LMNA underlies progressive cardiac conduction disease, atrial arrhythmias, and sudden death. *Circ Cardiovasc Genet* 4(3):280–7.

McKusick, V.A. 1991. The defect in Marfan syndrome. *Nature* 352(6333):279–81.

Medeiros-Domingo, A., T. Kaku, and D.J. Tester. 2008. SCN4B-encoded sodium channel beta4 subunit in congenital long-QT syndrome. *Circulation* 116(2):134–42.

Merner, N.D., K.A. Hodgkinson, A.F. Haywood et al. 2008. Arrhythmogenic right ventricular cardiomyopathy type 5 is a fully penetrant, lethal arrhythmic disorder caused by a missense mutation in the *TMEM43* gene. *Am J Hum Genet* 82(4):809–21.

Micke, P., A. Ostman, J. Lundeberg, and F. Ponten. 2005. Laser-assisted cell microdissection using the PALM system. *Methods Mol Biol* 293:151–66.

Millat, G., B. Kugener, P. Chevalier et al. 2009. Contribution of long-QT syndrome genetic variants in sudden infant death syndrome. *Pediatr Cardiol* 30(4):502–9.

Mineharu Y., K. Inoue, S. Inoue et al. 2008. Association analyses confirming a susceptibility locus for intracranial aneurysm at chromosome 14q23. *J Hum Genet* 53(4):325–32.

Mohler, P.J., J.J. Schott, A.O. Gramolini et al. 2003. Ankyrin-B mutation causes type 4 long-QT cardiac arrhythmia and sudden cardiac death. *Nature* 421(6923):634–9.

Moore, T.J., B.M. Psaty, and C.D. Furberg. 1998. Time to act on drug safety. *JAMA* 279(19):1571–3.

Motulsky, A.G. 1957. Drug reactions enzymes, and biochemical genetics. *JAMA* 165(7):835–7.

Murray, C., C. McAlister, and K. Elliott. 2006. Use of fluorescence in situ hybridisation and laser microdissection to isolate male non-sperm cells in cases of sexual assault. *Int Cong Ser* 1288:622–4.

Musshoff, F., U.M. Stamer, and B. Madea. 2010. Pharmacogenetics and forensic toxicology. *Forensic Sci Int* 203(1–3):53–62.

Nahed, B.V., A. Seker, B. Guclu et al. 2005. Mapping a Mendelian form of intracranial aneurysm to 1p34.3–p36.13. *Am J Hum Genet* 76(1):172–9.

Nakayama, T., M. Soma, Y. Watanabe et al. 2003. Splicing mutation of the prostacyclin synthase gene in a family associated with hypertension. *Adv Exp Med* 525:165–8.

Neuvonen, A.M., J.U. Palo, and A. Sajantila. 2011. Post-mortem ABCB1 genotyping reveals an elevated toxicity for female digoxin users. *Int J Legal Med* 125(2):265–9.

Neve, B., M.E. Fernandez-Zapico, V. Ashkenazi-Katalan et al. 2005. Role of transcription factor KLF11 and its diabetes-associated gene variants in pancreatic beta cell function. *Proc Natl Acad Sci U S A* 102(13):4807–12.

Newton-Cheh, C., T. Johnson, V. Gateva et al. 2009. Genome-wide association study identifies eight loci associated with blood pressure. *Nat Genet* 41(6):666–76.

Njolstad, P.R., O. Sovik, A. Cuesta-Munoz et al. 2001. Neonatal diabetes mellitus due to complete glucokinase deficiency. *N Engl J Med* 344(21):1588–92.

Ojanperä, I., M. Kolmonen, and A. Pelander. 2012. Current use of high-resolution mass spectrometry in drug screening relevant to clinical and forensic toxicology and doping control. *Anal Biochem* 403:1203–20.

Ou, X-l, J. Gao, H. Wang et al. 2012. Predicting human age with bloodstains by sjTREC quantification. *PLoS ONE* 7(8).

Ozturk, A.K., B.V. Nahed, M. Bydon et al. 2006. Molecular genetic analysis of two large kindreds with intracranial aneurysms demonstrates linkage to 11q24–25 and 14q23–31. *Stroke* 37(4):1021–7.

Padmanabhan, S., O. Melander, T. Johnson et al. 2010. Genome-wide association study of blood pressure extremes identifies variant near UMOD associated with hypertension. *PLoS Genet* 6(10):e1001177.

Patel, H.H., L.L. Hamuro, B.J. Chun et al. 2010. Disruption of protein kinase A localization using a trans-activator of transcription (TAT)-conjugated A-kinase-anchoring peptide reduces cardiac function. *J Biol Chem* 285(36):27632–40.

Petersen, E.E., S.A. Rasmussen, K.L. Daniel, M.M. Yazdy, and M.A. Honein. 2008. Prescription medication borrowing and sharing among women of reproductive age. *J Womens Health* 17: 1073–80.

Pinzani, P., C. Orlando, and M. Pazzagli. 2006. Laser-assisted microdissection for real-time PCR sample preparation. *Mol Aspects Med* 27(2–3):140–59.

Pirmohamed, M., A.M. Breckenridge, N.R. Kitteringham, and B.K. Park. 1998. Adverse drug reactions. *BMJ* 316(7140):1295–8.

Pirmohamed, M., S. James, S. Meakin et al. 2004. Adverse drug reactions as cause of admission to hospital: Prospective analysis of 18 820 patients. *Br Med J* 329:15–19.

Plaster, N.M., R. Tawil, M. Tristani-Firouzi et al. 2001. Mutations in Kir2.1 cause the developmental and episodic electrical phenotypes of Andersen's syndrome. *Cell* 105(4):511–9.

Plengvidhya, N., S. Kooptiwut, N. Songtawee et al. 2007. PAX4 mutations in Thais with maturity onset diabetes of the young. *J Clin Endocrinol Metab* 92(7):2821–6.

Podgoreanu, M.V., and D.A. Schwinn. 2005. New paradigms in cardiovascular medicine: Emerging technologies and practices: Perioperative genomics. *J Am Coll Cardiol* 46(11):1965–77.

Popiolek, D.A., H. Yee, K. Mittal et al. 2006. Multiplex short tandem repeat DNA analysis confirms the accuracy of p57(KIP2) immunostaining in the diagnosis of complete hydatidiform mole. *Hum Pathol* 37(11):1426–34.

Popiolek, D.A., M.K. Prinz, A.B. West, B.L. Nazzaruolo, S.M. Estacio, and Z.M. Budimlija. 2003. Multiplex DNA short tandem repeat analysis. A useful method for determining the provenance of minute fragments of formalin-fixed, paraffin-embedded tissue. *Am J Clin Pathol* 120(5):746–51.

Pounder, D., and G.R. Jones. 1990. Post-mortem drug resictribution—a toxicological nightmare. *Forensic Sci Int* 45:253–63.

Preuss, M., I.R Konig, J.R. Thompson et al. 2010. Design of the Coronary ARtery DIsease Genome-Wide Replication And Meta-Analysis (CARDIoGRAM) study: A genome-wide association meta-analysis involving more than 22 000 cases and 60 000 controls. *Circ Cardiovasc Genet* 3(5):475–83.

Priori, S.G., C. Napolitano, and P.J. Schwartz. 1999. Low penetrance in the long-QT syndrome: Clinical impact. *Circulation* 99(4):529–33.

Priori, S.G., C. Napolitano, N. Tiso et al. 2001. Mutations in the cardiac ryanodine receptor gene (*hRyR2*) underlie catecholaminergic polymorphic ventricular tachycardia. *Circulation* (2):196–200.

Priori, S.G., S.V. Pandit, I. Rivolta et al. 2005. A novel form of short QT syndrome (SQT3) is caused by a mutation in the *KCNJ2* gene. *Circ Res* 96(7):800–7.

Raeder, H., S. Johansson, P.I. Holm et al. 2006. Mutations in the CEL VNTR cause a syndrome of diabetes and pancreatic exocrine dysfunction. *Nat Genet* 38(1):54–62.

Rafestin-Oblin, M.E., A. Souque, B. Bocchi, G. Pinon, J. Fagart, and A. Vandewalle. 2003. The severe form of hypertension caused by the activating S810L mutation in the mineralocorticoid receptor is cortisone related. *Endocrinology* 144(2):528–33.

Rampazzo, A., G. Beffagna, A. Nava et al. 2003. Arrhythmogenic right ventricular cardiomyopathy type 1 (ARVD1): Confirmation of locus assignment and mutation screening of four candidate genes. *Eur J Hum Genet* 11(1):69–76.

Ria, M., J. Lagercrantz, A. Samnegard, S. Boquist, A. Hamsten, and P. Eriksson. 2011. A common polymorphism in the promoter region of the *TNFSF4* gene is associated with lower allele-specific expression and risk of myocardial infarction. *PLoS One* 2011;6(3):e17652.

Richard, P., P. Charron, L. Carrier et al. 2003. Hypertrophic cardiomyopathy: Distribution of disease genes, spectrum of mutations, and implications for a molecular diagnosis strategy. *Circulation* 107(17):2227–32.

Robino, C., M.R. Barilaro, S. Gino, R. Chiarle, G. Palestro, and C. Torre. 2006. Incestuous paternity detected by STR-typing of chorionic villi isolated from archival formalin-fixed paraffin-embedded abortion material using laser microdissection. *J Forensic Sci* 51(1):90–2.

Roos, Y.B., G. Pals, P.M. Struycken et al. 2004. Genome-wide linkage in a large Dutch consanguineous family maps a locus for intracranial aneurysms to chromosome 2p13. *Stroke* 35(10):2276–81.

Rutherford, S., M.P. Johnson, R.P. Curtain, and L.R. Griffiths. 2001. Chromosome 17 and the inducible nitric oxide synthase gene in human essential hypertension. *Hum Genet* 109(4):408–15.

Sajantila, A., J.U. Palo, I. Ojanperä, C. Davis, and B. Budowle. 2010. Pharmacogenetics in medicolegal context. *Forensic Sci Int* 203(1–3):44–52.

Sallee, F.R., C.L. DeVane, and R.E. Ferrell. 2000. Fluoxetine-related death in a child with cytochrome P-450 2D6 genetic deficiency. *J Child Adolesc Psychopharmacol* 10(1):27–34.

Schwarze, U., J.A. Goldstein, and P.H. Byers. 1997. Splicing defects in the *COL3A1* gene: marked preference for 5′ (donor) spice-site mutations in patients with exon-skipping mutations and Ehlers–Danlos syndrome type IV. *Am J Hum Genet* 61(6):1276–86.

Secrest, A.M., D.J. Becker, S.F. Kelsey, R.E. Laporte, and T.J. Orchard. 2011. Characterizing sudden death and dead-in-bed syndrome in Type 1 diabetes: analysis from two childhood-onset Type 1 diabetes registries. *Diabetes Med* 28(3):293–300.

Seidl, S., R. Burgemeister, R. Hausmann, P. Betz, and T. Lederer. 2005. Contact-Free isolation of sperm and epithelial cells by laser microdissection and pressure catapulting. *Forensic Sci Med Pathol* 1(2):153–158.

Simone, N.L., R.F. Bonner, J.W. Gillespie, M.R. Emmert-Buck, and L.A. Liotta. 1998. Laser-capture microdissection: opening the microscopic frontier to molecular analysis. *Trends Genet* 14(7):272–6.

Simonetti, G.D., M.G. Mohaupt, and M.G. Bianchetti. 2011. Monogenic forms of hypertension. *Eur J Pediatr*.

Sivapalaratnam, S., M.M. Motazacker, and S. Maiwald et al. 2011. Genome-wide association studies in atherosclerosis. *Curr Atheroscler Rep* 13(3):225–32.

Smilowitz, N.R., B.A. Sampson, C.R. Abrecht, J.S. Siegfried, J.S. Hochman, and H.R. Reynolds. 2011. Women have less severe and extensive coronary atherosclerosis in fatal cases of ischemic heart disease: an autopsy study. *Am Heart J* 161(4):681–8.

Soares-Viera, J.A., A.E.C. Billerbeck, E. Sadayo et al. 2007. Y-STRs in forensic medicine: DNA analysis in semen samples of azoospermic individuals. *J Forensic Sci* 52(3):664–70.

Splawski, I., J. Shen, K.W. Timothy et al. 2000. Spectrum of mutations in long-QT syndrome genes. KVLQT1, HERG, SCN5A, KCNE1, and KCNE2. *Circulation* 102(10):1178–85.

Splawski, I., K.W. Timothy, N. Decher et al. 2005. Severe arrhythmia disorder caused by cardiac L-type calcium channel mutations. *Proc Natl Acad Sci U S A* 102(23):8089–96; discussion 6–8.

Steck, A.K., and W.E. Winter. 2011. Review on monogenic diabetes. *Curr Opin Endocrinol Diabetes Obes* 18(4):252–8.

Stoffers, D.A., J. Ferrer, W.L. Clarke, and J.F. Habener. 1997. Early-onset type-II diabetes mellitus (MODY4) linked to IPF1. *Nat Genet* 17(2):138–9.

Syrris, P., D. Ward, A. Asimaki et al. 2007. Desmoglein-2 mutations in arrhythmogenic right ventricular cardiomyopathy: a genotype–phenotype characterization of familial disease. *Eur Heart J* 28(5):581–8.

Syrris, P., D. Ward, A. Evans et al. 2006. Arrhythmogenic right ventricular dysplasia/cardiomyopathy associated with mutations in the desmosomal gene desmocollin-2. *Am J Hum Genet* 79(5):978–84.

Tang, Y., B. Sampson, S. Pack et al. 2011. Ethnic differences in out-of-hospital fatal pulmonary embolism. *Circulation* 123(20):2219–25.

Tester, D.J., A. Medeiros-Domingo, M.L. Will, C.M. Haglund, and M.J. Ackerman. 2012. Cardiac channel molecular autopsy: insights from 173 consecutive cases of autopsy-negative sudden unexplained death referred for postmortem genetic testing. *Mayo Clin Proc* 87(6):524–39.

Tester, D.J., and M.J. Ackerman. 2009. Cardiomyopathic and channelopathic causes of sudden unexplained death in infants and children. *Annu Rev Med* 60:69–84.

Thierfelder, L., H. Watkins, C. MacRae et al. 1994. Alpha-tropomyosin and cardiac troponin T mutations cause familial hypertrophic cardiomyopathy: a disease of the sarcomere. *Cell* 77(5):701–12.

Thompson, E.E., H. Kuttab-Boulos, D. Witonsky, L. Yang, B.A. Roe, and A. Di Rienzo. 2004. CYP3A variation and the evolution of salt-sensitivity variants. *Am J Hum Genet* 75(6):1059–69.

Tiso, N., D.A. Stephan, A. Nava et al. 2001. Identification of mutations in the cardiac ryanodine receptor gene in families affected with arrhythmogenic right ventricular cardiomyopathy type 2 (ARVD2). *Hum Mol Genet* 10(3):189–94.

Titford, M. 2005. The long history of hematoxylin. *Biotechnic Histochem* 80(2):73–8.

Tu, E., R.D. Bagnall, J. Duflou, and C. Semsarian. 2011. Post-mortem review and genetic analysis of sudden unexpected death in epilepsy (SUDEP) cases. *Brain Pathol* 21(2):201–8.

Tu, E., R.D. Bagnall, L. Waterhouse, J. Duflou, and C. Semsarian. 2011. Genetic analysis of hyperpolarization-activated cyclic nucleotide-gated cation channels in sudden unexpected death in epilepsy cases. *Brain Pathol.*

Ueda, K., Y. Hirano, Y. Higashiuesato et al. 2009. Role of HCN4 channel in preventing ventricular arrhythmia. *J Hum Genet* 54(2):115–21.

Ueda, K., C. Valdivia, A. Medeiros-Domingo et al. 2008. Syntrophin mutation associated with long QT syndrome through activation of the nNOS–SCN5A macromolecular complex. *Proc Natl Acad Sci U S A* 105(27):9355–60.

van der Voet, M., J.M. Olson, H. Kuivaniemi et al. 2004. Intracranial aneurysms in Finnish families: confirmation of linkage and refinement of the interval to chromosome 19q13.3. *Am J Hum Genet* 74(3):564–71.

Vandewoestyne, M., D. Van Hoofstat, F. Van Nieuwerburgh, and D. Deforce. 2009. Automatic detection of spermatozoa for laser microdissection. *Int J Legal Med* 123(2):169–75.

Vatta, M., B. Mohapatra, S. Jimenez et al. 2003. Mutations in Cypher/ZASP in patients with dilated cardiomyopathy and left ventricular non-compaction. *J Am Coll Cardiol* 42(11):2014–27.

Vaxillaire, M., M. Rouard, K. Yamagata et al. 1997. Identification of nine novel mutations in the hepatocyte nuclear factor 1 alpha gene associated with maturity-onset diabetes of the young (MODY3). *Hum Mol Genet* 6(4):583–6.

Verlaan, D.J., M.P. Dube, J. St-Onge et al. 2006. A new locus for autosomal dominant intracranial aneurysm, ANIB4, maps to chromosome 5p15.2–14.3. *J Med Genet* 43(6):e31.

Vionnet, N., M. Stoffel, J. Takeda et al. 1992. Nonsense mutation in the glucokinase gene causes early-onset non-insulin-dependent diabetes mellitus. *Nature* 356(6371):721–2.

Visser, M. et al. 2011. mRNA-based skin identification for forensic applications. *Int J Legal Med* 125(2):253–63.

Vossen, C.Y., J. Conard, J. Fontcuberta et al. 2005. Risk of a first venous thrombotic event in carriers of a familial thrombophilic defect. The European Prospective Cohort on Thrombophilia (EPCOT). *J Thromb Haemost* 3(3):459–64.

Vuori, E., J.A. Henry, I. Ojanpera et al. 2003. Death following ingestion of MDMA (ecstasy) and moclobemide. *Addiction* 98:365–8.

Wallace, C., M.Z. Xue, S.J. Newhouse et al. 2006. Linkage analysis using co-phenotypes in the BRIGHT study reveals novel potential susceptibility loci for hypertension. *Am J Hum Genet* 79(2):323–31.

Wang Q., S. Rao, G.Q. Shen et al. 2004. Premature myocardial infarction novel susceptibility locus on chromosome 1P34-36 identified by genomewide linkage analysis. *Am J Hum Genet* 74(2):262–71.

Wang, H., Z. Li, J. Wang et al. 2010. Mutations in *NEXN*, a Z-disc gene, are associated with hypertrophic cardiomyopathy. *Am J Hum Genet* 87(5):687–93.

Wang, L., C. Fan, S.E. Topol, E.J. Topol, and Q. Wang. 2003. Mutation of MEF2A in an inherited disorder with features of coronary artery disease. *Science* 302(5650):1578–81.

Wang, L., H.L. McLeod, and R.M. Weinshilboum. 2011. Genomics and drug response. *N Engl J Med* 364:1144–53.

Wang, X., M. Ria, P.M. Kelmenson et al. 2005. Positional identification of TNFSF4, encoding OX40 ligand, as a gene that influences atherosclerosis susceptibility. *Nat Genet* 37(4):365–72.

Wang, X.L., A.S. Sim, R.F. Badenhop, R.M. McCredie, and D.E. Wilcken. 1996. A smoking-dependent risk of coronary artery disease associated with a polymorphism of the endothelial nitric oxide synthase gene. *Nat Med* 2(1):41–5.

Watanabe, H., T.T. Koopmann, S. Le Scouarnec et al. 2008. Sodium channel beta1 subunit mutations associated with Brugada syndrome and cardiac conduction disease in humans. *J Clin Invest* 118(6):2260–8.

Watkins, H., D. Conner, L. Thierfelder et al. 1995. Mutations in the cardiac myosin binding protein-C gene on chromosome 11 cause familial hypertrophic cardiomyopathy. *Nat Genet* 11(4):434–7.

Watkins, H., H. Ashrafian, and C. Redwood. 2011. Inherited cardiomyopathies. *N Engl J Med* 364(17):1643–56.

Wong, S.H., C. Happy, D. Blinka et al. 2010. From personalized medicine to personalized justice: the promises of translational pharmacogenomics in the justice system. *Pharmacogenomics* 11(6):731–7.

Wong, S.H., M.A. Wagner, J.M. Jentzen et al. 2003. Pharmacogenomics as an aspect of molecular autopsy for forensic pathology/toxicology: does genotyping CYP2D6 serve as an adjunct for certifying methadone toxicity? *J Forensic Sci* 48(6):1406–15.

Xu, X., J. Yang, J. Rogus, C. Chen, and N. Schork. 1999. Mapping of a blood pressure quantitative trait locus to chromosome 15q in a Chinese population. *Hum Mol Genet* 8(13):2551–5.

Yamagata, K., H. Furuta, N. Oda et al. 1996. Mutations in the hepatocyte nuclear factor-4alpha gene in maturity-onset diabetes of the young (MODY1). *Nature* 384(6608):458–60.

Yang, Y., B. Liang, J. Liu et al. 2010. Identification of a Kir3.4 Mutation in congenital long QT syndrome. *Am J Hum Genet*.

Yang, Y., J. Li, X. Lin et al. 2009. Novel KCNA5 loss-of-function mutations responsible for atrial fibrillation. *J Hum Genet* 54(5):277–83.

Yang, Z., N.E. Bowles, S.E. Scherer et al. 2006. Desmosomal dysfunction due to mutations in desmoplakin causes arrhythmogenic right ventricular dysplasia/cardiomyopathy. *Circ Res* 99(6):646–55.

Yasuno, K., K. Bilguvar, P. Bijlenga et al. 2010. Genome-wide association study of intracranial aneurysm identifies three new risk loci. *Nat Genet* 42(5):420–5.

Zackrisson, A.L., B. Lindblom, and J. Ahlner. 2010. High frequency of occurrence of *CYP2D6* gene duplication/multiduplication indicating ultrarapid metabolism among suicide cases. *Clin Pharmacol Ther* 88(3):354–9.

Zaracostas, J. 2007. Misuse of prescription drugs could soon exceed that of illicit narcotics, UN panel warns. *BMJ* 334:444.

Zhang, L.P., B. Hui, and B.R. Gao. 2011. High risk of sudden death associated with a PRKAG2-related familial Wolff–Parkinson–White syndrome. *J Electrocardiol* 44(4):483–6.

Zubakov, D., F. Liu, M.C. van Zelm et al. 2010. Estimating human age from T-cell DNA rearrangements. *Curr Biol* 20:R970–1.

Genetic Genealogy in the Genomic Era

19

JAKE K. BYRNES
NATALIE M. MYRES
PETER A. UNDERHILL

Contents

19.1 Introduction	483
19.2 DNA Testing for Genealogy	485
19.3 Haploid Chromosome Testing	486
19.3.1 Y Chromosome Testing	486
19.3.1.1 Discovering Paternal Lineages	486
19.3.1.2 Y-STR Haplotypes and Recent Ancestry	488
19.3.1.3 Y Chromosome Haplogroups and Deep Ancestry	489
19.3.1.4 Future of Y Chromosome Testing	490
19.3.2 mtDNA Testing	490
19.3.2.1 Discovering Maternal Lineages	490
19.3.2.2 mtDNA Testing Options	491
19.3.3 Future of Haploid Testing	492
19.4 Autosomal Testing	492
19.4.1 Why Assay the Autosomes?	492
19.4.2 Meiosis—A Genealogical Double-Edged Sword	493
19.4.3 Genetic Genealogy with Autosomal DNA	495
19.4.4 Ancestral Origin Estimation	495
19.4.5 Relative Identification	500
19.4.6 Conclusions	502
19.5 Discussion	503
Acknowledgments	504
References	504

19.1 Introduction

Molecular methods traditionally used by the scientific community for studying medical genetics, human evolution and developing forensic applications have been recently adapted to support the emerging and rapidly growing consumer genetic genealogy industry. Genetic genealogy applies DNA testing methodologies to traditional genealogical research, particularly in cases where the paper trail for identifying ancestors has run out, a common problem for genealogists known as "hitting a brick wall." In such cases, DNA matching techniques may locate family lineages that were not identifiable within written historical records.

    Genealogy is the study of family history, which involves tracing one's family lineages back through time by documenting vital statistics for each ancestor such as dates and places of births, deaths and marriages. Genealogy has grown into a popular pursuit for millions of people worldwide facilitated in recent decades by the digitization of vital records made accessible via the Internet. Genealogy is currently ranked as the second most popular search topic on the Internet (Falconer 2012) and there are currently a substantial number of companies offering genealogical products, many of which include genetic analysis (Shriver and Kittles 2004). The primary factors that motivate individuals to engage in genealogy research involve the combined thrill of discovering new family information while also experiencing an enhanced sense of personal identity. Serious genealogists and hobbyists alike describe the research as "addictive" because there is always a new ancestor to discover and learn about.

Genealogical research depends on written historical records for determining family relationships, and reliance on a chain of documents back through time inevitably leads to brick walls for researchers. While there are numerous reasons for such barriers, two common causes are that records have been destroyed or they were never kept in the first place. For example, during the American Civil War the southern states experienced extensive record loss resulting from a scorched earth policy that was implemented by Union General William T. Sherman. Designed to spread fear and break the morale of the Confederate States by destroying property, Sherman's March to the Sea was a 1000-mile march across the south where soldiers burned property indiscriminately, including government buildings where vital records were stored. As a result, many genealogy researchers with ancestry from the southern states cannot extend these lineages beyond the late 1800s. Similarly, African Americans have a particularly difficult task when tracing ancestors beyond the late 1800's since many of their African ancestors before this time were enslaved, and only limited records if any were kept on slaves. Prior to emancipation, slaves were considered property and therefore they could not legally marry, vote, own land or participate in other activities recorded in vital records. Furthermore, the 1870 census was the first time African Americans were listed by name as opposed to simply being listed as a number, age and gender.

The idea of applying molecular methods to overcome otherwise impenetrable brick walls in genealogical research emerged in the 1990's with several highly publicized reports of long-standing family mysteries being solved with genetic evidence in cases where traditional methods were unsuccessful. Examples include mtDNA analysis conclusively disproving the claims by Anna Anderson that she was actually Anastasia Romanov, daughter of Tsar Nicholas II of Russia (Coble 2011), and Y chromosome analysis showing that Thomas Jefferson, or a close paternal relative, fathered the children of one of Jefferson's slaves, Sally Hemmings (Foster et al. 1998). Another high-profile Y chromosome study showed the African Bantu-speaking Lemba tribe, with an oral tradition of Jewish ancestry, carried the Cohen Modal Haplotype, a haplotype found at high frequencies in major Jewish populations, thereby supporting the Lemba's claim of being descended from a Jewish priesthood lineage (Thomas et al. 2000). From these and other studies, it became readily apparent that genetic testing was a powerful tool to discover unknown relationships in both the recent and distant past beyond the point where traditional records end.

During the same time, the cost of genetic testing dropped precipitously, enabling numerous Internet companies to emerge offering genetic testing for genealogical purposes. Companies initially focused on Y chromosome short tandem repeat (STR) testing for surname-based patrilineal studies. Soon however, product lines expanded to include mtDNA hypervariable region and whole mitochondrial genome sequence analysis for maternal line

Genetic Genealogy in the Genomic Era

research. The popularity of these haploid tests has resulted in large and growing haplotype databases comprising hundreds of thousands of genetic profiles. Genetic genealogy companies are now offering autosomal microarray-based testing options, in some cases providing medical risk assessments in addition to genealogical matching services. As the genetic genealogy market continues to expand and its DNA databases continue to grow in size and diversity, significant repositories of human genetic variation are being created that may serve as a valuable resource for numerous disciplines focused on the study of human genetics.

19.2 DNA Testing for Genealogy

Genealogy researchers typically turn to DNA testing for two primary reasons. First, they want to confirm a particular family connection or piece of family lore that is unsubstantiated by written records. Second, they are searching for new leads after hitting a brick wall. To address these challenges DNA tests can be purchased over the Internet from any one of several DNA testing companies. Typically the company mails a DNA collection kit to the client/genealogist. The DNA kit provides cheek swabs, mouth rinse, or more recently saliva tubes to collect a biological sample from which to isolate DNA. The recipient returns

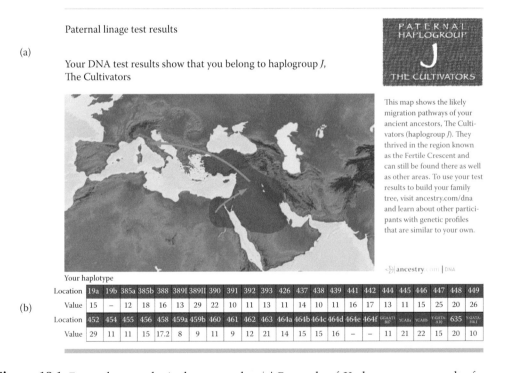

Figure 19.1 Example genealogical test results: (a) Example of Y chromosome results from a genetic genealogy company reporting the test subject belongs to Y chromosome haplogroup J. The map indicates the geographic region in which haplogroup J displays its highest frequencies and shows general dispersal/migratory routes. (b) The test subject receives his personal Y-STR haplotype, comprised of 46 Y-STR markers. The name of each STR marker is displayed in green boxes with the corresponding allele listed below in white boxes. The Y-STR haplotype can be used to search public and private Y chromosome databases.

the collection kit to the testing company and receives results several weeks later. The test results that the client receives typically include a personal DNA profile and information about deep (ancient) ancestry in the form of haplogroup membership or ethnicity estimation (Figure 19.1). The DNA profile is usually added to the company's DNA database and the client obtains access to online tools for comparing his/her genetic profile with others. In general, the primary objective of taking a DNA test is to compare one's DNA profile with those of other individuals to find close matches, often referred to as "genetic cousins," to identify a shared ancestor within the recent past (within the past 1000 years). Below we introduce the tools and techniques applied to genetic data to extract the most useful genealogical information.

19.3 Haploid Chromosome Testing

Until the recent introduction of autosomal testing the primary options available to genealogists were Y chromosome and mtDNA analyses. The haploid nature of these chromosomes makes them powerful systems for identifying common ancestors shared between individuals along specific ancestral lineages. Obvious differences exist between the Y chromosome and mtDNA genomes such as their accessible size for information content, 10 million versus 16 thousand nucleotides respectively, plus a higher overall mutation rate for the mtDNA resulting in more homoplasy (recurrent/revertant mutational events) within the mtDNA phylogenetic tree. Nonetheless, because these loci follow strict lineal inheritance patterns and are not subject to recombination and assortment, extant variation is due solely to mutations that have accumulated sequentially over time, thereby providing a molecular clock for inferring the number of generations separating two haploid genetic profiles or haplotypes. Additionally, because haploid chromosomes are not lost at each generation due to independent assortment, these chromosomes trace lineages both through recent history as well as into the very distant past. We discuss both haploid systems below, beginning with the Y chromosome.

19.3.1 Y Chromosome Testing

19.3.1.1 Discovering Paternal Lineages
In addition to being the first type of DNA test provided for genealogical purposes, Y chromosome testing remains a favorite tool among genealogists for researching paternal line ancestors. The Y chromosome is carried only by males and follows a strict paternal inheritance pattern, being passed from father to son with each successive generation (Figure 19.2a). This mode of inheritance mirrors the transmission of surnames in most western societies making the Y chromosome a particularly useful tool for genealogists (King and Jobling 2009).

The popularity of surname studies and Y chromosome databases demonstrates the utility of the Y chromosome for identifying relatedness along paternal lineages. There are currently thousands of surname studies underway which involve recruiting males who share the same surname or similar surname variants to undergo Y chromosome testing in order to identify the number of unique origins of their surname and to determine the individual lineages that converge within recent history. This enables related individuals to reconstruct the details of how they are connected to each other along their paternal lines and to collaborate

Genetic Genealogy in the Genomic Era

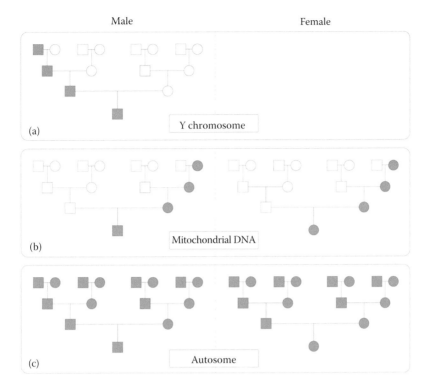

Figure 19.2 Utility of different genetic sources for genealogical research: This figure shows the familial inheritance patterns of different genetic material. Each panel (a–c) contains two family trees, one for a male descendant (left) and one for a female descendant (right) with contributing lineages highlighted in green. Male ancestors are represented as squares and female ancestors as circles. (a) The Y chromosome is inherited paternally and is only passed to male children. Therefore, it only reveals information about your paternal lineage and is only present in men. (b) Mitochondrial DNA is inherited maternally, but is passed on to both sexes. It reveals information about your maternal lineage and is useful for both men and women. (c) Autosomal DNA is inherited from both parents. Due to recombination and assortment, an individual may inherit autosomal DNA from any ancestor in the tree. It is also passed on to males and females making it useful for both males and females.

on extending their shared lineage further into the past. At the same time false leads are eliminated that may have been previously pursued, saving both time and money.

In addition to surname studies, Y chromosome databases are the largest and among the most frequently searched DNA databases. There are numerous public and private Y chromosome databases comprising tens to hundreds of thousands of Y chromosome haplotypes associated with various other types of data such as surnames, geographic locations and pedigrees. Genealogists often search Y chromosome databases when they are looking for new leads or attempting to solve problems caused by surname changes along paternal lineages. For example, surnames in Scandinavian countries change at each generation making paternal ancestry within these regions more difficult to trace. Similarly, many European American immigrants experienced surname changes when transiting through Ellis Island. Y chromosome analysis can circumvent confusing surname changes by identifying individuals within Y databases who have different surnames but who share closely related haplotypes.

19.3.1.2 Y-STR Haplotypes and Recent Ancestry

The most common type of Y chromosome testing involves assaying anywhere from 12 to 111 short tandem repeat (STR) loci located within the non-recombining portion of the Y chromosome. Different companies test many of the same Y-STR markers making Y-STR haplotypes relatively portable for comparison purposes. Y-STR markers are particularly useful to genealogists due to their multiallelic states and exceedingly high mutation rates which typically range between 10^{-3} and 10^{-4} mutations/marker/generation (Goedbloed et al. 2009). Such high variability translates into a very recent timeframe for identifying common ancestors.

The degree of relatedness between two Y chromosome haplotypes is quantified by calculating time to most recent common ancestor (TMRCA). The most common method used by genetic genealogy companies for estimating relatedness is a coalescence-based approach that produces a likelihood distribution of time to the MRCA (Walsh 2001). The method is based on the number of matching versus nonmatching Y-STR alleles observed between two Y haplotypes and the mutation rate of each marker included in the haplotype array. The more Y-STR markers that are used to assess relatedness the more precise the TMRCA calculation becomes and similarly, using locus-specific mutation rates when available, rather than average rates, increases TMRCA accuracy.

Unfortunately, genealogists easily misinterpret TMRCA by focusing on the maximum likelihood estimate rather than the likelihood distribution as a whole (Figure 19.3a). For example, if the most likely TMRCA is reported as six generations the genealogist may not recognize that a TMRCA of five or seven generations is almost as likely. To avoid this problem it is preferable to focus on the cumulative TMRCA distribution at specific confidence levels or percentiles when assessing relatedness between haplotypes (Figure 19.3b).

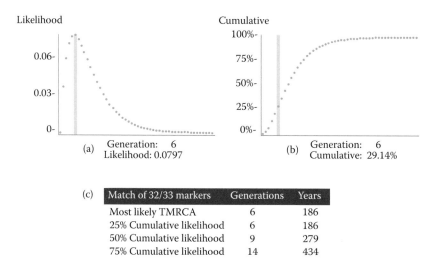

Figure 19.3 Time to most recent common ancestor (TMRCA) estimation from Y chromosome STR data: (a) Bayesian posterior distribution for time to most recent common ancestor between two 33-locus Y-STR haplotypes matching at 32 loci, using locus specific mutation rates and assuming an infinite alleles mutation model. (b) Cumulative distribution of the posterior TMRCA distribution from panel (a). (c) Summary of key values for assessing relatedness between the two haplotypes being compared.

19.3.1.3 Y Chromosome Haplogroups and Deep Ancestry

Although genealogists are fundamentally concerned with ancestry from the relatively recent past they have also shown intense interest in their deeper ancestral roots stretching back to over 100,000 years ago. Consequently, many Y chromosome tests include a component for Y haplogroup analysis to provide insight into deep ancestry. Haplogroup analysis is based on a robust and well-resolved Y chromosome phylogeny (Y chromosome gene tree) that delineates the sequential branching order and geographic distributions of genetic variants within global populations (Figure 19.4) (Karafet et al. 2008). Each branch on the tree represents a haplogroup defining mutation, typically a single nucleotide polymorphism (SNP). SNP markers have very low mutation rates (on the order of 10^{-8} mutations/generation) (Xue et al. 2009) and therefore are likely unique events when they occur. A haplogroup defining mutation is shared by all living males who are direct paternal line descendants of the ancestor who first carried the mutation. The Y chromosome tree currently includes hundreds of haplogroups and is continually expanding as new mutations are discovered. The International Society of Genetic Genealogy (ISOGG) maintains an up-to-date version of the Y chromosome tree on their website as a free resource to the genetic genealogy community (http://www.isogg.org). To determine a customer's haplogroup, genetic genealogy companies either directly assay haplogroup defining SNPs or they employ haplogroup prediction algorithms that use Y-STR marker data to infer haplogroup membership (Schlecht 2008).

The Y chromosome has a lower effective population size due to its haploid presence only in males, rendering the Y chromosome particularly sensitive to demographic forces such as bottlenecks, genetic drift and founder effects. As such, branches of the Y phylogeny are highly correlated with geography. Over the past decade numerous Y

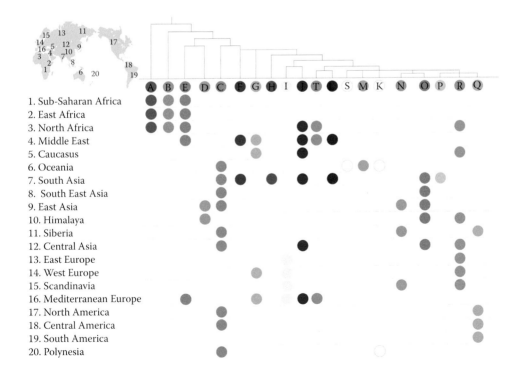

Figure 19.4 The phylogenetic relationships of the 20 major Y haplogroups and their typical geographic affiliations.

chromosome phylogeographic studies have been conducted on global populations that have contributed significantly to our understanding of the migration routes followed by modern humans to disperse out of Africa and spread across the globe. Hence, by providing a genealogist with his Y haplogroup membership, he can trace his paternal lineage within the Y chromosome phylogeny and understand how his ancient ancestors participated in global migrations.

19.3.1.4 Future of Y Chromosome Testing

The field of genetic genealogy is poised on the threshold of having the ability to more comprehensively profile a male individual's Y chromosome at the genome scale. Recently a pioneering study was published (Wei et al. 2013) that provides a possible didactic blueprint for how, in the not too distant future, genetic genealogists may be able to extract a much more detailed profile of an individual's constellation of single nucleotide variation on his Y chromosome. The research team reported the patterns of sequence variation detected across the same identical territory of approximately 9 million bp for each of 36 men of diverse continental ancestry in which eight of the 20 major haplogroups were represented. They listed the more than 6600 SNPs they detected in the sample set, that ranged from ancient haplogroup defining SNPs located deep in the tree to much younger (i.e., rarer) and more "private-like" SNPs located within some of haplogroups, especially those which had a larger membership sample size. Obviously, reliable knowledge of these rare recently evolved SNPs would be particularly valuable in a genealogical context. A representation of their phylogeny with the number of SNPs discovered within 3.2 Mb of sequence subjected to high resequencing coverage on each branch is shown in Figure 19.5. While the calibration of the rate at which SNPs accrue on the line of descent of a lineage is yet to be finalized, it is data like these, namely the number of accumulated SNPs on lineages which provide the opportunity to estimate both the entire time depth of the tree as well as the individual branches. Noteworthy are the comb-like branching patterns associated with the more numerous E and R haplogroup affiliated samples. Generally, in flawless data, all samples should have approximately the same overall number of mutations tracing back to the root of the phylogeny under a model of neutrality. The variability of overall lengths observed in this dataset can to a large extent be attributed to sequencing errors. While challenges associated with sequence errors, especially false negatives must be resolved in the future, sequence data similar to these demonstrate a path forward to more fully reveal the recent genealogical relationships between Y chromosome genomes.

19.3.2 mtDNA Testing

19.3.2.1 Discovering Maternal Lineages

Like the Y chromosome, mitochondrial DNA (mtDNA) follows a strict lineal inheritance pattern (Figure 19.2b). MtDNA is passed along the direct maternal line from mother to all her children. Hence, unlike the Y chromosome both males and females carry mtDNA but only females pass it to the next generation. Genealogists primarily use mtDNA testing as evidence to support or disprove a suspected maternal relationship rather than to identify a MRCA occurring in the recent past. The reason mtDNA is not used as frequently for this purpose is because the lower mutation rate of mtDNA, relative to that of Y-STRs, results in TMRCA values that predate the historical timeframe that is of most interest to genealogists.

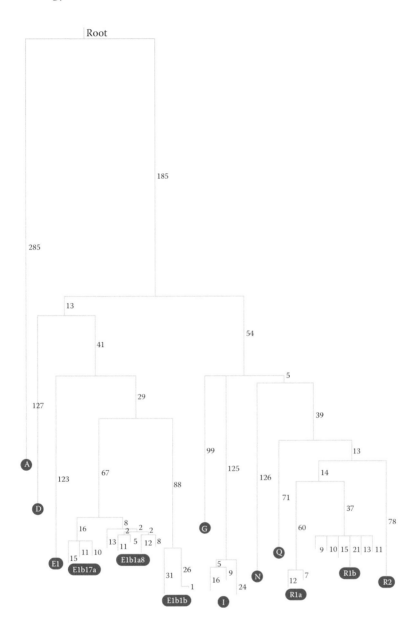

Figure 19.5 Gene tree of variants detected while resequencing identical 3.2-Mb regions in 36 Y chromosomes: Branch length is proportional to the number of mutations indicated along each branch. (Adapted from Wei, W., Ayub, Q., Chen, Y. et al., *Genet Res*, 23 (92), 388–395, 2013.)

19.3.2.2 mtDNA Testing Options

The most common type of mtDNA testing used for genealogical purposes involves sequencing one or more of the three hypervariable segments (HVS1, HVS2, HVS3) located within the mtDNA control region. With a mutation rate of 10^{-6} mutations/site/generation (Howell et al. 2003) the hypervariable segments display a significantly higher mutation rate than that of the coding region (Soares et al. 2009). By targeting the regions with the highest mutation rate, HVS testing captures the greatest amount of variability for the lowest cost, making the test affordable for consumers.

Although HVS testing is more common, DNA testing companies also offer whole mitochondrial genome sequencing in which the entire 16,569-bp circular genome is sequenced. While significantly more expensive, the whole genome test offers higher resolution for DNA matching and haplogrouping purposes. Similar to Y chromosome testing, mtDNA haplotypes are compiled into databases and often linked to genealogical information such as geographic locations and pedigree charts. Additionally, mtDNA haplogroup analysis is also included as part of most mtDNA tests to shed light on ancient maternal ancestry.

19.3.3 Future of Haploid Testing

The general trend in both Y and mtDNA testing is to continually test more markers to achieve greater resolution for matching purposes. By increasing the number of markers tested, rare alleles can be identified that connect populations and family groups on an ever finer scale. However, both the Y chromosome and mtDNA are limited in this capacity due their small sizes. Moving to autosomal testing offers a much greater opportunity to identify those rare alleles that have arisen over the recent past and are therefore of most use for genealogical purposes.

Numerous but rare sequence variants have arisen in the recent past as populations have experienced recent rapid growth (Keinan and Clark 2012). Furthermore, the probability that these newly arising mutations are passed on increases because they have not yet been lost by drift, purifying selection or rearranged by recombination. Analyses of a relatively small number of common variants can typically reveal continental (Lao et al. 2006) and often subcontinental patterning as well as recent admixture events in the offspring of parents of differing continental ancestry, especially as additional possible source populations are added to genealogy databases. While currently much more challenging from both an allele calling accuracy and database size perspective, the large reservoir of rare variants provides a potentially salient and powerful tool for detecting the distinguishing shared ancestry sought by genealogists ultimately at the familial level since the most recent and most heavily populated generations contribute the most recent *de novo* variation.

19.4 Autosomal Testing

19.4.1 Why Assay the Autosomes?

Both the Y chromosome and mtDNA reveal deep, detailed genealogical information that may provide important insights about personal ancestry. Uniparental inheritance patterns and a lack of recombination also make the interpretation of results in these haploid systems more straightforward. However, inferences based on Y and mtDNA will only be relevant to the paternal and maternal lineages respectively and these lineages only represent a small proportion of a complete family pedigree (Figure 19.2). More importantly, the history associated with the purely paternal and maternal lineages may be biased. In cases of sex biased migration, the Y chromosome and mitochondrial DNA may reveal different details about an individual's genealogical history. For example, the initial European colonization of the Americas in the 1500s primarily involved the migration of Southern European men to South and Central America (Mesa et al. 2000; Bryc et al. 2010). These men frequently married Native American women and living descendants of these couples retain admixed genetic ancestry having inherited genetic material from at least two distinct continental populations. Thus, many Y chromosome lineages present among Latinos in the Americas will be indicative of European ancestry. On the other hand, mtDNA analysis is more likely to suggest Native American ancestry. In any case, exclusively analyzing Y and mtDNA is unlikely to tell a complete story.

Fortunately, technological developments in the last decade have led to the opportunity to interrogate the entire human genome at greatly increased breadth. Although the rapid pace of next-generation sequencing innovation has made forensic genetics and genetic genealogy based on full genome sequences a possibility, this section will focus on what is presently possible with single nucleotide polymorphism genotyping microarrays, often called SNP chips, as this is currently the preferred technology in the industry. With genotyping microarrays, it is possible to assay millions of positions across not only the Y and mtDNA systems, but also within the autosomes. SNP chips assay alleles at a prescribed set of genomic sites known to be variable across multiple human populations. Since the Y and mtDNA make up less than 2% of the genome, a significantly more complete picture of genetic inheritance across the entire genome is gained by surveying autosomal sites as well. More importantly, as explained in the following section, the diploid nature of the autosomes makes it possible to inherit genomic segments from any ancestor in a pedigree.

19.4.2 Meiosis—A Genealogical Double-Edged Sword

Each human carries 22 pairs of autosomal chromosomes. One copy of each pair has been inherited maternally and the other paternally. During sexual reproduction, each parent transmits one copy of each autosome plus one sex chromosome (X or Y) through a haploid gamete to his or her child. Gametes are created by meiotic cell division, a process by which a single diploid cell replicates its DNA and then divides twice into four haploid cells. During Meiosis I autosomal DNA is replicated, undergoes recombination, and assorts independently into two haploid cells. The process of recombination allows genetic exchange between the non-sister chromatid copies of a chromosome. The result is the creation of mixed chromatids containing segments from both the paternal and maternal chromosome copies (Figure 19.6a). Following recombination, independent assortment partitions the chromosome copies so that each daughter cell gets only one copy of each chromosome. This partitioning occurs independently for each chromosome pair. After Meiosis I has completed, each daughter cell will complete a further division to separate sister chromatids into four haploid cells, and at least one of these cells will become a gamete. Both the process of recombination and the process of assortment have probabilistic aspects and this leads to the following facts that are specifically relevant for genetic genealogy:

1. An inherited autosome typically includes a mixture of genetic material from both copies of the autosome contained in the parent.
2. Autosomal inheritance is semi-conservative, meaning that a child's genome will only contain 50% of the genetic material present in each parent.
3. The mixture present in a transmitted autosome will likely be different for each inheritance event (i.e., two children will inherit different mixtures of each parent's autosomal chromosome copies unless they are monozygotic twins).

The first point implies that it is possible for an individual to inherit autosomal DNA from any ancestor in their pedigree (Figure 19.6b). While this suggests that the autosomes will provide much broader genealogical information, an additional consequence of recombination and independent assortment is genetic loss, described in point two. During meiotic

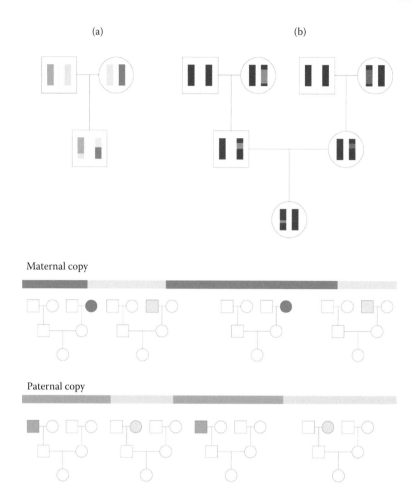

Figure 19.6 The effect of recombination and assortment on autosomal inheritance: (a) For every autosomal chromosome, a child gets one copy from each parent. The inherited copy from each parent is a mixture of each parents' two copies of that chromosome. On the scale of whole chromosomes, the breakpoints for these mixtures are distributed uniformly. (b) Through independent assortment following recombination, it is possible to inherit autosomal DNA from any ancestor (e.g., a paternal grandmother shown in blue). However, this assortment process also leads to the fact that any particular piece of DNA has a chance of being "lost" at each generation (red). (c) Contiguous blocks of the genome are inherited from the same ancestor, but each block may come from a different source, leading to an inheritance mosaic along each chromosome.

division, only one of the two mixed chromosome copies is transmitted into the haploid gamete. Because transmission is random, there is a fifty percent chance that a particular autosomal segment will not be transmitted to a child (Figure 19.6b). Since the process repeats each generation, the likelihood that a DNA segment is transmitted successfully from an ancestor to a descendant decreases with the number of generations between them. Point three highlights that each transmission event results in a unique mixture of parental autosomal DNA. This implies that while two siblings may each inherit 50% of their autosomal DNA from their mother, they are only expected to share 25% of that material with each other. Since they are also expected to share 25% of their paternally transmitted autosomal DNA, the total expected DNA sharing for full siblings is 50%. However, due to the stochastic nature of recombination this percentage can vary widely. Thus each autosome is a mosaic, made up of contiguous blocks of DNA each inherited from a single lineage in the pedigree (Figure 19.6c). The maternal copy contains segments inherited from lineages on the maternal side while the paternal copy will contain segments inherited from lineages on the paternal side.

19.4.3 Genetic Genealogy with Autosomal DNA

There are two primary avenues by which autosomal genetic analysis enhances genealogical research. First, it is possible to determine the historical populations that contributed significant amounts of genetic material to an individual. This is analogous to Y haplogroup analysis and mtDNA maternal lineage identification. By inferring ancestral genetic origins an individual may gain insight into deep family history. Second, similarly to Y STR haplotype matching, autosomal DNA can be used to identify relatives. By comparing the genetic content of two individuals, it is possible to infer not only that a relationship exists, but also to estimate how closely related two people are to one another. Identifying relatives creates opportunities to add these individuals to a family tree. Moreover, if an identified relative has recorded his or her family history, it can be used to help grow one's own family tree if a common ancestor can be identified.

19.4.4 Ancestral Origin Estimation

Estimating ancestral origins using autosomal DNA has recently become a very active area of research. Not only has this pursuit been successful for individuals alive today (Li et al. 2008; Nelson et al. 2008; Novembre et al. 2008; Winney et al. 2011; Reich et al. 2012) a surprising amount of knowledge has been gained from estimating ancestry in studies involving ancient DNA including specimens from Neanderthal (Green et al. 2010), extinct archaic humans (Reich et al. 2010), an extinct anatomically modern Palaeo-Eskimo (Rasmussen et al. 2010) and a 5300-year-old frozen mummy from the European Alps (Keller et al. 2012). As more genetic data is amassed across an ever-broadening range of modern human populations a number of powerful methods have been developed to use these data to answer questions regarding population structure and migration (Patterson, Price, and Reich 2006; Pritchard, Stephens, and Donnelly 2000; Alexander, Novembre, and Lang 2009; Lawson et al. 2012). While it should be noted that these methods have primarily been developed for demographic inference, they can often be easily adapted to infer ancestral origins using reference panels consisting of individuals with known ancestry. Though each approach may apply a unique algorithm, they all rely on genetic variation data, created primarily by population bottlenecks, mutation, genetic drift and selection.

Anatomically modern humans originated approximately 100–200 kya (thousands of years ago) in Eastern or Southern Africa, migrating out of sub-Saharan Africa 100 kya, first to the Near East (60–100 kya), then Europe and Asia (~40 kya), and finally the Americas (~15 kya) (Owens and King 1999; Underhill and Kivisild 2007; Henn, Cavalli-Sforza, and Feldman 2012). This migration proceeded through a serial-founder process in which each new population was seeded by a small sample of individuals from the neighboring larger ancestral group. Repetitive small sampling creates initial population differentiation due to differences in the frequency of genetic variants between the ancestral population and the seed sample. Following the establishment of distinct populations across the globe, historically limited migration has preserved the initial differentiation while mutation and genetic drift, the fluctuation in variant frequencies from generation to generation, has increased it. The methods developed for ancestral origin estimation can be broadly grouped into two classes based on whether they are designed to provide a single genome-wide ancestry estimate or localized ancestry estimate for each segment of the genome.

Though the first method for quantifying genetic variation within and between populations is probably due to Sewall Wright (Wright 1949), genome-wide ancestry estimation began in earnest with early work from Luca Cavalli-Sforza and colleagues, in which they demonstrated that patterns in human genetic variation are indicative of population history and can thus be used to identify ancestral origins (Cavalli-Sforza and Bodmer 1971). Cavalli-Sforza went on to be the first to apply principal components analysis (PCA) to genetic data and this is the inference method we consider first (Menozzi, Piazza, and Cavalli-Sforza 1978).

PCA is a mathematical method for performing an orthogonal transformation of a data matrix (Pearson 1901). The result is a matrix containing a set of linearly orthogonal vectors (principal components) ordered such that the first has maximal variance and each successive vector has the next largest variance. When using genotyping array data, PCA is applied to a data matrix containing I rows, each row representing an individual, and J columns, each column representing a SNP. Entry $g_{i,j}$ in the matrix represents the genotype of SNP j for individual i. Genotypes are codified as 0, 1, or 2 representing the count of a specific allele at the given site. Since PCA is an unsupervised method that simply returns a set of orthogonal vectors decreasingly ordered by variance, care must be taken in preparing data for PCA if the goal is to reveal population structure and infer origins. It is first important to "thin" the data set to reduce genotype correlation between alleles observed at neighboring sites known as linkage disequilibrium (LD). If the data are not thinned, it is possible that the components containing maximal variance will represent the information in a highly correlated set of variants in a single localized region of the genome. This type of variation is typically due to a genomic event that is specific to this region, such as a selective sweep, rather than capturing variation that is attributable to demographic forces like migration and differentiation that have a genome-wide affect on patterns of variation. It is also necessary to remove closely related individuals as these too may skew the transformation process. Finally, PCA is sensitive to the relative scaling of each variable, so each SNP column must be normalized to have mean zero and standard deviation of one. Following thinning, removal of related samples, and normalization we are typically guaranteed that the first few principal components will represent large-scale population structure. We can then make a scatter plot of the first two PCs in which each individual is given an x-coordinate based on PC1 and a y-coordinate based on PC2. Typically, we will see that the first two axes of the transformed matrix, the two accounting for maximal amounts of variance in the original genotype data, recapitulate the geographic origin of each sample (Figure 19.7a). Individuals from the same population will cluster together and population clusters themselves will tend to be in the proximity of their nearest geographic neighbors. Moreover, the Euclidean distance between pairs of points in this scatter plot is proportional to the genetic differentiation between them. For example, in Figure 19.7a, Greek samples ("G's" lower right) are more genetically similar to Italian samples ("I's" lower right) than Estonian samples ("E's" top center).

If we wish to infer ancestry for an individual of unknown origin, we first require a collection of reference samples, each with deep ancestry from a single point of origin such as those in Figure 19.7a. Given this reference PCA vector space, we simply project the new individual in to this space and infer the likely ancestral origin to be the population in which the new individual clusters. This approach works well for individuals with a single point of origin. However, for admixed individuals with recent ancestry from multiple source populations the problem is more complex. If an individual has a mixture of two ancestries,

Genetic Genealogy in the Genomic Era

they will be projected along the axis between these two population clusters at a position that is proportional to the relative amounts of each ancestry the individual's genome contains. For example, an individual with 50% Spanish ancestry and 50% Swedish ancestry would fall at the midpoint between these two clusters. In this case they would appear to fall within the French cluster. Thus without further information, PCA-based inference can be misleading for admixed individuals. The software package EIGENSTRAT represents the most developed application of PCA to ancestry inference (Price et al. 2006).

Another popular approach for genome-wide ancestral origin estimation is based on work by Jonathan Pritchard and colleagues (Pritchard et al. 2000). They developed a Bayesian model-based clustering algorithm called STRUCTURE that uses genome-wide genotyping data to infer ancestry. Given the number of expected population clusters in the data K and a genotype data matrix G where $g_{i,j}$ is the genotype for individual i at SNP j, the model consists of a $K \times J$ matrix P where $p_{k,j}$ is the frequency of one allele for population k at SNP j and an $I \times K$ matrix Q where $q_{i,k}$ is the proportion of ancestry individual i contains from population k. In the initial implementation the authors use a Markov chain Monte Carlo

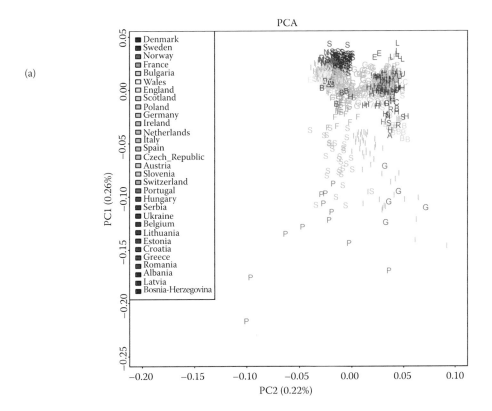

Figure 19.7 Ancestral origin inference from autosomal data: (a) Principal components analysis is used to project high dimensional data to a low dimensional space while still capturing the maximum amount of variation possible in the original data. In this case, we project a matrix of ~135,000 SNPs across ~1000 individuals of European ancestry down to a two dimensional space. The first two components of variation in this data reveal that individuals from the same geographic location tend to be more genetically similar. This is evidence of population structure and can be used to identify the ancestral sources of genetic information. (b) Results of running the clustering method Admixture with five clusters on ~100,000 SNPs for ~700 individuals. Each vertical bar represents an individual colored by proportions of ancestry belonging to each population cluster.

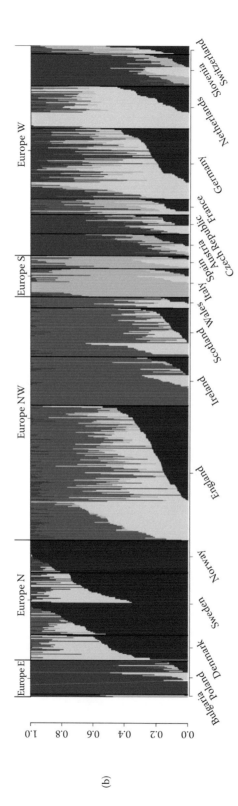

Figure 19.7 (Continued) Ancestral origin inference from autosomal data: (a) Principal components analysis is used to project high dimensional data to a low dimensional space while still capturing the maximum amount of variation possible in the original data. In this case, we project a matrix of ~135,000 SNPs across ~1000 individuals of European ancestry down to a two dimensional space. The first two components of variation in this data reveal that individuals from the same geographic location tend to be more genetically similar. This is evidence of population structure and can be used to identify the ancestral sources of genetic information. (b) Results of running the clustering method Admixture with five clusters on ~100,000 SNPs for ~700 individuals. Each vertical bar represents an individual colored by proportions of ancestry belonging to each population cluster.

approach to simultaneously estimate both *P* and *Q*. While the *P* matrix can be informative with regard to the architecture of populations contained in the sample, it is the *Q* matrix which provides an ancestral origin estimate for each sample. One obvious advantage with this approach is that q_i, is a numerical estimate of ancestry proportions whereas quantifying ancestry from PCA requires further extrapolation. More importantly, this method does not have difficulties distinguishing individuals of admixed ancestry from those with a single ancestral origin. It does however require both data thinning and removal of related samples much like PCA. Recently, a significantly faster model-based method called Admixture has become a standard clustering method for ancestry inference (Alexander, Novembre, and Lang 2009). Admixture is based on the likelihood model from STRUCTURE but uses a novel optimization approach to reduce computation time while returning comparable estimates of *P* and *Q*. An example of the estimated *Q* matrix for ~700 Europeans is provided in Figure 19.7b. In this case, *K* was set to five. As you can see, the majority of individuals are estimated to be composed of multiple ancestries. However, much like we observed in Figure 19.7a individuals from neighboring countries tend to show similar ancestry proportions. It is also worth noting that individuals from the same country often show a gradient in the proportion of a given ancestry. For example the "red" cluster in English samples varies from zero to nearly 50%. This gradient may be useful in further identifying ancestral origins. In the example above, the "red" cluster seems to represent Scandinavian ancestry and this suggests that English individuals with significant amounts of this ancestry may have ancestors from along the Eastern coast of England where Scandinavian admixture was historically more prevalent. Admixture also provides the opportunity to perform a semi-supervised clustering analysis in which a reference panel of single-origin individuals is provided and ancestry is estimated for only a subset of individuals of unknown origin.

As the density of genotyping arrays has continued to increase, it has become possible to provide ancestry estimates for each small segment along the genome. This is commonly known as local ancestry inference, where "local" is with respect to genomic position and numerous methods have been developed to estimate local ancestry (Sankararaman et al. 2008; Patterson et al. 2004; Price et al. 2009; Brisbin et al. 2012). While in theory this approach could be applied to any individual, in practice it is only useful for analyzing individuals with ancestry from multiple, significantly diverged populations. This is because the amount of information for distinguishing ancestral origins in each segment is small. Many methods attempt to alleviate this issue by using approaches such as hidden Markov models that allow for information sharing across neighboring segments, but the contributing source populations must still have a high degree of differentiation. Unlike the genome-wide ancestry methods presented above, local ancestry methods typically must be run in a supervised fashion and thus require reference data. Another difference is that nearly all of these methods require phased haplotype data rather than simple genotypes. A haplotype is the linear sequence of alleles observed across consecutive sites on the same chromosome copy. Since genotyping arrays simply return the unordered pair of alleles present at each site, we must first "phase" the data to recover the correct allele pairing from site to site. Local ancestry results tend to be highly dependent on the quality of the haplotype data used, so care must be taken in phasing both the reference panel and study samples. One notable exception to the phasing requirement is the algorithm HAPMIX which is arguably the gold standard in local ancestry inference (Price et al. 2009). While HAPMIX requires phased reference data, input sample data is unphased. Local ancestry analysis provides a high-resolution map of segments of the genome that originated from each of two or more populations. This information can often provide surprising details about one's origins. For example, while the average African-American genome is composed of approximately 80% West African segments and 20% European segments, there is significant variation in both the number and location of these segments. It might be enlightening to see what particular regions of a genome are derived from each source. For example, if many of the genes responsible for determining skin, hair and eye color lie within segments of a specific ancestral origin, it may be expected that an individual's appearance might be indicative of this origin. However, caution must be taken with these types of inferences as many phenotypic traits are polygenic and few are completely mapped such that we have a full understanding of how genotype controls phenotype. It has also been shown that the length distribution of segments of a given ancestry may be informative about when in the past the admixture event occurred. Since recombination acts to separate neighboring segments of DNA that might have been co-inherited from the same ancestor, it will also shorten contiguous segments of a given ancestry in admixed individuals over time. Thus, observed ancestry segments will tend to be shorter the more distant in time the admixture event was. As discussed in the next section, it is this pattern of coinheritance of neighboring segments that also provides information on relatedness.

19.4.5 Relative Identification

Ancestral origin inference can provide exciting clues to one's history, but the primary use of genetic data in genealogical research is in the discovery of relatives. Physical record databases including census data, military records and parish archives are typically the primary sources of information used to research and build one's family tree. However, the addition of genetic data provides a powerful independent source of information with which to confirm suspected relationships and discover new, previously unknown relatives.

In genetics, relative identification is accomplished by identifying pairs of individuals who share one or more common segments of DNA that are identical by descent (IBD). This implies that this pair of individuals is related and has inherited this segment of DNA from a common ancestor. More importantly, as with local ancestry both the amount of shared DNA and the number and length of shared segments varies with the number of generations back to the common ancestor. By deriving some simple quantities relating to expected IBD sharing proportions we can demonstrate the power of this approach for relative identification. The process of independent assortment in sexual reproduction implies that the probability that a particular position in the genome is shared IBD for a pair of individuals that are each g generations descended from a single common ancestor is $2^{-(2g-1)}$. This is also the expected proportion of the genome that will be shared IBD. If we wish to calculate the expected number of IBD segments, we simply need to multiply the probability of IBD by the expected number of independently inherited contiguous segments. Under simplifying assumptions, recombination can be thought to occur following a Poisson process each generation with a rate of one event per Morgan of sequence. A Morgan is a unit of genetic length, measuring the distance between two chromosome positions in which the expected number of crossover recombination events per meiosis is one. Thus the expected number of recombination events per generation is one multiplied by the length in Morgans, L, of the autosomal portion of the genome (for humans this is approximately 35.3 Morgans [McVean et al. 2004]). Since this process occurs independently each generation, we can use the fact that the sum of n independent Poisson processes each with a rate of r can be represented as a single Poisson process of rate nr. In our case, this implies that the expected number of recombinations between two individuals each g generations descended from a single common ancestor is $2Lg$. It is important to remember that the autosomal portion of the genome is already separated in to 22 independently inherited chromosomes, thus the final number of independently inherited autosomal segments is expected to be $(2Lg + 22)$ (Thomas et al. 1994; Huff et al. 2011). By multiplying this quantity by the chance that each segment has been inherited IBD, we see that the expected number of IBD segments, $N_s(g)$, in this case is $(2Lg + 22)2^{-(2g-1)}$. As g increases, this expectation is dominated by the $2^{-(2g-1)}$ term and thus for $g > 4$ the expected number of autosomal IBD segments is less than one (Table 19.1). This means that for distantly related individuals there is often a large probability that no IBD segments exist. The expected length of IBD segments is also related to the number of generations since common ancestry is due to the effect of recombination. Segment lengths, again measured in Morgans, tend to be exponentially distributed with a mean of length of $1/2g$, where g is again the number of generations back to the common ancestor for each individual. For example, for two individuals that are half first cousins (sharing one grandparental ancestor two generations in the past) the expected length of a shared segment is 0.25 Morgans, or 25 centiMorgans (cM). Similarly to the expected number of segments, the expected length of segments decreases with increasing g. Beyond $g > 10$, the expected segment length is only 5 cM and without very high-quality data it becomes very difficult to identify these segments.

While it may seem that IBD analysis is doomed to identify only recent shared ancestry, and only a small portion at that, this fact is alleviated by the sheer numbers of relatives each individual has for a given degree of relatedness. To calculate the expected number of relatives $N_r(g)$ an individual has due to common ancestry from g generations in the past we simply multiply the number of ancestor couples from generation g by the expected number of descendants each of these ancestral couples will produce in the current generation. This is equal to $2^{g-1} o^{g-1} (o - 1)$ where o is equal to the average number of offspring per generation (Henn et al. 2012). If we assume $o = 2.5$, which is not unreasonable for many human populations, we can see how this quantity grows (Table 19.1). What is surprising is the sheer number of relatives one expects at each level. Thus, although the expected number of IBD segments for a given pair of relatives is often far less than one, the number of relatives at a given generation of descent g implies that the chance of sharing an IBD segment with at least one individual who is related through an ancestor g generations actually increases with g (Table 19.1). While this

Table 19.1 Expected Identity by Descent and Numbers of Relatives by Degree of Relatedness

Relationship	Generations to Common Ancestor (g)	Expected # of Shared Autosomal Segments per Ancestor ($N_s(g)$) $(L*(2g) + 22)* 2^{-(2g-1)}$	Expected # of Relatives of This Degree ($N_r(g)$) $(2^{g-1}) * o^{(g-1)} * (o - 1)$	Expected # of IBD Segments from All Relatives of This Degree $2 * N_s(g) * N_r(g)$
1st Cousins	2	20.40	7.5	306.0
2nd Cousins	3	7.31	37.5	548.0
3rd Cousins	4	2.38	187.5	891.8
4th Cousins	5	0.73	937.5	1373.3
5th Cousins	6	0.22	4687.5	2039.8
6th Cousins	7	0.06	23,437.5	2953.7
7th Cousins	8	0.02	117,187.5	4197.1

Note: This table shows how the expected number of IBD segments (column 3) and relatives (column 4) vary by the degree of relationship (column 1). Column 4 represents the expected number of IBD segments inherited from all full relatives of a given degree. In this example $L = 35.3$ and $o = 2.5$.

demonstrates the power of using genetic data to identify relatives, the ability to successfully identify IBD segments is heavily dependent on data quality and the methods applied.

As with the ancestral origin inference, the identification of relatives using IBD analysis has been an active area of research for many years (Purcell et al. 2007; Browning and Browning 2010; Gusev et al. 2009; Browning and Browning 2011). While most methods rely on phased data as input, a few have been developed to identify long sequences of IBD consistent genotypes from unphased data. In this case, as long as two individuals share at least one allele at each position they are said to be IBD consistent. Methods that do not require phasing are fast, however minimizing false positives using only genotype data requires setting a fairly conservative threshold on the minimum number of consecutive IBD consistent genotypes one must observe before calling a segment IBD. This problem is worsened by the presence of linkage disequilibrium (LD) which may appear to increase the length of an IBD consistent segment due to the fact that consistency at one site will increase the probability of consistency at correlated neighboring sites. This implies that such methods will only have power to identify fairly large IBD segments and thus fairly recent common ancestry. One such example is the IBD method included as part of the PLINK analysis package (Purcell et al. 2007).

In contrast, significantly more sensitive methods can be developed which either take as input phased data or estimate IBD in conjunction with phasing. Many of these model-based methods are computationally expensive and thus only suitable for small data sets. However, two more recently developed methods of this type, GERMLINE and fastIBD, use hashing to regain some speed while maintaining improved sensitivity (Gusev et al. 2009; Browning and Browning 2011). Beginning with phased data, GERMLINE first identifies short, exact matching segments as anchor matches. From each anchor a matching segment may be extended using a more permissive "fuzzy" matching algorithm. This procedure allows for both phasing mistakes and genotyping error (Figure 19.8). Both are major sources of error in IBD analysis. A minimum segment length threshold is still applied, though it can be significantly shorter than that used for the unphased methods. The fastIBD method works in a similar fashion maintaining a probabilistic model of each individual's haplotypes and identifying candidate IBD segments as shared "rare" haplotypes. While a frequency threshold may appear to be distinct from the length threshold, they are actually much the same. As the length of the segment in question increases, the observed haplotype tends to become less common as more and more unique haplotypes are possible. Both of these methods can be run on data sets consisting of thousands of samples and provide accurate IBD segment identification of segments as small as 2–5 cM. Given this information, it is possible to identify even fifth cousins (six generations of separation) with some degree of certainty.

19.4.6 Conclusions

It is clear that autosomal analysis can add a tremendous amount of information to any genetic genealogical research project. Whereas the Y chromosome and mtDNA provide detailed information on single lineages, autosomal chromosomes undergo recombination and independent assortment during transmission and thus they may provide information about all ancestral lineages simultaneously. While this is a significant benefit, it comes at a cost, as genetic material is also lost through the same process each generation. Recombination and assortment also make the process of inference more difficult. A lack of recombination in haploid systems means that all samples can be placed on a single phylogenetic tree and the exact sequence of alleles observed implies only one possible history for each sample. On the other hand, the

Genetic Genealogy in the Genomic Era 503

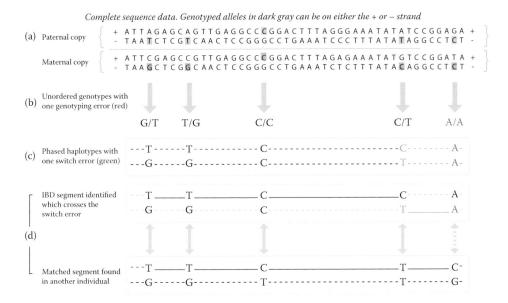

Figure 19.8 Identifying relatives through identity-by-descent (IBD) analysis: (a and b) Genotyping arrays provide unordered pairs of alleles observed from both copies of a chromosome assayed on one strand. Occasionally the assay returns erroneous genotypes (red). (c) These allele pairs must be phased to determine the underlying relationship between neighboring sites and to separate the chromosome copies. This process is imperfect and some switch errors (phasing continuity breaks) will remain (green). (d) Once phased, an individual's haplotypes can be compared to haplotypes from other individuals to identify long shared stretches that have been inherited identical by descent. IBD algorithms have been designed to allow for both genotyping and switch errors.

combination of alleles observed along an autosomal chromosome can be inherited from one or more ancestors and each segment may have a history that is effectively independent from neighboring regions and the effect of this fact must be accounted for in any analysis.

19.5 Discussion

The previous sections introduced and discussed the two primary genealogical questions on which we can gain insight via genetic analysis: haplogroup analysis or ancestral origin inference and STR haplotype matching or relative identification. From a broader perspective these are in fact the same question but on different timescales. In the context of ancestral origin inference, a population is a group of individuals with a somewhat distinct genetic signature due to mating within the group and isolation from other neighboring groups. In some sense this is also a family. On the other hand, a family can be thought of as a narrowly defined population, since members of a family share genetic material due to common ancestry and this material can be used to assign individuals to this group. Currently, the majority of methods for identifying ancestral origins are effective at distinguishing large-scale differences in populations that may be separated by many hundreds or thousands of years of evolution. On the other hand, our best IBD methods can identify relationships at a depth of roughly 5–15 generations which is approximately 150–450 years ago. Improvements in the methods for both of these types of analyses continue to close this gap, with the end goal that we can trace single DNA segments through space and time capturing the entire inheritance history from a large scale continental population in the distant past to a specific family in the last few generations. While this is obviously an extremely ambitious goal, active research in these areas is leading to progress. For more detail on these topics please see the recent review by Novembre and Ramachandran on ancestry inference and the recent review by Browning and Browning on IBD inference (Novembre and Ramachandran 2011; Browning and Browning 2012). Progress has not been limited to methods development. As mentioned in the Y testing discussion, new technologies in assaying genetic data are also enabling finer-scaled analyses. Most notably Next-Generation Sequencing (NGS) technologies are rapidly improving our catalog of global human variation. It should also be noted that the growth in this field is not limited to academic and casual users. In recent years, genealogical research companies such as Ancestry.com and Family Tree DNA have invested significant capital in expanding from Y and mtDNA analysis offerings to complete genome analysis demonstrating an understanding of the benefits of applying genetics to genealogy and a willingness to provide customers with the best tools to improve family history research.

Companies offering genetic testing can boast of delivering customers some truly amazing discoveries. Demand for these tests is due in part by an assumption that genetics can be used to "prove" the existence of a particular ethnic background in one's past, or "confirm" a relationship based on tenuous records. Similar to DNA forensics, the allure of using genetic analysis for genealogy is in part due to the perceived infallibility of the information. However, it is important to remember that although genetic analysis offers a powerful independent source of information for genealogical research, it too provides an incomplete, imperfect record of the past and is best used in conjunction with record-based research.

As interest in genetic genealogy continues to grow, companies offering genetic analysis are amassing large collections of individuals who have been sequenced or genotyped. For example, in 2011 23andMe announced it had more than 100,000 samples, while Ancestry DNA has surpassed the 100,000 sample mark in just 8 months of offering their autosomal ancestry test. Larger data sets not only improve the quality of results possible for genetic genealogy, but also open the door for many other research projects. 23andMe has already begun leveraging their data in numerous medical research collaborations. These large collections offer exciting new opportunities for research in forensics, medicine and basic biology, but with these opportunities come a host of new challenges in appropriate data usage, security, and customer protection.

Acknowledgments

We thank Jared Lewandowski and Josh Callaway for their graphic design expertise in creating the figures found in this chapter.

References

Alexander, D.H., Novembre, J., and Lange. K. 2009. Fast model-based estimation of ancestry in unrelated individuals. *Gen Res* 19(9): 1655–1664.
Brisbin, A., Bryc, K., Byrnes, J. et al. 2012. PCAdmix: principal components-based assignment of ancestry along each chromosome in individuals with admixed ancestry from two or more populations. *Hum Biol* 84(4): 343–364.
Browning, B.L., and Browning, S.R. 2011. A fast, powerful method for detecting identity by descent. *Am J Hum Gen* 88(2): 173–182.
Browning, S.R., and Browning, B.L. 2010. High-resolution detection of identity by descent in unrelated individuals. *Am J Hum Gen* 86(4): 526–539.

Browning, S., and Browning, B. 2012. Identity methods in human gene mapping. *Annu Rev Gen* 46(1).
Bryc, K., Velez, C., Karafet, T. et al. 2010. Genome-wide patterns of population structure and admixture among Hispanic/Latino populations. *Proc Natl Acad Sci U S A* 107(Supplement 2): 8954–8961.
Cavalli-Sforza, L.L., and W.F. Bodmer. 1971. *The Genetics of Human Populations.* W.H. Freeman, San Francisco (reprinted 1999 by Dover Publications).
Coble, M.D. 2011. The identification of the Romanovs: can we (finally) put the controversies to rest? *Invest Genet* 2(1): 1–7.
Falconer, B. 2012. We are family? *Bloomberg Businessweek* 30: 81–93.
Foster, E.A., Jobling, M.A., Taylor, P.G. et al. 1998. Jefferson fathered slave's last child. *Nature* 396(6706): 27–28.
Goedbloed, M., Vermeulen, M., Fang, R.N. et al. 2009. Comprehensive mutation analysis of 17 Y-chromosomal short tandem repeat polymorphisms included in the AmpFlSTR® Yfiler® PCR amplification kit. *Int J Leg Med* 123(6): 471–482.
Green, R.E., Krause, J., Briggs, A.W. et al. 2010. A draft sequence of the Neandertal genome. *Science* 328(5979): 710–722.
Gusev, A., Lowe, J.K., Stoffel, M. et al. 2009. Whole population, genome-wide mapping of hidden relatedness. *Gen Res* 19(2): 318–326.
Henn, B.M., Cavalli-Sforza, L.L., and Feldman, M.W. 2012. The great human expansion. *Proc Natl Acad Sci U S A* 109(44): 17758–17764.
Henn, B.M., Hon, L., Macpherson, J.M. et al. 2012. Cryptic distant relatives are common in both isolated and cosmopolitan genetic samples. *PloS ONE* 7(4): e34267.
Howell, N., Smejkal, C.B., Mackey, D.A., Chinnery, P.F., Turnbull, D.M., and Herrnstadt, C. 2003. The pedigree rate of sequence divergence in the human mitochondrial genome: there is a difference between phylogenetic and pedigree rates. *Am J Hum Genet* 72(3): 659–670.
Huff, C.D., Witherspoon, D.J., Simonson, T.S. et al. 2011. Maximum-likelihood estimation of recent shared ancestry (ERSA). *Gen Res* 21(5): 768–774.
Karafet, T. M., Mendez, F.L., Meilerman, M.B., Underhill, P.A., Zegura, S.L., and Hammer, M.F. 2008. New binary polymorphisms reshape and increase resolution of the human Y chromosomal haplogroup tree. *Gen Res* 18(5): 830–838.
Keinan, A., and Clark. A.G. 2012. Recent explosive human population growth has resulted in an excess of rare genetic variants. *Science* 336 (740): 740–743.
Keller, A., Graefen, A., Ball, M. et al. 2012. New insights into the Tyrolean Iceman's origin and phenotype as inferred by whole-genome sequencing. *Nature Comm* 3: 698.
King, T.E., and Jobling, M.A. 2009. What's in a name? Y chromosomes, surnames and the genetic genealogy revolution. *Trends Gen* 25(8): 351–360.
Lao, O., vanDuijn, K., Kersbergen, P., deKnijff, P., and Kayser, M. 2006. Proportioning whole-genome single-nucleotide-polymorphism diversity for the identification of geographic population structure and genetic ancestry. *Am J Hum Genet* 78(4): 680–690.
Lawson, D.J., Hellenthal, G., Myers, S., and Falush, D. 2012. Inference of population structure using dense haplotype data. *PLoS Genet* 8(1): e1002453.
Li, J.Z., Absher, D.M., Tang, H. et al. 2008. Worldwide human relationships inferred from genome-wide patterns of variation. *Science* 319(5866): 1100–1104.
McVean, G.A.T., Myers, S.R., Hunt, S., Deloukas, P., Bentley, D.R., and Donnelly, P. 2004. The fine-scale structure of recombination rate variation in the human genome. *Science* 304(5670): 581–584.
Menozzi, P., Piazza, A., and Cavalli-Sforza. L.L. 1978. Synthetic maps of human gene frequencies in Europeans. *Science* 201(4358): 786.
Mesa, N.R., Mondragón, M.C., Soto, I.D. et al. 2000. Autosomal, mtDNA, and Y-chromosome diversity in Amerinds: pre-and post-Columbian patterns of gene flow in South America. *Am J Hum Genet* 67(5): 1277–1286.
Nelson, M.R., Bryc, K., King, K.S. et al. 2008. The population reference sample, POPRES: a resource for population, disease, and pharmacological genetics research. *Am J Hum Genet* 83(3): 347–358.

Novembre, J., Johnson, T., Bryc, K. et al. 2008. Genes mirror geography within Europe. *Nature* 456(7218): 98–101.

Novembre, J., and Ramachandran, S. 2011. Perspectives on human population structure at the cusp of the sequencing era. *Annu Rev Genomics Hum Genet* 12: 245–274.

Owens, K., and King. M.C. 1999. Genomic views of human history. *Science* 286(5439): 451–453.

Patterson, N., Hattangadi, N., Lane, B. et al. 2004. Methods for high-density admixture mapping of disease genes. *Am J Hum Genet* 74(5): 979.

Patterson, N., Price, A.L., and Reich, D. 2006. Population structure and eigenanalysis. *PLoS Genet* 2(12): e190.

Pearson, K. 1901. LIII. On lines and planes of closest fit to systems of points in space. *London Edinburgh Dublin Philos Mag J Sci* 2(11): 559–572.

Price, A.L., Patterson, N.J., Plenge, R.M., Weinblatt, M.E., Shadick, N.A., and Reich, D. 2006. Principal components analysis corrects for stratification in genome-wide association studies. *Nat Genet* 38(8): 904–909.

Price, A.L., Tandon, A., Patterson, N. et al. 2009. Sensitive detection of chromosomal segments of distinct ancestry in admixed populations. *PLoS Genet* 5(6): e1000519.

Pritchard, J.K., Stephens, M., and Donnelly, P. 2000. Inference of population structure using multilocus genotype data. *Genetics* 155(2): 945–959.

Purcell, S., Neale, B., Todd-Brown, K. et al. 2007. PLINK: a tool set for whole-genome association and population-based linkage analyses. *Am J Hum Genet* 81(3): 559–575.

Rasmussen, M., Li, Y., Lindgreen, S. et al. 2010. Ancient human genome sequence of an extinct Palaeo-Eskimo. *Nature* 463(7282): 757–762.

Reich, D., Green, R.E., Kircher, M. et al. 2010. Genetic history of an archaic hominin group from Denisova Cave in Siberia. *Nature* 468(7327): 1053–1060.

Reich, D., Patterson, N., Campbell, D. et al. 2012. Reconstructing Native American population history. *Nature* 488(7411): 370–374.

Sankararaman, S., Sridhar, S., Kimmel, G., and Halperin, E. 2008. Estimating local ancestry in admixed populations. *Am J Hum Gen* 82(2): 290.

Schlecht, J., Kaplan, M.E., Barnard, K., Karafet, T., Hammer, M.F., and Merchant, N.C. 2008. Machine-learning approaches for classifying haplogroup from Y chromosome STR data. *PLoS Comput Biol* 4(6): e1000093.

Shriver, M.D., and Kittles, R.A. 2004. Genetic ancestry and the search for personalized genetic histories. *Nat Rev Genet* 5(8): 611–618.

Soares, P., Ermini, L., Thomson, N. et al. 2009. Correcting for purifying selection: an improved human mitochondrial molecular clock. *Am J Hum Genet* 84(6): 740–759.

Thomas, A., Skolnick, M.H., and Lewis, C.M. 1994. Genomic mismatch scanning in pedigrees. *Math Med Biol* 11(1): 1–16.

Thomas, M.G., Parfitt, T., Weiss, D.A. et al. 2000. Y chromosomes traveling south: the cohen modal haplotype and the origins of the Lemba— The "Black Jews of Southern Africa." *Am J Hum Genet* 66(2): 674–686.

Underhill, P.A., and Kivisild, T. 2007. Use of Y chromosome and mitochondrial DNA population structure in tracing human migrations. *Annu Rev Genet* 41: 539–564.

Walsh, B. 2001. Estimating the time to the most recent common ancestor for the Y chromosome or mitochondrial DNA for a pair of individuals. *Genetics* 158(2): 897–912.

Wei, W., Ayub, Q., Chen, Y. et al. 2013. A calibrated human Y-chromosomal phylogeny based on resequencing. *Genet Res* 23 92 0: 388–395.

Winney, B., Boumertit, A., Day, T. et al. 2011. People of the British Isles: preliminary analysis of genotypes and surnames in a U.K.-control population. *Eur J Hum Genet* 20(2): 203–210.

Wright, S. 1949. The genetical structure of populations. *Ann Hum Genet* 15(1): 323–354.

Xue, Y., Wang, Q., Long, Q. et al. 2009. Human Y chromosome base-substitution mutation rate measured by direct sequencing in a deep-rooting pedigree. *Cur Biol* 19(17): 1453–1457.

Law, Ethics, and Policy IV

DNA as Evidence in the Courtroom

20

DAVID H. KAYE
FREDERICK R. BIEBER
DAMIR PRIMORAC

Contents

20.1 Introduction	509
20.2 Legal Standards for Admitting Scientific Evidence	510
20.2.1 The General-Acceptance Standard	510
20.2.2 Scientific-Validity Standard	511
20.3 A Short History of Judicial Acceptance of DNA Evidence	512
20.3.1 Phase 1: Uncritical Acceptance	512
20.3.2 Phase 2: Challenges to Laboratory Methods and Population Genetics Models	513
20.3.3 Phase 3: Renewed Acceptance of VNTR Profiling	515
20.3.4 Phase 4: Acceptance of Polymerase Chain Reaction–Based Methods	516
20.4 New Methods and Unusual Applications	516
20.4.1 Mixtures and Low Template DNA	518
20.5 Presentation	518
20.5.1 Framing and Probabilities	518
20.5.2 Trawl Probabilities	520
20.6 Legal Procedures	521
20.7 Postconviction DNA Testing	522
References	524

20.1 Introduction

When judges and lawyers first heard that DNA profiling could identify the source of biological trace evidence such as blood or saliva, they were stunned. In 1988, a trial court in New York wrote that, "if DNA Fingerprinting works and receives evidentiary acceptance, it can constitute the single greatest advance in the 'search for truth' and the goal of convicting the guilty and acquitting the innocent since the advent of cross-examination." (*People v. Wesley*, 533 N.Y.S.2d 643, 644) (Sup. Ct. Albany Cnty. 1988). Considering that the legal profession's leading commentator on the law of evidence regarded cross-examination as "the greatest legal engine ever invented for the discovery of truth" (Wigmore 1974, § 1367), this is high praise indeed. The path to stable judicial acceptance of DNA identification evidence, however, was far from smooth (Aronson 2007, 270; Kaye 2010, 330). Today, as the science advances and new methods are introduced, the process of gaining legal acceptance continues.

This chapter outlines the legal principles that govern the admissibility of scientific evidence, surveys the history of legal challenges to forensic DNA typing, and identifies some current legal issues in the use of the technology in trials. The primary focus is on the legal rules and principles applicable in criminal trials in the United States, but variations in the British and continental legal systems also are mentioned. Section 20.1 outlines the legal standards for the admission of scientific evidence.* Section 20.2 gives an overview of the legal history that culminated in today's judicial receptivity toward DNA evidence. Section 20.3 identifies some of the established or emerging techniques or technologies that still raise admissibility objections. Section 20.4 describes some issues that arise in presenting the results at trial. Section 20.5 outlines some of the pretrial and trial procedures that require the participation of scientists and technicians who produce DNA evidence. Finally, Section 20.6 discusses the use of DNA testing to investigate claims of wrongful convictions years after the trial and normal appeals have concluded.

20.2 Legal Standards for Admitting Scientific Evidence

For centuries, the Anglo–American law did not distinguish one type of expert testimony from another. On the surface, a uniform standard governed the admission of the testimony of all qualified experts. The evidence had to be relevant and not too prejudicial or time-consuming, and it had to deal with matters not comprehensible to ordinary jurors without the assistance of an expert. A pristine statement of the position is that

> Any relevant conclusions which are supported by a qualified expert should be received unless there are other reasons for exclusion. Particularly, its probative value may be overborne by the familiar dangers of prejudicing or misleading the jury, unfair surprise, and undue consumption of time. (McCormick 1954)

Although this relevance–helpfulness requirement applies to all expert testimony, scientific and nonscientific alike, it need not have the same impact on all types of expert testimony. Scientific evidence tends to be time-consuming or difficult to understand. Courts fear that it comes cloaked in an aura of infallibility that leads jurors to give it more credence than it deserves. Consequently, *ad hoc* balancing of probative value and its counterweights can operate to exclude scientific evidence, especially if the science is not well established. Nevertheless, in practice, the relevance–helpfulness standard promotes "a generally laissez-faire approach to the admissibility of expert evidence" (Law Commission 2011, 16).

Given the perception that scientific evidence poses special problems, courts in the United States came to supplement the relevance–helpfulness standard with more specific rules that attend to the special features of scientific evidence. Two forms of additional scrutiny—a "general acceptance standard" and a "scientific validity standard"—typically are used.

20.2.1 The General-Acceptance Standard

The general acceptance standard made its debut in 1923, in the now celebrated case of *Frye v. United States*, 293 F. 1013 (D.C. Cir. 1923). James Alphonso Frye was charged with murder. He sought to introduce the testimony of a psychologist who had administered a systolic blood pressure test and was prepared to testify that Frye was truthful when he

---

* This section is an updated and condensed version of parts of Chapters 5 and 6 of Kaye, Bernstein, and Mnookin (2004).

denied committing the murder. The trial judge refused to allow the jury to hear this testimony. In affirming the trial court's rulings, the court of appeals wrote that exclusion was proper because *other* psychologists had yet to accept the expert's claim that he could verify honesty by measuring the speaker's blood pressure. The court wrote:

> Somewhere in this twilight zone [between the "experimental" and the "demonstrable"] the evidential force of the principle must be recognized, and while courts will go a long way in admitting expert testimony deduced from a well-recognized scientific principle or discovery, the thing from which the deduction is made must be sufficiently established to have gained general acceptance in the particular field in which it belongs (293 F. at 1014).

Concluding that the deception test lacked the requisite "standing and scientific recognition among physiological and psychological authorities" (ibid.), the court of appeals upheld the exclusion of the psychologist's testimony.

This "*Frye* standard" was not widely adopted at first, but in succeeding decades, it became the dominant rule in state and federal courts. It has been invoked to exclude from the courtroom polygraphy, graphology, hypnotic and drug-induced testimony, voice stress analysis, voice spectrograms, various forms of spectroscopy, infrared sensing of aircraft, blood alcohol tests, retesting of breath samples for alcohol content, polarized light microscopy, psychological profiles of battered women and child abusers, posttraumatic stress disorder as indicating rape, penile plethysmography as indicating sexual deviancy, astronomical calculations, blood group typing, DNA testing, therapy to recover repressed memories, and identifications based on ear prints and knife marks.

It may seem that several of these techniques should be admissible, and today a number of them are. (Others remain pseudoscience or insufficiently validated.) At the time of the decisions, however, the techniques were not so mature. Because the *Frye* test requires sufficient time for a method to be widely accepted as valid, it is fundamentally a conservative standard. The major advantage of the rule is that by looking to the views of the scientific community, it avoids having judges act like independent scientists in ascertaining whether the science in question is good enough for forensic use.

20.2.2 Scientific-Validity Standard

Since the early 1970s, in many jurisdictions, the *Frye* standard was subjected to critical analysis, limitation, modification, and finally, outright rejection. The adoption of the Federal Rules of Evidence intensified the retreat from *Frye*. These rules do not explicitly distinguish between scientific and other forms of expert testimony, and they do not mention general acceptance. As originally enacted, Rule 702 simply provided that "[i]f scientific, technical, or other specialized knowledge will assist the trier of fact to understand the evidence or to determine a fact in issue, a witness qualified as an expert by knowledge, skill, experience, training, or education, may testify thereto …." By 1990, a strong minority of jurisdictions had expressly repudiated *Frye* in favor of a "relevancy-plus" analysis that required a certain extra trustworthiness, accuracy, or fit beyond that needed to admit nonscientific testimony.

The trend culminated in a series of Supreme Court cases on scientific and expert testimony. The seminal case is *Daubert v. Merrell Dow Pharmaceuticals*, 509 U.S. 579 (1993). In *Daubert*, two young children born with deformed limbs and their parents sought damages against the manufacturer of Bendectin, a prescription drug taken by the boys' mothers to treat nausea and vomiting during pregnancy. The plaintiffs' case foundered when they were unable to point to any published epidemiological studies concluding that Bendectin causes limb reduction defects. The federal district court excluded the plaintiffs' evidence of causation—so-called structure–activity studies, *in vitro* or animal cell experiments, *in vivo* or live animal research, and an unpublished reanalysis of the epidemiological data. The court of appeals affirmed.

The Supreme Court held that in looking to general acceptance as indicated by peer-reviewed publications, the lower courts had applied the wrong standard. The Court proclaimed that the "austere [general acceptance] standard, absent from and incompatible with the Federal Rules of Evidence, should not be applied in federal trials" (ibid. at 589). Having jettisoned general acceptance as "the exclusive test for admitting expert scientific testimony," the Court announced that as the gatekeeper of evidence, "the trial judge must ensure that any and all scientific testimony or evidence admitted is not only relevant, but reliable" (ibid.). This *"evidentiary* reliability" presumes "scientific knowledge"—the proffered testimony must be "ground[ed] in the methods and procedures of science" (ibid. at 590). In a further elaboration, the Court suggested that this "reliability" determination "entails a preliminary assessment of whether the reasoning or methodology underlying the testimony is scientifically valid and … properly can be applied to the facts in issue" (ibid. at 592). This, in turn, depends on such things as (1) "whether it can be (and has been) tested," (2) "whether the theory or technique has been subjected to peer review and publication," (3) "the known or potential rate of error," (4) "the existence and maintenance of standards controlling the technique's operation," and (5) the "degree of acceptance within [a relevant scientific] community" (ibid. at 593–94).

In the wake of *Daubert*, some state courts (which are not bound by the federal rules of evidence) observed that they already used this version of the relevancy-plus standard, and a large number agreed to relax *Frye*'s insistence on general acceptance and follow in the Supreme Court's footsteps. Still others continued to follow *Frye*. The Law Commission of England and Wales has proposed that Parliament enact an evidentiary reliability standard that builds on *Daubert* (Law Commission 2011).

Forensic scientists, aware of these legal developments, sometimes try not only to publish the research necessary to validate a procedure and establish general acceptance, but also to present their opinions as to whether their work satisfies the legal standards (Koot, Sauer, and Fenton 2005; Perlin 2006). Whether scientific assessments should offer legal judgments has been questioned (Kaye 2005). In any event, it is the judges—based on their reviews of the scientific literature, previous decisions, the testimony of scientists, and other indicia of validity or general acceptance—who determine whether a particular technique is ready for courtroom use. The judicial response to DNA evidence illustrates the complications that can arise.

20.3 A Short History of Judicial Acceptance of DNA Evidence

20.3.1 Phase 1: Uncritical Acceptance

The first phase in the forensic use of DNA evidence was one of rapid and sometimes uncritical judicial acceptance (Kaye 2010). In this first wave of cases, expert testimony for the prosecution rarely was countered, and courts readily admitted restriction fragment length polymorphism (RFLP) findings. For example, in *Andrews v. State*, 533 So.2d 841 (Fla. Dist. Ct. App. 1988), the first DNA case to generate an opinion from an appellate court in the United States, a man wielding a straight razor had raped and robbed a woman in her home. Fingerprints on a window screen matched the defendant, and so did DNA obtained with a rape kit. The manager of forensic testing at Lifecodes Corporation reported that the percentage of the population with the same variable number tandem repeat (VNTR) alleles as those found in the semen sample was 0.0000012% (1/839,914,540). The defendant objected to this testimony, claiming that VNTR testing for identity was not sufficiently valid or generally accepted in the scientific community. But the trial court admitted the evidence, and the jury convicted.

The Florida District Court of Appeals affirmed the conviction. It concluded that

> The trial court did not abuse its discretion in ruling the test results admissible in this case. In contrast to evidence derived from hypnosis, truth serum and polygraph, evidence derived from DNA print identification appears based on proven scientific principles .... Given the evidence in this case that the test was administered in conformity with accepted scientific procedures so as to ensure to the greatest degree possible a reliable result, appellant has failed to show error on this point. (ibid. at 850–51)

This result was all but inevitable given the trial record. The state produced an impressive array of witnesses—including a molecular biologist from the Massachusetts Institute of Technology (MIT) and a geneticist with a degree from the University of Chicago. In response, the defendant produced no witnesses at all to support his view that the adaptation of gel electrophoresis to measure RFLP lengths to forensic testing was not generally accepted.

Andrews and similar cases as to the general acceptance or scientific validity of DNA testing soon become cited for the proposition that VNTR testing had achieved a level of maturity and accuracy that made it admissible in court. But courts relying on these early cases did not always attend to the fact that the opinions admitting the evidence were based on rather lopsided records. Even when defendants began to produce well-credentialed scientists who expressed doubts about the state of the technology, many courts treated *Andrews* and its brethren as proving that general acceptance was present—even though the earlier cases merely held that in the absence of adequate proof of a scientific controversy, trial courts could admit the DNA results.

20.3.2 Phase 2: Challenges to Laboratory Methods and Population Genetics Models

Within a few years, the early enthusiasm began to wane. Defendants pointed to problems at two levels—controlling the experimental conditions of the analysis and interpreting the results. Some scientists questioned certain features of the procedures for extracting and analyzing DNA used in forensic laboratories (Thompson and Ford 1990, 93). It became apparent that determining whether RFLPs in VNTR loci in two samples actually match can be complicated by measurement variability or by missing or spurious bands. A case in which these problems became prominent is *People v. Castro*, 545 N.Y.S.2d 985 (Sup. Ct. 1989). José Castro was charged in the Bronx in 1987 with the stabbing deaths of a pregnant woman and her 2-year-old daughter. The prosecution claimed that the woman's blood was found on Castro's watch. Castro said that the blood was his. Lifecodes, the first DNA laboratory in the United States to perform RFLP testing, concluded that the frequency of the VNTRs said to be present on the watch and in a sample of the victim's blood was 1 in 189 million in the U.S. Hispanic population. Richard Roberts, who was to receive the Noble prize a few years later, testified for the state of New York. The defense countered with, among others, Eric Lander, a mathematician-turned-biologist at MIT who became a leader in the Human Genome Project (Kolata 2012).

According to Lander (1992), "[i]n the pretrial hearing on the admissibility of the DNA evidence, numerous problems became evident" (also see Lander 1989). For example, Lifecodes reported that the samples both showed three bands when analyzed with a DXYS14 probe. (The DXYS14 probe can produce one to eight bands in a single individual's DNA.) However, every expert at the hearing agreed that there were two extra bands in the DNA from the watch. Lifecodes insisted that the extra bands were nonhuman contaminants with a pattern that showed that they could not have come from the DXYS14 locus. However, Howard Cooke of the Medical Research Council Human Genetics Unit in Edinburgh, who had discovered the locus, testified that there was no way to determine whether the extra bands were human on the basis of the pattern. Another problem with the reported match was that the bands at two autosomal loci fell outside the Lifecode's quantitative standard for declaring a match.

The trial court excluded the DNA evidence. The judge found that the principles of DNA testing were generally accepted, but wrote that "[i]n a piercing attack upon each molecule of evidence presented, the defense was successful in demonstrating to this court that the testing laboratory failed in its responsibility to perform the accepted scientific techniques and experiments" (ibid. at 996). Castro then pleaded guilty, and the judge imposed a lesser sentence than he would have had the DNA evidence been admitted.

The court's evaluation of Lifecodes' testing in this case was simplified by an unusual event. After reviewing the concerns that had been raised, Roberts proposed that the scientists for both parties meet without the lawyers. This unchaperoned tête-à-tête resulted in a joint statement concluding that "the DNA data in this case are not scientifically reliable enough to support the assertion that the samples ... do or do not match" (Lander 1992, 201).

Despite the defects in the laboratory work in *Castro* and a few other cases, most courts continued to find forensic RFLP-VNTR analyses to be generally accepted, and a number of states provided for admissibility of DNA tests by legislation (Melson 1992). The FBI convened a Technical Working Group on DNA Analysis Methods (TWGDAM) composed of forensic scientists and other experts to promulgate standards and protocols for forensic DNA testing and reporting. The effort has been depicted as part of a battle between the law enforcement establishment and the commercial laboratories to establish the standards for an emerging technology (Aronson 2007, 304). If so, the more cautious law enforcement approach prevailed, and the TWGDAM guidelines helped bring some standardization to the process of forensic VNTR-RFLP testing. Moreover, the pronouncements of a national group of experts—even one that was carefully picked by the FBI—helped persuade courts that the laboratory techniques used in forensic DNA analysis were generally accepted and scientifically valid.

However, a more sweeping attack on DNA profiling that began during this period led to a wave of cases in which many courts held that estimates of the probability of a coincidentally matching VNTR profile were inadmissible (Kaye 2010). These estimates relied on a simplified population-genetics model for the frequencies of VNTR profiles that treats each race as a large, randomly mating population. Some prominent scientists claimed that the applicability of the model had not been adequately verified and that it was inaccurate because ethnic or religious subgroups tend to mate preferentially among themselves (Lewontin and Hartl 1991). A heated debate on the significance of this population substructure spilled over from courthouses to scientific journals and convinced the supreme courts of several states that general acceptance was lacking. A 1992 report of the U.S. National Academy of Sciences (National Research Council 1992, 185) endorsed a more "conservative" computational method—dubbed "the ceiling principle"—as a compromise. The "ceiling principle" estimates the frequency of a DNA profile by combining the largest known allele frequencies from a potpourri of populations. The apparent need for a compromise seemed to undermine the claim of scientific acceptance of the less conservative procedure that was in general use (Kaye 2010).

Even as late as 1994, the defense in the televised case of *People v. Simpson* (referred to at the time as the "trial of the century") initially sought to exclude all the DNA evidence that seemed to tie the American football hero, O.J. Simpson, to the bloody murders of his ex-wife and another man. In the end, the evidence was admitted in such detail as to consume many days of the trial. Nevertheless, the defense overcame the evidence, largely by arguing that the large number of DNA matches resulted from unintentional cross-contamination in the laboratory and from police efforts to frame the defendant (Kaye 2010, 140–145).

20.3.3   Phase 3: Renewed Acceptance of VNTR Profiling

At this juncture, the legal history was poised to enter a third phase. In response to the population-genetics criticism and the 1992 National Academy of Sciences report came an outpouring of critiques of the report. The appeal of the "ceiling principle" lies largely in the prudential judgment that it protects the rights of defendants while permitting some use of incriminating DNA evidence. It is, in other words, more pragmatic than principled, and, strictly speaking, it is not guaranteed to give an upper bound on the true value of the population frequency. As such, scientists seeking a procedure that could be justified on purely scientific grounds roundly condemned it. One critic likened the exclusion of testimony about DNA matches by some courts to the Inquisition's attack on Galileo (Kaye 2010, 133).

To rebut the Academy's position on the value of the ceiling method, the FBI and other researchers published a series of new studies of the distribution of VNTR alleles in many populations. Relying on the burgeoning literature, a second National Academy panel concluded in 1996 that the usual method of estimating frequencies of VNTR profiles in broad racial groups was basically sound and suggested less drastic refinements (National Research Council 1996, 254). Moreover, in the period between the two NRC reports, an FBI geneticist and Eric Lander joined forces to write an article in *Nature* with the provocative title, "DNA Fingerprinting Dispute Laid to Rest" (Lander and Budowle 1994). Impressed with the public conversion of a prominent and respected critic to the view that the population genetics issues were no longer serious obstacles to admissibility, and reassured by the 1996 NRC report, courts began to regard concerns over population substructure as passé. Before long, the courts almost invariably returned to the earlier view that the statistics associated with VNTR profiling are generally accepted and scientifically valid both in major population groups and in subgroups (Kaye 2010).

20.3.4 Phase 4: Acceptance of Polymerase Chain Reaction–Based Methods

As results obtained with the polymerase chain reaction (PCR)–based methods entered the courtroom, it became necessary to ask whether these methods also rested on a solid scientific foundation or were generally accepted in the scientific community. The opinions quickly held that the current laboratory procedures for short tandem repeat (STR) typing (as well as earlier systems such as the HLA (human leukocyte antigen) DQ-α reverse dot-blot test)* satisfy these standards. Defense claims that DNA profiles should be inadmissible when generated using a manufacturer's proprietary, unpublished PCR primer sequences also were rejected (e.g., *State v. Kromah*, 657 N.W.2d 564 (Minn. 2003); *State v. Traylor*, 656 N.W.2d 885 (Minn. 2003)). The opinions also hold that the basic "product rule" for estimating the frequencies of genotypes in major populations groups is scientifically sound and generally accepted for the loci investigated in these tests. In sum, in little more than a decade, DNA typing made the transition from a novel set of methods for identification to a relatively mature and-well studied forensic technology (Aronson 2007; Kaye 2010).

20.4 New Methods and Unusual Applications

The early legal challenges to the admission of DNA evidence contributed to improvements in protocols and more extensive proficiency testing and stimulated research in population genetics and statistics. As further advances in DNA forensic technology are introduced, the process of technology transfer to the courts will continue. This involves both establishing that the science and technology is valid and reliable, and presenting scientific findings in a particular case to a lay judge or jury. This section identifies some of the established or emerging techniques or technologies that still raise admissibility objections. Section 20.5 then describes some problems that arise in presenting the results at trial.

As new loci and technologies are introduced, they too must be shown to be generally accepted, scientifically valid, and reliable, or at least helpful (depending on the jurisdiction). New STR loci and single nucleotide polymorphisms (SNPs) are finding forensic applications. Following suitable scientific publications establishing the ability of DNA analysts to detect particular alleles and to assess their significance in inferring identity, the admissibility of the newer loci or methods of detecting them will become an established legal practice.

For example, trial courts typically admit evidence of matching mitochondrial DNA sequences over objections based on *Daubert* or *Frye*, and a number of appellate courts have affirmed these rulings (Kaye and Sensabaugh 2011). However, relatively few published opinions have addressed the current conventions for declaring matches when the possibility of heteroplasmy is considered or the subtleties in estimating the population frequency of a sequence from existing forensic databases of mitotypes. Likewise, the ability

---

* Indeed, the first DNA evidence introduced in an American court was an early form of PCR-based DQα testing (Blake et al. 1992). However, the appellate opinion in the case makes no mention of any objection to this evidence in the prosecution for homicide of a couple who failed to provide food and medical care to an ailing 92-year-old man whom they had agreed to care for and whose bank account they looted. *Commonwealth v. Pestinikas*, 617 A.2d 1339 (Pa. Super. Ct. 1992).

to characterize Y-STRs has been amply demonstrated (e.g., Coble and Butler 2005), and the use of these haplotypes for identification in court has not generated significant controversy—notwithstanding limitations in the databases used in estimating random-match probabilities (Kaye 2010; Kaye and Sensabaugh 2011).

DNA from microbes, plants, and animals can sometimes link a defendant to a victim or a crime scene (Budowle et al. 2005; Kaye and Sensabaugh 2011). The laboratory methods and the genetics of the organisms in these cases have not posed insuperable obstacles to the proponents of the DNA evidence, but characterizing the implications of matching samples in quantitative terms has been more of a problem. In *State v. Bogan*, 905 P.2d 515 (Ariz. Ct. App. 1995), for example, a woman's body was found in the desert, near several Palo Verde trees. A detective found two seed pods in the bed of a truck that the defendant was driving before the murder. A biologist at the University of Arizona performed DNA profiling on this type of Palo Verde and testified that the two pods "were identical" and "matched completely with" a particular tree and "didn't match any of the [other] trees," and that he felt "quite confident in concluding that" the tree's DNA would be distinguishable from that of "any tree that might be furnished" to him (ibid. at 520). The jury convicted the defendant of murder, and jurors reportedly found this testimony very persuasive (Whiting 1993; Yoon 1993). The court of appeals held that this evidence satisfied the general-acceptance standard although very little quantitative data on the genetic variability of Palo Verde trees were available.*

On the other hand, the lack of knowledge (or proof) of population frequencies prompted an opinion holding that the trial court erred in a Washington murder case involving nonhuman DNA. In *State v. Leuluaialii*, 77 P.3d 1192 (Wash. Ct. App. 2003), the prosecution offered testimony of an STR match between DNA of a dog named Chief that had been shot along with its owner and a stain on a black jacket in a hotel room occupied by the defendants. The defendants objected, seeking a *Frye* hearing, but the trial court denied this motion and admitted testimony that included the report that "the probability of finding another dog with Chief's DNA profile was 1 in 18 billion [or] 1 in 3 trillion" (ibid. at 1196). The state court of appeals reviewed the scientific literature on canine STR identification and concluded that it was not sufficient to demonstrate general acceptance of the underlying computations. It remanded the case for a hearing on general acceptance, cautioning that "[b]ecause PE Zoogen has not yet published sufficient data to show that its DNA markers and associated probability estimates are reliable, we would suggest that other courts tread lightly in these waters and closely examine canine DNA results before accepting them at trial" (ibid. at 1201). Research since *Leuluaialii* has added to the literature on forensic canine population genetics (Eichmann et al. 2005; Halverson and Basten 2005).

---

* This issue had not been studied before. A blind trial showed that RAPD (random amplified polymorphic DNA) profiles correctly identified individual Palo Verde trees. The biologist who devised and conducted the experiments analyzed samples from the nine trees near the body and another 19 trees from across the county. He "was not informed, until after his tests were completed and his report written, which samples came from" which trees. Furthermore, unbeknownst to the experimenter, two apparently distinct samples were prepared from the tree at the crime scene that appeared to have been abraded by the defendant's truck. The biologist correctly identified the two samples from the one tree as matching, and he "distinguished the DNA from the seed pods in the truck bed from the DNA of all twenty-eight trees except" that one (*Bogan*, 905 P.2d at 521).

20.4.1 Mixtures and Low Template DNA

DNA samples with a pattern of alleles indicative of multiple contributors are more difficult to interpret than samples with a single profile. It is well known that complex mixtures with several minor contributors can lead to different interpretations from different examiners (Thompson 2009; Dror and Hampikian 2011). Interpretative guidelines and computer programs for deconvoluting the profiles of the contributors are available, and courts normally accept expert testimony about the profiles that seem to be present and statistics associated with them (Kaye and Sensabaugh 2011). Some courts also accept descriptions of the mixed stains without accompanying statistics (*Commonwealth v. Chmiel*, 30 A.3d 1111 [Pa. 2011]).

When sample size is very small, extra amplification cycles or other refinements in the analysis might generate useful information, but differentiating signal from noise in an electropherogram may be difficult. Defendants and their experts have argued that all attempts to discern alleles in an ill-defined category "low copy number" or "low template" DNA are unreliable and insufficiently validated to be admissible. In the handful of cases on this point, this view has not prevailed (Kaye, Bernstein, and Mnookin 2004; Kaye and Sensabaugh 2011).

With low-template DNA, software for mixture interpretation, new loci, and advanced technologies such as microfluidic devices that miniaturize and automate existing analytic procedures:

> [C]ertain basic questions will need to be answered [in court]. What is the principle of the new technology? Is it simply an extension of existing technologies, or does it invoke entirely new concepts? Is the new technology used in research or clinical applications independent of forensic science? Does the new technology have limitations that might affect its application in the forensic sphere? Finally, what testing has been done and with what outcomes to establish that the new technology is reliable when used on forensic samples? For next-generation sequencing technologies and microarray technologies, the questions may be directed as well to the bioinformatics methods used to analyze and interpret the raw data. Obtaining answers to these questions would likely require input both from experts involved in technology development and application and from knowledgeable forensic experts. (Kaye and Sensabaugh 2011, 150)

For both conventional and emerging methods, courts may want to assure themselves that the laboratory uses a detailed and validated protocol, that it has adhered to this protocol in the case before it (or can justify a departure from the usual protocol), that the laboratory work is fully documented, and that a written report sets forth the results fairly and completely so that it can be reviewed by other experts.

20.5 Presentation

20.5.1 Framing and Probabilities

Even if a form of DNA analysis is admissible in the abstract, the manner in which the results are characterized or framed may be legally problematic. Confusion often arises with regard to attempts to describe the (properly computed) probability of a random match. It is easy for the judge or jury to think that this number gives the probability that the match is random. The words are almost identical, but the probabilities can be quite different.

The random match probability is the probability that (A) the trace sample would have the identifying features possessed by the individual tested given that (B) the individual tested has been selected at random. In contrast, the probability that the match is random is the probability that (B) the individual tested has been selected at random given that (A) the trace sample has the identifying features possessed by the individual tested. In general, for two events A and B, the probability of A given B, $P(A \mid B)$, does not equal the probability of B given A, $P(B \mid A)$. The claim that $P(A \mid B) = P(B \mid A)$ is known as the fallacy of the transposed conditional, the inverse fallacy, or the prosecutor's fallacy.

To appreciate that the equation is fallacious, consider the probability that an individual picked at random from all graduate students has a full fellowship to support his or her study. Considering the huge number of students, this "random fellowship probability" is close to 0. At the same time, the probability that a person randomly selected from full graduate fellowship holders is a university student is 1. The "random fellowship probability" $P(\text{fellow} \mid \text{student})$ does not equal the transposed probability $P(\text{student} \mid \text{fellow})$. Likewise, the probability that a person randomly selected from all people who had the opportunity to commit a crime has the traits seen in the trace sample can be practically 0, whereas the probability that a person randomly selected from those people who have the traits seen in trace sample is the defendant can be far greater. Thus, the random match probability $P(\text{Match} \mid \text{Defendant not source})$ does not necessarily equal the transposed conditional probability $P(\text{Defendant not source} \mid \text{Match})$.

Although the logical fallacy in transposing the conditional seems obvious enough—after it has been pointed out—court opinions and transcripts are replete with it. No reported opinion in the United States has reversed a conviction merely because a validly computed probability was transposed,* but courts in other countries have, and a number of opinions of American courts have intimated that the mistake could constitute reversible error (Kaye, Bernstein, and Mnookin 2004).

Another source of confusion is the fallacious assumption that the relevant population in which to estimate a profile frequency or random-match probability is the defendant's racial or ethnic group (rather than all the groups from which the perpetrator of the crime might have come). In *Darling v. State*, 808 So.2d 145 (Fla. 2002), for instance, the Florida Supreme Court seemed to assume that simply because the defendant was from Bahama, a Bahamanian database was more appropriate than the Hispanic database—even though the crime was committed in the United States, not the Bahamas. Likewise, in *People v. Pizarro*, 3 Cal.Rptr.3d 21 (Cal. Ct. App. 2003), the state presented frequency estimates in two population groups—Hispanic and Caucasian. These were chosen because the defendant was said to be half-Caucasian and half-Hispanic, and there was no proof (other than that pointing to Pizarro) that the perpetrator of the crime was Hispanic. As such, the court of appeals reversed.†

---

* A federal court of appeals held that this error, committed by a DNA analyst, who then was a member of the Technical Working on DNA Analysis Methods, combined with other misstatements of probabilities, deprived the defendant in a rape case of due process of law. *Brown v. Farwell*, 525 F.3d 787 (9th Cir. 2008), rev'd sub nom. *McDaniel v. Brown*, 558 U.S. 120 (2010). The Supreme Court recognized the mistakes in the expert's statements but reversed, holding that the flawed testimony did not violate due process in manner claimed by the defendant. *McDaniel v. Brown*, 558 U.S. 120 (2010). The case is discussed in Kaye (2009a).
† Other reasons the Courts of Appeal gave for reversing were criticized (Kaye 2004) and then corrected in *People v. Wilson*, 136 P.3d 864 (Cal. 2006).

With the large number of loci now available for typing, DNA analysts have been willing to make source attributions—to conclude that the trace DNA must have come from a specific individual (or an identical twin). Although courts have perceived no impermeable legal barrier to source attributions (Kaye, Bernstein, and Mnookin 2004; Kaye and Sensabaugh 2011), a laboratory should have a well conceived policy as to when such inferences can be made (Scientific Working Group on DNA Analysis Methods 2010), and analysts should not testify that an expected number of matching individuals in a given population is the actual number (Kaye 2011).

20.5.2 Trawl Probabilities

In addition to the traditional use of trace evidence to show the presence of a known suspect at a crime scene, governments across the world have compelled many convicted offenders and arrested individuals to provide DNA samples for computer-searchable databases of their DNA profiles. These permanent databases help police to solve cases that have baffled them for decades and to catch previously convicted offenders who commit new crimes.

These database-trawl cases can be contrasted with traditional "confirmation cases" in which the defendant already was a suspect and the DNA testing provided additional evidence against him. In confirmation cases, statistics such as the estimated frequency of the matching DNA profile in various populations, the equivalent random-match probabilities, or the corresponding likelihood ratios generally can be used in court to indicate the probative value of the DNA match (Kaye, Bernstein, and Mnookin 2004).

In trawl cases, however, an additional question arises—does the fact that the defendant was selected for prosecution by trawling require some adjustment to the usual statistics? The legal issues are twofold. First, is a particular quantity—be it the unadjusted random-match probability or some adjusted probability—scientifically valid (or generally accepted) in the case of a database search? If not, it must be excluded under the *Daubert* (or *Frye*) standards. Second, is the statistic irrelevant or unduly misleading? If so, it must be excluded under the rules that require all evidence to be relevant and not unfairly prejudicial.

All statisticians agree that, in principle, the search strategy affects the probative value of a DNA match (Kaye 2009b). Most scholarship suggests that producing the suspect through a database search lowers the prior probability—the probability that the suspect is the source, as ascertained before considering the DNA results—and that it raises the likelihood ratio—the probability that the suspect is the source conditioned on the DNA evidence versus that probability without conditioning on the DNA evidence. It lowers the prior probability because, unlike the confirmation case, there normally was no stronger reason to suspect the subject of the "hit" than there was to suspect other persons in the database. Conversely, the database match raises the likelihood ratio primarily by eliminating all other individuals represented in the database.

Rather than address the hypothesis that the suspect is the source, however, frequentist statisticians emphasize the impact of the database match on the hypothesis that the database does not contain the source of the crime scene DNA. They ask how frequently searches of innocent databases—those for which the true source is someone outside the database—will generate cold hits. From this perspective, trawling is a form of data mining that requires a correction for this selection effect or ascertainment bias.

How to present the results of a database search most fairly remains debatable. The developing case law allows both adjusted and unadjusted statistics to be presented in the same case. This outcome is difficult to understand on logical or scientific grounds, but it might be justified for pragmatic reasons (Kaye 2009b).

20.6 Legal Procedures

The distinctive feature of the adversarial Anglo–American system of criminal justice is the use of a lay jury to find the facts and the judge to ensure that the opposing parties follow the rules for learning of the other side's evidence before trial and for presenting their evidence at the trial. The inquisitorial Continental system relies more on the judge to develop and assess the evidence. This difference in the role of the judge has several consequences for expert testimony. First, rules (such as those discussed in Section 20.1) that require judges to exclude evidence are less prominent in the civil law countries (Damaska 1997, 160). Second, instead of (or in addition to official expert reports) oral testimony from expert witnesses who are subject to cross-examination (rather than official expert reports) is often essential in common-law systems. Finally, party-retained experts as opposed to court-appointed ones dominate the common-law system, raising a significant risk of partisan or slanted testimony (Kaye, Bernstein, and Mnookin 2011). However, conflicts among experts occur in both systems (e.g., Povoledo 2011).

When judges must decide preliminary factual questions before admitting evidence, they may conduct pretrial or other hearings outside the presence of the jury. Often, proof regarding the collection, transport, and handling of evidentiary samples occurs only at trial, but disputes over whether methods for DNA analysis and associated statistics meet the threshold for scientific validity or acceptance have generated extended pretrial hearings (Kaye, Bernstein, and Mnookin 2011). At these hearings, concerns over the chain of custody of the evidence and the laboratory's adherence to appropriate testing protocols may be addressed as well.

Despite the oral tradition of imparting information to juries, laboratory reports are extremely important in common-law jurisdictions. They can be consulted to refresh a witness's recollection, they can document the chain of custody of the evidence, and they can be used in cross-examining a witness who departs from the conclusions expressed in a report. In addition, when a report is admitted as an exhibit, it becomes part of the trial record, which allows the jury to rely on the statements in it and to consult it during jury deliberations. Although a laboratory report offered into evidence to prove the facts recorded in it is technically hearsay, in many jurisdictions laboratory reports are admissible under an exception to the hearsay rule. In addition, even if the hearsay objection prevails and the report is not evidence in its own right, the expert who prepared the report or another expert may be permitted to rely on it as part of the basis for the testifying expert's own opinion (*Pendergrass v. State*, 913 N.E.2d 703, 709 [Ind. 2009]; Kaye, Bernstein, and Mnookin 2011, § 4.6).

Beyond the hearsay rule, the Constitution of the United States erects a further barrier to the use of a laboratory report as evidence against a defendant in a criminal case. In response to an early English practice of examining witnesses in private, without the defendant being present, and then introducing their testimony at the trial, a provision in the Bill of Rights guarantees criminal defendants the right to confront their accusers. In recent years, the Supreme Court has redefined the scope of the Confrontation Clause, particularly in its application to forensic-science laboratory reports. In *Melendez-Diaz v. Massachusetts*, 557 U.S. 305 (2009), the Supreme Court struck down the prosecutorial practice of introducing sworn "certificates of analysis" of the state Department of Public Health without calling as witnesses the analysts who performed the gas chromatography and recorded the results. The "certificates of analysis" in the case contained "only the bare bones statement that '[t]he substance was found to contain: Cocaine.' At the time of trial, petitioner did not know what tests the analysts performed, whether those tests were routine, and whether interpreting their results required the exercise of judgment or the use of skills that the analysts may not have possessed" (ibid. at 2537). The extent to which *Melendez-Diaz* permits the introduction of a DNA report—without the testimony of the experts who performed that analysis—through the testimony of another expert has been the subject of subsequent cases (*Williams v. Illinois*, 132 S.Ct. 2221 [2012]; *Bullcoming v. New Mexico*, 131 S.Ct. 2705 [2011]), but remains largely unresolved (Mnookin and Kaye 2013).

Regardless of the precise limits the Court discerns for constitutionally compelled confrontation, the principles underlying the rules about hearsay and confrontation make it more likely that a defendant who wishes to examine not just a written report, but the key laboratory analysts who examined the DNA samples, will have the opportunity to do so.

20.7 Postconviction DNA Testing

In addition to guiding police in their investigations and providing evidence for or against defendants at trials, DNA profiling has been instrumental in many postconviction inquiries into possible miscarriages of justice (National Commission on the Future of DNA Evidence 1999). In the United States, even with no national government-sponsored program of postconviction DNA testing in place, over 300 convicted men have been shown to be innocent.* Because DNA evidence is available in a thin slice of crimes—and only a fraction of those—the hundreds of "DNA exonerations" represent "the tip of the iceberg" (Garrett 2011, 262). Estimates of the incidence of wrongful convictions in the United States vary widely (e.g., Risinger 2007; Gross and O'Brien 2008), but studying the known cases shows how trials in a system that prides itself on "a Constitution that affords unparalleled protections against convicting the innocent" (*Herrera v. Collins*, 506 U.S. 390, 420 [1993] [O'Connor, J., concurring]) can convict the innocent along with the guilty.

Consider the case of Eddy Joe Lloyd. While hospitalized for mental illness, Lloyd wrote to police about the rape and murder of a 16-year-old girl in Detroit, Michigan. Police officers interrogated him in the hospital, suggesting that by confessing, he would help them "smoke out" the real perpetrator. When he said he did not know various details about the crime, officers supplied him with the location of the body and the type of jeans and earrings the victim was wearing. They had him sign a written confession and give a tape-recorded statement. The only forensic evidence at trial consisted of confirming the presence of semen and other biological matter on underwear used to strangle the victim and on some other items. The jury deliberated for less than an hour before convicting Lloyd of first-degree felony murder in 1985. Although Lloyd's attorney told the press that "If he's

---

* Not all DNA exonerations entail proof of actual innocence. In some instances, convicts have been freed because DNA analysis raised serious doubt about the soundness of the prosecution's original evidence.

not goofy, there's not a dog in Texas," Lloyd refused to argue insanity, insisting that he had never killed anyone. At sentencing, the judge lamented that he could not order a hanging. On appeal, an attorney appointed to represent Lloyd wrote that Lloyd should not be taken seriously because he was "guilty and should die."

Ten years later, from prison, Lloyd sought help from the Innocence Project, a law school clinic that represents inmates for whom DNA analysis might demonstrate innocence. Eventually, and with assistance from the prosecutor's office, students located physical evidence used at the trial. DNA testing conducted by a private laboratory and then confirmed by the Michigan State Crime Laboratory found a male DNA profile from sperm cells on the materials associated with the victim. The profile was not Eddie Joe Lloyd's. After serving 17 years in prison for a rape and murder he did not commit, Lloyd was exonerated. He died 2 years later (Innocence Project 2013).

Such exonerations face numerous obstacles (Kaye 2010, 185). The key physical evidence often is missing. Even when it is available, some officials are unwilling to release it for testing. The Supreme Court has not recognized a constitutional right to DNA testing—or even to be released from prison on a strong showing of actual innocence.* Almost all states have adopted special laws giving inmates a right to obtain DNA testing, but often the legislation applies to only a small subset of convicted individuals. Because the legal requirements and procedures for postconviction relief can be varied and complicated, we do not attempt to describe them here. However, many of the points we made about the use of DNA evidence at and before trial also pertain to postconviction hearings involving DNA testing for the purpose of demonstrating the innocence of convicted defendants after the process of trial and appeal has run its course.

Cases such as Lloyd's have stimulated much reflection, and some reform, with regard to other types of evidence as well. Indeed, in response to a civil rights lawsuit brought by Lloyd (reportedly settled for more than $4 million), the police department acceded to a consent decree requiring all interrogations in the most serious crimes be videotaped (Garrett 2011, 247). The New Jersey Supreme Court, citing the prevalence of faulty eyewitness identifications in cases of DNA exonerations, recently adopted sweeping new rules for trial judges to follow in admitting or excluding eyewitness testimony (*State v. Henderson*, 27 A.3d 872 [N.J. 2011]). Findings of false convictions in cases with scientific evidence of various kinds also helped convince Congress to request the National Academy Sciences to study ways to improve forensic science in America (National Research Council 2009, 328). With its now unquestioned power to convict the guilty and exonerate the innocent, DNA evidence has become a major force in the criminal justice system.

---

* In *District Attorney's Office for Third Judicial Dist. v. Osborne*, 557 U.S. 52 (2009), the Supreme Court avoided the issue of whether a prisoner has a right to be released upon a showing that he is probably innocent of the crime for which he was convicted after a fair trial. It did so in a 5-4 decision by reasoning that even if this right exists, a prisoner has no "free-standing" due process right to test the DNA from the scene of a rape after the conviction when (1) the convicted offender did not seek extensive DNA testing before trial even though it was available, (2) he had other opportunities to prove his innocence after a final conviction based on substantial evidence against him, (3) he had no new evidence of innocence (only the hope that more extensive DNA testing than that done before the trial would exonerate him), and (4) even a finding that he was not source of the DNA would not "conclusively" demonstrate his innocence.

References

Aronson, J. D. 2007. *Genetic Witness: Science, Law, and Controversy in the Making of DNA Profiling*. Piscataway, NJ: Rutgers University Press.

Blake, E., J. Mihalovich, R. Higuchi, P. S. Walsh, and H. Erlich. 1992. Polymerase chain reaction (PCR) amplification and human leukocyte antigen (HLA)-DQa oligonucleotide typing on biological evidence samples: Casework experience. *J Forensic Sci* 37:700–726.

Budowle, B., P. Garofano, A. Hellman et al. 2005. Recommendations for animal DNA forensic and identity testing. *Int J Legal Med* 119:295–302.

Coble, M. D., and J. M. Butler. 2005. Characterization of new miniSTR loci to aid analysis of degraded DNA. *J Forensic Sci* 50:43–53.

Damaska, M. R. 1997. *Evidence Law Adrift*. New Haven, CT: Yale University Press.

Dror, I. E., and G. Hampikian. 2011. Subjectivity and bias in forensic DNA mixture interpretation. *Sci Justice* 51:204–208.

Eichmann, C., B. Berger, M. Steinlechner, and W. Parson. 2005. Estimating the probability of identity in a random dog population using 15 highly polymorphic canine STR markers. *Forensic Sci Int* 151:37–44.

Garrett, B. L. 2011. *Convicting the Innocent: Where Criminal Prosecutions Go Wrong*. Cambridge, MA: Harvard University Press.

Gross, S. R., and B. O'Brien. 2008. Frequency and predictors of false conviction: Why we know so little, and new data on capital cases. *J Empirical Legal Stud* 5:927–962.

Halverson, J., and C. Basten. 2005. A PCR multiplex and database for forensic DNA identification of dogs. *J Forensic Sci* 50:352–363.

Innocence Project, Know the cases: Eddie Joe Lloyd. 2013. http://www.innocenceproject.org/Content/Eddie_Joe_Lloyd.php.

Kaye, D. H. 2004. Logical relevance: Problems with the reference population and DNA mixtures in *People v. Pizarro*. *Law Probability Risk* 3:211–220.

Kaye, D. H. 2005. The NRC bullet-lead report: Should science committees make legal findings? *Jurimetrics J* 46:91–105.

Kaye, D. H. 2009a. "False, but highly persuasive": How wrong were the probability estimates in *McDaniel v. Brown*? *Mich Law Rev First Impressions* 108:1–7. http://www.michiganlawreview.org/articles/false-but-highly-persuasive-how-wrong-were-the-probability-estimates-in-em-mcdaniel-v-brown-em (accessed February 24, 2013).

Kaye, D. H. 2009b. Rounding up the usual suspects: A legal and logical analysis of DNA database trawls. *North Carolina Law Rev* 87:425–503.

Kaye, D. H. 2010. *The Double Helix and the Law of Evidence*. Cambridge, MA: Harvard University Press.

Kaye, D. H. 2011. The expected value fallacy in *State v. Wright*. *Jurimetrics J* 52:1–6.

Kaye, D. H., D. E. Bernstein, and J. L. Mnookin. 2004. *The New Wigmore: A Treatise on Evidence: Expert Evidence*. New York: Aspen.

Kaye, D. H., D. E. Bernstein, and J. L. Mnookin. 2011. *The New Wigmore: A Treatise on Evidence: Expert Evidence*, 2nd ed. New York: Aspen.

Kaye, D. H., and G. Sensabaugh. 2011. Reference guide on DNA evidence. In *Reference Manual on Scientific Evidence*, 3rd ed., ed. National Research Council, Committee on the Development of the Third Edition of the Reference Manual on Scientific Evidence, 129–210. Washington, DC: National Academy Press.

Kolata, G. 2012. Power in numbers, *New York Times*, January 3 http://www.nytimes.com/2012/01/03/science/broad-institute-director-finds-power-in-numbers.html?pagewanted=all&_r=0 (accessed February 24, 2013).

Koot, M. G., N. J. Sauer, and T. W. Fenton. 2005. Radiographic human identification using bones of the hand: A validation study. *J Forensic Sci* 50:263–268.

Lander, E. S. 1989. DNA fingerprinting on trial. *Nature* 339:501–505.

Lander, E. S.1992. DNA fingerprinting: Science, law, and the ultimate identifier. In *The Code of Codes: Scientific and Social Issues in the Human Genome Project*, ed. D. J. Kevles and L. Hood, 191–210. Cambridge, MA: Harvard University Press.

Lander, E. S., and B. Budowle. 1994. DNA fingerprinting dispute laid to rest. *Nature* 27:735–738.

Law Commission. 2011. *Expert Evidence in Criminal Proceedings in England and Wales*. London: The Stationery Office. Also available online at http://www.official-documents.gov.uk/document/hc1011/hc08/0829/0829.pdf (accessed February 25, 2013).

Lewontin, R. C., and D. L. Hartl. 1991. Population genetics in forensic DNA typing. *Science* 254:1745–1750.

McCormick, C. T. 1954. *Handbook of the Law of Evidence*. St. Paul, MN: West.

Melson, K. E. 1992. Legal and ethical considerations. In *DNA Fingerprinting: An Introduction*, ed. L. T. Kirby, 189, 199–200. New York: Oxford University Press.

Mnookin, J., and D. Kaye. 2013. Confronting science: Expert evidence and the confrontation clause. *Supreme Court Rev* 2013:99–159.

National Commission on the Future of DNA Evidence. 1999. Postconviction DNA testing: Handling postconviction requests. Also available online at https://www.ncjrs.gov/pdffiles1/nij/177626.pdf (accessed February 25, 2013).

National Research Council. 2009. Committee on Identifying the Needs of the Forensic Sciences Community. *Strengthening Forensic Science in the United States: A Path Forward*. Washington, DC: National Academy Press.

National Research Council. 1996. Committee on DNA Forensic Science: An update. *The Evaluation of Forensic DNA Evidence*. Washington, DC: National Academy Press.

National Research Council. 1992. Committee on DNA Technology in Forensic Science. 1992. *DNA Technology in Forensic Science*. Washington, DC: National Academy Press.

Perlin, M. W. 2006. Scientific validation of mixture interpretation methods. In *Proceedings of the Seventeenth International Symposium on Human Identification*. Also available online at http://www.promega.com/products/pm/genetic-identity/ishi-conference-proceedings/17th-ishi-oral-presentations/ (accessed February 25, 2013).

Povoledo, E. 2011. Italian experts question evidence in Knox case. *New York Times*, June 30. http://www.nytimes.com/2011/06/30/world/europe/30knox.html (accessed February 24, 2013).

Risinger, D. M. 2007. Innocents convicted: An empirically justified factual wrongful conviction rate. *J Criminal Law Criminol* 97:761–806.

Scientific Working Group on DNA Analysis Methods. 2010. *SWGDAM Interpretation Guidelines for Autosomal STR Typing by Forensic DNA Testing Laboratories*. Also available online at http://www.fbi.gov/about-us/lab/codis/swgdam-interpretation-guidelines (accessed February 25, 2013).

Thompson, W. C. 2009. Painting the target around the matching profile: The Texas sharpshooter fallacy in forensic DNA interpretation. *Law Probability Risk* 8:257–276.

Thompson, W. C., and S. Ford. 1990. The meaning of a match: Sources of ambiguity in the interpretation of DNA prints. In *Forensic DNA Technology*, ed. M. A. Farley and J. J. Harrington, 93–152. Chelsea, MI: Lewis.

Whiting, B. 1993. Tree's DNA "fingerprint" splinters killer's defense. *Arizona Republic*, May 28.

Wigmore, J. H. 1974. *Evidence*. Rev. J. H. Chadbourn. Vol. 5. New York: Aspen.

Yoon, C. K. 1993. Forensic science–botanical witness for the prosecution. *Science* 260:894–895.

21 Some Ethical Issues in Forensic Genetics

ERIN D. WILLIAMS
DAVID H. KAYE

Contents

21.1 Introduction	527
21.2 General Concepts in Bioethics	528
21.2.1 Justice	528
21.2.2 Privacy and Confidentiality	528
21.2.3 Autonomy and Informed Consent	529
21.2.4 Beneficence	530
21.2.5 Utility	530
21.3 Ethical Issues in Acquiring DNA Samples	530
21.3.1 Crime-Scene Samples, and Shed or Abandoned DNA	530
21.3.2 Sampling with Consent	531
21.3.3 Acquiring DNA Samples from Medical Providers or Researchers	531
21.4 Law Enforcement DNA Databanks	533
21.5 Phenotypes and Racial Identifications from Genotypes	533
21.6 Identification of Remains	534
21.7 Ethics of Forensic Laboratory Reporting and Expert Testimony	535
Acknowledgments	536
References	536

21.1 Introduction

Identifying individuals, living or dead, from their genotypes can help to convict offenders or exonerate innocent suspects. It can confirm one's presence at a crime scene or one's place in a family tree. However, the extensive and sensitive nature of the genetic information locked in the coils of the DNA molecule also gives rise to ethical dilemmas.

This chapter surveys such issues. Section 21.1 offers an overview of the major concepts in bioethics, drawing some preliminary connections to issues that arise in forensic genetics. Section 21.2 addresses the ethics of acquiring DNA samples—from crime scenes and other locations and from the bodies of individuals. Section 21.3 outlines a range of ethical concerns with DNA databanks and databases. Section 21.4 concerns efforts to infer phenotypes from genotypes to assist in criminal investigations. Section 21.5 identifies ethical concerns that arise in identifying human remains, particularly in mass disasters or conflicts. Finally, Section 21.6 discusses the professional and ethical standard for reporting forensic laboratory results and testifying about them.

21.2 General Concepts in Bioethics

The key concepts in bioethics are justice, privacy and confidentiality, autonomy and informed consent, beneficence, and utility.

21.2.1 Justice

Justice refers to the allocation of burdens and benefits in a society. The types of justice pertinent in forensic genetics are retributive and distributive justice. Criminal law rests on theories of retributive justice. Competing theories, which we shall not pursue here, rely on different principles to ascertain which conduct should subject an offender to punishment and what punishment is appropriate for a particular offense and offender given the circumstances of the conduct and the nature, knowledge, and motivation of the criminal. For the purpose this chapter, we assume that the criminal code and its enforcement are generally just, so that the forensic scientist faces no ethical quandary in supplying information for criminal investigations and trials.

"[D]istributive justice refers to fair, equitable, and appropriate distribution [and] includes policies that allot diverse benefits and burdens such as property, resources, taxation, privileges or opportunities" (Beauchamp and Childress 2009, 241). To the extent that the criminal justice system burdens certain groups disproportionately (as compared to other groups) because the distribution of wealth in the society is fundamentally unfair (leading to a higher crime rate on the part of the poor), questions of comparative rather than individual, retributive justice arise (Feinberg 1973). How one balances claims of individual and comparative injustice is a difficult matter. In particular, complaints grounded in distributive or social justice have been raised in connection with law enforcement DNA databases in the United States and the United Kingdom because the profiles in these databases come disproportionately (compared to population percentages) from racial or ethnic minorities (Kaye and Smith 2003; Krimsky and Simoncelli 2011).

21.2.2 Privacy and Confidentiality

Privacy, broadly speaking, refers to freedom from unwanted observation, intrusion, or distribution of information about an individual. Confidentiality refers to information or items revealed in a protected relationship, such as that between and doctor and patient. Both privacy and confidentiality may involve control over how and when information is revealed, but confidentiality typically arises in a relationship of trust and refers to the duty of an individual entrusted with the information (e.g., a physician treating a patient or an attorney representing a client). Thus, in the medical arena maintaining the confidentiality of a patient's medical records is an important professional norm and the subject of a complex set of legal rules, but confidentiality typically is legally overridden in the pursuit of justice. For example, in the United States, the medical community may only disclose health and other genetic information in certain circumstances (Health Information Portability and Accountability Act, Pub. L. No. 104-191, 110 Stat. 1936 [1996]; Privacy Rule, 42 C.F.R. §§ 160.101-.552, 164.102-.106, 164.500-.534 [2002]; Genetic Information Nondiscrimination Act of 2008, Pub. L. No. 110-233, 122 Stat. 881). However, these laws permit disclosure of protected health information when required by other laws or for law enforcement purposes, including locating and identifying suspects of a crime (U.S. Department of Health and Human Services 2003). Likewise, in some jurisdictions, the physician–patient privilege for communications to one's doctor does not apply to defendants in criminal trials and certain other proceedings (Broun et al. 2013, sec. 104).

Research into human genetics and genomics and the application of the resulting knowledge in the health care system raises a plethora of privacy and confidentiality issues. Recent years have seen a veritable explosion of literature on the ethics of "biobanking" (Budimir et al. 2011). Traditionally, pathology tissue repositories have used samples collected from patients for a variety of purposes without securing the informed consent of the donors for each and every use (Korn 1999). Some physicians and ethicists have questioned this tradition (Clayton et al. 1995), and the liberal use and sharing of samples from repositories created for research into the genetics of diseases also has been challenged (Greely 1999). However, the norms that govern research biobanks do not necessarily apply without modification to law enforcement databanks (i.e., to military and law enforcement repositories of biological material maintained for the limited purpose of identification of remains or crime-scene samples). These samples were not obtained for therapeutic or research purposes on the basis of informed consent from the donors, and law enforcement personnel do not have a fiduciary duty to protect the interests of the suspects in criminal investigations (although they do have other ethical and legal duties toward suspects and witnesses) (Kaye 2000).

21.2.3 Autonomy and Informed Consent

The Kantian principle of autonomy respects persons and their choices. In the context of patient care and human experimentation, it means that "[e]very human being of adult years and sound mind has a right to determine what shall be done with his body…." (*Schloendorff v. Society of New York Hospital*, 105 N.E. 92, 93 [N.Y. 1914]). It is "consistent with the individualistic temper of American life, which emphasizes privacy and self-determination" (Pellegrino 1993, 10).

Informed consent protects autonomy by disclosing the pertinent risks and benefits of, and alternatives to, a particular course of action, allowing the individual to make a reasoned choice in light of his or her values and circumstances. Thus, "informed consent is based on respect for the individual and, in particular for each individual's capacity and right both to define his or her own goals and to make choices designed to achieve those goals" (President's Commission for the Study of Ethical Problems in Medicine and Biomedical and Behavioral Research 1982, 17). A surgeon, for example, generally may not perform a procedure—even when it clearly would benefit a patient—without the patient's informed content. *Schloendorff v. Society of New York Hospital*, 105 N.E. 92 (N.Y. 1914).

Yet, informed consent has a reduced bearing on actions in the law enforcement context. A suspect who refuses to submit to dental impressions authorized by a search warrant may be forced to undergo general anesthesia and nasal intubation (*Carr v. State*, 728 N.E.2d 125, 128 [Ind. 2000]; *cf. Winston v. Lee*, 470 U.S. 753, 760 [1985] "The reasonableness of surgical intrusions beneath the skin depends on a case-by-case approach, in which the individual's interests in privacy and security are weighed against society's interests in conducting the procedure.") The law routinely requires individuals to submit to much less invasive DNA sampling for investigative purposes. These accommodations result from conflicts among individuals' desires or acts. Defining the full and proper scope of autonomy, especially when there are conflicts between different autonomous agents, is an immensely difficult task.

21.2.4 Beneficence

Beneficence refers to the requirement that health care providers and medical researchers act to benefit others. One might argue that this derives from the physician's imperative to do no harm, but that would confuse beneficence with the more limited duty of nonmaleficence (Beauchamp and Childress 2009, 199). Achieving beneficence often requires a balance between the magnitude of the harm and the likelihood of the risk, especially in the research context (ibid., 223). Using tissue samples acquired for medical purposes to provide data on the distribution of the genotypes used in forensic identification, for example, should be conducted in a manner that protects the privacy interest of the sources of the samples (minimizing the psychosocial risks) and that benefits society as a whole.

21.2.5 Utility

Utilitarianism is a moral theory that evaluates acts or rules by their consequences. Roughly stated, it requires that we act to secure the greatest good for the greatest number of people. However, different utilitarians have different conceptions of what counts as "good" (ibid., 337). Utilitarianism differs from the right-based theories in the Kantian tradition in that the individual's rights are derived entitlements, respected only to maximize utility, and not because they are primary and intrinsic to the inherent worth of the individual. Thus, a common criticism of utilitarianism is that it permits (at least in principle) an individual's rights to be sacrificed for the general welfare. When indefeasible rights are not involved, there is general agreement that a utilitarian weighing of costs and benefits is crucial to developing good social policies.

With this catalog of general concepts and principles that should influence ethical judgments, we turn to a set of ethical issues that are apparent in forensic genetics. We start with the collection of DNA samples—from people and places.

21.3 Ethical Issues in Acquiring DNA Samples

21.3.1 Crime-Scene Samples, and Shed or Abandoned DNA

Collecting DNA from crime scenes and cadavers does not violate autonomy. It would infringe informational privacy if disease-related loci were tested and the results disseminated to employers or insurers, but the STR loci used in forensic identification are not known to cause disease or to have substantial correlations to sequences that affect health status (Kaye 2007). In some cases, DNA profiles from cadavers might reveal genetic conditions or relationships (or their absence, as in situations of "misattributed paternity") of which family members were not aware. This is a common occurrence in genetic research using pedigrees as well as in genetic counseling, where the obligations to individuals involved in (and consenting to) genetic tests as well as those affected by the testing have debated for some time (Juengst and Goldenberg 2008; Lucast 2007; Tozzo, Caenazzo, and Parker 2013).

DNA left at a crime scene is an example of abandoned DNA. An individual no doubt has some privacy interest in the secrecy of his travels, but it is implausible to contend that people have any right to prevent others from discovering, through trace evidence, that they were present in a particular place at some time in the past. A more appealing claim is that "genetic privacy" should encompass the genetic information in biological materials inadvertently left at ordinary places in the course of everyday life. Should a nosy neighbor be permitted to retrieve and analyze DNA from dental floss left in the trash to obtain possibly embarrassing information about genetic weaknesses? Whatever one thinks of a genetic privacy law that would protect individuals from such snooping (Joh 2011), the extraction of purely identifying information from the DNA that we routinely shed or abandon seems less problematic.

21.3.2 Sampling with Consent

Police often search individuals or their possessions with the ostensible permission of the individuals. The same is true of DNA sampling. In a rape case, for example, police may request "elimination samples" from spouses or sexual partners of the victim. They might ask a suspect to give a sample, telling him that he can establish his innocence by cooperating. Indeed, in exceptional cases, police have systematically collected samples from entire communities. The earliest "DNA dragnet" occurred in England in 1987, in Narborough (Wambaugh 1987). Investigating the rape and strangulation of two teenage girls, police requested blood samples from more than 4000 local men. Eventually, the killer was identified after he asked a coworker to give a sample in his place. Since then, DNA dragnets have been used with some success in Europe and the United States (Ripley et al. 2005; Maschke 2008).

DNA dragnets raise two sets of ethical and legal questions. The first is whether permission obtained in face-to-face encounters with the police, and under the pressure of becoming a suspect if one refuses to cooperate, is truly voluntary (Maclin 2008). The medical model of informed consent does not apply to encounters between the police and the public (*Schneckloth v. Bustamonte*, 412 U.S. 218 [1973]), but the threat of force can render consent to a search legally ineffective (cf. *Bumper v. North Carolina*, 391 U.S. 543 [1968]).

The second ethical issue is whether the police may keep the samples or profiles for later investigations of unrelated crimes. This should turn on the scope of the consent provided in the first instance (or whether the sample donor later agrees to the further use), and whether law enforcement practices require retention of all evidence collected as part of official investigations.

Whether the benefits of the practice outweigh the costs also has been questioned. Large-scale screening consumes considerable police resources and is an imposition on the many innocent people who are stopped and questioned. As a result, the practice is used only in the most serious cases, such as those of serial killings and rapes.

21.3.3 Acquiring DNA Samples from Medical Providers or Researchers

As noted in Section 21.1, ethical issues arise from the potential to use genetic information or samples collected for medical and other purposes for criminal investigations. DNA is collected for medical purposes, such as genetic-disease testing, and DNA resides in samples of tissue stored from medical procedures. It also can be collected for nonmedical purposes, such as direct-to-consumer testing, genealogy and ancestry identification, and nutrigenomics. It can even be collected by employers for monitoring exposure to toxins. Is it right to divert such samples and information to law enforcement uses? There can be no categorical answer to the question. Ethical propriety will depend on the nature of the law enforcement use, the facts of the case, and basis for the prior DNA collection. We will consider three examples.

First, a special act of the Swedish Parliament authorized access to a "neonatal database" to identify tsunami victims (Mund 2005). Although the humanitarian use was not contemplated originally, the data pertained to missing persons, most of whom were lost in the tsunami, and access to it was for the benefit of their families. Also, the use would not be likely to deter participation in the neonatal database. In this situation, the unanticipated use was benign to the DNA donors and likely to be consistent with their wishes if they had been informed of the situation.

Second, a case in Texas also involved neonatal samples, in the form of dried blood spots for a compulsory newborn screening program. Over the years, the Texas State Department of Health Services supplied some of these cards to medical researchers studying club foot, childhood cancer, and lead exposure. Five parents sued (Lewis et al. 2011), and the state promptly settled the case by agreeing to destroy millions of cards, to give parents clearer procedures to opt out of the storage of the cards, and to pay $26,000 in attorney's fees and costs. A journalist who read the public reports of the Health Department, then reported, confusingly, that nearly 800 of the samples had been provided to "the federal government to create a vast DNA database, one that could help crack cold cases and identify missing persons" (Ramshaw 2010). In reality, the cards—with personally identifying information removed—had gone to the United States Armed Forces DNA Identification Laboratory for research into the population frequencies of mitochondrial DNA types. In this case, then, there was even less of a risk to personal privacy than in the Swedish disclosure, but the use of the samples did not specifically benefit the individuals or families whose mitochondrial DNA was typed anonymously for statistical purposes.

The third example pertains to the apprehension of a notorious serial killer in the United States. Over a 30-year period, he killed at least 10 people, taunting the police in Wichita, Kansas, with letters signed BTK (for Bind, Torture, Kill). As police developed a case against Dennis Rader, they obtained a court order for a Pap smear his daughter had given years earlier at a university medical clinic. Investigators compared an STR (short tandem repeat) profile of the clinic specimen to that of DNA taken from several BTK crime scenes. A reverse paternity (statistical) analysis led detectives to conclude that the daughter, who neither knew of nor consented to the testing, was most likely the biological child of the killer. Rader confessed and pled guilty to 10 counts of murder (Nakashima 2008).

In this case, the interest of a manifestly innocent person—the daughter—in the confidentiality of the inchoate information—the Pap smear—yielded to the societal interest in catching the serial killer, but the propriety of this outcome is debatable. It would have been less invasive of the daughter's privacy to acquire a DNA sample from the suspect, surreptitiously if necessary. Although surreptitious sampling poses its own problems (see Section 21.2.1), it does not directly infringe the interests of third parties (such as the daughter). In addition, it could be argued that if medical information is used for criminal proceedings, patients will be less willing to allow their doctors to retain their tissue samples.

21.4 Law Enforcement DNA Databanks

Chapter 23 describes the use and expansion of law enforcement DNA databanks across the world. This development raises a plethora of issues:

- What crimes should trigger DNA collection? Only sex crimes? Only violent felonies? All felonies? Some or even all misdemeanors? Would a population-wide database be fairer than one based on encounters with the police or the courts?
- When should DNA samples be collected? After conviction? After a judicial finding of probable cause or an indictment? Upon custodial arrest? At birth (for a population-wide database)?
- Which loci should be tested to generate profiles for inclusion in the database?
- Who should have access to the samples and the profiles, and for what purposes?
- How long should such samples and profiles be retained?
- Should kinship matching between crime-scene and database samples be used to locate close relatives who might be the source of the crime-scene sample?

Resolving these issues requires balancing the public good (beneficence and utility) against distributive justice and "genetic privacy" concerns. Privacy protections in place in various databases include using profiles that are not associated with diseases, destroying samples once they no longer are needed, expunging profiles after certain time periods, and forbidding the release of samples for biological research (other than that related to the operation of the database or the interpretation of DNA matches).

21.5 Phenotypes and Racial Identifications from Genotypes

Although the STR loci used in forensic identification are not well suited for drawing inferences about physical traits or ancestral groups, geneticists can use other loci that provide clearer "ancestry and morphology information to infer a suspect's race and general appearance" (Ossorio 2006). Chapter 17 describes research into the genetics of eye, hair and skin color, height, and other physical features. Once inferences (with known uncertainties) of specific physical traits become available, it will be tempting to use them to guide investigations. Such publicly exposed traits cannot be considered private information. Furthermore, if it is acceptable to use the color of hair fibers found at a crime scene to help identify a criminal, why is it not acceptable to use the hair color as inferred from other biological material containing DNA?

The use of ancestry informative markers has been opposed, however, on the ground that it will reify race as a biologically meaningful category instead of the social construct and reflection of ancient migrations that it is (ibid.). There is also some concern that police authorities will not appreciate the limitations in inferring ancestry or phenotypes from crime-scene samples. Whether the risks warrant the suppression of characterizations of ancestry or race from DNA analysis depends on how much the practice would contribute to public misunderstanding and the value the descriptions would have in criminal investigations. Some commentators argue that "[p]olicymakers, judges, law enforcement officers, and civil rights activists alike must develop a more nuanced understanding of genetics" and note that "DNA ancestry tests and indirect molecular photofitting [could]… help remove the prejudicial blinders that investigators may have toward minorities" (Wagner 2009).

A very different issue is whether, in the courtroom or in reports to the police, experts should report population frequencies or likelihood ratios according to ethnicity or race. This practice is not an attempt to establish the race of an individual, but an effort to show that regardless of the subpopulation to which an accused offender might belong, it is improbable that the DNA match is coincidental. However, the same argument about reifying race has been made. According to one law professor, "the persuasive authority of DNA evidence and the reality of implicit prejudice call into question the legitimacy of using racialized DNA evidence" (Kahn 2009, 331).

21.6 Identification of Remains

This chapter has focused on ethical issues in forensic genetics as it has been applied in the criminal justice systems of individual countries. Ethical issues also arise in the identification of human remains (Kelly 2010; 43 C.F.R. § 10.11 [2010]). Many of them are well illustrated by the exhumation and analysis of remains from mass graves. Some of the largest of these in recent history were located in the former Yugoslavia. Exhumation and examination helped to give closure to loved ones and assisted in the prosecution of war criminals by the International Criminal Tribunal for the Former Yugoslavia (ICTY) (Williams and Crews 2003; Neuffer 2007).

How remains should be treated is complicated in the case of mass graves. To confound criminal prosecution, Bosnian Serbs used earth-moving equipment to exhume and redeposit remains from mass graves into secondary grave sites, and then mixed secondary sites into tertiary sites, disarticulating, commingling, mangling, and crushing the remains. Bodies needed to be reassembled. Protocols for the return of a body to family members and for making declarations of death could depend in part on the amount and type of remains located. For example, a body with a missing finger might merit one outcome, and a finger missing a body might merit another.

This commingling and reassembly of remains also affects families and states that wish to have remains returned and repatriated, or at least buried in accordance with their religious or cultural traditions. Some burial traditions maintain that rituals involving the dead body are necessary to ensure peace in the afterlife. Jewish and Muslim faiths require the burial of bodies including any body parts that may have been removed. The Hindu faith requires cremation. When only a sampling of remains from a grave can be identified, and when it may not be certain that all, or even the majority of bodies in a particular grave come from a particular state or tradition, the appropriate way to dispose of the commingled remains is not obvious. Ideally, the treatment of remains will respect the wishes of the deceased or their family members. But this ideal may be impossible to implement when the remains come from heterogeneous faiths or traditions. The best answer may be to enlist survivor groups in establishing, in advance, protocols that are respectful and fair to all involved by incorporating as many rituals as may be reflected by the composition of remains without giving undue priority to any one tradition.

21.7 Ethics of Forensic Laboratory Reporting and Expert Testimony

Expert reports convey useful information to the prosecution, the defense, and the court before trial. Indeed, in the United States, due process principles require the prosecution to disclose information that is "favorable to [the] accused" and "material either to guilt or to punishment" (*Brady v. Maryland*, 373 U.S. 83, 87 [1963]), as well as "evidence that the defense might have used to impeach the government's witnesses by showing bias or interest" (*United States v. Bagley*, 473 U.S. 667, 676 [1985]). Although the prosecution is responsible for disclosing all such information, even if it is in the hands of the law enforcement agency (*Kyles v. Whitley*, 514 U.S. 419, 437 [1995]), and the forensic scientist cannot control the actions of police or prosecutors, the expert may have an ethical obligation to avoid such practices as "preparation of reports containing minimal information in order not to give the 'other side' ammunition for cross-examination," "reporting of findings without an interpretation on the assumption that if an interpretation is required it can be provided from the witness box," and "omitting some significant point from a report to trap an unsuspecting cross-examiner" (Lucas 1989, 724; NIST 2012, 92).

At trial, expert witnesses use or impart specialized knowledge and information to assist the trier of fact to understand the evidence. Although an expert can control the content of laboratory notes and subsequent pretrial reports, during a trial, questioning by counsel and judges frames the expert's presentation. The expert may not simply decide what information to discuss but must answer the questions. This can create some tension between the goal of being complete and the need to be responsive (NIST 2012, 115). In resolving this tension, "ethical considerations and professional standards properly place a number of constraints on the expert's behavior" (Feinberg 1989, 161). One such constraint is "a requirement of candor. While an expert is ordinarily under no legal obligation to volunteer information, professional ethics may compel this ... when the expert believes that withholding information will change dramatically the picture that his ... analyses, properly understood, convey" (ibid., 161) Thus, the "guiding principles" proposed by the American Society of Crime Laboratory Directors/Laboratory Accreditation Board advise forensic scientists to "attempt to qualify their responses while testifying when asked a question with the requirement that a simple 'yes' or 'no' answer be given, if answering 'yes' or 'no' would be misleading to the judge or the jury" (ASCLD/LAB 2013, 3).

These precepts may seem more significant in the Anglo–American system of adversarial trials than in the Continental system, in which court-appointed experts are the rule rather than the exception, but the basic principles apply in all jurisdictions. Thus, the European Network of Forensic Science Institutes (2004, Standard I3) urges analysts to "deal with questions truthfully, impartially and flexibly in a language which is concise, unambiguous and admissible"; to "give explanations to specific questions in a manner that facilitates understanding by nonscientists"; to "consider additional information and alternative hypotheses that are presented to you"; to "consider and evaluate these and express relevant opinions taking into account the limitations on opinions which cannot be given without further examination and investigation"; and to "clearly differentiate between fact and opinion and ensure that the opinions you express are within your area of expertise." It is only natural that a forensic scientist who is convinced that his work should guide the court would wish to be as persuasive as possible, but persuasion should come from the clarity of the presentation and the quality and completeness of the scientific investigation.

The expert is not the decision maker; the forensic scientist's task is to educate the other participants in the proceedings so that the adjudication will use the scientific findings for what they are worth—no more, and no less.

Acknowledgments

The authors thank Nanette Elster, Joseph Montenegro, and anonymous reviewers for contributions to this chapter.

References

ASCLD/LAB. 2013. *ASCLD/LAB Guiding Principles of Professional Responsibility for Crime Laboratories and Forensic Scientists*. Version 1.2. http://www.ascld-lab.org/documents/AL-PD-1014.pdf (accessed March 8, 2013).

Beauchamp, T. L., and J. F. Childress. 2009. *Principles of Biomedical Ethics*. 6th ed. Oxford: Oxford University Press.

Broun, K. S., G. Dix, E. Imwinkelried et al. 2013. *McCormick on Evidence*. 7th ed. St. Paul, MN: West.

Budimir, D., O. Polašek, A. Marušić et al. 2011. Ethical aspects of human biobanks: A systematic review. *Croat Med J* 52(3):262–279.

Clayton, E. W., K. K. Steinberg, M. J. Khoury et al. 1995. Informed consent for genetic research on stored tissue samples. *JAMA* 274(22):1786–1792.

European Network of Forensic Science Institutes. 2004. Standing Committee for Quality and Competence. *Performance Based Standards for Forensic Science Pactitioners*. Also available at http://www.enfsi.eu/sites/default/files/documents/performance_based_standards_for_forensic_science_practitioners_0.pdf (accessed March 8, 2013).

Feinberg, J. 1973. *Social Philosophy*. Upper Saddle River, NJ: Pearson.

Feinberg, S. ed. 1989. *The Evolving Role of Statistical Assessments as Evidence in the Courts*. New York: Springer-Verlag.

Greely, H. T. 1999. Breaking the stalemate: A prospective regulatory framework for unforeseen research uses of human tissue samples and health information. *Wake Forest Law Rev* 34:737–766.

Joh, E. E. 2011. DNA theft: Recognizing the crime of nonconsensual genetic collection and testing. *Boston Univ Law Rev* 91:665–700.

Juengst, E. T., and A. Goldenberg. 2008. Genetic diagnostic, pedigree, and screening research. *The Oxford Textbook of Clinical Research Ethics*, ed. E. J. Emanuel, C. Grady, R. A. Crouch et al., 298–314. Oxford: Oxford University Press.

Kahn, J. 2009. Race, genes, and justice: A call to reform the presentation of forensic DNA evidence in criminal trials. *Brooklyn Law Rev* 74:325–374.

Kaye, D. H. 2000. Bioethics, bench, and bar: Selected arguments in Landry v. Attorney General. *Jurimetrics* 40:193–216.

Kaye, D. H. 2007. Please, let's bury the junk: The CODIS loci and the revelation of private information. *Northwest Univ Law Rev Colloquy* 102:70–81. http://www.law.northwestern.edu/lawreview/colloquy/2007/25/ (accessed March 8, 2013).

Kaye, D. H., and M. E. Smith. 2003. DNA identification databases: Legality, legitimacy, and the case for population-wide coverage. *Wisconsin Law Rev* 2003:414–459.

Kelly, R. L. 2010. Bones of contention. *New York Times*, December 13. http://www.nytimes.com/2010/12/13/opinion/13kelly.html (accessed March 8, 2013).

Korn, D. 1999. Genetic privacy, medical information privacy, and the use of human tissue specimens in research. In: *Genetic Testing and the Use of Information*, ed. C. Long, 16–84. Washington, DC: AEI Press.

Krimsky, S., and T. Simoncelli. 2011. *Genetic Justice: DNA Databanks, Criminal Investigations, and Civil Liberties.* New York: Columbia University Press.

Lewis, M. H., A. Goldenberg, R. Anderson, E. Rothwell, and J. Botkin. 2011. State laws regarding the retention and use of residual newborn screening blood samples. *Pediatrics* 127(4):703–712.

Lucas, D. 1989. The ethical responsibilities of the forensic scientist: Exploring the limits. *J Forensic Sci* 34:719–729.

Lucast, E. K. 2007. Informed consent and the misattributed paternity problem in genetic counseling. *Bioethics* 21(1):41–50.

Maclin, T. 2008. The good and bad news about consent searches. *McGeorge Law Rev* 39:27–82.

Maschke, K. J. 2008. DNA and law enforcement. In *From Birth to Death and Bench to Clinic: The Hastings Center Bioethics Briefing Book for Journalists, Policymakers, and Campaigns*, ed. M. Crowley, 45–50. Garrison, NY: The Hastings Center.

Mund, C. 2005. Biobanks–Data sources without limits? *Jusletter* 3 (October).

Nakashima, E. 2008. From DNA of family, a tool to make arrests. *Washington Post*, April 21. http://articles.washingtonpost.com/2008-04-21/news/36801213_1_familial-searches-dna-database-tania-simoncelli (accessed March 8, 2013).

Neuffer, E. 2007. Mass graves. In *Crimes of War 2.0: What the Public Should Know*, eds. A. Dworkin, R. Gutman, and D. Rieff, 238–240. New York: W. W. Norton & Company. Also available at http://www.crimesofwar.org/a-z-guide/mass-graves/(accessed March 8, 2013).

NIST. 2012. Expert Working Group on Human Factors in Latent Print Analysis. *Latent print examination and human factors: Improving the practice through a systems approach*, ed. D. H. Kaye. Gaithersburg, MD: National Institute of Standards and Technology. Also available online at http://www.nist.gov/customcf/get_pdf.cfm?pub_id=910745 (accessed March 8, 2013).

Ossorio, P. N. 2006. About face: Forensic genetic testing for race and visible traits. *J Law Med Ethics* 32(2):277–292.

Pellegrino, E. 1993. The metamorphosis of medical ethics: A 30-year retrospective. *JAMA* 269: 1158–1162.

President's Commission for the Study of Ethical Problems in Medicine and Biomedical and Behavioral Research. 1982. *Making Health Care Decisions: The Ethical and Legal Implications of Informed Consent in the Patient–Practitioner Relationship.* Vol 1. Washington, DC: Government Printing Office. Also available online at http://bioethics.georgetown.edu/pcbe/reports/past_commissions/making_health_care_decisions.pdf (accessed March 8, 2013).

Ramshaw, E. 2010. DSHS turned over hundreds of DNA samples to feds. *The Texas Tribune*, February 22. http://www.texastribune.org/texas-state-agencies/department-of-state-health-services/dshs-turned-over-hundreds-of-dna-samples-to-feds/ (accessed March 8, 2013).

Ripley A., T. Bates, M. Hequet, and R. Laney. 2005. The DNA dragnets. *Time*, January 16. http://www.time.com/time/magazine/article/0,9171,1018083-1,00.html (accessed March 8, 2013).

Tozzo, P., L. Caenazzo, and M. J. Parker. 2013. Discovering misattributed paternity in genetic counselling: Different ethical perspectives in two countries. *J Med Ethics*, doi:10.1136/medethics-2012-101062.

U.S. Department of Health and Human Services. 2003. *Summary of the HIPAA Privacy Rule.* Rev. ed. http://www.hhs.gov/ocr/privacy/hipaa/understanding/summary/privacysummary.pdf (accessed March 7, 2013).

Wagner, J. K. 2009. Just the facts, ma'am: Removing the drama from DNA dragnets. *NC J Law Technol* 11:51–101.

Wambaugh, J. 1987. *The Blooding.* New York: Bantam.

Williams, E. D., and J. D. Crews. 2003. From dust to dust: Ethical and practical issues involved in the location, exhumation, and identification of bodies from mass graves. *Croat Med J* 44(3): 251–258.

DNA in Immigration and Human Trafficking

22

SARA HUSTON KATSANIS
JOYCE KIM

Contents

22.1 Introduction	539
22.2 Relationship Testing in Immigration	540
22.2.1 Relationship Testing	540
22.2.2 Immigration Fraud and DNA	543
22.2.3 Rapid DNA Analysis	543
22.3 DNA Identification in Human Trade	544
22.3.1 Human Trade	544
22.3.2 Relationship Testing Strategies to Detect or Investigate Human Trafficking	545
22.3.3 DNA-PROKIDS	547
22.3.4 Dallas PDI	549
22.4 Ethical, Legal, and Social Considerations with DNA Identification	550
22.4.1 Defining "Family" in Immigration Procedures	550
22.4.2 Privacy of Genetic Information	550
22.4.3 Abuse of Power	551
22.4.4 Incidental Findings	551
22.4.5 Managing International Interoperable DNA Databases	552
22.4.6 Cultural Perspectives on Genetic Information	552
22.5 Summary	553
References	553

22.1 Introduction

The reliability of DNA as a crime-solving tool has opened a range of human genetic identification applications in government processes, civil cases, and for humanitarian purposes. Unlike other biometric identification tools, DNA can be used to establish biological relationships. As such, DNA identity applications are especially relevant to border control and migration policies to combat illicit intercountry adoptions and immigration fraud, and to identify victims of human trafficking.

Relationship testing as applied to missing persons is being adapted for use in migration policies and for humanitarian aid for live victims. Chapter 10 describes the application of genetic technologies to the identification of missing persons and victims of mass disasters. In addition, governments apply DNA testing to aid identification of military personnel. In the United States, for instance, all active duty and reserve military personnel provide DNA specimens upon their enlistment, reenlistment, or preparation for operational deployment.

In the wake of human rights atrocities and genocides, both governments and humanitarian organizations have sought DNA testing to identify postconflict remains and victims of war crimes (Holland et al. 1993; Wagner 2008; Pajnič, Pogorelc, and Balažic 2010) such as cases processed through the International Commission on Missing Persons (Davoren et al. 2007; Parsons et al. 2007; Huffine et al. 2001).

These applications are successful and valuable for identifying deceased persons, but the value of DNA for identifying victims while still alive remains underutilized. As government programs using DNA to determine identity expand, the forensic community is developing programs to tap this technology for the identification of human trafficking victims. At the same time, communities must grapple with the social implications of DNA collection of vulnerable populations and develop approaches that will protect these populations from additional harms.

In this chapter, we review the emerging and nascent applications of forensic DNA technologies to immigration procedures and human trafficking and explore the profound ethical, legal, and social implications of collecting DNA of noncriminals.

22.2 Relationship Testing in Immigration

22.2.1 Relationship Testing

As described in Chapter 10, DNA is useful for establishing not just identity, but also familial relationships. Relationship testing involves analysis of autosomal short tandem repeats (STRs) and statistical analysis of likelihood ratios to predict relatedness (Debenham 2006). With traditional autosomal STR testing, determining parent–child relationships is simplified by the fact that (barring mutation events) a child inherits one of each parent's alleles. Sibship and more distant relationships are also detectable using autosomal STRs, but more readily established using single-nucleotide polymorphisms (SNPs) to establish haplotype inheritance and ancestry informative markers (AIMs). In addition, Y-STRs and mitochondrial DNA (mtDNA) analysis can be useful for kinship analyses. mtDNA is inherited through the maternal lineage, passed from a mother to her children, and the Y chromosome is inherited through paternal lineage. As such, a man and his son will have nearly identical Y chromosomes, and a mother and her son or daughter will have nearly identical mtDNA sequence.

Whereas exact comparisons between two DNA profiles can predict likelihood of two samples coming from the same biological source, comparison of nonidentical samples can predict likelihood of a biological relationship based on how many markers are shared (Daiger and Chakravarti 1983). Private companies, hospitals, and academic laboratories provide relationship testing for more than 400,000 cases per year, primarily privately ordered paternity testing (Relationship Testing Program Unit 2008). Relationship testing is also applied in courts routinely for criminal cases of incest or paternity (e.g., sexual assault of a minor resulting in pregnancy) and in civil cases of paternity or maternity (e.g., to establish child custody or in inheritance cases).

Most relationship testing uses the likelihood ratio to estimate the probability of two genetic profiles if the two individuals are related versus unrelated, taking into account all of the typed markers (e.g., 13 in a CODIS [Combined DNA Index System] profile). The fraction of shared alleles expected between two biological relatives depends on how they are related (see Table 22.1). For example, in a parent–child relationship, the individuals share 50% of all alleles, with one common allele at each locus. In full sibling relationships, siblings also share on average 50% of all alleles, but not at each locus. The likelihood ratio can vary based on the frequency of each marker across the population, the number of loci compared in the profiles, and the kinship probabilities. The mathematical formula involves comparison of the probability of the genetic evidence under two propositions: that the two individuals are related and that they are unrelated (Butler 2005). The more shared rare alleles in a profile, the higher the likelihood ratio will be. The likelihood ratio is calculated as

$$\frac{\text{Probability (Biological Relationship)}}{\text{Probability (Unrelated)}}$$

Unlike other biometric identifiers (e.g., fingerprints, iris recognition, voice recognition), DNA analysis can confirm or refute claimed biological relationships. Thus, DNA plays a unique role in border security and immigration cases dependent on a claimed relationship of a petitioner to an applicant. Since the terror attacks of September 11, 2001, officials worldwide have sought more stringent requirements for travel documents. The incorporation of biometric technology into travel and other identity documents is likely to have a crucial impact on migration (Thomas 2006). A unique and permanent biometric profile ensures objective and efficient border controls, the question of identity being removed from the individual assessment of border guards to a neutral automated procedure (Thomas 2006). In the fight against illegal migration, biometric identifiers present a number of advantages. Biometric identification can be used to detect document falsification for immigration fraud (Wasem 2007). They also may help facilitate return procedures of failed asylum seekers. Under international law states cannot deport individuals without knowing their country of origin, yet many asylum seekers will have lost or destroyed their travel documents upon arrival. States of origin then may then use this uncertainty about the background of failed asylum seekers to justify a refusal to accept the return of a citizen (Thomas 2006). Asylum seekers can provide credible, immutable evidence of their claim, and traffickers may be hindered in their attempts to use false identities (Thomas 2006).

Proof of identity (POI) at the border includes three main "identity" attributes: biometric (e.g., fingerprints), attributed (e.g., your full name), and biographical (e.g., education or employment history). Traditionally, POI has relied on attributed and biographical data, but

Table 22.1 Probabilities of Sharing Alleles at Multiple Markers

Relationship	0 alleles	1 allele
Parent–child	0	1
Full siblings	1/4	1/2
Half siblings	1/2	1/2
Cousins	3/4	1/4
Uncle–nephew	1/2	1/2
Grandparent–grandchild	1/2	1/2

this approach is waning because of perpetrators' ability to compromise POI documents easily (Jamieson et al. 2008; Wenk 2011). Biometric attributes are becoming more important and are included on a number of different POI documentation.

In the United States, multiple programs apply biometric identification to border control efforts. For example, "e-passports," which have been issued by the United States since August 2007, facilitate digital face recognition of photographic identification and contain unique chip identification numbers and digital signatures (Goth 2008). The US-VISIT program, administered by the U.S. Department of Homeland Security, aids law enforcement and intelligence agencies involved in immigration and border security, including U.S. Customs and Border Protection, Department of State (DOS) consular officers, U.S. Citizenship and Immigration Services (USCIS), U.S. Immigration and Customs Enforcement, and the U.S. Border Patrol. US-VISIT conducts biometric identification and analysis, focusing on the collection and analysis of digital fingerprints and photographs from international visitors when they are issued U.S. visas or at ports of entry. Data collected and analyzed by US-VISIT may be used to identify illegal migrants or individuals who may pose a security risk to the United States (National Science and Technology Council 2008).

Several countries, in addition to the United States, have developed policies guiding the application of biometric identification in international travel. In Germany, passport applications have required a photograph and two fingerprints since November 2007 (Goth 2008). The European Union (EU) has proposed that all visitors entering the EU undergo biometric data analysis (Goth 2008). The UN Refugee Agency has applied biometric technologies to facilitate emergency supply distribution to refugees in Afghanistan and to combat human trafficking in Pakistan (Staff reporter 2004).

The United States, Canada, United Kingdom, France, Australia, Finland, Sweden, Norway, the Netherlands, Hong Kong Special Administrative Region of China, among other countries, have begun including DNA analysis and genetic relationship testing in border security measures. Typically, this involves confirmation of claimed biological relationships between immigrants and individuals who are already settled in the receiving country (Taitz, Weekers, and Mosca 2002). Because family reunification is a primary motivator in many immigration petitions, the authenticity of an applicant's family relationship claim may determine admission or visa-issuance decisions. The verification of family relationship claims through DNA analysis involves relationship testing of putative family members by private or government laboratories. The relationship testing statistical results are used to judge the accuracy of the claimed relationship (Taitz, Weekers, and Mosca 2002).

In 2000, the voluntary use of DNA in immigration cases was established in a memo from the U.S. Immigration and Naturalization Service (INS)—which is now USCIS—to its field offices. It had previously been established that the INS could require blood parentage testing (through Blood Group Antigen or Human Leukocyte Antigen). However, the 2000 memo states that there is no "statutory or regulatory authority to require DNA testing." In 2006, the USCIS Ombudsman recommended granting local offices the authority to require DNA testing in specific circumstances and launching a pilot program to require DNA testing in the immigration context (Esbenshade 2010).

In some cases, biometric markers can identify criminals, terrorists, or known traffickers attempting to travel under false identities. Since January 2009, the United States has required DNA collection for inclusion in CODIS from non-U.S. citizens arrested or detained under the authority of the United States since January 2009 (Public Law 106-386: Victims of Trafficking and Violence Protection Act of 2000).

The implementation of DNA testing into refugee petitioning highlights some of the major challenges of systematically relying upon DNA for immigration decisions. The UN and individual countries are exploring methods to document refugees petitioning for immigration (United Nations High Commissioner for Refugees 2008; Taitz, Weekers, and Mosca 2002). In the United States, refugee family reunification cases have been processed under the Priority 3 (P-3) Program since 2004. In 2008, DNA relationship testing was piloted as part of the P-3 Program in East Africa. The results of pilot DNA testing revealed high levels of fraud among East African P-3 refugees (Esbenshade 2010). As a result, applications under the P-3 Program were suspended (Bruno 2012). In 2010, the U.S. DOS passed new rules allowing DNA testing for international refugees seeking to reunify with their families in the United States (Holland 2011). The P-3 program resumed in 2012, requiring DNA confirmation of claimed relationships for the 20 countries admitted under the P-3 program (Bruno 2012).

22.2.2 Immigration Fraud and DNA

As the use of DNA relationship testing becomes common in immigration, it is necessary to understand the benefits and limitations to applying DNA technologies in international border security. Fraud in U.S. refugee and immigration programs has motivated the use of DNA analysis and other biometric measures to assess claimed biological relationships (Esbenshade 2010). Immigrant petitioners may falsify genetic relationship test results by providing DNA samples of related persons in place of unrelated persons. For example, distant blood relatives may claim a parent–child relationship in an attempt to accelerate or increase their chances of admission into a receiving country. In nations where polygamy is permitted, a male petitioner may claim his that younger wives are his daughters (Wenk 2011). Fraudulent DNA relationship testing may involve inadvertent or intentional "genotype recycling," wherein an individual repeatedly provides a biological sample under different names (Wenk 2010).

There are several ways for laboratories performing relationship testing to detect potential fraud. For one, a laboratory may suspect sample substitution when two or more people demonstrate identical DNA profiles. Discrepancies in paperwork and genetic results may be obvious, such as differences between the claimed sex and the genetic test results. But in ordinary cases, fraud may not be obvious or easily detectable unless the laboratory is specifically looking for discrepancies, such as comparing new profiles to a database of prior profiled cases. This is simple for fraud perpetrators to circumvent by using multiple laboratories for testing. Ideally, an interlaboratory DNA profile database would be optimal for detecting genotype recycling and attempted fraud.

To deter DNA fraud in immigration programs, authorities are developing approaches to maintain control and integrity of DNA collection processes. In some cases, this means better control over collection of DNA, such as requiring DNA collection at embassies and analysis by accredited commercial laboratories, whereas in other cases this means processing relationship testing through government-controlled laboratories.

22.2.3 Rapid DNA Analysis

Standard forensic DNA analysis involves multiple steps and can take hours to days to process, depending on the laboratory and equipment. Rapid DNA analysis (see also Chapter 16) processes profiles with a more efficient turnaround time from minutes to hours, depending on the system. Existing rapid DNA instruments process five to eight samples in 45–90 minutes (Lounsbury et al. 2013; Estes et al. 2012). Turnaround time is likely to further reduce with improved technology. Most rapid DNA systems use microfluidic technologies to isolate DNA, amplify loci, and detect STR alleles in a single integrated machine (Lounsbury et al. 2013; Estes et al. 2012; Foster and Laurin 2012; Laurin and Fregeau 2012; Hopwood et al. 2010). Instruments process reference samples from buccal swabs or bloodstain card punches.

This technology provides a range of opportunities for authorities to quickly evaluate a person's identity without detaining a person for an extended period. It may be used as on-site point-of-collecting processing of relationship tests without the need for transportation of specimens to an operational laboratory. It also permits comparisons of individuals for relationship testing without the need for databasing of DNA profiles, which may be problematic in some policies (such as DNA analysis of noncitizens). It is unclear whether DNA profiles resulting from rapid DNA technologies will adhere to current evidentiary standards of DNA profiling and whether profiles will be of sufficient quality for databasing national databases such as CODIS. As the technology develops, limitations may become apparent, such as the limited throughput or incompatibility with databasing quality standards. However, the ability to sample and DNA type an individual within minutes or hours has a host of applications, including anthropological and noncriminal applications.

Potentially, rapid DNA analysis may be integrated with kinship software for determining relationships quickly and on-site. This may be valuable at border crossings to confirm or refute claimed relationships and entry.

22.3 DNA Identification in Human Trade

22.3.1 Human Trade

The global trafficking of men, women, and children for reasons of prostitution and forced labor is a problem of international concern. The International Labor Organization estimates that 20.9 million individuals are enslaved at any given time around the world (International Labour Office 2012). Modern slavery includes involuntary domestic servitude, forced labor, child soldiers, forced marriages, and the sex trade. In the United States and much of Europe, adult and children trafficking victims are predominantly found in the sex trade, and child victims are often runaways, troubled, and homeless youth. Varying international definitions of "human trafficking" may include illicit intercountry adoption, the human organ trade, and child prostitution. Combating trafficking requires a cooperative approach of law enforcement, social services, and victim support groups, and effective communication and collaboration among stakeholders and agencies.

Under the United Nations Palermo protocols, trafficking in persons is defined as "the recruitment, transportation, transfer, harboring or receipt of persons, by means of the threat or use of force or other forms of coercion, of abduction, of fraud, of deception, of the abuse of power or of a position of vulnerability or of the giving or receiving of payments or benefits to achieve the consent of a person having control over another person, for the purpose of exploitation" (United Nations 2000). The United States' Trafficking Victims Protection Act (TVPA) comprises both sex trafficking, "the recruitment, harboring, transportation, provision, or obtaining of a person for the purpose of a commercial sex act" and labor trafficking, "the recruitment, harboring, transportation, provision, or obtaining of a person for labor or services, through the use of force, fraud, or coercion for the purpose of subjection to involuntary servitude, peonage, debt bondage, or slavery" (Public Law 106-386: Victims of Trafficking and Violence Protection Act of 2000). Outside of these definitions, humans are bought, sold, and exploited. Organ trade and illicit adoptions, for example, both can involve the purchase of an individual without consent. Intercountry adoptions have increased in the past decades, which have led to crime and fraud in some countries. In most countries, a child may be adopted from another country if he or she has no parents, or if both parents relinquish their rights, obligations, and claims to the child. Criminals and corrupt officials have developed creative ways to circumvent the orphan status requirement, either through theft of children, or through bribery and coercion of relinquishing parents. Mass disaster and war can increase children's vulnerability to displacement from their biological families. Displaced children may be illicitly offered for adoption out of their home countries.

22.3.2 Relationship Testing Strategies to Detect or Investigate Human Trafficking

Trafficking in persons remains a critical challenge to the management of migration and cross-border crime. High incidence of human trafficking along with advances in human genetic identification technologies present opportunities for international law enforcement agencies to apply DNA collection to detect trafficking in persons (Birchard 1998; Katsanis 2010; Sherwell 2008; Kim and Katsanis 2013). In addition to traditional police work and training to identify trafficking victims, identification may be accomplished by the comparison of DNA profiles from potential victims identified through law enforcement, health, and social services to reference databases of profiles. Systematic collection of DNA from family members of missing persons will enable the establishment of such reference databases for comparison with potential trafficking victim profiles.

Sharing and analyzing DNA profiles internationally presents multiple technical and administrative challenges. Governments must cooperate to ensure adequate communication as well as safely handling and exchanging genetic information to protect individuals' privacy (McCartney and Williams 2011). Moreover, the broad types and clandestine nature of the slave trade and trade in humans requires diverse approaches to investigate identity. Table 22.2 lays out potential approaches and early programs for using DNA to identify trafficking and human trade victims.

Some early adopters of DNA to identify live victims of trafficking have had varying success. In China, a nationwide DNA database program established in 2009 has used DNA collection and comparison between parents of missing children and abducted or homeless children. By 2012, the program had facilitated the reunification of more than 2300 children with their families (Yan 2013). A program in the United Arab Emirates facilitates the collection of DNA from children possibly trafficked as camel jockeys (Truong and Angeles 2005). In Guatemala, the Alba–Kenneth Warning System, enacted by the national Congress, requires DNA databanking for missing children cases to identify stolen children (Decreto Número 28-2010 2010).

In intercountry adoption fraud cases, DNA identification technologies can efficiently and accurately assess the veracity of false relationship claims. The standard processes for

Table 22.2 Potential Applications of DNA in Human Trade

Form of Human Trafficking or Trade	Potential DNA Application	Example Program
Illicit adoption	Confirm biological relationship of relinquishing parent to child Compare DNA of orphans to missing children database Collect DNA of intercountry adoptions for future identification purposes	USCIS required DNA for relinquishment cases in Guatemala (International Commission against Impunity in Guatemala (CICIG) 2010) DNA-PROKIDS (Eisenberg and Schade 2010)
Sex trafficking	Databank DNA for identification of sex workers at high risk of violent crime and trafficking Compare DNA of detained underage sex workers to missing children database	Dallas Prostitute Diversion Initiative (Felini et al. 2011)
Child soldiers	Postconflict family reunification of misplaced youth Collect DNA from regions at high risk of trafficking for future identification purposes	
Forced or bonded labor and migrant workers	Compare DNA of unidentified remains to missing persons database Compare DNA of alleged parents and children suspected to be trafficked Collect DNA from areas at high risk of trafficking for future identification purposes	United Arab Emirates program to identify trafficked camel jockeys (Truong and Angeles 2005)
Involuntary domestic servitude	Collect DNA of high-risk individuals issued diplomatic immigration visas	
Human organ trade	Databank DNA of donated human organs before transplantation for future identification purposes	

adopting a child from another country may involve DNA testing, particularly in the case of adopting a sibling of an existing adopted child. In such cases, authorities may require DNA testing to confirm this relationship (U.S. Citizenship and Immigration Services 2000). The number of international adoptions into the United States has accelerated since the 1990s and, in order to identify cases of adoption fraud, U.S. agencies began to require DNA testing as part of adoption procedures from Guatemala (see Box 22.1) and Vietnam and also recommended DNA testing in many other countries (U.S. Citizenship and Immigration Services 2008). Despite required DNA testing, reports of adoption fraud from Guatemala and Vietnam continued, and adoptions from both countries have been closed since 2008 (Joint Council on International Children's Services and National Council for Adoption 2007).

BOX 22.1   DNA TESTING IN GUATEMALA ADOPTIONS

As international adoptions become increasingly prevalent the risk of adoption fraud from the source countries increases, as was the case for Guatemalan adoptions, which rose over the course of a decade. From 2000 to 2008, nearly 30,000 Guatemalan children were placed for adoption with U.S. families, until adoptions were halted after suspicion of fraud (Joint Council on International Children's Services and National Council for Adoption 2007). Reports of fraud included birth mother coercion, fraudulent identities, and child theft (Siegal 2011; Birchard 1998; Katsanis 2010; Sherwell 2008).

DNA collection had been required for Guatemalan adoptions to demonstrate biological relatedness of a relinquishing mother to a child. Despite this, corrupt officials and bribery led to continued fraud, including substitutions of children at the DNA test, and forced consent. In response, the U.S. Embassy commenced a second DNA test before releasing the child to the adoptive parents. This brought to light several suspected cases of "genotype recycling" and possibly fraud at Guatemalan medical examiners' offices designated as DNA collection sites (International Commission against Impunity in Guatemala (CICIG) 2010; Corbett 2002; Groves 2009; Siegal 2011; Wenk 2011). Ultimately, several children were placed for adoptions despite the children being subject to investigations for kidnapping (Siegal 2011). One child was reunified with her family after being stolen more than a year prior and confirmed by DNA (Sherwell 2008; Siegal 2011).

To develop systematic approaches for identification of victims of crime and trafficked persons, academic centers are working alongside government authorities to collect and store DNA samples or profiles for identification of victims. Two such programs, DNA-PROKIDS (Program for Kids Identification with DNA Systems) and the Dallas Prostitute Diversion Initiative (PDI), facilitate prospective collection of noncriminals to identify trafficked children and postmortem high-risk victims of crime.

22.3.3 DNA-PROKIDS

DNA-PROKIDS, based in Granada, Spain, works internationally to promote DNA applications to combat human trafficking and adoption fraud. DNA-PROKIDS (www.dna-prokids.org) works with individual countries to establish DNA registries to facilitate the identification of trafficked children and reunification with their families and to apply scientific methodology to develop law enforcement cases and police intelligence. DNA-PROKIDS was initiated at the University of Granada (UGR, Granada, Spain) Genetic Identification Laboratory, which continues to serve as the headquarters. The University of North Texas Center for Human Identification (UNTCHI, Fort Worth, TX, USA) collaborates with UGR to develop and expand DNA-PROKIDS globally and provides DNA analysis for many countries that lack processing capabilities (Eisenberg and Schade 2010).

Countries may have samples collected through DNA-PROKIDS processed within their country or may send samples to UGR or UNTCHI for DNA profile generation and databasing. Samples are analyzed at UGR or UNTCHI using the same technical equipment and approach as forensic evidence. DNA data are administered separately from forensic evidence and secured in registries divided by country. Laboratory case reports include information on statistical analysis and likelihood ratio or exclusion. A centralized database has not been developed for DNA-PROKIDS, so specimen and profiles are stored locally. Participation in the DNA-PROKIDS program is usually initiated through a memorandum of understanding (MOU) between DNA-PROKIDS and a country's law enforcement, forensic institute, or government authorities. DNA-PROKIDS has established MOUs with several countries including Bolivia, El Salvador, Guatemala, Indonesia, Malaysia, Mexico, Nepal, Paraguay, Peru, Philippines, Thailand, and Sri Lanka (Eisenberg and Schade 2010). DNA-PROKIDS also accepts samples from regions where MOUs have not been signed, such as Nepal and a number of Indian states.

DNA-PROKIDS primarily distributes DNA sample collection kits, provides training and resources for law enforcement in participating countries, processes samples for countries requesting DNA analysis services, works with communities to encourage individuals to provide samples, and examines best practices for the establishment of international DNA registries (Eisenberg and Schade 2010). DNA-PROKIDS aims ultimately to facilitate the use and exchange of DNA technologies to combat human trafficking both within and between countries. To achieve this, they have identified aims for program implementation planned across three phases (currently in Phase 1): Phase 1—establish MOUs with a range of countries and ensure all sample collection protocols are followed properly within participating countries; Phase 2—receive high volumes of samples from a number of countries, generate data on sample collection and database searches, facilitate interaction between or sharing of databases, and encourage legislation on DNA applications to combat human trafficking; and Phase 3—operate a centralized database with the contributing countries, enact mandatory identification of all children given for adoption and all children without known families, and continue program implementation (personal communication Jose Lorente 2012). As of April 2012, more than 4200 samples had been processed, leading to hundreds of positive identifications and exclusions (Kim and Katsanis 2013).

The success of collaborative associations through DNA-PROKIDS is not always apparent. DNA-PROKIDS' experiences in Haiti (see Box 22.2) exposed a key program challenge: allowing collaborating institutions the autonomy to set their pace for sample collection while encouraging efficient resource allocation.*

BOX 22.2 DNA-PROKIDS IN POST-EARTHQUAKE HAITI

Displacement of thousands during natural disaster, war, and other mass disasters often increases individuals' vulnerability to trafficking. After the January 2010 earthquake in Haiti, for example, traffickers offered displaced children food, water, and shelter before smuggling them into the Dominican Republic, where they became victims of slavery, sexual exploitation, illegal adoption, and organized crime (Katsanis 2010). In response, aid organizations spurred efforts to rapidly identify and protect homeless children. Identifying the need to proactively deter child theft or misplacement after the 2010 earthquake in Haiti, DNA-PROKIDS provided substantial resources to Haitian authorities, including thousands of DNA collection kits, laptop computers with software for sample collection and data storage, scanners record information on sample collection cards, and equipment for instant photo documentation. Despite continuing reports of human trafficking, over the subsequent two years, DNA-PROKIDS received few if any samples for analysis from Haiti (Eisenberg and Schade 2010).

---

* The authors are grateful to Jose Lorente, Arthur Eisenberg, and Bruce Budowle for sharing insight into the DNA-PROKIDS program.

22.3.4 Dallas PDI

Sex workers are at an elevated risk for becoming victims of violent crime and trafficking. Because sex workers are rarely reported missing and typically do not carry identification, law enforcement often lack information necessary to pursue investigations and connect crimes involving sex workers (Quinet 2007). Sex work is considered a crime in most U.S. jurisdictions, but, in some instances, sex workers may be identified as trafficked victims. For example, the U.S. TVPA classifies any commercial sex act induced by force, fraud, or coercion or performed by minors as human trafficking (Public Law 106-386: Victims of Trafficking and Violence Protection Act of 2000). It is important for law enforcement to be able to distinguish victims of trafficking from consensual sex workers in order to avoid criminalization of trafficking victims.

The Dallas PDI (www.pdinewlife.org), established by the Dallas Police Department in 2007, is a prebooking program drawing effort from the Dallas County District Attorney's Office, City Attorney's Office, Courts, Dallas County Sheriff's Office, Dallas County Health and Human Services, Parkland Hospital, and more than 30 social service and faith-based organizations. In order to facilitate sex workers' permanent exit from the sex industry, PDI offers short-term residential treatment, including health care, counseling, and support for chemical dependency, as well as long-term support services to victims, such as offering alternative housing and career options. In addition to providing sex workers options for exit from the sex trade industry, the PDI endeavors to safeguard justice for those who may relapse after treatment.

The High Risk Potential Victims DNA Database (HRDNA) is the PDI's voluntary DNA collection program, in collaboration with the UNTCHI (http://hrdna.org). This DNA collection program establishes a databank of voluntary samples from sex workers completing the PDI program. The databank is intended to assist postmortem identification, allowing law enforcement to give closure to family members of sex worker victims of violent crimes, and could be used by law enforcement as a "cold association" to open or pursue a criminal investigation. Buccal samples for the HRDNA are collected by law enforcement from consenting individuals and stored unprocessed unless or until an investigation is opened to warrant DNA profiling. By 2013, hundreds of HRDNA samples had been collected and stored for this purpose. In cases that where a PDI participant is suspected to be missing, the stored sample may be processed and profiled to assist law enforcement with identification.

In addition to facilitating postmortem identification, the HRDNA could potentially be used to identify victims of human trafficking, particularly through routine collection of samples from detained underage sex workers. Matches between the HRDNA and profiles of family members reporting a missing child could identify cases in which sex workers have been trafficked, serve as evidence to prosecute cases of human trafficking, and facilitate family reunifications when appropriate. This untapped application remains under consideration as administrators balance the ethical and social implications of such a program.*

---

* The authors are grateful to Martha Felini, Arthur Eisenberg, and Bruce Budowle for sharing insight into the Dallas PDI program.

22.4 Ethical, Legal, and Social Considerations with DNA Identification

As this text attests, DNA reliably and efficiently identifies individuals and assesses biological relationships. But as law enforcement and government uses of DNA applications expand, particularly to the identification of living noncriminal citizens, some have questioned whether routine DNA collection could create a slippery slope to privacy and human rights violations. Ensuring the socially responsible use of genetic information is key in developing model policies to implement DNA testing.

22.4.1 Defining "Family" in Immigration Procedures

Many national and international policies and recommendations for immigration and border security emphasize the importance of family reunification. The 1989 Convention of the Rights of the Child describes a family as "the fundamental group of society and the natural environment for the growth and well-being of all its members and particularly children," emphasizing that a family should be "afforded the necessary protection and assistance so that it can fully assume its responsibilities within the community" (United Nations Office of the High Commissioner for Human Rights 1989). After conflict and disaster, family reunification is a central focus of efforts and governments' migration policies often favor cohesion of a family unit (Taitz, Weekers, and Mosca 2002). With the increased use of DNA relationship testing to support biological claims, immigration policies must take into consideration that families are social constructions, and not always biological entities. In some cultures, "family" may include a wide range of biological relatives as well as members who share religious, cultural, and social relationships (Parker, London, and Aronson 2013; Taitz, Weekers, and Mosca 2002). What may appear to a DNA analyst is a misattributed relationship may actually be a miscommunication of terms to define relationships in a family (Parker, London, and Aronson 2013).

22.4.2 Privacy of Genetic Information

The collection and use of DNA from noncriminals can raise questions of privacy and human rights violations, such as Fourth Amendment infringements in the United States (Henning 2010). Despite widespread agreement that DNA applications could critically aid law enforcement or government authorities, improve public security, and promote justice, privacy concerns have been raised that the government may be intruding on individuals' right to privacy (Kim and Katsanis 2013). One program to forcibly identify orphans of Argentina's "dirty war" raised questions about genetic privacy, citing concerns of valuing truth over privacy of the individuals concerned (Pertossi 2009). Although human rights activists intended to use DNA to identify hundreds of stolen children, some argued that the program was a privacy concern, particularly for children (now adults) unaware of their origins. Indeed, misattributed paternity can disrupt families and cause strife in communities peacefully living with the secrets of the past.

Developing policies ideally aim to balance individual privacy against public interest. Existing policies have not been resolved in most countries, with a range of court and legal opinions (Kaye 2013). The U.S. Supreme Court heard one case on arrestee collection

(*Maryland v. King*) in February 2013, deciding that collection from this population is no more an infringement on U.S. Fourth Amendment rights than dermatoglyphics.

22.4.3 Abuse of Power

In addition to privacy concerns, are concerns for abuse of power by authorities maintaining databanks of DNA specimens and DNA profile databases. Although a DNA profile includes only markers that do not contain personal or health-related information (e.g., disease susceptibility, behavioral predispositions) (Katsanis and Wagner 2013), a stored sample harbors a slew of identifying, personal, and medical information. A Canadian program similar to the Dallas PDI program to collect DNA from sex workers sparked controversy surrounding the intended use of the data and samples (Hainsworth 2010). The public expects clear guidelines on how samples are to be used and assurance from misuse. In some cases, informed consent of participants may be sufficient, whereas in scenarios where vulnerable populations are sampled, additional protections may be necessary.

Although many countries and states have developed policies to protect identifiable biological specimen, some argue that the protections are insufficient. Indeed, the transportation of human specimens may itself be considered human trafficking if the purpose for acquisition requires coercion or deception and the specimens are exploited in any way. Those developing policies and laws regarding DNA collection often emphasize the importance of protecting the privacy of profiled individuals, carefully outlining the intended uses of specimens and the resulting profiles, and providing provisions for voluntary consent when appropriate. It is also important to establish whether and how specimens and profiles can be expunged and how individuals may request expungement of their personal information.

In the United States, public opinion regarding the use of genetic testing in medicine versus law enforcement varies. For example, in 2007, a study revealed that, although more than 90% of Americans over the age of 18 support the use of genetic information for medical research, 54% of respondents reported that they had little or no trust in law enforcement to protect their genetic privacy (Kaufman et al. 2009). Response from outside the United States also reveals concerns about genetic privacy. When the U.S. government issued a notice of intent to collect DNA from immigrant detainees (28 CFR § 28.12), the Embassy of Canada formally commented that, "in Canada, the creation of biometric templates for indefinite storage attracts a much higher reasonable expectation of privacy than the initial collection of live photographs and fingerprints given the much greater potential for linking disparate information, tracking and surveillance, especially given advances in computer processing technology that allow for quick matching or linking of data from numerous sources" (Canadian Embassy 2008).

22.4.4 Incidental Findings

Any relationship testing has the potential of exposing hidden family relationships, such as nonpaternity or misattributed parentage. STR profiling alone has little chance of identifying medical traits that may be of import to the individual typed, but use of SNPs may lead to revelation of traits linked to particular markers. As well, use of AIMs in relationship testing may reveal family ancestries contrasting with family lore and oral history. To mitigate potential harms precipitated by such incidental findings, some laboratories have formal nondisclosure policies whereas others have no formal policies but weigh out the situations on a case-by-case basis (Parker, London, and Aronson 2013).

22.4.5 Managing International Interoperable DNA Databases

The development of global DNA databases could critically aid the detection and investigation of human trafficking. Global collaborations to share DNA data in human trafficking investigations present a slew of technical, legal, and administrative challenges. Possible interoperability of global DNA databases raises questions regarding international laws, policies, and guidelines on sample and profile storage as well as who maintains authority over data transferred between countries. Because an international antitrafficking DNA database may present the prospect of conducting searches across databases from different countries, it may become necessary to establish a standard approach to approving and conducting international DNA database searches. The establishment of a central database would simplify the processes, but authority for such a database of international import may be difficult to establish and justify (Kim and Katsanis 2013). A network of databases among cooperating countries could suffice, but parameters for searching and access to such an infrastructure will need to be carefully established and controlled by authorities without conflicts of interest. The development and use of international DNA databases should be governed by clear guidelines to prevent misuse and malfunction, and to assure security (Thomas 2006).

22.4.6 Cultural Perspectives on Genetic Information

As forensic genetics is applied to noncriminal purposes, and around the world, we must be mindful of the cultural dynamics of local communities. By and large, Western communities have embraced use of genomic information for crime solving and human rights investigations. This is more than evident in the U.S. and European cultural media. However, cultural and political differences in other communities may hold different attitudes toward use of DNA. Some communities may fear the use of DNA by authorities or even the process of sample collection. Other communities may oppose the evaluation of DNA for clues into ancestry or evolution that undermine their cultural beliefs. Even in regions where use of DNA is embraced for law enforcement purposes, there persists distrust of how the genetic information may be used (Kaufman et al. 2009). For example, differences in social or political context may shape individuals' concerns regarding DNA sample submission and fear of retribution for reporting suspicions or evidence of ongoing crimes.

Collection of DNA from noncriminals can involve populations who may be more vulnerable to coercion, such as missing children and their families or those whose legal standing is in question. Individuals approached by law enforcement or governmental authorities may believe they are required to provide samples as part of a criminal investigation. Moreover, even with informed consent, individuals may not fully understand the benefits, limitations, and potential risks associated with DNA collection and profiling, particularly when language or cultural barriers limit comprehension.

The predictive nature of genetic information, its relevance for family members, and its past use to support prejudice, heightens existing concerns for privacy and intrusion of authorities. In regions of conflict where trafficking and human trade are prominent, populations are often distrusting of authorities as well as foreigners. No matter how well intended, persuasion of groups of individuals to participate in DNA collection programs may be unwelcome. Policies for use of genetic information for human rights purposes must be mindful of the varying cultural perspectives.

22.5 Summary

DNA technologies provide an efficient and accurate approach to individual identification and assessment of biological relationship approaches valuable to the identification of victims of human trafficking and for immigration decisions. In order to protect the privacy of individuals and citizens included in DNA collection programs, especially collection from noncriminals and vulnerable populations, it is necessary to develop policies to guide the administration and quality control of DNA applications. Programs for DNA collection of immigrants and to prevent human trafficking should consider first whether DNA collection is a necessary component for identification or whether other biometric measures work sufficiently. If DNA is a valuable approach for the population and circumstances, then the program should establish parameters for what authority is most appropriate for collection of DNA while maintaining the integrity of the collected specimens. A DNA program should define the retention policies including how long specimens will be retained, who will have access to the specimens, and how specimens will be destroyed. Similar but distinct policies are necessary for the DNA profile and access to any DNA profile databases.

References

Birchard, K. 1998. Call for DNA testing for foreign adoptions. *Lancet* 352:9128.
Bruno, A. 2012. Refugee admissions and resettlement policies. Edited by Congressional Research Service. http://fpc.state.gov/documents/organization/187394.pdf (accessed March 25, 2013).
Butler, J. M. 2005. *Forensic DNA Typing*. 2nd ed. Burlington, MA: Elsevier Academic Press.
Canadian Embassy. 2008. Comment on FR Doc# E8-08339, May 18, 2008. http://www.regulations.gov/#!documentDetail;D = DOJ-OAG-2008-0009-1201 (accessed March 25, 2013).
Corbett, S. 2002. Where do babies come from? *New York Times*, July 16, 2002.
Daiger, S. P., and Chakravarti, A. 1983. Deletion mapping of polymorphic loci by apparent parental exclusion. *Am J Med Genet* 14(1):43–8.
Davoren, J., Vanek, D., Konjhodzic, R., Crews, J., Huffine, E., and Parsons, T. J. 2007. Highly effective DNA extraction method for nuclear short tandem repeat testing of skeletal remains from mass graves. *Croat Med J* 48(4):478–85.
Debenham, P. 2006. DNA fingerprinting, paternity testing and relationship (immigration) analysis. In *Encyclopedia of Life Sciences*. Chichester: John Wiley & Sons, Ltd.
Decreto Número 28-2010. Ley del Sistema de Alerta Alba-Keneth. El Congreso de la República de Guatemala. September 13, 2010.
Eisenberg, A., and Schade, L. 2010. DNA-PROKIDS: Using DNA technology to help fight the trafficking of children. *Forensic Magazine*.
Esbenshade, J. 2010. Special Report: An Assessment of DNA Testing for African Refugees. Edited by Immigration Policy Center, American Immigration Council. http://www.immigrationpolicy.org/special-reports/assessment-dna-testing-african-refugees (accessed March 25, 2013).
Estes, M. D., Yang, J., Duane, B. et al. 2012. Optimization of multiplexed PCR on an integrated microfluidic forensic platform for rapid DNA analysis. *Analyst* 137(23):5510–9.
Felini, M., Ryan, E., Frick, J., Bangara, S., Felini, L., and Breazeal, R. 2011. Prostitute Diversion Initiative annual report 2009–2010. http://www.pdinewlife.org/wp-content/uploads/2012/01/PDI-Annual-Report-2011_Final-RS.pdf (accessed March 25, 2013).

Foster, A., and Laurin, N. 2012. Development of a fast PCR protocol enabling rapid generation of AmpFlSTR(R) Identifiler(R) profiles for genotyping of human DNA. *Investig Genet* 3:6.

Goth, G. 2008. Biometrics Could Streamline Border Crossings. *Computing Now* 9(4), April 2008. http://www.computer.org/portal/web/computingnow/0210/theme/dso (accessed March 25, 2013).

Groves, M. 2009. Trafficking reports raise heart-wrenching questions for adoptive parents. *Los Angeles Times*, November 11, 2009.

Hainsworth, J. 2010. Sex workers question police DNA collection. *Xtra! News*, March 11, 2010.

Henning, A. C. 2010. Compulsory DNA collection: A Fourth Amendment analysis. February 16, 2010. Edited by Congressional Research Council. http://www.fas.org/sgp/crs/misc/R40077.pdf (accessed March 25, 2013).

Holland, E. 2011. Moving the virtual border to the cellular level: Mandatory DNA testing and the U.S. refugee family reunification program. *California Law Rev* 99:1635–82.

Holland, M. M., Fisher, D. L., Mitchell, L. G. et al. 1993. Mitochondrial DNA sequence analysis of human skeletal remains: Identification of remains from the Vietnam War. *J Forensic Sci* 38(3):542–53.

Hopwood, A. J., Hurth, C., Yang, J. et al. 2010. Integrated microfluidic system for rapid forensic DNA analysis: Sample collection to DNA profile. *Anal Chem* 82(16):6991–9.

Huffine, E., Crews, J., Kennedy, B., Bomberger, K., and Zinbo, A. 2001. Mass identification of persons missing from the break-up of the former Yugoslavia: Structure, function, and role of the International Commission on Missing Persons. *Croat Med J* 42(3):271–5.

International Commission against Impunity in Guatemala (CICIG). 2010. Decree 77-2007: Report on players involved in the illegal adoption process in Guatemala since the entry into force of the adoption law, December 1, 2010. http://cicig.org/uploads/documents/informes/INFOR-TEMA_DOC05_20101201_EN.pdf (accessed March 25, 2013).

International Labour Office. 2012. ILO 2012 Global Estimate of Forced Labour. Geneva, Switzerland. http://www.ilo.org/wcmsp5/groups/public/@-ed_norm/@-declaration/documents/publication/wcms_181921.pdf (accessed March 25, 2013).

Jamieson, R. W. D., Stephens, G., and Smith, S. 2008. Developing a conceptual framework for identity fraud profiling. Paper read at 16th European Conference on Information Systems, at Galway, Ireland. http://is2.lse.ac.uk/asp/aspecis/20080120.pdf (accessed March 25, 2013).

Joint Council on International Children's Services and National Council for Adoption. 2007. U.S. adoptions from Guatemala to halt as of January 1, 2008. *Adoption Social Work New York*, http://www.adoptionsocialworkny.com/2007/09/(accessed March 25, 2013).

Katsanis, S. 2010. Use DNA to stop child trafficking. *Toronto Globe and Mail*, February 23, 2010.

Katsanis, S. H., and Wagner, J. K. 2013. Characterization of the standard and recommended CODIS Markers. *J Forensic Sci* 58(Suppl 1):S169–72.

Kaufman, D. J., Murphy-Bollinger, J., Scott, J., and Hudson, K. L. 2009. Public opinion about the importance of privacy in biobank research. *Am J Hum Genet* 85(5):643–54.

Kaye, D. 2013. On the 'considered analysis' of collecting DNA before conviction. *Penn State Law* 60 (Legal Studies Research Paper No. 27-2012).

Kim, J., and Katsanis, S. 2013. Brave new world of human rights DNA collection. *Trends Genet* 29(6):329–32.

Laurin, N., and Fregeau, C. 2012. Optimization and validation of a fast amplification protocol for AmpFlSTR(R) Profiler Plus(R) for rapid forensic human identification. *Forensic Sci Int Genet* 6(1):47–57.

Lounsbury, J. A., Karlsson, A., Miranian, D. et al. 2013. From sample to PCR product in under 45 minutes: A polymeric integrated microdevice for clinical and forensic DNA analysis. *Lab Chip* 13(7):1384–93.

McCartney, C. W., and Williams, R. 2011. Transantional exchange of forensic DNA: Viability, legitimacy, and acceptability. *Eur J Criminal Policy Res* 17:305–22.

National Science and Technology Council. 2008. Biometrics in government post-9/11: Advancing science, enhancing operations, October 2008. Edited by Office of Science and Technology Policy, Executive Office of the President. http://www.fas.org/irp/eprint/biometrics.pdf (accessed March 25, 2013).

Pajnič, I. Z., Pogorelc, B. G., and Balažic, J. 2010. Molecular genetic identification of skeletal remains from the Second World War Konfin I mass grave in Slovenia. *Int J Legal Med* 124(4):307–17.

Parker, L. S., London, A. J., and Aronson, J. D. 2013. Incidental findings in the use of DNA to identify human remains: An ethical assessment. *Forensic Sci Int Genet* 7(2):221–9.

Parsons, T. J., Huel, R., Davoren, J. et al. 2007. Application of novel "mini-amplicon" STR multiplexes to high volume casework on degraded skeletal remains. *Forensic Sci Int Genet* 1(2):175–9.

Pertossi, M. 2009. Argentina forces dirty war orphans to provide DNA. *Associated Press*, November 20, 2009.

Public Law 106-386: Victims of Trafficking and Violence Protection Act of 2000. October 28, 2000.

Quinet, K. 2007. The missing missing: Toward a quantification of serial murder victimization in the United States. *Homicide Stud* 11:319–339.

Relationship Testing Program Unit. 2008. Annual report summary for testing in 2008. Edited by American Association of Blood Banks. http://www.aabb.org/sa/facilities/Documents/rtannrpt08.pdf (accessed March 25, 2013).

Sherwell, P. 2008. Guatemalan mother reunited with baby stolen and sold for adoption by U.S. couple. *The Sunday Telegraph*, July 26, 2008.

Siegal, E. 2011. *Finding Fernanda: Two Mothers, One Child, and a Cross-Border Search for Truth*. Cathexis Press.

Staff reporter. Pakistan: IOM developing strategy to counter human trafficking. 2004. *IRIN News Reports*, http://www.irinnews.org/fr/report/26502/pakistan-iom-developing-strategy-to-counter-human-trafficking (accessed March 25, 2013).

Taitz, J., Weekers, J. E., and Mosca, D. T. 2002. DNA and immigration: The ethical ramifications. *Lancet* 359(9308):794.

Taitz, J., Weekers, J. E. M., and Mosca, D. T. 2002. The last resort: Exploring the use of DNA testing for family reunitifcation. *Health Hum Rights* 6(1):20–32.

Thomas, R. 2006. Biometrics, international migrants and human rights. *Eur J Migration Law* 7(4): 377–411.

Truong, T., and Angeles, M. 2005. Searching for best practices to counter human trafficking in Africa: A focus of women and children, March 2005. Edited by United Nations Educational, Scientific and Cultural Organization. http://unesdoc.unesco.org/images/0013/001384/138447e.pdf (accessed March 25, 2013).

United Nations. 2000. Protocol to Prevent, Suppress and Punish Trafficking in Persons, Especially Women and Children, Supplementing the United Nations Convention Against Transnational Organized Crime. http://www.uncjin.org/Documents/Conventions/dcatoc/final_documents_2/convention_%20traff_eng.pdf (accessed March 25, 2013).

United Nations High Commissioner for Refugees. 2008. UNHCR note on DNA testing to establish family relationships in the refugee context, June 2008. http://www.unhcr.org/refworld/pdfid/48620c2d2.pdf (accessed March 25, 2013).

United Nations Office of the High Commissioner for Human Rights. 1989. Convention on the Rights of the Child. http://www.ohchr.org/EN/ProfessionalInterest/Pages/CRC.aspx (accessed March 25, 2013).

U.S. Citizenship and Immigration Services. 2000. Memorandum from Michael A. Pearson, Executive Associate Commissioner, Office of Field Operations to All Regional Directors, Director, International Affairs, FLETC/GLYNCO on Guidance on processing petitions for adopted alien children less than 18 years of age considered a "child" under the Immigration and Nationality Act through Public Law 1066-139. http://www.uscis.gov/files/pressrelease/adoptchi.pdf (accessed March 25, 2013).

U.S. Citizenship and Immigration Services. USCIS update: USCIS implements required DNA testing for Vietnamese adoptions, May 29, 2008. http://www.uscis.gov/portal/site/uscis/menuitem.5af9bb95 919f35e66f614176543f6d1a/?vgnextoid=49f0bfd9dd43a110VgnVCM1000004718190aRCRD&vg nextchannel=68439c7755cb9010VgnVCM10000045f3d6a1RCRD (accessed March 25, 2013).

Wagner, S. 2008. *To Know Where He Lies: DNA Technology and the Search for Srebrenica's Missing*. Berkeley: University of California Press.

Wasem, R. E. 2007. Immigration fraud: Policies, investigations, and issues. In *CRS Report for Congress*, ed. by R. E. Wasem. (accessed May 17, 2007).

Wenk, R. E. 2010. Sporadic genotype recycling fraud in relationship testing of immigrants. *Transfusion* 50(8):1852–3.

Wenk, R. E. 2011. Detection of genotype recycling fraud in U.S. immigrants. *J Forensic Sci* 56 (Suppl 1):S243–6.

Yan, Z. 2013. Database gives hope to abducted children. *China Daily USA*, January 22, 2013.

DNA Databases

23

CHRISTOPHER ASPLEN

Contents

23.1	Introduction	557
23.2	Volume of Databases Worldwide	559
23.3	United Kingdom and Wales	559
23.4	U.K. NDAD	560
	23.4.1 A Decrease in "Detections" or "Matches" in the Database	561
23.5	United Kingdom DNA Database and European Court of Human Rights Decision in *Marper*: What It Did and Did Not Say	561
23.6	United States—A Three-Tiered System	563
23.7	Tragedy in South Africa	564
23.8	Sharing DNA Data across Borders	566
23.9	European Union—Law for Connectivity	566
23.10	Interpol's Global DNA Gateway	566
23.11	Rapid DNA Technology	567
References		568

23.1 Introduction

Forensic DNA technology and forensic DNA databases are widely recognized as some of the most efficient and effective crime fighting tools available to law enforcement. It is, however, the combination of both the biological science of DNA and the computer science of databasing that makes DNA evidence so powerful at identifying and convicting the guilty while exonerating the innocent. Absent the "database" component in a crime fighting context, DNA technology fails to be an investigative tool that increases law enforcement's efficiency or effectiveness. Without the databasing component, DNA evidence can only be useful once police have identified a suspect, through the traditional investigative processes. DNA alone certainly makes a case better once police have performed their investigative role. However, when computer databasing technology is integrated into the investigative dynamic, the investigative process itself becomes significantly more effective and efficient. DNA becomes the driver of an investigation, rather than the goal of the investigation. This is especially true in cases where traditional investigative techniques have been unhelpful and when no suspect can otherwise be identified.

Every day, all around the world, hundreds of cases are solved by DNA evidence even when no viable suspect was previously identified by law enforcement. These cases are often our most heinous and violent and can be years or decades old. But increasingly DNA is being used to drive the investigation of cases, other than our most physically violent and dangerous. Law enforcement agencies around the world are beginning to appreciate the power of DNA databases to also solve property or "volume" crime such as burglary and car theft. As collection methods have improved, capacity has grown and turnaround times have shortened. Police have increasingly expanded the scope of their utilization of these databases, thus recognizing the highly recidivistic nature of property crimes and the positive cost/benefit analysis of solving volume crimes more quickly with DNA. DNA databases are quickly moving beyond violent crime to the much more voluminous context of property crimes.

Given the increasing potential of DNA evidence to solve crime through database utilization, more countries and law enforcement agencies around the world are creating and leveraging DNA databases (Hindmarsh and Prainsack 2010). It is not just the number of databases but rather the kinds of databases and the contextual utilization of them that is expanding. Ten years ago, the majority of forensic DNA databases around the world—and there were not many—were offender databases created under government authority, parliamentary, congressional, etc. The vast majority were found in North America and Europe where their construction was rather simplistic in that the entry criteria were established legislatively.

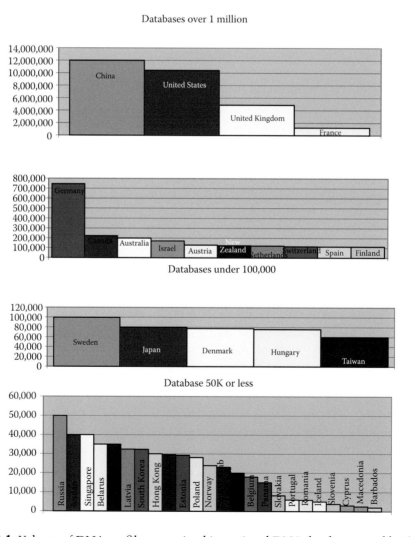

Figure 23.1 Volume of DNA profiles contained in national DNA databases worldwide.

DNA Databases																																			559

Since the creation of the first forensic DNA database in the United Kingdom, the entire history of government-established DNA databases is one of constant expansion. Upon seeing the effectiveness of the combination of DNA technology's scientific robust reliability and computer database algorithm development, no country has ever initiated a forensic DNA database with a certain volume of inclusion and then subsequently reduced the amount of DNA databasing work it required by law. Any country that has ever established a DNA database has only ever observed its benefits and ultimately expanded the entry criteria to include more classes and types of offenders. Even the European Court of Human Rights decision in the U.K. case *Marper* did not prevent the number and kinds of cases to be included in the U.K. database (Schellberg 2012) (Figure 23.1).

Even though the global landscape of DNA databasing legislation changes constantly, as of the writing of this chapter, 47 countries around the world have specifically passed legislation establishing a forensic DNA database for crime solving purposes (Schellberg 2012).*

Furthermore, at least 22 countries are actively considering legislation that would integrate DNA databases in their criminal justice systems (Schellberg 2012).†

23.2 Volume of Databases Worldwide

Although the volume of databases changes daily and probably by the thousands globally, the volume of DNA profiles is approximately 35 million records contained in forensic databases worldwide. Those countries with the largest databases are China, containing in excess of 12 million, the United States with more than 10.4 million, and the United Kingdom with more than 4.8 million (Schellberg 2012).

23.3 United Kingdom and Wales

Although it no longer has the largest database, England is widely recognized as having the most effective and efficient approach to the use of forensic DNA technology in the world. Since the establishment of the National DNA Database (NDNAD) on April 10, 1995, England has become a world leader in discovering innovative ways to use DNA to identify suspects, protect the innocent, and convict the guilty. DNA technology and DNA databasing have become central to the process of criminal investigation. The decision to integrate DNA technology so thoroughly and the subsequent success of the NDNAD can be attributed to three major factors: the political will of the Home Office, the original technical capability of the Forensic Science Service (FSS), and the operational desire of the police. The foundation of DNA-driven investigations in England and Wales is its expansive DNA

---

* Australia, Austria, Barbados, Belarus, Belgium, Brazil, Canada, Chile, China, Croatia, Cyprus, Denmark, Estonia, Finland, France, Germany, Hong Kong, Hungary, Iceland, Israel, Italy, Japan, Jordan, Kuwait, Latvia, Netherlands, New Zealand, Macedonia, Malaysia, Norway, Panama, Poland, Portugal, Qatar, Romania, Russia, Slovenia, Slovakia, Singapore, Spain, Sweden, Switzerland, Taiwan, United Arab Emirates, United Kingdom, United States. Source: Gordon Thomas Honeywell. Feb. 2012. Government Affairs, Tim Schellberg Presentation, American Academy of Forensic Sciences Meeting.
† India, Indonesia, Brazil, Pakistan, Nigeria, Mexico, Vietnam, Turkey, Thailand, South Africa, Ukraine, Argentina, Kenya, Peru, Saudi Arabia, Algeria, Morocco, Sri Lanka, Czech Republic, Ireland, Uruguay, Colombia.

database. Many factors, however, contribute to the success of the U.K.'s approach to forensic DNA applications.

England has a population of approximately 63.1 million. It has 43 municipal police forces with a total police personnel of more than 143,000 (Home Office Statistical Bulletin, March 31, 2010). Up until several years ago, 90% of forensic DNA analyses performed in England was performed by one, quasi-governmental agency, the FSS. The U.K. Government announced the closure of the FSS in December 2010, citing monthly losses of £2 million as justification. The FSS finally closed on March 31, 2012. In its place, the private DNA testing market has expanded, and now the United Kingdom relies entirely on privatized forensic DNA testing.

Averaging 2 million arrests per year, United Kingdom statistics show less than one-quarter of these arrests are first-time offenders. Furthermore, it has been shown that 85% of offenders receive their first conviction between the ages of 14 and 19 years, with the likelihood of a 14-year-old reoffending at 77% (Busher 2002). Government statistics also show 20% of the criminals commit 80% of crime and 60% of court appearances deal with only 21% of the offenders (Ibid., 21–25). All of these are factors strongly support a broad use of DNA databases, driven by a broad scope of inclusion.

23.4 U.K. NDAD

The FSS pioneered the use of DNA profiling in forensic science and set up the world's first national criminal intelligence DNA Database—launched in April 1995. The structure and use of the NDNAD, however, differs significantly from the structure and use of the local, state, and national DNA databases of the United States. In the United States, the National DNA Indexing System (NDIS) is based on a system in which convicted offender profiles from all U.S. states are included, but arrestee profiles can only be included from 24 states and searched against crime scene profiles. The U.K. NDNAD, however, is driven by the use of suspect profiles first and foremost. In England and Wales, police have powers to take and retain biological samples from anyone arrested for any recordable offence.

In April 2007, responsibility for the delivery of NDNAD services was transferred from the Home Office and the FFS to the National Police Improvement Agency (NPIA). The agency's role is to run the database operations, and maintain and ensure the integrity of the data. The NPIA is responsible for overseeing the NDNAD service, ensuring that it is operated in line with agreed standards. It is also responsible for accrediting all scientific laboratories that analyze DNA samples and oversee the contract for the operation and maintenance of the NDNAD (Table 23.1).

Table 23.1 National Policing Improvement Agency

All Police Forces including England, Wales, Scotland, and Northern Ireland as at March 31, 2012	April 2001–March 2012
Total number of crime scenes matching one or more subjects all offense types	409,715
Number of murder/manslaughter and attempted murder crime scenes matching one or more subjects	2595
Number of rape crime scenes matching one or more subjects	5729

Source: National DNA Database Statistics for England, Wales, Scotland, and Northern Ireland.

According to the NPIA, police forces in England, Wales, Scotland, and Northern Ireland have solved more than 409,000 crimes through the use of their DNA database.

According to the NDNAD Annual Report 2007/2009, approximately 500,000 offender profiles and 50,000 crime-scene profiles are entered into the database annually. That volume of offender and crime scene profiles results in more than 36,000 scene to offender hits per year for a 58.7% hit rate (National DNA Database Annual Report, 2007/2009).

23.4.1 A Decrease in "Detections" or "Matches" in the Database

Interestingly though, the NPIA acknowledges that the number of DNA detections (excluding "additional" detections) has fallen by 11% over a 5-year period, from 20,489 in 2003–2004 to 17,607 in 2008/2009. But this decline needs to be seen against the background of overall falling crime. In the same period, the total recorded crime in England and Wales fell by 22% from 6,013,759 in 2003/2004 to 4,703,814 in 2008/2009.

However, there has been an even more significant fall in those volume crime offenses where crime scenes are examined for DNA by police scenes of crime officers.

In 2008/2009,

- It was noted that 51.4% of all crime scenes examined by scenes of crime officers for DNA were burglary offenses. Over 5 years, burglary offenses dropped by 29% (from 820,013 to 581,397).
- Moreover, 20.4% of all crime scenes examined by scenes of crime officers for DNA were theft of and from vehicle offenses. Over 5 years, offenses against vehicles dropped by 40% from 985,006 to 592,117.

More relevant to the performance of the NDNAD in this discussion is the improvement in match rates. In 2002/2003, a new crime scene DNA profile being loaded to the DNA database had a 45% chance of matching a person's DNA profile. In 2008/2009, this had risen to almost 60%. To put this in perspective, should your house have been burgled and DNA was found, there was a 45% chance of the NDNAD identifying the donor of the DNA in 2002/2003. In 2010, there was just under a 60% chance of the donor being identified.

In other words, the rate of crime is falling in the United Kingdom and Wales at the same time that the likelihood of getting a match between crime scene and offender is rising. That is the exact crime-fighting statistical dynamic one would hope to see develop.

23.5 United Kingdom DNA Database and European Court of Human Rights Decision in *Marper*: What It Did and Did Not Say

Widely recognized as the most aggressive country in the world at including and keeping individuals' profiles in the forensic DNA database, the U.K. database provided for the inclusion and retention of profiles of individuals who, while arrested for or suspected of a particular crime, were subsequently never formally charged or convicted. It was the only country that retained profiles regardless of an individual's exoneration or exclusion from prosecution.

While tested in the Courts of the United Kingdom, and successfully upheld, the practice of retaining profiles of the "not convicted" was challenged in the European Court of Human Rights. Established to hear cases involving alleged violations of the 1950 European

Convention on Human Rights, the European Court of Human Rights sits in Strasbourg, France. Having considered the cases of *S. v. the United Kingdom* (2008) ECHR 1581 for several years, on December 4, 2008, the Court issued its "Grand Chamber Judgment" considering the issue of the "retention of fingerprints, cellular samples, and DNA profiles after criminal proceedings were terminated by an acquittal." The Court held unanimously (17 judges concurring) that there had been a violation of Article 8 (pertaining to a right of respect for private and family life) of the European Convention on Human Rights.

The decision stems from two separate cases originating in the United Kingdom. In the first case, "S" was an 11-year-old boy, arrested and charged with attempted robbery, who was acquitted 5 months later. Nonetheless, his DNA sample and fingerprints were taken, profiles were entered into the database, and the samples retained. The second case involved a charge of harassment in which the victim was Marper's partner. Marper and his partner reconciled and the case was dismissed before any trial was conducted. Again, however, the litigant's fingerprints and DNA were taken and held. In both cases, requests were made to have their DNA profiles and fingerprints removed from the databases and their biological samples destroyed. In both cases, those requests were denied by British authorities.

In its decision, the Court concluded that the retention of both cellular samples and DNA profiles amounted to an interference with the applicants' right to respect for their private lives, within the meaning of Article 8 § 1 of the Convention, concluding that fingerprints, DNA profiles, and cellular samples constituted personal data within the meaning of the Council of Europe Convention of 1981 for the protection of individuals with regard to automatic processing of personal data.

Recognizing that the practice of such retention was limited to England, Wales, and Northern Ireland, the court was concerned about what it considered to be the "indiscriminate" nature of the retention. It noted that the interests of the individuals concerned and the community as a whole in protecting personal data, including fingerprint, and DNA information, could be outweighed by the legitimate interest in the prevention of crime. And it accepted that the retention of fingerprint and DNA information pursued a legitimate purpose, namely, the detection, and therefore, prevention of crime. However, it found that the power of retention of the fingerprints, cellular samples, and DNA profiles of persons suspected but not convicted of offences, as applied in the case of the present applicants, failed to strike a fair balance between the competing public and private interests.

Although this decision was widely cited in news accounts around the world, it was often cited incorrectly. The European Court did not find a violation in the collection of DNA (or fingerprints) at arrest as was often cited by those opposing the taking of DNA from arrestees. The Court's decision was directed at the unlimited retention of those profiles once exonerated or deemed not prosecutable for any reason. It was as significant a decision in terms of what it allowed as it was in terms of what it disallowed. In fact, the Court recognized a legitimate purpose in collecting such data. Having had the opportunity to decry arrestee DNA sampling and profiling, it did not. As such, those objectors to the practice of arrestee testing should have a difficult time justifying their argument that such sampling and profiling is a human rights violation (*Maryland v. Alonzo Jay King*, 567 U.S. 2012).

It is instructive to understand the parameters of data protection and civil and human rights in various countries—particularly as we work toward a new dynamic of connectivity between countries in the fight against global crime and terrorism. Only then can we respect those parameters and encourage the continued or even accelerated flow of valuable information. It must be remembered, however, that those decisions are always made in jurisprudential, cultural, and historical contexts unique to each country. And ultimately, those decisions have great impact on the effectiveness of crime fighting technology.

For example, 10 years ago in the United States, the Federal Bureau of Investigation's Combined DNA Index System (CODIS) database was a much different database, with much different capabilities and applications than the current configuration. Legislatively established by the United States' Congress, the DNA Identification Act of 1994 created the NDIS, which served as a connection of U.S. state and local databases (DNA Identification Act of 1994). However, as originally crafted, the enabling legislation restricted the entry of offender profiles to only those profiles that came from individuals who were *convicted* of certain offenses.

However, over the past 8 years, since Federal legislation was passed allowing for the inclusion of arrestee profiles in the NDNAD, state governments throughout the United States have considered—and some passed—new laws allowing for the inclusion of arrestees in their state databases for ultimate searching in NDIS. Just as the U.K. law was challenged in the European Court of Human Rights, so the arrestee legislation is being challenged in State and Federal Courts throughout the United States.

At the time of publication of this book, 28 states in the United States have passed laws allowing for the taking of DNA from arrestees and for the entry of the profiles generated there from into the state NDIS database. Legal challenges have arisen in several states alleging that such collection from individuals not yet convicted violates the Fourth Amendment of the United States Constitution. Whereas the judicial systems of most "arrestee" states have found that to be a Constitutional use of governmental power, at the time of publication, two states—California and Maryland—have handed down appellate decisions declaring the practice to be a violation of the Constitution. Given the difference of opinions throughout the United States, and the fact that the NDIS is a Federal system intended to connect all the states together, the U.S. Supreme Court, at the time of writing this chapter, heard oral arguments regarding the issue of whether the U.S. Constitution allows for the taking of DNA at arrest (*Maryland v. Alonzo Jay King, Jr.*, 567 2012) (see author's note at end of chapter).

23.6 United States—A Three-Tiered System

DNA databasing in the United States is designed around a software system called CODIS. The CODIS databases exist at the local, state, and national levels. This tiered architecture allows crime laboratories to control their own data—each laboratory decides which profiles it will share with the rest of the country. NDIS is one part of CODIS—the national level—containing the DNA profiles contributed by federal, state, and local participating forensic laboratories. NDIS was implemented in October 1998. All 50 states, the District of Columbia, the federal government, the U.S. Army Criminal Investigation Laboratory, and Puerto Rico participate in NDIS.

The DNA Identification Act of 1994 (42 U.S.C. § 14132) authorized the establishment of this National DNA Index but has been amended several times to allow for the expansion of the database's utility (FBI 2013, National DNA Index System). The DNA Act and subsequent legislation specifies the categories of data that may be maintained in NDIS (convicted offenders, arrestees, legal, detainees, forensic [casework], unidentified human remains, missing persons and relatives of missing persons) as well as requirements for participating laboratories relating to quality assurance, privacy, and expungement.

In its original form, CODIS consisted of two indexes: the Convicted Offender Index and the Forensic Index. The Convicted Offender Index contains profiles of individuals convicted of crimes; state law governs which specific crimes are eligible for CODIS. (All 50 states have passed DNA legislation authorizing the collection of DNA profiles from convicted offenders for submission to CODIS.) The Forensic Index contains profiles developed from biological material found at crime scenes. Over the years, CODIS has added several other indexes, including an Arrestee Index, a Missing or Unidentified Persons Index, and a Missing Persons Reference Index.

For solving various crimes, CODIS searches the Forensic Index against itself and against the Offender Index. A Forensic to Forensic match provides an investigative lead that connects two or more previously unlinked cases. A Forensic to Offender match actually provides a suspect for an otherwise unsolved case. It is important to note that the CODIS matching algorithm only produces a list of candidate matches. Each candidate match is confirmed or refuted by a Qualified DNA Analyst. (To become qualified, a DNA analyst must meet specific education and experience requirements and undergo semiannual proficiency tests administered by a third party.)

As of January 2013, the NDIS contained more than 10,142,600 offender profiles, 1,362,800 arrestee profiles, and 472,500 forensic profiles. However, ultimately, the success of the CODIS program is measured by the crimes it helps to solve, not necessarily arrest or conviction. CODIS's primary metric, the "Investigation Aided," tracks the number of criminal investigations where CODIS has added value to the investigative process (FBI Laboratory Forensic Science Systems Unit 1998, https://www.ncjrs.gov/pdffiles1/nij/sl413apg.pdf). As of January 2013, CODIS had produced more than 200,300 hits assisting in more than 192,400 investigations.

In terms of what actual data are contained in the database, no personal information relating to the convicted offenders, arrestees, or detainees is stored in these DNA databases. No names or other personal identifiers of the offenders, arrestees, or detainees are stored using the CODIS software. The only information allowed to be stored is the following information and can be searched at the national level:

- The DNA profile—the numbers that designate the alleles at each of the various loci analyzed.
- The Agency Identifier of the agency submitting the DNA profile.
- The Specimen Identification Number—generally a number assigned sequentially at the time of sample collection. This number does not correspond to the individual's social security number, criminal history identifier, or correctional facility identifier.
- The DNA laboratory personnel associated with a DNA profile analysis.

23.7 Tragedy in South Africa

In the context of fighting crime with DNA technology, a country can possess the best, most advanced laboratory system in the world, the most rigorous quality assurance procedures, and send specialized crime scene analysts to every crime scene; however, these factors mean little if the law does not allow you to leverage the full potential of the technology and the evidence. Nowhere is this dynamic more tragically clear than in South Africa.

For more than 11 years, the government in South Africa has been considering legislation that would establish an offender DNA database in that country. Although the South African Police Service (SAPS) had created one of the most sophisticated and advanced automated systems for DNA analysis in the world, the system was ineffective because no legislation existed to allow police to include offenders in the database. The system was so impressive that the South Africans who built the system were invited by the United States Department of Justice to come to the United States and present on their system in 2002 (2002 DNA Initiative). South Africa was going to be a model, not only for Africa, but perhaps for the world. They had crime statistics that proved South Africa to be one of the most sexually violent places on the planet, and they had the capacity and technical sophistication to hit back hard. South Africa was going to prove the power of DNA like nowhere else.

Eleven years later, at the time of this publication, South Africa still has no legislation and no offender database. One of the most sophisticated forensic DNA analysis systems in the world fails to maximize its potential because the law prevents it from doing so. Eleven years after South Africa created one of the most important laboratory infrastructures in the world, the politicians in the South African Parliament have still failed to give police the legal authority to save literally thousands upon thousands of lives with DNA. Eleven years later South Africa, in contrast with more than 50 countries around the world, still has no legislation allowing for the establishment of a forensic DNA database.

South Africa is a strikingly beautiful country from its coastline at the Cape of Good Hope to Kruger National Park to the wine regions of Stellenbosch. It is also the economic anchor for sub-Saharan Africa. It has a technology portfolio that includes a nuclear weapons program (and the wisdom to subsequently dismantle it), a 2002 Noble Prize for work in microbiology, and the first human-to-human heart transplant ever performed. And most importantly, it is a country that engineered one of the most significant triumphs of human spirit and potential—the nonviolent elimination of apartheid.

However, according to the United Nations, South Africa also ranks second for murder and first for assaults and rapes per capita (The Eighth United Nations Survey on Crime Trends and the Operations of Criminal Justice Systems 2002). Fifty-two people are murdered every day there, and the number of rapes reported in a year is about 55,000. It is estimated that 500,000 rapes are committed annually in South Africa. In a 2009 survey, one in four South African men admitted to raping someone (June 2009. MRC Policy Brief, Understanding Men's Health and Use of Violence: Interface of Rape and HIV in South Africa). Even more insidious, South Africa has one of the highest incidences of child and baby rape in the world. It is a country where the beliefs exist that intercourse will cure or prevent HIV/AIDS and where child rape is used as a method of retaliation against someone else for a perceived wrong. Children are murdered and body parts used for "traditional" medicinal remedies. And in a country also cursed with epidemic rates of HIV/AIDS, rape takes on an exponentially tragic dimension (Jewkes et al. 2009).

What exacerbates the tragedy tenfold is the fact that, unlike many countries with the wisdom to implement DNA databases fully, South Africa already has all the other components necessary to leverage the power of DNA technology—the laboratory system, the finances, the education, and the commitment by police. All too often the issue is not political will or recognition of the literal lifesaving potential of DNA but rather a lack of resources—but not in South Africa. Everything else needed to be successful—to identify rapists and murderers, to protect children and the wrongly accused—exists in South Africa, except the law to make it a reality.

Note: On the day of submission of this book to the publisher, a DNA Bill (officially known as the Criminal Law [Forensic Procedures] Amendment Bill B09-2013) was introduced into Parliament for final review on May 8, 2013.

23.8 Sharing DNA Data across Borders

As more countries establish and expand their forensic DNA databases, the potential to successfully share data between countries has also expanded. Universally, countries continue to move toward the passage or expansion of legislative authority for forensic DNA databases. Even countries that lack DNA specific legislation have begun the process of databasing under other criminal justice authority. And as databases increase in size and effectiveness, the value of sharing data across borders becomes readily apparent.

23.9 European Union—Law for Connectivity

Increased capacity to share DNA profiles across borders has also been mirrored by an accelerated need for such connectivity. The advent of European unionization has made cross-border travel in Europe an unmonitored activity. People can move freely between countries absent the expectation that they will be questioned or checked against any database before entering the country. This applies not only to law-abiding citizens but also to those predisposed to commit crime. As such, an experienced criminal with an extensive criminal record can travel freely between countries, free from fear that his criminal record will follow him. This is particularly true in the case of DNA profiles contained in individual country databases. The most heinous child sexual predators, once paroled, can travel from the country that maintains his DNA profile across the border to a neighboring country where the power and potential of DNA is rendered useless. In other words, a perpetrator can be convicted in the United Kingdom (where DNA is used extensively to solve crime), then move to Spain (which maintains no offender database) and commit numerous crimes undetected for lack of sharing vital data. The weaker a country's database, the more attractive to the criminal.

23.10 Interpol's Global DNA Gateway

Interpol has also established the Interpol DNA Gateway, which is designed to facilitate the comparison of DNA profiles between Interpol member countries. INTERPOL maintains a database of DNA profiles accessible online to all member countries upon adoption of the charter governing its secure use. The database contains more than 117,000 DNA profiles submitted by 61 member countries and has recorded 51 international hits during 2011 (Interpol 2012).

A member state may submit DNA information for the inclusion on the DNA gateway at the General Secretariat via Interpol's secure telecommunications system, or as hardcopy and is based on the Interpol Standard Set of Loci (Interpol 2009). The system is indexed into the four main categories of crime scene samples, reference samples, missing persons, and unknown deceased. Importantly, Member States that submit DNA profiles to Interpol in accordance with their legislative requirements may limit access by naming specific countries to be prevented from accessing their information. Where member states are notified of a match, the contributing countries may then determine whether they release further information pertaining to the specific DNA profile.

International connectivity of forensic DNA databases is the next logical progression in the effort to maximize the crime fighting potential of DNA technology. Eventually, countries will possess the ability to use DNA to track down perpetrators anywhere in the world validating law enforcement's favorite credo, "You can run, but you can't hide."

23.11 Rapid DNA Technology

As this book goes to print, the most significant change in DNA technology since the development of PCR technology is progressing through an evaluation and validation process through the FBI and through the National Institute of Standards and Technology. Rapid DNA technology refers to the fully automated (hands-free) process of developing a CODIS Core STR profile from a reference sample buccal swab. The "swab in–profile out" process consists of automated extraction, amplification, separation, detection, and allele calling without human intervention. Currently, Rapid DNA instruments are being evaluated, and they maintain processing times (swab in–profile out) of between 60 and 90 minutes.

The development of Rapid DNA technology envisions several applications that will fundamentally change the application of forensic DNA technology, particularly in the context of DNA databases. Rapid instruments will ultimately be integrated into police booking stations enabling police to quickly develop profiles from suspect and arrestee samples. Once those profiles are developed, they will be searched against either CODIS or local databases to both ensure the correct identification of the person before the police and determine any linkages to other crimes. Similarly, immigration applications for Rapid DNA technology will improve the ability of immigration officials to identify illegal immigrants at both official and non-official points of entry into a given country. Rapid DNA analysis will also have significant applications for military fields of conflict, identifying casualties and enemies quickly.

In 2010, the FBI established a Rapid DNA Program Office to direct the development and integration of Rapid DNA technology for use by law enforcement. Working with the National Institute of Standards and Technology, the National Institute of Justice, the Department of Defense, the Department of Homeland Security, and other federal agencies, the Program Office is ensuring the coordinated development of this new technology among federal agencies. The Program Office also works with state and local law enforcement agencies and state bureaus of identification through the FBI's Criminal Justice Information Services Division Advisory Policy Board to facilitate the effective and efficient integration of Rapid DNA in the police booking environment.

At the time of publishing, several manufacturers have developed and commercialized instruments for validation by the FBI. The FBI is working with federal, state, and local CODIS laboratories and the Scientific Working Group on DNA Analysis Methods to test, evaluate, and validate the hands-free instruments for law enforcement use.

Author's Note

On June 11, 2013, the U.S. Supreme Court issued its most important ruling regarding forensic DNA databasing in the case of *Maryland v. King*, 569 U.S. ____ (2013). According to Supreme Court Justice Alito, it was also the Court's "most important criminal procedure case that this Court has heard in decades." (*Maryland v. King*, 569 U.S. ____ [2013] transcript p. 35 line 13). Although the constitutionality of taking DNA from an individual convicted of certain crimes had been established for many years, the issue of taking DNA from arrestees, before conviction, had not been settled. And although 28 states had passed arrestee legislation, the U.S. Supreme Court had not yet ruled on the issue of arrestee testing, a constitutional standard that would no doubt be more rigorous than the one for convicted offenders. Indeed, many states in the United States refused to pass such legislation until the Court had ruled on the issue.

In 2009, police arrested Alonzo Jay King, Jr. for first-degree assault. When, pursuant to Maryland law, King's DNA was obtained during the booking process and subsequently processed, they found it matched DNA evidence from a home invasion rape in which King was masked, held a gun to his elderly victim's head, and raped her several times. Relying on the match, Maryland charged and successfully convicted King of, among other things, first degree rape. A divided Maryland Court of Appeals overturned King's conviction, holding the collection of his DNA violated the Fourth Amendment because his expectation of privacy outweighed the State's interests. That decision immediately halted Maryland's collection of DNA from otherwise qualifying individuals—suspects arrested for violent crimes. Maryland applied for a stay of that judgment pending the Supreme Court's disposition of its petition for a writ of certiorari. The stay was granted, as was certiorari. In the case of *Maryland v. King*, by a vote of 5 to 4, the Court ruled, similar to application of fingerprint technology at arrest, that the securing of a biological sample at the time of arrest for identification purposes did not violate the U.S. Constitution.

In an opinion written by Justice Anthony Kennedy, the Court held, "When officers make an arrest supported by probable cause to hold for a serious offense and bring the suspect to the station to be detained in custody, taking and analyzing a cheek swab of the arrestee's DNA is, like fingerprinting and photographing, a legitimate police booking procedure that is reasonable under the Fourth Amendment" (*Maryland v. King*, 569 U.S. ____ [2013]). And while the issue of what constitutes a "serious" offense will have to be clarified in subsequent decisions, although not necessarily U.S. Supreme Court decisions, the issue of when police can take DNA samples is fundamentally settled. In doing so, the Supreme Court has empowered police to leverage the power of DNA technology earlier in the investigative process thus increasing its investigative effectiveness.

References

Busher, L. 2002. The use of the U.K. National DNA Database to support an intelligence led approach to the investigation of crime. *Journal of Forensic Medicine* 21–25.

CODIS, http://www.fbi.gov/about-us/lab/codis/ndis-statistics.

DNA Identification Act of 1994, codified at 42 U.S.C. § 14132.

DNA Initiative. 2002. *Transcripts of the Attorney General's Initiative on DNA Laboratory Backlogs (AGID-LAB) Working Group.* https://www.ncjrs.gov/pdffiles1/nij/238838.pdf.

European Convention on Human Rights. S. v. United Kingdom [2008] ECHR 1581. http://www.bailii.org/eu/cases/ECHR/2008/1581.html.
FBI Laboratory Forensic Science Systems Unit. 1998. Combined DNA Index System (CODIS). Hit Counting Guidelines (An Element of Performance Measurement) https://www.ncjrs.gov/pdffiles1/nij/sl413apg.pdf (access March 2013).
FBI. 2013. National DNA Index System, http://www.fbi.gov/about-us/lab/codis/codis-and-ndis-fact-sheet.
Bieber, F.R. 2006a. Turning base hits into earned runs: improving the effectiveness of forensic DNA data bank programs. Home Office Statistical Bulletin. 31 March 2010. *Police Service Strength England and Wales* 14/10. http://www.forensic.gov.uk/.
Bieber, F.R. 2006b. Turning base hits into earned runs: improving the effectiveness of forensic DNA data bank programs. Home Office Statistical Bulletin, 21–25.
Interpol. 2012. Interpol Fact Sheet: DNA Profiling. COM/FS/2012-02/FS-01.
Interpol. Second Edition 2009. Interpol Handbook on DNA Data Exchange and Practice: Recommendations from the Interpol DNA Monitoring Expert Group.
Maryland v. Alonzo Jay King, Jr., 567 U.S. _(2012).
National DNA Database Annual Report. 2007/2009. p. 11–15.
NPIA. http://www.npia.police.uk/en/13340.htm.
NPIA. http://www.npia.police.uk/en/14189.htm.
Jewkes, R., Y. Sikweyiya, R. Morrell, and K. Dunkle. 2009. MRC policy brief, understanding men's health and use of violence: Interface of rape and HIV in South Africa. http://www.mrc.ac.za/gender/violence_hiv.pdf.
Hindmarsh, R., and B. Prainsack eds. 2010. *Genetic Suspects: Global Governance of Forensic DNA Profiling and Databasing*. Cambridge: Cambridge Univ. Press.
The Eighth United Nations Survey on Crime Trends and the Operations of Criminal Justice Systems. 2002. United Nations Office on Drugs and Crime. Centre for International Crime Prevention. Last updated: April 1, 2003.
Schellberg T. 2012. Gordon Thomas Honeywell Government Affairs. American Academy of Forensic Sciences meeting.

Author Index

A

Abaz, J., 25
Abdel-Malek, Z., 436
Abdulkadir, S.A., 470
Abecasis, G.R., 441
Abrantes, D., 145
Abrecht, C.R., 458
Absher, D.M., 447, 495
Acar, E., 145
Accorsi, C.A., 374
Ackerman, M.J., 457, 460, 462, 463, 466
Acland, G.M., 321
Adamowicz, M., 283
Adams, D.E., 329
Adams, M.D., 5
Adams, R.L.P., 391
Adekola, B.A., 457
Adelson, D.L., 335
Adkins, W.K., 389
Agarwala, R., 325
Agellon, A.B., 115
Agronis, A., 319
Aguilar, P.V., 305
Aguirre, D., 142
Ahlner, J., 468, 469
Ahrens, P., 303
Aitichou, M., 305
Aitken, C., 246
Akagawa, H., 456
Akane, A., 107, 340
Akcali, C., 436
Akeson, M., 95
Akutsu, T., 389, 408, 409
Alacs, E., 317, 337, 340
Alamalakala, L., 358
Albarran, C., 320
Alberts, B., 391
Albornozm, J., 359
Al Dahouk, S., 302, 304
Alders, M., 455
Aldom, J.E., 305
Aler, M., 142
Alexander, D.H., 495, 499
Alford, R.L., 35
Al-Khaldi, S.F., 303
Allen, M., 417
Allen, R.C., 10
Allery, J.P., 389, 471
Allgeier, L., 46, 369, 370, 379
Almeida, A.M., 302
Almirall, J.R., 19
Alonso, A., 4, 141, 320
Alt, K.W., 141, 146, 148, 149
Altamura, B.M., 107
Altshuler, D.L., 441
Altukhova, V.V., 304
Alvarez, J.C., 287
Alvarez, M., 400
Alvarez-Iglesias, V., 418
Alverson, A.J., 86
Alves, C., 20, 142, 320
Alving, A.S., 463
Amendt, J.R., 354, 356, 358
Amendtand, J., 359
Ames, C., 174
Amigo, T., 141, 146, 148, 149
Amiott, B., 13, 38
Amiott, E.A., 328
Amorim, A., 17, 118, 124, 127, 136, 137, 138, 141, 142, 147, 148, 149, 154
Amory, S., 90
Andelinovic, S., 4
Andersen, M.M., 114
Anderson, G.S., 353
Anderson, M.W., 424
Anderson, R.E., 455, 532
Anderson, S.J., 86, 88, 92, 97, 138, 340, 426
Andersson, J.O., 86
Andersson, S.G., 86
Andjelinovic, Š., 4, 19, 373
Andreaggi, K.S., 135
Andrews, R.M., 18
Angeles, M., 545, 546
Angius, A., 456, 459
Angulo, R., 320
Angyal, M., 319
Anjos, M.J., 19
Anslinger, K., 340, 359, 471
Antcliff, J.F., 457
Antonellis, A., 457
Antonov, V.A., 304
Antzelevitch, C., 457
Ao, A., 90
Aouizerat, B.E., 456, 457, 458
Applegate, R.J., 456

Aquino, J., 142
Aranda, X.G., 144, 283
Arce, B., 287
Archibald, A.L., 321
Arking, D.E., 457, 461
Armitage, S.M., 321
Arnestad, M., 462
Arnheim, N., 86, 463
Arnold, J.J., 86, 461
Aronson, J.D., 509, 514, 516, 550, 552
Aronson, J.K., 465
Arvestad, L., 47
Asamura, H., 109, 140, 141
Asano, M., 387
Ashford, D.A., 294, 309
Ashkenazi-Katalan, V., 373, 374, 457
Ashley, M.V., 371, 372
Ashrafian, H., 461
Asimaki, A., 456
Asmonga, D., 472
Asper, M., 303
Athanasiadou, D., 141, 146, 148, 149
Athey, W., 126
Atiken, C.G.G., 239, 240, 243, 244, 245, 246
Atlas, R.M., 294, 309, 311
Attardi, G., 90
Atzei, R., 142, 147, 148, 149
Augustin, C., 145, 147, 148, 149, 151, 152, 153, 154, 358
Aurlien, D., 458, 463
Ayres, K.L., 321
Ayub, Q., 490, 491
Azeredo-Espin, A.M.L., 359
Azevedo, D.A., 143, 153
Aznar, J.M., 142

B

Baas, A.F., 456, 460
Babar, M.E., 143, 147, 149
Bacci, M., 141
Badenhop, R.F., 456, 459
Baechtel, F.S., 11
Baeta, M., 124, 127
Bagnall, R.D., 458, 463
Bahar, B., 391, 392
Bahring, S., 459
Bailey-Wilson, J.E., 280, 286
Bailliet, G., 127
Bajda, E., 178
Bajic, V.B., 115
Bakal, N., 231
Bakaric, P., 373
Baker, A., 340
Baker, D.J., 387
Balatbat, A.B., 305
Balažic, J., 540

Balding, D.J., 171, 178, 201
Balick, M.J., 376, 377
Balitzki-Korte, B., 340, 359
Ball, M., 495
Ball, S.L., 340, 356
Ballantyne, J., 115, 177, 280, 286, 392, 394, 395, 396, 399, 400, 401, 402, 404
Ballantyne, K.N., 115, 118, 119, 120, 121, 431, 432, 433, 438, 439, 442, 445
Ballinger, D.G., 431
Balog, R., 359
Balogh, M.K., 174
Balyan, H.S., 335
Bamshad, M., 447
Bandera, B., 145
Bangara, S., 546
Bankier, A.T., 86, 88, 92, 97, 340
Barac, L., 16
Barajas-Martinez, H., 457
Barash, M., 176
Barbaro, A., 145, 154
Barbosa, A.G., 143, 153
Bardai, A., 456, 458
Barendse, W., 321
Barilaro, M.R., 471
Barna, C., 283
Barnard, K., 127, 489
Barni, F., 387
Barns, S.M., 302
Barrell, B.G., 86, 88, 92, 97, 340
Barritt, S.M., 109
Barron, A.E., 94
Barry, P.L., 238
Barsh, G.S., 435
Barsheshet, A., 455
Baršic, B., 301, 307
Bartel, D.P., 401
Bartelink, E., 281
Bartoš, J., 6, 358
Bartsch, C., 340, 359
Baslamisli, F., 305
Basso, C., 465
Basten, C., 319, 329, 332, 517
Bastisch, I., 47
Bates, T., 531
Batzer, M., 329
Baudry, E., 358
Bauer, M., 392, 395, 467, 471
Baum, H.J., 112, 280, 286
Baur, M.P., 193
Baxter, E., 107
Baxter, S.J., 388
Bayer, B., 471
Bayley, H., 95
Beauchamp, T.L., 528, 530
Beck, J., 445
Beck, N., 146, 148

Author Index

Beck, T., 329
Becker, D.J., 145, 147, 148, 149, 461
Becker, K.A., 304
Becker-Ziaja, B., 303
Beckmann, J.S., 456, 459
Beffagna, G., 456
Begovac, J., 301, 307
Beighton, P., 460
Bekada, A., 141, 145
Beleza, S., 437, 440
Bellocq, C., 457
Belosludtsev, Y.Y., 359
Belson, M.G., 377
Bemiss, J.A., 340
Bender, K., 174
Benecke, M., 356, 360
Benenson, A.S., 301, 307
Benhamamouch, S., 141, 145
Benjeddou, M., 115
Bennett, E.A., 47
Bensasson, D., 329
Benschop, C.C.G., 178, 399, 408, 409
Benson, D.A., 341
Benson, G., 335
Benson, N., 329
Bentayebi, K., 145
Bentley, D.R., 500
Bentley, G., 422, 424, 426
Bento, A.M., 142
Benyamin, B., 435, 441
Berardi, G., 142
Berdos, P.N., 115
Beresford, L.G., 330
Berger, B., 107, 112, 146, 148, 149, 517
Berglund, E.C., 421
Berkelman, R., 311
Bernard, B.A., 447
Bernstein, D.E., 510, 518, 519, 520, 521
Berthelot, C., 47
Berthet, M., 457
Berti, A., 25
Bertilsson, L., 468, 469
Bestwick, M., 93
Betancor, E., 320
Betz, P., 471
Beus, A., 301
Bezzina, C.R., 456, 457, 458
Bianchetti, M.G., 459
Bibb, M.J., 86
Bieber, F.R., 196, 280, 283, 286, 471
Bienvenue, J.M., 471
Biesecker, L.G., 280, 286
Bigham, A., 440
Bijlenga, P., 456
Bilguvar, K., 456
Bill, M.R., 182, 192
Bille, T.W., 280

Billerbeck, A.E.C., 471
Billings, N.C., 330
Bing, D.H., 283
Bini, C., 141
Binladen, J., 339
Birchard, K., 545, 547
Bird, A.P., 405
Birney, E., 329
Birren, B., 329
Bittencourt, E.A., 142
Bjerregaard, P., 457
Blades, N., 308
Blake, E., 415, 516
Blanc, A., 471
Blanco-Verea, A., 124, 127
Bleeker, M., 358, 359, 372, 373, 374, 379
Bleyl, S.B., 457
Blinka, D., 467
Bloise, R., 455
Blundell, A.G., 337
Boakye, D.A., 359
Bobillo, C., 124, 126, 142
Bocchi, B., 459
Bodmer, W.F., 496
Boehme, P., 356
Bogdanovich, T., 303
Bogenhagen, D.F., 89, 92, 283
Bohr, V.A., 91
Božinovic, D., 307
Boll, K., 112
Bomberger, K., 284, 540
Bonella, P., 424
Bonne, G., 461
Bonner, R.F., 469
Boomsma, D., 431
Boonlayangoor, P.W., 245, 246
Boonseub, S., 340
Boquist, S., 456, 458
Bork, K.H., 311
Borowiec, M., 457
Borst, P., 86
Borsting, C., 154, 431, 432, 437, 439, 444
Bos, J.M., 457
Bostel, A., 303
Both, K., 107
Botkin, J., 532
Bouabdeallah, M., 145
Bouakaze, C., 436
Boucheix, C., 12
Boudjema, A., 141, 145
Bouhanna, P., 445
Bouloy, M., 305
Boumertit, A., 495
Bourke, M.T., 22, 369, 371, 373
Bowen, K.L., 280, 287
Bowen, M.D., 294, 302, 306, 309
Bowerman, D., 359

Bowles, K.R., 457
Bowles, N.E., 456
Brabetz, W., 145
Brandon, D.L., 377
Brandstatter, A., 416
Branicki, W., 432, 433, 436, 437, 439, 440, 444
Brauer, S., 115
Brauner, P., 176
Bravo, M.L., 142
Breazeal, R., 546
Breckenridge, A.M., 465
Breeze, R.G., 293, 294, 295, 301, 306, 307, 308, 311
Bregu, J., 44
Brenneman, R.A., 337
Brenner, C.H., 108, 112, 114, 118, 190, 193, 280, 286
Brennicke, A., 86
Brenyo, A., 455
Brettell, T.A., 19
Breur, H., 331
Briggs, A.W., 495
Brinkmann, B., 108, 112, 118, 120, 121, 141, 144, 218, 243, 245, 246
Brion, M., 124, 127, 422
Brisbin, A., 499
Bristol, N., 175
Brockschmidt, F.F., 447
Broman, K.W., 151, 200
Brookes, A.J., 431
Brookfield, J.F., 318
Brookmeyer, R., 308
Broudy, D.W., 455
Broun, K.S., 529
Brown, A.G., 374
Brown, B.L., 10
Brown, D.M., 321
Brown, J.F., 321
Brown, K.A., 407
Brown, T.A., 97
Brown-Harrison, M.C., 457
Browning, B.L., 502, 504
Browning, S.R., 502, 504
Bruck, H.A., 304
Brudnik, U., 433, 436, 440, 444
Brugada, R., 457
Brunak, S., 358
Bruni, I., 372, 376
Bruno, A., 543
Bryant, V.M., 374, 375
Bryc, K., 492, 495, 499
Buajordet, I., 465
Buckleton, J.S., 114, 118, 171, 172, 178, 190, 192, 193, 201
Buckley-Beason, 330
Budimir, D., 529
Budimlija, Z.M., 174, 281, 431, 432, 433, 437, 438, 439, 440, 442, 443, 444, 445, 471

Budowle, B., 10, 11, 115, 144, 174, 175, 280, 283, 286, 287, 293, 294, 295, 301, 306, 307, 308, 311, 320, 321, 323, 328, 329, 331, 338, 354, 358, 405, 406, 422, 430, 463, 465, 515, 517
Buel, E., 38
Buffalino, L., 107
Buffery, C., 195
Bulbul, O., 145
Bunker, R.K., 321
Buoncristiani, M.R., 228, 418
Burashnikov, E., 457
Burgemeister, R., 470, 471
Burger, G., 86
Burger, J.F., 174, 354
Burgoyne, L.A., 107
Burian, S.K., 340
Burke, A.P., 459
Burke, S., 92
Burnier, M., 456, 459
Burns, J.M., 340
Burroughes, T.R., 471
Burt, V.L., 459
Busam, D., 422
Busch, C., 303
Busher, L., 558
Butler, J.M., 5, 11, 19, 21, 32, 33, 38, 44, 108, 114, 118, 124, 140, 144, 145, 156, 158, 159, 160, 178, 183, 192, 232, 233, 280, 283, 286, 319, 333, 417, 418, 430, 517, 541
Bydon, M., 456
Byers, P.H., 456, 460
Byrnes, J., 499
Byun, H.M., 405

C

Cacció, S., 107
Caddy, B., 171, 174, 175
Caenazzo, L., 530
Cage, P.E., 318
Caglia, A., 108
Caine, L.M., 142, 145
Cairns, J., 468
Calandro, L.M., 109, 140
Calderon, J., 335
Caliebe, A., 114
Calloway, C.D., 19, 417, 418, 419
Calvert, V.S., 470
Cameron, C.E., 86
Campbell, D., 495
Campbell, R., 283
Campobasso, C.P., 283, 354, 358, 359, 360, 361
Caniglia, R., 337
Cannell, D.W., 447
Cao, L., 92, 93
Caraballo, G., 232

Author Index

Caragine, T., 175, 176, 178, 433, 437, 442, 443, 444, 445
Caratti, S., 141, 145
Carboni, I., 141
Cardaioli, E., 86
Cardoso, S., 142
Carlsson, B., 468
Carmody, G., 280, 286
Carnevali, E., 141
Carolino, I., 337
Carolino, N., 337
Carracedo, A., 108, 118, 127, 141, 142, 190, 318, 319, 333, 418, 422
Carrasco, F., 287
Carreno, R.A., 305
Carrier, L., 457
Carson, P.E., 463
Carter, J.M., 377
Cartinhour, S., 335
Carus, W.S., 308
Carvajal-Vergara, X., 461
Carvalho, G.R., 337
Carvalho, M., 142
Carvalho, R., 142
Caskey, C.T., 35
Caspers, H., 355, 360
Cassidy, B.G., 320
Castaneda, M., 147, 148, 149
Castelo, A.T., 335
Castillo, A., 142, 147, 148, 149
Castro, A., 141
Castro, J.A., 154
Catanesi, C.I., 147, 148, 149
Cattaneo, C., 354
Catts, E.P., 353
Cavalli-Sforza, L.L., 106, 122, 123, 495, 496
Cave, C.A., 280
Caviezel, F., 120
Ceccardi, S., 141
Celorrio, D., 142
Cerri, N., 112, 145
Cerrone, M., 462
Cervenka, V.J., 353
Cervini, M., 337
Cevario, S., 329
Chakraborty, R., 10, 35, 144, 283, 323
Chakravarti, A., 540
Chamberlain, V.F., 126
Champlot, S., 47
Chang, C.W., 109, 140
Chang, H.C., 337, 339
Chang, Y.Y., 143
Chang, Y.M., 107
Chang, Y.P., 456, 459
Chaplin, G., 436
Charron, P., 457

Chastain, T., 281
Chatel, S., 457
Chen, C., 402, 456, 459
Chen, F., 151
Chen, G.Q., 377
Chen, J.D., 145, 444
Chen, L., 457
Chen, Q., 457
Chen, W., 151, 436, 438, 444
Chen, X., 90
Chen, Y.F., 148, 149, 217, 490, 491
Chen, Z., 322
Cherf, G.M., 95
Cheswick, D., 283
Cheung, B.K., 389
Cheung, M.C., 171
Chevalier, P., 462
Chiang, H.L., 341
Chiarle, R., 471
Chiaroni, J., 122, 123
Childress, J.F., 528, 530
Chinnery, P.F., 18, 491
Chitayat, D., 468
Chizhikov, V., 303
Cho, N.S., 145
Chow, K.H., 305
Chowdhary, B.P., 321
Christensen, A.F., 47
Christiaans, I., 457
Chu, M.C., 302
Chu, Y., 86
Chun, B.J., 458
Chung, E., 305
Chung, S.K., 462
Chung, U., 109
Ciammaruconi, A., 303
Cicarelli, R.M.B., 142
Cítek, J., 337
Clark, A.G., 492
Clark, B., 339
Clark, S.H., 19
Clark, W.J., 339
Clarke, S., 337, 339
Clarke, W.L., 457
Clayman, C.B., 463
Clayton, D.A., 86, 89, 97, 283
Clayton, E.W., 529
Clayton, T.M., 174, 189
Clery, J.M., 359, 360
Cloete, K., 115
Cobb, J.E., 447
Coble, M.D., 18, 20, 21, 86, 89, 90, 97, 109, 124, 126, 140, 143, 144, 145, 153, 156, 158, 159, 160, 192, 418, 484, 517
Cohle, S.D., 465
Cole, T.J., 435, 441

Coletti, A., 141
Collins, K.A., 471
Collins, M.J., 354
Collins, P.J., 140, 144, 228
Collins-Morton, M.B., 360
Comey, C.T., 11, 329
Conard, J., 456
Concheiro, L., 318
Conklin, D., 44
Conneally, P.M., 280, 286
Conner, D., 457
Conquoz, R., 318
Convertini, P., 19
Cook, A.L., 438, 444
Cook-Deegan, R.M., 311
Coomber, N., 324, 325, 328
Cooper, G.M., 6
Copeland, W.C., 91
Coquoz, R., 47, 329
Corach, D., 108, 124, 126, 142
Corbett, S., 547
Cordeiro, J.M., 457
Corey, P.N., 465
Corfield, V., 458
Cormaci, P., 145
Cornelison, D.D.W., 171
Coronado, E., 44
Corrado, D., 465
Cortez, P., 183
Cortopassi, G., 86, 87
Costa, J.C., 142
Coto, I., 35
Cotton, E.A., 174
Cotton, R., 192
Cotton, W.R., 44
Council, N., 100
Courtois, G., 91
Courts, C., 402, 403
Cowell, R.G., 192
Cox, D.R., 431
Cox, M., 386, 387
Coyle, H.M., 367, 368, 369, 370, 371, 372, 373, 374, 375, 376, 377, 379
Craft, K.J., 373, 374
Crance, J.M., 305
Crawford, A.M., 321
Crawford, D.C., 294, 309
Creamer, J.I., 387
Cree, L.M., 92
Crespillo, M., 142
Crews, J.D., 284, 534, 540
Cronk, L.B., 457
Crotti, L., 462
Crubezy, E., 90, 436
Cuesta-Munoz, A., 457
Culver, M., 322
Cummings, J.H., 409
Curi, R.A., 337
Curic, G., 4, 244
Curran, J.M., 25, 174, 182, 192, 193, 201
Curtain, L.R., 456, 459
Curtain, R.P., 456, 459
Cutting, G.R., 456, 460
Cvetnic, Z., 303
Cybulska, L., 142
Cyr, D., 90
Cywinska, A., 340, 356
Czerneková, V., 337

D

Dadd, T., 436
Dafoe, B., 283
Dahl, M.L., 468
Daiger, S.P., 540
Dalen, I., 194
Damaska, M.R., 521
D'Amato, M.E., 115
D'Amico, M., 456, 459
Damon, I., 301, 302
Dando, M., 306, 311
D'Andrea, E., 318
Daniel, B., 387
Daniel, K.L., 466
Danielson, P.B., 93
DaPozzo, P., 86
Das, A., 305
Date Chong, M., 418
Daugharty, H., 304
David, V.A., 318, 321, 322, 323, 324, 325, 328, 329, 330
Davidson, E.M., 311
Davis, B., 322
Davis, C., 16, 465
Davison, S., 115
Davoren, J., 540
Dawnay, N., 317
Day, T., 495
Dayton, M.R., 331
Dean, L., 445
Dean-Nystrom, E.A., 303
Debenham, P., 540
Debue, K., 321
de Carli, L., 120
de Carvalho, E.F., 135, 142
Decher, N., 457
Decker, A.E., 144
DeClerck, G., 335
De Ferrari, F., 112, 145
Definis-Gojanovic, M., 4, 47
Deforce, D.L., 174, 471
De Forest, P.R., 367, 369, 373, 374, 376, 379
de Grandmaison, G.L., 108, 109
de Greef, C., 174

Author Index

Deichsel, D., 153
Dekairelle, A.F., 108, 109
de Knijff, P., 16, 114, 330, 440, 447, 492
De Leo, D., 142, 145, 146, 147, 148, 149
Dellefave, L., 457
Deloukas, P., 500
Del Pero, M., 141, 145
Delpon, E., 457
De Mattia, F., 372, 376
de Meo, P.D., 86
Demetrick, D.J., 470
den Dunnen, J.T., 447, 463
Denjoy, I., 457
De Paepe, A., 460
de Pancorbo, M.M., 141, 142, 146
De Santis, R., 303
Desiderio, M., 135, 142
de Sousa Lopes, S.C., 92
Dettlaff-Kakol, A., 108
De Ungria, M.C., 141, 146, 148, 149
DeVane, C.L., 468
de Waard, J.R., 358
Dib, C., 151
di Barletta, M.R., 457
Dick, L.K., 409
Diegoli, T.M., 89, 90, 135, 140, 143, 144, 145, 153, 156, 158, 159, 160
Dieltjes, P., 16
Dietrich, W., 321
Dietz, H.C., 456, 460
Di Francesco, L.M., 470
Džijan, S., 4, 244
Di Luise, E., 359, 360
Di Martino, D., 471
Dimick, G., 89, 91, 415, 416
Ding, C., 127
Dintzis, S.M., 470
Di Paolo, M., 455
Di Rienzo, A., 456, 459
Dissing, J., 339
Divall, G.B., 387
DiVella, G., 358, 359, 360, 361
Divne, A.M., 417
Dix, G., 529
Dixon, L., 109
DiZinno, J.A., 359, 360, 422
Dodds, K.G., 321
Doležel, J., 6
Dolezal, P., 86
Domenici, R., 141
Domingues, C., 142
Dominguez, A., 358
Dominko, T., 90
Donaldson, A.E., 408, 409
Donger, C., 457
Donley, M.A., 175
Donnai, D., 13

Donnelly, P., 495, 497, 500
Doody, M.J., 95
Dooley, J.S., 302
Doom, T.E., 192, 193
dos Santos, A.K., 142, 154
dos Santos, N.P., 154
dos Santos, S.E., 142
Downes, T.J., 14
Dowton, M., 340, 356
Drabkin, H.A., 457
Draus-Barini, J., 432, 437, 439, 444
Dravid, V.P., 305
Drayton, M.L., 176
Drebler, J., 153
Dressler, J., 115
Driggers, W.J., 91
Dring, L.G., 463
Drobnic, K., 231
Drogemüller, C., 330
Dror, I.E., 518
Drori, O., 339
Druid, H., 468
D'Souza, S.L., 461
Duane, B., 544
Dube, M.P., 456
Dubey, B., 337
Ducasse, N., 194
Duceman, B., 280, 286
Dudková, Z., 337
Duewer, D.L., 416
Duffy, D.L., 436, 437, 444
Duflou, J., 458, 463
Dufort, D., 90
Duim, B., 304
Dujon, B., 340
Duncalf, N., 471
Duncan, A.E., 445
Dunkle, K., 565
Dupuy, B.M., 150
Durbin, R.M., 441
Duverneuil, C., 108, 109
Duvnjak, S., 303

E

Ebbesen, J., 465
Ebensperger, C., 107
Ebihara, H., 303
Eckstein, P.E., 357
Ede, A.J., 321
Edelmann, J., 17, 115, 136, 137, 144, 145, 146, 147, 148, 149, 150, 151, 152, 153, 154
Edghill, E.L., 457
Edson, S.M., 47
Edwards, I.R., 465
Edwards, M., 440
Edwards, S.V., 340

Egeland, T., 150, 151, 194, 244
Egyed, B., 47, 318, 319
Ehler, E., 115
Ehrenreich, L., 115
Ehricht, R., 304
Ehrig, T., 470
Eiberg, H., 431, 432, 436, 437, 440, 444
Eichelbaum, M., 463
Eichmann, C., 516
Eisenberg, A.J., 10, 144, 174, 280, 283, 286, 546, 547, 548
Eisenmenger, W., 471
Eizirik, E., 322, 330
Eldar, M., 457
Elgh, F., 305
Elias, S., 356
Eliopoulos, G.M., 301
Ellard, S., 457
Ellegren, H., 321
Elliot, D.A., 387
Elliott, K., 471
Ellis, J.A., 445
El-Mostaqim, D., 143, 147, 148, 149
Elsmore, P., 174
Emerson, K., 357
Emes, A., 195
Emmert-Buck, M.R., 469
Emodi, G.P., 337
Encheva, V., 95
Endicott, P., 445
Engelthaler, D.M., 301
Englebrecht, J., 356
Entrala, C., 287
Epplen, J.T., 107, 108
Erickson, A.M., 329
Erickson, D.L., 371, 372
Erickson, M.A., 30
Eriksen, P.S., 183, 200
Erikssen, J., 465
Eriksson, N., 445
Eriksson, P., 456, 458
Erler, A., 108
Erlich, H.A., 328, 417, 418, 419, 421, 426, 516
Erlich, R.L., 426
Ermini, L., 491
Esbenshade, J., 542, 543
Eshoo, M.W., 302
Espina V., 476
Estacio, S.M., 281, 471
Estes, M.D., 544
Estrada, K., 434, 435, 441
Evans, A., 456
Evans, J.J., 330
Evans, M.J., 460
Evett, I., 100
Evett, I.W., 192, 195, 318
Ewing, M.M., 118

F

Fabbri, E., 337
Fach, P., 302
Fagart, J., 459
Faggioni, G., 303
Fahrenkrug, S.C., 335
Falamingo, R., 283
Falconer, B., 484
Falkenberg, M., 93
Falooona, F., 33
Falush, D., 495
Fan, C., 456, 458
Fan, W., 302
Fang, R.N., 107, 110, 115, 118, 119, 121, 488
Fang, S.G., 340
Fantin, D., 331
Farb, A., 459
Farhi, A., 459
Farmen, R.K., 183
Fattorinil, P., 107
Faure, C., 305
Faustino, L.P., 141, 146, 148, 149
Favaloro, E.J., 460
Fedele, C.G., 302
Federico, A., 86
Feinberg, J., 528
Feinberg, S., 535
Feldman, M.W., 495
Felini, L., 546
Felini, M., 546
Feng, J., 335
Fenicia, L., 302
Fenton, T.W., 512
Fernandez, I., 141
Fernandez-Zapico, M.E., 457
Ferrance, J.P., 471
Ferreira, S., 337
Ferreira da Silva, I.H., 143, 153
Ferreira, P.B., 337
Ferrell, R.E., 468
Ferrer, J., 457
Ferri, G., 141
Fields, B.S., 304
Fierer, N., 407
Fierro, M.F., 354
Figueiredo, M.G., 337
Filippini, G., 142, 145, 147, 148, 149
Fimmers, R., 120, 121, 193
Finch, C.E., 391, 392
Findlay, I., 172
Finlay, J.E., 435
Finnebraaten, M., 176
Fischbeck, K.H., 150
Fischer, A., 304
Fisher, A.B., 24, 253
Fisher, D.E., 435

Fisher, D.L., 18, 89, 329, 540
Fisher, D.R., 24, 253
FitzSimmons, N., 317, 337, 340
Flanagan, C.L., 463
Flanagan, L., 283
Flanagan, N., 109
Flanagan, S.E., 457
Fleming, M.A., 321
Fleming, R.I., 392, 395, 396, 404, 408, 409, 410
Flinter, F.A., 392
Florin, A.B., 359
Floyd, J.I., 471
Foger, M., 340
Fogt, F., 471
Fomarino, S., 16
Fontcuberta, J., 456
Foran, D.R., 356
Ford, M.J., 337
Ford, S., 513
Fordyce, S.L., 391
Foreman, L.A., 195
Forman, L., 280, 286
Formichi, P., 86
Forster, L., 92, 177
Forster, P., 92
Foster, A., 544
Foster, E.A., 120, 484
Foster, J.E., 357
Fourney, R.M., 280, 287, 321
Fournier, P.E., 304
Foxall, P.A., 140, 228
Fracasso, T., 141
Franceschini, N., 459
Frangoulidis, D., 304
Frank, W.E., 118
Frappier, R., 283
Frascione, N., 387
Frayling, T.M., 433, 435, 441
Frazer, K.A., 431
Frazier, R., 172
Fredlake, C.P., 94
Freeman, K., 86
Fregeau, C.J., 280, 287, 321, 544
Freitas, N.S., 154
French, C.E., 390, 391
Freschi, A., 142
Frick, J., 546
Fridez, F., 47, 318, 329
Friedlaender, J.S., 440
Frijters, A., 372, 373, 374, 379
Froede, R.C., 354
Frost, P., 436
Frudakis, T.N., 436
Frumkin, D., 172, 405
Fu, K., 92
Fuerst, P.A., 115
Fujimori, S., 109

Fujimoto, A., 445
Fukushima, H., 109, 140, 141
Funayama, M., 145
Funke-Kaiser, H., 456, 459
Furberg, C.D., 465
Furuta, H., 457

G

Gabriel, M.N., 19, 415, 416
Gaedigk, A., 468
Gaensslen, R.E., 217, 367, 369, 373, 374, 376, 379, 387, 388, 389
Gaillard, O., 445
Gajadhar, A.A., 358
Galaverni, M., 337
Galimberti, A., 372, 376
Gallus, G.N., 86
Gama, L.T., 337
Gamazo, C., 303
Gandolfi, B., 330
Gao, B.R., 457
Gao, G.R., 335
Gao, J., 463
Garcia, B., 142
Garcia, J.E., 337
Garin, D., 305
Garnica, W.T., 319
Garofano, P., 320, 321, 328, 331, 338, 354, 517
Garrett, B.L., 522, 523
Gaskin, Z., 436
Gaspari, M., 93
Gaspari, Z., 335
Gasparini, F., 145
Gateva, V., 456
Gaundry, E., 354
Gauvin, J., 400
Gawahara, H., 6
Ge, J., 16, 144
Geary, L.A., 330
Geberth, J.V., 24
Gefridese, L.A., 175
Geigl, E.M., 47
Geigl, J.B., 172
Geisterfer-Lowrance, A.A., 457
Gelfand, D.H., 33
Geller, D.S., 459
Geller, F., 445
George, A.L., Jr., 457
George, D.A., 319
Georges, A., 317, 337, 340
Geppart, M., 151, 153
Geppert, M., 115, 124, 127
Gerace, T.A., 378
Gergov, M., 467
Gerull, B., 456
Geserick, G., 120

Gessner, R., 457
Gharehbaghi-Schnell, E., 395
Ghosh, T., 142
Giampaoli, S., 25, 408, 409
Gilbert, D.A., 321
Gilbert, D.N., 301
Gile, M., 305
Giles, R.C., 471
Gill, J.R., 455
Gill, P.D., 14, 44, 47, 89, 97, 107, 108, 112, 118, 171, 172, 174, 178, 182, 189, 190, 192, 193, 195, 200, 201, 240, 329, 342, 417, 421, 430, 447, 471
Gillespie, J.W., 469
Gilmore, S.R., 372, 373, 374, 379
Ginestra, E., 359, 360
Gino, S., 141, 471
Ginther, C., 89
Giolitti, A., 141
Giraldo, P., 408
Girard, C.A., 457
Giuffré, G., 471
Gjerstad, L., 458, 463
Glesmann, L.A., 147, 148, 149
Gloyn, A.L., 457
Glurich, I., 460
Gock, S.B., 468
Goedbloed, M., 107, 110, 115, 118, 119, 121, 488
Goff, M.L., 353, 360
Goldberg, J.D., 171
Goldberg, S.M., 422
Goldenberg, A., 530, 532
Goldenberg, I., 455
Goldstein, J.A., 456, 460
Golze, M., 223
Gomes, I., 17, 141, 142, 144, 153
Gomes, V., 124, 127, 142
Gomez, A., 142
Gomez, J., 140
Gong, M., 456, 459
Gonzabay, C., 335
Gonzales, M., 456, 459
Gonzales, R.A., 320
Gonzalez, F.J., 463
Gonzalez-Lamuno, D., 141, 146, 148, 149
Goodwin, W., 11, 12, 38, 44
Goossens, B., 172
Gopalakrishnakone, P., 377
Gornik, I., 4, 244
Gornjak Pogorelc, B., 540
Goth, G., 542
Goyette, P., 457, 461
Graefen, A., 495
Graffy, L.A., 245, 246
Grahn, R.A., 330
Gramolini, A.O., 457

Graner, A., 335
Grange, T., 47
Grant, E.K., 460
Graur, D., 6, 7
Gray, D., 387
Gray, M.W., 86
Graziosil, G., 107
Greco, C., 337
Greely, H.T., 529
Green, P.M., 392
Green, R.E., 495
Greenberg, B., 353
Greene, C.M., 294, 309
Greenspoon, S.A., 176
Greilhuber, J., 6
Gretarsdottir, S., 456, 460
Grgicak, M.C., 44
Gribouval, O., 456, 459
Grieve, C.M., 303
Griffin, S., 172
Grimes, E.A., 387
Grinberg, L.M., 308
Grivell, L.A., 86
Grooms, K., 13, 38
Gross, A.M., 109, 115, 283
Gross, S.R., 522
Groves, M., 547
Grow, C.C., 302
Grškovic, B., 47
Grzybowski, G., 337
Grzybowski, T., 145
Guclu, B., 456
Gudbjartsson, D.F., 436
Gugic, D., 47
Gui, J., 458
Guilaine, J., 321
Guillaume, V., 305
Guisti, A.L., 10
Guisti, A.M., 10
Gunn, A., 24, 25, 354, 358
Gunther, S., 303
Guo, Y., 433, 435, 441
Gupta, P.K., 335
Gupta, S.K., 339, 340
Gurney, S.M.R., 92
Guruju, M., 456, 459
Gusev, A., 502
Gusmão, L., 17, 108, 118, 124, 127, 136, 137, 138, 141, 142, 143, 147, 148, 149, 153, 154, 190, 200, 338, 340, 341
Gussak, I., 457
Gustincich, S., 107
Gutman, G.A., 144
Gyapay, G., 151
Gyllensten, U., 90
Gyllenstrand, N., 359

H

Haas, C., 19, 392, 393, 395, 396, 399, 404, 411
Habener, J.F., 457
Haddrath, O., 340
Hadfield, T.L., 303
Hadi, S., 11, 12, 38, 44
Haff, L., 445
Haglund, C.M., 463, 466
Hahn, S., 171
Hainsworth, J., 551
Haissaguerre, M., 457
Hajibabaei, M., 340
Hajjar, I., 459
Haldar, D., 86
Haley, C.S., 321
Halkjaer-Knudsen, V., 311
Hall, L., 445
Hall, M., 354
Hall, T.A., 95
Hallenberg, C., 244, 245, 246
Hallwachs, W., 340
Halperin, E., 499
Halverson, J.L., 319, 329, 332, 517
Hammer, M.F., 118, 122, 123, 126, 127, 489
Hammond, H.A., 35
Hampikian, G., 518
Hamsten, A., 456, 459
Hamuro, L.L., 458
Han, J., 436, 444
Handyside, A.H., 171
Haned, H., 190, 194, 200, 201
Hanlin, E., 460
Hanneken, S., 445
Hanner, R.H., 340
Hansen, A.J., 339, 391
Hansen, J.E., 311
Hansen, L., 436, 437, 444
Hanson, E.K., 19, 115, 399, 400, 401, 402, 463
Hansson, O., 176
Hao, H., 143
Happy, C., 467
Haque, I., 337
Hara, M., 457
Harada, H., 457
Harbison, S., 392, 395, 396, 404, 408, 409, 410
Harding, A.E., 150
Hares, D.R., 36, 37
Harmelin, A., 172
Harrap, S.B., 445
Harris, H.A., 254
Hartl, D.L., 329, 515
Hashiyada, M., 145
Haskell, N.H., 353
Hass, M., 303
Hateley, C.A., 87, 91

Hattangadi, N., 499
Hauge, X.Y., 322, 325
Hauser, E., 303
Hausman, R.E., 6
Hausmann, R., 471
Hauswirth, W.W., 89, 92
Havey, M.J., 86
Hawksworth, D.L., 375
Hawley, W.A., 359
Hayakawa, M., 387
Hayashi, J., 90
Hayes, J., 358, 359, 360, 361
Haylock, R.W., 302
Haywood, A.F., 456
Hayworth, R., 471
Hazlewood, L., 335
He, X., 377
Healy, E., 436, 440
Hearing, V.J., 434, 435
Hebert, P.D.N., 340, 341, 356
Hecht, W., 338, 340, 341
Heddema, E.R., 304
Hedman, M., 115, 121, 145
Heegaard, E.D., 311
Heidel, M., 153
Heinrich, B., 287
Heiss, G., 459
Heit, J.A., 460
Helgadottir, A., 456, 458
Hellenthal, G., 495
Hellmann, A., 320, 321, 328, 331, 338, 354, 517
Hemachudha, T., 305
Hemenway, J., 369, 370, 379
Hemleben, V., 46
Henderson, D.A., 308
Henderson, M.S., 431, 432, 433, 436, 437, 438, 439, 440, 444
Henn, B.M., 495, 501
Hennessey, M.J., 281
Hennessy, G.J., 457
Hennessy, J.K., 228
Hennessy, J.R., 457
Hennessy, L.K., 140, 228
Henning, A.C., 540
Henriksen, G.M., 223
Henrique-Silva, F., 337
Henry, J.A., 466
Hensley, L.E., 301, 302, 306
Hequet, M., 531
Hering, S., 17, 144, 145, 146, 147, 148, 149, 150, 151, 152, 153, 154
Heron, A.J., 95
Herrmann, B., 177
Herrnstadt, C., 491
Hert, D.G., 94
Heuser, A., 456

Hewitt, G.M., 329
Heyer, E., 16
Higashiuesato, Y., 457, 458
Higgins, B., 89, 91, 415, 416
Higuchi, R., 322, 328, 417, 421, 422, 424, 426, 516
Hilbert, P., 456, 459
Hill, C.R., 178, 183, 430
Hill, D.S., 471
Hill, R.B., 455
Hillmer, A.M., 445
Himmelberger, A.L., 319
Hindmarsh, R., 558
Hinds, D.A., 431
Hirano, Y., 457, 458
Hirsch, C.S., 280
Hjort, B.B., 108, 109
Hlaca, D., 301
Ho, M., 217
Hobson, D., 175
Hochman, J.S., 458
Hochmeister, M.N., 387, 389
Hockenberry, T.L., 328
Hockwald, R.S., 463
Hodgkinson, K.A., 456
Hoffmann, B., 457
Hoffmaster, A.R., 294, 302, 306, 309
Hofreiter, M., 422
Hogendoom, K., 356
Hogers, R., 358, 359, 372, 373, 374, 379
Hoglund, B., 422, 424, 426
Hoh, J., 456, 459
Hohoff, C., 141, 218, 243, 245, 246
Holcomb, C.L., 424
Holland, C.A., 89, 97, 280, 340
Holland, E., 543
Holland, M.M., 18, 89, 91, 93, 94, 95, 96, 97, 98, 99, 280, 283, 329, 339, 417, 421, 540
Holliday, R., 405
Holm, P.I., 457
Holmes, B.H., 339
Holmgren, P., 468
Holmlund, G., 151, 244
Holzgreve, W., 171
Hon, L., 501
Honein, M.A., 466
Hong, K., 457
Hong, L.Z., 330
Hong, Y., 35
Hoorfar, J., 303
Hopgood, R., 89, 417, 421
Hopwood, A.J., 140, 339, 387, 544
Horii, T., 92, 93
Horikawa, Y., 457
Hornes, M., 358, 359, 372, 373, 374, 379
Horrocks, M., 374
Horsman-Hall, K.M., 174
Horvath, G., 145

Hoste, B., 108, 109
Hottel, H.E., 302
Hotzel, H., 304
Hou, M.Y., 304
Houck, M.M., 24, 25
Howard, C., 373, 379
Howard, R., 95
Howell, N., 18, 491
Howitt, T., 240
Hradecká, E., 337
Hsieh, H.M., 337, 339, 340, 341, 379
Hu, D., 457
Huang, D., 108, 146, 148, 149
Huang, L.H., 337, 339, 340, 341
Huang, T., 91
Huber, C.G., 95
Hubert De Pauw, I.P., 174
Huchette, S., 91
Hudlow, W.R., 228
Hudson, K.L., 551, 552
Huel, R., 540
Huff, C.D., 501
Huffine, E., 284, 540
Huffine, E.F., 283
Huggins, J., 301, 302
Hugh-Jones, M., 308
Hughs, D., 329
Huhne, J., 144
Hui, B., 457
Hummel, S., 177
Hundertmark, T., 147, 148, 151, 153
Hung, M.N., 304
Hunt, S., 500
Hunt, T., 391
Hunter, F.F., 340
Hurles, M.E., 7
Hurth, C., 544
Hussain, M., 143, 147, 149
Hutz, M.H., 142
Huynen, L., 340
Hwa, H.L., 143
Hysi, P.G., 463

I

Ibekwe, A.M., 303
Ibrahim, M.S., 305
Ichida, F., 457
Ickes, C.E., 463
Idle, J.R., 463
Ikejiri, M., 456
Ilina, M.V., 455
Illescas, M.J., 142
Illingworth, R., 405
Im, S., 436
Imes, D.L., 330
Immel, U., 151

Author Index

Imwinkelried, E., 529
Ingelman-Sundberg, M., 468
Inman, K., 10
Inoue, H., 388, 391, 392
Inoue, K., 456
Inoue, S., 456
Introna, F., Jr., 358, 359, 360, 361
Inturri, S., 141
Irwin, D., 340
Irwin, J.A., 18, 86, 90, 93, 109, 415
Ishida, T., 106, 445
Ishida, Y., 330
Issel-Tarver, L., 89
Itakura, Y., 145
Itzkovitz, S., 172
Ivanov, P.L., 18, 89, 91, 96, 97, 98, 329, 339
Ivanova, N.V., 340
Iwasaki, N., 457

J

Jablonski, N.G., 436
Jackson, G., 246
Jackson, I.J., 436, 440
Jacob, G.F., 388
Jacob, H.J., 321
Jaghø, R., 183
Jahan, N., 302
Jahrling, P.B., 301, 302, 306
Jakobs, S., 93
Jakubowska, J., 395, 399, 408, 409, 410
James, S.H., 256, 318
Jamieson, R.W.D., 542
Janica, J., 145, 146, 148, 149
Jannetto, P.J., 468
Janssens, A.C., 436
Janzen, D.H., 340, 371, 372
Jaqiello, G., 90
Jayashree, B., 335
Jeffreys, A.J., 4, 32, 318
Jennings, K.C., 356
Jentzen, J.M., 468
Jenuth, J.P., 92
Jeong, C.K., 141
Jernigan, J.A., 294, 309
Jewkes, R., 565
Jhala, U.S., 457
Jia, X., 426
Jiang, J., 305
Jimenez, S., 457
Jin, L., 35
Jin, M., 468
Jin, Q., 86
Jin, S., 127
Jobling, M.A., 7, 16, 106, 120, 430, 447, 471, 484, 486
Jochens, A., 114
Joe, B., 456, 459
Jogayya, K.N., 337
Joh, E.E., 531
Johansson, I., 468
Johansson, M., 321
Johansson, P., 305
Johansson, S., 457
Johnson, C.V., 19
Johnson, E.D., 471
Johnson, J., 422
Johnson, M.P., 456, 459
Johnson, S.A., 391, 392
Johnson, T., 456, 495
Johnson, W.E., 322, 330
Johnston, E., 331
Jones, D.A., 323
Jones, F.A., 371, 372
Jones, G.D., 374, 375
Jones, G.R., 466
Jones, M.K., 174, 183, 390
Jones, R.P., 458
Jordan, G.W., 305
Josefsson, A., 90
Jouan, A., 305
Ju, Z., 335
Juengst, E.T., 530
Jung, M., 145, 147, 148, 149, 392
Junge, A., 120
Junttila, M.J., 457, 461
Jurka, J., 335
Just, R.S., 18, 86, 89, 135, 415
Juusola, J., 392, 394, 395, 396, 404

K

Kaelin, C., 330
Kahl, G., 335
Kahlhofer, C., 304
Kahn, J., 534
Kaida, D., 19
Kakoi, H., 6
Kaku, T., 457
Kalis, S., 151
Kalow, W., 463
Kalpana, D., 142
Kan, Y.W., 171
Kaneda, H., 90
Kang, S.-C., 109
Kanki, T., 93
Kann, M., 459
Kanthaswamy, S., 319, 320, 321, 331, 333, 334, 354
Kantor, P.F., 455
Kaplan, M.E., 127, 489
Karafet, T.M., 122, 123, 127, 489, 492
Karczynski, S.L., 174
Karlsson, A., 544
Karlsson, A.O., 244
Karsch-Mizrachi, I., 341

Kashyap, V.K., 387
Kass, R.S., 457
Kass, S., 457
Kasten, F.H., 294
Katsanis, S.H., 36, 37, 545, 547, 548, 550, 551, 552
Katsnelson, A., 441
Katz, H., 321
Kaufman, D.J., 551, 552
Kaye, D.H., 509, 510, 512, 514, 515, 516, 517, 518, 519, 520, 521, 522, 523, 528, 529, 530, 550
Kaye, S., 389
Kayser, M., 16, 108, 114, 115, 120, 121, 123, 124, 125, 127, 128, 330, 431, 432, 433, 436, 438, 439, 440, 442, 445, 447, 463, 492
Kearney, V.A., 115
Kearney, V.F., 126
Keerl, V., 115, 119, 120, 121
Kehle, J., 303
Kehler, J.S., 330
Keil, W., 121
Keim, P., 293, 294, 295, 301, 302, 306, 307, 308, 311
Keinan, A., 492
Kelemen, O., 19
Keller, A., 495
Kelly, B.P., 457
Kelly, R.L., 534
Kelmenson, P.M., 456, 458
Kelsey, S.F., 461
Kennedy, B., 284, 540
Kenny, E.E., 440
Kent, E.J., 387
Kepron, C.A., 455
Kersbergen, P., 447, 463, 492
Ketchum, M., 354
Key, C.R., 455
Keyser, C., 90, 436
Khoury, M.J., 529
Kidd, J.R., 42
Kidd, K.K., 280, 286, 443
Kiesler, K.M., 42
Kiialainen, A., 421
Kijima-Suda, I., 6
Killeen, G.F., 359
Kim, J.D., 456, 459, 545, 548, 550, 552
Kim, J.H., 322
Kimmel, G., 499
Kimpton, C.P., 14, 44, 97, 107, 329
Kimura, A., 391, 392, 457
Kimura, R., 445
Kimura, S., 463
Kimura, T., 171
King, J.L., 405, 406
King, K.S., 392, 495
King, M.C., 89, 495
King, M.P., 90
King, T.E., 124, 486

Kinoshita, A., 445
Kipps, A.E., 388
Kircher, M., 495
Kirchheiner, J., 467
Kirkham, A., 182, 201, 240
Kirsch, G.E., 457
Kis, Z., 145
Kish, P.E., 256
Kisilevsky, A.E., 174
Kitchener, A.C., 341, 342
Kitpipit, T., 340, 342
Kitteringham, N.R., 465
Kittler, R., 108
Kittles, R.A., 436, 437, 440, 484
Kivisild, T., 495
Kjaer, K.W., 436, 437, 444
Klaassen, S., 457, 461
Klaric, I.M., 16
Kleiber, M., 140, 284
Klein, R., 90
Klein, S.B., 416
Kleinjung, F., 304
Kline, M.C., 144, 232, 416
Klintschar, M., 144
Kloep, F., 153
Kloosterman, A.D., 447, 463
Klotzbach, H., 356
Klussmann, S., 304
Knijff, P., 463
Knols, B.G., 359
Knop, A., 143
Knowler, J.T., 391
Knudsen, S., 356
Knutsen-Heitmann, I., 176
Koban, E., 337
Kobayashi, K., 140, 141
Kocher, T.D., 340
Kochl, S., 112
Koc-Zorawska, E., 145, 146, 148, 149
Kohler, N., 337, 339
Kohlmeier, F., 392
Koide, S., 456, 458
Koki, G., 440
Kolata, G., 514
Kolmonen, M., 466
Kolodji, Y., 95
Komaru, A., 90
Konig, I.R., 456, 458
Konjhodzic, R., 540
Kontadakis, K., 47, 318
Koopmann, T.T., 457
Kooptiwut, S., 457
Koot, M.G., 512
Kopke, K., 456, 459
Kopp, K., 30, 329
Koppelkamm, A., 391, 392

Koren, G., 468
Korn, D., 529
Korpelainen, H., 335, 375
Korschineck, I., 395
Koski, A., 466, 467, 468
Koskinen, M.T., 331
Kossarek, L.M., 321
Kostamo, K., 335, 375
Kostov, Y., 304
Kotchen, T.A., 459
Kott, T., 337
Kovacevic, L., 143
Kraft, P., 436, 444
Krane, C.M., 192, 193
Krause, D., 17, 146, 151
Krause, J., 495
Krawczak, M., 17, 114, 115, 118, 136, 137, 146, 151
Krenke, B.E., 13, 38, 109, 138, 140, 232, 283
Kreskas, M., 107
Kress, W.J., 371, 372
Krettek, R., 356, 358, 359
Krimsky, S., 528
Kristinsson, R., 93
Kriz, B., 306
Krüger, C., 120
Kruglyak, L., 431
Kruse, M., 458
Kubacka, I., 18
Kubat, M., 4, 244
Kucuktas, H., 335
Kuczius, T., 304
Kugener, B., 462
Kugiyama, K., 456, 458
Kuhlisch, E., 17, 146, 149, 151, 152
Kuhnau, W., 136, 137
Kuiper, M., 372, 373, 374, 379
Kuivaniemi, H., 456
Kukat, C., 93
Kull, S., 302
Kumar, A., 456, 459
Kumar, D.R., 460
Kumar, N., 335
Kunich, J.C., 353
Kunkel, T.A., 91
Kupiec, T., 436, 437, 439, 440, 444
Kuriger, J.K., 109
Kuroda-Kawaguchi, T., 105
Kuroiwa, T., 93
Kurosaki, Y., 303
Kurosu, A., 147, 148, 149
Kurushima, J.D., 330
Kuske, C.R., 302
Kutranov, S., 177
Kuttab-Boulos, H., 456, 459
Kuwayama, R., 340
Kwon, B.K., 141, 146, 148, 149

L

Ladd, C., 22, 46, 196, 283, 367, 368, 369, 372, 374, 379
LaFountain, M.J., 38
Lagace, R.E., 140
Lagercrantz, J., 456, 458
Lahat, H., 457
Lahti, R., 465
Laipis, P.J., 89, 92
Laleli-Sahin, E., 468
Lambert, C., 471
Lambert, D.M., 340
Lambert, J.A., 195
Lancaster, R., 463
Lancia, M., 141
Lander, E.S., 329, 514, 515
Landers, J.P., 471
Landstrom, A.P., 457
Lane, B., 499
Laney, R., 531
Lang, B.F., 86
Lange, K., 495, 499
Lango Allen, H., 433, 434, 435, 441
Langston, A.A., 321
LaNier, T., 46, 369, 370, 379
Lanktree, M.B., 433, 435, 441
Lao, O., 124, 126, 431, 432, 433, 438, 439, 442, 445, 492
Laporte, R.E., 461
Lappas, N.T., 389
Lareu, M.V., 124, 127, 416
Laroucau, K., 304
Larsson, M., 434
Larsson, N.G., 93
LaRue, B.L., 405, 406
La Spada, A.R., 150
Latorra, D., 11
Lauc, G., 4, 21, 244
Launiainen, T., 465, 466
Laurie, C., 339
Laurin, N., 544
Lauritzen, S.L., 192
Lautsch, S., 146, 148
Lawrence, J.B., 19
Lawson, D.J., 495
Lazarevic, V., 407
Lazarou, J., 465
Leader, D.P., 391
Leal, N.C., 302
Leamon, J.H., 94
Leat, N., 115
LeBlance, H., 354
Lechpammer, M., 471
Leclair, B., 280, 286, 287
Lederer, T., 471
Lednev, I.K., 387, 390

Le Donne, M., 471
LeDoux, S.P., 91
LeDuc, J.W., 301, 302
Lee, C.H., 46
Lee, C.L., 374
Lee, F.K., 377
Lee, H.C., 22, 46, 196, 254, 267–268, 271–272, 273, 367, 368, 369, 370, 371, 372, 373, 376, 379
Lee, H.S., 283
Lee, H.Y., 109, 141, 145, 405
Lee, J.C.I., 143, 339, 340, 341
Lee, N.H., 456, 459
Lee, S.S., 141, 146, 148, 149
Lee, Y.L., 445
Leeder, S.J., 468
Le Faou, A., 305
Lefeuvre, A., 305
Legler, M.M., 201
Lehman, H., 463
Lei, M., 458
Leibelt, C.S., 140, 144, 228
Leister, D., 340
Leite, F.P., 142
Leng, W., 86
Leonard, D., 174
Ler, S.G., 377
Leren, T.P., 458, 463
Le Scouarnec, S., 457
Lessig, R., 115
Lettre, G., 434, 435, 441
Levenstien, M.A., 456, 459
Levinson, G., 144
Levo, A., 467, 468
Lewis, C.M., 501
Lewis, K., 301
Lewis, M.H., 532
Lewis, S.E., 93
Lewontin, R.C., 515
Leys, R., 356
Li, C.T., 145
Li, H., 335
Li, J.Z., 447, 458, 495
Li, L., 145
Li, M., 91
Li, Q.G., 305
Li, S., 143
Li, W.H., 6, 7
Li, Y., 447, 448, 495
Liang, B., 457
Liang, Y., 401
Liang, Y.H., 459
Liao, S.P., 341
Liberty, A.A., 109
Lieberman, K.R., 95
Liew, C.W., 455
Lightowlers, R.N., 18
Lignitz, E., 141, 146, 148, 149

Likic, V., 86
Lilly, J.W., 86
Lim, E.J., 145
Lim, S.K., 115
Lima, G., 145
Lin, G., 217
Lin, J.H., 337, 339
Lin, J.Y., 435
Lin, P.S., 304
Lin, W.Y., 374
Lin, X., 458
Lin, Y.R., 337, 339
Linacre, A.M.T., 11, 12, 38, 44, 145, 156, 158, 159, 160, 171, 174, 175, 317, 321, 337, 338, 339, 340, 341, 342
Lincoln, S.E., 321
Lindahl, K.F., 90
Lindahl, T., 391
Lindblad-Toh, K., 329
Lindblom, B., 151, 469
Lindenbergh, A., 395, 396, 398, 399, 439, 442, 446
Lindgreen, S., 447, 448, 495
Lindler, L.E., 302
Lindner, I., 147, 148, 149
Lindpaintner, K., 456, 459
Lindstrom, L., 89, 91, 415, 416
Linhares, I.M., 408
Linton, L.M., 329
Linville, J.G., 356, 359, 360
Liotta, L.A., 469
Lipman, D.J., 341
Lipovich, L., 335
Lipp, G., 460
Lisic, M., 307
Liss, B., 171
Lithgow, T., 86
Litt, M., 322, 325
Little, W.C., 458
Liu, C.L., 379
Liu, F., 431, 432, 437, 438, 439, 445, 446, 463
Liu, J., 457
Liu, Q.L., 143, 148, 149
Liu, S.C., 19
Liu, T., 86
Liu, X., 456, 459
Liu, Z., 335
Lo, C.F., 337, 339
Lobry, J.R., 194
Lodeiro, M.F., 86, 93
Loffler, S., 145
Lohmueller, K., 201
London, A.J., 550, 552
London, B., 457
Long, Q., 107, 489
Longley, M.J., 91
Lopes, C.R., 337
Lopez, J.V., 329, 340

Author Index

Lopez Camelo, J.S., 127
Lopez-Goni, I., 303
Lord, W.D., 354, 359
Loreille, O.M., 90, 97
Lorente, J.A., 287
Lorente, M., 287
Lottanti, L., 141
Louis, E.E., Jr., 337
Lounsbury, J.A., 544
Lowe, A., 174, 183
Lowe, J.K., 502
Loyo, M.A., 232
Lu, D.J., 143, 148, 149
Lu, H.L., 148, 149
Lubahn, D.B., 150
Lubeck, P.S., 303
Lubenow, H., 400, 401, 402
Lubes, B., 90
Lucas, D., 535
Lucast, E.K., 530
Luchini, D., 455
Lucy, D., 174
Luczak, S., 145
Ludes, B., 436
Luebke, K., 359
Luider, J., 470
Lukashova, A., 335
Lunderberg, J., 47, 318, 470
Lundkvist, A., 305
Luo, H.B., 145
Lupi, R., 86
Lutz, S., 89, 97
Lutz-Bonengel, S., 145
Lv, D.J., 148, 149
Lynnerup, N., 339
Lyons, L.A., 330
Lyons, S.R., 303

M

Maaskant, P., 399
MacCormick, J.M., 462
Machado, F.B., 151
Machin, G., 431
Machray, G.C., 335
Mack, B., 471
Mackey, D.A., 491
MacKinnon, G., 281
Maclin, T., 531
MacPherson, J.M., 358, 445, 501
MacRae, C., 457
Madea, B., 120, 402, 403, 465, 466
Madi, T., 405
Maestrale, G.B., 456, 459
Maglia, G., 95
Mahgoub, A., 463
Main, F., 244
Maiwald, S., 458
Malecki, M.T., 457
Malicdem, M.T., 228
Malladi, V.S., 319
Malsom, S., 109
Malyarchuk, B.A., 145
Mameli, A., 174
Mandrekar, M.N., 30, 329
Mandrekar, P.V., 283
Manfredi, G., 90
Manganiello, S.D., 407
Mani, A., 456, 458
Maninchedda, G., 445
Mannucci, A., 107, 339
Manolescu, A., 456, 458
Manta, F.S.N., 135, 142
Manunta, P., 456, 459
Marchi, E., 281
Marchiarullo, D., 471
Marchuk, L., 391
Marcikic, M., 4, 244
Mardis, E.R., 421
Margiotta, G., 141
Marienfeld, J., 86
Marino, M.A., 93, 95
Marjanovic, D., 8, 9, 10, 15, 16, 22, 25, 143, 231, 253, 254
Markotic, A., 301
Markovic, D., 456, 459
Maroni, G., 7
Marquardt, M.L., 457
Marshall, A.R., 337
Marshall, J.A., 305
Marsman, R.F., 458
Marthandan, N., 359
Martin, N., 431
Martin, P., 4
Martina, P.F., 147, 148, 149
Martinez, A., 335
Martinez, M.J., 301, 302, 306
Martinez, R.E., 142
Martins, J.A., 142
Martins, W., 335
Marušic, A., 529
Marvan, R., 115
Maschke, K.J., 531
Mascia, M.B., 337
Masibay, A.S., 389
Massetti, S., 141
Masuda, R., 329, 340
Matheucci, E., Jr., 337
Mathewes, R.W., 374, 375
Matise, T.C., 151
Matsumoto, L., 89
Matthias, P., 456
Mattila, A.M., 319, 321, 331, 333, 334
Maxzud, K., 142

May, P.E., 321
Maybruck, J.L., 115
Mayr, W.R., 12
Mazza, J.J., 460
Mazzeo, E., 145
McAlister, C., 109, 471
McAlpine, P.J., 12
McCartney, C.W., 545
McCord, B.R., 140
McCord, B.R., 283
McCormick, C.T., 510
McCouch, S., 335
McCredie, R.M., 456, 459
McCulley, C.H., 462
McCurdy, L.D., 95
McDonald, A., 174
McElroy, A.K., 305
McEvoy, B.P., 435, 437, 440, 441
McEwen, J.E., 35, 38
McEwing, R., 317, 337
Mcgonagle, J.J., 328
McGrath, S., 302
McGuire, S.M., 337
McKenna, W.J., 456
McKeon, T.A., 377
McKinstry, D.M., 19
McKinstry, M.B., 19
McKusick, V.A., 460
McLeod, H.L., 464
McNally, E.M., 457
McQuillan, M.R., 91, 94, 95
McVean, G.A.T., 500
Mead, S., 445
Medeiros, R., 142
Medeiros-Domingo, A., 457, 463, 466
Medina-Acosta, E., 151
Meeusen, S., 93
Meganathan, P.R., 337
Mehdi, H., 457
Meiland, H.C., 178
Meilerman, M.B., 122, 123, 489
Meinking, T.L., 359
Melander, O., 456
Mellersh, C.S., 321
Mellmann, A., 303
Melo, A.C., 302
Melson, K.E., 512
Melton, L.J., 460
Melton, T., 89, 91, 93, 96, 97, 339, 340, 415, 416
Meltzer, M.I., 301
Mendez, F.L., 122, 123, 489
Mengel-From, J., 431, 432, 436, 437, 439, 440, 444
Menotti-Raymond, M.A., 318, 321, 322, 323, 324, 325, 328, 329, 330
Menozzi, P., 496
Merchant, N.C., 127, 489
Merck, M.D., 337

Mercuri, A.M., 374
Mercuri, E., 461
Merner, N.D., 456
Merriweather, A., 359
Merriwether, D.A., 440
Mesa, N.R., 492
Messmer, T.O., 304
Metzger-Boddien, C., 441, 448
Metzker, M.L., 439, 446
Meyer, A., 340
Meyer, H.J., 277, 304
Meyer, M., 422
Meyer, R.F., 294, 302, 306, 309
Meyers, C., 283
Mgone, C.S., 440
Michael, M., 146, 149, 150, 154
Michaels, G.S., 89
Michalec, M., 457
Michalek, W., 335
Micka, K.A., 328
Micke, P., 470
Mickiewicz, P., 174
Mieusset, R., 471
Mihalovich, J., 415, 516
Mijares, V., 147, 148, 149
Mikhailova, E., 95
Mikkelsen, A., 436, 437, 444
Mikulasovich, R., 175, 176, 178
Milbrandt, J., 470
Mildenhall, D.C., 374, 375
Millar, C.D., 340
Millar, J.D., 301
Millat, G., 462
Miller, M.T., 254, 367, 368, 369, 370, 373, 376, 379
Miller Coyle, H., 46, 367, 368, 369, 370, 371, 372, 373, 374, 377, 379
Milne, S.C., 283
Miltner, E., 90
Minaguchi, K., 143, 145
Minch, F., 106
Mineharu, Y., 456
Miranian, D., 544
Miskelly, G.M., 387
Mitani, T., 340
Mitchell, A.A., 183, 194, 202, 431, 432, 433, 437, 438, 439, 440, 443, 444, 445
Mitchell, L.G., 18, 89, 329, 540
Mitchell, R.J., 107
Mittal, K., 471
Mnookin, J.L., 510, 518, 519, 520, 521, 522
Modi, A., 47
Moellering, R.C. Jr., 301
Moffat, A.C., 376, 377
Mogensen, H.S., 108, 109, 183, 200
Mohapatra, B., 457
Mohaupt, M.G., 459
Mohler, P.J., 457

Mohr, D.N., 460
Molero, F., 302
Mondragón, M.C., 492
Monson, K.L., 10, 323
Montagna, S., 16
Montagnon, D., 436
Montali, E., 374
Montelius, K., 151
Montgomery, G.W., 436
Moonsamy, P.V., 424
Moore, J.E., 318
Moore, M.K., 340
Moore, T.J., 465
Moreno, L.I., 404
Moreno, R.D., 90
Moretti, T.R., 430
Morgan, D.G., 391, 392
Morin, P.A., 322
Morling, N., 145, 147, 148, 149, 154, 431, 432, 436, 437, 439, 440, 444
Moroney, J.F., 304
Morrell, R., 565
Morrison, T.B., 329
Morse, S., 293, 294, 295, 301, 306, 307, 308, 311
Mortari, N., 337
Mortera, J., 192
Morton, D.B., 318
Mosca, D.T., 542, 543, 550
Moss, A.J., 455
Mostad, P.F., 151, 194, 244
Motazacker, M.M., 458
Motti, J.M.B., 127
Motulsky, A.G., 463
Moustafa, I.M., 86
Mrsic, G., 47
Muchir, A., 461
Mukabana, W.R., 359
Mukerjee, S., 142
Mukherjee, M., 142
Mulero, J.J., 109, 140, 228
Muller, K., 90
Muller, S., 303
Mullikin, J.C., 322, 329
Mumford, B.R., 457
Mund, C., 532
Muniec, D.S., 91, 95, 96
Murphy, W.J., 322
Murphy-Bollinger, J., 551, 552
Murray, C., 174, 183, 471
Murray, J.C., 151
Murtaza, M., 433, 435, 441
Murthy, S.K., 470
Musone, S.L., 456, 457, 458
Musshoff, F., 465, 466
Muzzio, M., 127
My, D., 174
Myers, J.R., 389

Myers, K.M., 303
Myers, S.R., 495, 500

N

Nadeem, A., 143, 147, 149
Nagy, M., 120
Nahed, B.V., 456
Nakamura, Y., 143, 145
Nakanishi, H., 26, 407, 409, 410
Nakashima, E., 532
Nakayama, T., 456, 459
Nalca, A., 302
Nambiar, P., 145
Nan, H., 436, 444
Napolitano, C., 457
Nasidze, I., 91
Natanson, L., 337, 339
Naue, J., 145
Nava, A., 456
Nazzaruolo, B.L., 281, 471
Neale, B., 502
Negredo, A., 302
Neilson, J.R., 7
Nelson, K., 89, 91, 339, 340, 415, 416
Nelson, L.A., 340, 356
Nelson, L.S., 376, 377
Nelson, M.R., 495
Neubauer, H., 302
Neuenfeld, Y., 459
Neuffer, E., 534
Neuhaus, T., 456, 459
Neuhuber, F., 144
Neuman, C., 201
Neuvonen, A.M., 115, 469
Neve, B., 457
Newby, D.T., 303
Newhouse, S.J., 456, 459
Newton, C., 373, 374
Newton-Cheh, C., 456
Ng, T.K., 305
Nguyen, D., 91
Nickchen, K., 467
Nickerson, D.A., 431
Niederstätter, H., 18, 93, 95, 107, 112, 340
Nielsen, M., 436, 437, 444
Niemcunowicz-Janica, A., 145, 146, 147, 148, 149
Niemela, M., 456
Niezgoda, S., 280, 286
Nilsen, G.B., 431
Nilsson, M., 415
Nilsson, P., 305
Nishi, T., 147, 148, 149
Njolstad, P.R., 457
Nonneman, D., 335
Nordby, J.J., 318
Nordstrom, H., 305

Nori, A., 457
Norton, H.L., 436, 437, 440
Nothnagel, M., 151, 153
Novembre, J., 495, 499, 504
Nunez, C., 124, 127
Nunez, L., 457
Nunnari, J., 93
Nussbaumer, C., 395
Nuzzo, F., 120
Nyberg, B., 391
Nybom, H., 335

O

Obata, M., 90
Oberacher, H., 95
O'Brien, B., 522
O'Brien, S.J., 318, 322, 323, 324, 325, 328, 329, 330
O'Callaghan, J.E., 18, 415
O'Connell, K., 183, 202
Oda, N., 457
Odriozola, A., 140, 142
Oechslin, E., 457, 461
Oefner, J.P., 16
Oehmichen, M., 392
O'Fallon, W.M., 460
Ogden, R., 317, 333, 337, 338
Oguzturun, C., 145
O'Hanlon, K.A., 91, 94, 95
Ohashi, J., 445
Ohgaki, K., 93
Ohgane, J., 405
Ohmori, T., 26
Ojanperä, I., 465, 466, 467, 468
Oki, T., 109
Okii, Y., 340
Okinaka, R.T., 302
Olaisen, B., 150
Olasagasti, F., 95
Old, J.B., 389
Oldroyd, N.J., 228
Oliveira, R.N., 142
Olivo, P.D., 92
Olofsson, J., 108, 109
Olson, J.M., 456
Ommen, S.R., 457
Omoe, K., 304
Ooie, Y., 90
Oota, H., 106
Orchard, T.J., 461
Orihuela, Y., 174
Orlando, C., 470
Örnemark, 223
Orrego, C., 416
Ortiz, J., 335
Osborne, J., 302
Oshida, S., 145
Ossorio, P.N., 533
Ostell, J., 341
Osterziel, K.J., 457
Ostman, A., 470
Ota, M., 109, 140, 141
Ott, J., 456, 459
Ou, X.-l., 463
Outerbridge, C.A., 330
Overall, A.D.J., 321
Owens, J.D., 373, 374
Owens, K., 495
Oxborough, R.J., 318
Oya, M., 387
Ozaltin, E., 435
Ozawa, T., 340
Ozder, M., 337
Ozkan, E., 337
Ozturk, A.K., 456

P

Paabo, S., 340, 391
Pack, S., 456, 460
Packer, C., 321
Pádár, Z., 47, 318, 319
Padmanabhan, S., 456
Paessler, S., 305
Paic, F., 9
Pajnič, I.Z., 540
Pakstis, A.J., 42
Palacios, G., 303
Palestro, G., 471
Palha, T.D., 115, 142
Palha Tde, J., 142
Pallotti, F., 90
Palmbach, T.M., 46, 254, 367, 368, 369, 370, 372, 373, 374, 376, 379
Palmer, J.D., 86
Palmisano, I., 436
Palo, J.U., 115, 145, 465, 469
Pals, G., 456
Palstra, R.J., 436
Pamjav, H., 145, 147, 148, 149
Pandit, S.V., 457
Pandya, A., 107, 471
Paneto, G.G., 142
Pang, B.C., 389
Panicke, L., 337
Pank, M., 337, 339
Pant, P.V., 436
Paoletti, D.R., 192, 193
Papadopoulou, L.C., 90
Pape, T., 358, 359, 360, 361
Pardo, M., 303
Paredes, M., 142
Parfitt, T., 484
Park, B.K., 465

Park, J.E., 457
Park, L.K., 337
Park, M.J., 109, 141
Park, S.M., 399
Park, S.W., 145
Parker, L.S., 550, 552
Parker, M.J., 530
Parkin, E.J., 115, 124
Parra, E., 436, 437, 440
Parson, W., 86, 95, 107, 112, 146, 148, 149, 340, 354, 418, 517
Parsons, T.J., 18, 20, 87, 89, 90, 91, 93, 95, 96, 97, 98, 99, 280, 286, 329, 339, 418, 421, 540
Pasino, S., 141, 145
Passarino, G., 16
Patch, A.M., 457
Patel, H.H., 458
Patel, M.M., 377
Pathak, D., 455
Patterson, N.J., 495, 497, 499
Pattinson, J.K., 171
Patzelt, D., 392, 395, 471
Pauly, D., 302
Pavlinic, D., 4, 244
Pawlowski, R., 108
Payne, B.A., 87, 91
Pazoki, R., 456, 458
Pazzagli, M., 470
Peabody, A.J., 318
Peakall, R., 372, 373, 374, 379
Pearson, E.R., 457
Pearson, G.S., 306
Pearson, K., 496
Pecon-Slattery, J., 322
Pegoraro, K., 340
Peixe, C., 142
Pelander, A., 466
Peleman, J., 372, 373, 374, 379
Pellegrino, E., 529
Pène, L., 194
Penketh, R.J., 171
Pennisi, E., 10
Pepinski, W., 145, 146, 147, 148, 149
Peppin, L., 337
Pereira, R., 17, 142, 143, 144, 153, 154
Pereira, S.L., 339
Pereira, V., 142, 145, 147, 148, 149, 154
Perez, A., 142
Perez, F., 335
Perez, J., 194
Perez, T., 357
Perez, W.A.M., 337
Pericic, M., 16
Perlin, M.W., 201, 512
Perrin, D.G., 455
Perrot, A., 457
Pertossi, M., 550

Petek, M.J., 47
Peterman, O., 145, 147, 148, 149
Peters, G., 304
Petersen, E.E., 466
Petersen, J.M., 302
Petersmann, H., 141, 146, 148, 149
Peterson, A.C., 92
Peterson, J.L., 217
Petretto, E., 456, 459
Petricevic, Š., 373
Petterson, T.M., 460
Peyrefitte, C., 305
Pflueger, S.M., 330
Philips, M.B., 293, 294, 308
Phillips, C., 416, 445
Piazza, A., 496
Picard, C.J., 356, 358
Picardi, E., 86
Piccirillo, R., 436
Pico, A., 142, 147, 148, 149
Picornell, A., 145
Piercy, R., 329
Piko, L., 89
Pilli, E., 47
Pinchinet, R., 195
Pinheiro, F., 145
Pinheiro, M.F., 142
Pinheiro, R., 154
Pinkerton, N., 459
Pinon, G., 459
Pinto, N., 17, 136, 137, 138
Pinzani, P., 470
Pirastu, N., 445
Pirie, A.A., 174
Pirmohamed, M., 465
Pizzamiglio, M., 174
Planz, J.V., 144, 283
Plaster, N.M., 457
Plate, I., 146, 147, 148, 151, 153
Plenge, R.M., 497
Plengvidhya, N., 457
Pneuman, A., 433, 437, 442, 443, 444, 445, 448
Podesta, J.S., 331
Podgoreanu, M.V., 455
Poetsch, M., 141, 143, 146, 147, 148, 149
Poglio, A., 145
Pojskic, N., 143, 231
Polanskey, D., 422
Polašek, O., 529
Pollevick, G.D., 457
Polley, D., 174
Pomeranz, B.H., 465
Ponten, F., 470
Pontes, L., 145
Pontius, J.U., 322, 329
Popiolek, D.A., 471
Popovic, M., 47

Pospiech, E., 432, 437, 439, 444
Postma, A.V., 458
Pot, J., 372, 373, 374, 379
Poulter, M., 445
Pounder, D., 466
Povoledo, E., 521
Povolný, D., 356, 358
Powell, M.C., 175
Powell, W., 335
Power, D.A., 408, 409
Prachar, J., 92
Prainsack, B., 558
Pras, E., 457
Prasad, P., 335
Prata, M.J., 142, 154
Prenger, V.L., 93, 95
Presciuttini, S., 141
Preuss, J., 466
Preuss, M., 456, 458
Price, A.L., 495, 497, 499
Prieto, J.M., 318
Prieto, L., 20
Primorac, Da., 12, 56
Primorac, Dr., 4, 6, 8, 9, 10, 12, 15, 16, 19, 21, 22, 25, 35, 56, 244, 253, 254, 415, 416
Prinz, M.K., 16, 112, 142, 144, 153, 174, 175, 176, 178, 281, 283, 433, 437, 442, 443, 444, 471
Priori, S.G., 455, 457, 462
Pritchard, J.K., 495, 497
Probst, S., 457, 461
Procházková, H., 337
Prodi, D.A., 445
Prosser, R., 90
Provan, J., 335
Prufer, K., 422
Prusak, B., 337
Pruvost, M., 47
Psaty, B.M., 465
Pueschel, K., 356
Pugh, E., 280, 286
Pugh, J.E., 405
Punna, R., 335
Purcell, S., 502
Purschke, W.G., 304
Pusey, A.E., 321
Pyeritz, R.E., 456, 460

Q

Quan, L., 148, 149
Quarino, L., 389
Quattro, J.M., 340
Quickenden, T.I., 387
Quinet, K., 549
Quinones, I., 174
Quintans, B., 416
Quirke, T.P., 172

R

Rabbach, D., 13, 38
Racic, I., 303
Radam, G., 106
Radhakrishnan, J., 456, 458
Radi, E., 86
Radko, A., 337
Radstrom, P., 303
Radtke, F., 304
Raeder, H., 457
Rafestin-Oblin, M.E., 459
Rah, H.C., 330
Raimondi, E., 142
Rakha, A., 143
Rakotomavo, N., 143
Ralf, A., 123, 124, 125, 127, 128
Ramachandran, S., 504
Ramalho-Santos, J., 90
Ramallo, V., 127
Rameckers, J., 177
Rampazzo, A., 456
Ramshaw, E., 532
Rand, S., 218, 243, 245, 246
Randall, B.B., 354, 389
Randi, E., 337
Rao, G., 456, 458
Rao, P.N., 471
Raoult, D., 304
Raper, A.B., 388
Rasanen, I., 465, 466
Rasmussen, M., 447, 448, 495
Rasmussen, S.A., 466
Rasooly, A., 303, 304
Ratnasingham, S., 340, 341
Rayimoglu, G., 145
Raymond-Menotti, M., 47
Read, A., 13
Rebaca, K., 142
Rebecca, L.R., 19
Rechavi, G., 172
Redd, A.J., 115, 118
Reddy, A.G., 107
Redman, J.W., 144, 416
Redwood, C., 461
Reeder, D.J., 140, 228
Reefhuis, J., 294, 309
Rees, B., 388
Rees, J.L., 436, 440
Rehout, V., 337
Reich, D., 495, 497
Reich, K.A., 245, 246
Reichenberger, F., 456, 459
Reid, T.M., 144
Reijans, M., 358, 359, 372, 373, 374, 379
Reiner, A.P., 459
Reisinger, E.C., 302

Author Index

Reiter, C., 354
Relman, D.A., 301, 302
Remigio, E., 340
Ren, Z., 143, 145
Repenning, A., 141, 146, 148, 149
Replogle, J., 422
Reppy, J., 311
Reshef, A., 176
Resque, R.L., 154
Reynolds, H.R., 458
Reynolds, R.L., 415, 416
Reynolds, S.L., 86
Rhydderch, J.G., 337
Ria, M., 456, 458
Riancho, J.A., 141, 142, 146, 147, 148, 149, 153
Riazuddin, S., 143, 147, 148, 149
Riazuddin, S.A., 143, 147, 148, 149
Ribeiro-Dos-Santos, A., 115, 142
Ribeiro-Rodrigues, E.M., 115, 142, 154
Ricchetti, M., 340
Ricci, U., 112, 141
Rice, D.W., 86
Rice, S.M., 340
Richard, M.L., 109, 140, 232, 283, 399
Richard, P., 457
Richards, A.L., 305
Richards, J.B., 445
Richly, E., 340
Ripley A., 531
Risinger, D.M., 522
Rittner, C., 141, 146, 148, 149, 318
Ritz-Timme, S., 354
Rivolta, I., 457
Roberto, F.F., 303
Roberts, K.A., 415
Roberts, P., 246
Robertson, J., 317, 337, 340, 372, 373, 374, 379
Robin, E.D., 89
Robino, C., 141, 145, 471
Roby, R.K., 18, 89, 91, 96, 97, 98, 140, 228, 339
Roccella, E.J., 459
Rocchi, A., 141
Rochat, S., 47, 329
Rochholz, G., 354
Rodig, H., 115, 145, 147, 148, 149, 153
Rodrigues, E.M., 142
Rodrigues, R.A., 359
Rodriguez, J.J., 340
Roe, B.A., 456, 459
Roeder, A.D., 174, 396, 399
Roewer, L., 108, 114, 115, 120, 121
Rogalla, U., 145
Rognum, T.O., 462
Rogus, J., 456, 459
Rohrer, G.A., 335
Rolf, B., 121, 144, 340, 358, 471
Romano, C., 359, 360

Romero, C., 303
Roney, C.A., 47, 359
Roos, Y.B., 456
Roser, C., 303
Ross, P., 445
Rothberg, J.M., 94
Rothwell, E., 532
Rouard, M., 457
Rougé, D., 471
Rousseau, D., 92
Roussel, V., 91
Rowe, D.W., 19
Rowlands, D., 47
Rubinoff, D., 358
Rudan, I., 21
Rudan, P., 16
Rudin, N., 10, 201
Rufenacht, S., 330
Ruitberg, C.M., 144
Russell, M.W., 457
Rustgi, S., 335
Rutherford, S., 456, 459
Rutty, G.N., 174
Ryan, E., 463, 546

S

Saad, Y., 456, 459
Sabir, M.F., 143
Sabule, A., 141
Sacher, F., 457
Sachidanandam, R., 448
Sachse, K., 304
Sadayo, E., 471
Sadlock, J., 90
Saferstein, R., 24
Saffitz, J.E., 456
Sahin, O., 337
Saiki, R.K., 33
Sajantila, A., 16, 115, 121, 465, 466, 467, 468, 469
Sakai, H., 109, 140, 141
Sakamoto, Y., 456
Sakurada, K., 388, 399, 400
Salas, A., 142, 416, 445
Saldanha, G., 283
Saleh, S.S., 305
Salisbury, B.A., 445
Sallee, F.R., 468
Samejima, M., 145
Sammons, S., 302
Samnegard, A., 456, 458
Sampson, B.A., 456, 458, 460, 464
Sampson, K.J., 457
Samuels, D.C., 92
Sanches, A., 337
Sanchez, J.J., 154, 431, 432, 437, 439, 444, 445, 447
Sanchez, K., 232

Sanchez-Diz, P., 124, 127, 142
Sanchez-Seco, M.P., 302
Sandberg, R., 7
Sande, M.A., 301
Sandhu, H.S., 294, 309
Sani, I., 112
Sankararaman, S., 497
Sannes-Lowery, K.A., 95
Sano, N., 90
Sansone, M., 16
Santos, A.K., 142
Santos, F.R., 107
Santos, M.R, 127
Santos, S., 115, 142
Santovito, A., 141, 145
Sanudo, C., 141, 146, 148, 149
Saravo, L., 359, 360, 471
Sasaki, M., 171
Satkoski, J.A., 319
Satoh, M., 93
Sauer, J., 356, 358
Sauer, N.J., 512
Saunier, J.L., 18, 93
Sauvage, F., 194
Savolainen, P., 47, 318
Sayers, E.W., 341
Scarpetta, M.A., 176
Schade, L., 546, 547, 548
Schaefer, M., 91
Schäffer, A.A., 330
Schanfield, M.S., 11, 38
Scharf, S., 33
Scharnhorst, G., 320, 321
Schatten, G., 90
Schellberg, T., 559
Scheneider, P.M., 319, 333
Scherczinger, C.A., 22, 369, 371, 373
Scherer, S.E., 456
Schier, J.G., 377
Schierenbeck, K.A., 375, 376
Schiff, A.F., 389
Schiffner, L., 176, 178
Schill, W.B., 388
Schlecht, J., 127, 489
Schleinitz, D., 145
Schlotterer, C., 44
Schmaljohn, C., 305
Schmidt, A., 223
Schmidt, S.C., 389, 448
Schmidtke, J., 136, 137
Schmidt-Küntzel, A., 330
Schmidt-Traub, H., 457
Schneider, P.M., 140, 141, 146, 148, 149, 150, 190, 318, 392
Schneider-Poetsch, T., 19
Scholz, H.C., 302, 304
Schon, E.A., 90
Schonberg, A., 91

Schönwald, S., 301, 307
Schork, N., 456, 459
Schoske, R., 144
Schott, J.J., 457
Schroeder, H., 356
Schroeder, R., 91
Schubert, E., 304
Schulz, H., 456, 459
Schulze-Bahr, E., 458
Schumacher, G.F., 389
Schupp, J.M., 304
Schur, B.C., 468
Schurenkamp, M., 141, 218, 243, 245, 246
Schütt, S., 356, 358
Schutz, H.W., 354
Schutzer, S.E., 293, 294, 295, 301, 306, 307, 308, 311
Schwartz, M.B., 38, 90, 96
Schwartz, P.J., 457
Schwarze, U., 456, 460
Schwenke, P.L., 337
Schwinn, D.A., 455
Scoles, G.J., 358
Scordi-Bello, I.A., 455
Scott, D.L., 174
Scott, J., 551, 552
Scott, P., 47
Scurrah, K.J., 445
Sebestyen, J.A., 176
Sebillon, P., 457
Secrest, A.M., 461
Seda, P., 359
Seidl, S., 471
Seielstad, M.T., 106
Sejrsen, B., 339
Seker, A., 456
Seki, S., 107
Sekizawa, A., 171
Semeonoff, R., 318
Semino, O., 16
Semmler, T., 303
Semsarian, C., 458, 463
Sensabaugh, G.F., 322, 421, 516, 517, 518, 520
Seo, Y., 318
Serapion, J., 335
Serpico, C., 47
Serra, A., 142
Sesardic, I., 175
Settheetham-Ishida, W., 106
Setzer, M., 392
Severin, D.D., 302
Sevilla, A., 461
Sewell, J., 174
Shadick, N.A., 497
Shaler, B., 112
Shaler, R.C., 280, 281, 286, 387
Shapiro, E., 172
Sharma, A.K., 142

Author Index

Sharma, S., 335
Sharma, V.K., 303
Shastry, B.S., 431
Shaw, K., 175
Sheffield, V.C., 151
Sheikh, P.A., 337
Shekar, S.N., 436
Shelton, L.M., 328
Shen, J., 457
Shen, M., 148, 149
Shen, Q., 456, 458
Shen, Y., 140, 283
Sherry, S.T., 280, 286
Sherwell, P., 545, 547
Shi, S., 108
Shih, R.D., 376, 377
Shin, K.-J., 109, 141, 145, 146, 148, 149
Shinyashiku, F., 322
Shiono, H., 107
Shiota, K., 405
Shirley, N., 46, 369, 370, 379
Shitara, H., 92, 93
Shivji, M., 337, 339
Shojo, H., 26
Shore, G.D., 337
Shoubridge, E.A., 90, 92
Shows, T.B., 12
Shrestha, S., 200
Shriver, M.D., 437, 440, 484
Sibille, I., 108, 109
Siegal, E., 545
Siegel, J.A., 24, 25
Siegfried, J.S., 458
Sijen, T., 399
Sikora, M., 440
Sikweyiya, Y., 565
Silva, D.A., 135, 142
Silva, F.S., 142, 143, 153, 337
Silva, J., Jr., 305
Silva, R.H.A., 142
Silverstein, M.D., 460
Sim, A.S., 456, 459
Sim, J.E., 145
Simerly, C., 90
Simon, J.D., 436
Simoncelli, T., 528
Simone, A., 471
Simone, N.L., 469
Simonetti, G.D., 459
Simonetti, S., 90
Simonsen, Bo, 244, 245, 246
Simonson, T.S., 501
Sinelnikov, A., 201
Singh, L., 107, 339, 340
Singh, R., 335
Sistonen, J., 467
Sitaram, A., 436

Siu, B.L., 462
Sivapalaratnam, S., 458
Sjöqvist, F., 468
Skaletsky, H., 105
Škaro, V., 19
Skawronska, M., 145, 146, 147, 148, 149
Skelton, S.K., 304
Skoda, R.C., 463
Skoda, S.R., 358
Skolnick, M.H., 501
Skurnik, M., 303
Slack, F.J., 7
Slickers, P., 304
Sligar, S., 305
Sloan, D.B., 86
Slooten, K., 201
Slota, E., 337
Smalling, B.B., 319
Smejkal, C.B., 491
Smerick, J., 175
Smialek, J., 459
Smilowitz, N.R., 458
Smirnov, I., 445
Smith, B.C., 89
Smith, D.R., 322, 329
Smith, M.E., 528
Smith, M.A., 340
Smith, P.J., 177
Smith, R.L., 463
Smith, S., 542
Soares, P., 491
Soares-Viera, J.A., 471
Sobrino, B., 124, 127, 422
Socransky, S.S., 407
Soldat, J., 337
Soltyszewski, I., 145, 146, 148, 149
Soma, M., 456, 459
Somers, G.R., 455
Song, F., 405
Songtawee, N., 457
Sood, P., 401
Soto, I.D., 490
Souque, A., 459
Sousa, C.O., 337
Sovik, O., 457
Soysal, M.I., 337
Sozer, A., 280, 286
Spahn, P., 356
Spahr, H., 93
Spannbrucker, N., 463
Sparkes, R., 44, 189
Sparkes, R.L., 195
Spear, T.F., 319
Speed, W.C., 42
Speicher, M.R., 172
Spencer, C.E., 201
Sperling, F.A.H., 357, 358, 359, 360

Spichenok, O., 431, 432, 433, 437, 438, 439, 440, 443, 444, 445
Spicic, S., 303
Spinetti, I., 141
Spitaleri, S., 359, 360
Splawski, I., 457
Splettstoesser, W.D., 304
Sprecher, C.J., 13, 16, 38, 140
Spriggs, A.C., 47, 359
Sridhar, S., 497
Stacey, S.N., 434
Staiti, N., 471
Štambuk, S., 373
Stamer, U.M., 465
Stanglein, T.W., 328
Stanhope, M., 337, 339
Staples, T., 47
Staron, N., 463
Staub, R.W., 471
Steck, A.K., 462
Steighner, R.J., 93, 95
Steinberg, K.K., 529
Steincke, B., 463
Steinke, D., 339
Steinlechner, M., 107, 112, 340, 517
Steinmann, B., 460
Stenersen, M., 150
Stenzel, U., 422
Stephan, D.A., 456
Stephen, A.M., 409
Stephens, D.S., 294, 309
Stephens, G., 542
Stephens, J.C., 322, 445
Stephens, M., 495, 497
Stephens, R.M., 329
Stern, M.C., 11
Sternbach, G., 308
Stevens, J.R., 356, 358, 359
Stoddart, D., 95
Stoffel, M., 457, 502
Stoffel, S., 33
Stoffers, D.A., 457
Stokowski, R.P., 436
Stoneking, M., 91, 106
Stoney, D., 195
St-Onge, J., 456
Störmann, B., 302
Stover, M.L., 19
Stradmann-Bellinghausen, B., 141, 146, 148, 149
Strathdee, S.A., 200
Strauch, H., 106
Struycken, P.M., 456
Stuart, S.M., 19, 418, 419
Studer, R., 107
Sturk, K.A., 109
Sturm, R.A., 434, 436, 437, 444
Sturner, W.Q., 462
Stuve, L.L., 431
Su, Y., 143
Subramanian, S.V., 435
Sugano, Y., 147, 148, 149
Sugiyama, S., 456, 458
Sulaiman, I.M., 302
Sulem, P., 436
Sullivan, K.M., 89, 91, 95, 96, 107, 339, 417, 421
Sun, D., 337
Sun, H.Y., 115, 148, 149
Sun, R., 143
Sun, S., 304, 322
Sutlovic, D., 47, 373
Sutovsky, P., 90
Sutton, T.P., 256
Suzuki, I., 436
Swango, K.L., 228
Sweeney, A., 244
Swinfield, G., 124
Sykora, P., 91
Syndercombe Court, D., 145
Syrris, P., 456
Syvanen, A.C., 421
Szabady, B., 201
Szczerbinska, A., 436, 444
Szepietowska, I., 174
Szibor, R., 17, 136, 137, 140, 144, 146, 147, 148, 149, 150, 151, 152, 153, 154
Sztankoová, A., 337

T

Tabbada, K.A., 141, 146, 148, 149
Taberlet, P., 172
Tachezy, J., 86
Tada, A., 436
Tahir, M.A., 143, 147, 149
Taitz, J., 542, 543, 550
Tajima, A., 456
Takada, A., 303
Takahama, S., 90
Takeda, J., 457
Takiff, H., 232
Takken, W., 359
Tamarin, H.R., 5, 19
Tamariz, J., 175, 176, 178, 183, 202
Tan, C.G., 294, 309
Tan, J., 440
Tandon, A., 499
Tang, H., 445, 495
Tang, J., 359
Tang, K., 302
Tang, S., 91
Tang, W.M., 145, 147, 148, 149
Tang, X., 456, 459
Tang, Y.J., 135, 142, 305, 456, 460
Tanigawa, G., 457

Author Index

Tanriverdi, S., 305
Tanyeli, A., 305
Tao, N., 335
Taplin, D., 359
Tarditi, C.R., 330
Tariq, M.A., 143, 147, 148, 149
Tark, S.H., 305
Tarone, A.M., 356
Taroni, F., 239, 240, 243, 244, 245, 246
Taubøll, E., 458, 463
Tautz, D., 44, 321
Tavares, C.C., 142
Tawil, R., 457
Taya, C., 90
Taylor, G.P., 455
Taylor, G.R., 171, 174, 175
Taylor, P.G., 120, 482
Teitelbaum, S., 445
Tekaia, F., 340
Telmon, N., 471
Temenak, J.J., 305
Temnykh, S., 335
Tenorio, A., 302
Tereba, A., 138, 283
Terravainen, T., 436, 440
Terrill, M., 44
Tester, D.J., 457, 460, 462, 463, 466
Tetzlaff, S., 147, 148, 149
Thacker, C.R., 145
Thalheimer, A., 471
Thamboo, T.P., 388
Thangaraj, K., 107, 339
Thein, S.L., 4, 32, 318
Thibaut, S., 445
Thiel, T., 335
Thiele, K., 145
Thiene, G., 465
Thierfelder, L., 457
Thody, A.J., 436, 440
Thomas, A., 388, 501
Thomas, M.G., 436, 484
Thomas, R., 541, 552
Thomas, W.K., 340
Thompson, E.E., 456, 459
Thompson, J.M., 118
Thompson, J.R., 456, 458
Thompson, R., 457
Thompson, W.C., 239, 240, 242, 243, 244, 245, 246, 513, 518
Thomsen, A.T., 244, 245, 246
Thomson, J., 95, 177
Thomson, N., 491
Thorleifsson, G., 456, 458, 460
Thorp, H.H., 391
Thurber, A.E., 438, 444
Thyagarajan, D., 90
Tian, F., 337

Tie, J., 145
Tierney, A.A., 283
Tilford, C.A., 105
Tillmar, A.O., 145, 151
Timken, M.D., 228
Timothy, K.W., 457
Timpson, N.J., 440
Tishkoff, S.A., 437, 440, 442
Tiso, N., 456, 457
Titford, M., 471
Tiwawech, D., 106
Tkachenko, G.A., 304
To, K.Y., 145, 147, 148, 149
Tobe, S.S., 317, 337, 339, 340, 341, 342
Todd-Brown, K., 502
Togan, I., 337
Tokiyasu, T., 340
Tom, B.K., 319, 321, 331, 333, 334
Tomas, C., 145, 147, 148, 149, 154
Tomaso, H., 302, 304
Tomsey, C.S., 232
Toni, C., 141
Topol, E.J., 456, 458
Topol, S.E., 456, 458
Torre, C., 141, 145, 471
Torres, A., 303
Torres, R.A., 337
Torres, T.T., 359
Toscanini, U., 142
Toth, G., 335
Towner, J.S., 303
Tozzo, P., 530
Tresniowski, A., 331
Trevisan Grandi, G., 374
Tristani-Firouzi, M., 457
Troeger, C., 171
Troelsen, J., 436, 437, 444
Truc, P., 359
Trump, D., 458
Truong, T., 543, 544
Tsai, L.C., 337, 339, 340, 341, 379
Tsang, C.S., 47, 358
Tschentscher, F., 143, 147, 148, 149
Tsipouras, P., 460
Tsubata, S., 457
Tsuji, A., 456
Tu, E., 458, 463
Tucker, V.C., 140
Tuji, T., 391, 392
Tully, G., 47, 329
Tully, L.A., 93, 95
Tung, J.Y., 447
Turnbaugh, P.J., 407
Turnbull, D.M., 18, 491
Turrina, S., 142, 145, 146, 147, 148, 149
Tvedebrink, T., 183, 200
Tyler-Smith, C., 7, 106, 107, 115, 471

U

Uchida, A.U., 86, 93
Uchigasaki, S., 145
Ueda, K., 457, 458
Ullah, O., 147, 148, 149
Ulland, M.M., 109
Underhill, P.A., 16, 122, 123, 487, 495
Unnasch, T.R., 359
Unseld, M., 86
U'Ren, J.M., 304
Urquhart, A.J., 14, 174

V

Valdivia, C., 457
Valente, C., 142
Valenzuela, R.K., 431, 432, 434, 436, 437, 438, 439, 440, 444
Valeriani, F., 25
Vallejo, G., 388
Vallone, P.M., 42, 89, 124, 126, 144, 416, 430
Valverde, J.L., 142
Valverde, P., 436, 440
van Asch, B., 154, 320
van Daal, A., 174, 422
Van de Goor, L.H.P., 335
van de Lee, T., 358, 359, 372, 373, 374, 379
Vandenberg, N., 391
van der Beek, C.P., 178
van der Gaag, K., 117, 118
van der Voet, M., 456
Vandewalle, A., 459
Van de Walle, M.J., 92
Vandewoestyne, M., 471
van Duijn, K., 431, 432, 433, 436, 437, 439, 440, 442, 444, 445, 446, 447, 463, 492
Vanek, D., 115, 540
Van Ert, M.N., 304
Van Etten, R.A., 86
van Haeringen, W.A., 335
van Hannen, E.J., 304
van Hoofstat, D.E., 174, 471
VanHouten, B., 91
Van Niewerburgh, F., 471
van Oorschot, R.A.H., 107, 174, 183, 390, 391
van Oven, M., 123, 124, 125, 127, 128, 431, 432, 433, 438, 439, 442, 445
van Renterghem, P., 174
van Rotterdam, B.J., 302
Van Wormhoudt, A., 91
van Zelm, M.C., 463
Vargas, C., 142, 147, 148, 149
Varkony, T.A., 457
Varshney, R.K., 335
Vatta, M., 457
Vaughn, M., 174
Vaxillaire, M., 457
Velez, C., 492
Vella, C.M., 328
Venter, J.C., 5
Vergote, G., 388
Verlaan, D.J., 456
Verma, S.K., 340
Vermeulen, M., 107, 110, 115, 117, 118, 121, 488
Vernet, G., 305
Verrelli, B.C., 437, 440
Versage, J.L., 302
Verzeletti, A., 112, 145
Viatchenko-Karpinski, S., 457
Viculis, L., 109, 140, 232, 283
Vidal, R.L., 147, 148, 149
Vigne, J.D., 321
Villablanca, F.X., 340
Villalobo, E., 303
Villanueva, E., 287
Villard, E., 457
Villems, R., 16
Vinci, F., 283
Vingerling, J.R., 431, 432, 437, 444, 446
Vionnet, N., 457
Virkler, K., 387, 390
Virmani, R., 459
Virtanen, V., 335, 375
Visser, M., 399, 436, 463
Vissing, J., 90, 96
Vittinghoff, E., 456, 457, 458
Vo, T., 373, 374
Voglmayr, H., 6
Volckaert, F.A., 335
Volgyi, A., 145, 147, 148, 149
Volksone, V., 141
Vollrath, O., 151, 153
von Beroldingen, C.H., 322, 421
von Eiff, C., 304
von Wurmb-Schwark, N., 143
Vos, P., 358, 359, 372, 373, 374, 379
Vossen, C.Y., 456
Votano, S., 145
Voynarovska, K., 175
Vuori, E., 465, 466, 467, 468

W

Wacharapluesadee, S., 305
Wachter, L.L., 323, 325, 329
Wada, H., 456
Wade, N., 308
Wadhams, M.J., 18, 89, 91, 96, 97, 98, 339
Wagner, J.K., 6, 7, 36, 37, 534, 551
Wagner, M.A., 468
Wagner, S., 540
Wai, T., 90
Waite, E.R., 354

Author Index

Walberg, M.W., 86
Walker, D.H., 308
Walker, J.A., 329
Wall, R., 356, 358
Wallace, C., 456, 459
Wallace, D.C., 86, 91
Wallman, J.F., 340, 356
Walsh, A., 280, 286
Walsh, B., 118, 488
Walsh, K.A., 374
Walsh, M.H., 431, 432, 433, 436, 437, 438, 439, 440, 444
Walsh, P.S., 328, 415, 516
Walsh, S., 431, 432, 433, 438, 439, 440, 442, 445, 446
Walsh, S.J., 25
Walter, R.B., 335
Wambaugh, J., 5, 531
Wan, Q.H., 340
Wand, D., 136, 137
Wang, E., 305
Wang, H., 456, 458, 463
Wang, L., 456, 458, 464
Wang, Q., 107, 456, 458, 489
Wang, T., 458
Wang, X.L., 456, 458, 459
Wang, Z., 402, 403
Wangat, Y.Y., 145
Ward, D., 456
Ward, J., 372, 374
Ward, R.D., 339
Warhurst, D.C., 305
Warren, J.H., 471
Warren, W.C., 335
Warshauer, D.H., 390
Wasem, R.E., 541
Wasser, S.K., 339
Wasserstrom, A., 172, 405
Watanabe, A., 171
Watanabe, H., 457
Watanabe, Y., 456, 459
Waterhouse, L., 458, 463
Waterston, R.H., 329
Watkins, H., 457, 461
Watson, M.A., 470
Watt, P.M., 303
Waye, J., 32
Webb, J.L., 387
Webb, J.M., 340
Weber, J.L., 151, 321
Webster, K.A., 305
Wee, A., 388
Weedn, V.W., 18, 89, 91, 96, 97, 98, 339
Weedon, M.N., 433, 435, 441
Weekers, J.E., 542, 543, 550
Wegener, R., 147, 148, 149
Wehner, D.H., 46
Wei, W., 490, 491
Weigt, L.A., 371, 372

Weil, R., 359
Weinblatt, M.E., 497
Weinshilboum, R.M., 464
Weir, B.S., 114, 118, 192, 286, 328
Weis, J.J., 329
Weising, K., 335
Weiss, D.A., 484
Weiss, G., 115
Weissbach, L., 153
Weissenbach, J., 151
Weisser, H.J., 89, 97
Weissman, D., 448
Wells, J.D., 356, 361, 358, 359, 360
Wells, M.C., 335
Wenk, R.E., 542, 543, 547
Wenstrup, R.J., 460
Werren, J.H., 358
West, A.B., 471
West, D.M., 305
Westerhof, W., 437
Westerkofsky, M., 303
Westgard, J.O., 238
Wetmur, J.G., 445
Wetton, J.H., 47, 359
Wharton, D., 90
Whelton, P., 459
Whitaker, J.P., 174, 183, 189
Whitaker, N., 337
White, R.L., 151
Whitehead, P.H., 387, 388
Whitehouse, C.A., 302
Whitfield, J.B., 340
Whiting, B., 517
Whitworth, T., 357
Wichert, B., 330
Wichter, T., 456
Wickenheiser, R.A., 174
Wiegand, P., 90, 140, 146, 148, 149, 284
Wier, B.S., 100
Wiersema, J., 281
Wigmore, J.H., 509
Wilcken, D.E., 456, 459
Will, K.W., 357
Will, M.L., 463, 466
Willerslev, E., 339
Williams, D.W., 358
Williams, E.D., 534
Williams, R., 545
Williams, T., 424
Willott, G.M., 195, 389
Willuweit, S., 114
Wilson, A.C., 90, 340, 391
Wilson, E.M., 150
Wilson, G.L., 91
Wilson, I.J., 87, 91
Wilson, J.H., 391
Wilson, M.R., 359, 360, 422

Wilson, V., 4, 32, 318
Wilson III, D.M., 91
Wilson-Wilde, L., 317, 340
Wiltshire, P.E., 371, 372, 374, 375
Winney, B., 495
Winter, W.E., 462
Witherspoon, D.J., 501
Witkin, S.S., 408
Witonsky, D., 456, 459
Wittig, H., 89, 97
Wittwer, C.T., 329
Wohlhueter, R.M., 302
Wojas-Pelc, A., 432, 433, 436, 437, 439, 440, 444
Wolanska-Nowak, P., 436, 440, 444
Wold, B.J., 171
Wolfe, J., 107
Wolff, K., 335
Wolff, M.S., 445
Wollstein, A., 115, 117, 118, 119, 120, 121, 433, 440, 463
Wong, R., 89
Wong, S.H., 467, 468
Wong, T.H., 436, 440
Wooding, S., 445
Woodley, C.M., 340
Woodley, N.E., 340
Work, T.S., 86
Wraxall, 388
Wright, C.T., 86
Wright, S., 496
Wu, C., 392
Wu, J.H., 339, 340
Wu, W., 143
Wu, X.J., 143
Wu, X.L., 148, 149
Wu, X.Y., 148, 149
Wu, Y., 86
Wulfkuhle, J.D., 470
Wurdack, K.J., 371, 372
Wurm, C.A., 93
Wurmbach, E., 433, 437, 442, 443, 444, 448
Wyler, L.S., 337
Wysocka, J., 142

X

Xiao, X., 330
Xu, X., 456, 459
Xu, Z., 115
Xue, M.Z., 456, 459
Xue, Y., 107, 115, 489

Y

Yagi, S., 405
Yakes, F.M., 91
Yale, S.H., 460
Yam, W.C., 305
Yamagata, K., 457
Yamaguchi, Y., 434, 435
Yampolska, O., 308
Yan, Q., 335
Yan, Z., 543
Yang, C.Y., 340, 341
Yang, F., 86
Yang, J., 86, 435, 441, 456, 459, 544
Yang, L., 456, 459
Yang, M., 304
Yang, N., 196
Yang, Q., 146, 148, 149
Yang, R., 146, 148, 149
Yang, W.I., 145
Yang, Y., 457, 458
Yang, Z., 456
Yasuno, K., 456
Yazdy, M.M., 466
Ye, B., 457
Ye, Y., 145
Yee, H., 471
Yeoman, F., 240, 241
Yokota, M., 389
Yonekawa, H., 90
Yoo, S.Y., 145
Yoon, C.K., 517
Yoshida, M., 19, 340
Yoshimura, S., 340
Yoshiura, K., 447
Yu, C., 146, 148, 149
Yu, Y., 115
Yuan, X., 447
Yuen, K.Y., 305
Yuhki, N., 329, 330, 340

Z

Zabeau, M., 372, 373, 374, 379
Zabek, T., 337
Zackrisson, A.L., 469
Zalan, A., 145, 147, 148, 149
Zaloumis, S.G., 445
Zaracostas, J., 466
Zarrabeitia, A., 141, 146, 148, 149
Zarrabeitia, M.T., 140, 141, 142, 146, 147, 148, 149, 153
Zdelar-Tuk, M., 303
Zegura, S.L., 122, 123, 489
Zehner, R., 356, 358, 359
Zeller, M., 46
Zelson-Mundorff, A., 174, 281
Zemlak, T.S., 340
Zeng, X.P., 143, 145
Zhai, J., 303
Zhang, D., 329, 457
Zhang, H., 456, 459

Author Index

Zhang, L.P., 457
Zhang, S.H., 145
Zhang, X., 90
Zhang, Y., 335, 337
Zhang, Z., 19
Zhao, H., 148, 149
Zhao, S.M., 145
Zhao, Z.Z., 436, 437, 444
Zheng, X., 143
Zhinaula, M., 335
Zhivotovsky, L.A., 97
Zhong, X.Y., 171
Zhu, C., 108
Zhu, F., 143
Zhu, Y., 143
Zhuo, S., 86
Zilles, K., 392
Zimmer, E.A., 371, 372
Zimmer, T., 458
Zimmermann, B., 20
Zinbo, A., 284, 540
Zomorodipour, A., 86
Zozaya-Hinchliffe, M., 408
Zrnec, D., 47
Zubakov, D., 391, 392, 393, 395, 399, 401, 402, 403, 404, 463
Zuniga, S.B., 439, 442, 446
Zyga, A., 337

Subject Index

Page numbers followed by f and t indicate figures and tables, respectively.

A

Abandoned DNA, 531
Abrin, 377
Abuse of power, 551
Access control system (ACS), 215
Accreditation
   of forensic science providers, 223
   standardization, and quality assurance bodies, 212t–213t
Acid phosphatase, 24
   presumptive test, 262
Aconitine, 377
Adenine, 10
Adenosine triphosphate (ATP), 8
Admixture, 499
ADR. See Adverse drug reaction (ADR)
Adverse drug reaction (ADR), 465
African Americans, 484
   ancestary, 484
   eye color predictors in, 442t, 443
   genome, 500
   in pulmonary embolism (PE), 460
African haplogroups, 122
AGAT repeat, 153
Alleged parent, 68
Allele-drop, 44
Allele drop-in and drop-out, 174, 190, 198, 199, 200t
Allele(s), 9, 14, 15, 35, 68, 76, 193f
   frequencies, 155t–156t, 157t–158t, 159t–160t. See also X chromosome STR, forensic application of
   ladder, 43
   sharing, 108
Alternate Light Source (ALS units), 262, 263f
Aluquant® human DNA quantification system, 30. See also DNA quantification
Amelogenin (AMEL), 15, 107
American Academy of Forensic Sciences (AAFS), 211t
American Caucasian database, 99, 100. See also Caucasians
American Civil War, 484
American National Standards Institute (ANSI), 213t
American Society of Crime Laboratory Directors (ABCLD), 213t
American Society of Crime Laboratory Directors/Laboratory Accreditation Board, 535
Amino acids, 19
AmpFLSTR Identifiler, 362
AmpFLSTR® Identifiler™ PCR Amplification Kit, 38, 39f. See also Polymerase chain reaction (PCR) methods
AmpFlSTR® NGM™ PCR Amplification Kit, 41–42, 41f. See also Polymerase chain reaction (PCR) methods
AmpFLSTR® Profiler Plus®, 144
AmpFLSTR® Yfiler PCR Amplification Kit, 16
Amplified fragment length polymorphisms (AFLP), 356, 371, 372–373, 373f. See also Forensic botany
AmpliFLP™ D1S80 PCR Amplification Kit, 35
Amplitype Polymarker, 35
Amylase, 263
Analytical threshold (AT), 44
Ancestral origin estimation, 495–500
Ancestry informative markers (AIM), 21, 533
Anderson, Anna, 484
Andrews v. State, 513
Angiotensin receptor 1 (*AGTR1*), 459
Anglo-American system, of adversarial trials, 535
Animal DNA, forensic analysis of, 46–47
Animal forensic cases, 319
Animal forensic science, 331
Animal testing, 342
Annealing T studies, 230
Antemortem data, 280, 284, 287
Anthrax (*Bacillus anthracis*), 296t, 302t
Antibody–antigen reaction, 25
Aortic aneurysm rupture, 460
Argus X-8 kit, 151
Arrythmogenic right ventricular cardiomyopathy (ARVC), 461. See also Cardiomyopathy
Article 8, of European Convention on Human Rights, 562
Assign® ATF 454 software, 424
Association of Forensic Quality Assurance Managers (AFQAM), 212t
Automated sequencing, 45, 46
Autonomy, 529–530

Autopsy. *See also* Postmortem toxicogenetics
  cardiovascular diseases. *See* Cardiovascular diseases
  channelopathies, 462
  defined, 453
  diabetes mellitus, 461–462
  intracranial hemorrhage, 461
  medico-legal, 464–466
  metabolic diseases, 462
  negative, 462–463
  positive, 455–462
  SUDEP, 462–463
Autosomal short tandem repeat testing, 135, 138f
Autosomal testing, 492–503
  ancestral origin estimation, 495–500
  meiosis, 493–495, 494f
  relative identification, 500–502
Autosomes, 9
Avery, O., 5
AVRC. *See* Arrythmogenic right ventricular cardiomyopathy (ARVC)
Azoospermia, 25

B

Barcode for Life Consortium (BOLD), 340
Base pair (bp), 10
Basic Local Alignment Search Tool (BLAST), 335, 341
Bayesian formula, 69, 70
Bayes' theorem, 100
Beneficence, 530
Bilginturan's disease, 459
Bioethics
  autonomy, 529–530
  beneficence, 530
  concepts in, 518
  confidentiality, 528–529
  informed consent, 529–530
  justice, 528
  privacy, 527–529
  utility, 530
Biological evidence, determination of. *See also* DNA (deoxyribonucleic acid), biological sources of
  blood, 23–24
  feces, 26
  saliva, 25–26
  semen, 24–25
  urine, 26
  vaginal body fluid, 25
Biological evidence collection, 22–23
Biological semiconductor (BSC) transducer, 304t
Biological warfare, 293–294
Biological weapons, banned, 294
Biological Weapons Convention, 294
Biosafety and biosecurity, 310–311. *See also* Bioterrorism and forensic microbiology

Bioterrorism, 377. *See also* Toxicology
  defined, 377
Bioterrorism and forensic microbiology
  biological warfare, 293–294
  biosafety and biosecurity, 310–311
  bioterrorism, defined, 293
  classification of agents, 294–295, 296t–301t
  history of, 293–294
  microbial agents in bioterrorism, 294–295
  microbial forensic protocols and practices, 295, 301, 302t–305t, 306
  microbial forensics, defined, 293
  outbreak of infectious disease, 306–307
  suspicious infectious diseases outbreaks, 307–310
Bite mark evidence, 255. *See also* Sample collection from victim/suspect
Blood, 23–24. *See also* DNA (deoxyribonucleic acid), biological sources of
Blood collection methods
  dried blood stains, 260–261, 261f
  liquid blood samples, 261–262
  physiological fluid stains, 263–264
  seminal stains, 262, 263f
Blood evidence. *See also* Physical evidence (collection/preservation)
  confirmatory blood test, 258–260, 259f
  presumptive blood tests, 256–258, 257f, 258f
Blood stain pattern analysis, 264–265. *See also* Physical evidence (collection/preservation)
Bloodstains on clothing, 255f
Bluestar, 260
Botanical evidence, 368
Botulism (*Clostridium botulinum* toxin), 296t, 302t
Bovine forensic DNA testing, 335–337, 336t. *See also* Forensic animal DNA analysis
Brentamine Fast Blue B reagent, 25
Brucellosis (*Brucella* species), 298t, 303t

C

Caddisflies (case study), 360
Canine forensic DNA testing, 330–335, 332t, 333f, 334f. *See also* Forensic animal DNA analysis
Canine forensic genetic testing kit, 334f
Capillary electrophoresis, 109
Cardiac arrhythmia, 460
Cardiac conduction system, 462
Cardiomyopathy, 459–461
  genotype-phenotype correlation in, 461
  MYH7 gene, 461
Cardiovascular diseases. *See also* Autopsy; Sudden natural death
  aortic aneurysm rupture, 460
  cardiomyopathy, 460–461
  coronary artery atherosclerosis, 455, 458–459

Subject Index

deep venous thrombosis, 460
hypertensive cardiovascular disease, 459
pulmonary embolism (PE), 460
Carnivores, 354t
Carryover contamination, 243. *See also* Contamination
Case studies
cross-contamination, 242–243
error in forensic DNA analysis, 239, 241, 242, 244–245
forensic animal DNA analysis, 322
physical evidence collection, 266–276, 268f, 269f, 272f, 273f, 275f, 276f
Castro, José, 513–514
Catalytic tests, 387
Cat hair roots, 329
Caucasians. *See also* American Caucasian database
distribution of HV+ Mitotypes in, 420
HVI/II sequence in, 418
Cause of death (CoD). *See also* Autopsy
adverse drug reactions and, 465
classified, 465
defined by WHO, 465
medico-legal investigation, 465–466
postmortem toxicogenetics and, 466–469
standard procedures for, 465
Cavalli-Sforza, Luca, 496
Ceiling principle, 515
Centers for Disease Control (CDC), 294, 330
Central nervous system, 462–463
Centre d'Etudes du Polymorphisme Humaine (CEPH), 151
Cepheid Smart Cycler, 31, 31f
Certified Reference Material (CRM), defined, 234
Channelopathies, 462
Chelex®100, 27, 28. *See also* DNA isolation methods
Chemical confirmatory blood tests, 387
Chemical oxidation–reduction reaction, 257
Chemical reagents, 24
Chemiluminescence, 30
China, DNA databases in, 559
Christmas Tree Stain (CTS) method, 25
Chromosome Consortium Phylogenetic tree, 122, 123f
Chromosomes, 6, 9, 9f
and genes, 8–9, 9f. *See also* Human genetics
Classical confirmatory blood tests, 387
Cleaning agents, 256
Clopper–Pearson confidence interval calculation, 112
Close-up photographs, 254
Clothing/personal items as evidence, 256. *See also* Sample collection from victim/suspect
Codeine, 468
CODIS. *See* Combined DNA Index System (CODIS) software
Codon, 11

Cody's Law, 331
Collaborative exercises (CE), 220
Collaborative Testing Service (CTS), 246
College of American Pathologists (CAP), 246
Colorimetric method, 30
Color indicator, 24
Combined DNA Index System (CODIS), 563–564
Convicted Offender Index, 564
Forensic Index, 564
Combined DNA Index System (CODIS) software, 15, 35, 36, 59
markers, 36t, 37t
Combined parents not excluded (CRPNE), 78, 79
Combined paternity index (CPI), 68–69. *See also* Forensic identification
Combined probability of exclusion (CPE), 196
Combined probability of inclusion (CPI), 196
Combined random female not excluded (CRFNE), 75
Combined random man not excluded (CRMNE), 71
Commercial X STR multiplexes, 145t
Competitive reverse transcription-PCR, 302t
Complex repeats, 14
Computer software, 373
Confidentiality, 527–529
Confirmatory blood test, 258–260, 259f. *See also* Blood evidence
Constitution, United States, 521, 563
Contamination, 47, 231, 240–245. *See also* Error in forensic DNA analysis
carryover, 243
cross-contamination of DNA samples, 244
definition, 240
errors in interpretation of DNA results, 243–244
with foreign DNA, 356
inadvertent, 240
mixed samples, 243
Continental system, 535
Contributors. *See also* Forensic DNA mixtures
known, 181, 199
number of, 192–194, 193f, 194f
unknown, 199, 200
unrelated, 199
Control region (CR), 87, 88
Control samples, 374
Convention of the Rights of the Child (1989), 550
Convention on International Trade in Endangered Species of Wild Flora and Fauna (CITES), 318, 338, 338f, 339f
Convicted Offender Index, 59, 564
Cooke, Howard, 514
Coronary artery atherosclerosis, 455, 458–459. *See also* Cardiovascular diseases
gender differences in, 458
genomic risks, 458–459
Costs and benefits of QA implementation, 237–238

Counting method, 99
Court of Appeals, Maryland, 568
Courtroom, scientific evidence in. *See* Trials, scientific evidence in
Crime scene documentation, 369
Crime scene investigation, 368
Crime scene reconstruction. *See also* Physical evidence (collection/preservation)
    conjecture, 265
    data collection, 265
    hypothesis formation, 265–266
    testing, 266
    theory formation, 266
Cross-contamination
    case study, 242–243
    of DNA samples, 244. *See also* Contamination
Crown Prosecution Service (CPS), 176
Cryptosporidium parvum, 305t
Cultural perspectives, on genetic information, 552–553
Curly hair, 447
Cynomya cadaverina, 361
CYP2D6, 463, 468
    genotyping, 469
Cytochrome *c* oxidase I (COI), 340, 341, 361
Cytochrome P450 (CYP), 463, 464
Cytoplasmic chromosome, 18
Cytosine, 10

D

Dallas Police Department, 549
Dallas Prostitute Diversion Initiative (PDI), 549
Databases. *See* DNA databases
Data collection, 280
Daubert v. Merrell Dow Pharmaceuticals, 511–512
Debrisoquine, 463
Deconvolution, 196
Deep sequencing, 94, 95
Deep venous thromboembolism (VTE), 460
Defense hypothesis, 197
Deficiency paternity, 120
Denaturation, 33
Denaturing gradient gel electrophoresis (DGGE), 93
Denaturing high-performance liquid chromatography (dHPLC), 93
Denver convention, 15
Deoxyribonucleic acid (DNA). *See* DNA
Deoxyribose, 19
Dermicidine (DCD), 400
Detritivores, 354t
Developmental validation, 222, 227, 228–230. *See also* Validation
Diabetes mellitus, 431, 461–462
Dideoxynucleotide (ddNTP), 45
Differential DNA methylation, 405
Diffusion reactions, 24

Dilated cardiomyopathy, 460–461. *See also* Cardiomyopathy
Diploids, 6
Disaster victim identification (DVI), 120–122. *See also* Y chromosome in forensic evidence
Displacement loop (D-loop), 87
Disputed parent, 68
Disputed parentage, Y and X markers in, 76, 77t. *See also* Forensic identification
District Attorney's Office for Third Judicial Dist. v. Osborne, 523
DNA (deoxyribonucleic acid), 10, 10f. *See also* Human genetics
    analysis of, 210t, 356
    collection kit, 485, 486
    collection program, of Dallas PDI, 549
    double gelix, 11f
    evidence. *See* Scientific evidence
    evidence in court, 58–59. *See also* Forensic DNA analysis and statistics
    exonerations, 522, 523
    extraction, 27f, 89–90, 214, 356
    fingerprinting methodology, 318
    fraud, immigration, 543
    identification techniques, 282
    laboratory layout, 215f
    mitochondrial, 18–19
    relationship testing. *See* Relationship testing
    research, 5–6
    sample. *See* Sample acquisition, ethical issues in
    sequencing, 356
    variability of, 11–12, 11f, 12f
DNA (deoxyribonucleic acid), biological sources of, 21–22
    biological evidences, determination of, 23–26
    blood, 23–24
    feces, 26
    saliva, 25–26
    sample collection/storing, 22–23
    semen, 24–25
    urine, 26
    vaginal body fluid, 25
DNA (deoxyribonucleic acid), transfer of
    casual contact, 183
    direct/indirect contact, 183
    hand washing, 183
DNA Advisory Board (DAB), 206, 328
    versus ISO 17025, 223, 224t. *See also* Forensic DNA typing and quality assurance
DNA (deoxyribonucleic acid) databases, 98–99, 320
    CODIS, 563–564
    European Union and, 566
    global collaborations on sharing, 552
    human trafficking and, 552
    Interpol DNA Gateway, 566–567
    NDNAD, 559, 560–561
    overview, 557–559

sharing data across borders, 566
South Africa and, 564–566
United Kingdom. *See* United Kingdom
United States, 559, 560, 563–564
volume of, 559
"DNA Fingerprinting Dispute Laid to Rest," 515
DNA Identification Act of 1994 (US), 563
DNA (deoxyribonucleic acid) individualization. *See also* Forensic botany
 amplified fragment length polymorphisms (AFLP) methodology, 372–373, 373f
 short tandem repeat (STR), 373–374
DNA IQ System, 28
DNA (deoxyribonucleic acid) isolation methods, 26–28, 27f
 Chelex®100, 28
 DNA IQ System, 28
 FTA®, 28–29
 with organic solvents, 26–28
 Qiagen DNA isolation, 28
DNA-PROKIDS, 547–548
 in Haiti, 548
 MOUs with different countries, 548
DNA (deoxyribonucleic acid) quantification
 Aluquant®, 30
 hybridization (slot-blot) method, 30
 quantitative real-time PCR (QRT-PCR), 31–32, 31f
 spectrophotometry method, 29
 "yield" gel method, 29–30
DNase, 47
DNA (deoxyribonucleic acid) testing. *See also* Relationship testing
 autosomal testing, 492–503
 genealogy. *See* Genealogy
 haploid testing. *See* Haploid chromosome testing
 meiosis, 492–495, 494f
 mtDNA testing, 490–492
 postconviction, 522–523
 relative identification, 500–502
 Y chromosome testing, 486–490
DNA (deoxyribonucleic acid) variation measurement
 polymerase chain reaction (PCR), 33, 34f
 restriction fragment length polymorphism, 32
Documentation, 233–234. *See also* Forensic DNA typing and quality assurance
Dog Bite Law, 331
Dried blood stains, 260–261, 261f. *See also* Blood collection methods
Drop-out and drop-in, allelic, 174, 190, 198, 199, 200t
Drug efficacy, 467–468, 467t
Drug enforcement, 378. *See also* Toxicology
Drug intoxication
 poor metabolizing genotypes and, 468
 ultrarapid metabolizing genotypes and, 468–469

Drugs, adverse effects of, 465
D1S80, 35
D21S11 allele, 43
DSI-Semen™ kit, 405, 406f
Due process principles, 535
DVI. *See* Disaster victim identification (DVI)
DXS101, 155t–156t
DXS6789, 155t–156t
DXS6795, 160t
DXS6803, 160t
DXS7130, 160t
DXS7132, 160t
DXS7423, 159t
DXS7424, 159t
DXS8378, 160t
DXS9902, 159t
DXS10074, 153, 159t
DXS10079, 155t–156t
DXS10101, 157t–158t
DXS10103, 155t
DXS10134, 157t–158t
DXS10135, 153, 155t–156t
DXS10146, 153, 157t–158t
DXS10147, 159t
DXS10148, 155t–156t
DXYS14 probe, 514

E

Ear wax, 447
EDNAP Forensic mtDNA Population Database (EMPOP), 20
Electron microscopy, 362
Electropherograms, 13f, 17f, 125f, 325f, 398f
 of AmpFlSTR Identifiler Plus PCR amplification kit, 39f
 of AmpFlSTR NGM PCR amplification kit, 41f
 of Powerplex ES X PCR amplification kit, 40f
Electrophoresis, 24, 32
Electrospray ionization mass spectrometry (ESI-MS) approach, 95
Emery Dreifuss dystrophy, 461
Emulsion PCR (emPCR) technique, 94, 422
Encyclopedia of DNA Elements (ENCODE), 10
Endosymbiont, 86
England, DNA database in. *See* United Kingdom, DNA database in
Enhancement reagents, 257
Enzyme-linked immunosorbent assay (ELISA), 25, 389
Epithelial cells and tissues, 264. *See also* Stains from physiological fluid
Epsilon toxin of *Clostridium perfringens*, 298t, 303t
Error in forensic DNA analysis, 238–246. *See also* Forensic DNA typing and quality assurance
 case studies, 239, 241, 242, 244–245
 contamination, 240–245

mishandling of samples, 239
rates in forensic DNA testing, 245–246
Errors in DNA result interpretation, 243–244
Escherichia coli, 33, 92
Eskimo culture of Greenland, 447–448
Ethical issues
   acquiring DNA samples, 530–532
   bioethics. *See* Bioethics
   expert testimony, 535–536
   laboratory reporting, 535–536
   law enforcement DNA databanks, 533
   overview, 527
   phenotypes, 533–534
   racial identifications, 533–534
   remains identification, 534
Ethnicity, in pulmonary embolism (PE), 460
Ethylenediaminetetraacetic acid (EDTA), 27
Ethylene vinyl acetate thermoplastic film, 469
EU Council recommendation, 207
Eukaryotes, 6, 19
Europe, QA in, 207, 208t, 209, 210t, 211t–212t, 213t
European Convention on Human Rights, 561–562
European Court of Human Rights, 561–563
European DNA legislation, 207. *See also* Forensic DNA typing and quality assurance
European DNA Profiling Group (EDNAP), 19, 42, 393, 394t
European Network of Forensic Science Institutes (ENFSI), 40, 207, 208, 208t, 211t, 535
European PT Information System (EPTIS) database, 225
European Standard Set (ESS), 207
Evidence. *See* Scientific evidence
Evidence collection, 368–370. *See also* Forensic botany
Exclusionary power (EP), 71
Exonerations. *See* DNA (deoxyribonucleic acid) exonerations
Exons, 19
Expert testimony, 510, 535–536
Extrachromosomal DNA (plasmid), 6
Eye color prediction from DNA, 437–439
Eye-color-predictor-1, 445, 446
   validation of, 442, 442t
Eye-color-predictor-2
   validation of, 443, 444t
Eye color predictors, 436–437, 438, 439t

F

Fallacious assumption, 519
Family, in immigration procedures, 550
Family Assistance Center (FAC), 280. *See also* Missing persons identification by DNA
Fatal dog attack (case study), 332

Fecal material, 254, 264. *See also* Sample collection from victim/suspect; Stains from physiological fluid
Feces, 26, 409–410, 409t. *See also* DNA (deoxyribonucleic acid), biological sources of; Forensic tissue identification, nonhuman DNA/RNA for
Felid forensic DNA testing, 321–330. *See also* Forensic animal DNA analysis
Fetal and neonatal blood, 388. *See also* Forensic tissue identification, classical tests for
FFPET (formalin fixed paraffin embedded tissues) slide, 470f
Filtration, 177
Fingerprints, 318
Finnzyme Canine Genotypes™ 1.1 Multiplex STR Reagent kit, 334f
Finnzymes, 337
Florida District Court of Appeals, 513
Fluidigm Access Array system, 424
Fluorescence techniques, 45
Fluorescent detection technology, 43
Fluorescent properties, 260
Fluorophore, 32
Food safety threats, 298t, 303t
Forensic analysis
   of animal DNA, 46–47
   of plant DNA, 46
Forensic animal DNA analysis
   animal suspects/victims/witnesses, 319
   bovine forensic DNA testing, 335–337, 336t
   canine forensic DNA testing, 330–335, 332t, 333f, 334f
   case studies, 322, 323f, 324t, 332, 333–335
   categories, 319
   felid forensic DNA testing, 321–330
   forensic typing system for genetic individualization, 324–328, 325f, 326t–327t
   interlaboratory proficiency testing program, 321
   lack of commercially available forensic DNA kits, 320
   Locard's Exchange Principle, 318
   nonhuman DNA analysis, 318
   overview, 317–321
   technical inability, 320
   validation studies of cat multiplex, 328–330
   wildlife forensic DNA testing, 337–342, 338f, 339f
Forensic botany
   bioterrorism, 377
   defined, 367
   DNA individualization, 372–373, 373f
   drug enforcement, 378
   evidence collection, 368–370
   location of body, 375–376

Subject Index

microscopy, 370–371
missing persons, 375–376
mycology, 374
palynology, 374
plants as biological evidence, 367
species identification, 371–372, 371f
toxicology, 376–379
Forensic DNA analysis, NGS for, 426
sensitivity for. *See* Polymerase chain reaction (PCR) product detection
Forensic DNA analysis and statistics. *See also* Forensic identification
DNA evidence in court, 58–59
genetic and statistical principles, 56
Hardy–Weinberg equilibrium (HWE), 56–58
human remains, identification of, 77–81
linkage equilibrium, 58
parentage testing, 56
Forensic DNA assays, validation of, 445–446
Forensic DNA mixtures
case study, 201
contributors, number of, 192–194, 193f, 194f
IFSG recommendations on mixture analysis, 190
mixture ratios, 195, 195f
statistics, 196–200, 200t
SWGDAM guidelines on, 191–192
types of results, 189
Forensic DNA typing, 238
Forensic DNA typing and quality assurance
costs and benefits, 237–238
DNA Advisory Board (DAB) *versus* ISO 17025, 223, 224t
documentation, 233–234
error, potential for, 238–246
European DNA legislation, 207
ISO 17025–based quality management system (QMS), 214–216, 215f
overview, 205–206
proficiency testing (PT) and collaborative exercises, 216, 217t, 218t, 219f, 220f, 221f, 222t, 223
PT provider selection, 223, 225
quality assurance (QA) in Europe, 207, 208t, 209, 210t, 211t–212t, 213t
quality assurance (QA) in US, 206
reference materials, 234–237, 236f
validation, 225–233
Forensic entomology, 353
Forensic entomology, DNA-based methods in
case studies, 360–361, 362–363
human DNA extracted from insects, 358–362
human mtDNA, 362
insect DNA analysis methods, 356–359
insect mtDNA, 361
insects in forensics, 355t
invertebrates in forensics, 354t

overview, 353–356
Forensic identification, 59–60, 61t. *See also* Forensic DNA analysis and statistics
combined paternity index (CPI), 68–69
correction for substructuring, 60–61
individualization and identification, 61–62
low copy number (LCN) DNA testing, 64–65, 64t
maternity testing, 74, 75–76
mitochodrial DNA (mtDNA) testing, 66
mixed sample analysis, 62–64
motherless paternity testing, 72–74, 73t, 74t, 75t
mutations, effect of, 74
parentage testing, 67–68
parentage testing with mixed populations, 76
paternity index (PI), 68–69
probability of paternity, 69–71
random man not excluded (RMNE), 71–72
X chromosome STR testing, 66–67, 67t
Y and X markers in disputed parentage, 76, 77t
Y-STR analysis, 65–66
Forensic identification, traditional, 281
Forensic Index, 59, 564. *See also* Combined DNA Index System (CODIS)
Forensic insect specimen identification, 358
Forensic predictors, validation of, 442–444
Forensic Quality Services (FQS), 213t
Forensic Science Service (FSS), 172, 176, 201, 240, 319
UK, 559, 560
Forensics mixtures
and mtDNA, 421
and NGS, 422
Forensic Specialties Accreditation Board (FSAB), 213t
Forensic Statistical Tool (FST), 183, 202
Forensic tissue identification, classical tests for
fetal and neonatal blood, 388
menstrual blood, 387–388
nasal blood, 388
parturient blood, 388
peripheral blood, 386–387
saliva, 388–389
semen, 389
skin, 390–391, 390t
vaginal secretion, 389–390
Forensic tissue identification, DNA methylation for, 404–407, 409t
Forensic tissue identification, nonhuman DNA/RNA for
feces, 409–410, 409t
mouth-expirated blood, 407–408
saliva, 407
vaginal secretion, 408–409
Forensic tissue identification, RNA-based
miRNA markers for, 401–403, 402t, 403f
mRNA markers, 392–393, 393f, 394t

multiplex mRNA systems, 394–399, 396t–397t, 398f
stability of RNA, 391–392
stand-alone mRNA assays, 399–401
unresolved issues, 404–405
Forensic training for nurses, 262
Forensic typing system for genetic individualization (cat sample), 324–328, 325f, 326t–327t. *See also* Forensic animal DNA analysis
Forensim, 201
Fourth Amendment, of US Constitution, 563
Fraud, immigration. *See* DNA (deoxyribonucleic acid) fraud, immigration
Frye, James Alphonso, 510–511
Frye test, 180
Frye v. United States, 510–511
FTA®, 28–29. *See also* DNA isolation methods
Fungi in criminal investigations, 375

G

Gametes, 493
Gametic disequilibrium, 58
Gastric fluid, 264. *See also* Stains from physiological fluid
GATA165B12, 159t
GATA172D05, 153, 160t
GATA31E08, 153, 160t
GEDNAP (German DNA Profiling) certificate, 218f
mitochondrial DNA testing, 220f
stain analysis, 219f
Gel electrophoresis, 373
Gender differences, in coronary artery atherosclerosis, 458
Genealogy, 483–485
autosomal testing, 492–503
defined, 483–484
DNA testing, 485–486, 485f
haploid testing. *See* Haploid chromosome testing
meiosis, 493–495, 494f
mtDNA testing, 490–492
overview, 483–485
relative identification, 500–502
Y chromosome testing, 486–490
GeneChip® Human Mitochondrial Resequencing Array, 418
Gene expression, 393f
Gene flow, 57
GeneMapper electropherograms, 334f
General acceptance standard, 510–511
Genetic Analyzers, 40
Genetic bottleneck theory, 92
Genetic diversity, 10–11. *See also* Human genetics
Genetic drift, 57
Genetic genealogy. *See* Genealogy

Genetic Identification Laboratory at University of Granada, Spain, 547
Genetic information
cultural perspectives on, 552–553
privacy of, 550–551
Genetic markers, 228, 229
Genetic testing, 338
Geneva Protocol, 294
Genome Analyzer HiSeq and MiSeq, 421
454 genome sequencing technology, 422–424
1000 Genomes Project, 441
Genome structure, 6–8, 7t, 8f. *See also* Human genetics
Genome-wide association (GWA) studies, 435
Genome-wide risk, of sudden cardiac death, 461
Genotyping, 140
German Stain Commission, 63
GERMLINE, 502
Glanders (*Burkholderia mallei*), 298t, 304t
Ground zero, 281f
GS FLX, 421
GS Junior, 421
Guanine, 10

H

Hair color prediction, 440
Hair thickness, 447
Haiti, DNA-PROKIDS in, 548
Hand washing, 183
Hantavirus, 295, 301t
Hantavirus pulmonary syndrome (HPS), 307
Haplogroups, 122, 123f, 124, 125f
from haplotypes, 126–127
Y-SNP, 122–126, 123f, 125f
Haploid chromosome testing
future, 492
mtDNA testing, 490–492
overview, 484–485, 486
Y chromosome testing, 486–490
Haploids, 6
Haplotype(s), 113t, 114, 153
frequencies, 140, 150
HapMap consortium, 448
HAPMIX, 499
Hardy-Weinberg equilibrium (HWE), 56–58, 81, 140, 194, 196. *See also* Forensic DNA analysis and statistics
Heglostix, 24
Height prediction, 440
Heme, 257
Herbariums, 369
Heteroplasmy, 18, 425
mtDNA, 91–96. *See also* MtDNA (Mitochodrial DNA), forensic aspects of
Heterozygosities of cat STR multiplex, 326t–327t
Heterozygote, 9

Heterozygous studies, 232
High Risk Potential Victims DNA Database (HRDNA), 549
High template (HT) DNA, 175, 180
Hispanics, 443
HIV, nonintentional transmission of, 308
HLA 454 assay for mixture analysis, 424
HLA-DQA1 test, 417
Homicide (case study), 333
Homologous chromosomes, 9
Homozygote, 9
Homozygous markers, 150
Hong Kong Government Laboratory, 223
HPRTB, 153, 159t
HRDNA. See High Risk Potential Victims DNA Database (HRDNA)
Human chorionic gonadotropin (fÀhCG) genes, 397
Human DNA extracted from insects, 359–362. See also Forensic entomology, DNA-based methods in
Human DNA recovered from mosquitoes, 361–362
Human fetal hemoglobin (HbF), 388
Human genetics
    chromosomes and genes, 8–9, 9f
    DNA, 10, 10f
    DNA, variability of, 11–12, 11f, 12f
    genetic diversity, 10–11
    genome structure, 6–8, 7t, 8f
    mitochondrial DNA (mtDNA), 18–19
    RNA profiling, 19
    sex chromosomes, 15–17, 15f, 17f
    SNP molecular markers, 19–21, 20f
    STR markers, structure/nomenclature, 12–15, 13f, 14f
Human hypervariable region II (HVII), 360
Human identification (HID), 339, 367, 369
Human Identity Trade Association (HITA), 212t
Human/insect mtDNA analysis from maggots, 360
Human mtDNA, 362. See also Forensic entomology, DNA-based methods in
Human remains, identification of, 77–81, 534. See also Forensic DNA analysis and statistics
Human trafficking, 544–549
    Dallas PDI and, 549
    defined, 544–545
    DNA-PROKIDS and, 547–548
    relationship testing, 545–547
Human tyrosine hydroxylase, 15
Human Y-Chromosome DNA Profiling Standard Certificate of Analysis for SRM 2395, 237
Humeri with same DNA profile, 286f
Humic acid inhibition, 229
HV+ HaploArray, 418–419
    and mtDNA, 418
HV+ Mitotypes, distribution of, 420–421
HVS. See Hypervariable segments (HVS) testing

Hybridization (slot-blot) method, 30, 31. See also DNA quantification
Hydrogen bonds, 33
Hydrogen peroxide, 24
Hypertension. See Hypertensive cardiovascular disease
Hypertensive cardiovascular disease, 459
Hyperthermia, 377
Hypertrophic cardiomyopathy, 461. See also Cardiomyopathy
Hypervariable (HV) regions, mtDNA, 88f, 89
Hypervariable repeats, 14
Hypervariable segments (HVS) testing, 491–492

I

IBD. See Identity-by-descent (IBD)
Identity-by-Descent (IBD), 115, 500–502, 501t
Identity-by-State (IBS), 115
Illegal trade in endangered species, 338f, 339f. See also Wildlife forensic DNA testing
Immigration
    DNA fraud, 543
    family in, defining, 550
    rapid DNA analysis, 543–544
    relationship testing, 540–543
Immunoassay procedure, 258
Inadvertent contamination, 240. See also Contamination
Inbreeding, 57
InDels, 20
Indonesia, 223
Infectious disease outbreaks, 306–307. See also Bioterrorism and forensic microbiology
Informed consent, 529–530
Infrared (IR) laser, 469, 470f
Insect DNA analysis methods, 356–360. See also Forensic entomology, DNA-based methods in
Insect mtDNA, 360. See also Forensic entomology, DNA-based methods in
Insects in forensics, 355t
InterAmerican Accreditation Cooperation (IAAC), 213t
Intercolor balance, 232
Interlaboratory proficiency testing program, 321
Internal Lane Standard (ILS), 38
Internal proficiency tests, 217
Internal validation, 222, 232–233. See also Validation
International Association for Identification (IAI), 212t
International Human Genome Sequencing Consortium 2001, 5
International Laboratory Accreditation Cooperation (ILAC), 209, 212t
    guidelines and procedures, 213t

International Organization for Standardization (ISO), 212t
  defined, 207
International Panel of Microsatellites for Cattle Parentage Testing, 337
International Society for Animal Genetics (ISAG), 336t, 337
International Society for Forensic Genetics (ISFG), 14, 20, 63, 144, 189, 209, 210t, 340
  certificate-paternity testing interlaboratory comparison, 221f
  recommendations on mixture analysis, 190. See also Forensic DNA mixtures
International Society for Forensic Haemogenetics (ISFH), 144
International Society of Genetic Genealogy (ISOGG), 489
International standards, 209
International System of Gene Nomenclature in 1987, 12
INTERPOL, 212t
Interpol DNA Gateway, 566–567
Intoxication. See Drug intoxication
Intracolor balance, 232
Intracranial hemorrhage, 461
Intragenomic selection hypothesis (selfish DNA), 6
Introns, 11, 19
Invertebrates in forensics, 354t
Investigator® Argus X-12, 139f, 152f
Investigator ESSplex Plus Kit/Investigator IDplex Plus Kit, 42. See also Polymerase chain reaction (PCR) methods
Ion Torrent®, 421
"IrisPlex," 439, 440. See also Eye color predictors
ISO 17025, 222t, 223
ISO 9001 and ISO 17025 (differences), 222t
ISO 17025-based quality management system (QMS), 214–216, 215f. See also Forensic DNA typing and quality assurance
ISOGC. See International Society of Genetic Genealogy (ISOGG)
Isolated cell, 469

J

Judicial acceptance, of scientific evidence
  laboratory methods, challenges to, 513–515
  polymerase chain reaction (PCR)–based methods, 516
  uncritical acceptance, 512–513
  VNTR profiling, 515
Junk DNA, 6
Justice, 528

K

Kastle–Mayer reaction, 24
Kastle–Meyer phenolphthalein test, 386

Kcna1 gene, 463
KCNH2 gene, 463
Kennedy, Anthony, 568
King, Alonzo Jay, Jr., 568
Kinship analysis, 284

L

Laboratory Accreditation Board, 213t
Laboratory of Genomic Diversity (LGD), 322
Laboratory reporting, ethics of, 535–536
Laboratory reports, as scientific evidence, 521–522
Lab Retriever, 201
Lactobacillus crispatus, 408, 410
Lander, Eric, 514
Laser microdissection techniques (LMT), 469–471
Law Commission of England and Wales, 512
Law enforcement DNA databanks, 533. See also DNA databases
Leaf litter, 375
Left ventricular hypertrophy, 459
Legal standards, for admitting scientific evidence
  general acceptance standard, 510–511
  overview, 510
  scientific-validity standard, 511–512
Length heteroplasmy (LH), 92
LEOPARD syndrome, 461
Leucocrystal Violet, 257
Leucomalachite green, 260
Lifecodes Corporation, 513, 514
Likelihood ratio (LR), 60, 98, 189, 193
  case study, 181–183
LikeLTD, 201
Linear Array, 425
Linkage disequilibrium (LD), 58, 150, 151, 496, 502
Linkage equilibrium, 58. See also Forensic DNA analysis and statistics
Liquid blood samples, 261–262. See also Blood collection methods
Liquid urine, 254. See also Sample collection from victim/suspect
Lloyd, Eddy Joe, case of, 522–523
LMT. See Laser microdissection techniques (LMT)
Local ancestry analysis, 499–500
Locard's Exchange Principle, 318. See also Forensic animal DNA analysis
LoComatioN (software tool), 201
Locus descriptions for canine genotypes, 332t
Low copy number (LCN) DNA profiling
  case studies, 172, 173, 175–176, 178–179, 179–180, 181–182, 184
  LT-DNA procedural modifications, 175, 176, 177f
  LT-DNA testing, forensic applications of, 173–174
  overview, 171–172
  statistical considerations, 181, 182–183
  technical considerations, 174–175

Subject Index

Low copy number (LCN) DNA testing, 64–65, 64t. *See also* Forensic identification
Low template (LT) DNA, 518
    procedural modifications, 175, 176, 177f
    testing, 171, 172, 173–174. *See also* Low copy number DNA profiling
LRMix, 201
LRP6 gene, 458
LT-DNA testing. *See* Low template (LT) DNA testing
Luminol, 24, 260
    test, 387

M

Maggots (case study), 360–361
Malaysia, 223
Male lineage differentiation, 117f
Markov chain Monte Carlo (MCMC) method, 201
Marper case decision, of European Court of Human Rights, 562
Maryland Court of Appeals, 568
Maryland v. King, 568
Mass disaster victim identification by DNA. *See* Missing persons identification by DNA
Mass fatality events, 277. *See also* Missing persons identification by DNA
Maternal inheritance, 90, 91, 96, 339
Maternity testing, 74, 75–76. *See also* Forensic identification
Matrix-assisted laser desorption/ionization–time of flight (MALDI-TOF) mass spectrometry, 127
Matrix metallopeptidase 7 gene (*MMP7*), 399
Maxam-Gilbert (M-G) method, 45
Maximum likelihood method, 194
Medico-legal autopsy, 465–466
MEF2A gene, 458
Meiosis, 150, 493–495, 494f
Melanin, 435–437, 447
Melanogenesis, 440
Melendez-Diaz v. Massachusetts, 521–522
Melioidosis (*Burkholderia pseudomallei*), 299, 304t
Memorandum of understanding (MOU) DNA-PROKIDS, 547–548
Mendel, G., 5
Mendelian inheritance, principles of
    independent assortment, 56
    monohybrid crossing, 56
    segregation, 56
Mendelian laws, 68
Menstrual blood, 387–388. *See also* Forensic tissue identification, classical tests for
Mentype Argus X-8 kit, 66
Messenger RNA (mRNA), 19
Messenger RNA (mRNA) markers, 392–393, 393f, 394t. *See also* Forensic tissue identification, RNA-based

in Multitissue multiplex PCR assays, 396t–397t
Methylation-sensitive restriction endonuclease PCR (MSRE-PCR) technology, 405
Microbial forensic protocols and practices, 295, 301, 302t–305t, 306. *See also* Bioterrorism and forensic microbiology
Microbial forensics, 306
    defined, 293
Microfluidic technologies, 544. *See also* Rapid DNA analysis
Microgenomics, 469–471
MicroRNA (miRNA) markers, 401–403, 402t, 403f. *See also* Forensic tissue identification, RNA-based
Microsatellite(s), 35, 335
    markers, 337
    typing methods, 359
Microscopy, 370–371, 371f
Minifiler kit (Life Technologies), 21
Minimal haplotype (minHt), 65, 108
Minisatellite(s), 35, 373
MiniSTR molecular markers, 20, 21
Mishandling of samples, 239. *See also* Error in forensic DNA analysis
Mislabelling of blood sample, 239
Missing persons, 375–376. *See also* Forensic botany
Missing persons identification by DNA
    Family Assistance Center (FAC), 280
    mass fatality events, 277
    morgue operations, 279, 279f
    search and recovery, 278, 278f
    World Trade Center Remains Identification Project, 280–289, 281f–289f
Mitochodrial DNA (mtDNA). *See* mtDNA
Mitochondrial genome, loci on, 340–341
Mitochondrion, 6, 8, 8f
    and mtDNA genome structure, 85–89, 87f, 88f. *See also* MtDNA (Mitochondrial DNA), forensic aspects of
Mixed sample contamination, 243
Mixture ratios, 195, 195f
Mixture studies, 228, 230
Molecular autopsy. *See also* Autopsy
    defined, 454
Molecular biology techniques for bioterrorism agent detection, 302t–303t
Molecular markers
    miniSTR, 20, 21
    SNP, 19
Monogenic forms, of hypertension, 459. *See also* Hypertensive cardiovascular disease
Monohybrid crossing, 56
Morgan of sequence, 500
Morgue operations, 279, 279f. *See also* Missing persons identification by DNA
Motherless paternity testing, 72–74, 73t, 74t, 75t. *See also* Forensic identification

Mouth-expirated blood, 407–408. *See also* Forensic tissue identification, nonhuman DNA/RNA for
MPR. *See* Multidrug resistance protein (MPR)
mtDNA (Mitochondrial DNA), 6, 18–19. *See also* Forensic entomology, DNA-based methods in; Human genetics
   control region, 88f
   and forensics mixtures, 421
   genome, 87
   human, 361
   and HV+ HaploArray, 418
   hypervariable (HV) segments, 88f, 89
   insect, 360
   linear array technology, 417
   in species testing, 339–340. *See also* Wildlife forensic DNA testing
   testing, 66. *See also* Forensic identification
   typing, 135
mtDNA (Mitochondrial DNA), forensic aspects of
   application of, 96–97
   copy number, 89–90
   genetic variability and random match probabilities, 97–101
   heteroplasmy, 91–96
   mitochondrion and genome structure, 85–89, 87f, 88f
   mtDNA inheritance, 90–91
mtDNA 454 GS assay for mixture analysis, 424–426
mtDNA (Mitochondrial DNA) testing, 484, 490–492
   hypervariable segments (HVS) testing, 491–492
   maternal lineages, 490
   options, 491–492
Multidisciplinary approach, 287
Multidrug resistance protein (MPR), 464
Multilocus variable-number of tandem (MLVA), 302t
Multiplex ID (MID) tags, 422–425
Multiplex mRNA systems, 394–399, 396t–397t, 398f. *See also* Forensic tissue identification, RNA-based
Multiplex PCR assays, 141t–143t
Multiplex STR systems, 35–38, 36t, 37t. *See also* Polymerase chain reaction (PCR) methods
Mutation rates, 144, 146t–147t, 148t–149t, 150. *See also* X chromosome STR, forensic application of
Mutations, 330
   effect of, 74. *See also* Forensic identification
Mycology, 374. *See also* Forensic botany
MYH7 gene, 461
Myocardial infarction, 458

N

Nasal blood, 388, 399. *See also* Forensic tissue identification, classical tests for; Stand-alone mRNA assays

Nasal mucous, 255. *See also* Sample collection from victim/suspect
National Cancer Institute, 322
National Center for Biotechnology Information (NCBI), 341
National DNA Database (NDNAD), 559
   arrestee DNA profiles in, 563
   detections and matches, decline in, 561
   NPIA and, 560
   vs. NDIS of United States, 560
National DNA Index System (NDIS), 59, 563, 564
   vs. NDNAD of United Kingdom, 560
National Forensic Science Technology Center (NFSTC), 212t
National Institute of Justice (NIJ), 324
National Institute of Standards and Technology (NIST), 212t
National Police Improvement Agency (NPIA), 560, 560t
National Research Council (NRC), 206, 240, 309
Nature, 448
NDIS. *See* National DNA Index System (NDIS)
NDNAD. *See* National DNA Database (NDNAD)
Negative autopsy
   cardiac conduction system, 462
   central nervous system, 462–463
Neonatal blood, 400. *See also* Stand-alone mRNA assays
Neutralist hypothesis, 6
New Jersey Supreme Court, 523
New York City Office of Chief Medical Examiner (NYC OCME)
   experience, 280–289, 281f–289f
   morgue operations, 279f
NGS (Next Generation Sequencing), 417, 421–422
   for forensic DNA analysis, 426
   forensics applications of, 421–422
   and forensics mixtures, 422
Nicholas II of Russia, Tsar, 484
Nipah virus, 295, 301t
NIST SRM (National Institute of Standards and Technology standard reference material), 235–237, 236f. *See also* Reference materials (RM)
Noncoding DNA sequence, 10
Nonexcluded probability of paternity, 71
Nonhuman DNA analysis, 318. *See also* Forensic animal DNA analysis
Nonprotein-coding genomic elements (ncDNA), 6, 7t
Nonrecombining portion of Y (NRY), 105, 106
Nonspectrophotometric method, 29
Northern blot hybridization, 402
NPIA. *See* National Police Improvement Agency (NPIA)
Nuclear DNA (nucDNA), 86
Nuclear mitochondrial insertions (NUMT), 340